THE ELEMENTS

Element	Symbol	Atomic Number	Relative Atomic Mass*
Actinium	Ac	89	(227)
Aluminum	Al	13	26.98
Americium	Am	95	(243)
Antimony	Sb	51	121.8
Argon	Ar	18	39.95
Arsenic	As	33	74.92
Astatine	At	85	(210)
Barium	Ba	56	137.3
Berkelium	Bk	97	(247)
Beryllium	Be	4	9.012
Bismuth	Bi	83	209.0
Bohrium	Bh	107	(267)
Boron	B	5	10.81
Bromine	Br	35	79.90
Cadmium	Cd	48	112.4
Calcium	Ca	20	40.08
Californium	Cf	98	(249)
Carbon	C	6	12.01
Cerium	Ce	58	140.1
Cesium	Cs	55	132.9
Chlorine	Cl	17	35.45
Chromium	Cr	24	52.00
Cobalt	Co	27	58.93
Copper	Cu	29	63.55
Curium	Cm	96	(247)
Darmstadtium	Ds	110	(281)
Dubnium	Db	105	(262)
Dysprosium	Dy	66	162.5
Einsteinium	Es	99	(254)
Erbium	Er	68	167.3
Europium	Eu	63	152.0
Fermium	Fm	100	(253)
Fluorine	F	9	19.00
Francium	Fr	87	(223)
Gadolinium	Gd	64	157.3
Gallium	Ga	31	69.72
Germanium	Ge	32	72.61
Gold	Au	79	197.0
Hafnium	Hf	72	178.5
Hassium	Hs	108	(277)
Helium	He	2	4.003
Holmium	Ho	67	164.9
Hydrogen	H	1	1.008
Indium	In	49	114.8
Iodine	I	53	126.9
Iridium	Ir	77	192.2
Iron	Fe	26	55.85
Krypton	Kr	36	83.80
Lanthanum	La	57	138.9
Lawrencium	Lr	103	(257)
Lead	Pb	82	207.2
Lithium	Li	3	6.941
Lutetium	Lu	71	175.0
Magnesium	Mg	12	24.31
Manganese	Mn	25	54.94
Meitnerium	Mt	109	(268)
Mendelevium	Md	101	(256)

Element	Symbol	Atomic Number	Relative Atomic Mass*
Mercury	Hg	80	200.6
Molybdenum	Mo	42	95.94
Neodymium	Nd	60	144.2
Neon	Ne	10	20.18
Neptunium	Np	93	(244)
Nickel	Ni	28	58.70
Niobium	Nb	41	92.91
Nitrogen	N	7	14.01
Nobelium	No	102	(253)
Osmium	Os	76	190.2
Oxygen	O	8	16.00
Palladium	Pd	46	106.4
Phosphorus	P	15	30.97
Platinum	Pt	78	195.1
Plutonium	Pu	94	(242)
Polonium	Po	84	(209)
Potassium	K	19	39.10
Praseodymium	Pr	59	140.9
Promethium	Pm	61	(145)
Protactinium	Pa	91	(231)
Radium	Ra	88	(226)
Radon	Rn	86	(222)
Rhenium	Re	75	186.2
Rhodium	Rh	45	102.9
Roentgenium	Rg	111	(272)
Rubidium	Rb	37	85.47
Ruthenium	Ru	44	101.1
Rutherfordium	Rf	104	(263)
Samarium	Sm	62	150.4
Scandium	Sc	21	44.96
Seaborgium	Sg	106	(266)
Selenium	Se	34	78.96
Silicon	Si	14	28.09
Silver	Ag	47	107.9
Sodium	Na	11	22.99
Strontium	Sr	38	87.62
Sulfur	S	16	32.07
Tantalum	Ta	73	180.9
Technetium	Tc	43	(98)
Tellurium	Te	52	127.6
Terbium	Tb	65	158.9
Thallium	Tl	81	204.4
Thorium	Th	90	232.0
Thulium	Tm	69	168.9
Tin	Sn	50	118.7
Titanium	Ti	22	47.88
Tungsten	W	74	183.9
Uranium	U	92	238.0
Vanadium	V	23	50.94
Xenon	Xe	54	131.3
Ytterbium	Yb	70	173.0
Yttrium	Y	39	88.91
Zinc	Zn	30	65.41
Zirconium	Zr	40	91.22
		112**	(285)
		114	(289)
		116	(292)

*All relative atomic masses are given to four significant figures. Values in parentheses represent the mass number of the most stable isotope.

**The names and symbols for elements 112, 114, and 116 have not been chosen.

ARIS™ *Assessment, Review and Instruction System*

ONLINE STUDY & HOMEWORK MATERIALS

I M P O R T A N T : Following are instructions to access online resources to support your McGraw-Hill textbook

The URL associated with your text is:
http://www.mhhe.com/bauer

Option 1: **ARIS LOGIN.** Your instructor may use **ARIS** as a homework and assessment tool. If so, you must register in order to ensure your assignments are recorded into your instructor's gradebook.

Option 2: If your instructor is **NOT** using **ARIS** as a homework and assessment tool, you are welcome to access the material on the site without registering. Simply go to the URL listed above. You are free to access these materials for your own self-study.

ARIS LOGIN. *To Register you need:*

1. **Section Code:** Provided by your instructor.

2. **Registration Code:** Provided in the gray scratch-off area below.

3. **URL:** Go to the URL listed at the top of this card and follow the directions for creating an ARIS account.

Scratch off for registration code
This registration code can be used by one individual and is not transferable.

I M P O R T A N T : The registration code printed above can only be used once to create a unique student account. Students do not need a registration code to access the content of the site. Students choosing "ARIS Login" must login each time they visit the site in order for their grades to be saved to their instructor's gradebook.

Higher Education

T/A Bauer/Birk/Marks: A Conceptual Introduction to Chemistry
ISBN-13: 978-0-07-326270-3
ISBN-10: 0-07-326270-6

A CONCEPTUAL INTRODUCTION TO CHEMISTRY

Richard C. Bauer
Arizona State University

James P. Birk
Arizona State University

Pamela S. Marks
Arizona State University

Mc
Graw
Hill **Higher Education**

Boston   Burr Ridge, IL   Dubuque, IA   Madison, WI   New York   San Francisco   St. Louis
Bangkok   Bogotá   Caracas   Kuala Lumpur   Lisbon   London   Madrid   Mexico City
Milan   Montreal   New Delhi   Santiago   Seoul   Singapore   Sydney   Taipei   Toronto

Higher Education

A CONCEPTUAL INTRODUCTION TO CHEMISTRY

Published by McGraw-Hill, a business unit of The McGraw-Hill Companies, Inc., 1221 Avenue of the Americas, New York, NY 10020. Copyright © 2007 by The McGraw-Hill Companies, Inc. All rights reserved. No part of this publication may be reproduced or distributed in any form or by any means, or stored in a database or retrieval system, without the prior written consent of The McGraw-Hill Companies, Inc., including, but not limited to, in any network or other electronic storage or transmission, or broadcast for distance learning.

Some ancillaries, including electronic and print components, may not be available to customers outside the United States.

This book is printed on acid-free paper.

1 2 3 4 5 6 7 8 9 0 DOW / DOW 0 9 8 7 6 5

ISBN-13   978-0-07-285768-9
ISBN-10   0-07-285768-4

ISBN-13   978-0-07-310723-3 (Annotated Instructor's Edition)
ISBN-10   0-07-310723-9 (Annotated Instructor's Edition)

Publisher: *Thomas D. Timp*
Managing Developmental Editor: *Shirley R. Oberbroeckling*
Senior Developmental Editor: *Donna Nemmers*
Senior Developmental Editor: *Joan M. Weber*
Senior Marketing Manager: *Tamara L. Good-Hodge*
Lead Project Manager: *Joyce M. Berendes*
Senior Production Supervisor: *Sherry L. Kane*
Lead Media Project Manager: *Judi David*
Senior Media Producer: *Jeffry Schmitt*
Designer: *Laurie B. Janssen*
Cover/Interior Designer: *Rokusek Design*
(USE) Cover Image: Frozen grass, *Canada, British Columbia, the Kootenay's, Rod Currie, Stone Collection/GettyImages*
Senior Photo Research Coordinator: *John C. Leland*
Photo Research: *David Tietz*
Supplement Producer: *Tracy L. Konrardy*
Compositor: *Precision Graphics*
Typeface: *10.5/12 Times Roman*
Printer: *R. R. Donnelley Willard, OH*

The credits section for this book begins on page C-1 and is considered an extension of the copyright page.

Library of Congress Cataloging-in-Publication Data

Bauer, Richard C., 1963 Nov. 24–
    A conceptual introduction to chemistry / Richard C. Bauer, James P. Birk, Pamela S. Marks. — 1st ed.
    p. cm.
    Includes index.
    ISBN 978-0-07-285768-9 — ISBN 0-07-285768-4 (acid-free paper)
    1. Chemistry—Textbook. I. Birk, James P. II. Marks, Pamela. III. Title.

QD33.2.B38   2007
540—dc22   2005053865
                                                                                                CIP

www.mhhe.com

*To Trey for your support
and patience; and in loving
memory of my mother who, with
my father, instilled in me the value
of pursuing my academic interests.*
—*Rich Bauer*

*To my wife, Kay Gunter,
who encouraged me through
battles with blank pages and
shared the joys of completed
chapters; and in memory of my
parents, Albert and Christine Birk,
who taught me to love books
enough to see blank pages as a
worthwhile challenge.*
—*Jim Birk*

*To my husband Steve,
for his love and encouragement.*
—*Pam Marks*

About the Authors

Richard Bauer, Pamela Marks, and James P. Birk

Richard Bauer was born and raised in Saginaw, Michigan and completed his B.S. degree in chemistry at Saginaw Valley State University. While pursuing his undergraduate degree he worked at Dow Chemical as a student technologist. He pursued Masters and Ph.D. degrees in Chemistry Education at Purdue University under the direction of Dr. George Bodner. After Purdue, he spent two years at Clemson University as a visiting assistant professor.

Dr. Bauer presently serves as General Chemistry Coordinator at Arizona State University, where he has implemented an inquiry-based laboratory program. Dr. Bauer has taught Introductory and General Chemistry courses for 12 years, and also teaches a Methods of Chemistry Teaching course. He is especially fond of teaching Introductory Chemistry because of the diversity of students enrolled. In addition to general chemistry lab development, Dr. Bauer has interests in student visualization of abstract, molecular-level concepts; TA training; and methods of secondary school chemistry teaching. In addition to his scholarly interests, he plays the piano, sings, and directs choirs.

James P. Birk is Emeritus Professor of Chemistry and Biochemistry and a faculty member in the Center for Research on Education in Science, Math, Engineering, and Technology at Arizona State University. Born in Cold Spring, Minnesota, he received a B.A. degree in chemistry from St. John's University (Minnesota) and a Ph.D. in physical chemistry from Iowa State University. After a post-doctorate at the University of Chicago, he started his academic career at the University of Pennsylvania, where he was appointed to the Rhodes-Thompson Chair of Chemistry. Initially doing research on mechanisms of inorganic reactions, he switched to research on various areas of chemical education after moving to Arizona State University as Coordinator of General Chemistry. Dr. Birk's teaching responsibilities have been in General Chemistry, Introductory Chemistry, Chemistry for Engineers, Inorganic Chemistry, Methods of Teaching Chemistry, and graduate courses on Inorganic Reaction Mechanisms, Chemical Education, and Science Education. He has received several teaching awards, including Awards for Distinction in Undergraduate Teaching, Teaching Innovation Awards, the National Catalyst Award, and the President's Medal for Team Excellence. He has been a feature editor for the Journal of Chemical Education, editing the columns: Filtrates and Residues, The Computer Series, and Teaching with Technology. Recent research has focused on visualization (such as Dynamic Visualization in Chemistry and The Hidden Earth), on inquiry-based instruction, and on misconceptions (Chemistry Concept Inventory).

Pamela Marks is currently a Senior Lecturer at Arizona State University, where her main focus has been teaching Introductory Chemistry, General Chemistry, and Elementary Organic Chemistry for the past ten years. Prior to this, she coordinated the general chemistry laboratory program at the College of St. Benedict and St. John's University in Minnesota. Previous publications include multimedia-based general chemistry educational materials on CD. She received her B.A. in chemistry at St. Olaf College in 1984 and her M.A. in inorganic chemistry at the University of Arizona in 1988. She spends her free time with her husband, Steve, and her three children, Lauren, Kelsey, and Michael.

Brief Contents

Contents

Preface

As instructors of introductory chemistry, our lectures are significantly different from traditional lecture presentations in many ways. Beginning with the first week of classes and continuing through the rest of the semester, we follow a sequence of topics that allows us to explain macroscopic phenomena from a molecular perspective. This approach places emphasis on conceptual understanding over algorithmic problem solving. To help students develop conceptual understanding, we use numerous still images, animations, video clips, and live demonstrations. Roughly a third of each class period is devoted to explaining chemical phenomena from a conceptual perspective. During the remaining time, students work in groups to discuss and answer conceptual and numerical questions.

Our desire to create a conceptually-based text stems from our own classroom experience, as well as from educational research about how students learn. This book is grounded in educational research findings that address topic sequence, context, conceptual emphasis, and concept-embedded numerical problem solving. Throughout the text, we have made an effort to relate the content to students' daily lives and show them how chemistry allows us to understand phenomena—both simple and complex—that we encounter on a regular basis. Students' initial exposure to chemical concepts should be in the realm of their personal experience, to give context to the abstract concepts we want them to understand later. This text presents macroscopic chemical phenomena early, and uses familiar contexts to develop microscopic explanations.

This textbook is designed for the freshman level, introductory chemistry course that does not have a chemistry prerequisite, and is suitable for either a one-semester course or a two-semester sequence. The book targets introductory courses taken by non-physical science majors who may be in allied health, agriculture, or other disciplines that do not require the rigor of a science-major's general chemistry course, or for students fulfilling university liberal arts requirements for science credits. In addition, students who lack a strong high school science background often take the course as a preparation for the regular general chemistry sequence.

FEATURES OF THIS TEXT

Learning theory indicates that we should start with the concrete, macroscopic world of experience as the basis for developing student understanding of abstract, microscopic concepts. This textbook follows a **topic sequence** typically found in traditional general chemistry texts. That is, macroscopic ideas about chemical behavior are discussed *before* descriptions of abstract, microscopic concepts associated with electronic structure. The macroscopic ideas that begin chapters or sections are grounded in real-life experiences. Where appropriate, the macroscopic to molecular-level progression of ideas is carried over to topic sequence within individual chapters or sections, in addition to the general sequence of chapters.

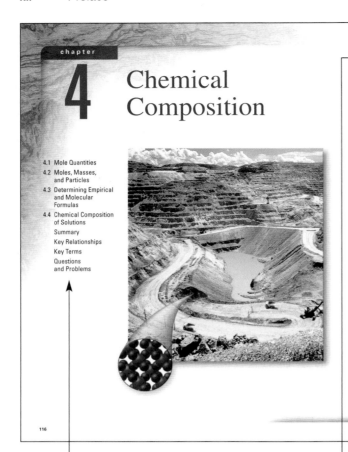

Copper plays an important role in our daily lives. To see how, let's follow Julio, a college student, through part of his day. Bzzz! The radio alarm goes off and awakens Julio. The circuits in the radio and the wires that connect it to the wall outlet contain copper. Julio leaps out of his bed, which has a brass frame. (Brass is a homogeneous mixture of copper and zinc.) He opens the bathroom door and turns on the light. The doorknob and hinges are made of brass. The light switch contains copper contacts, and the base of the lightbulb is made of copper. He turns on the water to wash his face. The faucets on the sink are brass-coated with nickel and chromium, and copper pipes carry water through the house to his bathroom. Examine your bedroom and bathroom. Do you see any other objects made of copper or brass?

Julio goes to the kitchen and pours orange juice into a glass. His juice is cold because copper tubes circulate coolant in his refrigerator. For breakfast, Julio cooks oatmeal in a copper-bottomed pot. After his meal, he grabs his car keys, which are made of nickel-plated brass, and drives to campus in a car that contains wiring, cables, windings, tubing, and other components made of copper—a total of about 50 lb (23 kg) of it. Julio drops some coins into the parking meter before dashing off to his chemistry class. The coins contain copper. After class, Julio drops by the computer lab to check his e-mail. The computer contains many copper connections. It is connected to the Internet with copper cables.

Julio's life seems to include copper at every turn. In fact, during his lifetime Julio will make use of over 1500 lb (680 kg) of copper metal. Will Julio ever stop to ask himself where all this copper comes from? Think about this yourself, but consider some other common metal such as iron. Where do you encounter iron in your daily life? How much do you depend on it? Where does it come from?

Most of the copper found in nature occurs as part of a mineral. A *mineral* is a naturally occurring element or compound that has a distinct chemical composition and solid form. The most common minerals of copper contain sulfide, oxide, or silicate along with other metals (Figure 4.1). Most copper is taken from open-pit mines in Chile or the southwestern United States (primarily Arizona, Nevada, Utah, and New Mexico), but a few northern states, including Montana and Michigan, mine some too. Copper comes out of the mines as ore, which has relatively small amounts of copper minerals embedded in rock. (Although the minerals contain large amounts of copper, the ore often contains only small amounts of minerals, so it often contains less than 1% copper by mass. In other words, 100 kg of ore may yield as little as 1 kg of copper.) At the mine, the ore is blasted loose from the ground and then lifted with large electric shovels into trucks that can hold as much as 300 tons (272 metric tons) of ore. The trucks haul the ore from the pits and dump it into a series of crushers and mills that break the large pieces into smaller ones.

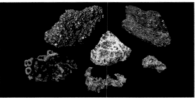

Figure 4.1
Copper is found mostly as various sulfide, oxide, and silicate minerals.

Each chapter begins with a chapter opening outline and an **opening vignette** that personalizes the content by telling a story about chemical phenomena encountered by students. These applications help students see how chemistry relates to their daily lives.

The chapter then offers some guiding questions typical of inquiry instruction. These **Questions for Consideration** serve as a guide in topic development through the chapter. **Margin notes** contain further explanations and chemical applications, combined with visuals, to help students conceptualize lessons.

We believe that an introductory chemistry textbook should maintain a focus on *chemistry,* rather than on math. Students' interest must be captured early in the semester if they're going to persevere in the class. Early in this text we introduce chemical reactions from macroscopic perspectives. A general fundamental knowledge of chemical behavior on a macroscopic level facilitates further development of molecular-level ideas, such as atomic structure.

Questions for Consideration

4.1 How can we count the number of atoms in a sample of a material? The number of molecules? The number of formula units?
4.2 How can we know the number of molecules in a sample of a material?
4.3 How can we use the masses of elements in a compound to determine its chemical formula?
4.4 How can we express the composition of a solution?

Math Tools Used in This Chapter

Scientific Notation (Math Toolbox 1.1)

Units and Conversions (Math Toolbox 1.3)

Although we cannot see infrared radiation, we can sometimes feel its effects as heat. All objects emit infrared radiation. The wavelength at which an object radiates most intensely depends on its temperature. The cooler the object, the longer the wavelength. We can detect infrared radiation using special cameras and film that detect differences in temperature. In the picture, the yellow areas are the warmest and the blue and black areas are the coldest.

Like charges repel. Opposite charges attract.

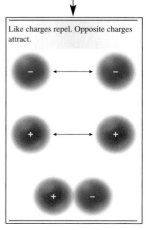

We believe that the best approach to incorporating math involves development of associated math on an as-needed basis with emphasis on concepts that problems are trying to illustrate. This text integrates need-to-know mathematical ideas that are important to chemists into conceptual discussions. Thorough math reviews are provided in **math toolboxes** that are referenced within appropriate sections of the text and placed at the end of the relevant chapter.

These toolboxes are referenced at the beginning of the chapters, where appropriate, in the **Math Tools Used in This Chapter** section. As problem solving is developed within the text, emphasis is placed on the underlying concepts, letting the numerical solutions emerge from conceptual understanding. Numerical-type problems often ask students to estimate answers and to consider the physical meaning of calculated quantities.

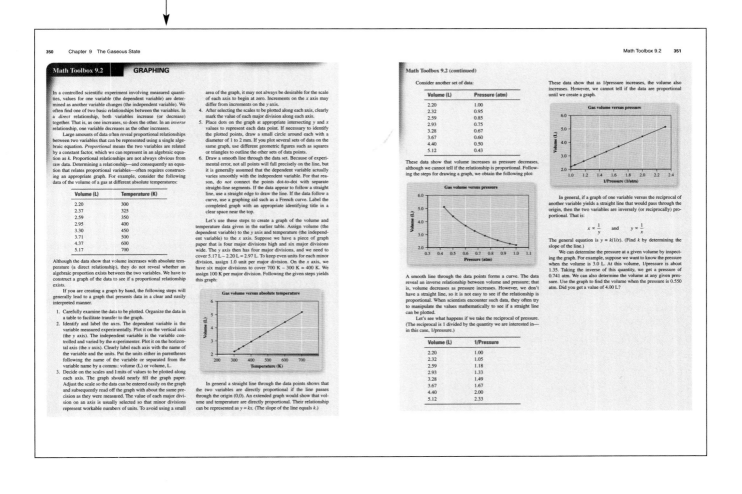

EXAMPLE 6.1   Mole-Mole Conversions

If 1.14 mol of CO_2 was formed by the combustion of C_3H_8, how many moles of H_2O were also formed?

$$C_3H_8(g) + 5O_2(g) \longrightarrow 3CO_2(g) + 4H_2O(g)$$

Solution:

We know the number of moles of C_3H_8 and we want to know the moles of CO_2:

The relationship we use to convert from moles of C_3H_8 to moles of CO_2 is the mole ratio we get from the balanced equation:

First we must ensure that the equation is balanced. Yes it is, so the coefficients in the equation give mole relationships between CO_2 and H_2O which can be written in two ways:

$$\frac{3 \text{ mol } CO_2}{4 \text{ mol } H_2O} \quad \text{and} \quad \frac{4 \text{ mol } H_2O}{3 \text{ mol } CO_2}$$

We multiply the known moles of CO_2 by the ratio that will cancel the old units (mol CO_2) and introduce the new units (mol H_2O):

$$\text{mol } H_2O = 1.14 \text{ mol } CO_2 \times \frac{4 \text{ mol } H_2O}{3 \text{ mol } CO_2} = 1.52 \text{ mol } H_2O$$

Notice that the units cancel properly.

Practice Problem 6.1

Pure methanol is used as a fuel for all race cars in the Indy Racing League and in the Championship Auto Racing Teams. It is used because methanol fires are easier to put out with water than the fires of most other fuels. The balanced equation for the combustion of methanol is

$$2CH_3OH(l) + 3O_2(g) \longrightarrow 2CO_2(g) + 4H_2O(g)$$

Given that 1.00 gal of methanol contains 94.5 mol, how many moles of oxygen gas will react with 1.00 gal of methanol?

Further Practice: 6.11 and 6.12

EXAMPLE 9.2   Graphical Relationship of Volume and Pressure for a Gas

The piston shown in the figure represents starting conditions for a helium gas sample. Suppose the volume and pressure correspond to point A on the graph. If the pressure increases by a factor of 2, what point along the curve corresponds to the new volume and pressure conditions?

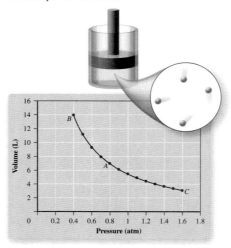

Solution:

Inspection of the graph indicates that when the pressure doubles, the volume is halved. Point A, the starting conditions of the gas in the piston, is about 0.8 atm in pressure and approximately 7 L in volume. If the pressure is doubled to 1.6 atm, the volume falls to about 3.5 L. This corresponds to point C.

Practice Problem 9.2

The atomic-level image in the figure represents the atoms present in a microscopic volume under the starting conditions defined by point A. Draw a picture to represent the atoms in the same microscopic volume at point C.

Further Practice: 9.21 and 9.22

The problem-solving approach used in this text is supported by worked **example boxes** that contain the following steps: question(s), solution, practice problems, and further practice. Roadmaps are also included to help guide students to problem solutions.

Problem solving in chemistry is much more than algorithmic number crunching. It involves applying principles to solve conceptual as well as numerical problems. Conceptual problems are those that require students to apply their understanding of concepts instead of applying an algorithm. This text emphasizes the underlying concepts when discussing numerical problems within in-chapter worked examples. Many end-of-chapter problems also emphasize conceptual problem solving.

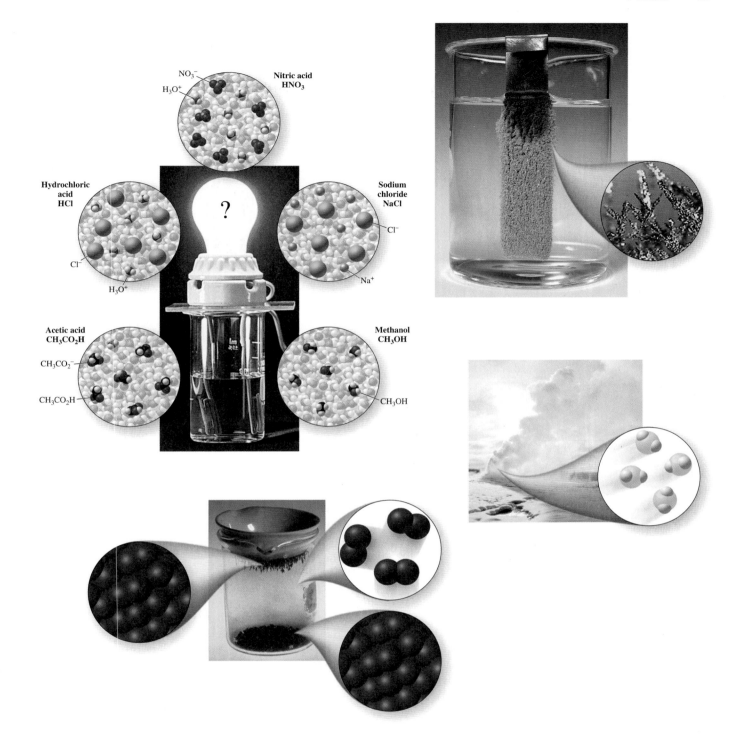

A conceptual understanding of chemistry requires students to visualize molecular-level representations of macroscopic phenomena, as well as to connect macroscopic and molecular-level understandings to symbolic representations. To help students connect verbal descriptions to molecular-level representations, this book has an extensive **art program.** You'll notice many examples of zoomed art where pictures or other macroscopic images have close-ups that show the particular phenomena at a molecular level.

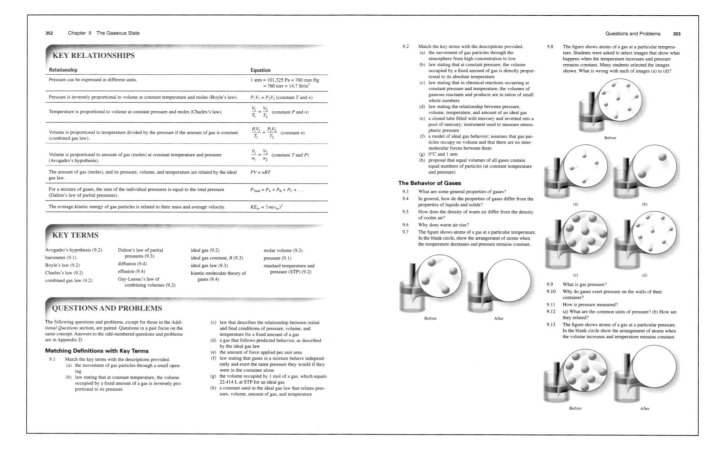

There are several other features of this textbook that support student learning. End-of-chapter materials include a **summary, math toolboxes** (when appropriate), **key terms list,** and **key relationships.** Each chapter has extensive **end-of-chapter problems** that range in difficulty and conceptual/quantitative emphasis. The questions are sorted by section and are paired, with odd-numbered answers appearing in the appendix. There are also **vocabulary identification questions** at the beginning of the end-of-chapter problems, as well as many questions involving interpretation of molecular-level images.

Students who enroll in an introductory chemistry course often take an associated lab. Most of the experiments these students conduct involve working with solutions. To enhance this lab experience, a brief **introduction to solution behavior** appears early in the textbook (Chapter 4). This early introduction will allow students to better understand what they experience in the lab, as well as understand the multitude of solutions we encounter on a daily basis.

SUPPLEMENTS FOR THE INSTRUCTOR

Annotated Instructor's Edition The Annotated Instructor's Edition is a guide to all pertinent information for integrating media and extra content into your lecture. Application icons are located throughout the textbook and within the end-of-chapter problems to denote:

 Teaching Tips These are points instructors may wish to emphasize, or applications that may be used to illustrate a topic.

 Teaching Support Activities These activities are simple and easy to incorporate into classroom discussion.

 Using Technology Relevant multimedia applications, along with their sources, are noted for incorporating into your classroom presentations.

 Demonstrations High quality demonstrations are noted for instructor use.

 Misconceptions Common misconceptions held by students are noted, so instructors can address these during lecture.

 Animations These animations appear within the ARIS website to support this text at www.mhhe.com/bauer, and also appear on the McGraw-Hill Chemistry Animations DVD for instructors.

 DCM (Digital Content Manager) Each figure, photo, or table that is available on the Digital Content Manager CD-ROM is marked by a DCM icon, along with the folder information for easy integration into your lecture presentation.

 Transparency/DCM This icon alerts you to figures that are available as a transparency, as well as in digital form.

Digital Content Manager Electronic art at your fingertips! This cross-platform CD-ROM provides you with artwork from the text in multiple formats. You can easily create customized classroom presentations, visually based tests and quizzes, dynamic content for a course website, or attractive printed support materials. Available on this CD are the following resources:

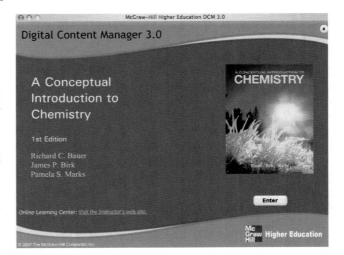

- **PowerPoint Lecture Outlines** Ready-made presentations combining art and lecture notes cover all of the chapters in the text. These lectures can be used as is or customized by you to meet your specific needs.
- **Art Library and Photo Library** Full-color digital files of all the illustrations and many of the photos in the text can be readily incorporated into lecture presentations, exams, or custom-made classroom materials.
- **TextEdit Art Library** A number of illustrations from each chapter have been made into editable art, so that you may custom tailor them for your classroom presentation needs.

- **Worked Examples Library and Tables Library** Access the worked examples and tables from the text in electronic format for inclusion in your classroom resources.

Instructor's Testing and Resource CD-ROM Instructors will save valuable time by utilizing the Instructor's Testing and Resource CD-ROM. This cross-platform CD-ROM includes the Instructor's Solutions Manual, which is comprised of all answers for the textbook's end-of-chapter problems, and a Test Bank of questions for each chapter, which can be used for exams or homework assignments. This computerized Test Bank utilizes testing software to allow you to quickly create customized exams by sorting questions by format, editing existing questions or adding new ones, and scrambling questions for multiple versions of the same test. Test questions are also available in Word format on this CD-ROM.

The following items may accompany this text. Please consult your McGraw-Hill representative for policies, prices, and availability, as some restrictions may apply.

The **Instructor's Solutions Manual** contains complete, worked-out solutions for all end-of-chapter problems in the text. This Manual can be accessed within the password-protected Instructor Edition of ARIS, the textbook website that accompanies this text, and is also available on the Instructor's Testing and Resource CD-ROM.

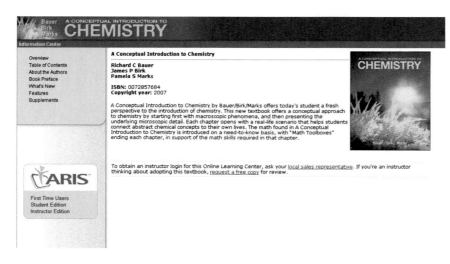

ARIS McGraw-Hill's Assessment, Review, and Instruction System for *A Conceptual Introduction to Chemistry* is a complete electronic homework and course management system, designed for greater ease of use than any other system available. Instructors can create and share course materials and assignments with colleagues with a few clicks of the mouse. Instructors can also edit questions, import their own content, and create announcements and due dates for assignments. ARIS has automatic grading and reporting of easy-to-assign homework, quizzing, and testing. Once a student is registered in the course, all student activity within McGraw-Hill's ARIS is automatically recorded and available to the instructor through a fully integrated gradebook that can be downloaded to Excel. This book-specific website is found at www.mhhe.com/bauer.

eInstruction McGraw-Hill has partnered with eInstruction to provide the revolutionary *Classroom Performance System* (CPS), to bring interactivity into the classroom. CPS is a wireless response system that gives the instructor and students immediate feedback from the entire class. The wireless response pads are essentially remotes that are easy to use and engage students. CPS allows you to motivate student preparation, interactivity, and active learning so you can receive immediate feedback and know what students understand. A text-specific set of questions, formatted for both CPS and PowerPoint, is available via download from the Instructor area of the ARIS textbook website at www.mhhe.com/bauer.

Over 300 animations are available within the **Chemistry Animations DVD.** Instructors can easily view the animations and import them into PowerPoint to create multimedia presentations.

A set of 150 **printed transparencies** feature key color images and tables from the text to assist instructors with classroom projection needs.

SUPPLEMENTS FOR THE STUDENT

The **Student Solutions Manual** contains detailed solutions and explanations for all odd-numbered problems in the text.

ARIS McGraw-Hill's Assessment, Review, and Instruction System for *A Conceptual Introduction to Chemistry* is available to students and instructors using this text. This user-friendly program allows students to complete their homework online, as assigned by their instructors. This text-specific website offers quizzing and animations for further chapter study, and can be found at www.mhhe.com/bauer.

ChemSkill Builder *ChemSkill Builder,* McGraw-Hill's powerful electronic homework system, gives you the tutorial practice you need to master concepts covered in your chemistry course. ChemSkill Builder contains more than 1500 algorithmically generated questions as well as interactive exercises, quizzes, animations, and study tools matched to each chapter of the text. A record of your work is maintained in an online gradebook so that your homework scores can be easily viewed.

ACKNOWLEDGMENTS

This project was a team effort, and we would like to thank all those who were personally involved with us as we developed this textbook. We extend a special thank you to Faith Brynie for her editing expertise, in smoothing out the differences in our writing styles and offering thoughtful suggestions for improvement. We also wish to acknowledge our ancillary authors who worked with us to make a cohesive learning package to support the textbook. Ellen Yezierski of Grand Valley State University researched and carefully prepared the annotations that appear in the Annotated Instructor's Edition of this text; Kirk Kawagoe of Fresno City College diligently worked and reworked solutions as he prepared the end-of-book Answers to Odd-Numbered Problems, and also the Instructor's Solutions Manual and Student Solutions Manual; Estelle Lebeau of Central Michigan University authored questions for instructor use with the eInstruction program; Carol Martinez of Albuquerque TVI Community College wrote the many and varied test bank questions that accompany this text; and Clarissa Sorenson of Albuquerque TVI Community College crafted the instructors' PowerPoint Lecture Outlines.

An important part of every text is accuracy. Although the authors accept the responsibility for any factual errors that fall within these pages, they wish to acknowledge the following individuals for their assistance in providing accuracy checking services for the text and ancillary products:

Jon Booze, *Accumedia Publishing Services*

Csillia Duneczky, *Johnson County Community College*

William Lumbley, *Indiana University*

Permider Sandhu, *Bellevue Community College*

Marcia Gillette, *Indiana University–Kokomo*

Jodi Kawagoe

Deborah McCool, *Penn State–Altoona*

To our friends at McGraw-Hill we extend the warmest appreciation. Vice President and Director of Marketing, Kent Peterson and Senior Developmental Editor,

Joan Weber got us started on this project. We especially thank Joan for her expert guidance in the initial review process. Senior Developmental Editor, Donna Nemmers took over as developmental editor when Joan moved on to pursue other projects, and has been invaluable in coordinating and organizing many of the activities associated with this project. Lead Project Manager, Joyce Berendes guided us through production and kept us on schedule during those times when we felt rather overwhelmed. We especially appreciated Joyce's remarkable eye for details and her great sense of humor. A special thank you to our coach, Publisher, Thomas Timp, who provided the pep talks we needed when the end seemed far, far from sight. We truly appreciate Thomas's support for us and our ideas as he guided this project along to completion. We also wish to thank Tami Hodge, the marketing manager of our project, who provided insights on faculty interest for preparatory chemistry. Finally, we wish to acknowledge our families. They helped us make difficult decisions regarding design, layout, art, and grammar. More importantly, they provided the support we needed to complete this long project.

We are especially grateful for the support of the Board of Advisors. This select group of chemical educators helped to shape this text with their careful review and commentary throughout the development of this text.

Board of Advisors

John R. Allen, *Southeastern Louisiana University*
Margaret R. Asirvatham, *University of Colorado, Boulder*
Claudia M.S. Hein, *Diablo Valley College*
Kirk T. Kawagoe, *Fresno City College*
Dennis N. Kevill, *Northern Illinois University*
Estelle L. Lebeau, *Central Michigan University*
Carol Anne Martinez, *Albuquerque TVI Community College*
Elsa C. Santos, *Colorado State University*
Thomas E. Sorensen, *University of Wisconsin–Milwaukee*
Marie Villarba, *Glendale Community College*
Sidney H. Young, *University of South Alabama*

We wish to thank the following people for their careful review and constructive comments during the manuscript development process, and for attending focus groups aimed at improving the final printed product. Your assistance has been invaluable to us.

Reviewers

Mamta Agarwal
  Chaffey College
John R. Allen
  Southeastern Louisiana University
Jeffrey R. Appling
  Clemson University
Margaret R. Asirvatham
  University of Colorado, Boulder
Joseph Bariyanga
  University of Wisconsin–Milwaukee
Joseph L. Barnes
  Pasadena City College
Bal Barot
  Lake Michigan College
Shay Bean
  Southwestern Illinois College
Christine V. Bilicki
  Pasadena City College

Sean R. Birke
  Jefferson College
Maria Bohorquez
  Drake University
Simon Bott
  University of Houston
David A. Boyajian,
  Palomar College
Bob Boykin
  Bossier Parish Community College
Bryan F. Breyfogle
  Southwest Missouri State University
Aaron D. Brown
  Los Angeles City College
Tim Burch
  Milwaukee Area Technical College
Fern M. Caka
  Utah Valley State College

Fernando A. Camou
Glendale Community College
Jeffrey S. Carver
Illinois Valley Community College
Douglas S. Cody
Nassau Community College
Ken Comer
Edison Community College
Sravanthi C. Cornell
*Albuquerque Technical & Vocational
Institute*
Mapi M. Cuevas
Santa Fe Community College
Robb Culp
Fresno City College
Judy Dirbas
Grossmont College
Csilla Duneczky
Johnson County Community College
Bill Durham
University of Arkansas
P. Mark Ebner
Gloucester County College
Bill Edinger
Santa Ana College
Evelyn S. Erenrich
Rutgers University
Karl Lee Essenburg
Grand Valley State University
Mary Evenson
St. Cloud State University
Nancy Faulk
Blinn College/Bryan Campus
C.W. Finley
Penn State University-New Kensington
Sheree J. Finley
Alabama State University
Vicki Flaris
Bronx Community College
John W. Francis
Columbus State Community College
David L. Frank
California State University–Fresno
Nancy Gardner
California State University–Long Beach
Steve Gentemann
Southwestern Illinois College
Marta E. Goicoechea-Pappas
Miami Dade College
Barry H. Gump
California State University-Fresno
Steve Gunther
*Albuquerque Technical Vocational
Institute*
Pauline Hamilton
*Community College of Baltimore
County–Essex Campus*
Kathleen A. Harter
Community College of Philadelphia
Claudia M.S. Hein
Diablo Valley College

Chu-Ngi Ho
East Tennessee State University
Paul A. Horton
Indian River Community College
Kamal Z. Ismail
*Bronx Community College–City
University of New York*
T. G. Jackson
University of South Alabama
Charles Jaffe
West Virginia University
Anton W. Jensen
Central Michigan University
Todd M. Johnson
Weber State University
Kirk T. Kawagoe
Fresno City College
Dennis N. Kevill
Northern Illinois University
Thomas D. Kim
Youngstown State University
Edith Preciosa Klingberg
University of Toledo
Silvia Kölchens
Pima Community College
Anne E. Kondo
Indiana University of Pennsylvania
Matt Koutroulis
Rio Hondo College
Martha J. Kurtz
Central Washington University
Richard Lavallee
Santa Monica College
Estelle L. Lebeau
Central Michigan University
Rosette Lewis
St. Augustine
Gerhard Lind
Metropolitan State College of Denver
Richard B. Lomneth
University of Nebraska–Omaha
Boon H. Loo
Towson University
Rudy L. Luck
Michigan Technological University
William D. Lumbley
Indiana University–Bloomington
Charmaine B. Mamantov
University of Tennessee–Knoxville
Larry A. Manno
Triton College
Carol Anne Martinez
TVI Community College
Diana Mason
University of North Texas–Denton
Johanna Mazlo
*University of North Carolina–
Greensboro*
Frank R. Milio
Towson University

Randy Miller
California State University, Chico
John A. Milligan
Los Angeles Valley College
Kathy Mitchell
St. Petersburg College
Villa Mitchell
Lipscomb University
Alice J. Monroe
St. Petersburg College
Michelle B. More
Weber State University
Thomas L. Neils
Grand Rapids Community College
William J. Nguyen
Santa Ana College
Jeffrey Paradis
*California State
University–Sacramento*
Amy J. Phelps
Middle Tennessee State University
H. Dale Pigott
Victoria College
Robert Potter
University of South Florida
Elizabeth Pulliam
Tallahassee Community College
Douglas E. Raynie
South Dakota State University
Melinda S. Ripper
Butler County Community College
William R. Robinson
Purdue University
Lynette Rushton
*South Puget Sound Community
College*
Victor Ryzhov
Northern Illinois University
Perminder Sandhu
Bellevue Community College
Elsa C. Santos
Colorado State University
Jerry L. Sarquis
Miami University

Douglas J. Sawyer
Scottsdale Community College
Jamie L. Schneider
Winona State University
James Selzler
Allan Hancock College
Mary C. Setzer
University of Alabama–Huntsville
William N. Setzer
University of Alabama–Huntsville
Clarissa Sorensen
Albuquerque TVI Community College
Bobby Stanton
University of Georgia
Susan E. Swope
Penn State University
Vicente Talanquer
University of Arizona
David Tanis
Grand Valley State University
Jacquelyn A. Thomas
Southwestern College
Anthony P. Toste
Southwest Missouri State University
Cyriacus Chris Uzomba
Austin Community College
Ramaiyer Venkatraman
Jackson State University
Marie Villarba
Glendale Community College
Jerry Walsh
*University of North Carolina–
Greensboro*
John Weide
Mesa Community College
Servet M. Yatin
Quincy College
Ellen Yezierski
Grand Valley State University
Sidney H. Young
University of South Alabama

A CONCEPTUAL INTRODUCTION TO CHEMISTRY

1

Matter and Energy

Anna and Bill are enrolled in an introductory chemistry course. For their first assignment, the professor has asked them to walk around campus, locate objects that have something to do with chemistry, and classify the things they find according to characteristics of structure and form.

Anna and Bill aren't sure what they should do, but they begin their trek at the bookstore. They spot a fountain, a large metallic sculpture, a building construction site, and festive balloons decorating the front of the store. They notice water splashing in the fountain and a heap of coins that have collected at the bottom. The metallic sculpture has a unique color and texture. At the building construction site they notice murals painted on the wooden safety barricade. Through a hole in the fence, they see a construction worker doing some welding.

Bill and Anna make a list of the things that attracted their attention and start trying to classify them. Inspecting the fountain, they notice that it appears to be composed of pebbles embedded in cement. As water circulates in the fountain, it travels in waves on the water's surface. The coins in the fountain, mostly pennies, vary in their shininess. Some look new, with their copper color gleaming in the bright sunshine. Others look dingy, brown, and old. The metal sculpture has a unique, modern design, but it's showing signs of age. A layer of rust covers its entire surface. Anna and Bill decide to classify the sculpture as a metal, like the coins in the fountain. They also conclude that the water, pebbles, and concrete in the fountain are not metals.

As they approach the construction site, Anna and Bill examine the painted mural. Through the peephole in the mural, they see a mess of gravel, cinder blocks, metallic tubes for ductwork, steel beams, and copper pipe. They add more nonmetals and metals to their list. A welder is joining two pieces of metal. Sparks are flying everywhere. Anna and Bill wonder what is in the sparks. Since the sparks are so small and vanish so rapidly, they don't know how to classify them.

As they continue their walk, they pass the intramural fields and the gym where they see students using tennis rackets, baseball bats, bicycles, and weight belts. They wonder how they will classify these items. For lunch, Bill and Anna buy pizza. They sip soft drinks from aluminum cans. They settle on a bench to enjoy their lunch in the sunshine and watch students playing volleyball in a sandpit. "Mustn't forget the sunscreen," Bill says, pulling a tube from his backpack. After lunch, they hurry off to an afternoon class. On the way, they notice a variety of vehicles on campus. Some are gasoline-powered cars and buses, but others have signs on them saying they operate on alternative fuels. Trucks lumber by, exhaust fumes spewing from their tailpipes. Bill and Anna feel the hoods of parked cars. Some are still warm from their engine's heat.

How are Bill's and Anna's observations related to chemistry? What characteristics have they identified that they can use for classification purposes? They have started their classification with metals and nonmetals. What other categories should they devise?

Now it's your turn. Make a list of things relevant to chemistry in the location where you are reading this. How will you classify the things on your list? What characteristics will you use to organize the items into categories? Most important, why bother to classify things at all?

In this chapter we will explore some answers to these questions. As you learn what chemistry is, you'll begin to develop explanations for how substances look, change, and behave.

Questions for Consideration

1.1 What characteristics distinguish different types of matter?
1.2 What are some properties of matter?
1.3 What is energy and how does it differ from matter?
1.4 What approaches do scientists use to answer these and other questions?

Math Tools Used in This Chapter

Scientific Notation (Math Toolbox 1.1)

Significant Figures (Math Toolbox 1.2)

Units and Conversions (Math Toolbox 1.3)

1.1   MATTER AND ITS CLASSIFICATION

All the *things* that Anna and Bill observed on campus are examples of matter. The fountain, the metal sculpture, the construction site, the balloons outside the bookstore, the exhaust fumes from buses, the pizza they had for lunch, even Bill and Anna themselves—all are matter. **Matter** is anything that occupies space and has mass. **Mass** is a measure of the quantity of matter. The interaction of mass with gravity creates weight, which can be measured on a scale or balance.

Some of Bill's and Anna's observations, however, were not of matter. Sunlight, the light from welding, and the heat of automobile engines are not matter. They do not occupy space, and they have no mass. They are forms of energy. *Energy* is the capacity to move an object or to transfer heat. We'll discuss energy in Section 1.3, but for now, let's focus on matter.

All of Anna's and Bill's observations are relevant to chemistry, because chemistry is the study of matter and energy. Since the entire physical world is matter and energy, chemistry would be an overwhelming subject of study if we did not classify phenomena in manageable ways. Anna and Bill used characteristics like shininess and hardness when they decided some materials were metals and others were not. Let's explore some other characteristics that can be used to classify matter.

Composition of Matter

One way to classify matter is by its chemical composition. Some types of matter always have the same chemical composition, no matter what their origin. Such matter is called a **pure substance** or more briefly, a *substance*. A pure substance has the same composition throughout and from sample to sample. It cannot be separated into components by physical means. For example, muddy water is not a pure substance because dirt can be removed from the water by filtration.

Some pure substances can be observed. For example, the aluminum in Anna's soda can is pure. It is not combined with any other substances, although it is coated with plastic and paint. Consider also the sandpit where Bill and Anna watched the volleyball game. The sand is not a pure substance, but if we removed all the dirt, minerals, and other contaminants, it would be the pure substance, silica, which is one kind of sand (Figure 1.1). Grains of silica differ in size, but they all have the same chemical composition, which can be determined in the laboratory.

In contrast to pure substances, other materials are mixtures. A **mixture** consists of two or more pure substances and may vary in composition. The fountain, for example, is made from a mixture of gravel, concrete, and pebbles. Even the water in the fountain is not a pure substance since small amounts of gases and minerals are dissolved in it. Like sand, however, it could be made pure if all the other substances were removed.

Are there any things where you are now that might be pure substances? Actually, pure substances are rare in our world. Most things are mixtures of some kind. Pure substances are found most often in laboratories where they are used to determine the properties and behavior of matter under controlled conditions.

Elements   All matter consists of pure substances or mixtures of substances. Pure substances, in turn, are of two types: elements and compounds. An **element** is a substance that cannot be broken down into simpler substances *even by a chemical*

Figure 1.1
Sand is composed of a mineral, silica. It contains the elements silicon and oxygen in specific proportions.

reaction. For example, suppose we first purified the water in a fountain to remove contaminants. Then we used a chemical process called *electrolysis* to separate it into its component elements. Water can be broken down by chemical means into hydrogen and oxygen, as shown in Figure 1.2, so water is not an element. The hydrogen and oxygen, however, are elements. We cannot break them down into any simpler substances using heat, light, electricity, or any chemical process. We can convert them into more complex substances, but not into simpler ones.

Elements are the building blocks of all matter. The many examples of matter that we use, see, and read about are all built up of only about 100 different elements in different combinations. The elements that are not isolated from natural sources on Earth have been synthesized by scientists. Some are so unstable that they have only a fleeting existence, including those that have not yet been formally named. Of the 109 elements that have been given names, 83 can be found in natural substances and in sufficient quantity to isolate. To classify elements, chemists use a *periodic table*, like that shown in Figure 1.3. The elements in each column, called *groups* or *families* of elements in the periodic table, share <u>similar characteristics, or *properties*</u>.

Elements are generally classified into two main categories: metals and non-metals. Generally, a **metal** can be distinguished from a **nonmetal** by its luster (shini-ness) and ability to conduct electricity (electrical conductivity). Copper, aluminum, iron, and other metals are good conductors of electricity. Nonmetal elements, such

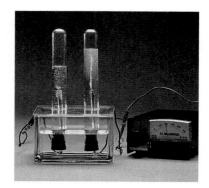

Figure 1.2
When electric current is passed through water, the water decomposes into the elements hydrogen and oxygen.

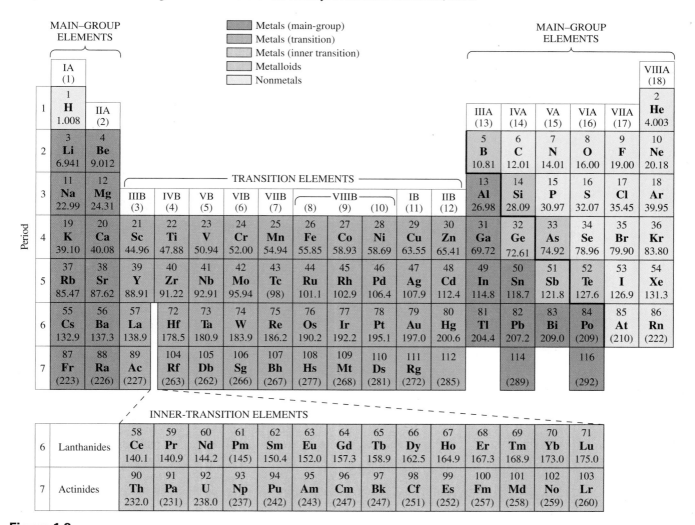

Figure 1.3
The periodic table organizes the known elements according to their properties. The letters are symbols for the names of the elements.

Figure 1.4

Some elements. Which of these are metals?

as carbon (in the form of diamond), chlorine, and sulfur, normally are not. Note the difference in appearance of the elements shown in Figure 1.4. Not all elements fit neatly into such categories. In Chapter 2 we'll discuss elements that have properties somewhere between metals and nonmetals.

EXAMPLE 1.1   Metals and Nonmetals

Which of the elements pictured are metals? Why do you think so?

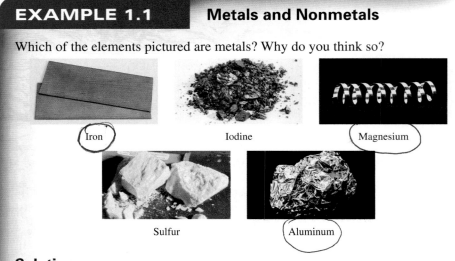

Solution:

Notice that three—iron, aluminum, and magnesium—have a luster; that is, they shine. They are metals. The other two have a dull surface. They are not metals. If you could handle and test the substances, you could use other properties such as electrical conductivity to distinguish between metals and nonmetals.

Practice Problem 1.1

Identify the nonmetals in Figure 1.4. Explain the characteristics you considered in making your decision.

Further Practice: 1.9 and 1.10

TABLE 1.1	Symbols of Selected Elements				
English Name	Original Name	Symbol	English Name	Original Name	Symbol
Copper	Cuprum	Cu	Potassium	Kalium	K
Gold	Aurum	Au	Silver	Argentum	Ag
Iron	Ferrum	Fe	Sodium	Natrium	Na
Lead	Plumbum	Pb	Tin	Stannum	Sn
Mercury	Hydragyrum	Hg	Tungsten	Wolfram	W

To avoid having to write out the name of an element every time we refer to it, we use a system of symbols. An **element symbol** is a shorthand version of an element's longer name. Often, the symbol is one or two letters of the element's name (C for carbon, He for helium, Li for lithium). The first letter is uppercase, and the second letter, if present, is lowercase. When the names of two elements start with the same two first letters (magnesium and manganese, for example), the symbol uses the first letter and a later letter to distinguish them (Mg for magnesium, Mn for manganese).

For a few elements, the symbols are based on their Latin names or on names from other languages. These are listed in Table 1.1. Some recently synthesized elements have been named for famous scientists. Others have not been given permanent names. You'll find a list of the modern names and symbols on the inside front cover of this book.

EXAMPLE 1.2   Element Symbols

Potassium is a soft, silver-colored metal that reacts vigorously with water. Write the symbol for the element potassium.

Solution:

The symbol for potassium is K. In the periodic table, potassium is element 19 in group (column) IA (1) of the periodic table.

Practice Problem 1.2

(a) Lead is a soft, dull, silver-colored metal. Write the symbol for the element lead.
(b) The symbol for a common element used to make jewelry is Ag. What is the name of this element?

Further Practice: 1.17 and 1.18

Compounds   A **compound,** sometimes called a *chemical compound,* is a substance composed of two or more elements combined in definite proportions. A compound has properties different from those of its component elements. For example, iron pyrite can be broken down into its component elements, iron and sulfur, but its characteristics are different from both (Figure 1.5). Anna and Bill saw many compounds that can be chemically separated into their component elements. Sand is a compound of silicon and oxygen. Water, as discussed earlier, is composed of hydrogen and oxygen. The cheese on their pizza contains many complex compounds, but each of the compounds contains carbon, hydrogen, oxygen, nitrogen, and a few other elements.

Figure 1.5
Iron pyrite is composed of the elements iron and sulfur.

Chemists represent compounds with formulas based on the symbols for the elements that are combined in the compound. (Chemical formulas are not the same as the mathematical formulas that may be familiar to you, such as $A = \pi r^2$ for the area of a circle.) A **chemical formula** describes the composition of a compound, using the symbols for the elements that make up the compound. Subscript numbers show the relative proportions of the elements in the compound. For example, water is known to consist of one unit of oxygen and two units of hydrogen. This compound is represented by the formula H_2O. Sodium chloride, the chemical compound commonly called table salt, contains equal portions of the elements sodium and chlorine. Its formula is therefore NaCl. We will discuss formulas in detail in Chapter 3.

Graphite leaves a mark similar to that made by dragging a rod of lead along a surface, so it was called lead. A hardness number indicates the relative amounts of graphite and clay in a pencil lead. A number 2 pencil is fairly soft, while a number 6 pencil is quite hard. Which has more graphite?

Mixtures   Some forms of matter, such as pencil lead, do not have the same composition in every sample. (Pencil lead isn't the element lead. It is a *mixture* of graphite and clay.) A mixture consists of two or more elements or compounds. It is possible to separate mixtures into their component pure substances. The separation can be done physically, using procedures such as grinding, dissolving, or filtering. Chemical processes are not needed to separate mixtures.

We can illustrate the difference between pure substances and mixtures by looking at salt water. Water that has been purified is a pure substance that is composed of hydrogen and oxygen, always in the same proportions. Salt water, on the other hand, is water mixed with salt and many other substances in varying proportions. For example, the Great Salt Lake in Utah is approximately 10% salt, while the Dead Sea is about 30% salt. In either case, we can readily separate salt from water by evaporating the water (Figure 1.6).

Figure 1.6
Water from the Great Salt Lake is diverted into large ponds. The water evaporates, leaving solid salt behind.

Mixtures differ in uniformity of composition. A **homogeneous mixture** has a uniform composition throughout and is often called a **solution.** Most solutions that we commonly encounter are composed of compounds dissolved in water. They are often clear. For example, a well-mixed sample of salt water prepared in a kitchen is uniform in appearance. The salt dissolved in it is invisible. Furthermore, any microscopically small portion of the sample would have the same composition as any other. The particles in the mixture might not be arranged in exactly the same pattern, but each sample, regardless of size, would have the same components in the same proportions.

A mixture that is not uniform throughout—a mixture of salt and pepper, for instance—is a **heterogeneous mixture.** Different samples have their components present in different proportions. Which of the things that Bill and Anna had for lunch is a homogeneous mixture? Which is heterogeneous? How about your own lunch? How can you tell?

We have considered a number of classes and subclasses of matter: mixtures, solutions, heterogeneous mixtures, pure substances, compounds, elements, metals, and nonmetals. A method for classifying matter into these categories is outlined in Figure 1.7. Note in the figure that yes or no answers to several questions distinguish one type of matter from another. First, we ask if the material can be separated physically. If so, then it is a mixture. If not, it must be a pure substance. If this substance can be decomposed (broken down into simpler substances) by chemical reactions, it is a compound. If it cannot, it is an element.

Not all solutions are liquids. For example, consider air that has been filtered to remove suspended solid particles. Filtered air is a gaseous solution containing a mixture of primarily oxygen and nitrogen gases, along with several other gases in lesser quantities. Solid solutions also exist and are called alloys. For example, the 14-carat gold used in rings is a solution of gold, silver, and copper.

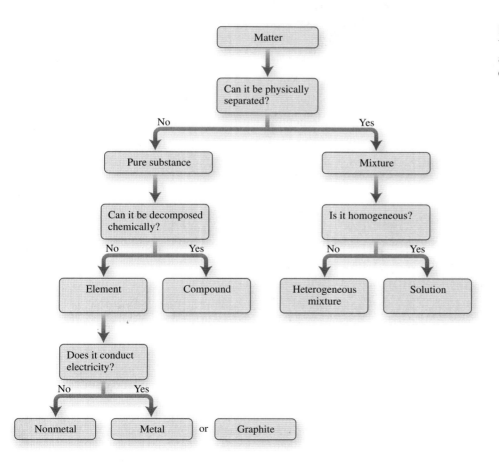

Figure 1.7
We can classify matter by answering the short series of questions in this flowchart.

EXAMPLE 1.3   Elements, Compounds, and Mixtures

Which of the following pictures represent pure substances?

Solution:

The copper on the outside of the coin and the helium inside the balloons are pure substances. (However, the helium and balloons considered together provide an example of a mixture.)

Practice Problem 1.3

Which of the pictures represent mixtures? Which are heterogeneous? Which are homogeneous?

Further Practice: 1.23 and 1.24

Representations of Matter

Chemists and other scientists view the world on several different levels. So far we have considered matter on a macroscopic scale. That is, we've discussed matter and phenomena we can see with our eyes. But simple observation is limited. Sometimes we cannot classify things merely by looking at them as Anna and Bill did. What do we do then? Chemists try to make sense of the structure of matter and its behavior on a scale that is much, much smaller than what we can see with our eyes.

Consider the copper pipe at the construction site, for example. If we could enlarge the tiniest unit that makes up the pipe, what would we see? Experimental evidence tells us copper is made up of discrete, spherical entities that all appear to be identical (Figure 1.8). Chemists identify these entities as atoms. An **atom** is the smallest unit of an element that has the chemical properties of that element. For example, we can imagine the helium inside a balloon as many, many atoms of helium. In Figure 1.9, each sphere represents a single helium atom. Similarly, if we could magnify the structure of water, we would find two small hydrogen atoms bound separately to a single larger oxygen

Figure 1.8
A copper pipe consists of a regular array of copper atoms.

Figure 1.9
Helium atoms are present inside the balloon.

Figure 1.10
Molecules containing hydrogen and oxygen make up the water in the fountain.

Although chemists generally use color coding to distinguish between atoms of different elements in representations, the atoms themselves do not have colors. Macroscopic samples of matter may have color, but these colors do not usually match those used to represent atoms. In accurate representations, the sizes of the spheres change to reflect the relative differences in size of atoms of different elements.

atom. Such a combination of elemental units is a **molecule.** Molecules are made up of two or more atoms bound together in a discrete arrangement. Several molecules of water are shown in Figure 1.10, where the central red sphere represents an oxygen atom and the two smaller, white spheres stand for hydrogen atoms. (Some compounds do not exist as molecules. We will discuss them in Chapter 3.)

In addition to molecules of compounds, molecules can also be formed by the combination of atoms of only one element. For example, as shown in Figure 1.11, the oxygen we breathe consists of molecules of two oxygen atoms joined together.

Chemists use many different ways to represent matter. Some are shown in Figure 1.12. Element symbols with subscripts represent a ratio of elements in a compound. One example is Figure 1.12B. To describe how the atoms are attached to one other, chemists often use lines and element symbols as shown in 1.12C. In Figure 1.12D spheres represent the atoms, and sticks show how they are connected. Figure 1.12E represents how the atoms fit together and their relative sizes. Macroscopic, molecular-level, and symbolic representations like these all have their advantages, and sometimes one is more convenient than another. You'll use them all as you progress through this course.

Figure 1.11
Oxygen molecules are made up of two interconnected oxygen atoms.

EXAMPLE 1.4 Representations of Matter

(a) Which of these images best represents a mixture of elements?

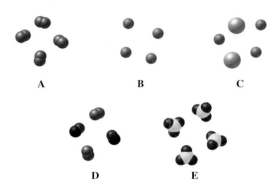

A   B   C

D   E

(b) If image A represents nitrogen, write its formula.

Solution:

(a) There are two mixtures represented in the images. Since the spheres (representing atoms) in image C have different colors and sizes, we can conclude that image C is a mixture of two elements. Image D is also a mixture, but it is a mixture of an element and a compound.
(b) The formula of the substance represented in image A is N_2. Note that two atoms are connected in the molecule.

Practice Problem 1.4

(a) Which of the images represents an element that exists as a molecule?
(b) If image E represents a compound of oxygen (red) and sulfur (yellow), what is its formula? (Write the symbol for sulfur first.)

Further Practice: 1.29 and 1.30

Animation: Three States of Matter
Online Learning Center, Animations

Chemistry Animations Library, Three States of Matter, Matter_States.rm

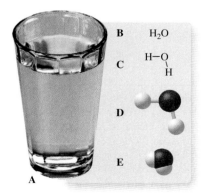

Figure 1.12
Different ways of representing water: (A) macroscopic, (B and C) symbolic, and (D and E) molecular.

States of Matter

Earlier we considered classification of matter based on composition. Let's look at a different way to classify matter: by its physical state. A **physical state** is a form that matter can take. The three most familiar to us are *solid, liquid,* and *gas.* Some substances, including some of those Anna and Bill observed, can be found in all three states under more or less ordinary conditions. Water, for example, can be a solid (ice), a liquid (flowing water), or a gas (water vapor) at environmental temperatures.

Other substances require extreme conditions to change from one state to another. For example, while carbon dioxide is a gas under normal conditions, it becomes a solid, called dry ice, at very low temperatures (Figure 1.13).

How do we know if a substance is in the solid, liquid, or gaseous state? Each state has characteristics that we can observe with our eyes and characteristics that are detectable or measurable at the molecular level. These characteristics are summarized in Table 1.2.

A **solid** has a fixed shape that is not related to the shape of the container holding it. When you place an iron pipe in a box, it does not change shape. Some solids can be made to change shape if enough force is applied. However, if you try to squeeze a solid to make it smaller, you'll fail. A solid cannot be compressed because its particles are arranged in a tightly packed, highly ordered structure that does not include much free space into which they might be squeezed. Note the closely packed particles in the solid state of iron shown in Figure 1.14.

A **liquid** is different from a solid in that it has no fixed shape. It takes the shape of the filled portion of its container, and it can be poured. The particles in a liquid are not arranged in ordered structures like those in a solid; they are free to move past one another. A liquid can be compressed slightly because its particles have a little free space between them. Note the differences between the liquid and solid states of iron shown in Figure 1.14.

A **gas** has no fixed shape; it adopts the shape of its container, expanding to fill available space completely. A gas is easily compressed. When squeezed, gases can undergo large changes in volume. The particles of a gas are widely separated with

Figure 1.13
Dry ice is the solid state of carbon dioxide. It converts from a gas to a solid at a very low temperature.

TABLE 1.2	**Characteristics of the Physical States of Matter**	
Solid	**Liquid**	**Gas**
Fixed shape	Shape of container (may or may not fill it)	Shape of container (fills it)
Its own volume	Its own volume	Volume of container
No volume change under pressure	Slight volume change under pressure	Large volume change under pressure
Particles are fixed in place in a regular array	Particles are randomly arranged and free to move about until they bump into one another	Particles are widely separated and move independently of one another

Figure 1.14
The liquid and solid states of iron.

Solid

Liquid

Figure 1.15
At the same temperature, a gas under high pressure has particles closer together than at low pressure.

Low pressure
Normal air

High pressure
Compressed air

Figure 1.16
Water condenses from a gas to a liquid on a cold surface. Air molecules are not shown.

Water vapor in
humid air

Condensed water
on glass

much empty space between them. When a gas is compressed, the amount of space between the particles is reduced. This happens when pressure is applied, such as when a bicycle tire is filled with air, as shown in Figure 1.15. Another characteristic of gases is that they move through space quickly. When Bill and Anna smelled the pizza they had for lunch, they were detecting particles that migrated as gases from the source of the food to their noses. When gases cool sufficiently, they become liquids. This occurs, for example, when water vapor in the air liquefies on the surface of a cold glass. Note the differences between the liquid and gaseous states of water shown in Figure 1.16.

TABLE 1.3	Symbols for Physical State	
Physical State	**Symbol**	**Example (bromine)**
Solid	(*s*)	$Br_2(s)$
Liquid	(*l*)	$Br_2(l)$
Gas	(*g*)	$Br_2(g)$
Aqueous (dissolved in water)	(*aq*)	$Br_2(aq)$

It is often convenient to show the physical state of a substance when representing it symbolically. For example, solid, liquid, and gaseous water can be represented as $H_2O(s)$, $H_2O(l)$, and $H_2O(g)$, respectively. The symbol (*aq*) represents an **aqueous solution,** a solution in which a substance is dissolved in water. A salt and water solution, for instance, can be written as $NaCl(aq)$. These symbols for physical state are shown in Table 1.3.

1.2 PHYSICAL AND CHEMICAL CHANGES AND PROPERTIES OF MATTER

Bill and Anna observed some of the *properties* of matter including *changes* in matter. Their observations could be either *qualitative,* based on some quality of the matter; or *quantitative,* based on a numerical value. When making qualitative observations, they described color, shape, texture, shininess, and physical state. Quantitative observations are different. They are numbers or measurements, and they must be carefully made and carefully reported.

Since quantitative data used to describe matter can involve both very large and very small numbers, it is often useful to express such numbers in *scientific* or *exponential notation.* Math Toolbox 1.1 (located at the end of this chapter) provides a review of this notation. In addition, it is necessary to express numbers in such a way as to indicate how accurately the value is known and how precisely it has been measured. The use of *significant figures* to properly express numerical values is presented in Math Toolbox 1.2.

Mass: 50 mg, 0.05 g, or 5×10^{-5} kg

Physical Properties

When reporting qualitative data, we can classify properties as either physical or chemical. When Bill and Anna observed the color, shape, texture, shininess, and physical state of things around them, they were noting their physical properties. A **physical property** is a characteristic that we can observe without changing the composition of a substance. Other examples of physical properties are odor, taste, hardness, mass, volume, density, magnetism, conductivity, and the temperatures at which a substance changes from one physical state to another. Let's take a close look at some of these properties: mass, volume, density, and temperature. These four properties are *quantitative;* they involve numerical values.

Mass   Recall that mass is a measure of the quantity of matter. We usually measure the mass of an object by weighing it on a balance. In chemistry, masses are often reported in units of grams (g). Large masses, like people or elephants, may be reported in units of kilograms (kg); and small masses, such as salt crystals or impurities in water, may be reported in units of milligrams (mg) or micrograms (µg) as shown in Figure 1.17. (Math Toolbox 1.3 summarizes the relationships among units such as these.) Sometimes the mass of something is reported in grams, but we might want to know the mass in another mass unit such as milligrams or kilograms.

Mass: 7×10^7 mg, 7×10^4 g, or 70 kg

Figure 1.17
A salt crystal has a mass of about 50 mg, while a person has a mass of about 70 kg.

Although we measure mass on a balance, mass and weight are not the same thing. Mass is the quantity of matter. Weight is the force exerted on a mass by gravity. (It is calculated by multiplying mass by the force of gravity.) Here astronaut Linda M. Godwin effortlessly holds up astronaut Jerry L. Ross aboard the orbiting space shuttle Atlantis. Ross's weight has changed from its value on Earth, but his mass has not.

We can easily convert a measurement from one unit to another if we know the relationship between the units. Example 1.5 shows how to convert between mass units. (See Math Toolbox 1.3 for more information on unit conversions.)

EXAMPLE 1.5   Units of Mass

Anna and Bill saw a pile of pennies in a fountain. They measured the mass of a penny and obtained a value of 2.50 g. However, they want to report its mass in units of milligrams (mg). How do they convert 2.50 g to mg?

Solution:

Most problems can be solved in more than one way. The general approach can be summarized by the following diagram:

The mass in grams has to be converted to the mass in milligrams. We need to find a relationship between these two quantities: 1 g = 1000 mg. Using this relationship, we get the following conversion:

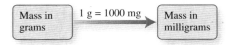

We will examine two approaches to this problem. In the first approach, we will use ratios as described in Math Toolbox 1.3. The relationship between milligrams and grams is

$$1000 \text{ mg} = 1 \text{ g}$$

The ratio of milligrams to grams is

$$\frac{1000 \text{ mg}}{1 \text{ g}}$$

Since the ratio of milligrams to grams is always the same, we can set up the following relationship between what we don't know and what we do know:

$$\frac{x \text{ mg}}{2.50 \text{ g}} = \frac{1000 \text{ mg}}{1 \text{ g}}$$

In this case we can solve for x by multiplying both sides of the equation by 2.50 g:

$$2.50 \text{ g} \times \frac{x \text{ mg}}{2.50 \text{ g}} = \frac{1000 \text{ mg}}{1 \text{ g}} \times 2.50 \text{ g}$$

$$x \text{ mg} = \frac{1000 \text{ mg}}{1 \text{ g}} \times 2.50 \text{ g}$$

$$= 2500 \text{ mg} \text{ or } 2.50 \times 10^3 \text{ mg}$$

Note that the gram units cancel to leave the appropriate mg units.

    Another way to solve this problem uses the dimensional analysis approach, which also uses the relationship between grams and milligrams. Consult Math Toolbox 1.3 for details. The ratio for converting between grams and milligrams is (1000 mg)/(1 g). We can also write the ratio in its inverted form, (1 g)/(1000 mg).

To convert from grams to milligrams, we can multiply 2.50 g by the ratio that will allow for cancellation of like units:

$$\text{Mass in mg} = 2.50 \; \cancel{g} \times \frac{1000 \; \text{mg}}{1 \; \cancel{g}} = 2500 \; \text{mg} \; \text{ or } \; 2.50 \times 10^3 \; \text{mg}$$

Practice Problem 1.5

The label on the can of soda that Anna had for lunch lists the contents as containing 35 mg of sodium. What is the mass of sodium in units of grams? In units of kilograms?

Further Practice: 1.45 and 1.46

Volume   **Volume** is the amount of space a substance occupies. We can determine the volume of a cube by measuring its length, width, and height and then multiplying them. For example, the volume of a cube that is 2.0 centimeters (cm) on each side is 8.0 cubic centimeters (cm^3):

$$\text{Volume of a cube} = \text{length} \times \text{width} \times \text{height}$$

$$\text{Volume} = 2.0 \; \text{cm} \times 2.0 \; \text{cm} \times 2.0 \; \text{cm} = 8.0 \; cm^3$$

The volumes of liquids are usually measured in units of liters (L) or milliliters (mL) as shown in Figure 1.18. One cubic centimeter is equal to 1 mL, so the volume of 8.0 cm^3 could also be reported as 8.0 mL. Larger volumes, such as big bottles of soda, are usually reported in liters. A 1-L bottle of soda contains 1000 mL. Example 1.6 shows how to convert between volume units.

Figure 1.18
Some 250-mL, 500-mL, and 1-L containers.

EXAMPLE 1.6   Units of Volume

For lunch, Anna and Bill had 12-ounce (oz) cans of soda. The volume of a 12-oz can of soda is 355 mL. What is its volume in units of liters?

Solution:

Again we will show two ways to solve this problem (see Math Toolbox 1.3). The following diagram summarizes both of them:

To convert volume in milliliters to volume in liters, we use the following relationship: 1000 mL = 1 L.

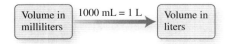

This is a unit conversion similar to the conversion we just did for mass. The ratio of milliliters to liters is

$$\frac{1000 \; \text{mL}}{1 \; \text{L}}$$

less dense ↑

0.93 —
1.00 —
1.01 —
1.03 —
1.13 —
1.32 —

more dense ↓

Figure 1.19
The densities of antifreeze, corn oil, dish detergent, maple syrup, shampoo, and water in g/mL are 1.13, 0.93, 1.03, 1.32, 1.01, and 1.00, respectively. Which layer is which substance?

Au

Al

Figure 1.20
Gold (Au) has a greater density than aluminum (Al) because gold has a greater mass per unit volume.

The following is the relationship between what we don't know and what we do know:

$$\frac{355 \text{ mL}}{x \text{ L}} = \frac{1000 \text{ mL}}{1 \text{ L}}$$

In this case we can solve for x, first by cross multiplying:

$$x \text{ L} \times 1000 \text{ mL} = 1 \text{ L} \times 355 \text{ mL}$$

We can solve for x by dividing both sides by 1000 mL:

$$\frac{x \text{ L} \times 1000 \text{ mL}}{1000 \text{ mL}} = \frac{1 \text{ L} \times 355 \text{ mL}}{1000 \text{ mL}}$$

$$x \text{ L} = \frac{1 \text{ L} \times 355 \text{ mL}}{1000 \text{ mL}}$$

$$= 0.355 \text{ L}$$

Note that the milliliter units cancel to leave the appropriate liter units.

We can also solve this problem by using the relationship between liters and milliliters as a conversion factor. The conversion factor for converting between liters and milliliters is (1000 mL)/(1 L) or (1 L)/(1000 mL). To convert from milliliters to liters, we can multiply 355 mL by the conversion factor that allows cancellation of like units:

$$\text{Volume in L} = 355 \text{ mL} \times \frac{1 \text{ L}}{1000 \text{ mL}} = 0.355 \text{ L}$$

Practice Problem 1.6

Anna and Bill saw some balloons outside the bookstore. The volume of gas inside one of the helium balloons was 4.60 L. What is the volume of gas in units of milliliters? In units of cubic centimeters?

Further Practice: 1.51 and 1.52

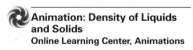
Animation: Density of Liquids and Solids
Online Learning Center, Animations

Chemistry Animations Library, Density of Liquids and Solids, Density.rm

These samples of metals have the same mass. Which has the greater density?

Density   The **density** of an object is the ratio of its mass to its volume. While mass and volume both depend on the size of the object or sample, density does not. Density is an unvarying property of a substance no matter how much of it is present. The densities of a few substances are listed in Table 1.4.

As Anna and Bill noted when they observed the fountain, a copper coin sinks in water. It sinks because copper (and the other metals in a penny) have a greater density than water. Conversely, air bubbles, just like other gases, rise to the top of water because gases are less dense than liquids. Oil floats on water for this same reason.

The density column in Figure 1.19 shows a variety of liquids with different densities. Which liquid has the greatest density? Which is the least dense?

If we compare *equal volumes* of two different substances, such as aluminum and gold, as shown in Figure 1.20, the substance with the greater mass has the greater density. How though can we compare densities if we do not have equal volumes? The mathematical relationship of mass, volume, and density reveals the answer:

$$\text{Density} = \frac{\text{mass}}{\text{volume}}$$

TABLE 1.4	Densities of Some Common Substances	
Substance	**Physical State**	**Density (g/mL)***
Helium	Gas	0.000178
Oxygen	Gas	0.00143
Cooking oil	Liquid	0.92
Water	Liquid	1.00
Mercury	Liquid	13.6
Gold	Solid	19.3
Copper	Solid	8.92
Zinc	Solid	7.14
Ice	Solid	0.92

*At room temperature and at normal atmospheric pressure, except gases at 0 degrees Celsius (°C) and water at 4°C.

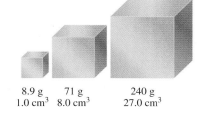

8.9 g   71 g   240 g
1.0 cm³ 8.0 cm³   27.0 cm³

Figure 1.21
The density of copper is 8.9 g/cm³. All three samples have the same ratio of mass to volume.

For example, a 1.0-cm³ sample of copper has a mass of 8.9 g. An 8.0-cm³ sample of copper has a mass of 71 g. A 27-cm³ sample of copper has a mass of 240 g. In all these samples (Figure 1.21), the mass of copper divided by its volume is 8.9 g/cm³. This is the density of copper. If we know the mass and volume of an object, we can determine its density by substituting directly into the density equation.

Additionally, if we know the density of a substance and its mass in our sample, we can determine its volume. For example, suppose we want to know the volume occupied by 100 g of copper. Should the volume be greater than or less than 100 cm³? There are many approaches to this problem. One way is to rearrange the density equation to solve for volume. Another way is to solve for the unknown volume in a set of equivalent ratios. Both of these methods are shown in Example 1.7.

EXAMPLE 1.7   **Density, Volume, and Mass**

What is the volume of 100.0 g of copper? The density of copper is 8.9 g/cm³.

Solution:

We need to carry out the following conversion:

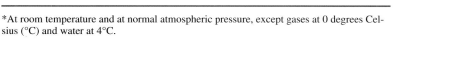

The relationship between mass and volume is given by density:

First, we rearrange the density equation to get volume on one side by itself:

$$\text{Density} = \frac{\text{mass}}{\text{volume}}$$

$$\text{Volume} = \frac{\text{mass}}{\text{density}}$$

A can of diet cola floats in water, but a can of regular cola sinks. Suggest a reason why. How can you use this information to quickly select your preferred type of soft drink from a cooler filled with ice water at a party?

Then, we substitute the known values of mass and density into the equation and solve for the value of volume:

$$\text{Volume} = \frac{100.0 \text{ g}}{8.9 \text{ g/cm}^3} = 11 \text{ cm}^3$$

In a second approach to this problem, consider that since the density of copper is always the same, the ratio of mass to volume is the same for both what we know and what we don't:

$$\frac{8.9 \text{ g}}{1 \text{ cm}^3} = \frac{100.0 \text{ g}}{x \text{ cm}^3}$$

Cross multiply to solve for x:

$$x \text{ cm}^3 = \frac{(1 \text{ cm}^3) \times (100.0 \text{ g})}{8.9 \text{ g}} = 11 \text{ cm}^3$$

Note that in both approaches, the gram units cancel to give the expected volume unit of cm^3.

Practice Problem 1.7

Solve the following problems using either approach.

(a) The density of pure gold is 19.3 g/cm³. What is the volume of 1.00 g of pure gold?

(b) 14-Carat gold is a homogeneous mixture of metals containing 58% gold by mass. The other 42% is a mixture of silver and copper. Silver and copper are both less dense than gold. Which of the following could be the mass of 1.00 cm³ of 14-carat gold: 16.0 g, 19.3 g, or 23.0 g?

Further Practice: 1.55 and 1.56

Water is unique among liquids because its solid form (ice) floats on its liquid form. This results from the relatively open structure adopted by water molecules in the solid state. What would happen to fish during the winter if ice were like other solids that sink in their liquid form?

Why do substances have different densities? Gases, in general, have very low densities because gas particles spread out and occupy large volumes. Metals tend to have high densities because their atoms pack together efficiently. Because ice floats on water, we can infer that water in its solid form must have a lesser density than water in its liquid form. Example 1.8 shows how to use molecular pictures to predict relative densities.

EXAMPLE 1.8   Explanations for Density

How do the molecular diagrams of ice and water help explain why ice is less dense than water?

Ice   Liquid water

Animation: Unique Properties of Water
Online Learning Center, Animations

Chemistry Animations Library, Unique Properties of Water, Water_Properties.rm

Solution:

In ice, the H_2O molecules have more space between them than in liquid water. The total volume occupied by a given number of molecules is greater. Because density is a ratio of mass to volume, the larger volume accounts for the lower density.

Practice Problem 1.8

Helium balloons rise in air, which is a mixture of oxygen and nitrogen molecules, so we know helium is less dense than air. Look at the molecular-level diagrams of helium and carbon dioxide. Predict whether a helium balloon rises or falls in an atmosphere of carbon dioxide.

less dense ↑

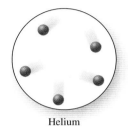   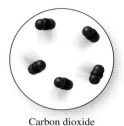

Helium   Carbon dioxide

Further Practice: 1.59 and 1.60

Temperature   Bill and Anna weren't happy with their lunches. The pizza was cold and their sodas were hot. When we make such comparisons, we are observing relative temperatures. **Temperature** is a measure of how hot or cold something is relative to some standard. We measure temperature with a thermometer.

In the United States, we often use the Fahrenheit scale to measure body temperature and air temperature. Fahrenheit is rarely used in science. Two other temperature scales are standard: the Celsius scale and the kelvin scale. The relationships between the three temperature scales, Fahrenheit (°F), Celsius (°C), and kelvin (K), are shown in Figure 1.22.

Another property of matter that is independent of sample size is the temperature at which the substance changes from one physical state to another. The *boiling point* is the temperature at which the liquid form of a substance changes to the gaseous form. At the *melting point,* the substance changes from a solid to a liquid.

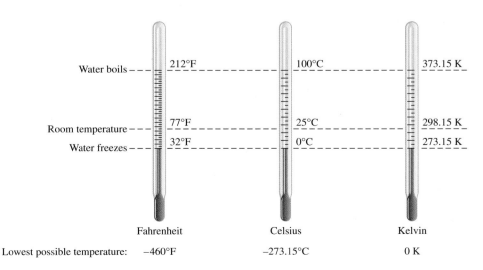

	Fahrenheit	Celsius	Kelvin
Water boils	212°F	100°C	373.15 K
Room temperature	77°F	25°C	298.15 K
Water freezes	32°F	0°C	273.15 K
Lowest possible temperature:	−460°F	−273.15°C	0 K

Figure 1.22
The Fahrenheit, Celsius, and kelvin temperature scales.

Temperatures are written differently for the different scales. While Celsius and Fahrenheit use the superscript ° to indicate degrees, the kelvin scale does not. The unit is written as K (the capital letter), but temperatures are measured in kelvins (lowercase).

Between these two temperatures, the substance is normally in its liquid state. For example, on the Celsius scale, the boiling point of water is 100°C. Water melts (or freezes, depending on its original state) at 0°C. On the kelvin scale, these values are 373.15 K and 273.15 K, respectively. On the Fahrenheit scale, they are 212°F and 32°F, respectively.

Note that on the kelvin scale, there are no negative values. It is an *absolute temperature scale* because its zero point is the lowest possible temperature observable in the universe. This value is absolute zero, which is equivalent to –273.15°C. Also note that the temperature increments on the kelvin scale are the same as those on the Celsius scale. The *difference* in temperature between the boiling point of water and the freezing point of water is 100 in both the Celsius (100°C – 0°C) and kelvin (373.15 K – 273.15 K) scales, while the difference is 180 on the Fahrenheit scale (212°F – 32°F). Because the temperature in kelvins is always 273.15° greater than the temperature in degrees Celsius, we can easily convert between them:

$$T_K = T_{°C} + 273.15$$

When converting between the Fahrenheit and Celsius scales, the calculation is more complicated because the degree increments are not equal:

$$T_{°F} = 1.8(T_{°C}) + 32$$

EXAMPLE 1.9   Units of Temperature

The melting point of copper is 1083°C. Above what temperature, in kelvins, is copper a liquid?

Solution:

Copper becomes a liquid above its melting point. In units of kelvin this temperature is

$$T_K = T_{°C} + 273.15$$

We substitute the value of the Celsius temperature into the expression and solve for the temperature in kelvins:

$$T_K = 1083 + 273.15 = 1356 \text{ K}$$

Practice Problem 1.9

(a) The boiling point of acetylene is –28.1°C. Below what temperature, in kelvins, is acetylene a liquid?
(b) The boiling point of helium is 4 K. Below what temperature, in degrees Celsius, is helium a liquid?
(c) Human body temperature is normally 98.6°F. What is this temperature on the Celsius and kelvin scales?

Further Practice: 1.63 and 1.64

Physical Changes

A process that changes the physical properties of a substance *without changing its chemical composition* is a **physical change.** For example, we can change liquid water to water vapor by heating it. This change from a liquid to a gas, called *boiling* or *vaporization,* is a physical change since both forms involve the same chemical substance, water (H_2O).

To represent such changes, we can refine the symbolic representations we developed for elements and for compounds. We write the chemical formula for the initial condition and composition of the matter we are considering, then an arrow, and finally the chemical formula for the final condition and composition. The arrow is used to show that a change has occurred and in which direction. Using this symbolism, the change of water from a liquid to a gas would be represented as

$$H_2O(l) \longrightarrow H_2O(g)$$

The molecular and symbolic representations in Figure 1.23 show that the water molecules do not themselves change, but their physical state does. All the processes that change water from one physical state into another are summarized in Figure 1.24.

Another example of a physical change is the separation of different substances in a mixture. For example, a magnet divides magnetic materials from nonmagnetic materials without changing their identities. A filter separates solid materials from liquid substances without changing either one chemically.

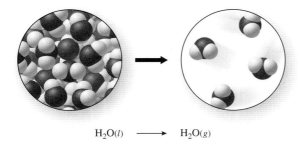

$$H_2O(l) \longrightarrow H_2O(g)$$

Figure 1.23
Molecular-level and symbolic representations of the evaporation of water.

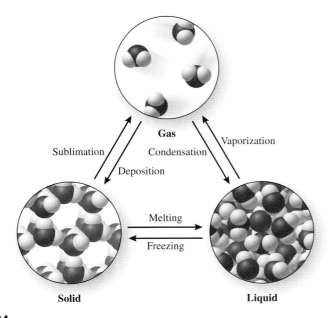

Figure 1.24
The physical states of solid, liquid, and gas can all change into one another either directly or by going through two changes of state. The names of these processes are shown here next to arrows that designate the direction of the change.

Chemical Changes

Remember the pennies in the fountain that Anna and Bill observed? Some were shiny and others looked dingy and brown. They might describe these less-shiny pennies as "tarnished." The pennies have undergone a **chemical change,** a process in which one or more substances are converted into one or more new substances. When pennies tarnish, some of the copper and zinc metal atoms in them combine with oxygen, forming compounds called metal oxides. The compounds are chemically different from either of the elements that formed them.

Suppose we clean a tarnished penny. Is the process a physical or a chemical change? It can be either. If you simply rub off the metal oxide coating with an eraser, the change is physical. Most penny collectors, however, prefer a chemical change that removes less metal. Rubbing ketchup on a penny is a great way to make it shiny. The vinegar in the ketchup reacts chemically with the metal oxides, freeing them from the surface of the penny. When the penny is rinsed, the result of the chemical change is easy to see.

Anna and Bill observed other examples of chemical change during their campus walk. When gasoline-powered cars burn fuel, a chemical change occurs. The gasoline reacts with oxygen to form carbon dioxide and water vapor. This chemical change releases the energy that runs the car. Chemical changes that involve burning are often accompanied by the release of energy. Anna and Bill also observed vehicles that run on alternative fuels. In hydrogen-powered vehicles, the hydrogen fuel combines with oxygen to form water vapor—and to release a lot of energy. The molecular-level and symbolic representations for this chemical change are shown in Figure 1.25. A chemical change is often called a **chemical reaction.** What are some examples of chemical reactions that you can observe around you?

Chemical Properties

The copper and zinc in a penny, the gasoline in a car, and the hydrogen in an alternative-fuel vehicle all share a common chemical property: They react with oxygen. However, they differ in how they react and what products they form. Only the latter two release sufficient energy rapidly enough to make their use as fuels possible.

A **chemical property** of a substance is defined by what it is composed of and what chemical changes it can undergo. For example, let's compare hydrogen and helium. Although they have similar physical properties (colorless gases, similar densities), their chemical properties are very different. While hydrogen reacts with many other elements and compounds (Figure 1.26), helium is considered *inert*. It has not yet been shown to react with any other element or compound.

Many metals combine with oxygen to form a metal oxide compound at the surface of the metal. When this occurs with iron, we call it *rust*.

Figure 1.25

A chemical change occurs when the atoms in H_2 and O_2 rearrange to form H_2O.

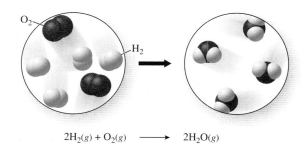

$$2H_2(g) + O_2(g) \longrightarrow 2H_2O(g)$$

Figure 1.26
The Hindenburg was a giant, rigid balloon filled with hydrogen gas. In 1937, it was destroyed when its hydrogen caught fire. Today, blimps are filled with the inert gas, helium.

EXAMPLE 1.10   Physical and Chemical Changes

Which of the following are physical changes and which are chemical changes?
(a) evaporation *physical property*
(b) burning methane gas to form carbon dioxide and water *chemical change*
(c) using a magnet to separate metal and plastic paperclips *physical change*
(d) rusting (conversion of iron to iron oxide) *chemical change*

Solution:

(a) Evaporation is a physical change because it involves only a change of state.
(b) Burning methane gas is a chemical change because new substances form.
(c) Separating components of a mixture is a physical change.
(d) Rusting is a chemical change because a new substance forms.

Practice Problem 1.10

Which of the following are physical properties and which are chemical properties?

(a) boiling point of ethanol *Phy*
(b) ability of propane to burn *Chem*
(c) tendency for silver to tarnish *Chem*
(d) density of aluminum *phy*

Further Practice: 1.69 and 1.70

Sometimes simple observation cannot tell us whether a change is chemical or physical. For example, bubbles appear when baking soda and vinegar mix. Bubbles also appear when water boils, but the change that produces the bubbles is different in these two cases. Baking soda and vinegar release bubbles because a chemical change takes place. They react to form carbon dioxide gas. However, when we warm water in a pan on the stove, small bubbles rise due to the release of dissolved air (mostly oxygen and nitrogen gas) from the water (before the water starts to boil). This process is only a physical change. If we could look at the nitrogen and

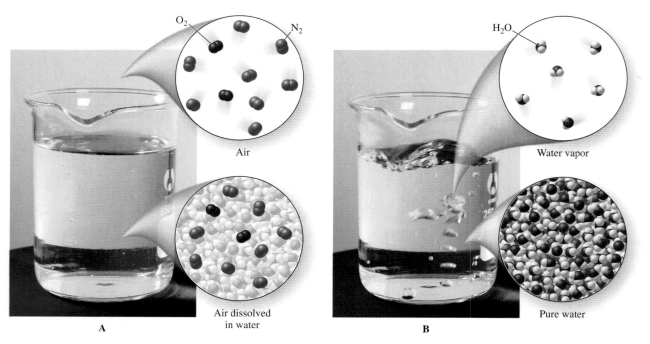

Figure 1.27
Water contains small amounts of dissolved nitrogen and oxygen gases. (A) When heated, these molecules go to the gaseous state in bubbles that rise to the surface. (B) When the water begins to boil, it no longer contains dissolved gases, and the bubbles contain gaseous water.

oxygen molecules, as shown in Figure 1.27A, we would see that they are the same whether they are dissolved in water or not. When these molecules are dissolved in water, they form a homogeneous mixture with it. Heating the water merely separates the oxygen and nitrogen molecules from the water molecules. If we continue to heat the water to boiling, larger bubbles form and then rise from the bottom, as shown in Figure 1.27B. These bubbles are gaseous water, or water vapor. The result of the physical change can be represented symbolically as

$$H_2O(l) \longrightarrow H_2O(g)$$

EXAMPLE 1.11   Physical and Chemical Changes

Do the following molecular-level images represent a chemical change or a physical change?

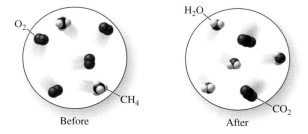

Before   After

Solution:

The substances after the change have a different composition than the substances before the change. Therefore, this is a chemical change.

Practice Problem 1.11

Do the following molecular-level images represent a chemical change or a physical change?

Before   After

Further Practice: 1.73 and 1.74

1.3   ENERGY AND ENERGY CHANGES

Physical and chemical changes involve energy. Energy is hard to define, but we see and feel evidence of it when something moves or changes temperature. At the construction site, Anna and Bill saw a worker pushing a wheelbarrow up a ramp. If released at the top of the ramp, the wheelbarrow would roll back down, converting energy from one form to another in the process. This release of energy is related to the spontaneous process of rolling down the ramp. (A spontaneous process is one that doesn't have to be forced to occur.) But returning the wheelbarrow to the top of the ramp is not spontaneous. It requires a continuous energy input. Similarly, chemical and physical changes are usually accompanied by energy changes. Some chemical reactions are spontaneous. They happen on their own. Others need a continuous energy input from an external source. Consider the reaction (Figure 1.28) of hydrogen and oxygen gases to form water vapor:

$$2H_2(g) + O_2(g) \longrightarrow 2H_2O(g)$$

This reaction is spontaneous and explosive. It releases a tremendous amount of energy. But the opposite reaction—the breakdown of water into hydrogen and oxygen gases—is not spontaneous. It occurs only if sufficient energy is continuously added, such as by passing electricity through liquid water. This process, called electrolysis, was shown in Figure 1.2, and can be described symbolically as follows:

$$2H_2O(l) \xrightarrow{\text{electrolysis}} 2H_2(g) + O_2(g)$$

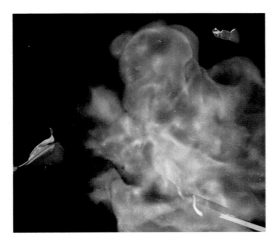

Figure 1.28
Hydrogen reacts explosively with oxygen to form water.

But what is energy? **Energy** is the capacity to do work or to transfer heat. **Work,** usually taken to mean mechanical work, occurs when a force acts over a distance. For example, work is done when the construction worker pushes the wheelbarrow up a ramp. Work is done when compressed gases, resulting from the combustion of a fuel, push the piston in the cylinder of an automobile engine. Not all reactions can be made to do work directly, but heat energy can be harnessed to do work. For example, boiling water produces steam, which turns the turbines in power plants. The turbines spin copper coils inside a magnetic field in a generator to produce an electric current.

Forms of Energy

Energy takes many different forms, and it can be converted from one form to another. Scientists describe two types of energy: kinetic energy and potential energy. **Kinetic energy** is the energy of motion. The wheelbarrow rolling down a ramp possesses kinetic energy. **Potential energy** is energy possessed by an object because of its position. Thus, the wheelbarrow resting at the top of the ramp has potential energy. If it did not contain this stored energy, it could not release energy when it rolled down the ramp. Any object in a position to be rolled, dropped, or otherwise allowed to move spontaneously has potential energy that will be converted to kinetic energy once the motion starts.

Other forms of energy—chemical, mechanical, electrical, and heat energy, for example—are really just forms of kinetic or potential energy. For example, chemical compounds can release chemical energy, the energy associated with a chemical reaction. Chemical energy is potential energy arising from the positions of the atoms and molecules in the compounds. A compound releases its potential energy when it undergoes a spontaneous chemical reaction that forms substances with less potential energy. For example, the explosive material TNT (trinitrotoluene) contains considerable potential energy that is released as kinetic energy when it decomposes. Chemical compounds can also have kinetic energy. Molecules move faster as the temperature rises. The motion of molecules or atoms is associated with heat energy, or the kinetic energy that increases with increasing temperature. The fast-moving gases produced by the explosion of TNT have high kinetic energy.

EXAMPLE 1.12   Molecular Motion and Kinetic Energy

Which of these two samples of argon gas has more kinetic energy?

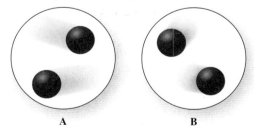

A   B

Solution:

The atoms that are moving faster have the greater kinetic energy. Thus, the atoms in A have the greater kinetic energy.

Practice Problem 1.12

Which of the two samples of argon gas is at a lower temperature?

Further Practice: 1.77 and 1.78

Electric energy is associated with the passage of electricity, generally through metals. Electric current passed through the filament of a lightbulb causes the metal to glow red and increases the motion of the atoms. This is a conversion of electric energy to kinetic energy. A lightbulb also gives off light energy. Nuclear energy involves both light and heat. Energy is released when one element is converted to another, as in a nuclear reactor or in the Sun, where hydrogen atoms fuse to form helium.

EXAMPLE 1.13   Forms of Energy

Identify examples of potential and kinetic energy in this picture.

Solution:

Anything that might move in the picture has potential energy. Kinetic energy is evident in the moving people and vehicles.

Practice Problem 1.13

Identify three additional forms of energy in the photograph.

Further Practice: 1.81 and 1.82

All these forms of energy can be converted into one another. For example, the welder at the construction site starts a gasoline engine that runs a generator that makes the electricity the welder uses to join two pieces of metal together. The chemical energy in the gasoline is converted to mechanical energy that turns the generator. The mechanical energy is converted to electrical energy by the generator. The electrical energy is converted to heat in an arc that is formed between the welding rod and the metal to be welded. This heat melts the metal and creates the weld. Some of the electrical energy is also converted to light. What energy conversions can you observe going on around you right now?

Units of Energy

Many packaged foods are labeled with the energy content. For example, a small box of raisins contains 130 Calories (Cal with a capital *C*). How much energy is this? Nutritionists and chemists use related, but not identical, units to measure energy. Chemists measure energy in units of *joules* (J) or *calories* (cal with a lower-case *c*). One calorie is the amount of heat energy needed to raise the temperature of 1 g of water by 1°C. A joule is smaller than a calorie:

$$4.184 \text{ J} = 1 \text{ cal}$$

A *kilojoule* (kJ), or 1000 J, is approximately the amount of energy that is emitted when a wooden kitchen match burns completely. The calorie used in chemistry should not be confused with the *Calorie* used by nutritionists (Figure 1.29), which is actually a kilocalorie (kcal), or 1000 cal.

Nutrition Facts

Serving Size: 1 cup (54g/1.9 oz.)
Servings Per Container: About 9

Amount Per Serving

Calories 190	Calories from Fat 10

	% Daily Value**
Total Fat 1g*	2%
Saturated Fat 0g	0%
Trans Fat 0g	
Cholesterol 0mg	0%
Sodium 0mg	0%
Potassium 180mg	5%
Total Carbohydrate 45g	15%
Dietary Fiber 6g	24%
Soluble Fiber 1g	
Insoluble Fiber 5g	
Sugars 7g	
Other Carbohydrates 32g	
Protein 5g	

Vitamin A 0%	•	Vitamin C 0%
Calcium 0%	•	Iron 8%

* Amount in cereal. One half cup of fat free milk contributes an additional 40 calories, 65mg sodium, 6g total carbohydrate (6g sugars), and 4g protein.

** Percent Daily Values are based on a 2,000 calorie diet. Your daily values may be higher or lower depending on your calorie needs.

	Calories:	2,000	2,500
Total Fat	Less Than	65g	80g
Sat. Fat	Less Than	20g	25g
Cholesterol	Less Than	300mg	300mg
Sodium	Less Than	2,400mg	2,400mg
Potassium		3,500mg	3,500mg
Total Carbohydrate		300g	375g
Dietary Fiber		25g	30g
Protein		50g	65g

Calories per gram:		
Fat 9 • Carbohydrate 4 • Protein 4		

INGREDIENTS: Organic Whole Grain Wheat, Organic Evaporated Cane Juice, Natural Flavor.

Figure 1.29

A nutritional label gives the energy content in Calories (or kilocalories) per serving.

EXAMPLE 1.14 Units of Energy

A can of cola from the United States contains 180 Cal, and a can containing the same amount of cola from Australia contains 900 J. Which contains more energy? Which is the diet cola?

Solution:

To answer this question, we must compare the energy in the two drinks using the same units. Thus, we must convert one value into the same unit as the other value. We will convert energy in Calories to energy in joules:

This is a unit conversion problem similar to those done earlier for mass and volume. We have no direct relationship between these two quantities, so we have to break this down into two steps. First note that the relationships between these two quantities are 1 Cal = 1000 cal and 1 cal = 4.184 J:

We convert 180 Cal to units of calories, using the concept that the relative values of cal and Cal remain constant:

$$\frac{x \text{ cal}}{180 \text{ Cal}} = \frac{1000 \text{ cal}}{1 \text{ Cal}}$$

Rearranging, we get

$$x \text{ cal} = 180 \text{ Cal} \times \frac{1000 \text{ cal}}{1 \text{ Cal}} = 180{,}000 \text{ cal}$$

As explained in Math Toolbox 1.1, numbers such as 180,000 are more conveniently expressed in scientific notation. The answer thus becomes 1.8×10^5 cal.

Now we can convert units of calories to units of joules using the ratio of these units:

$$\frac{x \text{ J}}{1.8 \times 10^5 \text{ cal}} = \frac{4.184 \text{ J}}{1 \text{ cal}}$$

Rearrange to solve for the energy of the soda in units of joules:

$$x \text{ J} = 1.8 \times 10^5 \text{ cal} \times \frac{4.184 \text{ J}}{1 \text{ cal}} = 7.5 \times 10^5 \text{ J}$$

This is considerably more energy than the 900 J found in the Australian cola, which must be the diet cola. Another way to solve the problem would be to convert 900 J to Calories and make the comparison. We could also have used conversion factors to reach the same conclusion.

Practice Problem 1.14

How many joules of energy are in a candy bar that contains 235 Cal?

Further Practice: 1.93 and 1.94

1.4   SCIENTIFIC INQUIRY

We began this chapter by describing some of the things that Anna and Bill saw around their campus. To classify the items, they observed similarities and differences in properties. They were making observations to help them understand nature.

Observation is one of the tools of scientific inquiry, but it is not the only one. The **scientific method** is an approach to asking questions and seeking answers that employs a variety of tools, techniques, and strategies. Although the scientific method is often explained as a series of steps and procedures, it is more accurately described as a way of looking at the world that differs from nonscience forms of inquiry. Scientists, like all humans, use intuition. They generalize about the world, sometimes with insufficient data. Chemists, especially, make inferences about atoms and molecules from data obtained from instruments that aren't quite capable of showing these tiny particles.

Scientists differ from professionals in nonscience disciplines in at least three important ways: (1) They test ideas by experimentation, (2) they organize their findings in particular (often mathematical) ways, (3) and they try to explain *why* things happen. Scientists use what is already known or believed about particular phenomena to gain insight into new observations from their experiments. Careful reasoning and insightful analogy are often employed, but sometimes intuition and luck play a part. Good scientists have an ability to couple objective scientific thinking with creative problem solving. In addition, a scientist must be curious enough to pursue the study of a seemingly trivial observation that can sometimes—albeit rarely—lead to a major advance in understanding.

Practicing scientists employ a variety of approaches to generate new knowledge or solve problems. Scientific inquiry generally includes observations, hypotheses, laws, and theories.

Observations

Scientific inquiry begins with ideas, knowledge, and curiosity in the minds of scientists. To look for answers to their questions, scientists collect data. Data may derive from the observation of a naturally occurring event or from deliberate experimentation. When experimenting, scientists set conditions, allow events to occur, and observe the result. This procedure permits scientists to examine events under controlled conditions not found in nature. In addition, scientists repeat one another's experiments and compare observations, thus checking the accuracy of their findings. A common experimental design involves the isolation of one factor at a time to determine which of many variables influences the outcome. The results of experiments may be qualitative (descriptive) or quantitative (numeric).

Consider this example. Suppose Anna and Bill want to experiment to find a way to clean the dull, brown coating from the pennies they saw in the fountain. After reading the results of previous studies, they think an acid might be a good agent to use. They try several, always being sure to work under the same conditions of temperature, light, ventilation, degree of tarnish, and so on. The pennies look a little cleaner after being placed in acetic acid, but the dull coating remains unaffected. With nitric acid, a reaction occurs. A blue-green liquid and a red gas form. The tarnish disappears, but so do the pennies (Figure 1.30). Hydrochloric acid appears to work best because the surface tarnish vanishes quickly, leaving the penny intact. However, Anna and Bill notice that one of the pennies appears to react with the hydrochloric acid to form bubbles. After a time, this penny begins to float (Figure 1.31). Upon closer examination of this penny, they notice that bubbles are coming from scratches on its surface. Their descriptions of the behavior of pennies in acids are observations made from deliberate and controlled experimentation.

Figure 1.30
Copper reacts vigorously with nitric acid to form copper nitrate in solution and gaseous nitrogen dioxide.

Figure 1.31
When some pennies were cleaned with hydrochloric acid, one began to bubble and eventually floated.

Hypotheses

To organize and correlate multiple observations and sets of collected data, scientists propose hypotheses. A **hypothesis** is a tentative explanation for the properties or behavior of matter that accounts for a set of observations and can be tested. Because hypotheses are usually starting points in the explanation of natural phenomena, they normally lead to further experimentation. In practice, hypotheses are intuitive guesses that may be based on small amounts of data. Often a hypothesis is modified repeatedly in light of the results of additional experimentation. The building of scientific knowledge involves a cyclic interplay between observations and the making and testing of hypotheses (Figure 1.32).

Consider the experiments with pennies and acids. Earlier, Bill and Anna observed that one of the pennies reacted with hydrochloric acid to form a colorless gas. That gas formed along scratches on the penny's surface. Their observation might lead them to form a hypothesis: Pennies react with hydrochloric acid to form bubbles of a gas because they have scratches that make the reaction faster. How might they test this hypothesis? They could scratch several pennies, dropping them into hydrochloric acid and checking for gas bubble formation. Suppose they performed this experiment with 10 pennies (Figure 1.33) and found that 7 of them formed bubbles at the scratches, but 3 did not. They would have to conclude that their hypothesis was only partially correct. Why? Because some of the pennies did not form bubbles as predicted by their hypothesis. Therefore, they might think further, asking themselves what other characteristics of pennies could influence the outcome.

While all pennies appear the same, close observation reveals detectable differences. For example, mints located in different cities produce pennies. The first letter of the city name appears just below the year in which the penny was minted (Figure 1.34). A *D* is stamped on pennies minted in Denver; *S* is used for San Francisco; and pennies minted in Philadelphia have no identifying mark. Given this new information, Anna and Bill might state a new hypothesis: Pennies react with hydrochloric acid to form bubbles if they are scratched *and* if they come from a

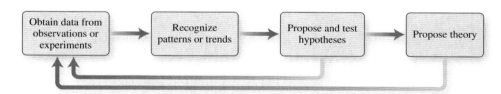

Figure 1.32
Observations and hypotheses are linked in a cyclic fashion during development of scientific knowledge.

Figure 1.33
Some of the pennies react with hydrochloric acid at the location of the scratches, while others do not.

specific mint—perhaps because that mint formulates its pennies differently. However, examination of the pennies from the previous experiment indicates that the mint location doesn't matter. Some scratched pennies from all the mints react and some don't. The amended hypothesis is incorrect.

As Bill and Anna set out to amend their hypothesis again, they realize they must isolate and assess one variable at a time. If they try to test too many ideas at once, they'd end up in a muddle of effects with several possible causes. So, while it's clear that scratches are necessary for bubbling to occur, some other variable or variables must be involved. As a next step, Anna and Bill reexamine the pennies from their previous experiment. They find that all the reactive pennies were minted in 1984 or more recently, while all the nonreactive pennies were minted in 1982 or before. This observation leads them to another amended hypothesis: Scratched pennies minted in 1982 or before do not react with hydrochloric acid to form bubbles, while scratched pennies minted since 1984 do—perhaps because the metal used to make the pennies was changed between 1982 and 1984.

How could they test their modified hypothesis? First, they must make sure that all old and new pennies respond in the same way as their original 10. They must then rule out any effects of chance by testing dozens, maybe even hundreds, of pennies. Assuming they get the results their hypothesis predicted, they may go on to cut some pennies open, perhaps verifying with direct observation that the insides of new and old pennies contain different materials.

Such findings would allow them to predict the behavior of pennies in hydrochloric acid, but they would not have a complete explanation for the chemical behavior. They would need to carry the process of scientific inquiry even further.

Figure 1.34
Pennies are marked with the year and city where they were minted.

Laws

When the behavior of matter is so consistent that it appears to have universal validity, we call this behavior a *law*. A scientific law describes the way nature operates under a specified set of conditions. For example, in the late eighteenth century, observations of the amounts of materials consumed and produced in chemical reactions led to the formulation of the *law of conservation of mass*. This law states that the mass of products obtained from a chemical reaction equals the mass of the substances that react. Every known chemical reaction that has been studied follows this law. For example, we could measure the mass of a penny and the mass of the nitric acid before allowing the reaction in Figure 1.30 to proceed. Upon completion of the reaction, we would find that the mass of the blue-green liquid plus the mass of the red gas is equal to the masses of our starting materials.

When nuclear reactions were discovered—such as the fusion reactions that power our Sun—scientists realized that processes that release very large amounts of energy do not conserve mass. To accommodate this new information, the *law of conservation of mass* had to be modified to become the *law of conservation of mass and energy.*

Theories

Earlier we developed a hypothesis about the behavior of pennies in hydrochloric acid. In the Laws section we described an example of a scientific law. Hypotheses and laws only describe *how* nature works, not *why. Theories* explain why observations, hypotheses, or laws apply under many different circumstances. For example, the *atomic theory,* which we will discuss in Chapter 2, explains many aspects of the behavior of matter, including the law of conservation of mass. Theories often employ mathematical or physical models that, if correct, explain the behavior of matter. Like hypotheses, a theory fits known observations. If new facts become known, theories may have to be modified or amended.

Let's return to the behavior of pennies in hydrochloric acid. Anna and Bill need to propose a theory that explains the behavior of all the pennies. Their Internet and library research reveals that before 1982 all pennies were minted from a copper alloy containing 5% zinc. After 1984 all pennies were copper-coated disks of zinc. The newer pennies contain approximately 97.5% zinc. If Anna and Bill can find out how pure zinc and pure copper react with hydrochloric acid, they might develop an

Figure 1.35
Zinc, but not copper, reacts with hydrochloric acid, producing hydrogen gas.

explanation for the chemical behavior of pennies in the presence of hydrochloric acid. A test with pieces of copper and zinc (Figure 1.35) indicates that zinc does indeed react with hydrochloric acid to release bubbles, while copper does not. They now have an explanation for the behavior of pennies in hydrochloric acid that is consistent with all the relevant observations. They can even explain why some of the pennies float. If the zinc is completely removed by reaction with hydrochloric acid, the copper shell that remains may fill with gas and rise to the surface of the liquid.

Scientific Inquiry in Practice

There is a perception that scientists have wild hair, dress funny, wear pocket protectors, lack social graces, and work in isolation in a laboratory. This stereotype is often perpetuated in movies, but is it real? Do you know any scientists who fit this description? Perhaps you do, but the stereotype is best left to the movies. Scientists are people, and they vary as much in their appearance, personalities, and preferences as people in any other vocation (Figure 1.36). Although many scientists do indeed work independently in the lab, research groups who share common interests do most scientific work collaboratively. Sometimes they make discoveries that follow long periods of painstaking work. At other times, new insights arise rapidly through serendipity. These are accidental, fortunate breakthroughs, such as Alexander Fleming's discovery of penicillin or Henri Becquerel's discovery of radioactivity. Such serendipitous events happen only occasionally, and they happen to individuals who can recognize their importance. Both Fleming and Becquerel noticed anomalies that less skilled observers might have overlooked. As Louis Pasteur stated, "In the fields of observation, chance favors only the prepared mind."

While most of scientific inquiry proceeds through hypothesis testing, this is not the only form the scientific method takes. A new approach, especially useful in the search for new drugs, is *combinatorial chemistry*. A series of related chemical compounds is systematically prepared and tested for effectiveness in disease treatment. Many different combinations are tried, using techniques involving miniaturization, robotics, and computer control. The compounds are screened as possible candidates for drug action. For example, if a drug is needed that binds to a particular enzyme to produce a biological effect, then the various compounds would be added to that enzyme to see if they do bind. If so, additional testing would be carried out.

An alternate approach involves the mixing of many chemicals, producing many different products in one container. These are then tested, either in the mixture or after separation. With current techniques, it is possible for one laboratory to synthesize and test as many as 100,000 new compounds in a month.

Figure 1.36
Chemists are a diverse group of people who work in a variety of environments.

SUMMARY

In this chapter we showed how things can be classified as matter or energy. We also demonstrated that matter can be classified in different ways based on its characteristics. Matter can be classified as an element, a compound, or as a mixture of one or both. Elements can be further classified as metals or nonmetals, and mixtures can be further classified as homogeneous mixtures (solutions) or heterogeneous mixtures.

Elements and compounds are commonly represented using element symbols and chemical formulas. Molecules can be either elements or compounds. Models can be constructed using spheres for atoms and combinations of spheres for molecules.

Matter can also be classified by its physical state, whether solid, liquid, or gas. Each substance has a set of unique physical and chemical properties. Some physical properties, such as mass, density, or boiling point, derive from measurements using appropriate units.

A change in a substance can be classified as either physical or chemical. In a physical change, the identity of the substance(s) remains unchanged. When a new substance is formed, the change is chemical, and the process of change is often called a chemical reaction.

The energy of an object is a combination of its kinetic and potential energy. Energy can change from one form to another, and it can be transferred as heat. Energy changes, often in the form of heat, accompany chemical reactions.

The scientific method includes a variety of methods of inquiry employed by scientists. Observation, or collection of data, is a necessary part of the scientific method. It frequently involves the design of carefully controlled experiments. Hypotheses are tentative explanations for the results of experiments. Laws result when observations appear to have universal validity, and theories are explanations for laws.

Math Toolbox 1.1   SCIENTIFIC NOTATION

Many numbers used in chemistry are either very large, such as 602,200,000,000,000,000,000,000 atoms in 12 g of carbon, or very small, such as 0.0000000001 centimeter per picometer (cm/pm). It is very easy to make mistakes with such numbers, so we express them in a shorthand notation called exponential or scientific notation. The numbers used in the examples are written as $(6.022 \times 10^{23}$ atoms$)/(12$ g C$)$ and 1×10^{-10} cm/pm in scientific notation. In this form, the numbers are easier to write, and it is easier to keep track of the position of the decimal point when carrying out calculations.

A number in scientific notation is expressed as $C \times 10^n$, where C is the *coefficient* and n is the *exponent*. The coefficient C is a number equal to or greater than 1 and less than 10 that is obtained by moving the decimal point the appropriate number of places. The exponent n is a positive or negative integer (whole number) equal to the number of places the decimal point must be moved to give C. For numbers greater than 1, the decimal point is moved to the left and the exponent is positive:

$$523 = 5.23 \times 10^2$$

For numbers smaller than 1, the exponent is negative, since the decimal point must be moved to the right:

$$0.000523 = 5.23 \times 10^{-4}$$

Such operations can be carried out more easily with a scientific calculator. Methods of use vary with different brands and models, so users should always consult their manual for specific instructions. However, the general approach is as follows. First enter the coefficient, including the decimal point. Then press the appropriate button for entering the exponent followed by the value of the exponent. The button is generally labeled EE, EXP, or 10^x. If the exponent has a negative value, press the change-sign button before entering the value of the exponent. Don't confuse this button with the subtraction button. The change-sign button is usually labeled +/– or (–).

When you carry out mathematical operations with numbers written in scientific notation, you need to know how to work on the exponential part of the number. You can usually do this with your calculator, but you should always check whether an answer from a calculator makes sense. To do this, verify the calculation

Math Toolbox 1.1 (continued)

by hand or make an estimate in your mind. The following instructions show how to work without a calculator. You should also perform these manipulations with your calculator to assure yourself that you can do them all.

Notice that in several of the examples, the coefficient is understood to be 1. That is, 1.0×10^4 is the same as 10^4. Since 1 multiplied or divided by 1 still equals 1, the coefficient does not need to be written.

To multiply exponential notation numbers, multiply the coefficients and add the exponents:

$$10^4 \times 10^7 = 10^{(4+7)} = 10^{11}$$
$$10^3 \times 10^{-5} = 10^{(3+-5)} = 10^{-2}$$
$$10^{-4} \times 10^4 = 10^{(-4+4)} = 10^0 = 1$$
$$(2 \times 10^{-4}) \times (4 \times 10^4) = (2 \times 4) \times 10^{(-4+4)} = 8 \times 10^0 = 8$$

To divide numbers written in exponential notation, divide the coefficients and subtract the exponents:

$$\frac{10^6}{10^4} = 10^{(6-4)} = 10^2$$

$$\frac{10^{-3}}{10^{-8}} = 10^{(-3--8)} = 10^5$$

$$\frac{8 \times 10^6}{2 \times 10^4} = \frac{8}{2} \times 10^{(6-4)} = 4 \times 10^2$$

To raise an exponential number to a power, raise both the coefficient and the exponent to the power by multiplying:

$$(10^4)^3 = 10^{(4 \times 3)} = 10^{12}$$
$$(10^{-2})^{-3} = 10^{(-2 \times -3)} = 10^6$$
$$(2 \times 10^4)^3 = (2)^3 \times 10^{(4 \times 3)} = 8 \times 10^{12}$$

To extract the root of an exponential number, take the root of both the coefficient and the exponent:

$$(10^8)^{1/2} = 10^{(8/2)} = 10^4$$
$$(10^{27})^{1/9} = 10^{(27/9)} = 10^3$$
$$(4 \times 10^8)^{1/2} = (4)^{1/2} \times 10^{(8/2)} = 2 \times 10^4$$

If the exponent is not evenly divisible by the root, we rewrite the number so the exponent is evenly divisible:

$$(10^7)^{1/2} = (10 \times 10^6)^{1/2} = 10^{1/2} \times (10^6)^{1/2} = 3.16 \times 10^3$$

To add or subtract numbers in exponential notation, first express both numbers with identical exponents. To do so, shift the decimal point in one of the numbers so its exponential part is the same as that of the other number. Then simply add or subtract the coefficients and carry the exponential part through unchanged. For example, suppose we want to add the following quantities:

$$5.2 \times 10^4 + 7.0 \times 10^3$$

We first change one number so they both have the same exponents:

$$52. \times 10^3 + 7.0 \times 10^3$$

Then we can add the two coefficients:

$$(52. + 7.0) \times 10^3$$

If necessary, we shift the decimal point to return the number to proper scientific notation:

$$59. \times 10^3 = 5.9 \times 10^4$$

Try the following examples. For (c) through (f) carry out the operations and report the answers in scientific notation.

  (a) Write 453,600 in scientific notation.
  (b) Write 0.00052 in scientific notation.
  (c) $(4.0 \times 10^6) \times (1.5 \times 10^{-3})$
  (d) $(2.0 \times 10^{-2})/(4.0 \times 10^3)$
  (e) $(3.0 \times 10^4)^2$
  (f) $4.3 \times 10^2 + 6.90 \times 10^3$

Did you get the following results?

  (a) 4.536×10^5
  (b) 5.2×10^{-4}
  (c) 6.0×10^3
  (d) 5.0×10^{-6}
  (e) 9.0×10^8
  (f) 7.33×10^3

Math Toolbox 1.2   SIGNIFICANT FIGURES

Numbers used in chemistry can be placed into two categories. Some numbers are *exact*. They are established by definition or by counting. Defined numbers are exact because they are assigned specific values: 12 in = 1 ft, 2.54 cm = 1 in, and 10 mm = 1 cm, for example. Numbers established by counting are known exactly because they can be counted with no errors. Since defined and counted numbers are known precisely, there is no uncertainty in their values.

Other numbers are not exact. They are numbers obtained by measurement or from observation. They may also include numbers resulting from a count if the number is very large. There is always some uncertainty in the value of such numbers because

they depend on how closely the measuring instrument and the experimenter can measure the values.

Precision and Accuracy

Uncertainty in numbers can be described in terms of either precision or accuracy. The *precision* of a measured number is the extent of agreement between repeated measurements of its value. *Accuracy* is the difference between the value of a measured number and its expected or correct value. If repetitive measurements give values close to one another, the number is precise, whether or not it is accurate. The number is accurate only if it is close to the true value.

Math Toolbox 1.2 (continued)

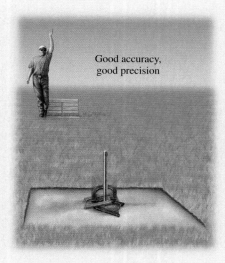

Good accuracy,
good precision

Poor accuracy,
good precision

Poor accuracy,
poor precision

We usually report the precision of a number by writing an appropriate number of significant figures. *Significant figures* in a number are all the digits of which we are absolutely certain, plus one additional digit, which is somewhat uncertain. For example, if we measure the height of a line on a graph calibrated with a line every 10 cm, we might obtain a value between 10 and 20 cm, which we estimate to be 18 cm. We are certain of the 1 in 18 cm, but not of the 8. Both these digits are considered significant; that is, they have meaning. Unless we have other information, we assume that there is an uncertainty of at least 1 unit in the last digit.

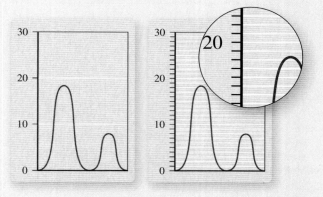

With a graph calibrated more finely, in centimeters, we see that the line height is indeed somewhat more than 18 cm. The line appears to reach about three-tenths of the distance between the 18 cm and 19 cm marks, so we can estimate the height at 18.3 cm. The number of significant figures is three, two of which are certain (1 and 8) and one of which is somewhat uncertain (3).

Determining the Number of Significant Figures

Every number represents a specific quantity with a particular degree of precision that depends on the manner in which it was determined. When we work with numbers, we must be able to recognize how many significant figures they contain. We do this by first remembering that nonzero digits are always significant, no matter where they occur. The only problem in counting significant figures, then, is deciding whether a zero is significant, using the following rules:

- A zero alone in front of a decimal point is not significant; it is used simply to make sure we do not overlook the decimal point (**0**.2806, **0**.002806).
- A zero to the right of the decimal point but before the first nonzero digit is simply a place marker and is not significant (0.**00**2806).
- A zero between nonzero numbers is significant (28**0**6, 0.0028**0**6).
- A zero at the end of a number and to the right of the decimal point is significant (0.002806**0**, 2806.**0**).

Math Toolbox 1.2 (continued)

- A zero at the end of a number and to the left of the decimal point (2806**0**) may or may not be significant. We cannot tell by looking at the number. It may be precisely known, and thus significant, or it may simply be a placeholder. If we encounter such a number, we have to make our best guess of the intended meaning.

The following table summarizes the significant figures in the numbers we just considered. The digits that are significant are highlighted.

Number	Count of significant figures
0.2806	4
0.002806	4
2806	4
0.0028060	5
2806.0	5
280600	4 or 5 or 6

To avoid creating an ambiguous number with zeros at its end, we write the number in scientific notation (see Math Toolbox 1.1) so that the troublesome zero occurs to the right of the decimal point. In this case, it is simple to show whether the zero is significant (2.8060×10^4) or not (2.806×10^4). The power of ten, 10^4, is not included in the count of significant figures, since it simply tells us the position of the decimal point.

Try this example: Determine the number of significant figures in 0.060520 and in 5020.01. Did you get the values 5 and 6?

Significant Figures in Calculations

We must also be concerned about the proper expression of numbers calculated from measured numbers. Calculators will not do this for us, so we have to modify their output. The rules for determining the proper number of significant figures in mathematical manipulations of measured numbers are simple. To determine the proper number of significant figures in the answer to a calculation, we consider only measured numbers. We need not consider numbers that are known exactly, such as those in conversions such as 1 foot (ft) = 12 inches (in). In the following examples, we will not use units so that we can focus on the manipulation of the numbers. Remember, however, that a unit must accompany every measured number.

Multiplication and Division

In a multiplication or division problem, the product or quotient must have the same number of significant figures as the least precise number in the problem. Consider the following example:

$$2.4 \times 1.12 = ?$$

The first number has two significant figures, and the second has three. Since the answer must match the number with the fewest significant figures, it should have two:

$$2.4 \times 1.12 = 2.7 \quad \text{(not 2.69)}$$

Try the following examples. Express the answers to the following operations with the proper number of significant figures:

(a) 5.27×3.20 (c) $6.0/2.9783$
(b) $1.5 \times 10^6 \times 317.832$ (d) $(2.01)^3$

Did you get the following results?

(a) 16.9   (c) 2.0
(b) 4.8×10^8 (d) 8.12

Note that powers and roots are just special cases of multiplication and division. Thus, the answer has the same number of significant figures as the number operated upon.

Addition and Subtraction

A sum or difference can only be as precise as the least precise number used in the calculation. Thus, we round off the sum or difference to the first uncertain digit. If we add 10.1 to 1.91314, for instance, we get 12.0, not 12.01314, because there is uncertainty in the tenths position of 10.1, and this uncertainty carries over to the answer. In this case, it is not the number of significant figures that is important but rather the place of the last significant digit. Consider another example:

$$0.0005032 + 1.0102 = ?$$

The first number has four significant figures and the second number has five. However, this information does not determine the number of significant figures in the answer. If we line these numbers up, we can see which digits can be added to give an answer that has significance:

$$\begin{array}{r} 0.0005032 \\ + 1.0102 \\ \hline 1.0107\cancel{032} \end{array}$$

The digits after the 5 in the first number have no corresponding digits in the second number, so it is not possible to add them and obtain digits that are significant in the sum. The answer is 1.0107, which has five significant figures, dictated by the position of the last significant digit in the number with the greatest uncertainty (1.0102).

Try the following examples. Express the answers with the proper number of significant figures:

(a) $24 + 1.001$ (c) $428 - 0.01$
(b) $24 + 1.001 + 0.0003$ (d) $14.03 - 13.312$

Did you get the following results?

(a) 25   (c) 428
(b) 25   (d) 0.72

Rounding Off Numbers

When a number contains more digits than are allowed by the rules of significant figures, we drop the digits after the last significant figure, using the following procedure:

- If the first digit being dropped is less than 5, leave the last significant figure unchanged. (For example, rounding to three significant figures, 6.073 becomes 6.07, and 6073 becomes 6.07×10^3.)

Math Toolbox 1.2 (continued)

- If the first digit being dropped is greater than 5 or is 5 followed by digits other than zero, increase the last significant figure by 1 unit (rounding to three significant figures, 6.077 becomes 6.08, and 60,751 becomes 6.08×10^4).
- If the first digit being dropped is 5 followed only by zeros or by no other digits, then increase the last significant figure by 1 unit if it is odd (6.075 becomes 6.08) but leave it unchanged if it is even (6.085 becomes 6.08).

Try the following examples. Round each of the following numbers to three significant figures:

- (a) 3245
- (b) 12.263
- (c) 0.001035
- (d) 312,486
- (e) 312,586

Did you get the following results?

- (a) 3245 becomes 3.24×10^3
- (b) 12.263 becomes 12.3
- (c) 0.001035 becomes 0.00104
- (d) 312,486 becomes 3.12×10^5
- (e) 312,586 becomes 3.13×10^5

Math Toolbox 1.3 — UNITS AND CONVERSIONS

All experimental sciences are based on observation. To be useful to the experimenter and to others, observations often must involve making measurements. Measurement is the determination of the size of a particular quantity—the number of nails, the mass of a brick, the length of a wall. Measurements are always defined by both a *quantity* and a *unit,* which tells what it is we are measuring.

But what system of units do we use? Most countries use the metric system, while the United States still uses primarily the English system. The English system is based on various units that are not related to one another by a consistent factor. Common units of length, for example, are inches, feet (12 in), yards (3 ft), and miles [1760 yards (yd)]. The metric system, on the other hand, uses units that are always related by a factor of 10, or by some power of 10. Units of length are centimeters (10^{-2} m), decimeters (10 cm or 10^{-1} m), meters, and kilometers (1000 m). Scientists in all countries use the metric system in their work.

The metric system is convenient to use because its units are related by powers of 10. Converting between related units is simply a matter of shifting a decimal point. The metric system adds a further convenience. It does not use arbitrary names like the English system, but rather, it defines *base units* of measure, and any multiple or fraction of these base units is defined by a special prefix.

Prefix	Factor	Symbol
giga	10^9	G
mega	10^6	M
kilo	10^3	k
deci	10^{-1}	d
centi	10^{-2}	c
milli	10^{-3}	m
micro	10^{-6}	μ
nano	10^{-9}	n
pico	10^{-12}	p

Metric Base Units and Derived Units

Although the metric system of prefixes can be used with any base unit, most scientists use SI units (from the French, *Système Internationale*). The seven base units are listed here.

Unit	Symbol	Quantity
meter	m	Length
kilogram	kg	Mass
second	s	Time
ampere	A	Electric current
kelvin	K	Temperature
mole	mol	Amount of substance
candela	cd	Luminous intensity

All other units are based on these, either by addition of a prefix or by combination. A number of commonly used units derived from the base units by multiplication and division are given special names in SI. An example is the unit of energy, the joule. It is force times distance, so 1 J is defined as 1 newton (N), the unit of force, times 1 m. Force is mass times acceleration, so the newton itself is a derived unit equal to 1 kg times 1 m, divided by 1 s^2. Thus, in base units, the formula for a joule is 1 J = 1 kg m^2 s^{-2}. In this formula, s^{-2} means $1/s^2$.

Conversion of Units

Measurements must be interpreted, often by mathematical manipulation of the data. This manipulation often involves converting one set of units into another using relationships between them. Two approaches to such conversions are *ratios* and *dimensional analysis*. In both cases, an analysis of the units provides clues to the correct solution of the problem.

Math Toolbox 1.3 (continued)

Consider a problem that can be solved almost automatically: "Convert 30 min into hours." You probably gave the answer without thinking: 30 min is one-half hour (0.50 h). But how did you know this answer? Let's systematize the problem-solving process. First—even if we don't always consciously think about it—we must decide what the problem is asking for. This problem requires converting a number of minutes into the corresponding number of hours. Next, we must know the number of minutes in 1 h. From this information, write a mathematical expression that shows the equivalence of two quantities having different units:

$$1 \text{ h} = 60 \text{ min}$$

We can use this expression in two ways to solve this problem. First we will describe the ratio approach.

Ratio Approach

In this approach, we use a known relationship to compare with our unknown relationship. We know how many minutes make up 1 h, but we want to know how many hours are in 30 min. From the known relationship, we develop a ratio of equivalent quantities that is equal to 1 and that has different units in the numerator and denominator. To develop a ratio to convert minutes into hours, we use the equivalent quantities and divide 1 h by 60 min.

$$\frac{1 \text{ h}}{60 \text{ min}}$$

The ratio for our unknown relationship should equal the known ratio as long as we use the same units, so we can set the unknown ratio equal to the known ratio:

$$\frac{x \text{ h}}{30 \text{ min}} = \frac{1 \text{ h}}{60 \text{ min}}$$

We can cross multiply to solve for *x:*

$$x \text{ h} \times 60 \text{ min} = 1 \text{ h} \times 30 \text{ min}$$

We can then solve for *x* by dividing both sides by 60 min:

$$\frac{x \text{ h} \times \cancel{60 \text{ min}}}{\cancel{60 \text{ min}}} = \frac{1 \text{ h} \times 30 \cancel{\text{ min}}}{60 \cancel{\text{ min}}}$$

We then carry out the mathematical operations:

$$x \text{ h} = \frac{1 \text{ h} \times 30 \cancel{\text{ min}}}{60 \cancel{\text{ min}}} = 0.5 \text{ h}$$

Notice that the units of minutes cancel out, leaving the units of hours, which is the unit called for in the problem.

Dimensional Analysis

We also use the ratio of equivalent quantities in the dimensional analysis approach. In this approach however, we get to the form of the manipulated ratio more quickly. We multiply the known quantity by the ratio so that units cancel and we get the unknown quantity with the desired units:

$$\text{Time in h} = 30 \text{ min} \times \frac{1 \text{ h}}{60 \text{ min}} = \frac{30 \times 1 \times \cancel{\text{min}} \times \text{h}}{60 \cancel{\text{min}}} = 0.50 \text{ h}$$

Notice how the starting units (minutes) cancel out, leaving the desired units (hours) to accompany the calculated number. In the dimensional analysis approach, units in the numerator and denom-

inator of a fraction are treated exactly the same as numbers—canceled out, multiplied, divided, squared, or whatever the mathematical operations demand. We set up the conversions so that desired units are introduced and beginning units cancel out.

Now reexamine the conversion just developed. From the same equivalence expression, 1 h = 60 min, we can derive another ratio by dividing 60 min by 1 h:

$$\frac{60 \text{ min}}{1 \text{ h}}$$

What happens when the original quantity is multiplied by this conversion factor?

$$\text{Time in h} = 30 \text{ min} \times \frac{60 \text{ min}}{1 \text{ h}} = \frac{30 \times 60 \times \text{min} \times \text{min}}{1 \times \text{h}}$$

$$= 1800 \frac{\text{min}^2}{\text{h}}$$

This answer is *clearly wrong* because the units make no sense. There are always two possible ratios of the known equivalent quantities, but only one gives the proper units in the answer. Which ratio is appropriate to use depends on how the problem is stated. But in every case, using the proper ratio cancels out the old units and leaves intact the desired units. Using the improper ratio results in nonsensical units (like min^2/h). This is a clear warning that the ratio was upside down or otherwise incorrect.

An Approach to Problem Solving

The processes just discussed can be generalized by the following diagram:

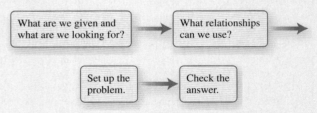

Consider the following problem: A liquid fertilizer tank has a volume of 6255 ft^3. How many gallons of liquid fertilizer can fit into this tank?

1. Decide what the problem is asking for.

First, read the problem carefully. If you are not sure what a term means, look it up. If it's necessary to use an equation to solve the problem, be sure you understand the meaning of each symbol. Look for clues in the problem itself—words or phrases such as *determine, calculate, what mass, what volume,* or *how much.* After deciding what quantity the problem is asking for, write the units in which this quantity must be stated. In the fertilizer tank example, we want to find the volume of the tank in units of gallons from units of cubic feet.

2. Decide what relationships exist between the information given in the problem and the desired quantity.

If necessary, recall or look up equivalence relationships (such as 1 h = 60 min) inside the back cover. This information may not be given in the problem itself. In the fertilizer tank example, we need to convert volume in cubic feet to volume in gallons, so we need an equivalence relationship between those two quantities. The expression is 1 ft^3 = 7.481 gal.

Math Toolbox 1.3 (continued)

The additional information that describes the relationship between two quantities won't always be a direct equivalence. A series of equivalence expressions and their derived conversion factors may be needed. Sometimes a mathematical equation is required to express the equivalence between two quantities [e.g., $T_{°C} = (T_{°F} - 32)/1.8$]. At other times, some chemical principle may have to be applied.

A word of caution: Just as a problem may contain less information than you need to solve it, it may also contain more information than you need. Never assume that you must use all the information. Examine all information critically and reject any that is not pertinent.

3. Set up the problem logically, using the relationships decided upon in step 2.

Starting with the relationships you obtained in step 2, develop the ratios needed to arrive at the final answer. Be sure to set up the ratios so that the old units cancel out and the desired units are introduced in the appropriate positions.

In the fertilizer tank example, the ratio is $(7.481 \text{ gal})/(1 \text{ ft}^3)$. The conversion is as follows:

$$\text{Volume in gal} = 6255 \ \cancel{\text{ft}^3} \times \frac{7.481 \text{ gal}}{1 \ \cancel{\text{ft}^3}} = 46{,}790 \text{ gal}$$

If you need a series of ratios to find the final answer, it is often helpful to "map out" the route you will follow to get there. Suppose you need to know how many centimeters are in 1.00 mile (mi). This conversion can be summarized by a diagram:

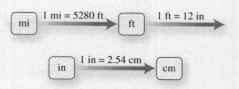

Unfortunately you can only find a table that gives the number of feet in a mile, the number of inches in a foot, and the number of centimeters in an inch. The diagram to get from miles to centimeters might look like this:

The problem setup would follow this progression:

$$\text{Distance in cm} = 1.00 \ \cancel{\text{mi}} \times \frac{5280 \ \cancel{\text{ft}}}{1 \ \cancel{\text{mi}}} \times \frac{12 \ \cancel{\text{in}}}{1 \ \cancel{\text{ft}}} \times \frac{2.54 \text{ cm}}{1 \ \cancel{\text{in}}}$$

$$= 1.61 \times 10^5 \text{ cm}$$

4. Check the answer to make sure it makes sense, both in magnitude and in units.

This step is just as important as the others. You must develop an intuitive feeling for the correct magnitude of physical quantities. Suppose, for example, you were calculating the volume of liquid antacid an ulcer patient needed to neutralize excess stomach acid, and the result of your calculation was 32 L. Now, units of liters looks reasonable, since the liter is a valid unit of volume; but it should be obvious that something is wrong with the numerical answer. Since a liter is about the same volume as a quart, this patient would need to drink 8 gal of antacid! The source of the problem is probably an arithmetic error, or possibly omission of the metric prefix *milli-* (10^{-3}) somewhere, since 32 mL would be an appropriate volume to swallow. In any event, a result such as this one should be recognized as unacceptable and the setup and calculations checked for errors.

In the fertilizer tank example, the units cancel properly, so the answer should be correct, if there are no arithmetic errors. Since there are about 7.5 gal in 1 ft³, the answer should be greater than the volume in cubic feet by something less than a factor of 10. The answer is indeed within this expected range.

Try the following example. If a laser beam fired from the moon takes 1.30 s to reach Earth, what is the distance in meters between the moon and Earth? Light travels in a vacuum at a speed of 3.00×10^{10} cm in 1 s or 3.00×10^{10} cm/s. Did you get 3.90×10^8 m as your answer?

KEY RELATIONSHIPS

Relationship	Equation
The density of an object is the ratio of its mass and volume.	Density = mass/volume
The absolute temperature in kelvins is offset from the Celsius temperature by 273.15°.	$T_K = T_{°C} + 273.15$
The Fahrenheit temperature has degrees that are 1.8 times as large as Celsius degrees. The freezing point of water is set at 32°F and 0°C.	$T_{°F} = 1.8(T_{°C}) + 32$

KEY TERMS

aqueous solution (1.1)

atom (1.1)

chemical change (1.2)

chemical formula (1.1)

chemical property (1.2)

chemical reaction (1.2)

compound (1.1)

density (1.2)

element (1.1)

element symbol (1.1)

energy (1.3)

gas (1.1)

heterogeneous mixture (1.1)

homogeneous mixture (1.1)

hypothesis (1.4)

kinetic energy (1.3)

liquid (1.1)

mass (1.1)

matter (1.1)

metal (1.1)

mixture (1.1)

molecule (1.1)

nonmetal (1.1)

physical change (1.2)

physical property (1.2)

physical state (1.1)

potential energy (1.3)

pure substance (1.1)

scientific method (1.4)

solid (1.1)

solution (1.1)

temperature (1.2)

volume (1.2)

work (1.3)

QUESTIONS AND PROBLEMS

The following questions and problems, except for those in the *Additional Questions* section, are paired. Questions in a pair focus on the same concept. Answers to the odd-numbered questions and problems are in Appendix D.

Matching Definitions with Key Terms

1.1   Match the key terms with the following descriptions.
   (a) a measure of the quantity of matter
   (b) a characteristic of a substance involving the possible transformations that the substance can undergo to produce a new substance
   (c) a combination of two or more substances that can be separated by physical means
   (d) a pure substance that cannot be broken down into simpler stable substances in a chemical reaction
   (e) the capacity to do work or to transfer heat
   (f) a characteristic of a substance that can be observed without changing its composition
   (g) the physical state in which matter has no characteristic shape but takes the shape of the filled portion of its container
   (h) the ratio of the mass of a substance to its volume
   (i) a mixture with uniform composition
   (j) the physical state of matter characterized by a fixed shape and low compressibility

1.2   Match the key terms with the following descriptions.
   (a) the smallest particle of an element that retains the characteristic chemical properties of that element
   (b) a change in which substances are converted into new substances that have compositions and properties different from those of the original substances
   (c) anything that occupies space and is perceptible to the senses
   (d) a substance composed of two or more elements combined in definite proportions
   (e) a combination of atoms of one or more elements
   (f) a process characterized by changes only in the physical properties of a substance, not in its composition

   (g) the physical state in which matter has no fixed shape or volume but expands to fill its container completely
   (h) the energy possessed by an object because of its position
   (i) a tentative explanation for the properties or behavior of matter that accounts for a set of observations and can be tested
   (j) the energy possessed by an object because of its motion

Matter and Its Classification

1.3   How would you classify the following items observed by Bill and Anna?
   (a) water in a fountain containing dissolved dye
   (b) a copper pipe
   (c) the contents of a balloon after blowing it up by mouth
   (d) a slice of pizza

1.4   How would you classify the following items observed by Bill and Anna?
   (a) sand in a volleyball court
   (b) a baseball bat made entirely of aluminum
   (c) the contents of a balloon filled from a helium tank
   (d) a glass filled with a soft drink

1.5   Which of the following are examples of matter?
   (a) sunlight
   (b) gasoline
   (c) automobile exhaust
   (d) oxygen gas
   (e) iron pipe

1.6   Which of the following are not examples of matter?
   (a) light from a fluorescent bulb
   (b) sand
   (c) wheelbarrow rolling down a ramp
   (d) helium balloons
   (e) heat from a welding torch

1.7   How are elements distinguished from compounds?

1.8   How are homogeneous mixtures distinguished from heterogeneous mixtures?

1.9   List characteristics of metals.

1.10   List characteristics of nonmetals.

1.11   Name the following elements.
       (a) Ti   (b) Ta   (c) Th   (d) Tc   (e) Tl

1.12   Name the following elements.
       (a) C   (c) Cr   (e) Cu   (g) Cs
       (b) Ca   (d) Co   (f) Cl

1.13   Name the following elements.
       (a) B   (b) Ba   (c) Be   (d) Br   (e) Bi

1.14   Name the following elements.
       (a) S   (b) Si   (c) Se   (d) Sr   (e) Sn

1.15   Name the following elements.
       (a) N   (c) Mn   (e) Al
       (b) Fe   (d) Mg   (f) Cl

1.16   Name the following elements.
       (a) Be   (c) Ni   (e) Ti
       (b) Rb   (d) Sc   (f) Ne

1.17   What are the symbols for the following elements?
       (a) iron   (d) gold
       (b) lead   (e) antimony
       (c) silver

1.18   What are the symbols for the following elements?
       (a) copper   (d) sodium
       (b) mercury   (e) tungsten
       (c) tin

1.19   A chemical novice used the symbol Ir to represent iron. Is this an acceptable symbol for the element? If not, what is the correct symbol for iron?

1.20   A chemical novice used the symbol SI to represent silicon. Is this an acceptable symbol for the element? If not, what is the correct symbol?

1.21   The symbol NO was used by a student to represent nobelium, an unstable, synthetic element. Is this an acceptable symbol for the element? If not, what is the correct symbol?

1.22   A student used the symbol CO to represent cobalt, an element found in vitamin B_{12}. Is this an acceptable symbol for the element? If not, what is the correct symbol?

1.23   Classify each of the following as a pure substance, a homogeneous mixture (solution), or a heterogeneous mixture: hamburger, salt, soft drink, and ketchup.

1.24   Classify each of the following as a pure substance, a homogeneous mixture (solution), or a heterogeneous mixture: sand, boardwalk, ocean, and roller coaster.

1.25   Elemental hydrogen normally exists as two hydrogen atoms bound together. Write the formula for this molecule, and draw a picture to represent how it might look on a molecular level.

1.26   Elemental chlorine normally exists as two chlorine atoms bound together. Write the formula for this molecule, and draw a picture to represent how it might look on a molecular level.

1.27   This image is a representation for a compound containing nitrogen and oxygen. Write the formula for this compound.

1.28   This image represents a compound containing phosphorus and chlorine. Write the formula for this compound.

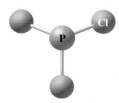

1.29   Which of the images represents a mixture of an element and a compound?

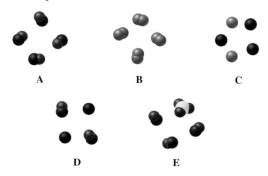

1.30 Which of the images in Question 1.29 represents a pure substance that is a compound?

1.31 Classify each of the following as an element or a compound.
(a) O_2 (c) P_4 (e) NaCl
(b) Fe_2O_3 (d) He (f) H_2O

1.32 Classify each of the following as an element or a compound.
(a) hydrogen gas (d) nitrogen dioxide
(b) water (e) aluminum chloride
(c) salt (f) neon

1.33 Under normal conditions, mercury is a liquid. Draw molecular-level pictures of what mercury atoms might look like in the liquid and solid states.

1.34 Under normal conditions, bromine is a liquid. Draw molecular-level pictures of what bromine, Br_2, might look like in the liquid and gaseous states.

1.35 What type of matter expands to fill its container and can be compressed to a smaller volume?

1.36 What type of matter is composed of particles that do not move past one another?

1.37 Identify the physical state of each of the following elements from their symbols.
(a) $Cl_2(g)$ (b) $Hg(l)$ (c) $C(s)$

1.38 Identify the physical state of each of the following compounds from their symbols.
(a) $NaCl(s)$ (b) $CH_3OH(l)$ (c) $CO_2(g)$

1.39 What physical state is represented in this diagram?

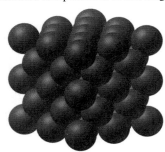

1.40 Draw a picture of the gaseous state of the substance shown in Question 1.39.

1.41 How might you symbolically represent a homogeneous mixture of oxygen, O_2, and water?

1.42 Why does the symbol $H_2O(aq)$ make no sense?

Physical and Chemical Changes and Properties of Matter

1.43 At the beginning of the chapter, Anna and Bill made many observations about what they saw. Some of their observations included color, texture, and shininess. Are these physical or chemical properties?

1.44 At the beginning of the chapter, you were asked to classify processes like a truck running on gas or welding pieces of metal. After reading the chapter, would you classify these processes as physical or chemical changes?

1.45 A slice of Swiss cheese contains 45 mg of sodium.
(a) What is this mass in units of grams?
(b) What is this mass in units of ounces (oz)? (16 oz = 453.6 g)
(c) What is this mass in pounds (lb)? (1 lb = 453.6 g)

1.46 A package of Swiss cheese has a mass of 0.340 kg.
(a) What is this mass in grams?
(b) What is this mass in ounces (oz)? (16 oz = 453.6 g)
(c) What is this mass in pounds (lb)? (1 lb = 453.6 g)

1.47 A grain of salt has a mass of about 1.0×10^{-4} g. What is its mass in the following units?
(a) milligrams
(b) micrograms
(c) kilograms

1.48 If a dog has a mass of 15.2 kg, what is its mass in the following units?
(a) grams
(b) milligrams
(c) micrograms

1.49 If you drank 1.2 L of a sports drink, what volume did you consume in the following units?
(a) milliliters
(b) cubic centimeters
(c) cubic meters

1.50 If the volume of helium in a balloon is 145 cm^3, what is its volume in the following units?
(a) milliliters
(b) liters
(c) cubic meters

1.51 If the length, width, and height of a box are 8.0 cm, 5.0 cm, and 4.0 cm, respectively, what is the volume of the box in units of milliliters and liters?

1.52 If a cubic box (all sides the same length) has a volume of 1.0 L, what is the length of each side of the box?

1.53 A slice of cheese has a mass of 28 g and a volume of 21 cm^3. What is the density of the cheese in units of g/cm^3 and g/mL?

1.54 Two stones resembling diamonds are suspected of being fakes. To determine if the stones might be real, the mass and volume of each were measured. Both stones have the same volume, 0.15 cm^3. However, stone A has a mass of 0.52 g and stone B has a mass of 0.42 g. If diamond has a density of 3.5 g/cm^3, could the stones be real diamonds? Explain.

1.55 If the density of a sugar solution is 1.30 g/mL, what volume of this solution has a mass of 50.0 g?

1.56 The density of a certain type of plastic is 0.75 g/cm^3. If a sheet of this plastic is 10.0 m long, 1.0 m wide, and 1 cm thick, what is its mass?

1.57 Why do liquids have greater densities than gases?

1.58 When a balloon filled with air is heated, the balloon increases in volume. Does the density of the air in the balloon increase, decrease, or remain the same?

1.59 A piece of plastic sinks in oil but floats in water. Place these three substances in order from least density to greatest density.

1.60 What special molecular-level feature of ice explains why ice floats in water?

Ice   Liquid water

1.61 Acetone, a component of some types of fingernail polish, has a boiling point of 56°C. What is its boiling point in units of kelvin?

1.62 The boiling point of liquid nitrogen is 77 K. What is its boiling point in units of degrees Celsius?

1.63 What is the difference in temperature between the boiling point of water and the freezing point of water in each of the following temperature scales?
  (a) Celsius scale
  (b) kelvin scale
  (c) Fahrenheit scale

1.64 If the temperature of a cup of coffee decreases from 60.0°C to 25.0°C, what is the decrease in temperature in units of degrees Celsius and kelvin?

1.65 Does the boiling point of a substance depend on how much of this substance you have?

1.66 Does the melting point of a substance depend on how much of this substance you have?

1.67 Identify each of the following as a physical property or a chemical property.
  (a) mass
  (b) density
  (c) flammability
  (d) resistance to corrosion
  (e) melting point
  (f) reactivity with water

1.68 Identify each of the following as a physical property or a chemical property.
  (a) boiling point
  (b) reactivity with oxygen
  (c) resistance to forming compounds with other elements
  (d) volume

1.69 Identify each of the following as a physical change or a chemical change.
  (a) boiling acetone
  (b) dissolving oxygen gas in water
  (c) combining hydrogen and oxygen gas to make water
  (d) burning gasoline
  (e) screening rocks from sand
  (f) the conversion of ozone to oxygen,
      $2O_3(g) \longrightarrow 3O_2(g)$

1.70 Identify each of the following as a physical change or a chemical change.
  (a) condensation of ethanol
  (b) combining zinc and oxygen to make the compound zinc oxide
  (c) dissolving sugar in water
  (d) burning a piece of paper
  (e) combining sodium metal with water, producing sodium hydroxide and hydrogen gas
  (f) filtering algae from water

1.71 Write a symbolic representation and a molecular-level representation for the change that occurs during condensation of chlorine, Cl_2.

1.72 Write a symbolic representation and a molecular-level representation for the process of freezing oxygen, O_2.

1.73 Do the changes shown in this diagram represent a physical or chemical change?

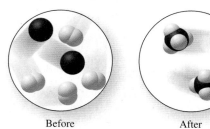

Before   After

1.74 Draw a picture that shows CH_4 condensing from a gas to a liquid. Does this picture represent a physical or a chemical change?

Energy and Energy Changes

1.75 Anna and Bill saw a construction worker welding pipe. Classify the forms of energy they were observing.

1.76 Bill and Anna watched students playing volleyball in the sunshine. Classify the forms of energy they were observing.

1.77 Which of these two samples of carbon dioxide gas has more kinetic energy? Explain your answer.

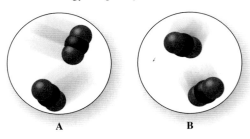

A   B

1.78 Which of these two samples of methane gas is at a higher temperature? Explain your answer.

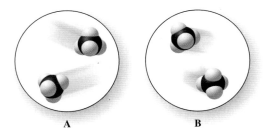

A   B

1.79 Give examples of potential energy that you can find in your room.

1.80 Give examples of kinetic energy that you can find in your room.

1.81 Distinguish between different types of energy shown in the following picture.

1.82 Distinguish between different types of energy shown in the following picture.

1.83 Describe how some common types of energy can change from one type to another.

1.84 If energy cannot be created, what is the source of the energy that is released when gasoline is burned?

1.85 Consider water falling into a fountain. How would you classify the transformation of energy in this process in terms of kinetic and potential energy?

1.86 Consider propane fuel being burned to power a vehicle. How would you classify the transformation of energy in this process in terms of kinetic and potential energy?

1.87 Convert an energy of 526 cal to units of joules.

1.88 Convert an energy of 225 cal to units of joules.

1.89 Convert an energy of 145 kJ to units of calories.

1.90 Convert an energy of 0.675 kJ to units of calories.

1.91 Convert an energy of 876 J to units of Calories.

1.92 Convert an energy of 2430 J to units of Calories.

1.93 Sodium cyclamate, $NaC_6H_{12}NSO_3$, was a popular nonsugar sweetener until banned by the Food and Drug Administration. Its sweetness is about 30 times that of sugar (or sucrose, $C_{12}H_{22}O_{11}$). The energy that can be supplied by ingestion of each compound is 16.03 kJ/g for sodium cyclamate and 16.49 kJ/g for sucrose. What is the energy saving in Calories resulting from using 1.00 g of sodium cyclamate in place of 30.0 g of sucrose?

1.94 In some areas, gasoline is formulated to contain 15% ethanol during some seasons. If gasoline releases 11.4 kcal/g upon combustion, but ethanol releases only 7.12 kcal/g, how much less energy in kJ/g is provided by the mixture than by pure gasoline?

Scientific Inquiry

1.95 Explain how a hypothesis is used in scientific research.

1.96 Explain the difference between a hypothesis and a theory.

1.97 Classify each of the following as observation, hypothesis, law, or theory.
(a) Bad luck results from walking under a ladder.
(b) Oil floats on water.
(c) Oil floats on water because it is less dense.
(d) Wood burns.

1.98 Classify each of the following as observation, hypothesis, law, or theory.
(a) When wood burns, oxygen is consumed.
(b) Heavier-than-air objects always fall toward the center of the Earth.
(c) Matter is composed of atoms.
(d) Crime rates increase when the moon is full.

1.99   You observe a piece of balsa wood floating on water and propose the hypothesis that all wood floats on water because wood is less dense than water. Suggest experiments to test this hypothesis.

1.100   You observe coins in a fountain and propose the hypothesis that fountains are built with coins in them. Suggest experiments to test this hypothesis.

Additional Questions

1.101   The density of air in a balloon is less at high altitudes than at low altitudes. Explain this difference.

1.102   Rank the following measurements in order from smallest to largest: 1.0×10^{-4} m, 2.0×10^{-5} m, 3.0×10^{-6} km, 4.0×10^{2} mm, 0.0 m, 1.0 m.

1.103   If you have a sample of zinc and a sample of copper, and both have the same mass, which has the greatest volume?

1.104   If the temperature in a room increases from 20.0°C to 30.0°C, what is the temperature change in units of kelvin?

1.105   Give the symbols for the following noble gas elements:
   (a) helium   (c) argon   (e) xenon
   (b) neon   (d) krypton   (f) radon

1.106   Give the symbols for potassium and phosphorus.

Atoms, Ions, and the Periodic Table

When Anna and Bill went looking for things related to chemistry (in Chapter 1), they classified nearly everything they saw as matter. Now Anna, who is writing a history paper, wants to know how our modern ideas about the atom arose. She learned in a philosophy class that the ancient Greeks thought the world was made up of four kinds of matter: earth, air, fire, and water. They called them elements. How did we come to understand matter as being composed of atoms of the different elements in the periodic table?

As Anna sets out to learn more, she discovers that the concept of the atom dates back to 450 BC. The Greek philosopher Democritus believed that there was a limit to how far matter could be broken down into smaller pieces. A piece of sand, for example, could be split into smaller pieces, and those pieces into even smaller pieces. But it made no sense to Democritus that the piece of sand could be broken down into smaller pieces forever.

Democritus gave the smallest components of matter the name *atomos* (*atom* in English), which is a Greek word meaning "unbreakable." Democritus thought that individual atoms were too small to see, but when combined together they made up everything that was visible. Although not all the philosophers of the day agreed with him, Democritus suggested that each "element" (earth, air, fire, and water) might be made up of different types of atoms.

After Democritus, some embraced the idea and tried to convince others, but the concept of the atom did not become popular until centuries later. In the 1700s, chemists realized that they must look for elements by breaking down matter into simpler substances. Once they had something that could not be broken down any further, they called it an *element*. About 30 different elements had been discovered by the end of the 1700s. They included most of the common metals, such as gold, silver, and copper; less common metals such as nickel, cobalt, and uranium; and two of the gaseous elements in air, nitrogen and oxygen.

Throughout the 1700s, few chemists believed that elements were made of atoms. The experiments of many chemists, however, provided evidence for the atomic nature of matter. A French chemist, Antoine Lavoisier, discovered in 1782 that when one substance changes into another, no mass is gained or lost. Another French scientist, Joseph Proust, showed that elements combine in specific mass ratios to form new substances. For example, when forming water, hydrogen and oxygen always combine in a mass ratio of 1:8.

Anna learns that the science and math teacher John Dalton offered the first convincing argument for the existence of atoms. Dalton realized Lavoisier's and Proust's findings were best explained by assuming that matter is made up of atoms. In 1808, Dalton published a book in which he described his views on atoms. So persuasive were his arguments that the idea came to be called *Dalton's atomic theory*.

Anna can't help but wonder why Dalton got the credit for the atom instead of Democritus. Their views were very similar. The difference is that Democritus and other philosophers had no scientific evidence to support their ideas. Their concepts were merely opinions of what seemed right. Dalton, however, pored over hundreds of experimental results and showed how they could be explained by assuming that atoms exist.

As Anna continues her research, she learns that the acceptance of Dalton's atomic theory by the scientific community paved the way for our current understanding of the atom—the *modern model of the atom*. Many additional questions pop into Anna's head now. How are atoms of different elements, such as carbon and hydrogen, different from one another? Do they have different masses? Are atoms composed of even smaller particles?

Anna isn't the only one who's finding that chemistry relates to other subjects. Anna's roommate Megan, a nutrition major, is quick to realize that the human body is composed of matter, and that everything we are made of is just different combinations of elements. Although we are about 99% carbon, hydrogen, and oxygen,

small amounts of many other elements are essential to proper functioning. Many of these are called *essential minerals.* Minerals are crucial to the growth and production of bones, teeth, hair, blood, nerves, and skin, not to mention the enzymes and hormones that living cells and tissues need to function. The essential minerals that we need in the greatest quantity are calcium, phosphorus, potassium, sodium, chlorine, magnesium, and iron. We get them from the foods we eat and drink. For example, calcium, which makes bones strong, is found in significant amounts in milk products. Potassium, an important regulator of cell fluids, is found in many fruits and vegetables. Iron, a building block for the oxygen-carrying hemoglobin in our blood, comes from fruits, vegetables, and red meat.

Megan notices that many of the minerals are classified as metals on the periodic table. Are we ingesting *metals* in the foods we eat? Generally, no. We take them in as components of compounds in which a metal such as iron exists as an ion. An *ion* is an atom with an electrical charge. The properties of metal ions are different from the properties of pure metal elements. Ions are important to the chemical processes that occur in our bodies. After learning about ions as nutrients, Megan begins to wonder how ions differ from atoms and why they have different properties.

In this chapter we will learn about the atoms that compose the elements and about the similarities and differences among them. We will answer Anna's and Megan's questions and many of your own.

Questions for Consideration

2.1   What evidence suggests that matter is composed of atoms?

2.2   How does the composition of different atoms differ?

2.3   How do ions differ from the atoms of elements?

2.4   How can we describe the mass of the atoms of an element?

2.5   How does the periodic table relate to the structure and behavior of atoms?

Math Tools Used in This Chapter

Scientific Notation (Math Toolbox 1.1)

Significant Figures (Math Toolbox 1.2)

2.1   DALTON'S ATOMIC THEORY

As you will learn in Chapter 15, mass is not conserved in reactions that involve the nucleus of the atom. In nuclear fusion reactions that occur in the Sun, for example, matter is converted into energy. Nuclear reactions were not known in Lavoisier's and Dalton's time.

In Anna's research, she learned that before the nineteenth century, many believed that matter was continuous—that is, that it could be divided infinitely (Figure 2.1). That notion began to change in 1808, when John Dalton (Figure 2.2) published his atomic theory. Dalton based his argument on experimental evidence that had begun accumulating about three decades earlier. For example, Antoine Lavoisier's experimental results led to the **law of conservation of mass,** which he published in 1787. His experiments showed that no measurable change in mass occurs during a chemical reaction. The mass of the products of a reaction always equals the mass of the reacting substances.

When we observe chemical changes, it may seem in some cases that the law of conservation of mass does not apply. After wood is burned in a bonfire, for example, there is less mass left in the fire pit. When an iron nail rusts, the rusted nail has a greater mass than the original nail. However, if we carry out each of these reactions in a closed container and find the mass of the closed container and its contents before and after the change, the mass does not change. With an open container, the gases produced when wood burns escape into the atmosphere, so they are not included in the final measurement. The gases consumed when a nail rusts are not included in the

Figure 2.1
A microscope image does not show the particle composition of the copper pipe. It seems continuous. We now know that copper metal, and all matter, is composed of particles called atoms.

Figure 2.2
John Dalton published his atomic theory in 1808. Dalton was not known as a researcher. Instead, he had a gift for developing theories from available experimental data.

original mass. The example in Figure 2.3 shows conservation of mass when a reaction occurs in a closed system. Lavoisier did experiments like these in which he was careful not to let any matter enter or escape.

Joseph Proust did similar experiments, mostly reacting metals with oxygen. He found that the oxygen content of the products was always fixed at one or two values, rather than showing a range of all possible values. His findings, published between 1797 and 1804, led to the **law of definite proportions.** This law states that all samples of the same compound always contain the same proportions by mass of the component elements. For example, pure water is always composed of oxygen and hydrogen with a mass ratio of 8 g of oxygen for every 1 g of hydrogen. Conversely, when water is broken down into its elements using electricity, oxygen and hydrogen are formed in this same mass ratio of 8:1.

Figure 2.3
When sodium carbonate is placed in a solution of hydrochloric acid, a reaction forms carbon dioxide gas. Because the carbon dioxide is not allowed to dissipate into the atmosphere, its mass is included in the final mass in the experiment shown here. How would the result be different if the container were open?

Figure 2.4

Two H_2 molecules combine with one O_2 molecule and rearrange to form two H_2O molecules. A chemical reaction is simply the rearrangement of atoms into new combinations.

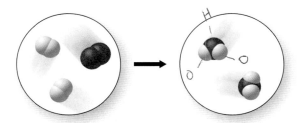

Dalton reasoned that the law of conservation of mass and the law of definite proportions could be explained only if matter was composed of atoms. The following postulates summarize **Dalton's atomic theory:**

1. All matter is composed of exceedingly small, indivisible particles, called atoms.
2. All atoms of a given element are identical both in mass and in chemical properties. However, atoms of different elements have different masses and different chemical properties.
3. Atoms are not created or destroyed in chemical reactions.
4. Atoms combine in simple, fixed, whole-number ratios to form compounds.

A chemical reaction, according to Dalton's atomic theory, is a rearrangement of atoms into a new combination, resulting in the formation of one or more new chemical substances (Figure 2.4). Although Dalton's theory has been modified over the past 200 years, it still provides the basis for understanding how atoms are the building blocks of matter. As with most theories, atomic theory has been amended as new evidence has been found. We now know that some of Dalton's postulates are not quite correct. The first statement, for example, is not strictly true. Atoms are composed of even smaller particles, called *subatomic particles*. The second statement is not quite accurate either. As you will learn in Section 2.2, atoms of a given element actually can vary in mass.

A light microscope cannot resolve images smaller than the wavelength of the light used to examine the object.

Still, Dalton's central idea has stood the test of time and experimentation—although for nearly two centuries, no one had a way to see the atoms that Dalton proposed. Because atoms are so small, they cannot be seen using an optical microscope. However, with the invention of the scanning tunneling microscope (STM) in 1981, scientists can now observe what we interpret to be single atoms on the surface of a material (Figure 2.5).

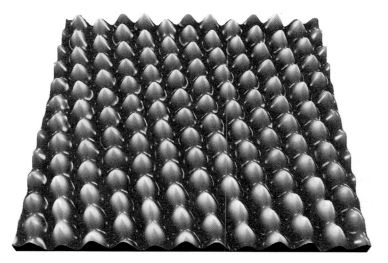

Figure 2.5

Image of atoms on the surface of gold captured by a scanning tunneling microscope (STM).

2.2 STRUCTURE OF THE ATOM

When reading nutritional labels, Megan noticed that a food supplement contained both iron and selenium. She wondered how the atoms of these elements differ. Scientists have also asked this question about atoms in general and found ways to probe their composition. They have studied the smaller particles that actually make up atoms.

Subatomic Particles

Atoms are not the small, indivisible particles Dalton envisioned. They consist of even smaller particles that together make up all atoms. A **subatomic particle** is a smaller particle found inside an atom. More than 40 are known, but only three have importance in chemical behavior. These subatomic particles are the *proton, neutron,* and *electron.* It was not until the late 1800s that researchers developed instruments and techniques capable of detecting subatomic particles and revealing their numbers and arrangements in atoms.

The existence of the **electron,** a negatively charged subatomic particle, was demonstrated by J. J. Thomson in 1897. He conducted a series of experiments with cathode-ray tubes (Figure 2.6). In a partially evacuated cathode-ray tube, a voltage is applied by connecting each end of the tube to a battery. Electricity then flows from one end of the tube to the other in the form of a ray. The invisible rays can be observed when they cause certain materials coated on the glass to glow. Thomson found that, in a magnetic or electric field, the rays bent toward a positively charged plate and were deflected away from a negatively charged plate outside the tube. He knew that like electrical charges repel each other and opposite charges attract each other. The bending of the beam toward the positive plate (and away from the negative plate) showed that the beam was composed of negatively charged particles. Thomson showed that the rays had a negative electrical charge no matter what material was used for the source of the rays. This result indicated that the rays were composed of identical, negatively charged particles common to all matter. We call these particles electrons. Thomson was also able to determine the charge-to-mass ratio of the electron from such experiments.

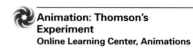

Animation: Thomson's Experiment
Online Learning Center, Animations

Chemistry Animations Library, Thomson's Experiment, Thomson_Expt.mov

Animation: Cathode-Ray Tube
Online Learning Center, Animations

Chemistry Animations Library, Cathode-Ray Tube, 01_Cathode_Ray_Tube.swf

Cathode-ray tubes (CRTs) are the fundamental components of television picture tubes and computer monitors. The screen contains chemical compounds that glow when struck by fast-moving electrons. Different chemicals that glow different colors provide a color picture.

How did the discovery of electrons make it necessary to change one of Dalton's postulates?

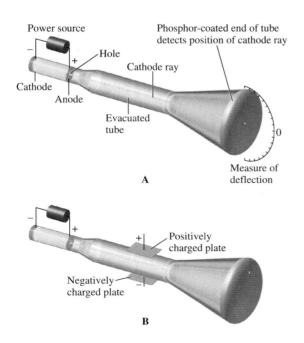

Figure 2.6
J. J. Thomson's experiments with cathode-ray tubes led to the discovery of the electron. (A) In a normal cathode-ray tube, the cathode ray travels in a straight path when there is no external field. (B) When Thomson applied an external electric field to the cathode-ray tube, the cathode ray bent toward the positive plate. It was known that like charges repel each another and opposite charges attract each other. The bending of the beam toward the positive plate (and away from the negative plate) indicated that the beam was composed of negatively charged particles.

Figure 2.7
In Millikan's oil-drop experiment, the electric field strength required to suspend an oil droplet was dependent upon the number of extra electrons on it. These experiments allowed Millikan to determine the charge of a single electron and calculate its mass.

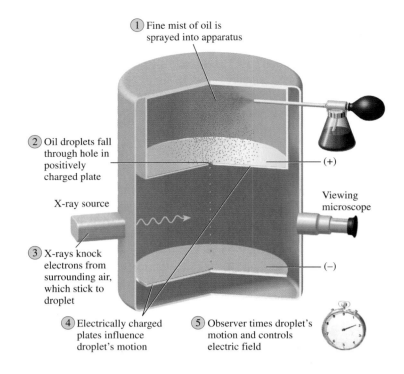

1. Fine mist of oil is sprayed into apparatus

2. Oil droplets fall through hole in positively charged plate

X-ray source

3. X-rays knock electrons from surrounding air, which stick to droplet

(+)

Viewing microscope

(−)

4. Electrically charged plates influence droplet's motion

5. Observer times droplet's motion and controls electric field

Animation: Millikan Oil Drop
Online Learning Center, Animations

Chemistry Animations Library, Millikan Oil Drop, 02_Millikan_Oil_Drop.swf

Building on these results, Robert Millikan experimented with oil droplets in an effort to measure the strength of the negative charge on an electron (Figure 2.7). When he exposed the droplets to radiation, they took on an electrical charge. By measuring the magnitude of the electric field necessary to cause the droplets to hang suspended in air, Millikan determined that the charge on an electron is -1.6022×10^{-19} coulombs (C) (*coulomb* is a unit of electrical charge). He then calculated the mass of an electron to be 9.1094×10^{-28} g from its charge and the charge-to-mass ratio determined by Thomson. Since even the lightest atoms have a mass greater than 10^{-24} g, an electron contributes only a small part to the mass of an atom. In fact, the mass of an electron is 1836 times less than the mass of one hydrogen atom, the lightest of all the elements. Because of the electron's small mass, it was originally thought that thousands of them must lie inside a single hydrogen atom. We now know there is only one.

The discovery of the electron stimulated many more experiments in search of other subatomic particles. Since atoms are electrically neutral, scientists reasoned that atoms must contain positively charged particles to counter the negatively charged electrons. The positively charged particle, called a **proton,** has a charge equal in magnitude to the electron but opposite in sign, $+1.6022 \times 10^{-19}$ C. To be electrically neutral, an atom must have equal numbers of protons and electrons. To make it easier to deal with electrical charges in matter, we usually express the charges as a multiple of the charge of an electron or of a proton, instead of in units of coulombs. Expressed in this way, the charge of an electron is 1−, and the charge of a proton is 1+.

The Nuclear Atom

How might protons and electrons be arranged in an atom? Thomson's model of atomic structure, called the "plum pudding" model, assumed that protons and electrons were evenly distributed throughout the atom (Figure 2.8). Ernest Rutherford designed an experiment to test this model, and his associate, Hans Geiger, carried it out. The experiment involved bombarding thin gold foil with alpha particles. *Alpha particles* were known at the time as positively charged

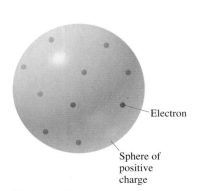

Electron

Sphere of positive charge

Figure 2.8
Thomson's model suggested that electrons in the atom might be embedded in a sphere of positive charge, like raisins in plum pudding.

particles thousands of times greater in mass than electrons. (Today we know them as helium atoms that have lost their electrons.) According to the plum pudding model, none of the alpha particles should have been affected by the dispersed bits of positive and negative charge in the gold atoms. They should have zipped right through the gold foil, and most did. However, some were deflected slightly, and a few actually bounced backwards, as shown in Figure 2.9. The result was quite unexpected. It was as if you fired a bullet at a sheet of tissue paper and it came back and hit you! The deflection of these massive alpha particles suggested that most of the mass of the atom had to be concentrated in a positively charged core, which Rutherford called the **nucleus.** The electrons, he reasoned, had to be dispersed in the large volume outside of the nucleus. The large electron space was the area penetrated by most of the alpha particles. Only if an alpha particle came close enough to the incredibly dense nucleus would it be deflected from its original path. The alpha particles that hit the nucleus of a gold atom head-on were deflected backwards.

Rutherford's experiment was the basis for the *nuclear model of the atom* (Figure 2.10), developed in 1907. The model suggested that the nucleus contains the protons and most of the mass of the atom. The electrons exist outside the nucleus in what is often called an "electron cloud."

The diameter of the nucleus is about 10^{-14} m, and the diameter of the atom is about 10^{-10} m. These relative sizes are comparable to a flea in the center of a domed stadium. The mass of a proton, 1.6726×10^{-24} g, is nearly the same as the mass of a hydrogen atom. The proton is equal in charge (but opposite in sign) to the electron, but about 1840 times greater in mass.

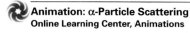

Animation: α-Particle Scattering
Online Learning Center, Animations

Chemistry Animations Library, α-Particle Scattering, 3Rutherford_Expt.exe

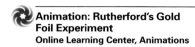

Animation: Rutherford's Gold Foil Experiment
Online Learning Center, Animations

Chemistry Animations Library, Rutherford's Gold Foil Experiment, 2Rutherford_Expt.rm

Although a penny contains about 1×10^{22} atoms, most of the penny is empty space. Why?

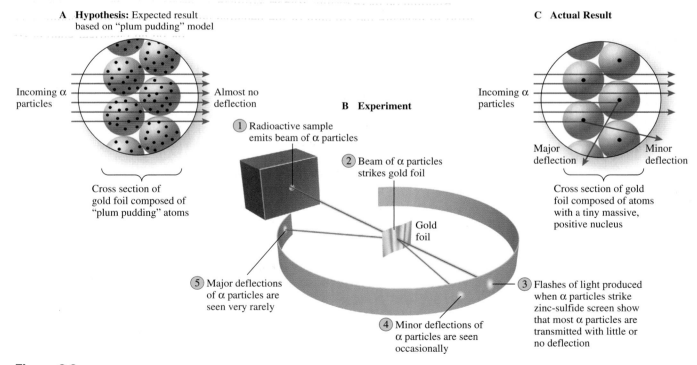

A **Hypothesis:** Expected result based on "plum pudding" model

Incoming α particles

Almost no deflection

Cross section of gold foil composed of "plum pudding" atoms

B **Experiment**

1 Radioactive sample emits beam of α particles

2 Beam of α particles strikes gold foil

Gold foil

5 Major deflections of α particles are seen very rarely

4 Minor deflections of α particles are seen occasionally

C **Actual Result**

Incoming α particles

Major deflection

Minor deflection

Cross section of gold foil composed of atoms with a tiny massive, positive nucleus

3 Flashes of light produced when α particles strike zinc-sulfide screen show that most α particles are transmitted with little or no deflection

Figure 2.9
Rutherford's gold foil experiment led to the nuclear model of the atom. A beam of positively charged alpha particles was directed through a thin layer of gold. A zinc-sulfide screen detected the alpha particles by producing a flash of light upon contact. (A) According to the plum pudding model, all alpha particles should have penetrated straight through the gold atoms. (B) In the experiment, many of the alpha particles penetrated straight through the atom, but some were deflected. (C) The nuclear model of the atom explains the experimental results. The positively charged protons are packed tightly together in the very center of the nucleus. When the nucleus is in the path of an alpha particle, the alpha particle is deflected from its original path.

Figure 2.10
In the nuclear model of the atom, protons and neutrons are located in a tiny nucleus at the center of the atom. The space outside the nucleus is occupied by the electrons.

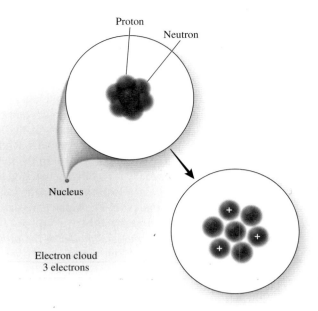

These experiments helped Rutherford and other scientists understand the structure of the atom better, but they could not account for the entire mass of the atom. Most atoms other than the hydrogen atom have masses that are at least twice the sum of the masses of the protons and electrons they contain. For example, calcium contains 20 protons and 20 electrons. Together, their mass is 3.3471×10^{-23} g. Yet a calcium atom has a mass that is nearly twice this value (6.6359×10^{-23} g).

To account for the extra mass, Rutherford hypothesized the neutron. A **neutron** is an uncharged particle in the nucleus of the atom. Because of the electrical neutrality of the neutron, it was difficult to find. It was not until 1932 that James Chadwick, a scientist working with Rutherford, did experiments that detected the neutron. The mass of the neutron was determined to be 1.6749×10^{-24} g, slightly greater than the mass of a proton. The properties of the electron, proton, and neutron are summarized in Table 2.1. All the subatomic particles in the table are important to the nuclear model of the atom as we understand it today (Figure 2.10).

In Chapters 7 and 8 you will learn how electrons relate to chemical reactivity. In Chapter 15, you will learn how neutrons relate to nuclear radioactivity. In this chapter, we will focus on how protons determine the identity of an atom, neutrons help determine its mass, and electrons determine its charge.

Isotopes, Atomic Number, and Mass Number

Recall that Anna was wondering how atoms of carbon and hydrogen differ. What is it about an atom of an element that distinguishes it from an atom of some other element? The same scientists who identified the subatomic particles found an answer to this question. They learned, for example, that all hydrogen atoms contain just one proton, and that any atom containing just one proton is a hydrogen atom. In a similar fashion, any atom that contains two protons is a helium atom. An atom with three protons is a lithium atom, and so on. The number of protons in an atom's

TABLE 2.1	Subatomic Particles		
Particle	Mass (g)	Actual Charge (C)	Relative Charge
Electron	9.1094×10^{-28}	-1.6022×10^{-19}	1−
Proton	1.6726×10^{-24}	$+1.6022 \times 10^{-19}$	1+
Neutron	1.6749×10^{-24}	0	0

nucleus determines the identity of that element. The number of protons in the nucleus of each atom of an element is the **atomic number** of that element. On the periodic table in this book, elements are shown with the atomic number just above the element symbol, as shown for the element gold (Au) in Figure 2.11.

How can we determine the number of electrons and neutrons in an atom? Atoms are electrically neutral. This means that the number of electrons in an atom equals the number of protons—the atomic number. For example, the atomic number of gold (Au) is 79, so an atom of gold has 79 protons and 79 electrons. Not all atoms of an element contain the same number of neutrons, however. An **isotope** of an element is an atom that contains a specific number of neutrons. For example, most hydrogen atoms contain no neutrons. Their nucleus consists only of a single proton. Not all hydrogen atoms are alike, however. Some contain one neutron and some even have two. All three are isotopes of hydrogen. They are still hydrogen atoms because they have one proton, but their nuclei differ in the number of neutrons (Figure 2.12). The isotopes of an element have essentially identical chemical properties, but their physical properties, such as melting point and boiling point, may differ slightly.

One way to distinguish between isotopes is by their *mass number*. The **mass number** (A) of an isotope is the sum of the number of protons (Z) and the number of neutrons (N) in its nucleus:

$$\text{Mass number} = \text{number of protons} + \text{number of neutrons}$$
$$A = Z + N \qquad M = P + N$$

The three isotopes of hydrogen have different mass numbers ($A = 1, 2,$ and 3) because they have different numbers of neutrons, also called the **neutron number** ($N = 0, 1,$ and 2). The mass number is not an actual mass; it is a count of the number of particles in the nucleus.

Example 2.1 shows how you can determine the atomic number and mass number for an isotope of an element by looking at an atomic-level representation.

Figure 2.11
The atomic number of each element is indicated on the periodic table in this book just above the element symbol.

How does the existence of isotopes make it necessary to change one of Dalton's postulates?

Specific isotopes have many applications, especially in medical technology. They are important in medical imaging, testing, and treatment. Iodine-131, for example, is used to diagnose thyroid problems and to destroy thyroid tumor cells.

EXAMPLE 2.1   Determining Atomic Number and Mass Number

For the atom represented in the following diagram,

(a) Determine the number of protons and neutrons.
(b) Identify the atomic number and the element.
(c) Determine the mass number for this isotope.

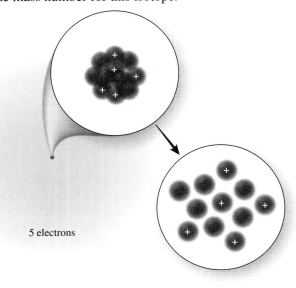

5 electrons

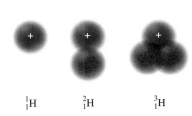

$^{1}_{1}\text{H}$ $^{2}_{1}\text{H}$ $^{3}_{1}\text{H}$

Figure 2.12
Isotopes of hydrogen vary in the number of neutrons. In these representations, the blue spheres with a plus in them represent protons. The red spheres represent neutrons. How many neutrons are in each isotope? What do the letters and numbers below each diagram represent?

Solution:

(a) The protons and neutrons make up the nucleus. The protons are positively charged. There are five protons. The neutrons have no charge. There are six neutrons.
(b) The atomic number, the number of protons, is 5. From the periodic table we see that the element with atomic number 5 is boron.
(c) The mass number, the sum of the protons and neutrons, is 11.

Practice Problem 2.1

For the atom represented in the following diagram,

(a) Determine the number of protons and neutrons.
(b) Identify the atomic number and the element.
(c) Determine the mass number for this isotope.

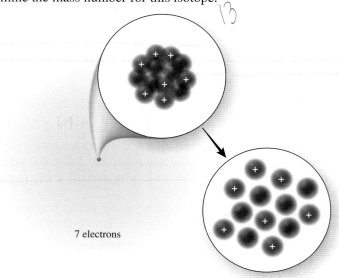

7 electrons

Further Practice: 2.29 and 2.30

As shown by Example 2.2, you can determine the number of each type of sub-atomic particle when given both the atomic number and the mass number.

EXAMPLE 2.2   Numbers of Subatomic Particles

The only stable isotope of naturally occurring fluorine has a mass number of 19. How many protons, electrons, and neutrons are in a fluorine atom?

Solution:

The periodic table shows the atomic number of fluorine is 9. Because the atomic number is equal to the number of protons, the number of protons in a fluorine atom is 9. The number of electrons in an atom equals the number of protons, so fluorine has 9 electrons. The number of neutrons can be determined from the known mass number and known number of protons. The mass number, the total number of protons and neutrons, is 19. There are 9 protons, so there must be 10 neutrons to give a mass number of 19.

Practice Problem 2.2

A rare isotope of carbon has a mass number of 14. How many protons, electrons, and neutrons are in this isotope of carbon?

Further Practice: 2.31 and 2.32

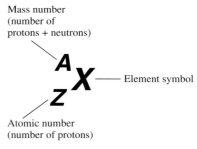

Figure 2.13
An isotope symbol consists of the element symbol with a subscript to denote the number of protons and a superscript to denote the sum of the number of protons and neutrons, the mass number.

The mass number and atomic number of an isotope are often represented by the notation called an **isotope symbol,** as shown in Figure 2.13. In an isotope symbol, X is the element symbol shown on the periodic table, A is the mass number, and Z is the atomic number (number of protons). For example, the isotope symbol for the isotope that contains one proton and no neutrons in its nucleus is ^1_1H, while hydrogen with one proton and one neutron is represented as ^2_1H (see Figure 2.12). The symbol for carbon with six protons and eight neutrons is $^{14}_6C$. Because each element can only have one atomic number Z, sometimes the atomic number is left off the isotope symbol. For example, the isotope of carbon with eight neutrons can also be represented as ^{14}C. Other common representations use the name of the element followed by the mass number (carbon-14), or the element symbol followed by the mass number (C-14).

EXAMPLE 2.3   Writing Isotope Symbols

Write two representations for the following isotope.

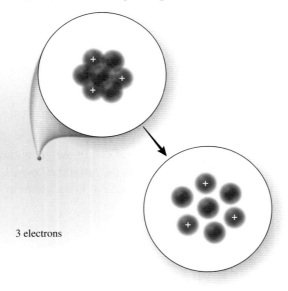

3 electrons

Solution:

There are three protons, so the atomic number is 3, corresponding to the element lithium (Li). There are four neutrons, so the mass number is 7, the sum of the protons and neutrons. This isotope can be represented as ^7_3Li, ^7Li, lithium-7, or Li-7.

Practice Problem 2.3

Write two representations for the following isotope.

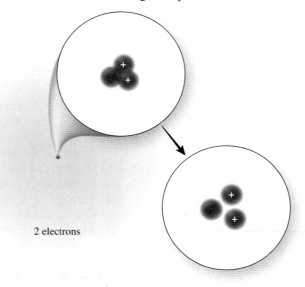

2 electrons

Further Practice: 2.33 and 2.34

As shown by Example 2.4, you can determine the number of neutrons in an atom from its isotope symbol.

$e = p \rightarrow n$

EXAMPLE 2.4   Interpreting Isotope Symbols

Determine the number of neutrons in each of the following isotopes.

(146)

(a) $^{238}_{92}U$ (b) ^{23}Na (c) hydrogen-3

Solution: (12) H^{-3} (2)

The number of neutrons is equal to the difference between the mass number and the atomic number. If the atomic number is not given in the symbol, look on the periodic table above the element.

(a) 238 − 92 = 146 neutrons
(b) 23 − 11 = 12 neutrons
(c) 3 − 1 = 2 neutrons

Practice Problem 2.4

Determine the number of neutrons in each of the following isotopes.

(78)

(a) $^{131}_{53}I$ (b) ^{37}Cl (20) (c) carbon-13 (7)

Further Practice: 2.35 and 2.36

The isotope of hydrogen that has a mass number of 2 ($_1^2$H) has a special name, *deuterium*. Deuterium (sometimes represented by the symbol D) accounts for less than 1% of naturally occurring hydrogen. It is used frequently by chemists to observe specific hydrogen atoms in experiments. When deuterium (D$_2$) combines with oxygen, deuterated water (D$_2$O) forms. It behaves chemically just like ordinary water, but it has different physical properties. The photo in Figure 2.14 shows two ice cubes in water. One ice cube is made of H$_2$O and the other of D$_2$O. What property is different for these ice cubes?

Since one ice cube sinks and the other does not, we can conclude that they have different densities. Deuterated water (D$_2$O) is denser than H$_2$O because deuterium has two particles in its nucleus and hydrogen-1 only has one. In its solid state, its density is greater than one, so it sinks in water. For this reason D$_2$O is sometimes called *heavy water.*

Figure 2.14
Which ice cube contains deuterium (hydrogen-2)? How can you tell?

2.3   IONS

In the introduction, Megan was interested in the essential minerals that our bodies need. She noted that the minerals we consume in our food are mostly metal elements in their *ion* (electrically charged) form. She was especially interested in sodium because her doctor told her to reduce it in her diet. Dietary sodium, found in large quantities in snack foods and sports drinks, is also in an ion form. How are ions different from neutral atoms?

In a neutral (uncharged) atom, the number of electrons equals the number of protons, so the overall charge on the atom is zero. This is because the negative charge on each electron is equal in magnitude and opposite in sign to the positive charge on each proton. What happens when the protons and electrons are *not* equal? Then the overall charge is not zero. The overall charge is positive if the atom contains fewer electrons than protons. It is negative if there are more electrons than protons. When an atom contains more or less electrons than protons, it has a charge and is called an **ion.**

Many elements exist naturally as ions. The elements sodium and calcium, for example, are generally *not* found in nature as neutral atoms. Instead, they exist as ions—either combined with other ions in compounds or dissolved in water. Seawater, for example, contains dissolved sodium and calcium ions. Supplies of elemental sodium and calcium that contain neutral atoms are made from the naturally occurring ionic forms by chemical reactions. Some elements, such as gold, do exist naturally as neutral atoms but can be made into ions by chemical reactions.

Ions can be classified as *cations* or *anions*. A **cation** is a positively charged ion that contains fewer electrons than the number of protons in the nucleus. The overall charge is represented as a superscript to the right of the element symbol. An example is a calcium ion, Ca^{2+}, which contains 20 protons and only 18 electrons. An **anion** is a negatively charged ion that contains more electrons than the number of protons in the nucleus. An example is Cl$^-$, which contains 17 protons and 18 electrons.

Although many ions exist naturally, they can also be formed from their neutral atoms. When a magnesium atom loses two electrons, it forms a Mg^{2+} cation. When a nitrogen atom gains three electrons, it forms an N^{3-} anion. Figure 2.15 shows the formation of cations and anions from neutral atoms. The charge on the ion is the net charge that results from unequal numbers of protons and electrons. As shown in Example 2.5, an ion symbol lets us determine the number of protons and electrons in the ion.

Ions do *not* form by gaining or losing protons or neutrons. Changing the number of protons would change the identity of the element! This does not occur under conditions other than nuclear reactions.

Figure 2.15
In an ion, the numbers of protons and electrons are not equal. (A) The cation Mg^{2+} forms when a magnesium atom loses two electrons. (B) The anion N^{3-} forms when a nitrogen atom gains three electrons.

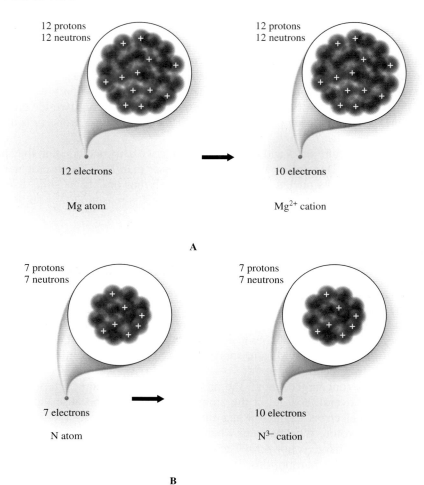

12 protons
12 neutrons

12 electrons

Mg atom

12 protons
12 neutrons

10 electrons

Mg^{2+} cation

A

7 protons
7 neutrons

7 electrons

N atom

7 protons
7 neutrons

10 electrons

N^{3-} cation

B

EXAMPLE 2.5   Ions

How many protons and electrons compose the following ions? Identify each as a cation or anion.

(a) Na^+ (b) O^{2-} (c) Cr^{3+}

Solution:

(a) The number of protons is determined from the atomic number. The atomic number of sodium is 11, so there are 11 protons in Na^+. It has a positive charge because it does not have enough electrons to balance the positive charges from the protons. A 1+ charge shows that there is one less electron than the number of protons. So its total number of electrons must be 10. The Na^+ ion is a cation because it is positively charged.

(b) The atomic number for oxygen is 8, so there are 8 protons. A negative charge results when there are more electrons than protons. Since the charge is 2−, there are 2 more electrons than protons, so there are 10 electrons. The O^{2-} ion is an anion because it is negatively charged.

(c) The atomic number of chromium is 24, so there are 24 protons in Cr^{3+}. A 3+ charge indicates that there are 3 fewer electrons than the number of protons, so there are 21 electrons. The Cr^{3+} ion is a cation because it is positively charged.

Practice Problem 2.5

How many protons and electrons compose the following ions? Identify each as a cation or anion.

(a) F^- (b) Mg^{2+} (c) N^{3-}

Further Practice: 2.43 and 2.44

If we know the number of protons and electrons in an ion, we can write an ion symbol for it. If we also know the number of neutrons, we can write a combination isotope-ion symbol by writing an isotope symbol and adding the overall charge as the right-hand superscript. This is shown in Example 2.6.

EXAMPLE 2.6   Writing Isotope Symbols for Ions

Write the isotope symbols for ions that have the following numbers of protons, neutrons, and electrons.

(a) 35 protons, 44 neutrons, and 36 electrons
(b) 13 protons, 14 neutrons, and 10 electrons
(c) 47 protons, 62 neutrons, and 46 electrons

Solution:

(a) The atomic number (number of protons) is 35 indicating the element symbol is Br. The mass number is the sum of the protons and neutrons, which is 79. The charge is 1– because there is one more electron than protons. The symbol is $^{79}_{35}Br^-$.
(b) The atomic number is 13, so the element symbol is Al. The mass number is 27. The charge is 3+ because there are three fewer electrons than protons. The symbol is $^{27}_{13}Al^{3+}$.
(c) The atomic number is 47, so the element is silver (Ag). The mass number is 109. The charge is 1+ because the number of electrons is one less than the number of protons. The symbol is $^{109}_{47}Ag^+$.

Practice Problem 2.6

Write the isotope symbols for ions that have the following numbers of protons, neutrons, and electrons.

(a) 16 protons, 18 neutrons, and 18 electrons
(b) 11 protons, 12 neutrons, and 10 electrons
(c) 20 protons, 20 neutrons, and 18 electrons

Further Practice: 2.45 and 2.46

2.4 ATOMIC MASS

How can we describe the mass of an atom of an element? Although single atoms cannot be weighed on a balance, modern techniques such as *mass spectrometry* (Figure 2.16) can be used to determine individual atomic masses accurately. For example, the mass of a single hydrogen-1 atom is 1.67380×10^{-24} g. The mass of a carbon-12 atom is 1.99272×10^{-23} g, about 12 times the mass of a hydrogen-1 atom. Numbers as small as these are difficult to remember and use. It is more convenient to think of a carbon-12 atom as being about 12 times the mass of a hydrogen-1 atom.

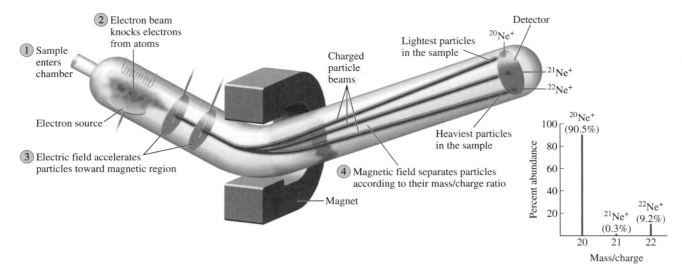

Figure 2.16

A mass spectrometer measures the masses of individual atoms. Electrons are removed from atoms (or molecules) to produce ions that are accelerated through a magnetic field. The degree of bending in the path of the ions is related to the mass and the charge of the ions. Mass spectrometry is also used to determine the relative amounts of different isotopes in a sample of an element.

For this reason, scientists have devised a method for expressing masses of atoms in a more convenient way. They use the atomic mass unit (amu). This mass scale uses carbon-12 (^{12}C), the most abundant isotope of carbon, as the standard to which all other atoms are compared. Carbon-12 is assigned an atomic mass of exactly 12 atomic mass units, or 12 amu. One **atomic mass unit (amu)** is equal to one-twelfth the mass of a carbon-12 atom:

$$1 \text{ amu} = \tfrac{1}{12} \times \text{mass of 1 C-12 atom} = 1.6606 \times 10^{-24} \text{ g}$$

Atomic mass units make it easy to compare masses of atoms. For example, if a carbon atom has a mass 12 times that of a hydrogen atom, then the mass of the hydrogen atom is about 1 amu. A hydrogen-2 atom is one-sixth the mass of the carbon-12 atom, so it has a mass of approximately 2 amu. A hydrogen-3 atom is one-fourth the mass of a carbon atom, so its mass is approximately 3 amu.

Note that there are three different, naturally occurring isotopes of hydrogen, each with a different mass. How do we describe the mass of the atoms in a sample of an element if it is composed of isotopes of different masses? From mass spectrometry (Figure 2.16), we can find the mass of each isotope and measure how much of each is present in a sample. This allows us to take a *weighted average.* The average of the mass of the individual isotopes, taking into account the naturally-occurring relative abundance of each, is the **relative atomic mass** of the element. The relative atomic mass of hydrogen, to four significant figures, is 1.008 amu. The relative atomic mass of hydrogen is closer to that of hydrogen-1 than the other isotopes because the hydrogen-1 isotope is present in the largest abundance (99.99%) in the Earth's crust and atmosphere. The relative atomic mass of carbon is 12.01 amu because carbon-12 is most abundant (98.93%), with smaller amounts of carbon-13 and carbon-14.

To find the relative amounts of each isotope present in a sample using a mass spectrometer, samples of an element from many parts of the world are analyzed. The percent abundance of each isotope is determined and an average is taken. The relative abundances from many different locations are similar.

On the periodic table on the inside front cover of this book, the relative atomic mass of each element is listed below its element symbol. Elements such as hydrogen and carbon have relative atomic mass values very close to the mass of a particular isotope. Others, like silver (Ag), have relative atomic mass values that are averages based

The term *relative atomic mass* is used by the International Union of Pure and Applied Chemistry (IUPAC), the worldwide governing body that determines official names in chemistry. Relative atomic mass is used to describe the weighted average of the atomic masses of the isotopes that compose an element as it is found in nature. This term is synonymous with the older term *atomic weight,* which is still commonly used.

on significant contributions from more than one isotope. For example, natural silver has a relative atomic mass of 107.9 amu (Figure 2.17). At first it may seem that natural silver is composed mostly of the isotope ^{108}Ag. However, silver has been found to consist of two naturally occurring isotopes: 51.82% ^{107}Ag and 48.18% ^{109}Ag. The exact mass of ^{107}Ag is 106.9051 amu, and the exact mass of ^{109}Ag is 108.9048 amu. To obtain the relative atomic mass of silver, we calculate the weighted average using the masses of Ag-107 and Ag-109. To do this, we multiply the mass of each isotope by its relative abundance, expressed in decimal form (percent divided by 100). This gives the mass contribution from each isotope. Summing the mass contributions gives the weighted average, which is the relative atomic mass listed on the periodic table:

Isotope mass × abundance = mass contribution from isotope

^{107}Ag 106.9051 amu × 0.5182	= 55.40 amu	
^{109}Ag 108.9048 amu × 0.4818	= 52.47 amu	

107.87 amu (relative atomic mass of Ag)

The relative atomic mass of silver listed on the periodic table is 107.9 amu, rounded to four significant figures.

If only two major isotopes comprise an element, we can usually determine which isotope is most abundant by referring to the relative atomic mass of the element on the periodic table. This is shown in Example 2.7.

47
Ag
107.9

Figure 2.17
The relative atomic mass of each element is indicated on the periodic table in this book just below the element symbol.

See Math Toolbox 1.2 for a review of significant figures.

EXAMPLE 2.7   Relative Atomic Mass

Naturally occurring chlorine consists of ^{35}Cl (34.9689 amu) and ^{37}Cl (36.9659 amu). Which isotope is most abundant?

Solution:

The relative atomic mass of chlorine is 35.45 amu as shown on the periodic table. Since the relative atomic mass is closer to the mass of ^{35}Cl (34.9689 amu) than the mass of ^{37}Cl (36.9659 amu), the ^{35}Cl isotope must be most abundant. (The actual percentages are 75.77% ^{35}Cl and 24.23% ^{37}Cl.)

Practice Problem 2.7

The copper mined from the Earth's crust consists of ^{63}Cu (62.93 amu) and ^{65}Cu (64.93 amu). Which isotope is most abundant?

Further Practice: 2.67 and 2.68

Example 2.8 gives you practice calculating the relative atomic mass for an element given the masses and percent abundances of its isotopes.

EXAMPLE 2.8   Calculating Relative Atomic Mass

The element magnesium is composed of three isotopes, magnesium-24, magnesium-25, and magnesium-26. The mass and percent abundance of each are listed in the following table. Calculate the relative atomic mass of magnesium, and compare your calculated value to the value in the periodic table.

Isotope	Mass (amu)	Natural Abundance (%)
^{24}Mg	23.985	78.99
^{25}Mg	24.986	10.00
^{26}Mg	25.983	11.01

Solution:

To determine the mass contribution of each isotope, multiply its mass by its percent abundance in decimal form. To find the relative atomic mass, total the mass contributions of the isotopes:

^{24}Mg   23.985 amu $\times$ 0.7899 = 18.95 amu

^{25}Mg   24.986 amu $\times$ 0.1000 = 2.500 amu

^{26}Mg   25.983 amu $\times$ 0.1101 = 2.861 amu

24.31 amu   (calculated relative atomic mass)

The calculated relative atomic mass, 24.31 amu, is the same as that on the periodic table to four significant figures.

Practice Problem 2.8

The element lithium is composed of two isotopes, lithium-6 and lithium-7. The mass and percent abundance of each is listed in the following table. Calculate the relative atomic mass of lithium, and then compare your answer to the value listed on the periodic table.

Isotope	Mass (amu)	Natural Abundance (%)
^{6}Li	6.01512	7.590
^{7}Li	7.01600	92.41

Further Practice: 2.69 and 2.70

Note that the mass of an isotope, as measured by mass spectrometry, is similar in value to its mass number. A magnesium-24 atom, for example, has a mass of 23.985 amu. Its mass number, 24, is a whole number that is close to its atomic mass. Remember that the mass number is not an actual mass; it is a count of the number of particles in the nucleus. The mass of an atom must be measured. Because the mass number and the atomic mass of an isotope are always similar in value, for convenience we often approximate the mass of an isotope as a whole number that is also its mass number. The approximate mass of a magnesium-24 atom, therefore, is about 24 amu. Its more exact mass (23.985 amu) is measured experimentally and is provided in published data tables.

We use relative atomic mass when we refer to an "average atom" of an element, or a group of atoms. For example, the mass of 100 magnesium atoms can be determined by multiplying 100 atoms by the relative atomic mass of magnesium, 24.31 amu per atom, to get a total mass of 2431 amu. What is the mass of 1000 lithium atoms?

2.5 THE PERIODIC TABLE

As Anna continued her historical research, she read about the Russian scientist Dmitri Mendeleev. He developed and published the basic arrangement of the periodic table between 1869 and 1871 (Figure 2.18), decades before the discovery of protons. Mendeleev did not arrange the elements in order of atomic number. He arranged the 63 known elements in order of increasing relative atomic mass, and he grouped elements with similar properties into columns and rows so that the properties of the elements varied in a regular pattern (periodically). Once he arranged all the elements that were known, he predicted the existence and properties of three elements that were unknown at the time: gallium (Ga), scandium (Sc), and germa-

			Ti = 50	Zr = 90	? = 180
			V = 51	Nb = 94	Ta = 182
			Cr = 52	Mo = 96	W = 186
			Mn = 55	Rh = 104,4	Pt = 197,4
			Fe = 56	Ru = 104,4	Ir = 198
		Ni = Co = 59		Pd = 106,6	Os = 199
H = 1			Cu = 63,4	Ag = 108	Hg = 200
	Be = 9,4	Mg = 24	Zn = 65,2	Cd = 112	
	B = 11	Al = 27,4	? = 68	Ur = 116	Au = 197?
	C = 12	Si = 28	? = 70	Sn = 118	
	N = 14	P = 31	As = 75	Sb = 122	Bi = 210?
	O = 16	S = 32	Se = 79,4	Te = 128?	
	F = 19	Cl = 35,5	Br = 80	J = 127	
Li = 7	Na = 23	K = 39	Rb = 85,4	Cs = 133	Ti = 204
		Ca = 40	Sr = 87,6	Ba = 137	Pb = 207
		? = 45	Ce = 92		
		?Er = 56	La = 94		
		?Yt = 60	Di = 95		
		?In = 75,6	Th = 118?		

Figure 2.18
Mendeleev based his original periodic table on the elements that were known at the time. He used it to predict the existence of other, then unknown, elements. The numbers accompanying the element symbols are relative atomic mass values that were known at the time.

nium (Ge). He also placed two pairs of elements (Co/Ni and Te/I) in an order that did not match their relative atomic masses. Using what he thought was the correct order based on the properties of these elements, Mendeleev thought that their calculated relative atomic mass values must be in error. We now know that the periodicity of the properties of the elements occurs in the order of atomic number (number of protons), not relative atomic mass.

Classification of Elements

The modern version of the **periodic table,** in which all the known elements are arranged in columns and rows to emphasize periodic properties, is shown in Figure 2.19 and on the inside cover of this book. In the periodic table, the elements are in the order of their atomic number (number of protons) from smallest to largest. There are many ways to use the classification system the periodic table provides. One way is to look at elements in the same vertical column. Because they have similar properties, we call them a **family** or a **group.** Each group has a Roman numeral (I through VIII) and a letter (A or B) associated with it. A newer system (shown in parentheses here) uses only Arabic numbers (1 through 18) to designate each group.

A horizontal row of elements in the periodic table is a **period.** Elements in the same period have properties that tend to vary in a regular fashion. Periods are labeled with numbers (1 through 7). Figure 2.20 shows a picture of elements in period 3 and in group VA (15). Note the gradual variation in physical appearance of these elements.

Look for the stair-step line that begins at boron and moves down the periodic table. This line separates the *metals* (to its left) from *nonmetals* (to its right). (The properties of metals and nonmetals were discussed in Chapter 1.) A **metalloid,** or *semimetal,* is an element that has physical properties resembling a metal but chemical reactivity more like that of a nonmetal. The metalloids lie along the stair-step line. Some metals, nonmetals, and metalloids are shown in Figure 2.21.

Any element in one of the eight groups labeled with the letter A is a **main-group element,** or a *representative element.* An element in any of the 10 groups labeled with the letter B is a **transition metal.** The 14 members of the *lanthanide* and *actinide* series of elements are usually placed on separate lines at the bottom of the table to conserve space, but they belong immediately after elements La (Z = 57) and Ac (Z = 89), respectively. Any lanthanide or actinide element is an **inner-transition metal.**

Several of the groups have descriptive names that are frequently used instead of their group numbers. Any member of group IA (1) except hydrogen is an **alkali metal.** Any element in group IIA (2) is an **alkaline earth metal.** An element in group VIIA (17) is a **halogen.** On the far right are the group VIIIA (18) elements. An element in this group is called a **noble gas.** These names remind us that elements in a group display similar properties. The alkali metals, for example, are said

A - main grp element

B - transition metal

lanthanide ⎫ inner-
actinide ⎬ tansitional
metal

Animation, Properties of Alkali and Alkaline Earth Materials
Online Learning Center, Animations

Chemistry Animations Library, Properties of Alkali and Alkaline Earth Materials, GroupI_And_II_rxn.rm

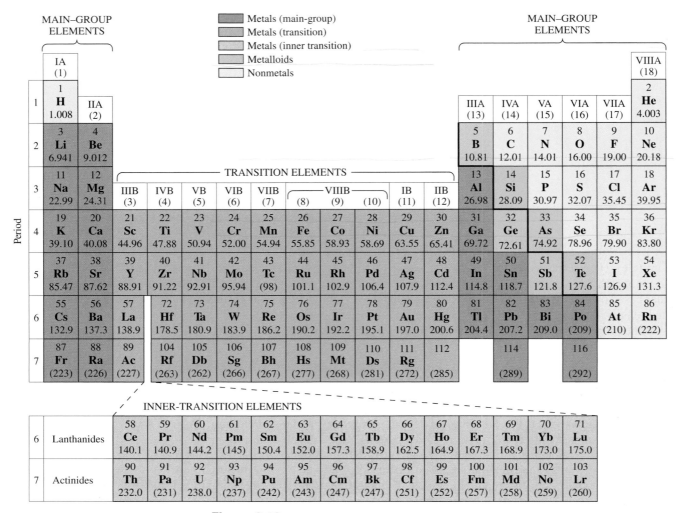

Figure 2.19
The periodic table helps us to classify elements in a variety of ways.

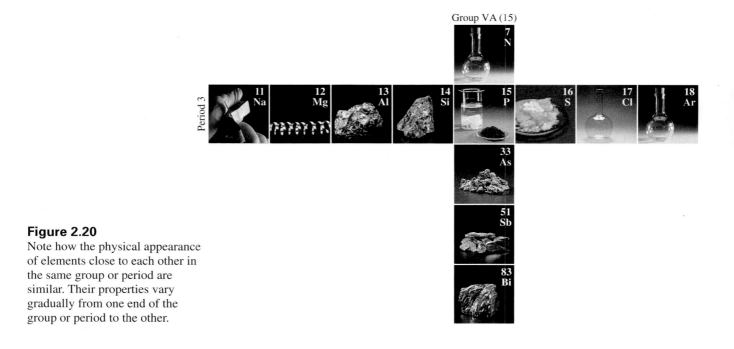

Figure 2.20
Note how the physical appearance of elements close to each other in the same group or period are similar. Their properties vary gradually from one end of the group or period to the other.

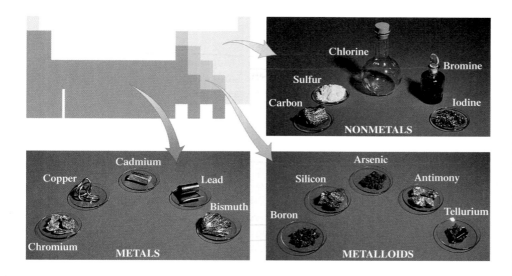

Figure 2.21
The metals, on the left side of the periodic table, have a shiny appearance. They are malleable and ductile (can be pulled into thin wires). They conduct heat and electricity. The nonmetals, on the right side of the periodic table, are gases, liquids, or powder or crystalline solids. They are insulators (nonconductors of heat and electricity). The metalloids have properties between the metals and the nonmetals.

Figure 2.22
Potassium, an alkali metal, reacts violently with water, producing flammable hydrogen gas.

to be reactive. This means they react readily with many other elements and compounds, including water (Figure 2.22). The alkaline earth metals are less reactive than the alkali metals but more reactive than most of the transition metals.

The halogens are similar in several ways. One is that they occur as elements in the form of diatomic molecules. A **diatomic molecule** is a molecule consisting of two atoms. When we represent the halogens as elements, we use a subscript to show that they are diatomic molecules: F_2, Cl_2, Br_2, I_2. Other elements that exist naturally as diatomic molecules are hydrogen (H_2), oxygen (O_2), and nitrogen (N_2). The elements that occur as diatomic molecules are shown in Figure 2.23.

The noble gases are unique in that they exist naturally in their elemental form as single atoms. They are said to be inert; that is, they do not react chemically with other elements or compounds. The noble gases were not known until the late 1800s, although many of them rank as highly abundant elements in the universe and in the Earth's atmosphere. As you will learn later, the unique stability of the noble gases helps us understand the chemical reactivity of other elements.

EXAMPLE 2.9   Classification of Elements

Classify each of the following elements by group number, group name (if applicable), and period, and as a metal, nonmetal, or metalloid.

(a) sodium   (b) silicon   (c) bromine   (d) copper

Solution:

(a) Na is in group IA (1), the alkali metal group, and in period 3, and is a metal.
(b) Si is in group IVA (14) and in period 3, and is a metalloid.
(c) Br is in group VIIA (17), the halogen group, and in period 4, and is a nonmetal.
(d) Cu is in group IB (11), a transition metal group, and in period 4, and is a metal.

Practice Problem 2.9

Identify the element that is described.

(a) the element that is in group VA (15) and in period 2, and is a nonmetal
(b) the element that is a noble gas and in period 3, and is a nonmetal
(c) the element that is an alkaline earth metal and in period 4
(d) the element that is in group IB (11), is a transition metal, and is in period 5

Further Practice: 2.75 and 2.76

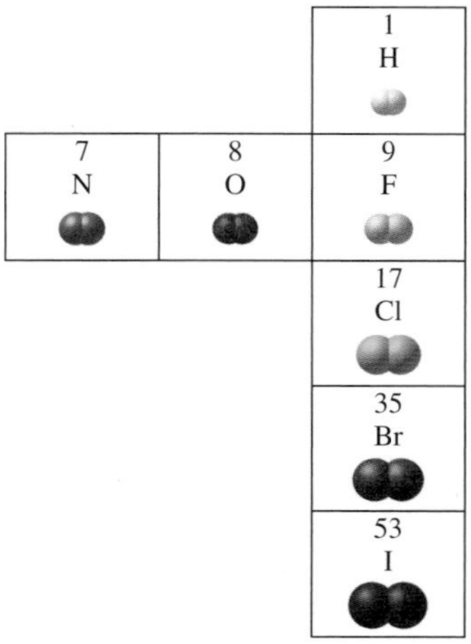

Figure 2.23
Seven elements occur as diatomic molecules.

The alkali metals (group IA) got their name because they react with water to form an alkaline (basic) solution. The word *alkali* comes from Arabic and means "ashes," the material that was originally used to make alkaline solutions. The name for the alkaline earth metals (group IIA) is derived from their presence in metal oxides, which were once called *earths*.

Astatine (At), a halogen, is found in very small amounts in the Earth's crust. It is radioactive and very short lived. Since astatine has properties most similar to iodine, scientists believe astatine also exists as diatomic molecules.

Although no compounds of noble gases exist naturally, a few stable compounds of krypton and xenon have been prepared. A few compounds of argon are known, but they are stable only at low temperatures. This picture shows one synthetic compound of a noble gas, XeF_2, adhering to the interior of the reaction vessel.

Ions and the Periodic Table

In Section 2.3, we saw that ions have charges because they have a different number of electrons than there are protons in the nucleus. Megan noticed that some of the essential nutritional elements are ions that have the same charge. For example, sodium and potassium both exist as ions with a 1+ charge. Similarly, calcium and magnesium exist as 2+ ions. She found out that the pattern is no coincidence. The position of an element in the periodic table helps us to predict the charge on its ion. To understand this trend, let's start by looking at the elements that do not form ions, the noble gases.

The noble gases are the most stable—the least reactive—of all the elements. Their stability is associated with the number of electrons they contain. To achieve a similar stability, many atoms of the main-group elements gain or lose electrons to form ions with the same electron count as the nearest noble gas. The nonmetals usually *gain* electrons to form anions that have a noble gas electron count. Most main-group metals *lose* electrons to form cations that have a noble gas electron count. For example, a sodium atom, with 11 electrons, forms a Na^+ ion by losing one electron. When it does, it has the same number of electrons as neon, 10. How many electrons does a magnesium atom lose when it forms an ion? When oxygen forms O^{2-} ions, each atom, with eight electrons, gains two electrons for a total of 10, like neon. How many electrons will a nitrogen atom gain to form an ion?

Because elements in the same group have an electron count that differs from that of the nearest noble gas by the same number, elements in the same group often form ions of the same charge. Like sodium, all group IA (1) elements lose one electron to form a cation with a charge of 1+. Group IIA (2) elements lose two electrons to form cations with a charge of 2+. Aluminum atoms of group IIIA (3) lose three electrons to form 3+ ions. Group VIIA (17) elements gain one electron to form anions with a charge of 1–. Group VIA (16) elements gain two electrons to form anions with a charge of 2–. Nitrogen and phosphorus of group VA (15) gain three electrons to form anions with a charge of 3–. The charges for ions that can be predicted from the periodic table are labeled on the abbreviated periodic table in Figure 2.24. Most of the other elements form two or more different ions of different charges. The charges of these ions cannot be predicted from the periodic table.

EXAMPLE 2.10   Predicting Charges on Ions

Write the symbol for the ion that each of the following elements is predicted to form.

(a) magnesium
(b) bromine
(c) nitrogen

Solution:

These ions can be predicted by their positions in the periodic table.

(a) Magnesium is in group IIA (2), so it will lose two electrons to form Mg^{2+}.
(b) Bromine is in group VIIA (17), so it will gain one electron to form Br^-.
(c) Nitrogen is in group VA (15), so it will gain three electrons to form N^{3-}.

Practice Problem 2.10

Write the symbol for the ion that each of the following elements is predicted to form.

(a) lithium
(b) sulfur
(c) aluminum

Further Practice: 2.95 and 2.96

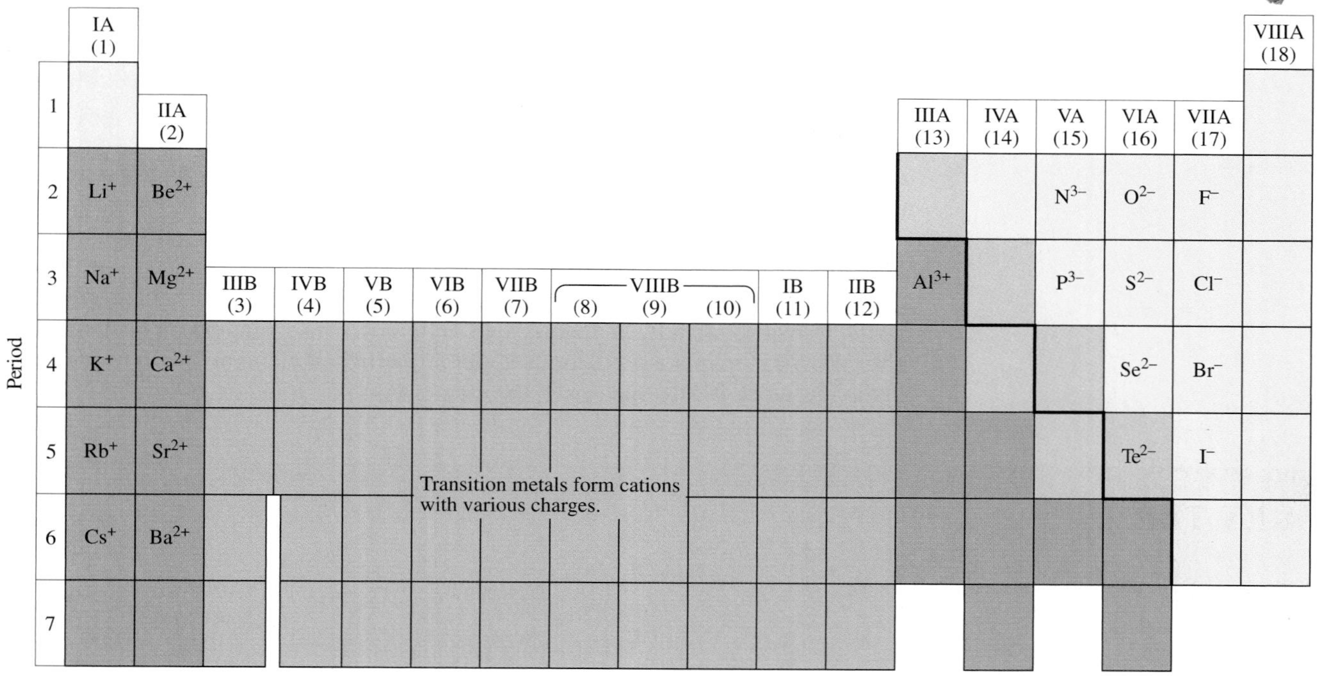

Period	IA (1)	IIA (2)	IIIB–IIB (3)–(12)	IIIA (13)	IVA (14)	VA (15)	VIA (16)	VIIA (17)	VIIIA (18)
1									
2	Li^+	Be^{2+}				N^{3-}	O^{2-}	F^-	
3	Na^+	Mg^{2+}	Transition metals form cations with various charges.	Al^{3+}		P^{3-}	S^{2-}	Cl^-	
4	K^+	Ca^{2+}					Se^{2-}	Br^-	
5	Rb^+	Sr^{2+}					Te^{2-}	I^-	
6	Cs^+	Ba^{2+}							
7									

Figure 2.24

Many of the main-group elements form ions that have charges that can be predicted by their positions in the periodic table. The ions shown here are sometimes called common ions because these elements rarely form ions of other charges.

In our bodies, ions usually exist dissolved in body fluids. Ions can also exist as components of compounds. In Chapter 3 you will learn about these *ionic compounds*. You will learn about their properties and how to name them. You will also learn how to determine charges on ions that cannot be predicted from the periodic table.

SUMMARY

The acceptance of atomic theory by the scientific community in the early 1800s paved the way for our current understanding of matter and its chemical and physical properties. Dalton's atomic theory was based on experimental evidence that supported what became the law of conservation of mass and the law of definite proportions. Atoms are the building blocks of matter, and the differences among elements result from differences in their atoms. During ordinary chemical reactions, atoms rearrange to form new substances, but atoms are neither created nor destroyed.

Today's understanding of subatomic particles began with experiments done in the early 1900s. Protons and neutrons compose the nucleus of the atom and contribute most to its mass. The mass number of an atom is the sum of the protons and neutrons in its nucleus. Protons are positively charged, and the number of protons is the atomic number of an element. Electrons are negatively charged particles that exist in the large space outside the nucleus. In a neutral (uncharged) atom, the number of protons and the number of electrons are equal. Atoms of the same element have the same number of protons. Isotopes of an element differ in the number of neutrons (and therefore mass number). Because isotope symbols always include the mass number, the number of neutrons in an isotope can be determined.

Ions contain unequal numbers of protons and electrons and have a charge. Ions are formed by the gain or loss of electrons from a neutral atom. When electrons are gained, negatively charged ions called anions form. Cations are positively charged ions formed when electrons are lost from a neutral atom.

The mass of an atom is usually reported in atomic mass units (amu). Masses of atoms have numerical values that are very close to the mass number for the isotope. The relative atomic mass for an element is listed on the periodic table in this book under the element symbol. It is the weighted average of the atomic masses of the naturally occurring isotopes.

The periodic table provides specific information about the elements. Elements with similar properties occur together in vertical columns called groups. We can also classify elements as metals, nonmetals, and metalloids; or as main-group elements, transition metals, lanthanides, or actinides. The arrangement of the periodic table allows us to make predictions about properties of elements, such as the charge on the ion when it is formed from the element.

KEY TERMS

alkali metal (2.5)

alkaline earth metal (2.5)

anion (2.3)

atomic mass unit (amu) (2.4)

atomic number (2.2)

cation (2.3)

Dalton's atomic theory (2.1)

diatomic molecule (2.5)

electron (2.2)

family (2.5)

group (2.5)

halogen (2.5)

inner-transition metal (2.5)

ion (2.3)

isotope (2.2)

isotope symbol (2.2)

law of conservation of mass (2.1)

law of definite proportions (2.1)

main-group element (2.5)

mass number (2.2)

metalloid (2.5)

neutron (2.2)

neutron number (2.2)

noble gas (2.5)

nucleus (2.2)

period (2.5)

periodic table (2.5)

proton (2.2)

relative atomic mass (2.4)

subatomic particle (2.2)

transition metal (2.5)

QUESTIONS AND PROBLEMS

The following questions and problems, except for those in the *Additional Questions* section, are paired. Questions in a pair focus on the same concept. Answers to the odd-numbered questions and problems are in Appendix D.

Matching Definitions with Key Terms

2.1   Match the key terms with the descriptions provided.
    (a) an electrically neutral subatomic particle
    (b) a law stating that the mass of the substances produced in a reaction equals the mass of the substances that reacted
    (c) a positively charged subatomic particle
    (d) a member of one of the A groups of elements in the periodic table
    (e) the average mass of an atom of an element, taking into account the masses and abundances of all the naturally occurring isotopes
    (f) the sum of the protons and neutrons in an atom
    (g) an atom of an element with a specific number of neutrons
    (h) an ion with a positive charge
    (i) any particle found within an atom
    (j) any element in group IA (1)
    (k) a chart of all the known elements ordered by increasing atomic number and arranged in columns and rows to emphasize similar properties

2.2   Match the key terms with the descriptions provided.
    (a) a member of one of the B groups of elements in the periodic table
    (b) a law stating that all samples of a pure compound always contain the same proportions of the component elements
    (c) a negatively charged subatomic particle
    (d) an ion with a negative charge
    (e) a neutral particle consisting of two bound atoms
    (f) any element in group VIIIA (18)
    (g) a set of elements in the same horizontal row in the periodic table
    (h) the number of protons in an atom
    (i) the unit of mass used to describe single atoms or small numbers of atomic particles which is equal to one-twelfth the mass of a carbon-12 atom
    (j) a set of elements in the same vertical column in the periodic table
    (k) any element in group IIA (2)
    (l) the central core of the atom which contains the protons and neutrons, and most of the mass of the atom

Dalton's Atomic Theory

2.3   Which laws did Dalton use to argue that matter consists of atoms?

2.4 What modern technique allows us to "see" the surface of atoms?

2.5 How does Dalton's atomic theory describe atoms of different elements?

2.6 How does Dalton's atomic theory explain the law of conservation of mass?

2.7 How does Dalton's atomic theory explain the law of definite proportions?

2.8 How does Dalton's atomic theory explain the formation of compounds from elements?

2.9 Does the following diagram represent a chemical reaction that obeys the law of conservation of mass? Explain your answer.

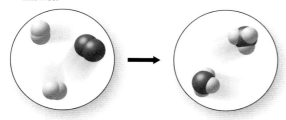

2.10 Does the following diagram represent a chemical reaction that obeys the law of conservation of mass? Explain your answer.

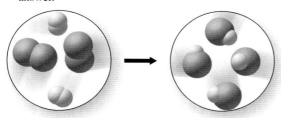

Structure of the Atom

2.11 Which experiment showed that all atoms contain negatively charged particles with masses far less than the mass of a hydrogen atom?

2.12 What information about the structure of the atom did Rutherford's gold foil experiment provide?

2.13 Which subatomic particles are negatively charged?

2.14 Which subatomic particles are positively charged?

2.15 Draw a nuclear model of a helium atom with a mass number of 4 (helium-4). Show the location of the nucleus, protons, neutrons, and electrons. Indicate the number of each type of subatomic particle.

2.16 Draw a nuclear model of a hydrogen atom with a mass number of 3 (hydrogen-3). Show the location of the nucleus, protons, neutrons, and electrons. Indicate the number of each type of subatomic particle.

2.17 Which subatomic particle has approximately the same mass as a proton?

2.18 Which of the following best describes about how many times greater in mass a proton is than an electron: 200, 2000, or 20,000?

2.19 Explain why the mass of a carbon atom is about twice the mass of the protons in a carbon atom. What accounts for the difference in mass?

2.20 Which subatomic particle was discovered last? Why was it difficult to detect?

2.21 What is the atomic number for each of the following elements?
(a) hydrogen
(b) nitrogen
(c) mercury

2.22 How many protons are in each atom of each of the following elements?
(a) helium
(b) neon
(c) gold

2.23 The number of which subatomic particle determines the identity of an element?

2.24 How do isotopes of an element differ?

2.25 What information do you need to determine the atomic number of an atom?

2.26 What information do you need to determine the mass number of an atom?

2.27 Which of the following are always the *same* for atoms of the same element?
(a) mass number
(b) atomic number
(c) neutron number
(d) mass of an atom

2.28 Which of the following are *different* for isotopes of an element?
(a) mass number
(b) atomic number
(c) neutron number
(d) mass of an atom

2.29 Given the composition of the nuclei, what are the atomic number, neutron number, and mass number of each isotope of hydrogen shown?

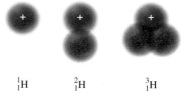

$_1^1H$ $\qquad$ $_1^2H$ $\qquad$ $_1^3H$

2.30   Identify the element from the atom shown. What are the atomic number, neutron number, and mass number of this isotope?

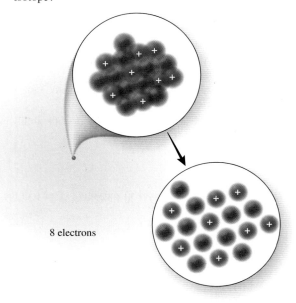

8 electrons

2.31   How many protons, neutrons, and electrons are in each of the following?
   (a)   an oxygen atom with a mass number of 15
   (b)   a silver atom with a mass number of 109
   (c)   a chlorine atom with a mass number of 35

2.32   How many protons, neutrons, and electrons are in each of the following?
   (a)   a hydrogen atom with a mass number of 1
   (b)   a magnesium atom with a mass number of 26
   (c)   a lithium atom with a mass number of 6

2.33   What is the isotope symbol for atoms that contain the following numbers of subatomic particles?
   (a)   1 proton, 1 electron, and 2 neutrons
   (b)   4 protons, 4 electrons, and 5 neutrons $^{13}_{7}N^{3-}$ ←
   (c)   15 protons, 15 electrons, and 16 neutrons

2.34   What is the isotope symbol for atoms that contain the following numbers of subatomic particles?
   (a)   2 protons and 1 neutron
   (b)   47 protons and 62 neutrons
   (c)   82 protons and 125 neutrons

2.35   How many protons and neutrons are in an atom represented by the following?
   (a) $^{56}_{26}Fe$
   (b) ^{39}K
   (c)   copper-65

2.36   How many protons and neutrons are in an atom represented by the following?
   (a) $^{11}_{5}B$
   (b) ^{68}Zn
   (c)   iodine-127

Ions

2.37   How do an atom and an ion of the same element differ?

2.38   What changes when an ion forms from an atom, or when an atom forms from an ion?

2.39   What forms when a neutral atom undergoes each of the following changes?
   (a)   It gains one electron.
   (b)   It loses two electrons.

2.40   What forms when an ion with a 1+ charge undergoes each of the following changes?
   (a)   It gains one electron.
   (b)   It loses two electrons.

2.41   Write the symbol for the ion that is formed after each of the following changes. Identify each as a cation or an anion.
   (a)   A copper atom loses two electrons.
   (b)   An iodine atom gains one electron.

2.42   Write the symbol for the ion that is formed after each of the following changes. Identify each ion as a cation or an anion.
   (a)   A sulfur atom gains two electrons.
   (b)   A silver atom loses one electron.

2.43   How many protons and electrons are found in each of the following?
   (a) Zn^{2+}
   (b) F^-
   (c) H^+

2.44   How many protons and electrons are found in each of the following?
   (a) P^{3-}
   (b) Al^{3+}
   (c) O^{2-}

2.45   Complete the following table.

Isotope Symbol	Number of Protons	Number of Neutrons	Number of Electrons
$^{37}_{17}Cl^-$	17	20	18
$^{25}_{12}Mg^{2+}$	12	13	10
	7	6	10
$^{40}Ca^{2+}$	20	20	18

2.46   Complete the following table.

Isotope Symbol	Number of Protons	Number of Neutrons	Number of Electrons
$^{81}_{35}Br^-$			
	38	50	36
	1	1	2
$^1H^+$			

2.47   What element has 10 electrons when it forms a cation with a 1+ charge?   Na

2.48   What element has 10 electrons when it forms an anion with a 2– charge?

2.49   What element has 27 electrons when it forms a cation with a 2+ charge?

2.50   What element has 46 electrons when it forms a cation with a 1+ charge?

2.51 How do $^7\text{Li}^+$ and ^6Li each differ from a neutral lithium-7 atom? Which differs from lithium-7 by the greatest mass?

2.52 How do $^{79}_{35}\text{Br}^-$ and $^{81}_{35}\text{Br}$ differ in their numbers of subatomic particles? Which should have the greater mass?

2.53 Approximately how much greater in mass is a D_2 molecule than an H_2 molecule?

2.54 Approximately how much greater in mass is a D_2O molecule than an H_2O molecule? (Assume the oxygen atoms have a mass number of 16.)

2.55 Approximately how much greater in mass is a krypton-80 atom than an argon-40 atom?

2.56 Approximately how much greater in mass is a magnesium-24 atom than a carbon-12 atom?

Atomic Mass

2.57 What is the basis for the atomic mass unit (amu) scale?

2.58 What is the mass of a carbon-12 atom on the amu scale? Is this value exact or approximate?

2.59 What is the *approximate* mass, in atomic mass units, of the following isotopes?
(a) ^2_1H
(b) $^{238}_{92}\text{U}$

2.60 Estimate the combined mass (in atomic mass units) of ten cobalt-59 atoms.

2.61 Why do we use the amu mass scale instead of the gram mass scale when discussing masses of atoms?

2.62 What is the relationship between grams and atomic mass units? How many atomic mass units are in 1 g?

2.63 What is the difference between the mass of an atom and the mass number of an atom?

2.64 Which isotope has exactly the same mass (in atomic mass units) as its mass number?

2.65 How is the mass of an individual atom determined? How do you determine the mass number of an atom?

2.66 Which of the following is most likely the mass of a nickel-62 atom: 62.0000000 amu, 62.495654 amu, 61.5871338 amu, or 61.9283461 amu. Why?

2.67 Naturally occurring calcium is composed of two isotopes: calcium-40 and calcium-44. Which isotope is most abundant?

2.68 Naturally occurring silicon is composed of just one isotope. Write its symbol.

2.69 An unknown element (X) discovered on a planet in another galaxy was found to exist as two isotopes. Their atomic masses and percent abundances are listed in the following table. What is the relative atomic mass of the element?

Isotope	Mass (amu)	Natural Abundance (%)
^{22}X	21.995	75.00
^{20}X	19.996	25.00

2.70 Suppose that the isotope abundance of magnesium on another planet has been determined to be different from that on Earth as shown in the following table. Calculate the relative atomic mass of magnesium on this planet.

Isotope	Mass (amu)	Natural Abundance (%)
^{24}Mg	23.985	20.00
^{25}Mg	24.985	20.00
^{26}Mg	25.983	60.00

2.71 What is the mass in amu of 1000 boron atoms?

2.72 What is the mass in amu of 1000 mercury atoms?

2.73 Which contains more atoms, 2500 amu of boron atoms or 2500 amu of mercury atoms?

2.74 A sample of pure silver and a sample of pure gold have the same mass. Which contains the greatest number of atoms?

The Periodic Table

2.75 Identify which of the elements Br, K, Mg, Al, Mn, and Ar can be classified in each of the following ways.
(a) alkali metal
(b) halogen
(c) transition metal
(d) alkaline earth metal
(e) noble gas
(f) main-group element

2.76 Identify which of the elements He, Zn, Pb, I, Ca, and Na can be classified in each of the following ways.
(a) alkali metal
(b) halogen
(c) transition metal
(d) alkaline earth metal
(e) noble gas
(f) main-group element

2.77 Name the element that is a halogen in period 2.

2.78 Name the element that is an alkaline earth metal in period 6.

2.79 What element is in group IVB (4) and in period 4?

2.80 What element is in group IVA (14) and in period 2?

2.81 Identify each of the following elements as metal, nonmetal, or metalloid.
(a) calcium
(b) carbon
(c) potassium
(d) silicon

2.82 Identify each of the following elements as a metal, nonmetal, or metalloid.
(a) phosphorus
(b) chromium
(c) arsenic
(d) sodium

2.83 Identify the following elements as a main-group element, transition metal, lanthanide, or actinide.
(a) calcium
(b) carbon
(c) lead
(d) uranium
(e) cobalt

2.84 Identify the following elements as a main-group element, transition metal, lanthanide, or actinide.
(a) phosphorus
(b) chromium
(c) plutonium
(d) barium
(e) radon

2.85 In which group of the periodic table do all the elements occur as diatomic molecules?

2.86 Seven elements occur as diatomic molecules. Write molecular formulas for each.

2.87 Which of the following elements does *not* occur as a diatomic molecule: nitrogen, fluorine, or neon?

2.88 Which of the following elements does *not* occur as a diatomic molecule: sulfur, hydrogen, or oxygen?

2.89 In which group of the periodic table do all the elements exist naturally as gases of uncombined atoms?

2.90 The noble gases are sometimes called "inert gases." Why? Why do they not form ions?

2.91 The ions of many of the main-group elements have the same number of _____ as the noble gas nearest to them in the periodic table.

2.92 In which group of the periodic table do all the elements not form ions?

2.93 Identify the groups of the periodic table in which all the elements form ions of the charge indicated.
(a) 1+ cations
(b) 2+ cations
(c) 1– anions
(d) 2– anions

2.94 For each group listed, identify the charge on the ions that they form.
(a) group VIA (16)
(b) halogens
(c) alkali metals
(d) group IIA (2)

2.95 For each element listed, write the symbol for the ion it forms.
(a) sodium
(b) oxygen
(c) sulfur
(d) chlorine
(e) bromine

2.96 For each element listed, write the symbol for the ion it forms.
(a) nitrogen
(b) phosphorus
(c) magnesium
(d) potassium
(e) aluminum

Additional Questions

2.97 When iron rusts, a compound forms with the formula Fe_2O_3. When iron nails rust, should their mass increase, decrease, or remain the same? Explain your answer.

2.98 Is the law of conservation of mass obeyed in the reaction shown? Explain.

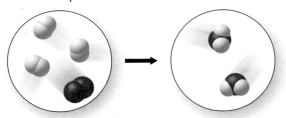

2.99 A 100-g sample of zinc sulfide contains 67.1 g zinc and 32.9 g sulfur. If a 1.34-g sample of zinc is heated with excess sulfur, 2.00 g of zinc sulfide form. Show how these data are in agreement with the law of definite proportions.

2.100 A 100-g sample of HgO contains 92.6 g mercury and 7.4 g oxygen. What mass of mercury atoms and what mass of oxygen atoms are contained in any 100-g sample of HgO?

2.101 What properties of electrons contributed to why they were the first subatomic particles to be discovered?

2.102 What experimental evidence led to each of the following advances in our understanding of atomic structure?
(a) discovery of the electron and its charge-to-mass ratio
(b) determination of the mass and charge of the electron
(c) the nuclear model of the atom

2.103 Which isotope has a mass number of 60 and an atomic number of 28?

2.104 Why is it impossible for an element to have a mass number less than its atomic number?

2.105 How many protons and neutrons are in a potassium-39 atom?

2.106 What number of each type of subatomic particle is found in most H^+ ions?

2.107 Explain why the relative atomic mass of cobalt is greater than that of nickel, although the atomic number of nickel is one greater than that of cobalt.

2.108 Which element has an isotope that has a mass approximately three times greater than its most abundant isotope?

2.109 Some tables list relative atomic mass to as many significant figures as can be reported accurately. The relative atomic mass of sulfur, 32.066 amu, is known to five significant figures, while that of fluorine, 18.9984032 amu, is known to nine significant figures. Natural samples of sulfur contain three isotopes, but natural samples of fluorine contain only one. Explain why their relative atomic masses differ in the number of significant figures reported.

2.110 How many atoms are in each of the following samples of carbon-12?
(a) 120 amu
(b) 12,000 amu
(c) 7.22×10^{24} amu

2.111 Naturally occurring iodine is composed of only one isotope. What is its mass number?

2.112 How many atoms are in each of the following samples of naturally occurring iodine? (Iodine has only one naturally occurring isotope.)
(a) 127 amu
(b) 12,700 amu
(c) 7.22×10^{24} amu

2.113 A sample of pure carbon and a sample of pure iodine have the same mass. Which has the greatest number of atoms?

2.114 Two equal-volume balloons contain the same number of atoms. One contains helium and one contains argon. Comment on the relative densities of the gases in these balloons.

2.115 Naturally occurring boron comprises two isotopes, boron-10 and boron-11. The atomic mass of boron-10 is 10.013 amu. The atomic mass of boron-11 is 11.009 amu. Which of the following is the best estimate of the percent abundance of each isotope of boron? Why?

50.0% boron-10 and 50.0% boron-11

20.0% boron-10 and 80.0% boron-11

80.0% boron-10 and 20.0% boron-11

95.0% boron-10 and 5.0% boron-11

5.0% boron-10 and 95.0% boron-11

2.116 Sodium metal reacts with water to form the compound sodium hydroxide and hydrogen gas. What is the formula for the hydrogen gas? Would you expect a similar reaction when potassium is added to water? How about copper or silver? Explain why.

2.117 Bromine is a reddish-brown liquid at room temperature. Write the formula that represents liquid bromine.

2.118 Nitrogen and oxygen are the main components of the air we breathe. Write formulas to represent these gases.

2.119 Which element is in group IA (1) but is not an alkali metal?

2.120 Which element is the only nonmetal in group IVA (14)?

Chemical Compounds

Some students taking a chemistry course went on a canoe trip to a local river. They collected water samples from several places along the river for analysis back in their lab. As they paddled downstream, two of the students, Jeff and Megan, began talking about all the different kinds of water they encounter in their daily lives: bottled, mineral, mountain spring, seltzer, tonic, carbonated, lake, ocean, pond, hard, soft, and more. In their precollege science courses, Jeff and Megan recalled learning that water is one of the most important and abundant substances on the Earth's surface and that it supports the existence of all living things on this planet. As a liquid, it occurs in the rivers, lakes, and oceans that cover over 70% of the Earth's surface. It is also found underground. Groundwater provides about three-fourths of the water used by cities in the United States. Water also exists on Earth in solid form as snow and ice at the polar ice caps and in other cold regions. It is found in the atmosphere as gaseous water vapor and clouds.

The orange juice Jeff and Megan drank as they collected samples along the river is mostly water. The salads they ate for lunch are mostly water, too; and because the day was warm and humid, water condensed on the outside of their iced tea glasses. Before they left for the trip, they showered and brushed their teeth using water, and their bathrooms were cooled by air-conditioning systems that circulate water. When they rode the bus to the canoe launch site, the vehicle released water into the air from its exhaust pipe. The water formed as a product of combustion in the engine. As they paddled the canoe, Megan and Jeff perspired. "Even our sweat is mostly water," Megan joked.

To a chemist, pure water exists as H_2O molecules in which two hydrogen atoms are connected to one oxygen atom (Figure 3.1). In this form, it is colorless, odorless, and tasteless. In nature, water is rarely pure because it has the unique ability to dissolve so many other substances. One of Megan and Jeff's assignments was to analyze their water samples for various materials—suspended and dissolved—that had been carried down the river.

To understand where those materials come from, let's consider how water makes its way from a cloud over Minnesota to the surface of the Earth, across the land to the Mississippi River, down the river to New Orleans and the Gulf of Mexico, and ultimately out into the Atlantic Ocean. The water picks up many different substances as it travels, such as minerals and organic matter that will later be absorbed or ingested by plants and animals. Falling through the air as rain, it dissolves gases from the atmosphere. As it percolates through the ground and runs down streambeds, it dissolves a variety of minerals and gases. As it flows in smaller rivers that feed the Mississippi, it gathers more minerals. Finally it becomes "salty" as it mixes with seawater in the Gulf of Mexico. Water from the surface enters the atmosphere as puddles and ponds evaporate. Evaporations from rivers, lakes, and oceans also contribute to atmospheric water. Transpiration (water loss) from plants adds water to the atmosphere, too. In the air, gaseous water condenses to form clouds, beginning the process all over again (Figure 3.2).

Figure 3.1
Pure water, to a chemist, consists of two hydrogen atoms attached to an oxygen atom.

The human body is about 70% water.

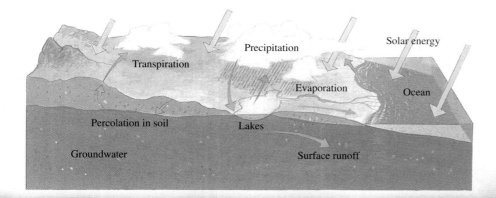

Transpiration
Precipitation
Solar energy
Evaporation
Ocean
Percolation in soil
Lakes
Groundwater
Surface runoff

Figure 3.2
Water is transported from place to place on Earth in a series of processes called the water cycle. As water flows across the land or falls through the air, many substances end up suspended or dissolved in it.

A Circuit complete   B Circuit not complete   C Circuit complete

Figure 3.3
(A) When electricity flows around a complete circuit, the bulb lights. (B) Snip any wire and the circuit is broken. (C) When the ends of the snipped wire are placed in a solution of a substance that contains ions, the bulb glows again. This is a simple procedure for testing the conductivity of a substance dissolved in water.

How does *dissolving* differ from *melting*?

Too much nitrogen and phosphorus, in the form of nitrates and phosphates, in water—from fertilizers, sewage, or livestock runoff—stimulate the growth of algae. These algal growth spurts, or blooms, soon die off. Bacteria decompose the dead plants, consuming oxygen in the process. The reduction of dissolved oxygen in the water kills fish and other aquatic organisms.

From rain cloud to ocean, water is always in the form of a solution, with water acting as the solvent. Numerous substances dissolved in the water provide nutrients to plants and animals. For example, many of the minerals needed by living organisms are present in water as ions. They include ions of iron, calcium, potassium, and magnesium. The ability of water to dissolve other substances is very important. Many of the topics we discuss throughout this book involve chemical behavior of substances dissolved in water.

When they return to the lab, Megan and Jeff will test their river water for substances of biological and chemical origin. Some are normal, harmless substances, but others that can enter the water as municipal, industrial, and agricultural wastes can be harmful. They endanger living things and contaminate the human water supply. For example, some of the bacteria that grow in polluted water cause diseases among animals and humans if the water is consumed. Some kill trees and crops. In addition, overgrowths of algae in polluted water can damage aquatic ecosystems by removing dissolved oxygen from it.

Jeff and Megan will perform several tests on their samples, including one to determine the water's electrical conductivity. They will use a simple device like that shown in Figure 3.3 to see if substances dissolved in the river water allow a complete electric circuit to operate. Notice in Figure 3.3B the snipped wire that leads down from the lightbulb. If the snipped wires were to touch, the circuit would be complete and the bulb would glow. Another way of completing the circuit is to immerse the wires in a solution that conducts electricity. For an electric current to pass through a solution, ions must be present. Jeff and Megan's conductivity experiments will tell them if chemical compounds have dissolved in the river water to form ions.

Many of the substances that dissolve in water can be classified into categories of chemical compounds. The categories vary greatly in their physical and chemical properties. In this chapter we'll examine the characteristics and behavior of these compounds and look at different ways of naming them.

Questions for Consideration

3.1 How do ionic compounds differ from molecular compounds?
3.2 What kinds of ions are in ionic compounds?
3.3 What do formulas for ionic compounds represent?

3.4 How are ionic compounds named?
3.5 What do formulas for molecular compounds represent and how are they named?
3.6 What are some common acids and bases and how are they named?
3.7 How can naming different kinds of compounds be compared?

3.1   IONIC AND MOLECULAR COMPOUNDS

Megan and Jeff analyzed their river water samples and determined that the water conducted an electric current. To investigate further, they tested several different materials in their lab. Consider the three images shown in Figure 3.4. When the electrical conductivity of pure water is tested, the lightbulb does not glow. Pure water is not a good enough conductor to complete the circuit, and neither is a solution of sugar in water. However, when electricity is passed through a sodium chloride, NaCl, solution, the bulb glows brightly. For electric current to flow through an aqueous solution, ions must be present. NaCl provides them, while pure water and sugar water do not—at least not in sufficient quantities to carry the electric current. A substance like NaCl that releases ions when dissolved in water is an **electrolyte** (Figure 3.5). One that does not, such as sugar, is a **nonelectrolyte.** Electrolyte solutions conduct electricity while nonelectrolyte solutions do not.

We can be more specific by distinguishing among strong electrolytes, weak electrolytes, and nonelectrolytes. Consider the solutions shown in Figure 3.6. Which compounds appear to break apart as ions in solution? Do some of the compounds produce more ions than others? The process by which a compound dissolves in water to produce ions is called *dissociation* or *ionization* depending on the type of compound dissolving. Because a **strong electrolyte** dissociates extensively in water, it conducts electricity well. Notice in Figure 3.6 that sodium chloride, NaCl, and hydrochloric acid, HCl, dissociate extensively. They are good conductors of electricity. Other substances, such as acetic acid, CH_3CO_2H, dissociate only partially in water. Such a substance does not conduct electricity well and is called a **weak electrolyte.** The solution containing methanol, CH_3OH, does not conduct electricity. Methanol does not dissociate into ions at all. It is a nonelectrolyte.

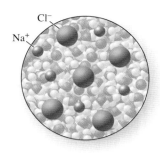

Figure 3.5
Sodium chloride (NaCl) is an electrolyte because it separates into ions when dissolved in water.

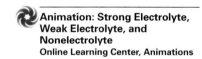

Animation: Strong Electrolyte, Weak Electrolyte, and Nonelectrolyte
Online Learning Center, Animations

Chemistry Animations Library, Strong electrolyte, weak electrolyte, and nonelectrolyte, 07_Strong_Weak_Nonelectrolytes.

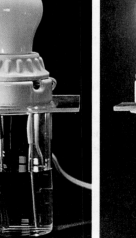

| **A** Pure water | **B** Aqueous sugar solution | **C** Aqueous NaCl solution |

Figure 3.4
When electrical conductivity is tested with a lightbulb apparatus as shown in Figure 3.3, (A) pure water and (B) a sugar solution do not appear to complete the circuit, but (C) a sodium chloride solution does—the bulb glows.

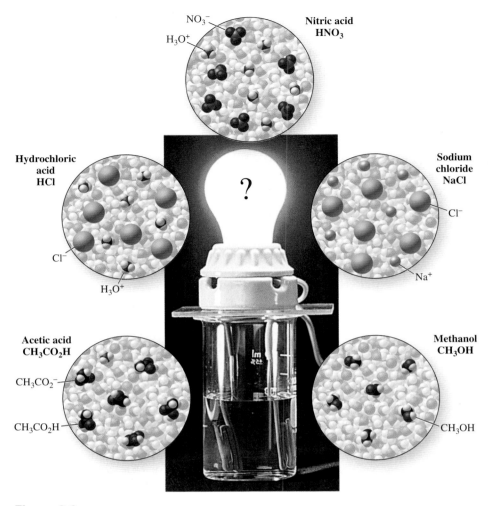

Nitric acid
HNO₃

NO_3^-

H_3O^+

Hydrochloric
acid
HCl

Cl^-

H_3O^+

Sodium
chloride
NaCl

Cl^-

Na^+

Acetic acid
CH₃CO₂H

$CH_3CO_2^-$

CH_3CO_2H

Methanol
CH₃OH

CH_3OH

Figure 3.6
Inspect the molecular-level images for each of the solutions. If any of these substances dissociate into ions, separate particles of single atoms (or atom groups) are visible. Which of the compounds dissociate completely in solution? Which only partially dissociate? Which do not dissociate at all? Note that the acids do not simply dissociate, but form H_3O^+ ions in solution.

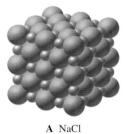

A NaCl

B CO₂

C O₂

Figure 3.7
Sodium chloride, NaCl, is an ionic compound. Carbon dioxide, CO₂, is a molecular compound. Elemental oxygen, O₂, is a molecular element. Note: The structures are not drawn to the same scale.

Many experiments, including those conducted by Jeff and Megan, have shown that, in general, compounds can be classified into two categories. An **ionic compound** (sometimes generically referred to as a *salt*) consists of oppositely charged cations and anions in proportions that give electrical neutrality. Ionic compounds are easily identified because they usually consist of ions from a *metal* with ions from a *nonmetal*. Sodium chloride, which is table salt, and is composed of Na^+ and Cl^- ions, is an example of an ionic compound (Figure 3.7A). A **molecular compound,** on the other hand, is composed of atoms from two or more *nonmetals*. It exists as a discrete unit of atoms held together in a molecule. Carbon dioxide (CO_2) is an example of a molecular compound (Figure 3.7B), and O_2 is a molecular element (Figure 3.7C).

EXAMPLE 3.1 Ionic and Molecular Compounds

Based on their formulas, which of the following compounds are ionic?

(a) KCl
(b) CO_2
(c) CaO
(d) CCl_4

Solution:

We can determine if a compound is ionic by looking at the elements that compose it. An ionic compound is usually composed of ions from a metal and a nonmetal. Two of the compounds meet this criterion, KCl and CaO.

Practice Problem 3.1

Which of the compounds listed in the example are molecular?

Further Practice: 3.7 and 3.8

To see how ionic and molecular compounds differ, let's first consider their appearance and what happens to them in aqueous solution. Ionic compounds that dissolve in water separate into the independent ions that make them up. When a compound composed of ions separates into its component ions in water, the process is described as dissociation. In contrast, molecular elements and compounds generally retain their molecular structure. The atoms contained in the molecule usually stay together when the substance dissolves in water.

Notice in Figure 3.8 the ordered arrangement of cations and anions in the ionic solid sodium chloride. This ordered arrangement is one type of solid structure called a *crystal structure* or *crystal lattice*. The strong attraction between the oppositely

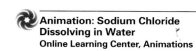

Animation: Sodium Chloride Dissolving in Water
Online Learning Center, Animations

(PMM View) from Chemical Education Research Group, Department of Chemistry, Iowa State University
www.chem.iastate.edu/group/Greenbowe/sections/projectfolder/flashfiles/thermochem/solutionSalt.html

Figure 3.8
In solid NaCl, the ions are arranged in a crystal lattice. Once sodium chloride dissolves in water, the ions of sodium and chlorine separate and become surrounded by water molecules.

Like charges repel. Opposite charges attract.

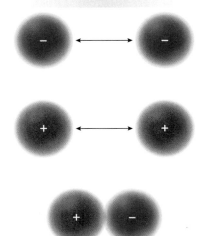

Ozone is actually a triatomic oxygen molecule, O_3. The ozone layer within the stratosphere (the atmospheric layer that extends from 10 to 50 km above the Earth) helps protect living things on the surface by absorbing much of the harmful ultraviolet radiation in sunlight.

charged ions in an ionic compound—along with the pattern in which they arrange themselves—gives ionic solids their unique properties. They are solids at room temperature, have very high melting points, and are brittle.

The crystal lattice of ionic solids breaks down when they dissolve in water. The solid dissociates into ions. Compare the molecular-level image of solid sodium chloride in Figure 3.8 to the representation of it dissolved in water. How do the ions of sodium and chlorine differ in the pictures?

We represent the process of sodium chloride dissolving with the equation

$$NaCl(s) \xrightarrow{\text{H}_2\text{O}} Na^+(aq) + Cl^-(aq)$$

where the H_2O written above the arrow indicates that water participates in the process but is not permanently changed. The (aq) after Na^+ and Cl^- indicates that these ions are in aqueous solution.

In contrast to ionic compounds, most molecular compounds remain as molecules when dissolved in water. When not in solution, molecular compounds can be gas, liquid, or solid at room temperature. They have much lower melting and boiling points than ionic compounds. Carbon dioxide (CO_2) is an example of a molecular compound. Because carbon dioxide is composed of molecules of uncharged atoms, not ions, its appearance and behavior are different from those of an ionic compound (Figure 3.9). Molecular elements also differ from ionic compounds. For example, elemental oxygen is diatomic; that is, it exists as two oxygen atoms bound together in a molecule, represented as O_2. When it dissolves in water, the oxygen atoms remain together as a molecule (Figure 3.10).

Other differences between ionic and molecular compounds are summarized in Table 3.1. To explain these differences, we can compare the molecular-level structures of ionic and molecular compounds. For example, compare solid NaCl to solid NH_3 in Figure 3.11 and imagine what is involved when these two compounds melt. The process of melting for any substance requires the separation of

Figure 3.9
While molecular compounds like carbon dioxide are composed of molecules, not ions, some can also form solids. At room temperature, CO_2 readily goes from the solid phase directly to a gaseous state. (Note that other gas particles are not shown in the molecular-level image of the gaseous state).

TABLE 3.1	Properties of Ionic and Molecular Compounds Compared	
Ionic Compounds	**Molecular Compounds**	
Crystalline solid	Gas, liquid, or solid	
Hard, brittle solid	Soft solid	
Very high melting point	Low melting point	
Very high boiling point	Low boiling point	
High density	Low density	
Strong electrolyte in aqueous solution	Weak electrolyte or nonelectrolyte in aqueous solution	
Electrical conductivity good when compound is molten	Electrical conductivity poor in pure form	

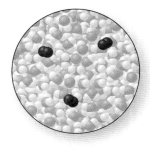

Figure 3.10
Many molecular substances such as oxygen, O_2, remain as intact molecules when dissolved in water.

the components of the compound. In the case of NaCl, the cations separate from the anions when the salt melts. The melting point of NaCl is 801°C. When NH_3 melts, individual molecules of NH_3 separate from each other. Because more heat energy is required to overcome the strong attraction between oppositely charged ions in ionic compounds, their melting points are much higher than those of molecular compounds. The melting point of NH_3 is –77.7°C. The low melting point of NH_3 indicates that there is not a strong attraction between the molecules when they are in the solid state.

Figure 3.11
Strong attractive forces between oppositely charged cations and anions must be overcome for solid NaCl to melt. Molecules of NH_3 in the solid state are not held together by oppositely charged particles. Instead, weak forces between the NH_3 molecules hold them together in the crystal. Note: The structures are not drawn to the same scale.

EXAMPLE 3.2   Properties of Ionic and Molecular Compounds

Which would you expect to have the lower melting point, KCl or CCl_4?

Solution:

Ionic compounds consist of oppositely charged ions that have a strong force of attraction between them. These compounds require significantly higher temperatures for melting. It would take a lot more energy to break apart the ions of KCl than to separate CCl_4 molecules from each other, so CCl_4 should have the lower melting point.

Practice Problem 3.2

Which would you expect to have the higher boiling point, KCl or CCl_4?

Further Practice: 3.9 and 3.10

In Example 3.2, the compounds listed are composed of only two elements. They are binary compounds. A **binary compound** is a compound containing atoms or ions of only two elements. Many chemical compounds are binary. Some ionic compounds contain atoms of two different nonmetals that form a single, binary ion. You'll learn more about these special ions in Section 3.2.

3.2   MONATOMIC AND POLYATOMIC IONS

When Jeff and Megan tested their samples from the river, they discovered a lot of sodium, magnesium, and iron in the water. These elements were not present in the form of metals. They were dissolved in the water as cations. In addition, Megan and Jeff found several anions. When they completely evaporated all the water from a sample by boiling it off, a variety of compounds remained as a solid residue in the flask. Most of them were ionic compounds.

Monatomic Ions

To better understand the composition of ionic compounds, let's look more closely at the kinds of ions that make them up. In Chapter 2 we described how the charges on the ions of some main-group elements can be predicted from their positions in the periodic table. For example, find magnesium and sulfur in the periodic table. Based on their positions, what charge would you expect for their ions? A magnesium atom has two more electrons than an atom of its nearest noble gas, neon. A magnesium ion will have the same number of electrons as a neon atom, which is two less than the magnesium atom. We represent that ion as Mg^{2+}. In contrast, a sulfur atom contains two fewer electrons than the nearest noble gas, argon. A sulfur ion will have the same number of electrons as an argon atom, which is two more than a sulfur atom. A sulfur atom becomes negatively charged with two extra electrons, represented as S^{2-}. The ions of magnesium and sulfur are monatomic. A **monatomic ion** is an ion of a single atom.

Several common monatomic ions are shown in Figure 3.12. Many ions of the main-group elements, those on the far left or far right of the periodic table, can be predicted from their positions. However, the charges of ions for elements in the middle, especially the transition metals, cannot be predicted from their group numbers. These elements often form more than one kind of ion.

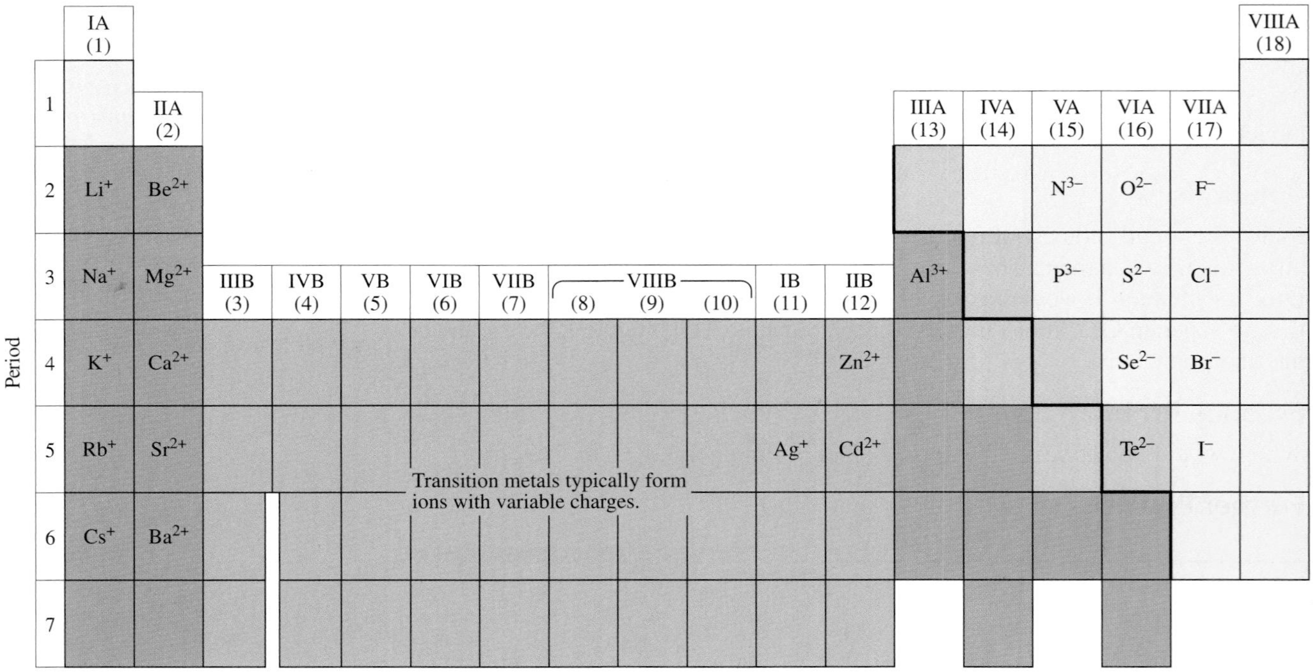

Figure 3.12
The charges on the common monatomic ions of many main-group elements correlate with the elements' positions in the periodic table. Charges on the other elements are not easily predicted from their positions. A few of the transition metals exhibit only one charge.

TABLE 3.2	Common Monatomic Ions					
Ion	Name of Ion	Ion	Name of Ion	Ion	Name of Ion
O^{2-}	oxide ion	Al^{3+}	aluminum ion	K^+	potassium ion
Na^+	sodium ion	Mg^{2+}	magnesium ion	F^-	fluoride ion
N^{3-}	nitride ion	Cl^-	chloride ion		

Nomenclature is a system of naming. To distinguish one chemical entity from another, unique names are assigned to all elements, ions, and compounds. Just as elements have unique names and symbols, ionic compounds and ions have their own names—and a special set of rules for correct naming. Several common monatomic ions are listed in Table 3.2. Try to determine some of the rules that were used to name them.

Did you figure out how to name ions? Did you determine when the *–ide* ending is used and how to use it? Now try to apply your rules in Example 3.3.

EXAMPLE 3.3   Formulas and Names of Ions

Based on their positions in the periodic table, predict the charge, write the formula, and name the ions that the following elements are expected to form.

(a) lithium
(b) sulfur

Solution:

(a) An atom of lithium is expected to lose one electron to have the same number as helium. Its ion is represented with the formula Li^+ and is named *lithium ion.*
(b) Sulfur is expected to gain two electrons to have the same number as argon. The representation for the ion of sulfur is S^{2-} and is named *sulfide ion.*

Practice Problem 3.3

Based on their positions in the periodic table, predict the charge, write the formula, and name the ions that the following elements are expected to form.

(a) barium
(b) bromine

Further Practice: 3.11 and 3.12

If you had trouble establishing rules for naming ions, the following summary may help. Monatomic anions are named as the first part of the element name (called the root of the name) and *-ide* is added as a suffix. For example, S^{2-} is known as sulfide ion. Monatomic cations are not given special endings to their root names, so we refer to them as an ion of their element name; for example, Na^+ is known as *sodium ion.*

Polyatomic Ions

Not all ions are monatomic. Some are composed of multiple atoms. A **polyatomic ion** is an ion containing two or more atoms, usually of more than one element. An example is the nitrate ion, NO_3^-. Nitrate ions are discrete units of three oxygen atoms and one nitrogen atom as represented in Figure 3.13. Although the formula

NO_3^-

Figure 3.13
In the nitrate ion, three oxygen atoms surround one nitrogen atom.

S^{2-}
Sulfide

SO$_3^{2-}$
Sulfite

SO$_4^{2-}$
Sulfate

Figure 3.14
Anions containing sulfur differ in the number of oxygen atoms attached to the sulfur. All have a 2– charge.

In a formula for a binary molecular compound or polyatomic ion, the element present in the least amount is usually central to the atoms of the other element. The central atom is usually listed first in the formula.

Many commercial and domestic water-softening products and devices are available for removing the Ca^{2+}, Mg^{2+}, and Fe^{3+} ions in hard water.

for nitrate ion might suggest that the oxygen atoms are attached to each other, they are actually attached to a central nitrogen atom. This polyatomic ion bears a charge of 1– because it has one more electron than the total number provided by the nitrogen and oxygen atoms.

The most common polyatomic ions are anions that contain oxygen attached to some other element. Such an anion is called an **oxoanion.** Oxoanions are typically combinations of oxygen with a nonmetal, although some contain metals. For a given nonmetal, there are often two common oxoanions. When this is the case, the two oxoanions differ in the number of oxygen atoms surrounding the nonmetal atom. For example, there are two known oxoanions of sulfur, both with a 2– charge. One contains a central sulfur atom with three oxygen atoms attached to it. The other is a central sulfur atom with four oxygen atoms. These oxoanions are represented in Figure 3.14 with the formulas SO$_3^{2-}$ (sulf*ite* ion) and SO$_4^{2-}$ (sulf*ate* ion).

Some water supplies contain a variety of aqueous ions, including sulfate, carbonate, phosphate, and chloride ions, as well as ions of calcium, magnesium, iron, and aluminum. Water containing the cations Ca^{2+}, Mg^{2+}, and sometimes Fe^{3+}, along with anions such as Cl$^-$, SO$_4^{2-}$, and HCO$_3^-$, is called *hard water.* When Megan and Jeff boiled away their water samples, they noticed a residue left on the glassware. They tested it and found that carbonate (CO$_3^{2-}$) ions were present. The hard water deposits were mostly metal carbonates, CaCO$_3$ and MgCO$_3$. They had noticed similar residues at home on cooking pans and drinking glasses from which water had boiled or evaporated (Figure 3.15A). Carbonate deposits also build up in pipes (Figure 3.15B). Although their formulas are more complex than those of binary compounds, CaCO$_3$ and MgCO$_3$ behave like binary ionic compounds. When CaCO$_3$ dissolves in water, Ca^{2+} and CO$_3^{2-}$ ions form (Figure 3.16). When the water is removed by evaporation or boiling, metal carbonates in solid form stay behind.

Nearly all polyatomic ions are negatively charged. An exception is the common polyatomic cation NH$_4^+$ (called ammonium ion). Like monatomic ions, polyatomic ions have their own rules for naming. Several common ones are shown in Table 3.3. Can you figure out the rules used to name them?

Now apply your rules to naming the polyatomic ions in Example 3.4.

A

B

Figure 3.15
(A) Hard water leaves behind a solid buildup of metal carbonates.
(B) Over time, solid buildup of carbonates can close off pipes.

Before   After

Figure 3.16
A solution of $CaCO_3$ contains Ca^{2+} and CO_3^{2-} ions. Boiling the water from the solution leaves behind solid deposits of calcium carbonate.

TABLE 3.3	Common Oxoanions
Formula of Ion	**Name of Ion**
NO_3^-	nitrate ion
NO_2^-	nitrite ion
SO_4^{2-}	sulfate ion
SO_3^{2-}	sulfite ion

EXAMPLE 3.4   Formulas and Names of Polyatomic Ions

There are two oxoanions of phosphorus. One has three oxygen atoms attached to a phosphorus atom and has a 3– charge. The other has the same charge and four oxygen atoms attached to a phosphorus atom. Write the formula for the phosphite ion.

Solution:

From Table 3.3 we can tell that for a particular nonmetal oxoanion, the one that contains fewer oxygen atoms gets an *–ite* ending. We represent the phosphite ion as PO_3^{3-}.

Practice Problem 3.4

Write the formula for the phosphate ion.

Further Practice: 3.23 and 3.24

If you had trouble establishing rules for naming oxoanions, the following summary may help. If there are only two oxoanions of an element, the one with the greater number of oxygen atoms is named by combining the root of the nonmetal

B BO_3^{3-} borate	C CO_3^{2-} carbonate	N N^{3-} nitride NO_3^- nitrate NO_2^- nitrite	O O^{2-} oxide O_2^{2-} peroxide	F F^- fluoride No oxoanions
	Si SiO_4^{4-} silicate	P P^{3-} phosphide PO_4^{3-} phosphate PO_3^{3-} phosphite	S S^{2-} sulfide SO_4^{2-} sulfate SO_3^{2-} sulfite	Cl Cl^- chloride ClO_4^- perchlorate ClO_3^- chlorate ClO_2^- chlorite ClO^- hypochlorite
		As As^{3-} arsenide AsO_4^{3-} arsenate AsO_3^{3-} arsenite	Se Se^{2-} selenide SeO_4^{2-} selenate SeO_3^{2-} selenite	Br Br^- bromide BrO_4^- perbromate BrO_3^- bromate BrO_2^- bromite BrO^- hypobromite
			Te TeO_4^{2-} tellurate TeO_3^{2-} tellurite	I I^- iodide IO_4^- periodate IO_3^- iodate IO_2^- iodite IO^- hypoiodite

Figure 3.17
Formulas and charges of monatomic anions and common oxoanions. Do you see any patterns?

element name and the ending -*ate*. The oxoanion with the lesser number of oxygen atoms uses the ending -*ite*. So sul*ate* is the sulfur oxoanion with the formula SO_4^{2-} and sul*ite* is the name of SO_3^{2-}. Formulas and charges for several of the most common oxoanions are shown in Figure 3.17. Do you see any patterns in the ions shown?

Some nonmetals form more than two oxoanions. For example, chlorine forms four known oxoanions. They are named as before with the –*ate* or –*ite* suffixes. However, the prefix *per-* is placed before the root of the element name and –*ate* is added to the end for the oxoanion that contains the most oxygens. The prefix *hypo–* and the ending –*ite* are used for the anion with the fewest oxygens. These, plus other polyatomic ions whose names do not fit regular patterns, are listed in Table 3.4.

TABLE 3.4   **Important Polyatomic Ions**

1– Ions		2– Ions	
nitrate	NO_3^-	chromate	CrO_4^{2-}
nitrite	NO_2^-	dichromate	$Cr_2O_7^{2-}$
bicarbonate		sulfate	SO_4^{2-}
(hydrogen carbonate)	HCO_3^-	sulfite	SO_3^{2-}
perchlorate	ClO_4^-	carbonate	CO_3^{2-}
chlorate	ClO_3^-	oxalate	$C_2O_4^{2-}$
chlorite	ClO_2^-		
hypochlorite	ClO^-	peroxide	O_2^{2-}
cyanide	CN^-	**3– Ions**	
hydroxide	OH^-	phosphate	PO_4^{3-}
		phosphite	PO_3^{3-}
acetate	$CH_3CO_2^-$	borate	BO_3^{3-}
permanganate	MnO_4^-	**1+ Ion**	
		ammonium	NH_4^+

Figure 3.18
Ionic compounds containing polyatomic ions such as calcium carbonates, CaCO₃, exist in a regular array of cations and anions. Notice the similarities between sodium chloride and calcium carbonate.

In their solid form, ionic compounds exist in an organized array of alternating positive and negative ions. Calcium carbonate, $CaCO_3$, the hard water residue that is also a major component of limestone, is another example of an ionic compound. Although its formula might suggest that it is nothing like sodium chloride, it does have similar properties. It is composed of positive and negative ions: the calcium ion Ca^{2+} and the carbonate ion CO_3^{2-}. Notice in Figure 3.18 that calcium carbonate forms an ordered array of alternating cations and anions similar to sodium chloride's crystal lattice structure. Instead of a monatomic anion, the anion in this case is CO_3^{2-}. As indicated by the solid structure, a polyatomic ion behaves as a unit.

3.3 FORMULAS FOR IONIC COMPOUNDS

Some of the ionic compounds that Megan and Jeff isolated from their river water sample were $MgCl_2$, $NaNO_3$, $CaSO_4$, and $FeCl_3$. What do these formulas represent? How are they determined? Whether in a solid or in solution, the ions that make up an ionic compound do not exist as molecules. Contrast sodium chloride and methane in Figure 3.19. The lattice structure of sodium and chloride ions is different from the molecular structure of CH_4. Methane exists as discrete molecules, each composed of one carbon atom and four hydrogen atoms.

If ionic compounds don't exist as molecules, then how can we write formulas for them? Combinations of oppositely charged ions, either in solid form or in solution, are present in quantities that result in a net charge of zero. That is, the number of positive charges equals the number of negative charges. Formulas for ionic compounds are written to reflect this fact. The sum of the positive charges must equal the sum of the negative charges:

$$\begin{array}{c} \text{Total positive charge} \\ \text{from cations} \end{array} + \begin{array}{c} \text{total negative charge} \\ \text{from anions} \end{array} = \text{zero net charge}$$

Based on their positions in the periodic table, we can predict that the charge on sodium in an ionic compound is 1+ and the charge on chlorine in an ionic compound is 1−. For the compound to be electrically neutral, the two ions have to be present in a one-to-one ratio, or Na_1Cl_1. When there is only one of a given atom or ion in a formula, the subscript 1 is understood, so the formula for sodium chloride can be written simply as NaCl.

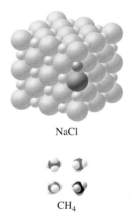

NaCl

CH₄

Figure 3.19
Ionic compounds such as NaCl do not exist as discrete molecules. The highlighted region of the NaCl structure is the formula unit for this compound. Molecular compounds such as CH_4 do exist as a discrete unit. Note: The structures are not drawn to the same scale.

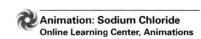
Animation: Sodium Chloride
Online Learning Center, Animations

Chemistry Animations Library, Sodium Chloride, 7Sodium_Chloride.rm, 8Sodium_Chloride.rm.

Compare this formula to the highlighted region of the sodium chloride structure in Figure 3.19. It has repeating units of one sodium ion and one chloride ion. Because an ionic compound does not exist as a molecule, its formula represents its smallest repeating unit. We call this repeating unit a **formula unit.** The formula unit of sodium chloride is NaCl. How do you know to put the sodium first? Generally, chemical names start with the name of the element that is farther to the left or farther down on the periodic table. Most often, the name or symbol of a metal is written first.

For ionic compounds, the formula represents the *ratio* of ions of each kind, not an exact count of ions. Consider another example, sodium sulfide. Sodium ion should be Na^+ and sulfide ion should be S^{2-} according to their positions in the periodic table. Two sodium ions are required to provide the 2+ charge needed to balance the 2– charge of the sulfide ion, so these ions will combine to form a compound with the formula Na_2S.

Formulas for ionic compounds containing polyatomic ions must also be written so that the sum of the positive charges from the cations equals the sum of the negative charges from the anions. For example, consider calcium carbonate (Figure 3.20). From the periodic table we know that the calcium ion has a charge of 2+, Ca^{2+}. From Table 3.4 we know carbonate ion has a charge of 2–, CO_3^{2-}. Therefore, the formula unit is written $CaCO_3$. The calcium appears first in both the name and the formula because it is the cation. Note that we cannot deduce the charge on a polyatomic ion from the periodic table. We must remember it or look it up.

Sometimes subscripts are needed to balance the charges on the cations and anions. For example, consider a compound containing calcium ion and nitrate ion. Based on its position in the periodic table we expect the calcium ion to carry a charge of 2+, Ca^{2+}. The nitrate ion is known to have a charge of 1–, NO_3^-. Two nitrate ions are required for every calcium ion to write a formula for the compound that is neutral: $Ca(NO_3)_2$. The parentheses around the formula for the polyatomic ion show that the subscript applies to the whole unit. If only one polyatomic ion is represented in a formula unit, the parentheses are not needed. For examples, see Figure 3.21.

Figure 3.20
The formula unit of calcium carbonate is $CaCO_3$.

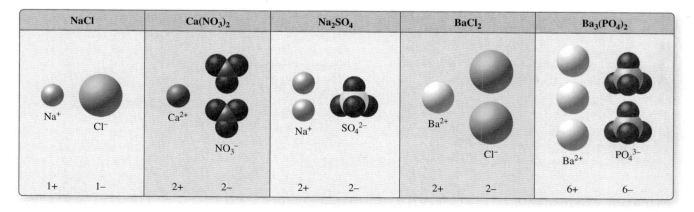

Figure 3.21
Ionic compounds are formed by combinations of cations and anions in proportions that lead to electrical neutrality. Note that the images represent building a formula unit of each compound.

EXAMPLE 3.5   Ionic Compounds

The given images represent aqueous solutions of ionic compounds. Match the images to the following formulas: $BaCl_2(aq)$, $Na_2S(aq)$, and $KBr(aq)$.

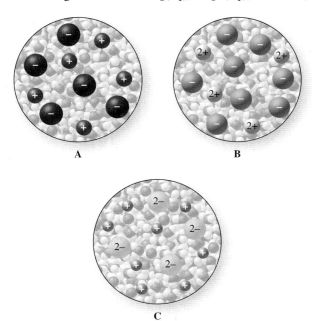

Solution:

Based on their positions in the periodic table, barium is expected to have a charge of 2+ and the chloride ion is expected to be 1−. Image B corresponds to $BaCl_2$. Based on its position in the periodic table, we expect sodium ion to have a charge of 1+. The sulfide ion has a charge of 2−. Image C corresponds to Na_2S. Based on their positions in the periodic table, the potassium ion is expected to have a charge of 1+ and the bromide ion is expected to be 1−. Image A corresponds to KBr. Notice that in all the images ions are present in appropriate ratios of cations and ions.

Practice Problem 3.5

Suppose an ionic compound containing magnesium and chloride ions is dissolved in water. Beginning with a representation for five magnesium ions, draw a picture of a solution containing these ions that is electrically neutral. You can omit the water molecules.

Further Practice: 3.27 and 3.28

In Example 3.5 we matched formulas with molecular-level representations. We can also write formulas from ion names. First, we determine if a polyatomic ion is present. Next, we determine the charge on the ions that make up the compound. If the ion is monatomic, the charge can often be predicted from the periodic table. If a polyatomic ion is present, we must recall its charge. Once we know the charges on the cations and anions in the compound, we can write a correct formula as Example 3.6 demonstrates.

EXAMPLE 3.6 **Formulas for Ionic Compounds**

Write the formulas for compounds containing the following ions:

(a) calcium ion and nitride ion
(b) barium ion and nitrate ion
(c) potassium ion and sulfate ion

Solution:

(a) Calcium ion and nitride ion are monatomic. Their charges can be determined from their positions in the periodic table. Calcium ion is expected to have a 2+ charge, and nitride ion is expected to have a 3– charge. Combining three calcium ions with two nitride ions yields a formula that has equal amounts of positive and negative charges: Ca_3N_2.

(b) Barium ion is a monatomic cation. Nitrate ion is a polyatomic anion. We can use the periodic table to predict the charge of the cation. The barium ion is expected to have a 2+ charge. From Table 3.4 we know that nitrate has a charge of 1–. It takes two nitrate ions for every one barium ion to give equal amounts of positive and negative charges: $Ba(NO_3)_2$.

(c) Potassium ion is a monatomic cation. Sulfate ion is a polyatomic anion. Based on its position in the periodic table, potassium ion is expected to have a 1+ charge. From Table 3.4 we know that the charge on sulfate ion is 2–. It takes two potassium ions for every one sulfate ion to yield equal amounts of positive and negative charges: K_2SO_4

Practice Problem 3.6

What are the formulas for all the compounds that can be formed by K^+, Fe^{3+}, Br^-, and SO_4^{2-}?

Further Practice: 3.29 and 3.30

3.4 NAMING IONIC COMPOUNDS

On their canoe trip, Megan and Jeff took along a packet of instant drink mix. As Jeff mixed the powder with water from his canteen, he noticed its list of ingredients (Figure 3.22). He asked Megan if she knew what any of the long chemical names meant. Although she didn't recognize them all, several names of compounds were familiar to her. Some were ionic compounds such as potassium iodide, copper sulfate, and magnesium oxide. Megan knew that they provide essential minerals the body needs for good health. Although she didn't know the formulas for some of the others, she recognized them as sources of other metal ions the body needs. They include ferrous fumarate, which contains iron, and sodium selenite, a source of selenium.

Jeff and Megan, like most new students to chemistry, wondered why chemical names have to be so complicated. Actually, they are not as complicated as they look at first. Chemical names have been established by applying a consistent set of rules that allow chemists to communicate accurately and efficiently. The systematic rules of nomenclature assure that each chemical compound has a unique name that distinguishes it from all other compounds. Formulas and names for several ionic compounds are listed in Table 3.5. Can you determine the rules used for naming them?

Figure 3.22
Here is a label from a protein drink mix. Do you recognize the names of any of the ingredients?

TABLE 3.5	Names of Some Ionic Compounds		
Formula	**Name**	**Formula**	**Name**
NaCl	sodium chloride	$Mg(NO_3)_2$	magnesium nitrate
$NaNO_2$	sodium nitrite	BaO	barium oxide
$MgCl_2$	magnesium chloride	Li_3N	lithium nitride

The systematic rules of nomenclature are established by the International Union of Pure and Applied Chemistry (IUPAC).

The names in the table follow from the rules explained in Sections 3.2 and 3.3. Apply them as you name the ionic compounds in Example 3.7.

EXAMPLE 3.7   Naming Ionic Compounds

Name the compounds given by the formulas (a) Na_2O and (b) $Ca_3(PO_4)_2$.

Solution:

(a) The first compound, Na_2O, is composed of cations and anions whose charges can be predicted from the periodic table. We state the name of the metal ion first, followed by the root of the name for the nonmetal with an *–ide* ending: *sodium oxide*.

(b) The charge on a calcium ion can be predicted from the periodic table. The polyatomic ion is PO_4^{3-}, named phosphate ion. The name of the compound is *calcium phosphate*.

Practice Problem 3.7

Name the compounds K_2O and $MgSO_3$.

Further Practice: 3.39 and 3.40

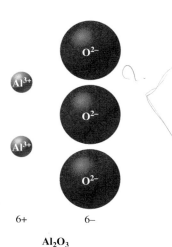

6+   6–

Al_2O_3

Figure 3.23
Because aluminum has a 3+ charge and oxygen has a 2– charge, the formula for aluminum oxide is Al_2O_3.

If you had trouble establishing rules for naming ionic compounds, the following summary may help. For binary ionic compounds, we give the name of the cation first, simply using its element name (or stating ammonium if NH_4^+ is present). To name the anion we add the suffix *–ide* to the root of the name of the element that corresponds to the anion. Generally for ionic compounds, a prefix indicating the atomic ratio is not used. (As we'll see later in Section 3.5, prefixes are used for molecular compounds.) Thus, NaCl is composed of Na^+ and Cl^- ions and is named sodium chloride. The compound with the formula $AlBr_3$ is composed of Al^{3+} ions and Br^- ions and is named aluminum bromide.

Note that names for ionic compounds do not directly communicate the relative number of ions in the formula unit. This is unnecessary because we can use the ionic charges for cations and anions predicted from the periodic table or recalled from a table such as Table 3.4 to write the formula unit. For example, the compound aluminum oxide contains an ion of an element that is a metal and an ion of an element that is a nonmetal, so we recognize it as ionic. Aluminum, in group IIIA (13), forms a 3+ ion. Oxygen, in group VIA (16), forms a 2– ion. To be electrically neutral, the formula for aluminum oxide must be Al_2O_3 (Figure 3.23).

Naming ionic compounds containing polyatomic ions is similar to naming binary ionic compounds. The cation is named first, followed by the name of the polyatomic ion. For example, the compound that contains the sodium ion (Na^+) and the sulfate ion (SO_4^{2-}) has the formula Na_2SO_4 and is called sodium sulfate. Some rules for naming ionic compounds are summarized in Figure 3.24.

Some metals can combine with a nonmetal to form more than one ionic compound. For example, copper can combine with chlorine to form CuCl and $CuCl_2$. In these two compounds, the copper has two different charges (1+ and 2+). Many of the metals, especially the transition metals, can exhibit more than one charge (Figure 3.25).

The names we assign to the two compounds of copper and chlorine must be unique so that one can be distinguished from the other. There are two ways of naming compounds of this kind. One is systematic and more current. The other uses common names that were developed—not quite so systematically—in past centuries.

Some formulas and systematic names for ionic compounds containing metals that can have more than one charge are shown in Table 3.6. Try to figure out the rules used to name them.

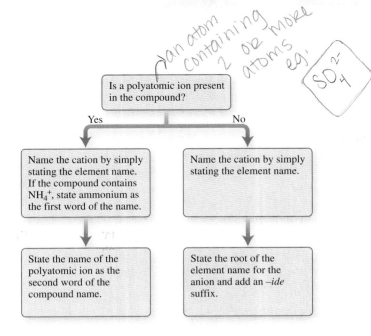

an atom containing 2 or more atoms eg.

SO_4^{2-}

Figure 3.24
A flowchart can be used to determine names for some ionic compounds.

Is a polyatomic ion present in the compound?

Yes → Name the cation by simply stating the element name. If the compound contains NH_4^+, state ammonium as the first word of the name. → State the name of the polyatomic ion as the second word of the compound name.

No → Name the cation by simply stating the element name. → State the root of the element name for the anion and add an –*ide* suffix.

$Na_2^+ SO_4^{-2}$

	IA (1)													IIIA (13)	IVA (14)	VA (15)	VIA (16)	VIIA (17)	VIIIA (18)

Periodic table with transition metal charges:

Period 4: Cr^{2+}, Cr^{3+} (VIB, 6); Mn^{2+}, Mn^{3+} (VIIB, 7); Fe^{2+}, Fe^{3+} (8); Co^{2+}, Co^{3+} (9); Cu^+, Cu^{2+} (IB, 11); Zn^{2+} (IIB, 12)

Period 5: Ag^+ (IB, 11); Cd^{2+} (IIB, 12); Sn^{2+}, Sn^{4+} (IVA, 14)

Period 6: Hg_2^{2+}, Hg^{2+} (IB, 11); Pb^{2+}, Pb^{4+} (IVA, 14)

Figure 3.25
This periodic table shows some metals that exhibit multiple charges. Notice that zinc, silver, and cadmium are exceptions. They exhibit only one charge.

Cr – Chromium
Mn – manganese
Fe – Iron
Co – cobalt
Cu – copper
Hg – mercury
Sn – tin
Pb – lead
S is an anion

TABLE 3.6	Ionic Compounds Containing Metals with Various Charges	
Compound	**Cation**	**Systematic Name**
$FeCl_2$	Fe^{2+}	iron(II) chloride
$FeCl_3$	Fe^{3+}	iron(III) chloride
Cu_2O	Cu^+	copper(I) oxide
CuO	Cu^{2+}	copper(II) oxide
$CuSO_4$	Cu^{2+}	copper(II) sulfate ← why –ate
SnO	Sn^{2+}	tin(II) oxide

Did you figure out that the Roman numerals represent charge? Try to use this rule to name the ionic compounds in Example 3.8.

EXAMPLE 3.8 **Naming Ionic Compounds Containing Metals with Variable Charges**

Name the compounds given by the formulas (a) $SnCl_2$ and (b) SnO_2.

Solution:

In both cases the charge on tin is not apparent from its position in the periodic table, so we'll have to deduce its charge in each compound from the anions.

(a) Based on its position in the periodic table, we predict that chloride ion has a 1– charge. For the compound to be electrically neutral, the tin must have a 2+ charge to counter the negative charges from the two chloride ions. We can now write the name for the compound by beginning with the metal name followed by the Roman numeral to represent its charge. Then we write the root of the nonmetal name for the anion with an –ide ending. The name of the compound is *tin(II) chloride*.

(b) Based on its position in the periodic table, we predict that oxide ion has a 2– charge. For the compound to be electrically neutral, the tin must have a 4+ charge to counter the four negative charges from the two oxide ions. We can now write the name for the compound by beginning with the metal name followed by the Roman numeral to represent its charge. Then we write the root of the nonmetal name for the anion with an –ide ending. The name of the compound is *tin(IV) oxide*.

Practice Problem 3.8

Name the compounds Cr_2S_3 and $FeSO_3$.

Further Practice: 3.43 and 3.44

Zinc, silver, and cadmium normally form only one ion, so Roman numerals are not used when naming their compounds.

If you had trouble establishing rules for naming ionic compounds containing metals that exhibit variable charges, the following summary may help. The systematic method used to name ionic compounds is called the *Stock system*. It uses Roman numerals in parentheses after the name of the metal to indicate the charge. Note that there is no space between the metal name and the parentheses containing the Roman numeral. For example, CuCl is named copper(I) chloride because the copper in this compound has a charge of 1+. The other compound of copper and chlorine, $CuCl_2$, is named copper(II) chloride because the copper has a 2+ charge. Rules for naming ionic compounds containing a cation that can exhibit variable charges are shown in Figure 3.26.

Constructing formulas from Stock system names is particularly straightforward, since the name gives the charge of the metal. For example, in chromium(III) oxide, the chromium has a charge of 3+. From its position in the periodic table, we assume oxygen has a charge of 2–. For the sum of the positive and negative charges to be equal, there must be three oxide ions for every two chromium ions, giving a formula of Cr_2O_3. To practice writing formulas from names, try Example 3.9.

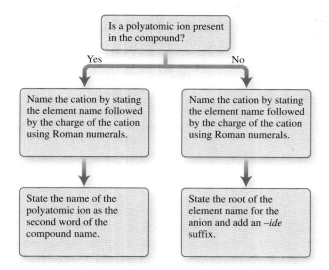

Figure 3.26
A flowchart can be used to determine names for ionic compounds that contain metals with variable charges.

EXAMPLE 3.9   **Formulas for Compounds Containing Metals That Exhibit Variable Charge**

Write the formulas for the compounds with the following names:

(a) iron(III) sulfide $Fe^{3+}_2 \; S^{2-}_3$
(b) cobalt(II) nitrate

Solution:

(a) The name tells us that iron has a 3+ charge in this compound, Fe^{3+}. Sulfide is the name of the monatomic ion of sulfur that, from its position in the periodic table, we expect to have a charge of 2–, S^{2-}. For the sum of the positive and negative charges to be equal, there must be three sulfide ions for every two iron(III) ions, giving a formula of Fe_2S_3.
(b) We know from the name that cobalt has a 2+ charge in this compound, Co^{2+}. Nitrate is the name of a polyatomic ion with a charge of 1–, NO_3^-. For the sum of the positive charges and the sum of the negative charges to be equal, there must be two nitrate ions for every one cobalt(II) ion, giving a formula of $Co(NO_3)_2$.

Practice Problem 3.9

Write the formulas for the compounds with the following names:

(a) copper(I) sulfate
(b) iron(II) oxide

Further Practice: 3.49 and 3.50

−ous (vs) −ic
↓   ↓
the lower   higher of
charge   2 charge

An older method of naming is still commonly used. In this method, the name of the metal is changed by adding a suffix, either -*ous* or -*ic*, to the root of the original Latin name for the metal. (See Chapter 4 for Latin names of some of the elements.) The suffix -*ous* indicates the lower of two charges, and the suffix -*ic* indicates the higher. For example, the two compounds of copper and chlorine described earlier have the formulas CuCl and $CuCl_2$. In the first one, the copper ion has a charge of 1+. This is the lower possible charge, so the compound is called cupr*ous* chloride.

TABLE 3.7		Names of Some Ionic Compounds Composed of Metals with Variable Charges	
Compound	**Ion**	**Stock Name**	**Old Name**
$FeCl_2$	Fe^{2+}	iron(II) chloride	ferrous chloride
$FeCl_3$	Fe^{3+}	iron(III) chloride	ferric chloride
Cu_2O	Cu^+	copper(I) oxide	cuprous oxide
CuO	Cu^{2+}	copper(II) oxide	cupric oxide
SnO	Sn^{2+}	tin(II) oxide	stannous oxide
SnO_2	Sn^{4+}	tin(IV) oxide	stannic oxide

The other compound, $CuCl_2$, contains copper ions with the higher possible charge, so it is called cup*ric* chloride. The older method is not often used in laboratories, because it applies only to metals that exhibit two charges. However, it is common on the labels of consumer products. Some examples of common names are listed in Table 3.7, along with the corresponding Stock system names. Example 3.10 asks you to employ the system of common names.

EXAMPLE 3.10   Common Names for Ionic Compounds Containing Metals That Exhibit More Than One Charge

(a) Using common nomenclature name the two compounds of iron and oxygen with the formulas FeO and Fe_2O_3.
(b) Write formulas for the compounds cuprous nitrate and cupric nitrate.

Solution:

(a) From its position in the periodic table we know that the oxygen ions in both compounds have a charge of 2–. In the first compound, FeO, the iron has a charge of 2+ to make the sum of the positive charges and the sum of the negative charges equal. In the second compound, the iron ion has a charge of 3+. The compound with the formula FeO contains the iron ion with the lower of two possible charges, so it is called *ferrous oxide*. The other compound, Fe_2O_3, contains the iron ion with the higher charge, so it is called *ferric oxide.*
(b) From Figure 3.25 we know that copper ions have two possible charges, 1+ and 2+. The name of the first compound, cuprous nitrate, indicates that it contains copper ions with the lower of two possible charges. This compound contains Cu^+ ions. Nitrate ion has the formula NO_3^-. For the sum of the positive charges and the sum of the negative charges to be equal, the formula for cuprous nitrate must be $CuNO_3$. The name of the second compound, cupric nitrate, indicates that it contains copper ions with the higher of two possible charges, so Cu^{2+} is present in this compound. For the sum of the positive charges and the sum of the negative charges to be equal, the complete formula must be $Cu(NO_3)_2$.

Practice Problem 3.10

(a) Using common nomenclature name the compounds with the formulas $Sn(SO_4)_2$ and $SnSO_4$.
(b) Write formulas for the compounds ferric sulfide and ferrous sulfide.

Further Practice: 3.51 and 3.52

3.5 NAMING AND WRITING FORMULAS FOR MOLECULAR COMPOUNDS

Among the substances Megan and Jeff found in their samples of river water were carbon dioxide and carbon tetrachloride. These are binary molecular compounds. The naming of such compounds follows rules that differ slightly from those used for naming ionic compounds. Several molecular compounds are listed in Table 3.8. Try to determine the rules used to name them.

Did you figure out how to name binary molecular compounds? Did you determine when and how to use prefixes? Now try to apply the rules to name the molecular compounds in Example 3.11.

TABLE 3.8	Names of Some Common Molecular Compounds		
Formula	**Name**	**Formula**	**Name**
CO	carbon monoxide	SO_3	sulfur trioxide
CO_2	carbon dioxide	N_2O_4	dinitrogen tetroxide
CCl_4	carbon tetrachloride	PF_5	phosphorus pentafluoride

EXAMPLE 3.11   Formulas and Names for Molecular Compounds

Name the compounds (a) P_4O_{10} and (b) NO_2.

Solution:

(a) Since P_4O_{10} is a compound containing two nonmetals, we consider it a molecular compound. Phosphorus is named first, because it occurs farther toward the left and down on the periodic table. Because there are four phosphorus atoms in the molecule, we add the prefix *tetra-*. Oxygen is named second by taking the root of its name and adding the suffix *–ide*. We must add the prefix *deca-* to the beginning of the element to indicate that there are 10 oxygen atoms in the molecule. The full name is *tetraphosphorus decoxide*. (When two vowels are adjacent, as in deca-oxide, the vowel at the end of the prefix is often dropped.)

(b) The second compound, NO_2, is also molecular because it contains elements that are nonmetals. Nitrogen is to the left of oxygen, so it is named first. Because there is only one atom of nitrogen in a molecule of the compound, the prefix *mono-* is omitted but implied. The prefix *mono-* is not used for the first element named in a molecular compound. Oxygen is named second by taking the root of its name and adding the suffix *–ide*. We add the prefix *di-* to the beginning of the element to indicate that there are two oxygen atoms in the molecule. The full name is *nitrogen dioxide*.

Practice Problem 3.11

Name the compounds P_4O_6 and N_2O_5.

Further Practice: 3.61 and 3.62

CO

CO_2

Figure 3.27
Carbon and oxygen form two common molecular compounds.

If you had trouble establishing rules for naming molecular compounds, the following summary may help. Because some nonmetals can combine to form more than one compound, we must identify the appropriate ratio of atoms in the molecule. For example, there are two common compounds containing carbon and oxygen (Figure 3.27).

SO₂

SO₃

Figure 3.28
These images represent two molecular compounds containing atoms of sulfur and oxygen.

The prefix *mono-* is generally not used except in carbon monoxide (CO) and nitrogen monoxide (NO).

The formulas SO_3 and SO_3^{2-} correspond to two very different chemical species. The first corresponds to a compound containing one sulfur atom and three oxygen atoms. The second represents a polyatomic ion that has a 2– charge.

TABLE 3.9		Common Greek Prefixes			
Prefix	**Number**	**Prefix**	**Number**	**Prefix**	**Number**
mono-	1	penta-	5	octa-	8
di-	2	hexa-	6	nona-	9
tri-	3	hepta-	7	deca-	10
tetra-	4				

One of them, CO, contains one atom each of carbon and oxygen. Its name is carbon monoxide. (The extra vowel is dropped from mono-oxide.) The other compound, CO_2, contains one atom of carbon and two atoms of oxygen. It is called carbon dioxide.

To name molecular compounds systematically we begin by stating the name of the element farther to the left and farther down on the periodic table. To the root of the second element's name we add an *–ide* ending. Unlike ionic compounds where the number of ions in the electrically neutral compound is not stated, we specifically identify the number of atoms in a molecular compound. To indicate this number we use Greek prefixes such as *mono-, di-, tri-,* and so on. (These and other prefixes are listed in Table 3.9.)

Let's consider a few compounds to clarify and practice the naming rules. First, consider a compound of sulfur and oxygen in which a molecule contains one sulfur atom and two oxygen atoms, SO_2 (Figure 3.28). Notice that sulfur is listed first in the formula. Similarly, when we name the compound, sulfur will be identified first. Since there is only one sulfur atom in the molecule, a prefix is not necessary. To the root of the name for the second element in the compound, we will add an *–ide* ending as we did for ionic compounds. We call this compound sulfur dioxide to distinguish it from other compounds containing sulfur and oxygen. Another such compound, SO_3, is shown in Figure 3.28. What is its name?

Let's consider a compound with the formula HCl. Using the rules just discussed, its name would be hydrogen monochloride. However, prefixes are not used, because there are no other compounds with this combination of elements. Its name is simply hydrogen chloride. This pattern of naming is similar for HF, HBr, and HI. Rules for naming binary molecular compounds are summarized in Figure 3.29.

Knowing the process by which compounds are named allows us to write formulas for molecular compounds from chemical names. Unlike the formula for an ionic compound, which reveals the simplest ratio of cations and anions in a formula unit, a **molecular formula** tells us the actual number of atoms present in a single molecule of a compound. The formula CO_2, for example, shows that each molecule is made up of one carbon atom and two oxygen atoms. Consider how molecular formulas are written in Example 3.12.

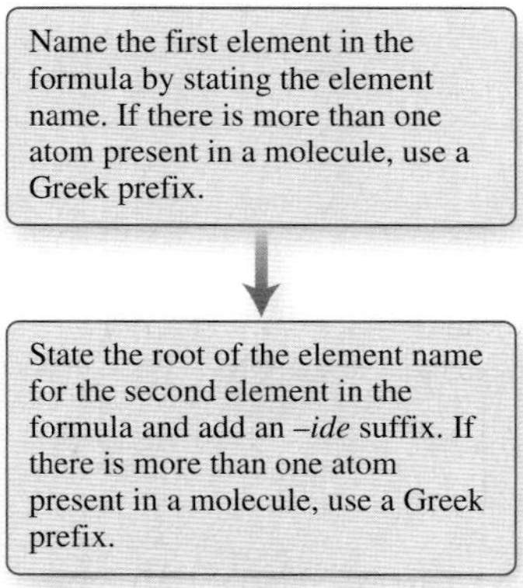

Name the first element in the formula by stating the element name. If there is more than one atom present in a molecule, use a Greek prefix.

↓

State the root of the element name for the second element in the formula and add an *–ide* suffix. If there is more than one atom present in a molecule, use a Greek prefix.

Figure 3.29
A flowchart can be used to determine names for binary molecular compounds.

EXAMPLE 3.12   Writing Formulas from Names for Molecular Compounds

What is the formula for a compound with the name chlorine dioxide?

Solution:

The first word indicates that chlorine is present in the compound and its symbol should be listed first in the formula. Since there is no prefix preceding its name, one atom is understood. The second element in the formula is oxygen. The prefix *di–* indicates that there are two oxygen atoms in a molecule of this compound. The formula is ClO_2.

Practice Problem 3.12

What is the formula for a compound with the name dinitrogen trioxide?

Further Practice: 3.63 and 3.64

Some binary molecular compounds are known by *trivial,* or nonsystematic, names. Such names often bear no direct relation to the compound's composition. For example, dihydrogen monoxide, H_2O, is commonly known by its trivial name, water. Ammonia, NH_3, and hydrogen peroxide, H_2O_2, are other examples.

In a water molecule, the oxygen atom is the central atom. The formula H_2O is an exception to listing the central atom first in a formula. Other exceptions are H_2S, H_2Se, and H_2Te.

3.6   ACIDS AND BASES

Megan and Jeff tested their river water samples using an instrument called a pH meter. It revealed that some of their samples tended to be acidic, while others were more basic. An **acid,** simply defined, is a substance that when dissolved in water provides hydrogen ions, H^+. When a compound that is not ionic provides ions when dissolved in water, the process is called *ionization.*

To understand the behavior of acids, consider hydrogen chloride, HCl. By itself this substance is a gas under normal conditions (room temperature and atmospheric pressure). It is a molecular compound that, like other acids, behaves quite differently from the molecular compounds discussed earlier. When gaseous HCl is bubbled into water, it ionizes, producing hydrogen ions, H^+, and chloride ions, Cl^-, (Figure 3.30).

In general, acids are compounds that usually contain a hydrogen atom that can be removed as an H^+ ion when the compound is placed in water. The process can be summarized for hydrogen chloride by the following equation:

$$HCl(g) \xrightarrow{\text{H}_2\text{O}} H^+(aq) + Cl^-(aq)$$

One important class of acids is the organic (carbon-based) compounds that contain the combination of atoms $-CO_2H$. They are called carboxylic acids. An example is acetic acid, CH_3CO_2H, the substance present in vinegar that gives it its sour taste.

The "acid" in acid rain comes from oxides of sulfur and nitrogen, released into the air as gases from the burning of coal and other fuels. The oxides react with water in the air to form sulfuric and nitric acids, which can damage grasslands, forests, and buildings and threaten the health of humans and other animals.

A hydrogen ion, H^+, does not actually exist in solution as an isolated entity. Instead, it is surrounded by several water molecules. An H^+ ion in solution is often represented as H_3O^+ (called *hydronium ion*) to show that the H^+ ion is really attached to water molecules.

We can represent the ionization process in a few ways:

$$HCl(g) \xrightarrow{\text{H}_2\text{O}} HCl(aq)$$

$$HCl(g) \xrightarrow{\text{H}_2\text{O}} H^+(aq) + Cl^-(aq)$$

$$HCl(g) + H_2O(l) \rightarrow H_3O^+(aq) + Cl^-(aq)$$

These expressions represent the same process, but the last is the most complete.

Figure 3.30
The molecular compound HCl ionizes to $H^+(aq)$ and $Cl^-(aq)$ when dissolved in water. $H^+(aq)$ is associated with one or more water molecules.

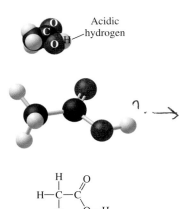

Figure 3.31

A class of organic (carbon-containing) compounds contains a –CO$_2$H group. These compounds are called carboxylic acids. The hydrogen atom attached to an oxygen atom is acidic; the others are not.

The structure of acetic acid is shown in Figure 3.31. Various organic acids are found in the human body and in many of the foods we eat. For example, citric acid is present in citrus fruits such as lemons, oranges, and grapefruit.

Not all compounds containing hydrogen are acids in water. When hydrogen combines with an element on the far right of the periodic table (excluding the noble gases), the compound is an acid. For example, HF and HCl are both acids, but methane, CH$_4$, and ammonia, NH$_3$, are not.

When writing formulas for acids, we generally place the hydrogen first and write (*aq*) after the formula to indicate that the compound is an acid when dissolved in water. For example, HCl(*g*) represents a molecular compound that exists as a gas and we call it hydrogen chloride. The formula HCl(*aq*) corresponds to a solution that's been prepared by dissolving gaseous hydrogen chloride in water and obtaining aqueous hydrogen ions and chloride ions. It's called hydrochloric acid.

Hydrogen combined with a polyatomic ion can also behave as an acid. For example, nitric acid (HNO$_3$) ionizes in water according to the equation

$$HNO_3(l) \xrightarrow{\text{H}_2\text{O}} H^+(aq) + NO_3^-(aq)$$

Many other compounds containing hydrogen and oxoanions are acids. Notice in Figure 3.32 that the hydrogen atoms are bound to oxygen atoms in the molecule. When the compound is added to water, the hydrogen atom can separate from the rest of the molecule as an aqueous H$^+$ ion.

A **base**, simply defined, is a substance that reacts with an acid in aqueous solution to form water. Most common bases either contain hydroxide ion, OH$^-$, or can provide OH$^-$ ions in solution. For example, sodium hydroxide, NaOH, is an active ingredient in chemical drain cleaners. It is a base that dissolves in water according to the equation

$$NaOH(s) \xrightarrow{\text{H}_2\text{O}} Na^+(aq) + OH^-(aq)$$

Ammonia (NH$_3$) used in various household cleaners, is another common base, but its behavior is not apparent from its formula. When gaseous ammonia is bubbled into water, a reaction occurs in which some of the ammonia molecules produce aqueous ammonium ions, NH$_4^+$(*aq*), and hydroxide ions, OH$^-$(*aq*) (Figure 3.33).

Many consumer products contain acids or bases. Some examples are shown in Figure 3.34.

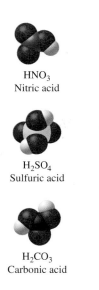

HNO$_3$
Nitric acid

H$_2$SO$_4$
Sulfuric acid

H$_2$CO$_3$
Carbonic acid

Figure 3.32

In oxoacids the acidic hydrogen atoms are attached to oxygen atoms.

Figure 3.33

Some ammonia, NH$_3$, molecules react with water to produce ammonium ions, NH$_4^+$, and hydroxide ions, OH$^-$.

TABLE 3.10   **Some Strong Acids**

Formula	Name	Formula	Name
HCl(*aq*)	hydrochloric acid	H_2SO_4(*aq*)	sulfuric acid
HNO_3(*aq*)	nitric acid	$HClO_4$(*aq*)	perchloric acid

TABLE 3.11   **Names of Some Common Acids**

Formula	Name	Formula	Name
HF(*aq*)	hydrofluoric acid	H_2SO_4(*aq*)	sulfuric acid
HCl(*aq*)	hydrochloric acid	H_2SO_3(*aq*)	sulfurous acid
HI(*aq*)	hydroiodic acid	$HClO_4$(*aq*)	perchloric acid
H_2S(*aq*)	hydrosulfuric acid	$HClO_3$(*aq*)	chloric acid
H_2CO_3(*aq*)	carbonic acid	$HClO_2$(*aq*)	chlorous acid
HNO_3(*aq*)	nitric acid	HClO(*aq*)	hypochlorous acid

Figure 3.34
Many consumer products exhibit acidic or basic properties. For example, window cleaners often contain ammonia. Sodas often contain phosphoric acid or citric acid. The next time you are in the grocery store, look at the labels on some products. How can you determine which contain acids or bases?

Acids and bases vary in the extent of their ionization or dissociation in water. Strong acids and bases ionize or dissociate completely. They are strong electrolytes. For example, hydrochloric acid is a strong acid used in many industrial and cleaning applications including some toilet bowl cleaners. It ionizes completely in water to produce ions, so in aqueous solution it is a strong electrolyte. Acetic acid, on the other hand, is present in vinegar and can be consumed safely. In water, relatively few of its molecules ionize. It is both a weak electrolyte and a weak acid. The behaviors of hydrochloric acid and acetic acid are represented in Figure 3.35.

You may encounter several strong acids while working in the laboratory. Some of them are listed in Table 3.10. These acids dissociate completely and are, therefore, strong electrolytes. They must be handled with care as they can burn skin and eyes and eat holes in clothing.

Like the nomenclature of ionic and molecular compounds, the naming of acids follows systematic rules. Table 3.11 shows the formulas and names for many common acids. Can you determine the rules for naming acids?

Did you figure out how to name binary acids? Did you determine how to name acids containing polyatomic ions? Now apply your rules to name the acids in Example 3.13.

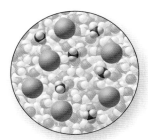
hydrochloric acid

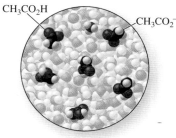
acetic acid

Figure 3.35
Hydrochloric acid dissociates completely. Most acetic acid molecules, on the other hand, remain intact.

EXAMPLE 3.13   **Naming Acids**

(a) Name the acid with the formula HBr(*aq*).
(b) The compound with the formula HNO_3(*aq*) is called nitric acid. What is the name for the acid with the formula HNO_2(*aq*)?

Solution:

(a) Since HBr(*aq*) contains just hydrogen and a nonmetal, it is a binary acid. If it were not in aqueous solution, it would be called hydrogen bromide. Since it is in aqueous solution, we can conclude that it is an acid. Binary acids have a *hydro-* prefix, the stem of the element name to which the hydrogen is attached, and an *-ic* suffix. The stem of the name derives from the name of the element hydrogen is attached to. In this case, the element bromine provides the stem *brom.* Assembling the prefix, the stem, and the suffix, along with the word *acid,* gives the name *hydrobromic acid.*

(b) In acids containing a nonmetal that can form two oxoanions, the one with the higher number of oxygen atoms is named with the root for the oxoanion with an *–ic* ending followed by the word *acid.* Thus $HNO_3(aq)$ is nitric acid as stated in the problem. The oxoanion with fewer oxygens is named with the root for the oxoanion name and an *–ous* ending. The name ends with the word *acid,* so $HNO_2(aq)$ is *nitrous acid.*

Practice Problem 3.13

(a) Name the compound $H_2Se(aq)$.
(b) An acid containing an oxoanion of phosphorus with the formula $H_3PO_3(aq)$ is named phosphorous acid. What is the name for the acid with the formula $H_3PO_4(aq)$?

Further Practice: 3.67 and 3.68

Perhaps the rules you came up with allowed you to complete Example 3.13 successfully. If you had trouble, the following summary for naming acids may help. Binary acids in solution are named with the prefix *hydro-* followed by the stem of the name of the nonmetal with the suffix *-ic* and the word *acid* attached. Thus, $HCl(aq)$ is named hydrochloric acid.

containing 2 or 1 atoms

Naming acids containing polyatomic ions requires modifying the name of the polyatomic ion. The prefix *hydro-* is not used in naming acids containing polyatomic ions. For example, the compound $H_2CO_3(aq)$, present in many carbonated beverages, contains the carbonate ion. In naming this acid, we remove the *–ate* ending from the name of the polyatomic ion and replace it with *–ic*. Then we add the term *acid* to complete the name. This acid is carbonic acid.

If there is more than one polyatomic ion for a given nonmetal, the names of the acids must distinguish among them. For example, there are two common polyatomic ions containing sulfur. They are the sulfate ion, SO_4^{2-}, and the sulfite ion, SO_3^{2-}. The acids of these ions have the formulas $H_2SO_4(aq)$ and $H_2SO_3(aq)$. In naming the acid containing the sulfate ion, the *–ate* ending is replaced with *–ic* and the term *acid* is added at the end. $H_2SO_4(aq)$ is sulfuric acid. Naming the acid containing sulfite ion requires us to replace the *–ite* ending with *–ous* and add the term *acid* to the end. Therefore, $H_2SO_3(aq)$ is called sulfurous acid. Rules for naming acids are summarized in Figure 3.36.

Figure 3.36
A flowchart can be used to determine names for acids.

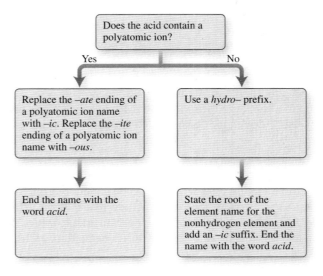

3.7   PREDICTING PROPERTIES AND NAMING COMPOUNDS

Jeff and Megan encountered a variety of compounds in their analysis of river water samples. At first they thought that some of the names for compounds were overly complicated, but once they learned the rules, they felt more comfortable. They also realized that by deducing the elements in a compound from its name or formula, they could predict many of its properties. For example, suppose they find out that a compound contains magnesium and chlorine and has the formula $MgCl_2$. Looking at the component elements in the periodic table, they see that the compound contains a metal and a nonmetal. They conclude that the compound is ionic and, therefore, that it probably has some characteristics in common with other ionic compounds. They predict that it might be a brittle solid with a very high melting point. And because it is an ionic compound, they feel justified in applying the appropriate naming rules and calling it magnesium chloride.

If you don't know which rule to apply to name a particular compound, the flowchart in Figure 3.37 provides some guidance. The questions are designed to help distinguish among ionic compounds, molecular compounds, and acids. For example, consider $Cr_2(SO_4)_3$. This compound contains a metal, so we follow the "Yes" branch in response to the first question. The metal exhibits more than one charge, so Roman numerals are necessary in its systematic name. The metal ion is chromium in the 3+ state, so the first part of the name will be chromium(III). The compound consists of the polyatomic ion SO_4^{2-}, which is the sulfate ion, so the name ends with sulfate. The name of this compound is chromium(III) sulfate.

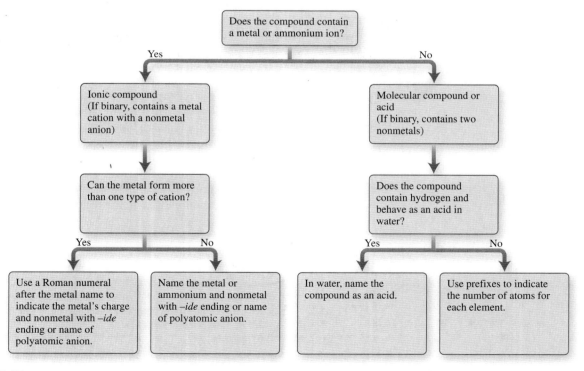

Figure 3.37
A flowchart can be used to determine which rules to apply in naming compounds.

EXAMPLE 3.14 — Using a Flowchart to Name Compounds

Name compounds with the formulas (a) CCl_4 and (b) $H_3PO_4(aq)$.

Solution:

(a) The compound CCl_4 is molecular because it is composed of two nonmetals. We list the carbon first, followed by the prefix *tetra-* to indicate four chlorine atoms. The name of this compound is *carbon tetrachloride.*

(b) The compound $H_3PO_4(aq)$ is an acid containing the phosphate ion. To name this compound we replace the *-ate* ending with *-ic* and add the word *acid.* This compound is *phosphoric acid.*

Practice Problem 3.14

Name compounds with the formulas PBr_3, $MgCl_2$, $H_2SO_4(aq)$.

Further Practice: 3.80 and 3.81

SUMMARY

In this chapter we described two broad categories of chemical compounds. Ionic compounds consist of a metal ion (or NH_4^+) combined with a nonmetal ion. Nonmetal ions can be monatomic or polyatomic. Molecular compounds consist of atoms of two or more nonmetals combined in a molecule. Ionic compounds differ from molecular compounds in their physical properties and molecular-level structure.

The ability of a chemical compound to dissociate or ionize in water to provide ions determines how well it conducts electricity. Compounds that dissociate extensively into ions are strong electrolytes. Examples include soluble ionic compounds, strong acids, and strong bases. Substances that dissociate only partially in water are weak electrolytes. Those that do not dissociate when dissolved are nonelectrolytes.

Formulas and names distinguish one compound from all others. The rules for naming compounds were established by international agreement. Formulas for molecular compounds indicate the number of atoms of each element in a molecule of the compound. Formulas for ionic compounds can only tell us the ratio of cations to anions in a compound.

Within the broad categories of ionic and molecular compounds, some chemical compounds can be described in terms of their acidic or basic properties. Acids are compounds that provide hydrogen ions in solution. Common bases are often compounds that provide hydroxide ions in solution and react with acids to form water.

Compounds are given unique names based on the elements present and how the compound behaves. The rules are summarized in the following table:

Type of Compound	Naming
Ionic	Cation named first followed by name of anion
Molecular	First atom in formula (the element farther down or to the left in the periodic table) named first with the second element named as if it were an anion. Greek prefixes are used to designate the number of atoms in the molecule.
Acids	Binary acids are named as *hydro-* followed by the root of the element name with an *-ic* suffix and the word *acid* placed at the end of the name. Acids containing polyatomic ions are named by taking the root of the polyatomic ion name, replacing *–ate* with *–ic* or replacing *–ite* with *–ous* and adding the word *acid* at the end.

KEY TERMS

acid (3.6)

base (3.6)

binary compound (3.1)

electrolyte (3.1)

formula unit (3.3)

ionic compound (3.1)

molecular compound (3.1)

molecular formula (3.5)

monatomic ion (3.2)

nonelectrolyte (3.1)

oxoanion (3.2)

polyatomic ion (3.2)

strong electrolyte (3.1)

weak electrolyte (3.1)

QUESTIONS AND PROBLEMS

The following questions and problems, except for the last section of *Additional Questions,* are paired. Questions in a pair focus on the same concept. Answers to the odd-numbered questions and problems are in Appendix D.

Matching Definitions with Key Terms

3.1   Match the key terms with the descriptions provided.
   (a) the smallest repeating unit in an ionic compound
   (b) a substance that dissociates or ionizes completely into ions in solution and conducts electricity well
   (c) a compound composed of two or more nonmetals that exists in discrete units of atoms held together
   (d) a compound that releases hydrogen ions when dissolved in water
   (e) a substance that retains its molecular identity when it dissolves and does not conduct electricity
   (f) an anion containing oxygen attached to some other element

3.2   Match the key terms with the descriptions provided.
   (a) a formula that indicates the actual number of atoms present in a molecule
   (b) a substance that dissociates partially into ions upon dissolving and conducts electricity to a slight extent
   (c) a compound consisting of cations and anions in proportions that give electrical neutrality
   (d) a substance that reacts with an acid to form water
   (e) a compound containing atoms or ions of two elements
   (f) an ion containing two or more atoms, such as NO_3^-

Ionic and Molecular Compounds

3.3   Identify the following compounds containing sulfur as representations of ionic or molecular compounds.

K_2S

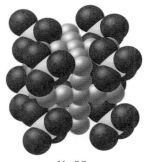

Na_2SO_4

SO_2

3.4   Identify the following compounds containing nitrogen as representations of ionic or molecular compounds.

Na$_3$N

KNO$_3$

NO$_2$

3.5   Identify the following combinations of elements in a compound as ionic or molecular.
(a) nitrogen and oxygen
(b) potassium and oxygen
(c) phosphorus and fluorine
(d) magnesium and chlorine

3.6   Identify the following combinations of elements in a compound as ionic or molecular.
(a) nitrogen and chlorine
(b) sulfur and oxygen
(c) calcium and chlorine
(d) sodium and nitrogen

3.7   Determine if the following formulas represent ionic or molecular compounds.
(a) PCl$_5$ (b) LiF   (c) BaCl$_2$ (d) N$_2$O$_5$

3.8   Determine if the following formulas represent ionic or molecular compounds.
(a) Al$_2$S$_3$ (b) P$_2$O$_5$ (c) MgBr$_2$ (d) K$_2$O

3.9   Which of the compounds LiF, CO$_2$, or N$_2$O$_5$ is expected to have the highest melting point?

3.10   Which of the compounds NaCl, CO, or KF is expected to have the lowest melting point?

Monatomic and Polyatomic Ions

3.11   Based on their positions in the periodic table, predict the charge, write the formula, and name the monatomic ions of the following elements.
(a) sodium   (b) potassium   (c) rubidium

3.12   Based on their positions in the periodic table, predict the charge, write the formula, and name the monatomic ions of the following elements.
(a) fluorine   (b) chlorine   (c) iodine

3.13   Based on their positions in the periodic table, predict the charge, write the formula, and name the monatomic ions of the following elements.
(a) calcium   (b) nitrogen   (c) sulfur

3.14   Based on their positions in the periodic table, predict the charge, write the formula, and name the monatomic ions of the following elements.
(a) aluminum   (b) barium   (c) phosphorus

3.15   The following image represents an oxoanion of nitrogen with a 1– charge. Write a formula for this ion and name it.

3.16   The following image represents an oxoanion of nitrogen with a 1– charge. Write a formula for this ion and name it.

3.17   Name the following polyatomic ions.
(a) SO$_4^{2-}$ (b) OH$^-$ (c) ClO$_4^-$

3.18   Name the following polyatomic ions.
(a) SO$_3^{2-}$ (b) CN$^-$ (c) ClO$_2^-$

3.19   Write formulas for the following ions.
(a) nitride   (b) nitrate   (c) nitrite

3.20   Write formulas for the following ions.
(a) chlorite   (b) chloride   (c) chlorate

3.21   Write formulas for the polyatomic ions with the following names.
(a) carbonate
(b) ammonium
(c) arsenate
(d) permanganate

3.22   Write formulas for the polyatomic ions with the following names.
(a) bicarbonate
(b) acetate
(c) periodate
(d) hypochlorite

3.23   This image is a representation for an oxoanion of sulfur with a 2– charge. Write the formula and name this ion.

3.24   This image is a representation for an oxoanion of nitrogen with a 1– charge. Write the formula and name this ion.

3.25   The iodate ion is an oxoanion of iodine bearing a negative one (1–) charge and containing three oxygen atoms. Write the formula for the iodate ion.

3.26 The phosphite ion is an oxoanion of phosphorus bearing a negative three (3−) charge and containing three oxygen atoms. Write the formula for the phosphite ion.

Formulas for Ionic Compounds

3.27 Suppose an ionic compound containing aluminum and chloride ions dissolves in water. Beginning with a representation for five aluminum ions, draw a picture of a solution that is electrically neutral.

3.28 Suppose an ionic compound containing magnesium and nitrate ions dissolves in water. Beginning with a representation for five magnesium ions, draw a picture of a solution that is electrically neutral.

3.29 Write the formula of the compound formed by each of the following sets of ions.
(a) Ba^{2+} and Cl^-
(b) Fe^{3+} and Br^-
(c) Ca^{2+} and PO_4^{3-}
(d) Cr^{3+} and SO_4^{2-}

3.30 Write the formula of the compound formed by each of the following sets of ions.
(a) Na^+ and N^{3-}
(b) Cs^+ and ClO_3^-
(c) Ti^{3+} and CO_3^{2-}
(d) NH_4^+ and S^{2-}

3.31 Identify the ions in the compounds represented in the following formulas.
(a) KBr
(b) $BaCl_2$
(c) $Mg_3(PO_4)_2$
(d) $Co(NO_3)_2$

3.32 Identify the ions in the compounds represented in the following formulas.
(a) NaBr
(b) $AlCl_3$
(c) $Ba_3(PO_4)_2$
(d) $Mn(NO_3)_2$

3.33 Iron forms two ions, Fe^{2+} and Fe^{3+}.
(a) What are the formulas for the compounds the two iron ions can form with oxide ion (O^{2-})?
(b) What are the formulas for the compounds the two iron ions can form with chloride ion (Cl^-)?

3.34 Two ions of chromium are Cr^{2+} and Cr^{3+}.
(a) What are the formulas for the compounds the two chromium ions can form with oxide ion (O^{2-})?
(b) What are the formulas for the two compounds the chromium ions can form with chloride ion (Cl^-)?

3.35 Sodium ion has a 1+ charge: Na^+. In combination with the sulfate ion, it forms the compound Na_2SO_4.
(a) What is the charge on the sulfate ion?
(b) What is the charge on a strontium ion if it forms the compound $SrSO_4$?

3.36 Magnesium has a 2+ charge: Mg^{2+}. In combination with the nitrate ion, it forms the compound $Mg(NO_3)_2$.
(a) What is the charge on the nitrate ion?
(b) What is the charge on a potassium ion if it forms the compound KNO_3?

3.37 The listed formulas are incorrect. Determine what is wrong with each and correct it.
(a) $NaCl_2$ (b) KSO_4 (c) Al_3NO_3

3.38 The listed formulas are incorrect. Determine what is wrong with each and correct it.
(a) MgCl
(b) $Na(SO_4)_2$
(c) $K_2(NO_3)_2$

Naming Ionic Compounds

3.39 Name the following ionic compounds.
(a) $MgCl_2$
(b) Al_2O_3
(c) Na_2S
(d) KBr
(e) $NaNO_3$ — nitrate ion
(f) $NaClO_4$ → perchlorate

3.40 Name the following ionic compounds.
(a) $BaCl_2$
(b) $(NH_4)_2S$
(c) MgO
(d) KNO_2
(e) $Mg_3(PO_4)_2$
(f) $KClO_2$

3.41 For which of the following compounds must we specify the charge on the metal when we name the compound: $MgSO_4$, $MnSO_4$, $CaCl_2$, $CoCl_2$, $AgNO_3$?

3.42 For which of the following compounds must we specify the charge on the metal when we name the compound: $Cr(NO_3)_2$, $Ba(NO_3)_2$, $AlBr_3$, NiS, $ZnCl_2$?

3.43 Name the following ionic compounds.
(a) Cu_2O
(b) $CrCl_2$
(c) $FePO_4$
(d) CuS

3.44 Name the following ionic compounds.
(a) $TiCl_3$
(b) $CoBr_2$
(c) Mn_2O_3
(d) $Fe_3(PO_4)_2$

3.45 Write formulas for the following ionic compounds.
(a) calcium sulfate
(b) barium oxide
(c) ammonium sulfate
(d) barium carbonate
(e) sodium chlorate

3.46 Write formulas for the following ionic compounds.
(a) potassium bicarbonate
(b) sodium oxide
(c) ammonium hydroxide
(d) magnesium bromide
(e) sodium hypochlorite

3.47 Given the following formulas, what is the charge on the metal? What is the name of each compound?
(a) $CoCl_2$
(b) PbO_2
(c) $Cr(NO_3)_3$
(d) $Fe_2(SO_4)_3$

3.48 Given the following formulas, what is the charge on the metal? What is the name of each compound?
   (a) $CoBr_3$
   (b) $CuCl_2$
   (c) Cr_2O_3
   (d) $FeSO_4$

3.49 Write formulas for the following ionic compounds.
   (a) cobalt(II) chloride
   (b) manganese(II) nitrate
   (c) chromium(III) oxide
   (d) copper(II) phosphate

3.50 Write formulas for the following ionic compounds.
   (a) copper(II) bromide
   (b) iron(III) nitrate
   (c) chromium(VI) oxide
   (d) copper(II) chlorate

3.51 What are the common names for $Fe(NO_3)_2$ and $Fe(NO_3)_3$?

3.52 What are the common names for Cu_2SO_4 and $CuSO_4$?

3.53 Complete the following table by writing the names and formulas for the compounds formed when the cations listed across the top combine with the anions listed along the side.

	Ca^{2+}	Fe^{2+}	K^+
Cl^-			
O^{2-}			
NO_3^-			
SO_3^{2-}			
OH^-			
ClO_3^-			

	Mn^{2+}	Al^{3+}	NH_4^+
Cl^-			
O^{2-}			
NO_3^-			
SO_3^{2-}			
OH^-			
ClO_3^-			

3.54 Complete the following table by writing names and formulas for the compounds formed when the cations listed across the top combine with the anions listed along the side.

	Ba^{2+}	Fe^{3+}	Na^+
Br^-			
S^{2-}			
NO_2^-			
SO_4^{2-}			
HCO_3^-			
ClO_4^-			

	Ni^{2+}	Cr^{3+}	Ag^+
Br^-			
S^{2-}			
NO_2^-			
SO_4^{2-}			
HCO_3^-			
ClO_4^-			

3.55 Complete the following table by writing formulas for the compounds formed when the cations listed across the top combine with the anions listed along the side.

	potassium	iron(III)	strontium
iodide			
oxide			
sulfate			
nitrite			
acetate			
hypochlorite			

	aluminum	cobalt(II)	lead(IV)
iodide			
oxide			
sulfate			
nitrite			
acetate			
hypochlorite			

3.56   Complete the following table by writing formulas for the compounds formed when the cations listed across the top combine with the anions listed along the side.

	sodium	chromium(II)	calcium
chloride			
sulfide			
nitrate			
sulfite			
hydroxide			
chlorate			

	ammonium	iron(III)	lead(II)
chloride			
sulfide			
nitrate			
sulfite			
hydroxide			
chlorate			

3.57   What is the formula for silver chloride?

3.58   What is the formula for zinc chloride?

Naming and Writing Formulas for Molecular Compounds

3.59   Write formulas for the molecular compounds represented by the following images.

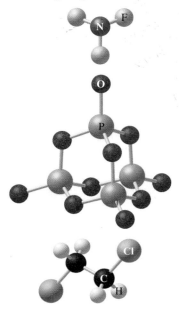

3.60   Write formulas for the molecular compounds represented by the following images.

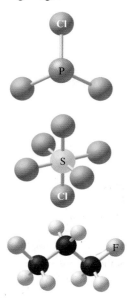

3.61   Name the following compounds.
   (a) PF_5
   (b) PF_3
   (c) CO
   (d) SO_2

3.62   Name the following compounds.
   (a) SO_3
   (b) N_2O_4
   (c) NO_2
   (d) CS_2

3.63   Write formulas for the following compounds.
   (a) sulfur tetrafluoride
   (b) tricarbon dioxide
   (c) chlorine dioxide
   (d) sulfur dioxide

3.64   Write formulas for the following compounds.
   (a) carbon disulfide
   (b) dinitrogen pentoxide
   (c) boron nitride
   (d) iodine heptafluoride

Acids and Bases

3.65   Which of the following images correctly represents phosphoric acid?

3.66   Which of the images shown in Question 3.65 corresponds to phosphorous acid?

3.67   Name the following acids.
   (a) HF(aq)
   (b) HNO₃(aq)
   (c) H₃PO₃(aq)

3.68   Name the following acids.
   (a) HI(aq)
   (b) HNO₂(aq)
   (c) H₃PO₄(aq)

3.69   Write formulas for the following acids.
   (a) hydrofluoric acid
   (b) sulfurous acid
   (c) perchloric acid

3.70   Write formulas for the following acids.
   (a) hydroiodic acid
   (b) sulfuric acid
   (c) hypochlorous acid

3.71   When nitric acid dissociates in water, what ions are present in solution?

3.72   When hydrobromic acid dissociates in water, what ions are present in solution?

Additional Questions

3.73   Determine what ions and how many of each are formed for every one formula unit that dissociates in water.
   (a) NaCl
   (b) MgCl₂
   (c) Na₂SO₄
   (d) Ca(NO₃)₂

3.74   Determine what ions and how many of each are formed for every one formula unit that dissociates in water.
   (a) KBr
   (b) Na₂O
   (c) K₃PO₄
   (d) Al(NO₃)₃

3.75   Indicate whether each of the following substances is an electrolyte or a nonelectrolyte.
   (a) NaOH(aq)
   (b) HCl(aq)
   (c) NaCl(aq)
   (d) C₁₂H₂₂O₁₁(aq) (sucrose solution)

3.76   Indicate whether each of the following substances is an electrolyte or a nonelectrolyte.
   (a) CH₂Cl₂(aq)
   (b) HNO₃(aq)
   (c) KBr(aq)
   (d) KOH(aq)

3.77   Many transition metals form more than one ion and are named using Roman numerals to designate the charge on the ion. Why are Roman numerals unnecessary when naming ionic compounds containing silver, zinc, or cadmium?

3.78   Determine what ions are present when each of the following dissociates in water.
   (a) HCl
   (b) NaOH
   (c) CaCl₂
   (d) KNO₃
   (e) Na₃PO₄

3.79   Write formulas, including charge, for each of the following polyatomic ions.
   (a) nitrate
   (b) sulfite
   (c) ammonium
   (d) carbonate
   (e) sulfate
   (f) nitrite
   (g) perchlorate

3.80   Name the following compounds.
   (a) KBr(s)
   (b) N₂O₅(g)
   (c) HBr(aq)
   (d) Na₂SO₄(s)
   (e) Fe(NO₃)₃(s)

3.81   Name the following compounds.
   (a) MgBr₂(s)
   (b) H₂S(g)
   (c) H₂S(aq)
   (d) CoCl₃(s)
   (e) KOH(aq)
   (f) AgBr(s)

3.82   Name the following compounds.
   (a) NH₃
   (b) H₂O
   (c) H₂O₂

3.83   Write formulas for each of the following compounds.
   (a) lead(II) chloride
   (b) magnesium phosphate
   (c) nitrogen triiodide
   (d) iron(III) oxide
   (e) calcium nitride
   (f) barium hydroxide
   (g) dichlorine pentoxide
   (h) ammonium chloride

3.84   Suppose you need some iron(III) chloride for an experiment. When you go to the stockroom there is one bottle labeled $FeCl_2$ and another with the formula $FeCl_3$. Which bottle should you select?

3.85   The active ingredient in baking soda is sodium bicarbonate (also called sodium hydrogen carbonate). What is the formula for this compound?

3.86   An active ingredient in toothpaste is tin(II) fluoride. What is the formula of this compound? What is the common name for this compound?

3.87   Calcium hypochlorite is a compound used in water treatment to kill disease-causing organisms. What is the formula of this compound?

3.88   Calcium hydrogen carbonate, magnesium chloride, and calcium sulfate can cause scale buildup in pipes. What are the formulas for these compounds?

3.89   Aluminum sulfate and calcium oxide are used in water treatment to form a compound that allows solid materials to be removed from water.

$$Al_2(SO_4)_3(aq) + 3CaO(s) + 3H_2O(l) \rightarrow 2Al(OH)_3(s) + 3CaSO_4(s)$$

   Based on their formulas, what substances in this chemical equation are molecular?

3.90   Consider the following formulas and names. Explain what is wrong with each name.
   (a) $FeBr_2$, iron dibromide
   (b) CS_2, copper(IV) sulfide
   (c) $Co(NO_3)_3$, cobalt trinitrite
   (d) $Mg(OH)_2$, magnesium dihydroxide
   (e) Cu_2O, dicopper(II) oxide

3.91   When solid copper metal is added to a solution of silver nitrate, solid silver and copper(II) nitrate form. Write formulas for all the substances involved in this process.

3.92   Indicate whether each of the following combinations of elements would form molecular or ionic compounds.
   (a) sodium and iodine
   (b) selenium and chlorine
   (c) oxygen and chlorine
   (d) chromium and oxygen

3.93   Write formulas for the following compounds.
   (a) ammonia
   (b) nitric acid
   (c) nitrous acid

3.94   A sports drink contains the ingredients sucrose, whey protein concentrate, fructose, maltodextrin, citric acid, sodium chloride, lecithin, magnesium oxide, vitamin E, and ferrous sulfate. Write formulas for the molecular and ionic compounds present in this product that can be derived from what you learned in this chapter.

3.95   Methane (CH_4) is not an acid, but CH_3CO_2H is. Given the following figures, which hydrogen is responsible for the acid properties of the second compound, acetic acid?

Methane   Acetic acid

3.96   The compounds CH_4 and NH_3 are not acids, but glycine is. Given the structure shown, which hydrogen is responsible for the acidic properties of this amino acid?

3.97   List similarities and differences between each of the following substances.

Note that these structures are not drawn to scale.

Chemical Composition

Copper plays an important role in our daily lives. To see how, let's follow Julio, a college student, through part of his day. Bzzz! The radio alarm goes off and awakens Julio. The circuits in the radio and the wires that connect it to the wall outlet contain copper. Julio leaps out of his bed, which has a brass frame. (Brass is a homogeneous mixture of copper and zinc.) He opens the bathroom door and turns on the light. The doorknob and hinges are made of brass. The light switch contains copper contacts, and the base of the lightbulb is made of copper. He turns on the water to wash his face. The faucets on the sink are brass-coated with nickel and chromium, and copper pipes carry water through the house to his bathroom. Examine your bedroom and bathroom. Do you see any other objects made of copper or brass?

Julio goes to the kitchen and pours orange juice into a glass. His juice is cold because copper tubes circulate coolant in his refrigerator. For breakfast, Julio cooks oatmeal in a copper-bottomed pot. After his meal, he grabs his car keys, which are made of nickel-plated brass, and drives to campus in a car that contains wiring, cables, windings, tubing, and other components made of copper—a total of about 50 lb (23 kg) of it. Julio drops some coins into the parking meter before dashing off to his chemistry class. The coins contain copper. After class, Julio drops by the computer lab to check his e-mail. The computer contains many copper connections. It is connected to the Internet with copper cables.

Julio's life seems to include copper at every turn. In fact, during his lifetime Julio will make use of over 1500 lb (680 kg) of copper metal. Will Julio ever stop to ask himself where all this copper comes from? Think about this yourself, but consider some other common metal such as iron. Where do you encounter iron in your daily life? How much do you depend on it? Where does it come from?

Most of the copper found in nature occurs as part of a mineral. A *mineral* is a naturally occurring element or compound that has a distinct chemical composition and solid form. The most common minerals of copper contain sulfide, oxide, or silicate along with other metals (Figure 4.1). Most copper is taken from open-pit mines in Chile or the southwestern United States (primarily Arizona, Nevada, Utah, and New Mexico), but a few northern states, including Montana and Michigan, mine some too. Copper comes out of the mines as ore, which has relatively small amounts of copper minerals embedded in rock. (Although the minerals contain large amounts of copper, the ore often contains only small amounts of minerals, so it often contains less than 1% copper by mass. In other words, 100 kg of ore may yield as little as 1 kg of copper.) At the mine, the ore is blasted loose from the ground and then lifted with large electric shovels into trucks that can hold as much as 300 tons (272 metric tons) of ore. The trucks haul the ore from the pits and dump it into a series of crushers and mills that break the large pieces into smaller ones.

Figure 4.1
Copper is found mostly as various sulfide, oxide, and silicate minerals.

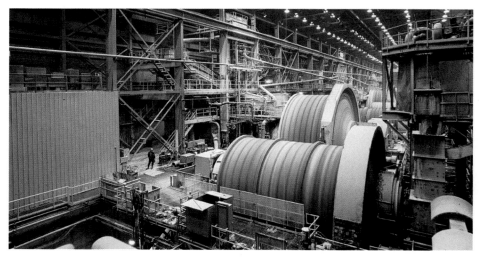

Figure 4.2
Copper ores are crushed in large rotating drums, called mills, in which either larger pieces of ore or steel balls are used to crush smaller pieces of ore.

Figure 4.3
Crushed ore is mixed with water, lime, and special reagents to make bubbles and to make copper minerals stick to the bubbles. Air is blown through the mixture and the copper minerals float to the surface, leaving the rest of the ore to sink to the bottom.

Two different processes are used to get the copper out of the ore. Which one is used depends on the mineral that the ore contains. Ores containing sulfide mineral, such as chalcopyrite, are first crushed to particles as small as grains of sand. This is done in large rotating mills (Figure 4.2) where 3-in-diameter steel balls grind the ore to a powder. The finely crushed ore is mixed with water and lime (CaO), air is blown into the container, and the mixture is agitated vigorously. The copper minerals float to the surface and are removed, leaving the denser rock behind (Figure 4.3). After drying, the copper minerals, containing about 25% to 30% copper (along with quantities of iron and silicon dioxide), are placed in a furnace heated by burning natural gas or coal (Figure 4.4). Oxygen is blown in and the copper minerals ignite, converting the sulfur to sulfur dioxide, which is trapped and used to make sulfuric acid. The elemental copper and iron separate and fall to the bottom of the furnace, where layers form. The top layer, which is poured off, is mostly molten iron and silicon dioxide. The bottom layer—containing copper, some iron, and other impurities—is transferred to another furnace, where iron and any remaining sulfur are burned off, leaving impure copper. This copper is poured into molds and cooled to make sheets. An electric current is used to transfer copper from the impure sheets to other sheets that are pure copper (Figure 4.5). Impurities fall to the bottom as sludge. Valuable elements such as silver, gold, platinum, palladium, antimony, selenium, and tellurium can be recovered from the sludge.

Ores containing oxides of copper are processed by a different method. They are crushed, but not as finely as are sulfide ores. The coarsely ground ore is placed in piles and sprayed with a dilute sulfuric acid solution. The acid converts the copper oxide to a solution of copper(II) sulfate. The solution is concentrated and purified, and then electrically plated onto copper sheets, much the same as the last step in the process used for sulfides.

Just as Julio gives little thought to where copper comes from, he's probably never wondered how its amounts are determined in ores, minerals, and chemical compounds. Yet for chemists in the past, such measurements were a big problem. How did they learn what other elements different copper compounds contain? How did they figure out the ratios of atoms in copper compounds so they could write formulas for them?

In this chapter we will investigate the composition of chemical compounds such as the minerals from which we recover copper metal. We will determine the formulas of compounds and learn how to relate these formulas to the amounts of the elements that compose them.

Figure 4.4
Impure copper is recovered from a series of furnaces used to burn off most of the sulfur, iron, and silicon.

Figure 4.5
Copper is purified by using electricity to transfer the metal from impure sheets into solution as copper(II) ions. The copper(II) ions are converted back to copper metal by electricity and deposited onto pure sheets.

Questions for Consideration

4.1 How can we count the number of atoms in a sample of a material? The number of molecules? The number of formula units?
4.2 How can we know the number of molecules in a sample of a material?
4.3 How can we use the masses of elements in a compound to determine its chemical formula?
4.4 How can we express the composition of a solution?

Math Tools Used in This Chapter

Scientific Notation (Math Toolbox 1.1)

Units and Conversions (Math Toolbox 1.3)

4.1   MOLE QUANTITIES

Let's consider the copper wires in Julio's computer. They started out as the copper sulfide mineral chalcopyrite. When dissolved in acid, chalcopyrite forms a blue solution that gives off a gas that smells like rotten eggs. This odor arises from the molecular compound hydrogen sulfide, which typically forms as a gas when an acid reacts with an ionic compound containing sulfide ions. We would like to describe the composition of hydrogen sulfide and write its chemical formula, but how can we know how many atoms of each component element are in it? If we could count the atoms of each element in a small sample of hydrogen sulfide, we could determine the relative numbers of each and deduce the chemical formula.

Consider the molecular-level view of hydrogen sulfide gas shown in Figure 4.6. We can identify individual molecules of hydrogen sulfide, with the white spheres representing hydrogen atoms and the yellow spheres representing sulfur atoms. Each molecule is made up of two hydrogen atoms and one sulfur atom. In a collection of molecules, such as the four molecules pictured in Figure 4.6, we can count eight hydrogen atoms and four sulfur atoms, giving a ratio of two hydrogen atoms per sulfur atom, the same ratio as in each molecule.

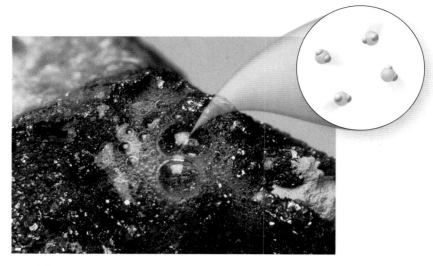

Figure 4.6
Hydrogen sulfide (H_2S) is given off when a sulfide ore reacts with acid.

We only have to go to the hardware or grocery store to find many examples of counting by weighing. We buy nails and walnuts by the pound rather than by the piece, for example.

Recall that a formula unit is the smallest repeating unit in an ionic compound. The composition of a formula unit is the same as the composition of the ionic compound.

Counting by units other than one piece is a common practice in other areas as well. We count eggs in units of a dozen, paper in units of a ream (500 sheets), paperclips by the gross (144), and soldiers by the battalion. It is simply more convenient to use a unit that counts by the group rather than by the individual.

We use the abbreviation mol for mole, but molecule should not be abbreviated.

Figure 4.7
The bumps in this image are atoms on the surface of a mica crystal. Mica is a mineral containing potassium and aluminum cations and silicate and hydroxide anions.

Unfortunately, we cannot see atoms and molecules in this way because they are too small. Although we can use a scanning tunneling microscope to "see" atoms and molecules on the surface of a solid, as shown in Figure 4.7, this technique reveals only atoms on a surface, not those beneath. Furthermore, because atoms are so small, a dust-sized piece of a solid contains over 10^{16} of each type of atom. We certainly wouldn't want to count that many atoms individually, even if we could. Fortunately, we don't have to see atoms to count them. Instead we can find the relative masses of the elements in a compound. From this, we can determine the 2:1 ratio of atoms and deduce the formula H_2S.

In Chapter 2, we saw that the average masses of the elements can be expressed as their *relative atomic masses.* For the elements in hydrogen sulfide, the relative atomic masses are 1.008 amu for hydrogen and 32.07 amu for sulfur. While relative atomic masses are very useful for comparing the masses of elements, atomic mass units are not useful to chemists weighing substances to use in chemical reactions like those for processing copper oxides. Common laboratory balances measure masses in grams, not atomic mass units. We cannot weigh one or even a few atoms, molecules, or formula units of a substance, since the mass of most atoms and common molecules is less than 10^{-21} g. To work with a mass of about 1 g, we have to measure a number of formula units greater than 10^{21}. For example, to get 1 g of H_2S, we would need almost 2×10^{22} molecules. To get an idea of how large this number is, consider a teaspoon of water. There are as many molecules of water in this teaspoon (about 1.7×10^{23}) as there are teaspoonfuls of water in all the oceans on Earth. Similarly, there are about as many SiO_2 formula units in one grain of quartz sand as there are sand grains on all the beaches on Earth. Because numbers this large are too unwieldy to work with easily, we need some way to measure the amount of a substance without counting the number of atoms, formula units, or molecules.

The solution is to use a unit called the mole. A **mole** contains 6.022×10^{23} atoms, molecules, ions, or formula units. This number, called **Avogadro's number,** is the same as the number of atoms in exactly 12 g of the isotope carbon-12, or ^{12}C. Figure 4.8 shows 1 mole (mol) of several different substances. Although 1 mol contains different volumes and different masses of different materials, it is always the same number of basic units (atoms, molecules, or formula units). For example, 1 mol of copper contains 6.022×10^{23} *atoms* of copper. One mol of O_2 contains 6.022×10^{23} *molecules* of O_2. The number of atoms, molecules, or formula units in a mole is always the same, so we can calculate multiples or fractions of moles easily. For example, 2 mol

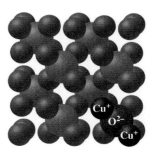

Figure 4.9
One formula unit of Cu_2O, shown highlighted, contains two Cu^+ ions and one O^{2-} ion.

Figure 4.8
Moles of various elements and compounds have different masses and volumes, but the same number of formula units, atoms, or molecules.

of copper contain $2 \times (6.022 \times 10^{23})$ or 1.204×10^{24} atoms, and $\frac{1}{2}$ mol of copper contains $\frac{1}{2} \times (6.022 \times 10^{23})$ or 3.011×10^{23} atoms. Also, we can easily convert between numbers of molecules and numbers of atoms. For example, 1 mol of O_2 contains $2 \times (6.022 \times 10^{23})$ or 1.204×10^{24} atoms of oxygen. With ionic solids, we have to count formula units. Thus, 1 mol of copper(I) oxide (the mineral cuprite) contains 6.022×10^{23} formula units of Cu_2O. How many ions does it contain? Each formula unit contains two Cu^+ ions and one O^{2-} ion, as shown in Figure 4.9, so we have two Cu^+ ions and one O^{2-} ion per formula unit for a total of three ions per formula unit. This gives us $2 \times (6.022 \times 10^{23})$ or 1.204×10^{24} Cu^+ ions plus 6.022×10^{23} O^{2-} ions for a total of $3 \times (6.022 \times 10^{23})$ or 1.807×10^{24} ions in each mole of Cu_2O.

EXAMPLE 4.1   Particles in 1 Mole

How many H_2S molecules are in exactly 1 mol of hydrogen sulfide (H_2S)? How many hydrogen atoms and sulfur atoms are there?

Solution:

Each mole of H_2S contains 6.022×10^{23} H_2S molecules. Since each molecule contains 2 H atoms and 1 S atom, each mole contains $2 \times (6.022 \times 10^{23})$ or 1.204×10^{24} H atoms and 6.022×10^{23} S atoms.

Practice Problem 4.1

How many N_2O_5 molecules are in exactly 1 mol of dinitrogen pentoxide (N_2O_5)? How many nitrogen atoms and oxygen atoms?

Further Practice: 4.11 and 4.12

The relative atomic mass is the average mass of atoms of an element in atomic mass units. This definition describes the average mass of one atom, but we can extend it to larger quantities of matter as well. Recall that the basis of the definition is that one atom of ^{12}C has a mass of exactly 12 amu. Avogadro's number has been defined so that the mass of 1 mol of ^{12}C (that is, 6.022×10^{23} atoms of ^{12}C) is exactly 12 g. Consequently, the mass of a basic particle of any substance in atomic mass units and the mass of 1 mol of the substance in grams have exactly the same numerical value. The definition of a mole can be restated, then, in terms of the mass of our reference, ^{12}C: *A mole is the amount of substance that contains as many basic particles (atoms, molecules, or formula units) as there are atoms in exactly 12 g of ^{12}C.* The term that describes the mass of 1 mol of a substance is **molar mass.**

Since the molar mass value for any atom, in units of grams per mole (g/mol), is numerically the same as the relative atomic mass value (in units of amu per atom), we can use the periodic table to determine the molar mass of any element or compound. For example, the average mass of 1 hydrogen atom is 1.008 amu, and the mass of 1 mol of hydrogen atoms is 1.008 g. The molar mass of a molecular or ionic substance has the same value as the sum of the relative atomic masses of its component elements. For example, the average mass of a molecule of H_2S is 34.07 amu, and its molar mass is 34.07 g/mol. Molar masses of molecules (or formula units) are obtained by adding the molar masses of the component elements, each weighted by the number of that atom in the molecule (or formula unit). For simplicity, we will abbreviate molar mass as *MM*.

EXAMPLE 4.2   Molar Masses of Substances

When copper sulfide ores are roasted in a furnace, sulfur dioxide gas (SO_2) forms. What is the molar mass of SO_2?

Solution:

The molar mass of a compound is the sum of the molar masses of its component elements—in this case, sulfur and oxygen. A molecule of sulfur dioxide contains one sulfur atom and two oxygen atoms, so we must multiply the molar mass of oxygen by a factor of 2. We can find the molar masses of sulfur and oxygen on a periodic table, multiply the molar mass of oxygen by 2, and find the sum:

$$\text{Mass of 1 mol of S} = 1 \text{ mol} \times 32.07 \text{ g/mol} = 32.07 \text{ g}$$
$$\text{Mass of 2 mol of O} = 2 \text{ mol} \times 16.00 \text{ g/mol} = \underline{32.00 \text{ g}}$$
$$\text{Mass of 1 mol of } SO_2 = 64.07 \text{ g}$$

The molar mass of SO_2 is 64.07 g/mol.

Practice Problem 4.2

Many of the minerals that yielded the copper used to make Julio's brass bed came out of the ground in ores of silicon dioxide (SiO_2). What is the molar mass of SiO_2?

Further Practice: 4.17 and 4.18

4.2 MOLES, MASSES, AND PARTICLES

The copper in Julio's light switch was processed from the mineral chalcopyrite. Suppose we dissolve a sample of chalcopyrite in acid. We get a blue solution, characteristic of copper(II) ions. The solution gives off the odor of rotten eggs, good evidence that sulfur is also present in the mineral. If we add ammonia to some of the solution, the color gets much more intense, confirming the presence of copper. We also see a brownish-red solid settling out of the solution, which indicates that iron is also present. The tests show that copper, sulfur, and iron are present in chalcopyrite, but how do we determine their amounts? One way is to experiment. Suppose careful separating and measuring revealed that a 3.67-g sample of chalcopyrite contained 1.27 g Cu, 1.12 g Fe, and 1.28 g S. While these numbers are useful, they do not describe all samples, only our original 3.67-g sample. If we examined a larger sample—for example, a sample with a mass of 8.33 g—the composition would be 2.88 g Cu, 2.54 g Fe, and 2.91 g S.

There is another convenient way to express composition, and it is valid for samples of any size. We can convert the mass of each component to a percentage of the total mass. The result is the **percent composition by mass** (or *percent composition*). It is an expression of the portion of the total mass contributed by each element. It is constant, no matter what the size of the sample. For example, to calculate the percent copper in our 3.67-g sample of chalcopyrite, divide the mass of copper by the mass of the sample and multiply this ratio by 100% (to convert to a percentage):

$$\% \text{ Cu} = \frac{1.27 \text{ g Cu}}{3.67 \text{ g sample}} \times 100\% = 34.6\% \text{ Cu}$$

Similar calculations give 30.5% Fe and 34.9% S. The sum of those percentages is 100% (Figure 4.10).

Let's examine this relationship more closely to see how it arises. Percent means "per hundred." You use percent frequently. For example, if you pay a sales tax, your total expenditure is greater than the purchase price. For example, a total cost of 107 cents may arise from merchandise costing 100 cents with a tax of 7 cents. Although sales tax is usually expressed as a percentage of the purchase price, what is the percentage of the total cost due to the tax? The percentage due to the tax is 6.54% of the total cost. If an item costs $30, you pay $32.10, including $2.10 in taxes, which again is 6.54% of the total cost. You pay $14.00 in taxes for every $200.00 purchased, which is 6.54% of the total cost of $214.00. Although those numbers sound different, they all have the same composition of 7 parts tax and 100 parts of purchase to every 107 parts of total cost. In a similar fashion, if we know the ratio of the mass of copper to the mass of the sample, we can easily find the unknown mass of Cu in a 100-g sample. Why? Because the ratio of the part to the whole is always the same:

$$\frac{1.27 \text{ g Cu}}{3.67 \text{ g sample}} = \frac{\text{mass Cu}}{100 \text{ g sample}}$$

If we rearrange the equation to solve for mass Cu, we get the following:

$$\text{Mass Cu} = \frac{1.27 \text{ g Cu}}{3.67 \text{ g sample}} \times 100 \text{ g sample} = 34.6 \text{ g Cu}$$

Since this is the mass of Cu per 100-g sample, the value is the % Cu:

$$\% \text{ Cu} = 34.6\%$$

In future calculations, we will use the one-step conversion summarized by the general equation

$$\% \text{ E (E is any element)} = \frac{\text{mass of E}}{\text{mass of sample}} \times 100\%$$

The percent composition by mass for any substance is always independent of the sample size. Use Example 4.3 to convince yourself that the percentages would be the same for an 8.33-g sample of chalcopyrite as for a 3.67-g sample.

Even though we cite the mass of each element using the element's symbol, our notation does not necessarily indicate that these elements are present as atoms. Rather, they may be present as monatomic or polyatomic ions.

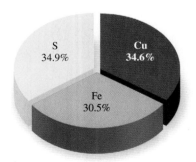

Figure 4.10
The percent by mass of the component elements must total 100%. How would this graph change if we doubled the amount of chalcopyrite under consideration?

EXAMPLE 4.3 — Percent Composition from Masses

What is the percent copper in an 8.33-g sample of chalcopyrite that contains 2.88 g Cu?

Solution:

Set up the equation for % Cu as the mass Cu divided by the mass of the sample, with the ratio multiplied by 100%:

$$\% \, Cu = \frac{2.88 \, g \, Cu}{8.33 \, g \, sample} \times 100\% = 34.6\% \, Cu$$

Note that the answer is the same percentage as for the smaller chalcopyrite sample. This is what we would expect since minerals, like other chemical compounds, have a constant composition throughout.

Practice Problem 4.3

What are the percent iron and the percent sulfur in an 8.33-g sample of chalcopyrite that contains 2.54 g Fe and 2.91 g S?

Further Practice: 4.27 and 4.28

Conversion between mass and moles is very common. For example, if we want to figure out how much of a substance is formed by a chemical reaction, we have to convert masses of starting materials to moles and then figure out the moles of the substance formed. Then we can determine the mass of that product.

We can also express the composition of a substance in terms of the number of moles of each element. This means converting the masses of Cu, Fe, and S into numbers of moles using the molar mass. For example, we know from the periodic table that 63.55 g of copper is the mass of 1 mol of copper. If we have 127.1 g of copper, which is 2×63.55 g, then we have 2 mol of copper. And if we have 31.77 g of copper, which is $\frac{1}{2} \times 63.55$ g, then we have $\frac{1}{2}$ mol of copper.

The number of moles of a substance is directly proportional to the mass of the substance. The ratio of mass to number of moles is always the same. It equals the molar mass, which is the mass of 1 mol. The 3.67-g sample of chalcopyrite contains 1.27 g Cu. Suppose we want to know how many moles of copper are in this amount. We can equate the ratio of 1.27 g of copper to the unknown number of moles of copper and the mass-to-mole ratio for 1 mol of copper (the molar mass):

$$\frac{1.27 \, g \, Cu}{x \, mol \, Cu} = \frac{63.55 \, g \, Cu}{1 \, mol \, Cu}$$

We rearrange this equation to solve for moles of copper:

$$x \, mol \, Cu = 1.27 \, \cancel{g \, Cu} \times \frac{1 \, mol \, Cu}{63.55 \, \cancel{g \, Cu}} = 0.0200 \, mol$$

The ratio approach will always work to solve problems such as these, but we can generally solve them more quickly by using dimensional analysis. For a review of the ratio approach and the dimensional analysis approach, see Math Toolbox 1.3. The conversion of mass to moles follows this pathway:

In dimensional analysis, we start with an expression that equates two equivalent quantities. Here, we know that 1 mol of Cu has a mass of 63.55 g. The equivalence expression is

$$1 \text{ mol Cu} = 63.55 \text{ g Cu}$$

We use this expression to create a ratio of the two quantities. The ratio can be in two forms:

$$\frac{63.55 \text{ g Cu}}{1 \text{ mol Cu}} \quad \text{and} \quad \frac{1 \text{ mol Cu}}{63.55 \text{ g Cu}}$$

We select the ratio that will cancel the old units (grams in this case) and introduce the new units (moles). We multiply the starting quantity (1.27 g Cu) by the ratio to obtain the moles of copper:

$$\text{mol Cu} = 1.27 \text{ g Cu} \times \frac{1 \text{ mol Cu}}{63.55 \text{ g Cu}} = 0.0200 \text{ mol Cu}$$

This equation is the same as the final equation developed in the ratio approach. We will use dimensional analysis for the remaining examples in this chapter, but remember that the same results can be obtained either way.

Example 4.4 demonstrates the dimensional analysis approach for calculating the moles of Fe and S in our sample of chalcopyrite.

EXAMPLE 4.4   Moles from Mass

How many moles of iron are in the 1.12 g of Fe found in our sample of chalcopyrite?

Solution:

From the periodic table, we obtain the molar mass of iron to use in an equivalence expression:

$$1 \text{ mol Fe} = 55.85 \text{ g Fe}$$

We use this expression to create possible conversion ratios:

$$\frac{55.85 \text{ g Fe}}{1 \text{ mol Fe}} \quad \text{and} \quad \frac{1 \text{ mol Fe}}{55.85 \text{ g Fe}}$$

Since we want to convert grams to moles, we select the second conversion ratio to carry out the conversion:

$$\text{mol Fe} = 1.12 \text{ g Fe} \times \frac{1 \text{ mol Fe}}{55.85 \text{ g Fe}} = 0.0201 \text{ mol Fe}$$

Practice Problem 4.4

How many moles of sulfur are in the 1.28 g of S found in our sample of chalcopyrite?

Further Practice: 4.29 and 4.30

So far we have calculated 0.0200 mol Cu, 0.0201 mol Fe, and 0.0399 mol S in the 3.67-g sample of chalcopyrite. We can calculate the moles of a compound in the same manner. We use the molar mass of the compound in the same way that we used the molar mass of an element to make the conversion from mass to moles. In addition, if we know the number of moles and need the mass, we can use the molar mass to create a ratio of mass to moles. Example 4.5 illustrates this situation.

EXAMPLE 4.5   Mass from Moles

While Julio puzzles over mole-to-mass conversions, a friend heats water in a copper kettle and makes him a cup of tea. He tells Julio he placed 0.0120 mol of table sugar (sucrose, $C_{12}H_{22}O_{11}$) in the tea. What mass of sugar has he added?

Solution:

We use the molar mass of sugar to convert moles to mass. First we obtain the molar mass from the molar masses of the atoms, each multiplied by the number of atoms of that element in the formula of the compound $C_{12}H_{22}O_{11}$:

$$\text{Mass of 12 mol of C} = 12 \text{ mol} \times 12.01 \text{ g/mol} = 144.1 \text{ g}$$
$$\text{Mass of 22 mol of H} = 22 \text{ mol} \times 1.008 \text{ g/mol} = 22.2 \text{ g}$$
$$\text{Mass of 11 mol of O} = 11 \text{ mol} \times 16.00 \text{ g/mol} = \underline{176.0 \text{ g}}$$
$$\text{Mass of 1 mol of } C_{12}H_{22}O_{11} \qquad\qquad = 342.3 \text{ g}$$

The molar mass of $C_{12}H_{22}O_{11}$ is 342.3 g/mol. Now we have an equivalence expression:

$$1 \text{ mol } C_{12}H_{22}O_{11} = 342.3 \text{ g } C_{12}H_{22}O_{11}$$

From this expression, we have two possible ratios for conversion between units:

$$\frac{342.3 \text{ g } C_{12}H_{22}O_{11}}{1 \text{ mol } C_{12}H_{22}O_{11}} \quad \text{and} \quad \frac{1 \text{ mol } C_{12}H_{22}O_{11}}{342.3 \text{ g } C_{12}H_{22}O_{11}}$$

We know moles and want to calculate mass, so we use the first of ratios:

$$\text{g } C_{12}H_{22}O_{11} = 0.0120 \text{ mol } C_{12}H_{22}O_{11} \times \frac{342.3 \text{ g } C_{12}H_{22}O_{11}}{1 \text{ mol } C_{12}H_{22}O_{11}}$$

$$= 4.11 \text{ g } C_{12}H_{22}O_{11}$$

This is about the amount of sugar in one teaspoonful.

Practice Problem 4.5

Julio's friend prefers an artificial sweetener such as aspartame ($C_{14}H_{18}N_2O_5$) in his tea. A packet of aspartame contains 40 mg of sweetener. How many moles of aspartame are in the packet? What mass of aspartame would give 0.0120 mol? How many packets is this?

Further Practice: 4.33 and 4.34

If we know how many moles of a substance are in a sample, we can calculate the number of atoms or molecules in this sample. For example, because we know the number of moles of each element in our 3.67-g sample of chalcopyrite, we can calculate the number of atoms it contains. Recall that we found 0.0200 mol Cu, 0.0201 mol Fe, and 0.0399 mol S in the sample. To get the number of atoms of Cu, use the relationship between moles and Avogadro's number:

$$1 \text{ mol} = 6.022 \times 10^{23} \text{ atoms}$$

We use this equivalence to set up the possible conversion ratios:

$$\frac{6.022 \times 10^{23} \text{ atoms}}{1 \text{ mol}} \quad \text{and} \quad \frac{1 \text{ mol}}{6.022 \times 10^{23} \text{ atoms}}$$

Since we must convert moles to number of atoms, we use the first ratio:

$$\text{Atoms Cu} = 0.0200 \text{ mol Cu} \times \frac{6.022 \times 10^{23} \text{ atoms Cu}}{1 \text{ mol Cu}} = 1.20 \times 10^{22} \text{ atoms Cu}$$

Similar calculations give us the number of iron and sulfur atoms: 1.21×10^{22} Fe atoms and 2.40×10^{22} S atoms. (If you need to review mathematical operations using scientific notation, see Math Toolbox 1.1.)

Suppose we do not know the number of moles of the substance, but we know its mass. To find the number of molecules, we first have to convert the mass to moles, using the molar mass. Then we find the number of molecules using Avogadro's number. The conversion between mass, moles, and number of molecules (or atoms or formula units) involves molar mass and Avogadro's number, and follows this pathway:

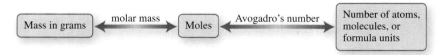

Any of these conversions can be carried out in either direction. Each equivalence expression (using molar mass or Avogadro's number) can be used to create two conversion ratios. One of them converts units left to right; the other converts units right to left.

EXAMPLE 4.6   Number of Molecules from Mass

A substance named Agorca M5640 (officially 5-nonyl salicylaldoxime) is used for concentrating extracted copper ore. Its molecular formula is $C_{16}H_{25}NO_2$. It has a molar mass of 263.4 g/mol. If you have a 150.0-g sample of Agorca M5640, how many molecules do you have?

Solution:

To convert from mass to number of molecules, we will need two conversion steps. First, we need equivalence expressions for each conversion, using the molar mass and Avogadro's number:

$$1 \text{ mol } C_{16}H_{25}NO_2 = 263.4 \text{ g } C_{16}H_{25}NO_2$$

$$1 \text{ mol } C_{16}H_{25}NO_2 = 6.022 \times 10^{23} \text{ molecules } C_{16}H_{25}NO_2$$

We use expressions to develop possible conversion ratios:

$$\frac{1 \text{ mol } C_{16}H_{25}NO_2}{263.4 \text{ g } C_{16}H_{25}NO_2} \quad \text{and} \quad \frac{263.4 \text{ g } C_{16}H_{25}NO_2}{1 \text{ mol } C_{16}H_{25}NO_2}$$

$$\frac{6.022 \times 10^{23} \text{ molecules}}{1 \text{ mol}} \quad \text{and} \quad \frac{1 \text{ mol}}{6.022 \times 10^{23} \text{ molecules}}$$

Following the pathway outlined earlier, we start with 150.0 g $C_{16}H_{25}NO_2$ and convert this quantity to moles. Then we convert moles to molecules. The first of each pair of conversion ratios accomplishes changes:

$$\text{Molecules } C_{16}H_{25}NO_2 = 150.0 \text{ g } C_{16}H_{25}NO_2 \times \frac{1 \text{ mol } C_{16}H_{25}NO_2}{263.4 \text{ g } C_{16}H_{25}NO_2}$$

$$\times \frac{6.022 \times 10^{23} \text{ molecules } C_{16}H_{25}NO_2}{1 \text{ mol } C_{16}H_{25}NO_2}$$

$$= 3.430 \times 10^{23} \text{ molecules}$$

Practice Problem 4.6

After working through this example, Julio develops a headache and takes an aspirin tablet. One 5-grain aspirin tablet contains 0.324 g of acetylsalicylic acid ($C_9H_8O_4$, molar mass = 180.2 g/mol). How many molecules of acetylsalicylic acid are in one tablet?

Further Practice: 4.39 and 4.40

4.3 DETERMINING EMPIRICAL AND MOLECULAR FORMULAS

We have seen how to express the composition of a substance in terms of masses, percent composition, moles, and atoms or molecules of the component elements. While these are all useful ways to express composition, none of them is very compact. In this section we will see how to convert such information into a chemical formula, which is an efficient way of representing the composition of a substance.

Empirical and Molecular Formulas

There are two types of chemical formulas that are related to the composition of a substance. An **empirical formula** expresses the simplest ratios of atoms in a compound. It is written with the smallest possible whole-number subscripts. It is not always the same as a molecular formula (review Section 3.3), which expresses the actual number of atoms in a molecule. The molecular formula is either the same as the empirical formula, or it is some multiple of the empirical formula. Thus, the molecular and empirical formulas for water, with two hydrogen atoms and one oxygen atom per molecule, are both H_2O. The molecular formula for hydrogen peroxide, which contains two hydrogen atoms and two oxygen atoms, is H_2O_2, but its empirical formula is HO. The difference between molecules of water and hydrogen peroxide is shown in Figure 4.11.

For some substances there is no useful molecular formula, because no specific molecule can be identified in the structure. Quartz sand, for example, is simply networks of silicon and oxygen atoms linked together as shown in Figure 4.12. The atoms occur in a 1:2 ratio, and the substance, silicon dioxide, is commonly represented by the empirical formula SiO_2. Similarly, formulas for ionic compounds are usually empirical formulas. As shown in Figure 4.13, a copper(II) oxide crystal contains equal numbers

H_2O

H_2O_2

Figure 4.11

For the water molecule, the molecular and empirical formulas are the same, H_2O. For hydrogen peroxide, the smallest ratio of hydrogen to oxygen atoms is 1:1, but the molecule actually contains two atoms of each, so the molecular and empirical formulas are different, H_2O_2 and HO, respectively.

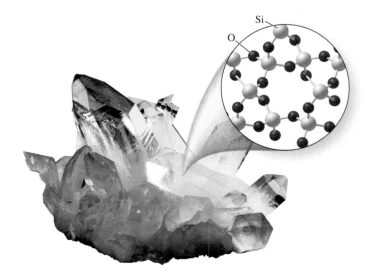

Figure 4.12

In quartz, there are two oxygen atoms for every one silicon atom, but there is no identifiable SiO_2 molecule. For such extended structures, there is no molecular formula, only an empirical formula.

Figure 4.13
In ionic solids, the empirical formula represents the smallest repeating unit in the crystal, the formula unit. These crystals of tenorite contain equal numbers of copper(II) ions and oxide ions, so the formula is CuO.

TABLE 4.1	Some Empirical and Molecular Formulas	
Substance	**Molecular Formula**	**Empirical Formula**
cyclopentane	C_5H_{10}	CH_2
cyclohexane	C_6H_{12}	CH_2
ethylene	C_2H_4	CH_2
hydrogen sulfide	H_2S	H_2S
calcium chloride	There is no molecular formula for an ionic compound.	$CaCl_2$

of copper(II) ions and oxide ions, but no molecules. For such ionic compounds, we represent the compound symbolically with its empirical formula.

Whenever possible, chemists like to use molecular formulas. Molecular formulas contain more information—not only the ratios of atoms but also the actual numbers of atoms in a molecule. Consider, for example, the substances benzene and acetylene. They have the molecular formulas C_6H_6 and C_2H_2, respectively (Figure 4.14). Both have the empirical formula CH. The empirical formula tells us only that their molecules contain carbon and hydrogen atoms in equal numbers, or in a 1:1 ratio. It does not distinguish benzene and acetylene. We need molecular formulas to tell them apart. Some other examples of empirical and molecular formulas are given in Table 4.1.

Examine Figure 4.15 and determine the empirical formula for each substance shown. To do this, find the smallest subscript and divide the other subscripts by its value. If the other subscripts are *evenly* divisible by the smallest one, then the division results in the empirical formula. Sometimes, if this fails, all the subscripts may be evenly divisible by some other number, so other choices have to be checked as well. If the subscripts are not evenly divisible by some number other than 1, the molecular formula and the empirical formula are the same.

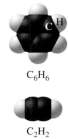

C_6H_6

C_2H_2

Figure 4.14
Benzene (C_6H_6) and acetylene (C_2H_2) have the same empirical formula, CH.

EXAMPLE 4.7   Empirical and Molecular Formulas

Consider the molecular formula $C_6H_3Cl_3$ (see Figure 4.15). What is the empirical formula of this compound?

Solution:

The subscripts in the formula are all evenly divisible by 3, the smallest subscript. Carrying out this division gives us $C_{6/3}H_{3/3}Cl_{3/3}$ or C_2HCl as the empirical formula. The molecular formula is three times the empirical formula.

Practice Problem 4.7

What is the empirical formula of each of the following compounds?

(a) CH_2Cl_2 (b) CH_3CO_2H (c) P_4O_6

Further Practice: 4.51 and 4.52

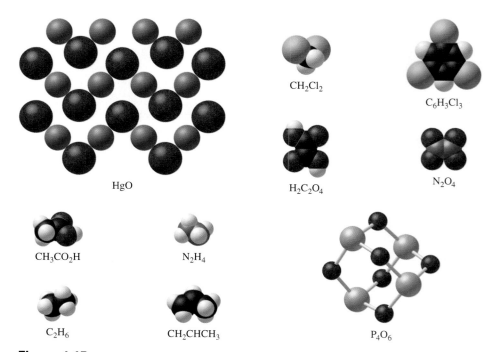

Figure 4.15
For which of these substances is the empirical formula the same as the molecular formula? Note: The structures are not drawn to the same scale.

Determining Empirical Formulas

We can now determine empirical formulas from data on the chemical composition of a compound. Recall, for example, our 3.67-g sample of chalcopyrite. It contained 1.27 g Cu, 1.12 g Fe, and 1.28 g S. We have converted the masses of the constituent elements to numbers of moles and numbers of atoms. Either of these sets of numbers can give us a chemical formula. Because the numbers are easier to manipulate, let's use the number of moles: 0.0200 mol Cu, 0.0201 mol Fe, and 0.0399 mol S. To write a chemical formula for chalcopyrite, we need only convert these values to whole numbers. Why? Because the relative number of moles of atoms in a sample of a compound must be the same as the relative number of atoms in one formula unit or one molecule of this compound. To get the relative number of atoms, divide the number of moles of each element by the number of moles of the element present in the least amount. In this example, copper is present in the least amount, so we need only divide the moles of Fe and S by the moles of Cu to reduce all the numbers to whole numbers:

$$\frac{0.0201 \text{ mol Fe}}{0.0200 \text{ mol Cu}} = \frac{1.01 \text{ Fe}}{1.00 \text{ Cu}}$$

$$\frac{0.0399 \text{ mol S}}{0.0200 \text{ mol Cu}} = \frac{2.00 \text{ S}}{1.00 \text{ Cu}}$$

Slight errors in the experimental determination of the masses can yield values that are close to whole numbers, but not exactly. In such cases, we can safely round the values to whole numbers.

When rounded, the whole number values indicate that there are 1 mol Cu, 1 mol Fe, and 2 mol S in 1 mol of chalcopyrite. From these results, we can write the empirical formula of chalcopyrite as $CuFeS_2$. The procedure and the result would be the same if we used the number of atoms of each element instead of the number of moles.

Empirical Formulas from Percent Composition

The copper in Julio's lights, refrigerator, bed, and computer came from different minerals, but the same procedures we used to determine the empirical formula for chalcopyrite can be used when working with any of the others. For compounds containing only two elements, A and B, the conversions from percent composition to empirical formula can be represented schematically as follows:

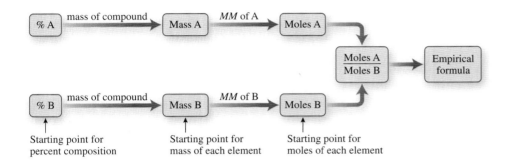

If the compound contains more than two elements, we add a line to the schematic for each additional element.

If we know the percent composition instead of actual masses, we must convert percent composition to relative masses before we calculate moles. The simplest approach is to use a hypothetical sample weight of exactly 100 g, so the percent composition is numerically equal to the mass of each component. We can actually pick any mass. The exact sample mass is not important since we calculate the ratio of moles in the sample, and this ratio is independent of the sample size.

EXAMPLE 4.8   **Empirical Formula from Percent Composition**

Determine the empirical formula for the mineral chalcocite, which has the percent composition 79.8% Cu and 20.2% S.

Solution:

If the sample of chalcocite has a mass of 100 g, the masses of copper and sulfur will be numerically equal to their percent composition values:

$$79.8\% \text{ of } 100 \text{ g is } 79.8 \text{ g} = \text{mass Cu}$$
$$20.2\% \text{ of } 100 \text{ g is } 20.2 \text{ g} = \text{mass S}$$

Next we convert the mass of each element into its number of moles. The equivalence expressions arise from the molar masses of the elements:

$$1 \text{ mol Cu} = 63.55 \text{ g Cu}$$
$$1 \text{ mol S} = 32.07 \text{ g S}$$

We convert these expressions to ratios to convert mass to moles:

$$\frac{63.55 \text{ g Cu}}{1 \text{ mol Cu}} \quad \text{and} \quad \frac{1 \text{ mol Cu}}{63.55 \text{ g Cu}}$$

$$\frac{32.07 \text{ g S}}{1 \text{ mol S}} \quad \text{and} \quad \frac{1 \text{ mol S}}{32.07 \text{ g S}}$$

Since we need to cancel grams of each element and calculate moles, the second ratio is used in each case:

$$\text{mol Cu} = 79.8 \ \cancel{\text{g Cu}} \times \frac{1 \ \text{mol Cu}}{63.55 \ \cancel{\text{g Cu}}} = 1.26 \ \text{mol Cu}$$

$$\text{mol S} = 20.2 \ \cancel{\text{g S}} \times \frac{1 \ \text{mol S}}{32.07 \ \cancel{\text{g S}}} = 0.630 \ \text{mol S}$$

Finally, divide by the smaller of the two mole quantities:

$$\frac{\text{mol Cu}}{\text{mol S}} = \frac{1.26 \ \text{mol Cu}}{0.630 \ \text{mol S}} = \frac{2.00 \ \text{Cu}}{1.00 \ \text{S}}$$

Since 2 mol of copper are present for each 1 mol of sulfur, the empirical formula is Cu_2S.

Practice Problem 4.8

Determine the empirical formula for the mineral covellite, which has the percent composition 66.5% Cu and 33.5% S.

Further Practice: 4.57 and 4.58

Empirical Formulas for Compounds Containing More Than Two Elements

For a compound that contains more than two elements, we follow the same procedure, but we add a series of conversions for each additional element.

EXAMPLE 4.9   Empirical Formulas for More Than Two Elements

The mineral malachite is a beautiful green color, often with swirling bands of different intensity. It is often used in jewelry because of its attractive appearance. An analysis of malachite gives the following composition: 57.48% copper, 5.43% carbon, 0.91% hydrogen, and 36.18% oxygen. Determine the empirical formula of malachite.

Solution:

Consider a 100-g sample of malachite, so the masses of each element are numerically the same as the percent composition. Then calculate the moles of each element in 100 g of the compound. We use the molar masses to create equivalence expressions and convert them to ratios for the conversions. As in Example 4.8, the correct ratio has moles divided by grams:

$$1 \ \text{mol Cu} = 63.55 \ \text{g Cu} \qquad \frac{1 \ \text{mol Cu}}{63.55 \ \text{g Cu}}$$

$$1 \ \text{mol C} = 12.01 \ \text{g C} \qquad \frac{1 \ \text{mol C}}{12.01 \ \text{g C}}$$

$$1 \ \text{mol H} = 1.008 \ \text{g H} \qquad \frac{1 \ \text{mol H}}{1.008 \ \text{g H}}$$

$$1 \ \text{mol O} = 16.00 \ \text{g O} \qquad \frac{1 \ \text{mol O}}{16.00 \ \text{g O}}$$

We multiply the mass of each element by the corresponding conversion ratio to get moles of this element:

$$\text{mol Cu} = 57.48 \ \cancel{\text{g Cu}} \times \frac{1 \ \text{mol Cu}}{63.55 \ \cancel{\text{g Cu}}} = 0.9045 \ \text{mol Cu}$$

$$\text{mol C} = 5.43 \ \cancel{\text{g C}} \times \frac{1 \ \text{mol C}}{12.01 \ \cancel{\text{g C}}} = 0.452 \ \text{mol C}$$

$$\text{mol H} = 0.91 \ \cancel{\text{g H}} \times \frac{1 \ \text{mol H}}{1.008 \ \cancel{\text{g H}}} = 0.90 \ \text{mol H}$$

$$\text{mol O} = 36.18 \ \cancel{\text{g O}} \times \frac{1 \ \text{mol O}}{16.00 \ \cancel{\text{g O}}} = 2.261 \ \text{mol O}$$

To calculate the mole ratios, first identify the element present in the smallest amount in moles. In this example, it's carbon. Then divide the moles of each of the other elements by the moles of carbon:

$$\frac{\text{mol Cu}}{\text{mol C}} = \frac{0.9045 \ \text{mol Cu}}{0.452 \ \text{mol C}} = \frac{2.00 \ \text{Cu}}{1.00 \ \text{C}}$$

$$\frac{\text{mol H}}{\text{mol C}} = \frac{0.90 \ \text{mol H}}{0.452 \ \text{mol C}} = \frac{1.99 \ \text{H}}{1.00 \ \text{C}}$$

$$\frac{\text{mol O}}{\text{mol C}} = \frac{2.261 \ \text{mol O}}{0.452 \ \text{mol C}} = \frac{5.00 \ \text{O}}{1.00 \ \text{C}}$$

When the mole ratios are rounded to the nearest whole number, we see that one formula unit has one carbon atom, two copper atoms, two hydrogen atoms, and five oxygen atoms. From this information, we can write the empirical formula $Cu_2CO_5H_2$. This formula is usually written as $Cu_2(OH)_2CO_3$, to show which ions are found in the mineral. Malachite contains two anions, hydroxide (OH^-) and carbonate (CO_3^{2-}). However, this cannot be deduced from the empirical formula.

Practice Problem 4.9

Another copper mineral is chrysocolla. An analysis of chrysocolla gives the following composition: 36.18% copper, 15.99% silicon, 2.29% hydrogen, and 45.54% oxygen. Determine the empirical formula of chrysocolla.

Further Practice: 4.59 and 4.60

Empirical Formulas with Fractional Mole Ratios

In Example 4.9, we calculated a mole ratio of 1.99, which we rounded to a value of 2. The decimal values will not always be that close to a whole number, so rounding off is not always appropriate. The ratios may, however, have fractional values that correspond to ratios of small whole numbers, such as 1.25 (5/4), 1.33 (4/3), 1.50 (3/2), and 1.67 (5/3). In such cases, we multiply each ratio by a small whole number to make all the subscripts whole numbers. For example, a compound of nitrogen and oxygen has the following mole ratio:

$$\frac{\text{mol O}}{\text{mol N}} = \frac{2.50 \ \text{mol O}}{1.00 \ \text{mol N}}$$

This corresponds to a whole-number ratio of 5/2, so we multiply each number by 2. There must be five oxygen atoms for every two nitrogen atoms in the compound, giving the formula N_2O_5.

EXAMPLE 4.10 **Empirical Formulas with Fractional Mole Ratios**

The copper mineral azurite has the deep-blue color azure. Azurite contains 55.31% copper, 6.97% carbon, 37.14% oxygen, and 0.58% hydrogen. Calculate the empirical formula of azurite.

Solution:

Calculate the moles of each element in 100 g of the compound, following the same procedure used in Examples 4.8 and 4.9:

$$\text{mol Cu} = 55.31 \ \text{g Cu} \times \frac{1 \ \text{mol Cu}}{63.55 \ \text{g Cu}} = 0.8703 \ \text{mol Cu}$$

$$\text{mol C} = 6.97 \ \text{g C} \times \frac{1 \ \text{mol C}}{12.01 \ \text{g C}} = 0.580 \ \text{mol C}$$

$$\text{mol H} = 0.58 \ \text{g H} \times \frac{1 \ \text{mol H}}{1.008 \ \text{g H}} = 0.58 \ \text{mol H}$$

$$\text{mol O} = 37.14 \ \text{g O} \times \frac{1 \ \text{mol O}}{16.00 \ \text{g O}} = 2.321 \ \text{mol O}$$

Now calculate the ratio of the moles of each element to the moles of the element present in the smallest amount (carbon):

$$\frac{\text{mol Cu}}{\text{mol C}} = \frac{0.8703 \ \text{mol Cu}}{0.580 \ \text{mol C}} = \frac{1.50 \ \text{Cu}}{1.00 \ \text{C}}$$

$$\frac{\text{mol H}}{\text{mol C}} = \frac{0.58 \ \text{mol H}}{0.580 \ \text{mol C}} = \frac{1.0 \ \text{H}}{1.00 \ \text{C}}$$

$$\frac{\text{mol O}}{\text{mol C}} = \frac{2.321 \ \text{mol O}}{0.580 \ \text{mol C}} = \frac{4.00 \ \text{O}}{1.00 \ \text{C}}$$

One of the ratios is not close to a whole number, but it corresponds to a ratio of whole numbers (3/2). We can multiply this ratio by 2 to get a whole number. Of course, we must also multiply all the other ratios by 2. Thus, for every 2 mol of C, we have 3 mol of Cu, 2 mol of H, and 8 mol of O. The empirical formula is $Cu_3C_2H_2O_8$. This formula is usually written $Cu_3(CO_3)_2(OH)_2$ to show the ions in the mineral.

Practice Problem 4.10

Shattuckite is a fairly rare copper mineral. It has the composition 48.43% copper, 17.12% silicon, 34.14% oxygen, and 0.31% hydrogen. Calculate the empirical formula of shattuckite.

Further Practice: 4.61 and 4.62

Molecular Formulas from Empirical Formulas

The molar mass of a compound can be determined experimentally by a variety of methods. We can also calculate the mass of 1 mol of formula units, using the empirical formula. Comparing the experimental and calculated masses allows us

Formula units   **Mass of formula units**

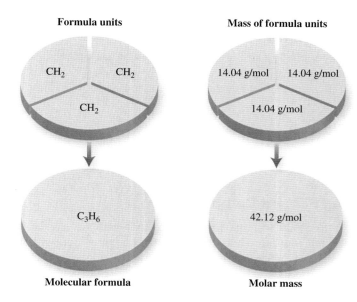

Figure 4.16
Three formula units of CH_2 make up one molecule of C_3H_6. The masses of the formula units (14.04 g/mol) make up 1 molar mass of 42.12 g/mol.

to determine and confirm the molecular formula of a compound. In other words, if the experimental molar mass is the same as the calculated one, the molecular formula is the same as the empirical formula. If the experimental molar mass is greater than that calculated from the empirical formula, then the molecular formula is some multiple of the empirical formula (Figure 4.16). This concept is applied in Example 4.11.

All compounds with the same empirical formula have the same percent composition. In such cases, we cannot distinguish between the compounds by comparing either the percent compositions or the empirical formulas.

EXAMPLE 4.11   Molecular Formulas

Many copper and brass cleaners contain oxalic acid, which consists of 26.68% carbon, 2.24% hydrogen, and 71.08% oxygen. It has a molar mass of about 90.0 g/mol. What is the molecular formula of oxalic acid?

Solution:

The empirical formula is obtained from the percent composition as in previous examples. You should verify for yourself that this turns out to be HCO_2. The molar mass is about 90.0 g/mol. If the empirical formula is also the molecular formula, then the molar mass of HCO_2 should be 90.0 g/mol. We calculate the molar mass of HCO_2 by summing the weighted molar masses of the constituent atoms:

$$
\begin{aligned}
\text{Mass of 1 mol of H} &= 1\ \text{mol} \times 1.008\ \text{g/mol} = & 1.008\ \text{g} \\
\text{Mass of 1 mol of C} &= 1\ \text{mol} \times 12.01\ \text{g/mol} = & 12.01\ \text{g} \\
\text{Mass of 2 mol of O} &= 2\ \text{mol} \times 16.00\ \text{g/mol} = & \underline{32.00\ \text{g}} \\
\text{Mass of 1 mol of } HCO_2 & & = & 45.02\ \text{g}
\end{aligned}
$$

The molar mass calculated for HCO_2 is 45.02 g/mol. The ratio of the experimental molar mass and the calculated molar mass is

$$
\frac{\text{experimental } MM}{\text{calculated } MM} = \frac{90.0\ \text{g/mol}}{45.02\ \text{g/mol}} = 2.00
$$

Since 90.0 g/mol is very close to twice the molar mass calculated from the empirical formula, the subscripts in the empirical formula must all be multiplied by 2, and the molecular formula must be twice the empirical formula: $H_2C_2O_4$.

Figure 4.17
Although they have similar formulas, cuprite, Cu_2O (*top*), and chalcocite, Cu_2S (*bottom*), contain different percentages of copper by mass.

Practice Problem 4.11

Potassium persulfate is a strong bleaching agent. It has a percent composition of 28.93% potassium, 23.72% sulfur, and 47.35% oxygen. The experimental molar mass is 270.0 g/mol. What are the empirical and molecular formulas of potassium persulfate?

Further Practice: 4.67 and 4.68

Determining Percent Composition

If we know a chemical formula for a compound, we can use the relationships between moles and mass to determine its percent composition without having to determine masses by experiment. For example, we might compare the copper content of two minerals, such as the cuprite (Cu_2O) and chalcocite (Cu_2S) shown in Figure 4.17, to determine which would yield more copper.

To simplify converting a chemical formula to a percent composition, we can assume a sample size of 1 mol. Then the mass of an element in 1 mol of the compound equals the number of moles of the element times the molar mass of the element:

The percent composition of a given element then equals the mass of this element divided by the molar mass of the compound (which is the mass of 1 mol) and multiplied by 100% to convert to a percentage:

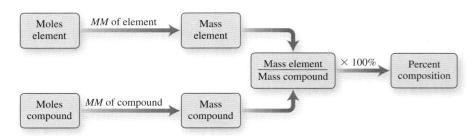

For example, there are 2 mol of copper in 1 mol of Cu_2O (cuprite). Thus, we can determine that 1 mol of the compound contains 2 mol $\times$ 63.55 g/mol or 127.1 g of copper. Adding the molar masses of the elements gives 143.1 g/mol for the molar mass of cuprite, so 1 mol has a mass of 143.1 g. The percent copper in the compound is thus

$$\% \, Cu = \frac{127.1 \text{ g}}{143.1 \text{ g}} \times 100\% = 88.82\% \, Cu$$

Since the percent composition of all the component elements must add up to 100%, there must be 11.18% oxygen in cuprite.

EXAMPLE 4.12   Percent Composition from Chemical Formula

Calculate the percent copper in chalcocite, Cu_2S.

Solution:

For chalcocite, 1 mol of the compound contains 2 mol × 63.55 g/mol or 127.1 g of copper. Adding the molar masses of the elements gives 159.2 g/mol for the molar mass of chalcocite. The percent copper in the compound is thus

$$\% \, Cu = \frac{127.1 \, g}{159.2 \, g} \times 100\% = 79.84\% \, Cu$$

Note that this mineral contains a lower percentage of copper than does cuprite (Cu_2O), even though each contains 2 mol of copper per mole of compound, because sulfur contributes a greater percentage to the mass of Cu_2S than oxygen contributes to the mass of Cu_2O.

Practice Problem 4.12

Calculate the percent copper in covellite, CuS.

Further Practice: 4.69 and 4.70

4.4   CHEMICAL COMPOSITION OF SOLUTIONS

Since chemical reactions between solids are normally slow, we usually carry out such reactions by first dissolving the compounds in a liquid to form a solution. For example, the acid leaching process used to treat oxide ores of copper produces a solution containing copper(II) sulfate. Which of the liquids that you encounter every day are solutions?

Recall from Chapter 1 that a *solution* is any mixture that is homogeneous at the molecular or ionic scale. In a solution, the substance that is dissolved is called the **solute** (usually present in a lesser amount), and the substance doing the dissolving is called the **solvent** (usually present in a greater amount). Figure 4.18 shows a solution of copper(II) sulfate (the solute) dissolved in water (the solvent).

Figure 4.18
In this system, dissolved copper(II) sulfate is present in a lesser quantity than water, so copper(II) sulfate is the solute and water is the solvent.

EXAMPLE 4.13   Solute and Solvent

From this picture of a solution containing $CuSO_4$ and water, identify the solute and the solvent.

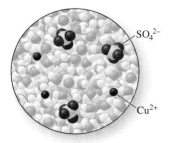

SO_4^{2-}

Cu^{2+}

Solution:

In this image, the red spheres represent Cu^{2+} ions. The yellow and red groups of spheres represent SO_4^{2-} ions. The pale red and white groups represent H_2O molecules. The number of Cu^{2+} and SO_4^{2-} ions is equal and is less than the number of H_2O molecules. Thus, the solute is $CuSO_4$, which yields Cu^{2+} and SO_4^{2-} ions in solution. The solvent is H_2O since it is present in the greater amount.

Practice Problem 4.13

In the following image of a solution containing hydrogen sulfide and water, identify the solute and the solvent.

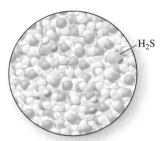

H_2S

Further Practice: 4.75 and 4.76

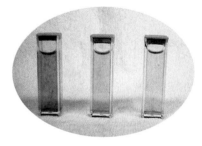

Figure 4.19
The color intensity of a colored solution decreases as the concentration of the solute decreases.

The terms *strong* and *weak* are used here in their everyday sense, not in their scientific sense. When we discuss strong and weak acids, we are not talking about their concentration, but about their ability to dissociate in solution.

Concentration

Solutions are homogeneous mixtures, but different solutions can contain varying amounts of solute and solvent. So how do we express the composition of a solution? One way is to describe its **concentration,** which is the relative amounts of solute and solvent in it. When comparing solutions, we can describe them as either dilute or concentrated. A **dilute solution** contains a relatively small amount of solute. In contrast, a **concentrated solution** contains a comparatively large amount of solute. These terms are helpful when we compare two solutions of different concentrations, indicating that one contains more or less solute than the other.

We see differences in concentration in everyday solutions. For example, when Julio brews a pot of tea, the color is more intense if the tea is "strong" (concentrated) than if it is "weak" (dilute). What are some other solutions that you see in your environment where color gives you an indication of the concentration? What other ways can you compare concentrations of solutions? The concentrations of sugar solutions could be compared by how slowly they pour or by their density. If you compared a teaspoonful of sugar dissolved in a cup of water to molasses or syrup (also sugar solutions), which would pour more slowly? Which would be denser? Taste is another way we compare concentration in our daily lives (but never in the laboratory), but it is not an accurate measure of concentration. Would you want the nurse in the hospital to taste the saline solution to see if it is the correct concentration before putting it into your IV?

A variety of experimental methods allows us to determine the concentrations of solutions. For example, if the solute is colored and the solvent is colorless, then the intensity of color in the solution is a measure of its concentration. Consider the aqueous solutions of copper(II) sulfate shown in Figure 4.19. The more copper(II) sulfate dissolved in the water, the more intense is the blue color due to the Cu^{2+} ions.

Figure 4.20 shows, at a molecular level, how the relative amounts of solute and solvent vary with the concentration. If you count the solute particles [copper(II) ions and sulfate ions] and compare them to the number of water

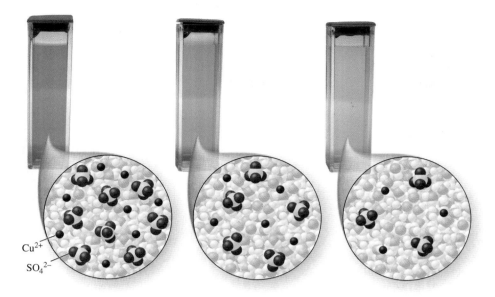

Cu^{2+}
$SO_4{}^{2-}$

Figure 4.20
Compare the concentrations of ions in the solution for these copper(II) sulfate solutions. Which has more copper(II) ions? More sulfate ions? Which solution has the greatest concentration?

molecules, you can see that as the solution becomes more dilute, the number of ions in a measured amount decreases and the number of water molecules in that same measured amount increases. Because the concentration of copper(II) ions decreases relative to the number of water molecules, the color of the blue solution fades.

Percent by Mass

Just as we could express the composition of a chemical compound as a percent by mass of each component, we can do the same for the composition of a solution. The percent by mass equals the mass of the solute divided by the mass of solution, multiplied by 100% to convert the ratio to a percentage:

$$\% \text{ mass} = \frac{\text{mass solute}}{\text{mass solution}} \times 100\%$$

For example, hydrochloric acid is sold in solutions that are 36% by mass HCl, with the remainder of the mass supplied by the solvent, water. The mass of solution is the sum of the masses of solute and solvent. This way of measuring concentration has two advantages. First, it does not depend on temperature. The volume may change with temperature, but the mass does not. Second, although the volume of a solution may be affected by the presence of solute, the percent mass is not affected. Thus, we don't need special glassware that measures exact volumes to prepare solutions accurately by mass.

Molarity

In addition to percent by mass, the concentration of a solution can be expressed in a variety of ways. One of the most common is molarity (*M*). The **molarity** of a solution is the number of moles of solute dissolved in 1 L of solution. The molarity of any solution can be calculated by dividing the moles of solute by the liters of solution:

$$\text{Molarity} = \frac{\text{moles of solute}}{\text{liters of solution}}$$

Animation: Making a Solution
Online Learning Center, Animations

Chemistry Animations Library, Making a Solution, Making_a_Solution.swf.

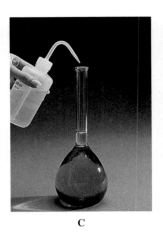

A   B   C   D

Figure 4.21
Solutions of known molarity are usually prepared in volumetric flasks, which are calibrated to contain a specific volume. (A) To prepare 250 mL of a 0.100 *M* solution of $CuSO_4 \cdot 5H_2O$, weigh 6.24 g (which is 0.0250 mol) of the solute. (B) Transfer it to the flask and add some solvent, water. Dissolve the solute by swirling. (C) Solvent is added with swirling until the bottom of the liquid surface just matches the marked line on the neck of the flask. (D) Then a stopper is inserted into the flask and the flask is inverted several times to thoroughly mix the contents. Convince yourself that this procedure will indeed result in a solution whose concentration is 0.100 *M*.

In Chapter 6 we will see how chemical reactions can be used to determine how much solute is present in a solution.

A solution having a volume of 1.000 L and containing 159.6 g (1.000 mol) of $CuSO_4$ is a 1.000 *M* copper(II) sulfate solution. If the liter of solution had half this much copper(II) sulfate, 79.8 g, its concentration would be 0.500 *M*. What would the concentration be if 79.8 g were contained in 0.500 L of solution?

Molarity is calculated from liters of *solution,* not liters of *solvent.* How large the difference is depends on the concentration of the solution. Dilute solutions contain only a small amount of solute. The volume of such solutions is very nearly the same as the volume of solvent. However, the difference can be significant for solutions in which the volume of the solvent is significantly less than the volume of the solution. To assure accuracy, solutions of the desired molarity are usually prepared in volumetric flasks, as shown in Figure 4.21.

If we know the mass of the solute and the volume of the solution, we can calculate the molarity of the solution. To begin, calculate the number of moles of solute contained in the solution. Then divide by the volume, in liters, of the solution.

EXAMPLE 4.14   Molarity of Solutions

A solution is prepared from 17.0 g of NaCl dissolved in sufficient water to give 150.0 mL of solution. What is the molarity of the solution? (The molar mass of NaCl is 58.44 g/mol.)

Solution:

We know the volume of solution, but we must find the moles of solute before we can calculate the molarity:

The moles of sodium chloride can be calculated from its mass using the molar mass, as we have done previously:

$$\text{mol NaCl} = 17.0 \text{ g NaCl} \times \frac{1 \text{ mol NaCl}}{58.44 \text{ g NaCl}} = 0.291 \text{ mol NaCl}$$

Convert the volume to units of liters:

$$\text{Volume (L)} = 150.0 \text{ mL} \times \frac{1 \text{ L}}{1000 \text{ mL}} = 0.1500 \text{ L}$$

Divide the number of moles by the volume in liters to get molarity:

$$\text{Molarity} = \frac{0.291 \text{ mol NaCl}}{0.1500 \text{ L solution}} = 1.94 \frac{\text{mol NaCl}}{\text{L solution}} = 1.94 \, M$$

Practice Problem 4.14

A solution is prepared from 22.5 g of H_2S dissolved in sufficient water to give 250.0 mL of solution. What is the molarity of the solution?

Further Practice: 4.83 and 4.84

We commonly express the concentration of a solution as a molarity of the compound. For example, a solution might contain 0.100 M $CuCl_2$. If we are interested in the concentrations of ions in a solution of this ionic compound, we must account for the fact that the solution contains twice as many chloride ions as copper(II) ions. For each formula unit of $CuCl_2$ in the solution, we have one Cu^{2+} ion and two Cl^- ions. For each mole of $CuCl_2$ in the solution, we have 1 mol of Cu^{2+} ions and 2 mol of Cl^- ions. Thus, the 0.100 M $CuCl_2$ solution contains 0.100 M Cu^{2+} and 0.200 M Cl^-.

If we know the volume and the molarity of a solution, we can calculate the mass of solute in that volume of solution. This is the reverse of the process illustrated in Example 4.14.

EXAMPLE 4.15   Mass from Molarity

A sample of copper ore is soaked in hydrochloric acid, giving a solution that contains some copper(II) chloride. If the solution has a concentration of $3.94 \times 10^{-6} \, M$ $CuCl_2$, what mass of $CuCl_2$ (in grams) is contained in 1500.0 L of the solution?

Solution:

To determine the mass of the solute, first determine the moles of solute by multiplying the molarity by the volume of solution in liters:

$$\boxed{\text{Molarity}} \xrightarrow{\text{volume}} \boxed{\text{Moles of solute}} \xrightarrow{MM} \boxed{\text{Grams of solute}}$$

$$\text{mol CuCl}_2 = 3.94 \times 10^{-6} \frac{\text{mol}}{\text{L}} \times 1500.0 \text{ L} = 5.91 \times 10^{-3} \text{ mol CuCl}_2$$

Many ionic compounds, especially those of the transition metals, contain not only a metal cation and a nonmetal anion or oxoanion, but also water molecules. Such compounds are called hydrates. The amount of water in a formula unit of the compound is shown by the formula of water preceded by the number of water molecules in a formula unit that is separated from the rest of the formula by a centered dot. An example is $CuSO_4 \cdot 5H_2O$, which, as can be seen in the photo, has different properties than $CuSO_4$. The blue material in the center of the $CuSO_4$ was obtained by adding water.

$CuSO_4$

$CuSO_4 \cdot 5H_2O$

Note that we expressed the molarity as mol/L in order to allow the units to cancel. Molar mass is then used to convert moles to mass:

$$g\ CuCl_2 = 5.91 \times 10^{-3}\ mol\ CuCl_2 \times \frac{134.5\ g\ CuCl_2}{1\ mol\ CuCl_2} = 0.795\ g\ CuCl_2$$

Practice Problem 4.15

Copper compounds such as bluestone are often used to control algae in fish-farming ponds. Bluestone is copper(II) sulfate pentahydrate, $CuSO_4 \cdot 5H_2O$, with a molar mass of 249.7 g/mol. A sample of pond water was found to have a concentration of 6.2×10^{-5} M copper(II) sulfate. If the pond has a volume of 1.8×10^7 L, what mass of bluestone did the farmer add to the pond?

Further Practice: 4.85 and 4.86

Finally, we can calculate the volume of solution required to make a solution of a specific concentration from a given amount of solute. Again, we need to use only conversions arising from the definition of molarity and the gram-to-mole conversion using molar mass.

EXAMPLE 4.16   Volume and Molarity

A solution of copper(II) acetate, $Cu(CH_3CO_2)_2$, is used as a green dye for textiles. We want to prepare a 0.150 M solution of copper(II) acetate, starting with 40.0 g of $Cu(CH_3CO_2)_2$. What should be the total volume of the solution?

Solution:

We need a two-step conversion to go from mass of solute to volume of solution:

$$\boxed{\text{Grams of solute}} \xrightarrow{MM} \boxed{\text{Moles of solute}} \xrightarrow{molarity} \boxed{\text{Volume of solution}}$$

Convert the mass of solute to moles of solute, using the molar mass (181.6 g/mol) to carry out the conversion:

$$mol\ Cu(CH_3CO_2)_2 = 40.0\ g\ Cu(CH_3CO_2)_2 \times \frac{1\ mol\ Cu(CH_3CO_2)_2}{181.6\ g\ Cu(CH_3CO_2)_2}$$

$$= 0.220\ mol\ Cu(CH_3CO_2)_2$$

Then using the molarity of the solution, convert moles of solute to volume of solution:

Volume of solution =

$$0.220\ mol\ Cu(CH_3CO_2)_2 \times \frac{1\ L\ solution}{0.150\ mol\ Cu(CH_3CO_2)_2} = 1.47\ L\ solution$$

Practice Problem 4.16

Copper(II) chlorate, $Cu(ClO_3)_2$, is used to treat fibers before dying them so the dye will adhere. Suppose we wish to prepare a 0.225 M solution of $Cu(ClO_3)_2$, starting with 1.89 g of copper(II) chlorate. What should be the total volume of the solution?

Further Practice: 4.87 and 4.88

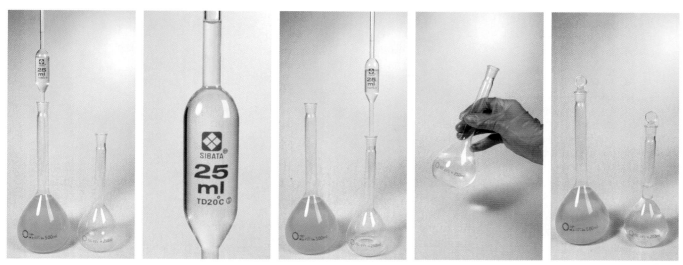

Figure 4.22
A pipet is used to accurately measure 25.00 mL of 1.000 M CuSO$_4$ solution and deliver it into a 250.0-mL volumetric flask. Water is added with swirling until the diluted solution fills the flask to the marked level. A stopper is placed in the flask, and the flask is inverted several times to completely mix the solution. The new solution has a concentration of 0.1000 M.

Figure 4.23
As water is added to a concentrated solution of copper(II) sulfate, the amount of water increases, leading to a lesser concentration of solute. The number of copper(II) ions and sulfate ions in a given volume of solution decreases, but the total number of copper(II) ions and sulfate ions in the flask is constant.

Concentrated solution   Water   Dilute solution

Dilution

To lower the concentration of a solution, we use the process of **dilution,** or adding more solvent to a solution, as depicted in Figure 4.22. In this process, the relative numbers of solute and solvent particles change. Adding more solvent increases the number of solvent particles and increases the volume. For a given amount of the original solution, the number of solute particles stays the same, but they are now spread out through a greater volume, so their concentration is less, as shown in Figure 4.23.

Suppose you taste your lemonade and find that it is too sweet. How might you make it less sweet? The easiest way is to add more water. Adding water to a solution to lessen its concentration is called dilution. Dilution is a simple way to reduce concentration, but how would you increase the concentration of a solution?

Rearranging an equation such as $M_{dil}V_{dil} = M_{con}V_{con}$ involves dividing both sides by one of the quantities. If we divide by V_{dil}, we get the following ratios:

$$\frac{M_{dil}\,\cancel{V}_{dil}}{\cancel{V}_{dil}} = \frac{M_{con}V_{con}}{V_{dil}}$$

Since the terms cancel on the left side of the equation, we get the following rearranged equation:

$$M_{dil} = \frac{M_{con}V_{con}}{V_{dil}}$$

We could divide by any of the three other variables as well, giving different versions of rearranged equations.

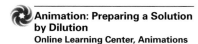

If we dilute a solution of known molarity, we can calculate the molarity of the diluted solution. The number of moles of solute contained in the volume of the more concentrated solution (V_{con}) is found by multiplying the molarity (M_{con}) of this solution by its volume in liters:

$$\text{Moles}_{con} = M_{con} \times V_{con}$$

Since we are adding only solvent, the moles of solute do not change upon dilution, so the moles of solute before and after dilution are equal:

$$\text{Moles}_{con} = \text{moles}_{dil}$$

The moles of solute in the diluted solution also equal the product of molarity (M_{dil}) and volume (V_{dil}) in liters, where the volume is the total volume after dilution:

$$\text{Moles}_{dil} = M_{dil} \times V_{dil}$$

Since the number of moles of solute is unchanged, we get an equation that relates the molarity of the diluted solution to the molarity and volume of the original, more concentrated, solution:

$$M_{dil}V_{dil} = M_{con}V_{con}$$

We can rearrange this equation to solve for the molarity of the diluted solution:

$$M_{dil} = \frac{M_{con}V_{con}}{V_{dil}}$$

Thus, if 25.00 mL of a 0.1000 M copper(II) sulfate solution is diluted to 250.0 mL, the concentration of the diluted solution is obtained by substituting the molarity of the concentrated solution and the two volumes, first converted to liters, into the equation

$$M_{dil} = \frac{0.1000\ M \times 0.02500\ \cancel{L}}{0.2500\ \cancel{L}} = 0.01000\ M$$

Variations on this calculation can be carried out as long as any three of the four variables in the equation are known.

EXAMPLE 4.17   Dilution

If 85.2 mL of 2.25 M $CuCl_2$ solution is diluted to a final volume of 250.0 mL, what is the molarity of the diluted solution of $CuCl_2$?

Solution:

The numbers of moles of copper(II) chloride are the same before and after dilution, so we can use the equation for dilution:

$$M_{dil} = \frac{M_{con}V_{con}}{V_{dil}}$$

Insert the values of the quantities and solve the equation for the molarity of the diluted solution:

$$M_{dil} = \frac{2.25\ M \times 0.0852\ \cancel{L}}{0.2500\ \cancel{L}} = 0.767\ M$$

Practice Problem 4.17

If 42.8 mL of 3.02 M H_2SO_4 solution is diluted to a final volume of 500.0 mL, what is the molarity of the diluted solution of H_2SO_4?

Further Practice: 4.93 and 4.94

SUMMARY

In this chapter you learned that metals are extracted from minerals and ores and are used to make many useful products. Before a metal is extracted from an ore, chemists often determine the amounts of elements in the compounds found in the ore. Because we cannot see the atoms in compounds to count them, we must use indirect measurements. In laboratory experiments, elements in a sample of a compound can be separated by chemical reactions and measured in terms of their mass. The average mass of one atom of an element is its relative atomic mass measured in atomic mass units.

As important as the relative atomic masses are, they are not practical for everyday use, because the amounts of a compound used in chemical processes contain huge numbers of atoms. To work with large numbers of atoms, molecules, or ions, we use another unit, the mole, defined as Avogadro's number (6.022×10^{23}) of particles. The mass of 1 mol of a substance is the molar mass of that substance. Molar masses and Avogadro's number can be used to convert between the mass of a sample of pure matter, the number of moles of the chemical substance, and the number of atoms or molecules in the sample.

The chemical composition of compounds can be summarized in terms of the masses of the component elements, the percent composition by mass, moles of each element, and atoms or ions of each element. Each of these can be used to determine a chemical formula that represents the composition of a compound.

Chemical formulas represent the composition of chemical compounds. An empirical formula provides the simplest ratios of the elements, while a molecular formula gives the actual numbers of atoms of each element in a molecule. Empirical formulas can be determined from experimentally measured masses of the component elements or from percent composition by mass. Molecular formulas can be determined from empirical formulas if the molar mass of the compound is known.

The composition of a solution, a homogeneous mixture, is described by its concentration—or the relative amounts of solute and solvent. Concentration is commonly expressed in terms of percent by mass or of molarity, the moles of solute per liter of solution. Dilute solutions (of lesser concentration) can be prepared from concentrated solutions (of greater concentration) by adding known amounts of solvent, a process known as dilution.

KEY RELATIONSHIPS

Relationship	Equation
Percent composition is expressed as the fraction of the sample mass due to a specific element, multiplied by 100 to convert to a percent.	$\% \text{ E (E is any element)} = \dfrac{\text{mass of E}}{\text{mass of sample}} \times 100\%$
Percent by mass composition of a solution is the fraction of the solution mass due to the solute, multiplied by 100 to convert to a percent.	$\% \text{ mass} = \dfrac{\text{mass solute}}{\text{mass solution}} \times 100\%$
The concentration of a solution in units of molarity is the ratio of moles of solute to liters of solution.	$\text{Molarity} = \dfrac{\text{moles of solute}}{\text{liters of solution}}$
When diluting a solution, the moles of solute in the solution remain unchanged. Moles are calculated by multiplying molarity by volume in liters. This product has the same value before and after dilution.	$M_{\text{dil}} V_{\text{dil}} = M_{\text{con}} V_{\text{con}}$

KEY TERMS

Avogadro's number (4.1)

concentrated solution (4.4)

concentration (4.4)

dilute solution (4.4)

dilution (4.4)

empirical formula (4.3)

molar mass (4.1)

molarity (4.4)

mole (4.1)

percent composition by mass (4.2)

solute (4.4)

solvent (4.4)

QUESTIONS AND PROBLEMS

The following questions and problems, except for those in the *Additional Questions* section, are paired. Questions in a pair focus on the same concept. Answers to the odd-numbered questions and problems are in Appendix D.

Matching Definitions with Key Terms

4.1   Match the key terms with the descriptions provided.
(a) the amount of a substance that contains the same number of atoms, molecules, or formula units as there are atoms in exactly 12 g of carbon-12 (^{12}C)
(b) the number of fundamental particles (atoms, molecules, or ions) in 1 mol of any substance
(c) a formula written with the simplest ratios of atoms or ions (the smallest whole-number subscripts)
(d) the substance being dissolved; usually, that component of a solution that is present in the lesser amount
(e) the number of moles of solute per liter of solution
(f) a solution that contains a relatively high concentration of solute

4.2   Match the key terms with the descriptions provided.
(a) the mass of 1 mol of any substance, in units of grams per mole
(b) an expression of the percent of the total mass due to each element in a compound
(c) the substance doing the dissolving; usually, that component of a solution that is present in the larger amount
(d) the relative amounts of solute and solvent in a solution
(e) a solution that contains a relatively small concentration of solute
(f) the process of adding more solvent to give a solution of lesser concentration

Mole Quantities

4.3   Write the formulas for the compounds described by the following compositions.
(a) twice as many hydrogen atoms as sulfur atoms
(b) 1.5 times as many oxygen atoms as nitrogen atoms
(c) one-half as many calcium ions as chloride ions

4.4   Write the formulas for the compounds described by the following compositions.
(a) three times as many hydrogen atoms as phosphorus atoms
(b) 2.5 times as many oxygen atoms as nitrogen atoms
(c) one-third as many aluminum ions as chlorate ions

4.5   What are the formulas of the following molecules?

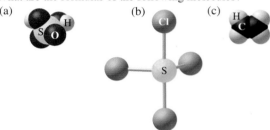

4.6   What are the formulas of the following molecules?

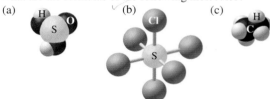

4.7   What is the molecular formula of these molecules?

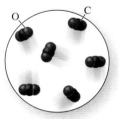

4.8   What is the molecular formula of these molecules?

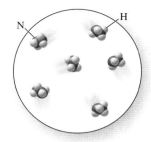

4.9   How many formula units are in $\frac{1}{2}$ mol of Cu_2S?

4.10   How many formula units are in 1.5 mol of $CuSO_4$?

4.11   How many sulfur atoms are in 0.2 mol of SO_2?

4.12   How many oxygen atoms are in 0.2 mol of SO_2?

4.13   How many calcium ions are in 1 mol of $CaCl_2$?

4.14   How many chloride ions are in 2 mol of $CaCl_2$?

4.15   Calculate the molar mass of each of the following substances.

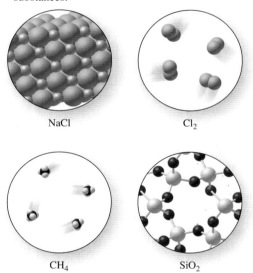

NaCl Cl_2

CH$_4$ SiO_2

Note: The structures are not drawn to the same scale.

4.16   Calculate the molar mass of each of the following substances. Note: The structures are not drawn to the same scale.

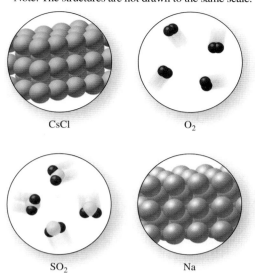

CsCl O_2

SO_2 Na

4.17   Calculate the molar mass of each of the following compounds.
   (a) Hg_2Cl_2 (c) Cl_2O_5
   (b) $CaSO_4 \cdot 2H_2O$ (d) $NaHSO_4$

4.18   Calculate the molar mass of each of the following compounds.
   (a) K_2SO_4 (c) $C_2H_4Cl_2$
   (b) $NiCl_2 \cdot 6H_2O$ (d) $KSbO_3$

4.19   In general, why must we weigh a sample in order to count the number of atoms it contains?

4.20   What is a mole? Why do chemists need to use the concept?

4.21   If 1.00 mol of LiCl has a mass of 42.394 g, what is the average mass of 1 LiCl formula unit in atomic mass units?

4.22   If one formula unit of $CuCl_2$ has an average mass of 134.5 amu, what is the mass of 1.00 mol of $CuCl_2$?

4.23   Calculate the molar mass of the following substances.
   (a) I_2 (b) $CrCl_3$ (c) C_4H_8

4.24   Calculate the molar mass of the following substances.
   (a) P_4 (b) CrO_2Cl_2 (c) CaF_2

4.25   If 2.01×10^{23} molecules of a substance have a mass of 12.0 g, what is the molar mass of the substance?

4.26   If the molar mass of a substance is 98.09 g/mol, what is the mass of 3.01×10^{23} molecules of the substance?

Moles, Masses, and Particles

4.27   A 4.55-g sample of limestone ($CaCO_3$) contains 1.82 g of calcium. What is the percent Ca in limestone?

4.28   A 3.75-g sample of limestone ($CaCO_3$) contains 1.80 g of oxygen and 0.450 g of carbon. What is the percent O and the percent C in limestone?

4.29   Calculate the number of moles in 10.0 g of the following substances.
   (a) $KHCO_3$ (c) Se
   (b) H_2S (d) $MgSO_4$

4.30   Calculate the number of moles in 100.0 g of each of the following compounds.
   (a) SO_2 (c) $BaSO_4$
   (b) Na_2SO_4 (d) $KAl(SO_4)_2$

4.31   Which element, Mo, Se, Na, or Br, contains the most moles of atoms in a 1.0-g sample?

4.32   Which element, Cu, P, Ag, or B, contains the least moles of atoms in a 5.0-g sample?

4.33   A chemical reaction requires 5.6 mol of zinc acetate, $Zn(CH_3CO_2)_2$. What mass of zinc acetate is needed?

4.34   A chemical reaction is used to produce 3.4 mol of copper(II) bicarbonate, $Cu(HCO_3)_2$. What mass of copper(II) bicarbonate is produced?

4.35   A sample of ammonia (NH_3) weighs 25.0 g. Calculate the following quantities.
   (a) moles of NH_3 (c) number of N atoms
   (b) number of NH_3 molecules   (d) moles of H atoms

4.36   A sample of sodium carbonate (Na_2CO_3) weighs 15.0 g. Calculate the following quantities.
   (a) moles of Na_2CO_3 (c) number of Na^+ ions
   (b) number of Na_2CO_3 (d) moles of CO_3^{2-}
       formula units   ions

4.37   Which of these substances has the most atoms per mole? Which has the least?

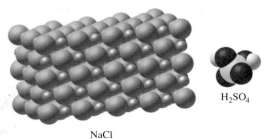

H_2SO_4

NaCl

SO_3

Na

Note: The structures are not drawn to the same scale.

4.38   Which of these substances has the most atoms per mole? Which has the least?

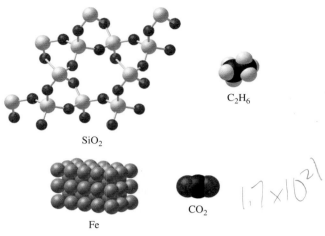

SiO_2

C_2H_6

Fe

CO_2

1.7×10^{21}

Note: The structures are not drawn to the same scale.

4.39   A raindrop weighs 0.050 g. How many molecules of water are in one raindrop?

4.40   A grain of sand weighs 7.7×10^{-4} g. How many formula units of silicon dioxide (SiO_2) are in one sand grain?

4.41   What is the mass of 2.4×10^{22} molecules of SO_2? 2.6

4.42   What is the mass of 3.0×10^{21} molecules of H_2SO_4?

4.43   Which compound, NH_3, NH_4Cl, NO_2, or N_2O_3, contains the most nitrogen atoms in a 25.0-g sample?

4.44   Which compound, NaCl, PCl_3, $CaCl_2$, or $HClO_2$, contains the most chlorine atoms (or ions) in a 100.0-g sample?

Determining Empirical and Molecular Formulas

4.45   You have two colorless gases, each made of sulfur and oxygen. If they have different percent compositions, can they be the same substance?

4.46   Describe some uses for the percent composition of a substance.

4.47   What is the difference between an empirical formula and a molecular formula?

4.48   Why do we normally use an empirical formula instead of percent composition to represent the composition of a substance?

4.49   Which of the following molecules have an empirical formula that is different from their molecular formula?

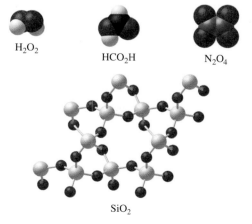

H_2O_2

HCO_2H

N_2O_4

SiO_2

Note: The structures are not drawn to the same scale.

4.50 Which of the following molecules have an empirical formula that is the same as their molecular formula?

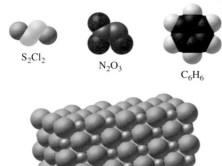

S_2Cl_2

N_2O_3

C_6H_6

NaCl

Note: The structures are not drawn to the same scale.

4.51 What is the empirical formula of each of the following compounds?
(a) P_4O_{10} (c) $PbCl_4$
(b) Cl_2O_5 (d) $HO_2CC_4H_8CO_2H$

4.52 What is the empirical formula of each of the following compounds?
(a) As_4O_6 (c) $CaCl_2$
(b) H_2S_2 (d) C_3H_6

4.53 Given the following molecular formulas, write the empirical formulas.
(a) $C_6H_4Cl_2$ (b) C_6H_5Cl (c) N_2O_5

4.54 Given the following molecular formulas, write the empirical formulas.
(a) N_2O_4 (b) $H_2C_2O_4$ (c) CH_3CO_2H

4.55 Which of the following compounds of nitrogen and oxygen have identical empirical formulas: N_2O, NO, NO_2, N_2O_3, N_2O_4, N_2O_5?

4.56 Which of the following compounds of carbon and hydrogen have identical empirical formulas: CH_4, C_2H_4, C_3H_6, C_4H_{12}, C_6H_6?

4.57 What are the empirical formulas of the compounds with the following compositions?
(a) 72.36% Fe, 27.64% O
(b) 58.53% C, 4.09% H, 11.38% N, 25.99% O

4.58 What are the empirical formulas of the compounds with the following compositions?
(a) 85.62% C, 14.38% H
(b) 63.15% C, 5.30% H; 31.55% O

4.59 Eugenol, a chemical substance with the flavor of cloves, consists of 73.19% C, 19.49% O, and 7.37% H. What is the empirical formula of eugenol?

4.60 One of the compounds in cement has the following composition: 52.66% Ca, 12.30% Si, and 35.04% O. What is its empirical formula?

4.61 The explosive trinitrotoluene (TNT) has the composition 37.01% C, 2.22% H, 18.50% N, and 42.27% O. What is the empirical formula of TNT?

4.62 Strychnine (rat poison) has the composition 75.42% C, 6.63% H, 8.38% N, and 9.57% O. What is the empirical formula of strychnine?

4.63 What information is needed to determine a molecular formula?

4.64 In what way does a molecular formula represent more information than an empirical formula?

4.65 A compound with the empirical formula CH_2O has a molar mass of approximately 90 g/mol. What is its molecular formula?

4.66 A gaseous compound has the empirical formula NO_2. If its molar mass is approximately 92 g/mol, what is its molecular formula?

4.67 If a compound has a molar mass of approximately 180 g/mol and a percent composition of 40.00% C, 6.72% H, and 53.29% O, what is the molecular formula?

4.68 A compound with the molar mass of approximately 142 g/mol has the composition 50.7% C, 9.9% H, and 39.4% N. What is its molecular formula?

4.69 What is the percent composition of each of the following compounds?
(a) NH_3 (b) $FeCl_2$ (c) Na_3PO_4 (d) KCl

4.70 What is the percent nitrogen in each of the following compounds?
(a) $NaNO_3$ (b) NH_4Cl (c) N_2H_4 (d) N_2O

4.71 Consider the following minerals as potential sources of copper metal. Which mineral contains the highest percentage of copper?
(a) chalcocite, Cu_2S (c) cuprite, CuO
(b) malachite, $Cu_2(CO_3)(OH)_2$ (d) azurite, $Cu_3(CO_3)_2(OH)_2$

4.72 Calculate the percent iron in the following iron minerals, which might be found in iron ores. Which contains the highest percentage of iron?
(a) wustite, FeO (c) magnetite, Fe_3O_4
(b) hematite, Fe_2O_3 (d) siderite, $FeCO_3$

Chemical Composition of Solutions

4.73 What is a solution? Give five examples of solutions that can be found in your home.

4.74 Distinguish between a solute and a solvent. Identify the solute in five solutions that can be found in a grocery store.

4.75 Identify the solute and solvent in this solution containing calcium chloride and water. Explain your answer.

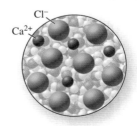

Cl^-

Ca^{2+}

4.76 Identify the solute and solvent in this solution containing sodium chloride and water. Explain your answer.

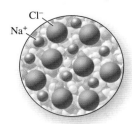

4.77 What is the difference between a dilute and a concentrated solution?

4.78 Give an example of two solutions that could be described as dilute and concentrated.

4.79 What relationship is described by concentration?

4.80 Define molarity.

4.81 Which of the following solutions of NaCl is more concentrated?

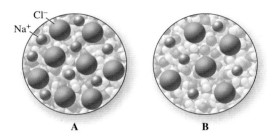

A   B

4.82 Which of the following solutions of $CaCl_2$ is more dilute?

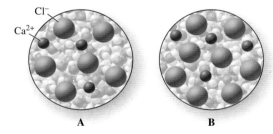

A   B

4.83 Calculate the molarity of each of the following solutions.
(a) 122 g of acetic acid (CH_3CO_2H) in 1.00 L of solution
(b) 185 g of sucrose ($C_{12}H_{22}O_{11}$) in 1.00 L of solution
(c) 70.0 g of hydrogen chloride (HCl) in 0.600 L of solution
(d) 45.0 g of potassium hydroxide (KOH) in 250.0 mL of solution

4.84 Calculate the molarity of each of the following solutions.
(a) 6.30 g HNO_3 dissolved in 255 mL of solution
(b) 49.0 g H_2SO_4 dissolved in 125 mL of solution
(c) 2.80 g KOH dissolved in 525 mL of solution
(d) 7.40 g $Ca(OH)_2$ dissolved in 200.0 mL of solution

4.85 Calculate the moles and the mass of solute in each of the following solutions.
(a) 250.0 mL of 1.50 M KCl
(b) 250.0 mL of 2.05 M Na_2SO_4

4.86 Calculate the moles and the mass of solute in each of the following solutions.
(a) 150.0 mL of 0.245 M $CaCl_2$
(b) 1450 mL of 0.00187 M H_2SO_4

4.87 Calculate what volume of the following solutions is required to obtain 0.250 mol of each solute.
(a) 0.250 M $AlCl_3$ (b) 3.00 M HCl

4.88 Calculate what volume of the following solutions is required to obtain 0.250 mol of each solute.
(a) 1.50 M H_2SO_4 (b) 0.750 M NaCl

4.89 How much water had to be added to 10.0 mL of the first solution to obtain the second solution?

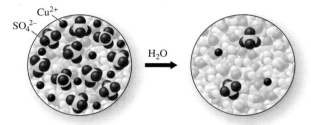

4.90 How much water had to be added to 50.0 mL of the first solution to obtain the second solution?

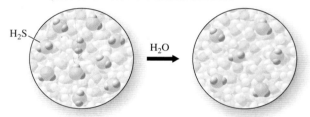

4.91 How much water must be added to 935.0 mL of 0.1074 M HCl to obtain a solution that is exactly 0.1000 M?

4.92 If you wish to prepare a 0.055 M solution of $NaNO_3$, to what volume would you have to dilute 25 mL of 3.0 M $NaNO_3$?

4.93 If 26.35 mL of 0.2473 M HCl is diluted to 250.0 mL, what is the concentration of the diluted solution?

4.94 What is the concentration of a solution prepared by diluting 35.0 mL of 0.150 M KBr to 250.0 mL?

Additional Questions

4.95 What is the mass of 0.100 mol of $Cu(OH)_2$?

4.96 How many (a) moles, (b) grams, and (c) atoms of vanadium are in 52.5 g of V_2O_5?

4.97 What is the average mass of one atom of argon in atomic mass units and in units of grams?

4.98   How many molecules are present in 15.43 g of butyl alcohol (C_4H_9OH)?

4.99   Calculate the number of moles of atoms and the number of atoms in the following quantities.
  (a) 36.1 g of argon
  (b) the 44.5-carat Hope diamond, which consists of carbon (1 carat = 0.200 g)
  (c) 2.50 mL of mercury with a density of 13.6 g/mL

4.100   What mass of iodine contains the same number of atoms as 50.0 g of chlorine?

4.101   A beaker contains 100.5 g of water. What mass of methanol (CH_3OH) must be added to the beaker for the resulting mixture to contain 2.00 mol of methanol per mole of water?

4.102   A gaseous compound has the composition 92.3% C and 7.7% H by mass. What is its empirical formula? If the molar mass of the compound is 78.1 g/mol, what is its molecular formula?

4.103   Tear gas has the composition 40.25% C, 6.19% H, 8.94% O, and 44.62% Br. What is its empirical formula?

4.104   A chemical procedure calls for 3.54 mol of dry ice (solid CO_2). What mass of CO_2 should we add?

4.105   How many oxygen atoms are contained in 10.00 g of H_3PO_4?

4.106   Calculate the mass percent of each element in Na_2CO_3.

4.107   How many moles of CO_2 are contained in the 960 g of CO_2 exhaled daily by an average person?

4.108   Para-aminobenzoic acid, often abbreviated PABA, is used in sunscreen formulations to prevent sunburn caused by ultraviolet radiation. PABA contains 61.30% C, 23.33% O, 10.22% N, and 5.15% H. What is the empirical formula of PABA?

4.109   Vanillin, the active component of the flavoring vanilla, has the composition 63.15% C, 5.30% H, and 31.55% O.
  (a) What is its empirical formula?
  (b) If the molar mass of vanillin is 152.14 g/mol, what is its molecular formula?

4.110   Monosodium glutamate ($NaC_5H_8NO_4$) is used extensively as a flavor enhancer. What is its percent composition?

4.111   If 5.00 g of Al is burned completely in excess oxygen gas, 9.45 g of an aluminum oxide is formed. What is the empirical formula of that oxide?

4.112   If 4.32 g Cu combines with 1.09 g of S to form 5.41 g of the bluish-black mineral chalcocite, what is the empirical formula of chalcocite?

5

Chemical Reactions and Equations

Janelle climbs the stairs to the second floor of the classroom building and begins her scheduled class. She goes to her workstation and selects two solutions, one of iron(II) sulfate, $FeSO_4$, and the other of potassium ferricyanide, $K_3Fe(CN)_6$. She mixes the two solutions, and a deep blue solid forms (Figure 5.1). She filters the mixture to isolate the solid and places it in an oven to dry. She then grinds the dried solid to a fine powder in oil. What course do you think Janelle is taking?

Meanwhile Antonio is at his workstation on the first floor of the classroom building. He selects an iron nail and places it in a solution of copper(II) sulfate, $CuSO_4$. Over a few minutes, the shiny piece of iron takes on a dull brown coating (Figure 5.2). He notices that the blue solution becomes less intense as the brown coating forms. He concludes that copper(II) ions have been removed from the solution. In which course do you think Antonio is enrolled?

Next door, Breanna drops pieces of a rock into beakers containing different concentrations of sulfuric acid. The rocks begin to give off bubbles of gas, which rise to the surface (Figure 5.3). She collects some of the gas and bubbles it through a solution of barium chloride ($BaCl_2$). A white solid forms. What course do you think Breanna is taking?

In the basement of the classroom building, Michelle takes some silicon dioxide (SiO_2) and heats it in a furnace to 1500°C. When it melts, she mixes it with some cobalt(II) oxide (CoO). After allowing the molten material to cool slowly, she obtains a deep blue solid. What course do you think Michelle is taking?

Elsewhere, Jennifer takes a ceramic pot and soaks it in an acetic acid solution (CH_3CO_2H). After an hour, she heats the solution to evaporate the acetic acid, leaving a solid residue. She adds some water and potassium iodide (KI), forming a yellow solid (Figure 5.4). She recognizes this solid as lead(II) iodide (PbI_2), indicating that the pot contains a lead compound. What course do you think Jennifer is taking?

If you think these students are taking a chemistry course, you are right about Jennifer and wrong about all the others. They are taking various art courses. Art and chemistry are intimately connected. Just as artists classify works of art as paintings, sculptures of stone or metal, stained glass, or ceramics, chemists classify reactions into different groups. Each of these students' experiments provides an example of a different art form and a different category of reactions.

Figure 5.1
When solutions of iron(II) sulfate and potassium ferricyanide mix, the insoluble blue compound $KFe[Fe(CN)_6]$ forms.

Figure 5.3
The calcium carbonate in marble reacts with sulfuric acid to release carbon dioxide gas.

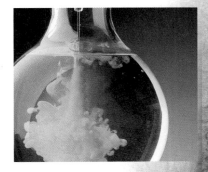

Figure 5.4
When potassium iodide is added to a solution containing lead(II) ions, the yellow solid lead(II) iodide forms.

Figure 5.2
When iron metal is placed in a solution of copper(II) sulfate, copper metal is deposited as a coating on the iron.

Around 1704, the color manufacturer Diesbach, working in Berlin, accidentally discovered the pigment known as Prussian blue (or Berlin blue). It was prepared by melting dried blood in the presence of potassium salts and adding iron(II) sulfate to an aqueous extract of the mixture. We now have less gory methods of preparing Prussian blue.

The solid that Janelle prepared is called Prussian blue. It has been used as a pigment since about 1700. Although many pigments were originally found as minerals in the Earth and may still be obtained in this way, today pure pigments are usually synthesized in a laboratory. Many are ionic compounds made by the kinds of reactions we will investigate in this chapter. When Janelle was grinding the blue powder with oil, she was making paint, which consists of pigments dispersed in a liquid, such as linseed oil, that will dry to form a film. When added to glass, the same transition metal oxides used as paint pigments produce stained glass. Michelle is making cobalt glass from silicon dioxide (sand) and cobalt(II) oxide.

Antonio is investigating the reaction between metals under corrosive conditions. His study will help him understand how to construct large metal structures to be displayed outdoors. An example is the Statue of Liberty. It was constructed of copper panels around an iron framework. In the corrosive marine environment of New York Harbor, the statue degraded over about a century. It eventually had to be restored because of damage that resulted from contact between the iron and the copper. We will examine the kinds of reactions responsible for corrosion in this chapter. Corrosion will be explored in more detail in Chapter 14.

Breanna is investigating the effect of acid rain on marble sculptures. Many old statues and architectural features, such as gargoyles, have corroded as a result of air pollution. Sulfur and nitrogen oxides released into the atmosphere by the burning of coal and other fuels react with water to form sulfuric and nitric acids. These acids react with the calcium carbonate in marble and limestone to release carbon dioxide gas. In this chapter we will examine some reactions of acids with other substances.

Jennifer's experiment is for chemistry class, but it is related to both art and human health. Before the dangers of lead poisoning were recognized, lead compounds were added to the glazing material on ceramic pots. They lowered the melting temperature of the glaze and made it form a glassy coating. Jennifer's test reveals whether the pot she is studying has lead in its glaze.

Today, different materials are used to obtain the color and finish of ceramic glazes. They don't contain lead, but many do contain transition metal ions. By controlling the oxygen concentration, or by adding other materials such as wood or charcoal, the heating process in a kiln causes an oxidation-reduction reaction. It changes the charge on the transition metal ions and, as a result, changes the color of the glaze. We will explore oxidation-reduction reactions in this chapter and examine them in detail in Chapter 14.

Questions for Consideration

5.1 What happens in a chemical reaction?
5.2 How do we know whether a chemical reaction takes place?
5.3 How do we represent a chemical reaction with a chemical equation?
5.4 How are chemical reactions classified? How can the products of different classes of chemical reactions be predicted?
5.5 How do we represent chemical reactions in aqueous solution?

5.1 WHAT IS A CHEMICAL REACTION?

Each of the students described in the introduction was carrying out some kind of chemical reaction. For example, when Antonio put iron metal into a solution of copper(II) sulfate, a chemical reaction occurred. What is a chemical reaction?

A chemical reaction is the conversion of one substance or set of substances into another. Any substance that we start with is a **reactant.** A new substance that forms during the reaction is a **product.** Products are different from reactants *only* in the *arrangement* of their component atoms. Chemical reactions neither destroy atoms nor create new atoms. They occur because the bonds that hold atoms together can be broken and rearranged. (You'll read more about chemical bonding in Chapter 8.)

Chemical reactions are not the same as nuclear reactions, in which new elements may form. You'll learn about those in Chapter 15.

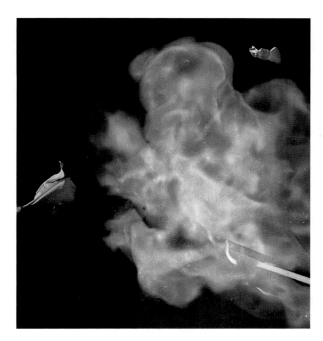

Figure 5.5
When ignited, hydrogen molecules and oxygen molecules in the air react explosively to form gaseous water.

Consider the reaction of hydrogen gas with oxygen gas, as shown in Figure 5.5. Hydrogen and oxygen both occur naturally as diatomic molecules. If they mix, they react slowly; but if ignited, the reaction is vigorous, even explosive. In either case, the reactants form the same product: gaseous water molecules, each containing two hydrogen atoms and one oxygen atom. Figure 5.6 shows the rearrangement of the atoms during the reaction. Note that this representation shows the reactant molecules separated from the product molecules, but they are actually separated by time, not distance. The reactants are in the container where the reaction occurs, and the products are in that same container after the reaction is complete.

Count the number of atoms of hydrogen and oxygen in the reactants and in the product. Has either number changed? No. In every chemical reaction the number of atoms of each element stays constant. This is consistent with the law of conservation of mass (see Chapter 2), which states that matter cannot be created or destroyed by ordinary chemical means.

Now consider the arrangement of atoms in Figure 5.6. What are oxygen atoms attached to in the reactants? In the products? How about hydrogen atoms? In the reactants, each hydrogen atom is attached to another hydrogen atom. Each oxygen atom is attached to another oxygen atom. In the products, two hydrogen atoms are attached to one oxygen atom. The number of atoms is the same, but their arrangement is different.

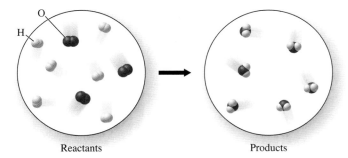

Reactants   Products

Figure 5.6
The hydrogen atoms from hydrogen molecules combine with the oxygen atoms from oxygen molecules to form gaseous water molecules.

EXAMPLE 5.1   Identify Reactant and Product

Breanna mixed a solution of sulfuric acid and solid calcium carbonate. Carbon dioxide gas bubbled out of the solution as the initial solid disappeared, leaving behind a deposit of calcium sulfate. Identify the reactants and products in this chemical reaction.

Solution:

Sulfuric acid and calcium carbonate are the starting substances, the reactants. They are converted into carbon dioxide and calcium sulfate, the products. (Another product is water, which can't be identified from this description.)

Practice Problem 5.1

Antonio placed a piece of iron in a solution of copper(II) sulfate. He obtained a deposit of copper metal and a solution containing iron(II) sulfate. Identify the reactants and products in this chemical reaction.

Further Practice: 5.3 and 5.4

5.2   HOW DO WE KNOW A CHEMICAL REACTION OCCURS?

Just as artists use paint and stone to create works of art, chemists use elements and compounds to create new substances. Both the artist and the chemist are constrained by the materials available to them. Both must work within the rules of nature regarding the properties and behaviors of their materials. Both rely on their materials to produce new creations, but is simply using materials enough? Is putting some paint on canvas art? Is mixing two substances together chemistry?

Consider the two beakers containing clear, colorless liquids shown in Figure 5.7. What happened when the two liquids mixed? Can you tell if a chemical reaction occurred? Just by looking, you can't. You started with two colorless liquids, and their mixture remains colorless. A rearrangement of atoms might or might not have taken place. There's no obvious way to tell.

Without an observable change in properties, the only way to know if a reaction has occurred is to separate the products and determine their chemical compositions as described in Chapter 4. Such methods are difficult and time-consuming, so we would like to have an easier way to know. Look at Figure 5.7 again. Suppose you touch the beaker containing the mixture and find that it is hot. Would you then have evidence that a chemical reaction occurred? Yes, a change in temperature often accompanies a chemical reaction. What are some other clues? Make a list of ideas, and then examine Figure 5.8 to see if any of the photos of chemical reactions match your list of clues.

Figure 5.7
Does a chemical reaction occur when these two colorless liquids mix? How can you tell?

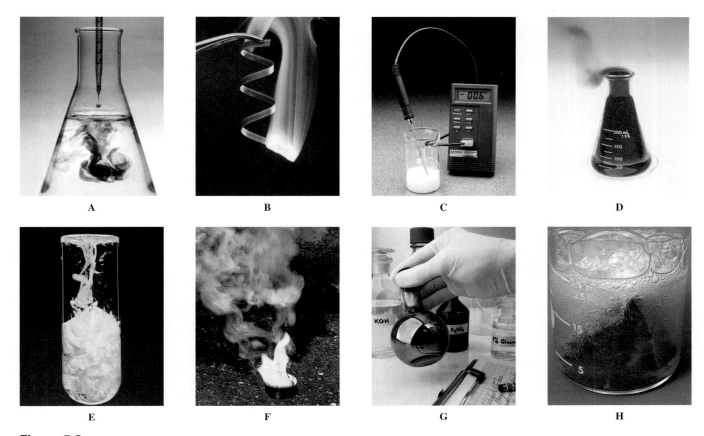

Figure 5.8
Do any of these photos of chemical reactions illustrate items on the list you made of nature's clues that a chemical reaction has occurred?

(A) $I_2(aq)$ + NaI(aq) + starch(aq)
(B) Mg(s) + $O_2(g)$
(C) Ba(OH)$_2 \cdot$8H$_2$O(s) + NH$_4$Cl(s)
(D) Cu(s) + HNO$_3(aq)$

(E) BaCl$_2(aq)$ + Na$_2$SO$_4(aq)$
(F) P$_4(s)$ + O$_2(g)$
(G) AgNO$_3(aq)$ + glucose(aq)
(H) CaCO$_3(s)$ + HCl(aq)

The clues most often used by chemists are the following:

- Change in color
- Production of light
- Formation of a solid (such as a *precipitate* in solution, smoke in air, or a metal coating)
- Formation of a gas (bubbles in solution or fumes in the gaseous state)
- Absorption or release of heat (sometimes appearing as a flame)

Did you come up with any clues that don't fit one of these categories?

> These clues are only indications that a reaction *may* have taken place. Physical changes can also cause such effects. For example, a temperature change might result from dissolving a compound, and a color change might arise from mixing two colored compounds.

5.3 WRITING CHEMICAL EQUATIONS

Describing chemical reactions in words is awkward and time-consuming. For this reason, chemists use a chemical equation to show what reactants are involved and what products are formed. A **chemical equation** is a symbolic representation of a chemical reaction.

To see how equations are written, let's look at the reaction shown in Figure 5.9. Two powders—gray aluminum metal (Al) and red iron(III) oxide (Fe$_2$O$_3$)—are mixed and poured into a ceramic container with a small hole in the bottom. A wick in the pile of powder is ignited. The powder smokes and then gives off a shower of sparks. As this spectacular display continues and then subsides, liquid metal runs

Animation: Thermite Reaction
Online Learning Center, Animations

Chemistry Animations Library, Thermite Reaction, Thermite_Rxn.rm.

Figure 5.9
When a powdered mixture of aluminum metal and iron(III) oxide is heated, it reacts vigorously to form liquid iron metal and aluminum(III) oxide. This reaction is called the thermite reaction.

When reading an equation, we replace the arrow with a word such as *yields, produces,* or *forms.*

Figure 5.10
The thermite reaction is used to make molten iron. In the device shown here, the liquid iron runs down from the crucible on top into the mold. Upon solidification, the iron joins two sections of railroad track.

out of the hole in the container. This liquid is molten iron. Left behind is aluminum oxide, which was also present in the smoke. This reaction is called the *thermite reaction.* It is sometimes used to weld pieces of iron or steel together, as in reinforcing rod (rebar) in concrete buildings or in the rails for high-speed railroad tracks (Figure 5.10).

A verbal description is one way to represent the thermite reaction, but as you can see, it is not very efficient. We might also draw pictures to show the changes that occur at the atomic level, but this is time-consuming and takes up a lot of space, too. The easier, shorter way is to use the symbols and formulas you've already learned to write and read.

To write a chemical equation, we first identify the reactants and write their chemical formulas with plus signs (+) between them. We place an arrow after them. The arrow stands for the process of converting one set of substances into another. Then we write the chemical formulas of the products, again with plus signs between them if there is more than one.

The physical state—solid (*s*), liquid (*l*), gas (*g*), or aqueous solution (*aq*)—is shown for each reactant and product. If heat must be supplied for the reaction to take place, or if some other special conditions are needed, we describe them above the arrow. Finally, the equation must be balanced. In a **balanced equation,** *the number of atoms of each element is the same in the products as in the reactants.* We indicate the relative number of molecules or formula units of reactants or products by *coefficients,* which are whole numbers that immediately precede the formula of each substance. If a coefficient has a value of 1, it is omitted, but understood to be there. The process of balancing an equation involves examining and modifying coefficients until the reactants and products contain the same numbers of the component elements. The coefficients are written as the smallest possible whole numbers. The coefficients are whole numbers because parts or fractions of atoms or molecules do not exist.

Let's work through writing and balancing the thermite reaction. In words, it looks like this:

$$\text{aluminum} + \text{iron(III) oxide} \xrightarrow{\text{heat}} \text{aluminum oxide} + \text{iron}$$

Using the rules developed in Chapter 3, we can substitute formulas for names in the equation. We should also add the physical state of each substance:

$$Al(s) + Fe_2O_3(s) \xrightarrow{\text{heat}} Al_2O_3(s) + Fe(l)$$

But the equation is not complete as written. The same number of atoms of each element must appear on both sides of the arrow. The equation must balance. In its present form, it does not. It has the following numbers of atoms:

Atoms in Reactants $Al(s) + Fe_2O_3(s)$		Atoms in Products $Al_2O_3(s) + Fe(l)$	
1 Al	●	2 Al	● ●
3 O	● ● ●	3 O	● ● ●
2 Fe	● ●	1 Fe	●

The number of oxygen atoms is balanced at 3 in the reactants and 3 in the products, but the other elements are not balanced. The reactants contain only one aluminum atom, but the products contain two. The reactants contain two iron atoms, but the products contain only one. To balance the equation, we must change coefficients until the reactants and products contain the same number of atoms of each element. *It is important to remember that you cannot change the subscript of an element in a compound to balance an equation.* For example, Al_2O_3 cannot be changed to AlO_3 to balance the number of aluminum atoms. Even if AlO_3 existed, it would be an entirely different compound. When the atoms of the reactants and products have not yet been equalized, the equation is called a *skeletal equation* or *unbalanced equation.*

To balance the skeletal equation for the thermite reaction, we need to do nothing about oxygen. It is already balanced. Note however that there are two aluminum atoms in the product Al_2O_3 but only one in the reactant Al. To balance the aluminum atoms, we can place the coefficient 2 in front of Al on the reactant side of the equation:

$$2Al(s) + Fe_2O_3(s) \xrightarrow{\text{heat}} Al_2O_3(s) + Fe(l)$$

This balances the aluminum atoms, but the iron atoms remain unbalanced. We can place the coefficient 2 in front of the product Fe to balance them:

$$2Al(s) + Fe_2O_3(s) \xrightarrow{\text{heat}} Al_2O_3(s) + 2Fe(l)$$

If we count the numbers of each element again, we see that they are now the same on both sides of the equation. The equation is balanced. Mass is conserved.

Atoms in Reactants 2Al(*s*) + Fe$_2$O$_3$(*s*)		Atoms in Products Al$_2$O$_3$(*s*) + 2Fe(*l*)	
2 Al	⚫⚫	2 Al	⚫⚫
3 O	⚫⚫⚫	3 O	⚫⚫⚫
2 Fe	⚫⚫	2 Fe	⚫⚫

To see how and why equations balance, use a model kit or gumdrops and toothpicks to build models of reactant molecules. Take them apart and rearrange them into product molecules. Use the smallest number of atoms possible to achieve conservation of mass. Reactions don't actually occur this way, but we want to emphasize the conservation of atoms in reactions as a means of balancing equations.

Let's consider this model approach with the reaction of oxygen (O_2) and methane (CH_4), a gas commonly used as a source of heat in the laboratory and in gas stoves (Figure 5.11). The products of the reaction are carbon dioxide (CO_2) and water (H_2O). All the substances are in the gaseous state. Let's start by writing a skeletal equation so we know which models we need:

$$CH_4(g) + O_2(g) \longrightarrow CO_2(g) + H_2O(g)$$

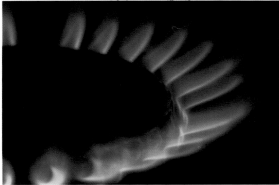

Figure 5.11
In a gas stove, methane reacts with oxygen to form carbon dioxide and water.

The equation can be written with the reactants in any order on the left side of the arrow and with the products in any order on the right side of the arrow. Any of the elements in the products can be used as a starting point for the balancing process.

If you do not have a model kit, you can make one with some gumdrops and toothpicks. Use different colors of gumdrops to represent different elements. Connect the gumdrops with toothpicks to represent molecules. Build several molecules of each reactant, but use only the minimum necessary to disassemble and build whole numbers of products.

The model for O_2 shows two bonds (sticks) holding the two atoms (spheres) together. In Chapter 8, we will explore the nature of bonds between atoms and we will see why two bonds are needed in this molecule.

We start with at least one molecule of each reactant:

We can disassemble our models to see what parts we have to work with to form products. We have 1 carbon atom, 4 hydrogen atoms, and 2 oxygen atoms. Note that we have not removed the original models from our diagram so we can keep track of their numbers:

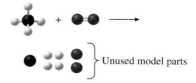

From the atoms we have available, we can make either 1 carbon dioxide molecule or 2 water molecules. Either approach will work. Let's pick the carbon dioxide to build first:

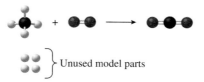

That leaves us with 4 extra hydrogen atom models. We have not formed any water molecules, although water is a product. According to the law of conservation of mass, all the reactant atoms must be completely consumed, so we know we need to make some water molecules from the leftover hydrogen atoms. But we need oxygen to do that, and the only way to get it is to add another oxygen molecule to the reactants. That gives us 2 additional oxygen atom models to work with.

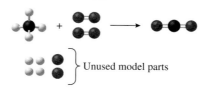

Now we have enough parts to build 2 water molecules from the 2 oxygen atom models and 4 hydrogen atom models. This leaves no unused atoms, so the equation must be balanced.

You can verify what your models show by counting the atoms of each element on each side of the equation. Since the numbers are equal, the equation is indeed balanced. Since we do not write coefficients of 1, the final version of the equation is

$$CH_4(g) + 2O_2(g) \longrightarrow CO_2(g) + 2H_2O(g)$$

The general approach to balancing equations is as follows:

- *Identify the reactants and products and write their correct formulas.* A molecular formula cannot be changed, because such a change would change the identity of the substance.
- *Write a skeletal equation including physical states.*
- *Change coefficients one at a time until the atoms of each element are balanced.* It is usually easiest to start by changing the coefficients for the substances that

To balance an equation, we need equal numbers of *atoms* of each element on each side of the equation, but not necessarily equal numbers of *molecules.* Consider the example $N_2(g) + 3H_2(g) \longrightarrow 2NH_3(g)$. In this reaction four molecules of reactants form two molecules of product. It is common for a reaction to yield more or fewer molecules than were initially present, although the total number of atoms always stays the same.

contain elements that occur *least* often in the equation. These substances often have the more complex formulas.

- *Make a final check by counting the atoms of each element on both sides of the equation.* They should be the same.

Before we start balancing equations using formulas, let's reexamine the equation we balanced using models. We can write a skeletal equation using molecular images:

We count atoms to see what coefficients need to be adjusted. The carbon atoms are balanced, but the hydrogen atoms and oxygen atoms are not. Adding another water molecule to the products balances the hydrogen atoms:

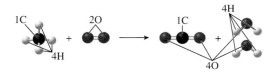

To balance the oxygen atoms, we add 1 more oxygen molecule to the reactants:

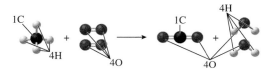

Now the numbers of each element are equal on both sides of the equation. The equation is balanced.

Balancing this equation was relatively simple, but sometimes changing a coefficient to balance one element unbalances another. In such cases, we must go back and change coefficients. Example 5.2 shows how.

EXAMPLE 5.2   Balancing Chemical Equations

Potassium chlorate is a white solid that decomposes when heated, forming solid potassium chloride and oxygen gas. Balance the equation that represents the reaction.

Solution:

Potassium chlorate is the only reactant. It has the formula $KClO_3$ and is in the solid state. The products are solid potassium chloride (KCl) and oxygen gas (O_2). The word equation gives the name of the reactant before the arrow and the names of the products, separated by a plus sign, after the arrow:

$$\text{potassium chlorate} \xrightarrow{\text{heat}} \text{potassium chloride} + \text{oxygen}$$

To get the skeletal equation, we substitute the formulas and their physical states for the names:

$$KClO_3(s) \xrightarrow{\text{heat}} KCl(s) + O_2(g)$$

Next we count atoms to see where to begin changing coefficients. Potassium and chlorine occur once on each side of the arrow and thus are balanced already, but there are 3 oxygen atoms on the left and 2 oxygen atoms on the right:

$$KClO_3(s) \xrightarrow{\text{heat}} KCl(s) + O_2(g)$$

Atoms in Reactants	Atoms in Products
1 K 1 Cl 3 O	1 K 1 Cl 2 O

We can balance the oxygen atoms by placing a 2 in front of $KClO_3$ and a 3 in front of O_2. Now there are 6 oxygen atoms on each side of the equation:

$$2KClO_3(s) \xrightarrow{\text{heat}} KCl(s) + 3O_2(g)$$

Atoms in Reactants	Atoms in Products
2 K 2 Cl 6 O	1 K 1 Cl 6 O

Now, however, the potassium and chlorine are no longer balanced. The left side of the equation has 2 of each, while the right side has only 1. We can rebalance the K and Cl atoms by placing a 2 in front of KCl on the product side of the equation:

$$2KClO_3(s) \xrightarrow{\text{heat}} 2KCl(s) + 3O_2(g)$$

A final count shows that this is a correct, balanced equation:

Atoms in Reactants	Atoms in Products
2 K 2 Cl 6 O	2 K 2 Cl 6 O

Practice Problem 5.2

When sodium metal is dropped into liquid water, the mixture releases gaseous hydrogen, leaving behind a colorless solution of sodium hydroxide. Balance the equation that represents this reaction.

Further Practice: 5.27 and 5.28

At intermediate stages, coefficients needed to balance an equation may not be whole numbers. Sometimes a temporary balancing with fractional coefficients makes balancing easier. But in the end, the coefficients must be whole numbers. Example 5.3 shows a method for balancing equations this way.

EXAMPLE 5.3   Balancing Chemical Equations with Fractional Coefficients

Ethane (C_2H_6) is a gaseous fuel that can be burned with oxygen. Balance the following skeletal equation for this reaction:

$$C_2H_6(g) + O_2(g) \longrightarrow CO_2(g) + H_2O(g)$$

Solution:

We can see that this equation is not balanced by counting atoms on each side of the equation:

$$C_2H_6(g) + O_2(g) \longrightarrow CO_2(g) + H_2O(g)$$

Atoms in Reactants	Atoms in Products
2 C 6 H 2 O	1 C 2 H 3 O

Since oxygen occurs in three substances, we should start by balancing the carbon and hydrogen. There are 2 carbon atoms on the left, but only 1 on the right, so we place a 2 in front of CO_2:

$$C_2H_6(g) + O_2(g) \longrightarrow 2CO_2(g) + H_2O(g)$$

Atoms in Reactants	Atoms in Products
2 C 6 H 2 O	2 C 2 H 5 O

There are 6 hydrogen atoms on the left and 2 on the right, so we place a 3 in front of H_2O:

$$C_2H_6(g) + O_2(g) \longrightarrow 2CO_2(g) + 3H_2O(g)$$

Atoms in Reactants	Atoms in Products
2 C 6 H 2 O	2 C 6 H 7 O

Now all we have left to balance is the oxygen. There are 4 oxygen atoms in 2 CO_2 molecules and 3 oxygen atoms in 3 H_2O molecules, making a total of 7 oxygen atoms in the products. Since we cannot balance the oxygen atoms in the reactants with a whole number unless we unbalance everything else, we can use a fractional coefficient, $\frac{7}{2}$ (or 3.5) temporarily:

$$C_2H_6(g) + \tfrac{7}{2}O_2(g) \longrightarrow 2CO_2(g) + 3H_2O(g)$$

Atoms in Reactants	Atoms in Products
2 C 6 H 7 O	2 C 6 H 7 O

The atoms of each element are equal on both sides of the equation, but the coefficients are not all whole numbers. To get whole-number coefficients, we must multiply all coefficients in the equation by 2:

$$2C_2H_6(g) + 7O_2(g) \longrightarrow 4CO_2(g) + 6H_2O(g)$$

This is a correct, balanced equation, as shown by the final atom counts.

Atoms in Reactants	Atoms in Products
4 C 12 H 14 O	4 C 12 H 14 O

Practice Problem 5.3

Liquid butane is often used as a fuel in cigarette lighters. When the pressure is released, butane evaporates and can burn with oxygen. Balance the following skeletal equation for the reaction of butane with oxygen gas.

$$C_4H_{10}(g) + O_2(g) \longrightarrow CO_2(g) + H_2O(g)$$

Further Practice: 5.37 and 5.38

Let's consider one final example that illustrates how to proceed if part of a reactant is unchanged in the products. This situation usually arises with ionic compounds in which the anion is an oxoanion or some other polyatomic anion. If the anion is unchanged in the products, it can be treated as a unit instead of as a collection of atoms that must be balanced individually.

EXAMPLE 5.4   Balancing an Equation with an Unchanged Oxoanion

When Antonio adds a piece of iron to a solution of copper(II) sulfate, the following reaction occurs:

$$Fe(s) + CuSO_4(aq) \longrightarrow Fe_2(SO_4)_3(aq) + Cu(s)$$

Balance this skeletal equation.

Solution:

Notice that the sulfate ion (SO_4^{2-}) occurs unchanged in the reactants and products. Its component elements, S and O, do not occur anywhere else in the reaction, so the sulfate ion can be treated as a unit. We start by counting atoms and sulfate ions.

$$Fe(s) + CuSO_4(aq) \longrightarrow Fe_2(SO_4)_3(aq) + Cu(s)$$

Atoms in Reactants	Atoms in Products
1 Fe 1 Cu 1 SO_4^{2-}	2 Fe 1 Cu 3 SO_4^{2-}

The copper atoms are balanced. Two iron atoms occur in the products, but only 1 in the reactants, so we can change the coefficient of Fe to 2:

$$2Fe(s) + CuSO_4(aq) \longrightarrow Fe_2(SO_4)_3(aq) + Cu(s)$$

Atoms in Reactants	Atoms in Products
2 Fe 1 Cu 1 SO_4^{2-}	2 Fe 1 Cu 3 SO_4^{2-}

The sulfate ions are not balanced. We can balance them by changing the coefficient of $CuSO_4$ to 3:

$$2Fe(s) + 3CuSO_4(aq) \longrightarrow Fe_2(SO_4)_3(aq) + Cu(s)$$

Atoms in Reactants	Atoms in Products
2 Fe 3 Cu 3 SO_4^{2-}	2 Fe 1 Cu 3 SO_4^{2-}

This change unbalanced the copper atoms, so we have to change the coefficient of Cu to 3:

$$2Fe(s) + 3CuSO_4(aq) \longrightarrow Fe_2(SO_4)_3(aq) + 3Cu(s)$$

Atoms in Reactants	Atoms in Products
2 Fe 3 Cu 3 SO_4^{2-}	2 Fe 3 Cu 3 SO_4^{2-}

The equation is now balanced, as indicated by the final counts.

Practice Problem 5.4

Antonio also reacts zinc metal with nitric acid. Balance the following skeletal equation for this reaction.

$$Zn(s) + HNO_3(aq) \longrightarrow Zn(NO_3)_2(aq) + H_2(g)$$

Further Practice: 5.39 and 5.40

5.4   PREDICTING CHEMICAL REACTIONS

In all the chemical reactions we have considered so far, the identities of the reactants and products are known. The identity of the reactants is easy enough, since we usually know what we have mixed. But how do we know what products will result? The answer could come from an experiment. We could carry out the reaction and then isolate and identify the products, or if the reaction has been investigated before, we could use or extend the experimental work of others.

We saw in Chapter 2 that the chemical properties of the elements are periodic. That is, elements in the same group in the periodic table undergo similar reactions. We can use this periodicity to predict the products of chemical reactions. For example, if we place a piece of lithium metal in liquid water, aqueous lithium hydroxide and hydrogen gas form (Figure 5.12):

$$2Li(s) + 2H_2O(l) \longrightarrow 2LiOH(aq) + H_2(g)$$

How would you know that's what forms?

The principle of periodicity suggests that the other alkali metals will also react with water to form a metal hydroxide and hydrogen gas. As we can see from the sequence of photographs in Figure 5.12, all these metals react with water. What differs is the vigor of the reaction. With the potassium reaction, the hydrogen always catches fire. For the alkali metals below potassium in the periodic table, the reaction with water is explosive.

In addition, atoms have only a limited number of ways to rearrange themselves. If we can classify reactions into types, we can determine what kinds of atomic rearrangements are theoretically possible for different combinations of chemical substances. For example, suppose we depict elements and compounds with letters and spheres:

Spheres A and B are elements, either monatomic or diatomic. Spheres C, D, E, and F might be atoms, monatomic ions, or polyatomic ions arranged to form molecules as shown. How many different types of chemical reactions can you identify by rearranging these atoms or groups of atoms? To keep this exercise from getting too complicated, use only one substance by itself, or any combination of two substances, but no more than two as your reactants and your products. When you have completed this exercise, read on.

To start, let's consider the case in which we have just one reactant. We probably have to heat the substance for something to happen. If we have an element, what could result? A monatomic element can't do anything because there is no way to rearrange a single atom. A diatomic element might break down into its component atoms. We won't encounter examples of such reactions, so we won't consider them further.

lithium   sodium   potassium   rubidium   cesium

Figure 5.12
The alkali metals all react with water to form the corresponding alkali metal hydroxide and hydrogen gas. The reactions differ only in their vigor.

If we heat the compound CD, it can break into smaller pieces, either atoms or smaller molecules, C and D:

Example: $NH_4Cl(s) \longrightarrow NH_3(g) + HCl(g)$

The compound EF could similarly break up into E and F.

Next let's consider combinations of two substances. Let's start with the elements. We can react element A with element B to form a compound:

Example: $Zn(s) + Cl_2(g) \longrightarrow ZnCl_2(s)$

We can combine an element with a compound to make a new compound:

Example: $H_2(g) + CO(g) \longrightarrow H_2CO(g)$

This type of combination could also happen between A and EF, between B and CD, and between B and EF. All four would be the same reaction type.

An element could also react with a compound to form a new compound and release an element from the original compound:

Example: $Zn(s) + CuCl_2(aq) \longrightarrow Cu(s) + ZnCl_2(aq)$

Depending on the identity of the component elements, the products might instead be the new compound CA and the released element D. This type of reaction could also happen with the compound EF, and element B could also react in this way with either compound CD or compound EF.

Finally let's consider the reaction between two compounds. They could combine to make a new compound:

Example: $CaO(s) + SO_2(g) \longrightarrow CaSO_3(s)$

Or two compounds could form two new compounds:

Example: $AgNO_3(aq) + KCl(aq) \longrightarrow AgCl(s) + KNO_3(aq)$

Did your list of reaction types match these? Did you notice any patterns in the reactants and products? In each of the examples, we had either one or two reactants and either one or two products. They could be either elements or compounds. Let's make a list of the possible combinations:

Reactants	Products
1 compound	2 elements (or smaller compounds)
2 elements or compounds	1 compound
1 element and 1 compound	1 element and 1 compound
2 compounds	2 compounds

This gives us just four different patterns of reactivity that we have to investigate. Chemists give them the following names: decomposition reactions, combination (or synthesis) reactions, single-displacement reactions, and double-displacement reactions. They are summarized in Table 5.1.

TABLE 5.1	Classes of Chemical Reactions		
Class	**Reactants**	**Products**	**Example**
Decomposition	1 compound	2 elements (or smaller compounds)	CD $\longrightarrow$ C + D
Combination	2 elements or compounds	1 compound	A + B $\longrightarrow$ AB
Single-displacement	1 element and 1 compound	1 element and 1 compound	A + CD $\longrightarrow$ C + AD
Double-displacement	2 compounds	2 compounds	CD + EF $\longrightarrow$ CF + ED

We will investigate examples of each class in the remainder of this section. At this point, we can place known chemical reactions into classes by considering the type and number of reactants and products.

EXAMPLE 5.5   Classes of Reactions

Classify each of the following as a decomposition, combination, single-displacement, or double-displacement reaction.

(a) $Na_2S(aq) + 2HCl(aq) \longrightarrow H_2S(g) + 2NaCl(aq)$
(b) $H_2(g) + CuO(s) \longrightarrow Cu(s) + H_2O(g)$
(c) $2Fe(s) + 3Cl_2(g) \longrightarrow 2FeCl_3(s)$
(d) $2Na_2O_2(s) \longrightarrow 2Na_2O(s) + O_2(g)$

Solution:

To classify the reactions, look for the patterns outlined in Table 5.1.

(a) This reaction involves two compounds as reactants and two compounds as products. This pattern is called double-displacement.
(b) This is a reaction of an element with a compound, producing a different element and compound. This is a single-displacement reaction.
(c) In this reaction, two elements combine to form a compound. This pattern is found in combination reactions.
(d) Here, one compound is converted into a simpler compound and an element. This is a decomposition reaction.

Practice Problem 5.5

Classify each of the following as a decomposition, combination, single-displacement, or double-displacement reaction.

(a) $NH_3(g) + HCl(g) \longrightarrow NH_4Cl(s)$
(b) $CuCl_2(aq) + Na_2S(aq) \longrightarrow CuS(s) + 2NaCl(aq)$
(c) $NiSO_3(s) \longrightarrow NiO(s) + SO_2(g)$
(d) $Ca(s) + PbCl_2(aq) \longrightarrow CaCl_2(aq) + Pb(s)$

Further Practice: 5.51 and 5.52

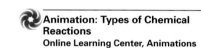

Animation: Types of Chemical Reactions
Online Learning Center, Animations

Chemistry Animations Library, Types of Chemical Reactions, Reaction_Types.rm.

Decomposition Reactions

When a compound undergoes a **decomposition reaction,** it breaks down into the elements of which it is composed or into simpler compounds. The compounds that form are predictable—usually they are common, stable, small molecules, especially gases. If a gas can form as a product, it nearly always does. Patterns for predicting the products of decomposition reactions can be formulated for a variety of compounds. Some of these patterns are listed with examples in Table 5.2.

Decomposition reactions may form more than two products in some cases. Similarly, combination reactions may involve more than two reactants.

TABLE 5.2	Decomposition Reactions That Occur When Compounds Are Heated

- *Oxides* and *halides* of the metals Au, Pt, and Hg decompose to the elements.

$$2HgO(s) \xrightarrow{\text{heat}} 2Hg(l) + O_2(g)$$

$$PtCl_4(s) \xrightarrow{\text{heat}} Pt(s) + 2Cl_2(g)$$

- *Peroxides* decompose to oxides and oxygen gas.

$$2H_2O_2(aq) \xrightarrow{\text{heat}} 2H_2O(l) + O_2(g)$$

- *Metal carbonates,* except those of the group IA (1) metals, decompose to metal oxides and carbon dioxide gas.

$$NiCO_3(s) \xrightarrow{\text{heat}} NiO(s) + CO_2(g)$$

- *Oxoacids* decompose in a similar way to form nonmetal oxides and water. Other compounds that contain the components of water frequently decompose with elimination of water. The other product is what is left after water is removed.

$$H_2CO_3(aq) \xrightarrow{\text{heat}} CO_2(g) + H_2O(l)$$

$$Ca(OH)_2(s) \xrightarrow{\text{heat}} CaO(s) + H_2O(g)$$

$$CuSO_4 \bullet 5H_2O(s) \xrightarrow{\text{heat}} CuSO_4(s) + 5H_2O(g)$$

- *Ammonium compounds* lose ammonia.

$$(NH_4)_2SO_4(s) \xrightarrow{\text{heat}} 2NH_3(g) + H_2SO_4(l)$$

A few binary compounds decompose to their constituent elements upon heating. For example, when red solid mercury(II) oxide is heated, it slowly decomposes. Gaseous oxygen fills the container, and droplets of liquid mercury settle on its cooler parts (Figure 5.13).

The oxide ions lose their negative charge as they are converted to oxygen atoms. The mercury(II) ions lose their positive charge to form mercury atoms. These changes are shown in Figure 5.14.

Many metal compounds containing oxoanions, such as carbonates, decompose upon heating. A common example is calcium carbonate.

$$CaCO_3(s) \xrightarrow{\text{heat}} CaO(s) + CO_2(g)$$

Reactions in which the ionic charges change are called *oxidation-reduction reactions.* In the decomposition of HgO, the charge of mercury changes from 2+ to 0 when Hg(l) is formed. The charge of oxygen changes from 2– to 0 when $O_2(g)$ is formed. We will explore oxidation-reduction reactions in more detail in Chapter 14.

Figure 5.13
When solid mercury(II) oxide (HgO) is heated, it decomposes into liquid mercury metal and oxygen gas.

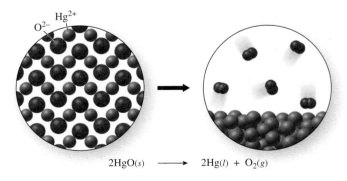

$$2HgO(s) \longrightarrow 2Hg(l) + O_2(g)$$

Figure 5.14
The regular array of mercury(II) ions and oxide ions breaks down when mercury(II) oxide is heated. The oxide ions form oxygen atoms, which combine to form gaseous oxygen molecules. The mercury(II) ions form mercury metal in the liquid state.

Figure 5.15
Limestone chunks, when heated by burning fuel in the large rotary kiln in the center of the photograph, decompose to form lime and carbon dioxide.

Portland cement has calcium oxide (lime) as one of its ingredients. The calcium oxide is made from limestone (calcium carbonate) by decomposition in large kilns at about 1900°C (Figure 5.15).

A number of compounds that contain water or the components of water—hydrates, hydroxides, and oxoacids—lose water when heated. Hydrates—compounds that contain water molecules—lose water to form **anhydrous** compounds, compounds free of molecular water. One example is hydrated calcium sulfate. It can be dehydrated by heating to form a powder:

$$CaSO_4 \cdot 2H_2O(s) \xrightarrow{heat} CaSO_4(s) + 2H_2O(g)$$

The anhydrous calcium sulfate can be mixed with water to use in making plaster for walls or casts, or as an artist's surface for painting.

EXAMPLE 5.6   Decomposition Reactions

Michelle needs some cobalt(II) oxide to make cobalt glass. She has the following cobalt compounds available to her: $CoCl_2 \cdot 6H_2O$, $CoCO_3$, CoS, and $Co(OH)_2$. Which of these compounds can make the cobalt(II) oxide she needs? What products would form if the other compounds were heated?

Solution:

In Table 5.2 we can see that Michelle could use $CoCO_3$ or $Co(OH)_2$. Metal carbonates lose carbon dioxide gas when heated, leaving a metal oxide:

$$CoCO_3(s) \xrightarrow{heat} CoO(s) + CO_2(g)$$

Metal hydroxides lose water to form metal oxides:

$$Co(OH)_2(s) \xrightarrow{heat} CoO(s) + H_2O(g)$$

Heating a hydrate results in the loss of gaseous water:

$$CoCl_2 \cdot 6H_2O(s) \xrightarrow{heat} CoCl_2(s) + 6H_2O(g)$$

We have seen no patterns that suggest that anything would happen to CoS upon heating.

Figure 5.16
Aluminum reacts with bromine to form aluminum(III) bromide.

Combination reactions in which two elements combine, or in which an element and a compound combine, are oxidation-reduction reactions, in which a change in charge occurs. For example, in the formation of $AlBr_3$ from Al and Br_2, the charge on aluminum changes from 0 to 3+, and the charge on bromine changes from 0 to 1–.

Nonmetals become more reactive the closer they are to the upper-right corner of the periodic table, ignoring the noble gases. The most reactive nonmetal is fluorine (F_2).

Practice Problem 5.6

Janelle wants to make paper pulp to be used in a printing project. She needs magnesium sulfite trihydrate ($MgSO_3 \cdot 3H_2O$) to bleach the paper pulp. This compound decomposes upon heating. Predict the products and write complete, balanced equations for any reactions.

Further Practice: 5.55 and 5.56

Combination Reactions

In a **combination reaction** (also called *synthesis*), two elements, an element and a compound, or two compounds react to produce a single compound. An important type of combination reaction involves two elements. *Most metals react with most nonmetals to form ionic compounds.* The products can be predicted from the charges expected for the cation of the metal and the anion of the nonmetal, as you learned in Chapter 2. For example, the silver-colored metal aluminum reacts with bromine (Figure 5.16) to form solid aluminum bromide. The product of this reaction can be predicted from the charges of the ions: 3+ for aluminum ion, and 1– for bromide ion:

$$2Al(s) + 3Br_2(l) \longrightarrow 2AlBr_3(s)$$

Similarly, *a nonmetal may react with a more reactive nonmetal to form a molecular compound.* For example, the yellow solid sulfur reacts with oxygen gas to form colorless, gaseous sulfur dioxide (Figure 5.17):

$$S_8(s) + 8O_2(g) \longrightarrow 8SO_2(g)$$

This compound has a pungent odor and is one of the substances that make acid rain.

A compound and an element may combine to form another compound if one exists with a higher atom:atom ratio. For example, carbon dioxide has a higher oxygen:carbon atom ratio than does carbon monoxide. Carbon monoxide forms when carbon (coal), hydrocarbons, or other carbon-containing materials are burned in the presence of insufficient oxygen. It is a colorless, odorless, poisonous gas.

$$2C(s) + O_2(g) \longrightarrow 2CO(g)$$

It reacts further with oxygen to form colorless, odorless carbon dioxide (Figure 5.18), which is a component of the atmosphere and of exhaled breath:

$$2CO(g) + O_2(g) \longrightarrow 2CO_2(g)$$

These changes are shown on a molecular level in Figure 5.19.

Two compounds may react to form a new compound. For example, calcium oxide, a white solid, reacts with carbon dioxide to form white solid calcium carbonate, which is found naturally in limestone, marble, and seashells:

$$CaO(s) + CO_2(g) \longrightarrow CaCO_3(s)$$

Reactions such as this happen during the hardening and aging of cement.

Figure 5.17
In a combination reaction, the elements sulfur and oxygen form sulfur dioxide.

Figure 5.18
Carbon monoxide burns in air to form carbon dioxide. This is a combination reaction between a compound and an element.

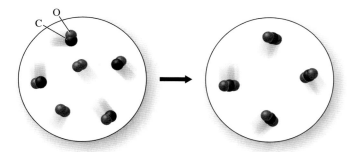

Figure 5.19
In the burning of carbon monoxide, oxygen atoms from oxygen molecules add to carbon monoxide molecules to form carbon dioxide molecules.

EXAMPLE 5.7   Combination Reactions

When pure calcium metal is exposed to the oxygen in air, a white coating appears on the surface. Predict the product and write a complete, balanced equation for this reaction.

Solution:

When two elements react, they form a compound. Since the elements are a metal and a nonmetal, the compound should be ionic. The metal should form a cation, and the nonmetal should form an anion. From their positions in the periodic table, we can predict that the charge on the calcium cation should be 2+ and that the charge on the oxide anion should be 2–. The product should be the compound formed from Ca^{2+} and O^{2-}, CaO. From this information, we can write the following word equation:

$$\text{calcium} + \text{oxygen} \longrightarrow \text{calcium oxide}$$

Using formulas, we get the following skeletal equation:

$$Ca(s) + O_2(g) \longrightarrow CaO(s)$$

Finally, following the usual method for balancing a skeletal equation, we get

$$2Ca(s) + O_2(g) \longrightarrow 2CaO(s)$$

Practice Problem 5.7

When magnesium metal burns in air, it forms two white compounds. One is the compound expected for the reaction of magnesium with oxygen in air, while the other, magnesium nitride (Mg_3N_2), results from a reaction with nitrogen in air. Predict the product of the reaction of magnesium with oxygen. Write complete, balanced equations for the reactions that form both products.

Further Practice: 5.59 and 5.60

Single-Displacement Reactions

In a **single-displacement reaction,** a free element displaces another element from a compound to produce a different compound and a different free element. An example is the thermite reaction between aluminum and iron(III) oxide, used in the production of iron metal for welding:

$$2Al(s) + Fe_2O_3(s) \xrightarrow{\text{heat}} Al_2O_3(g) + 2Fe(l)$$

In single-displacement reactions, the displaced element is often a metal, but it can be a nonmetal. Let's consider a common type of single-displacement reaction,

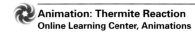

Animation: Thermite Reaction
Online Learning Center, Animations

Chemistry Animations Library, Thermite Reaction, Thermite_Rxn.rm.

Figure 5.20
Calcium reacts with water to form aqueous calcium hydroxide and gaseous hydrogen. An indicator that turns pink in the presence of a base was added to the solution. This solution is pink, indicating that hydroxide ions are produced in this reaction.

All single-displacement reactions are oxidation-reduction reactions. In the thermite reaction, aluminum displaces iron from its oxide compound. Aluminum changes its charge from 0 to 3+, while iron changes its charge from 3+ to 0.

one in which a highly reactive metal displaces hydrogen from water. For example, calcium reacts with cold water to form calcium hydroxide and hydrogen gas, as shown in Figure 5.20:

$$Ca(s) + 2H_2O(l) \longrightarrow Ca(OH)_2(aq) + H_2(g)$$

Some metals, such as magnesium, do not react with cold water, but react slowly with steam:

$$Mg(s) + 2H_2O(g) \xrightarrow{\text{heat}} Mg(OH)_2(aq) + H_2(g)$$

Metals that are still less reactive do not react with water at all, but they do displace hydrogen from acids. Cobalt is an example:

$$Co(s) + 2HCl(aq) \longrightarrow CoCl_2(aq) + H_2(g)$$

The reaction of metals with acids is sometimes called corrosion, especially if the result is undesirable. **Corrosion** is a process in which a solid is slowly eaten away. Metals that are even less reactive, such as copper, do not react with either water or acids in a single-displacement reaction.

How do we predict which metals react with water or acids? We can use an **activity series**, a list of the metals in order of their reactivity. *A more active element displaces a less active element from its compounds.* The most active elements, found at the top of the activity series, react with cold water to release hydrogen gas. The next most active elements react with steam to release hydrogen gas. All the elements more active than hydrogen in the series react with acids. None of the elements less active than hydrogen displace hydrogen gas from any compound. A short activity series is given in Figure 5.21.

Copper, mercury, and gold do not react with acids to release hydrogen gas, but they do react with nitric acid to form the metal nitrate and nitrogen dioxide gas. The reaction for copper is given by the following equation:

$$Cu(s) + 4HNO_3(aq) \longrightarrow$$
$$Cu(NO_3)_2(aq) + 2NO_2(g) + 2H_2O(l)$$

This is another example of an oxidation-reduction reaction.

EXAMPLE 5.8 — Reaction of Metals with Water and Acids

Antonio is asked to make a magnesium statue to be placed in the middle of a public fountain. Should he take the commission? Why or why not?

Solution:

From its position in the activity series, we know that magnesium reacts with steam to release hydrogen gas:

$$Mg(s) + 2H_2O(l) \xrightarrow{\text{heat}} Mg(OH)_2(aq) + H_2(g)$$

If splashed with water from the fountain and heated by the sun, the statue will corrode. Antonio should suggest a different metal.

Practice Problem 5.8

Tomatoes are acidic. Why don't we use iron cans coated with zinc to hold tomato sauce?

Further Practice: 5.63 and 5.64

The displacement of hydrogen from water or acids is just one type of single-displacement reaction. Other elements can also be displaced from their compounds. For example, copper displaces silver. If a piece of copper metal is placed in an aqueous solution of silver nitrate, silver metal is deposited on the surface of the copper (Figure 5.22). The copper is converted to an ionic copper(II) compound in solution:

$$Cu(s) + 2AgNO_3(aq) \longrightarrow Cu(NO_3)_2(aq) + 2Ag(s)$$

As shown in Figure 5.23, atoms on the surface of the copper metal develop a positive charge by losing electrons and go into solution as copper(II) ions. Silver ions in solution, upon contacting the piece of copper metal, lose their positive charge by gaining electrons to form silver atoms that stick to the copper's surface. The anions are not involved in the reaction and remain unchanged in the solution.

A single-displacement reaction was responsible for severe corrosion that occurred in the Statue of Liberty. In the damp environment of New York Harbor, the copper panels on the outside of the statue developed a coating of copper(II) oxide, copper(II) hydroxide, and copper(II) carbonate, while the iron used for the inner framework rusted to form iron(III) oxide. This brought copper metal in contact with iron(III) ions and iron in contact with copper(II) ions. Would either of these combinations undergo a single-displacement reaction? As the National Park

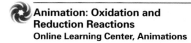

Animation: Oxidation and Reduction Reactions
Online Learning Center, Animations

Chemistry Animations Library, Oxidation and Reduction Reactions, 06_Oxidation_Reduxn_Rxns.swf.

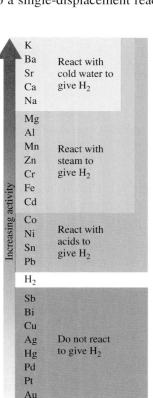

Figure 5.21
In this chemical activity series, the most active elements appear at the top. The notations to the right side indicate the reactivity of these elements with water or acids to release molecular hydrogen gas.

Figure 5.22
Copper metal reacts with aqueous silver nitrate to form silver metal and aqueous copper(II) nitrate. A microscopic view shows that the silver metal is deposited as crystals.

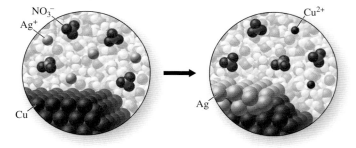

Figure 5.23
Copper atoms are converted to copper(II) ions, and silver ions are converted to silver atoms. Nitrate ions remain unchanged in solution.

Figure 5.24
The Statue of Liberty was closed between 1984 and 1986. It was rebuilt to remedy the effects of corrosion.

Service found out, the answer is yes. The iron was converted to iron(II) ions and the copper(II) ions were converted to copper metal:

$$Fe(s) + CuO(s) \longrightarrow Cu(s) + FeO(s)$$

Consequently, corrosion of the metal framework was accelerated by a factor of 1000, and the statue had to be rebuilt (Figure 5.24).

Will a single-displacement reaction occur with all combinations of metals and ionic compounds? No. Earlier we saw that if we place a piece of copper metal in a solution of silver nitrate, silver metal is deposited and copper goes into solution as copper(II) nitrate. But the opposite does not occur. If we place a piece of silver in a solution of copper(II) nitrate, nothing happens (Figure 5.25). If we have two metals and solutions of their ions, only one combination will undergo a single-displacement reaction. How can we tell which? We can use the activity series shown in Figure 5.21. The more active metal displaces the less active metal from its ionic compound. Thus, magnesium will displace iron from $FeCl_2$, but tin or copper will not.

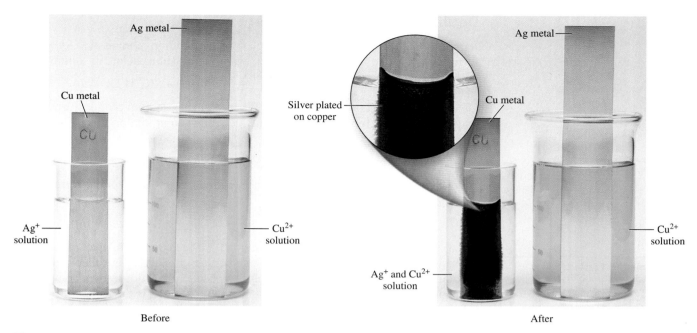

Ag metal

Cu metal

Ag⁺
solution

Cu²⁺
solution

Silver plated
on copper

Ag metal

Cu metal

Ag⁺ and Cu²⁺
solution

Cu²⁺
solution

Before

After

Figure 5.25
Although a single-displacement reaction occurs when a piece of copper metal is placed in a solution of silver nitrate, no reaction occurs when a piece of silver is placed in a solution of copper(II) nitrate. Copper is more active than silver.

EXAMPLE 5.9   Displacement of Metal Ions

Antonio has two pieces of metal, zinc and iron. He also has two aqueous solutions, one of zinc chloride and one of iron(II) chloride. Which metal will corrode when placed in one of the solutions? Which solution? Write a balanced equation for the reaction.

Solution:

A metal placed in a solution containing its own ions will not react. Thus, we must consider whether iron metal will displace zinc ions from solution or whether zinc metal will displace iron(II) ions from solution. To determine which reaction will occur, we consult the activity series. The more active metal will displace ions of the less active metal from solution. Figure 5.21 shows that zinc is more active than iron. Therefore, it will displace ions of iron from solution. The products of the reaction will be iron metal and zinc chloride. The word equation is

$$\text{zinc} + \text{iron(II) chloride} \longrightarrow \text{zinc chloride} + \text{iron}$$

Substituting the formulas gives us the following skeletal equation:

$$\text{Zn}(s) + \text{FeCl}_2(aq) \longrightarrow \text{ZnCl}_2(aq) + \text{Fe}(s)$$

Counting atoms of each element, we see that this is also the balanced equation for the reaction.

Practice Problem 5.9

Antonio has two other pieces of metal, cobalt and tin. He also has two aqueous solutions containing their ions, one of cobalt(II) sulfate and one of tin(II) sulfate. Which metal will corrode when placed in one of the solutions? Which solution? Write a balanced equation for the corrosion reaction.

Further Practice: 5.67 and 5.68

Double-Displacement Reactions

Jennifer mixed a solution of lead(II) acetate with a solution of potassium iodide and formed a yellow solid of lead(II) iodide, leaving potassium acetate in solution:

$$Pb(CH_3CO_2)_2(aq) + 2KI(aq) \longrightarrow PbI_2(s) + 2KCH_3CO_2(aq)$$

Other kinds of reactions are similar. When aqueous sodium sulfide reacts with hydrochloric acid, for example, it forms aqueous sodium chloride and hydrogen sulfide gas:

$$Na_2S(aq) + 2HCl(aq) \longrightarrow 2NaCl(aq) + H_2S(g)$$

When aqueous potassium hydroxide reacts with aqueous nitric acid, it forms water and aqueous potassium nitrate:

$$KOH(aq) + HNO_3(aq) \longrightarrow H_2O(l) + KNO_3(aq)$$

What do these reactions have in common? In each reaction, two compounds exchange parts to form two new compounds. Such a reaction is called a **double-displacement reaction.** In all double-displacement reactions, two compounds exchange ions or elements to form new compounds. What is the *driving force* that causes two compounds to exchange partners? In the first example, called a **precipitation reaction,** one product compound separates from the reaction mixture because it is insoluble. It forms a solid. In the second example, one product compound separates from the reaction mixture as an insoluble gas. In the third example, a stable molecular substance, water, is formed. In each case reactant ions are removed from the solution by forming a new substance—either a solid, a gas, or a stable molecular compound.

Not all combinations of two compounds can undergo a double-displacement reaction. Consider the compounds potassium chloride and nitric acid, for example. If they are dissolved in aqueous solution, they form the following ions: $K^+(aq)$, $Cl^-(aq)$, $H^+(aq)$, and $NO_3^-(aq)$. Nothing appears to happen. We see no sign of a reaction because no combination of the component ions can form either a compound that will separate from the solution or a stable molecular compound.

We will consider three classes of double-displacement reactions: precipitation, gas formation, and acid-base neutralization reactions. For each class we will examine how to determine whether a reaction will occur.

Precipitation Reactions   In precipitation reactions, reactants exchange ions to form a precipitate. A **precipitate** is an insoluble ionic compound that does not dissolve in water. *If a possible product is insoluble, a precipitation reaction should occur.* We can predict whether such a compound will form by consulting the solubility rules, some of which are listed in Table 5.3.

Suppose we mix colorless, aqueous solutions of barium chloride and sodium sulfate. As shown in Figure 5.26, this reaction produces a white precipitate. Since we started with solutions of the reactants, they must be soluble, so neither of these substances could be the insoluble solid. We have to identify which possible product would be insoluble. We start examining the possibilities by considering the products that could be made by exchanging the cations and anions present in the reactants. In this case, we must consider barium sulfate and sodium chloride as possible products. We then have to consider whether either or both is insoluble.

According to the solubility rules in Table 5.3, most sulfate compounds, sodium compounds, and chloride compounds are soluble. An exception is barium sulfate, which is insoluble. Sodium chloride, according to the rules, is soluble. Thus, the insoluble product must be barium sulfate. From this information, we can write the following skeletal equation:

$$BaCl_2(aq) + Na_2SO_4(aq) \longrightarrow BaSO_4(s) + NaCl(aq)$$

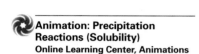

Animation: Precipitation Reactions (Solubility)
Online Learning Center, Animations

Chemistry Animations Library, Precipitation Reactions (Solubility), 3Precipitation Reactions.swf.

Animation: Barium Precipitation
Online Learning Center, Animations

Chemistry Animations Library, Barium Precipitation, 1Precipitation.avi.

Figure 5.26
When solutions of barium chloride and sodium sulfate mix, a precipitate of barium sulfate forms, leaving sodium chloride in the solution.

TABLE 5.3	Rules Used to Predict the Solubility of Ionic Compounds
Ions	**Rule**
Na^+, K^+, NH_4^+ (and other alkali metal ions)	Most compounds of alkali metal and ammonium ions are soluble.
NO_3^-, $CH_3CO_2^-$	All nitrates and acetates are soluble.
SO_4^{2-}	Most sulfates are soluble. Exceptions are $BaSO_4$, $SrSO_4$, $PbSO_4$, $CaSO_4$, Hg_2SO_4, and Ag_2SO_4.
Cl^-, Br^-, I^-	Most chlorides, bromides, and iodides are soluble. Exceptions are AgX, Hg_2X_2, PbX_2, and HgI_2. ($X = Cl$, Br, or I.)
Ag^+	Silver compounds, except $AgNO_3$ and $AgClO_4$, are insoluble. $AgCH_3CO_2$ is slightly soluble.
O^{2-}, OH^-	Oxides and hydroxides are insoluble. Exceptions are alkali metal hydroxides, $Ba(OH)_2$, $Sr(OH)_2$, and $Ca(OH)_2$ (somewhat soluble).
S^{2-}	Sulfides are insoluble. Exceptions are compounds of Na^+, K^+, NH_4^+ and the alkaline earth metal ions.
CrO_4^{2-}	Most chromates are insoluble. Exceptions are compounds of Na^+, K^+, NH_4^+, Mg^{2+}, Ca^{2+}, Al^{3+}, and Ni^{2+}.
CO_3^{2-}, PO_4^{3-}, SO_3^{2-}, SiO_3^{2-}	Most carbonates, phosphates, sulfites, and silicates are insoluble. Exceptions are compounds of Na^+, K^+, and NH_4^+.

We can balance this equation in the usual manner, giving the following balanced equation:

$$BaCl_2(aq) + Na_2SO_4(aq) \longrightarrow BaSO_4(s) + 2NaCl(aq)$$

Let's also consider this reaction on a molecular level (Figure 5.27). From the diagrams, we see that mixing the two solutions gives us a solution that contains four kinds of ions. After precipitation of the barium sulfate, the solution still contains the two ions, $Na^+(aq)$ and $Cl^-(aq)$, that do not form an insoluble compound. These ions do not undergo a chemical change.

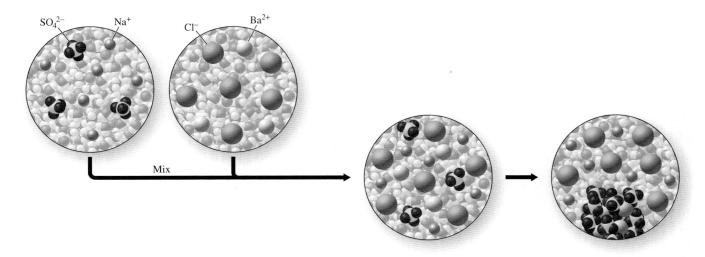

Figure 5.27
In double-displacement reactions, one pair of ions combines to form a solid, a gas, or an undissociated molecule, leaving the other pair of ions in solution.

EXAMPLE 5.10 Precipitation Reactions

Janelle finds a recipe for the preparation of the pigment chrome yellow. She dissolves lead(II) nitrate and potassium chromate in separate volumes of water, and then mixes the two solutions. She obtains a yellow precipitate. Identify the precipitate and write a balanced equation for the reaction she carried out.

Solution:

The starting solutions contain $Pb^{2+}(aq)$, $NO_3^-(aq)$, $K^+(aq)$, and $CrO_4^{2-}(aq)$. The solubility rules say that most compounds of potassium are soluble and all nitrates are soluble, so the precipitate must be formed between lead(II) ions and chromate ions. By matching the positive and negative charges to get electrical neutrality, we can determine the formula of the precipitate to be $PbCrO_4(s)$. The solution contains the ions $K^+(aq)$ and $NO_3^-(aq)$. The skeletal equation for this reaction is

$$Pb(NO_3)_2(aq) + K_2CrO_4(aq) \longrightarrow PbCrO_4(s) + KNO_3(aq)$$

The balanced equation is

$$Pb(NO_3)_2(aq) + K_2CrO_4(aq) \longrightarrow PbCrO_4(s) + 2KNO_3(aq)$$

Practice Problem 5.10

Jennifer mixes solutions of cadmium(II) nitrate and sodium sulfide and obtains an orange precipitate that is the pigment cadmium orange. Identify the precipitate, and write a balanced equation for the reaction she carried out.

Further Practice: 5.77 and 5.78

Gas Formation Reactions In a similar type of reaction, the formation of an insoluble (or only slightly soluble) gas provides the driving force for a reaction. Thus *a double-displacement reaction should occur if an insoluble gas would be formed.* All gases are soluble in water to some extent, but only a few are highly soluble. The common very soluble gases are $HCl(g)$ and $NH_3(g)$. Sulfur dioxide (SO_2) is fairly soluble, with about 3 mol dissolving in a liter of cold water. All other gases, generally binary molecular compounds, are sufficiently insoluble to provide a driving force if they are formed as a reaction product. For example, many sulfide compounds, such as the ZnS used in some television screens, will react with acids to form gaseous hydrogen sulfide:

$$ZnS(s) + 2HCl(aq) \longrightarrow ZnCl_2(aq) + H_2S(g)$$

Sometimes insoluble gases do not form directly from a double-displacement reaction. Instead they form when an unstable product of a double-displacement reaction decomposes. For example, calcium carbonate reacts with hydrochloric acid to form calcium chloride and carbonic acid:

$$CaCO_3(s) + 2HCl(aq) \longrightarrow CaCl_2(aq) + H_2CO_3(aq)$$

Carbonic acid (H_2CO_3) is an unstable substance. It readily decomposes to form water and carbon dioxide (Figure 5.28), as described in Table 5.2:

$$H_2CO_3(aq) \longrightarrow H_2O(l) + CO_2(g)$$

The net reaction is

$$CaCO_3(s) + 2HCl(aq) \longrightarrow CaCl_2(aq) + H_2O(l) + CO_2(g)$$

Figure 5.28
When an acid reacts with a metal carbonate, bubbles of carbon dioxide gas form.

Such reactions are responsible for the acid rain damage to marble statues (composed of $CaCO_3$) that Breanna was investigating. Acid rain erodes marble and limestone, pitting their surfaces and diminishing the detail of carvings and castings. In the air,

Figure 5.29
A sandstone statue sculpted in 1702 was photographed in 1908 (*left*) and again in 1969 (*right*). The damage resulted from the reaction of acid rain with the calcium carbonate that binds the sand particles together in sandstone.

oxides of nitrogen and sulfur, released from motor vehicles and some industrial processes, form nitric and sulfuric acids, which fall to Earth with the rain. As shown in Figure 5.29, exposure of marble or sandstone statues to acid rain leads to loss of some of the calcium carbonate of which they are made. Many old statues and architectural features have eroded over time because of air pollution.

EXAMPLE 5.11   Gas-Forming Reactions

Magnesium sulfite ($MgSO_3$) is used as a bleach to make wood pulp for making paper. A gas is released when this compound is mixed with acid. Identify the gas, and write a balanced chemical equation for the reaction of magnesium sulfite with hydrochloric acid.

Solution:

Sulfites react with acid to give sulfur dioxide gas. Initially, we have a double-displacement reaction:

$$MgSO_3(s) + 2HCl(aq) \longrightarrow MgCl_2(aq) + H_2SO_3(aq)$$

In a sufficiently concentrated solution, the sulfurous acid then undergoes a decomposition reaction:

$$H_2SO_3(aq) \longrightarrow H_2O(l) + SO_2(g)$$

The overall reaction forms a metal compound with the anion present in the acid, along with molecular water and sulfur dioxide gas:

$$MgSO_3(s) + 2HCl(aq) \longrightarrow MgCl_2(aq) + H_2O(l) + SO_2(g)$$

Practice Problem 5.11

The reaction of barium sulfide with sulfuric acid is used to prepare barium sulfate. Barium sulfate, combined with zinc sulfide and zinc oxide, is used to make the white pigment lithopone. Write a balanced equation for the reaction of barium sulfide with sulfuric acid.

Further Practice: 5.79 and 5.80

Metal oxides can behave like bases when dissolved in water because they react with water to form metal hydroxides. For example, the equation for the reaction of calcium oxide with water is

$$CaO(s) + H_2O(l) \longrightarrow Ca(OH)_2(aq)$$

Acid-Base Neutralization Reactions   A **neutralization reaction** is a double-displacement reaction of an acid and a base. Recall from Chapter 3 that acids are compounds that release hydrogen ions, and bases are compounds that neutralize acids. They react with the hydrogen ions released by acids. The most common bases are hydroxide compounds of metals. Normally, *an acid reacts with a base to form an ionic compound and water.* The formation of stable water molecules from hydrogen ions and hydroxide ions is the driving force for neutralization reactions. A common laboratory example of an acid-base neutralization is the reaction between hydrochloric acid and sodium hydroxide:

$$HCl(aq) + NaOH(aq) \longrightarrow NaCl(aq) + H_2O(l)$$

EXAMPLE 5.12   **Acid-Base Neutralization Reactions**

Upset stomachs are sometimes treated with the antacid Milk of Magnesia™, which is a suspension of magnesium hydroxide in water. Stomach acid is hydrochloric acid. Write a balanced equation for the reaction that occurs when Milk of Magnesia™ mixes with stomach acid.

Solution :

The reactants are $Mg(OH)_2(s)$ and $HCl(aq)$. A reaction between two compounds is usually a double-displacement reaction. Since we can recognize $Mg(OH)_2$ as a base and HCl as an acid, this must be a neutralization reaction. The products can be predicted by exchanging partners—Mg^{2+} with Cl^- and H^+ with OH^-, giving $MgCl_2$ and H_2O. From the solubility rules, we see that $MgCl_2$ should be soluble. Writing the skeletal equation gives

$$Mg(OH)_2(s) + HCl(aq) \longrightarrow MgCl_2(aq) + H_2O(l)$$

The complete, balanced equation is

$$Mg(OH)_2(s) + 2HCl(aq) \longrightarrow MgCl_2(aq) + 2H_2O(l)$$

Practice Problem 5.12

Calcium oxide is the white powder, lime. When added to water, it makes slaked lime, which is a solution of the base calcium hydroxide. If sulfuric acid is added to slaked lime, calcium sulfate and water form. Write a complete, balanced equation for the reaction between sulfuric acid and calcium hydroxide.

Further Practice: 5.81 and 5.82

Combustion Reactions

Any reaction that involves oxygen molecules as a reactant and that rapidly produces heat and flame is a **combustion reaction.** We see a flame when a large amount of energy in the form of heat is released very rapidly. The burning of a fuel, such as logs in a fireplace or methane in a gas stove, is a common combustion reaction. Combustion reactions are also used to heat kilns for firing pottery. Many other substances can undergo combustion, as shown in Figure 5.30.

Both metallic and nonmetallic elements can undergo combustion. In such instances, the reactions can also be classified as combination reactions. The following are examples of both combustion and combination reactions:

$$Mg(s) + O_2(g) \longrightarrow MgO(s)$$

$$S(s) + O_2(g) \longrightarrow SO_2(g)$$

If insufficient oxygen is present, combustion of a hydrocarbon produces carbon monoxide instead of carbon dioxide. Because carbon monoxide is poisonous, a barbecue grill with charcoal briquettes should never be used indoors.

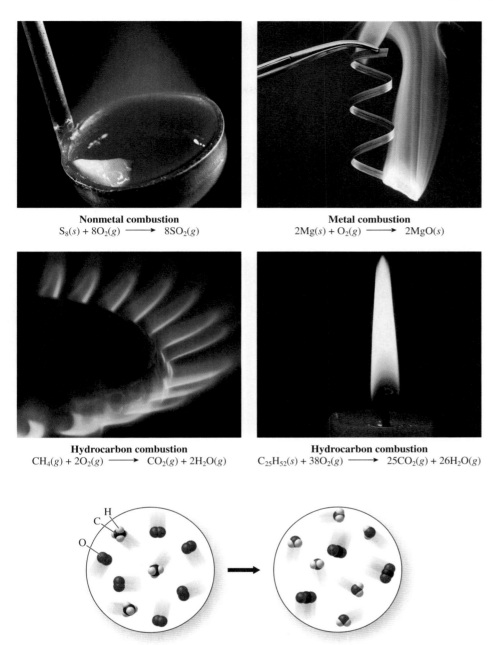

Nonmetal combustion
$$S_8(s) + 8O_2(g) \longrightarrow 8SO_2(g)$$

Metal combustion
$$2Mg(s) + O_2(g) \longrightarrow 2MgO(s)$$

Hydrocarbon combustion
$$CH_4(g) + 2O_2(g) \longrightarrow CO_2(g) + 2H_2O(g)$$

Hydrocarbon combustion
$$C_{25}H_{52}(s) + 38O_2(g) \longrightarrow 25CO_2(g) + 26H_2O(g)$$

Figure 5.30
Many substances can undergo combustion reactions. Which of these reactions could also be classified as combination reactions?

Figure 5.31
In combustion reactions, oxygen combines with other elements in the other reactant to form oxides. In this example, the combustion of methane produces carbon dioxide and water.

Look back at the section on combination reactions and see which of the examples there can also be classified as combustion reactions.

The most common combustion reactions involve hydrocarbons as one of the reactants. *Hydrocarbons* are chemical compounds made only of hydrogen and carbon. Propane (C_3H_8) is a hydrocarbon used as a fuel for barbecue grills. The combustion of propane is described by the following equation:

$$C_3H_8(g) + 5O_2(g) \longrightarrow 3CO_2(g) + 4H_2O(g)$$

When plenty of oxygen is available for the reaction, the hydrogen in hydrocarbons forms water vapor, and the carbon forms carbon dioxide, as shown in Figure 5.31.

Hydrocarbon derivatives such as alcohols and sugars participate in combustion reactions. If elements other than hydrogen, oxygen, and carbon are reactants in a combustion reaction, they are converted to their oxides.

EXAMPLE 5.13   Combustion Reactions

Antonio uses an oxyacetylene torch to cut some pieces of metal. The heat from the torch comes from the combustion of acetylene, $C_2H_2(g)$. Write a balanced equation for the reaction.

Solution:

Acetylene reacts with molecular oxygen to form carbon dioxide and water. The skeletal equation is

$$C_2H_2(g) + O_2(g) \longrightarrow CO_2(g) + H_2O(g)$$

Balancing this equation results in

$$2C_2H_2(g) + 5O_2(g) \longrightarrow 4CO_2(g) + 2H_2O(g)$$

Practice Problem 5.13

Shellac, used as a transparent coating on wood and other materials, is a suspension of a tree sap (resin) in methanol, $CH_3OH(l)$. The mixture is highly combustible. Write an equation to describe the combustion of methanol.

Further Practice: 5.85 and 5.86

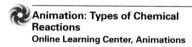

Animation: Types of Chemical Reactions
Online Learning Center, Animations

Chemistry Animations Library, Types of Chemical Reactions, Reaction_Types.rm.

Figure 5.32
Mixing solutions of lead nitrate and potassium chromate produces the yellow solid lead(II) chromate and a solution of potassium nitrate.

5.5   REPRESENTING REACTIONS IN AQUEOUS SOLUTION

A chemical equation is a shorthand way of showing the changes that occur in a chemical reaction. Many of the reactions mentioned in this chapter occur in solution. We have written them as if the reactants were molecular compounds, but in many cases, they are not. To be accurate in representing reactions that occur in solution, we should represent soluble ionic compounds, as well as strong acids and bases, by the formulas of their component ions. For example, $Pb(NO_3)_2(aq)$ can be represented as $Pb^{2+}(aq) + 2NO_3^-(aq)$. From Chapter 3 recall that ionic compounds dissolved in water break apart into their component ions. They are electrolytes. Insoluble solids are not electrolytes.

Consider the reaction of aqueous solutions of lead(II) nitrate and potassium chromate, both of which are ionic compounds and strong electrolytes. Their reaction is used to prepare the pigment chrome yellow, which is lead(II) chromate, leaving an aqueous solution of potassium nitrate, as shown in Figure 5.32. Earlier, we wrote the following equation for this reaction:

$$Pb(NO_3)_2(aq) + K_2CrO_4(aq) \longrightarrow PbCrO_4(s) + 2KNO_3(aq)$$

An equation in this form is often called a **molecular equation,** because it represents the substances as if they existed as molecules (or formula units) in solution rather than dissociated into ions. The following **ionic equation** represents the substances more appropriately, since both reactants are soluble strong electrolytes that dissociate into ions in solution:

$$Pb^{2+}(aq) + 2NO_3^-(aq) + 2K^+(aq) + CrO_4^{2-}(aq) \longrightarrow$$
$$PbCrO_4(s) + 2K^+(aq) + 2NO_3^-(aq)$$

Note that potassium ions and nitrate ions occur on both sides of the equation in equal numbers. They are called **spectator ions** because they do not participate in the reaction. Spectator ions need not be written in the final form of a balanced equation. They can be eliminated from both sides. Thus the reaction can be represented as follows:

$$Pb^{2+}(aq) + CrO_4^{2-}(aq) \longrightarrow PbCrO_4(s)$$

Such an equation, called the **net ionic equation,** includes only those ions that are involved in the reaction.

Metal ions of group IA (1) always form soluble compounds, so they are always spectator ions in double-displacement reactions. Look back at Figure 5.27. Identify the spectator ions in that molecular diagram.

EXAMPLE 5.14   Net Ionic Equations

Consider the reaction between colorless aqueous solutions of silver nitrate ($AgNO_3$) and potassium bromide (KBr), which gives a pale yellow precipitate and a colorless solution. The precipitate is used extensively in photographic films and papers. Identify the precipitate and write molecular, ionic, and net ionic equations for the reaction.

Solution:

Both compounds are strong electrolytes, so they dissociate into the ions $Ag^+(aq)$, $NO_3^-(aq)$, $K^+(aq)$, and $Br^-(aq)$ in solution. Possible formulas for the precipitate, then, are $AgNO_3$, KBr, AgBr, and KNO_3. The first two choices can be eliminated because we started with solutions of $AgNO_3$ and KBr. If they could have precipitated as solids, they would have done so before mixing. That leaves AgBr and KNO_3. The solubility rules (Table 5.3) indicate that KNO_3 should be soluble and AgBr should be insoluble. Thus we can write the molecular equation

$$AgNO_3(aq) + KBr(aq) \longrightarrow AgBr(s) + KNO_3(aq)$$

Since $AgNO_3$, KBr, and KNO_3 are all strong electrolytes in solution, we can write their formulas as the separate ionic formulas, giving the following ionic equation:

$$Ag^+(aq) + NO_3^-(aq) + K^+(aq) + Br^-(aq) \longrightarrow AgBr(s) + K^+(aq) + NO_3^-(aq)$$

Finally we note that NO_3^- and K^+ occur on both sides of the equation, so they are spectator ions. They can be eliminated, giving us the following net ionic equation:

$$Ag^+(aq) + Br^-(aq) \longrightarrow AgBr(s)$$

Practice Problem 5.14

Barium sulfate, used in the white pigment lithopone, can be prepared by mixing solutions of barium chloride and sodium sulfate. Write balanced molecular, ionic, and net ionic equations for the reaction.

Further Practice: 5.101 and 5.102

SUMMARY

In chemical reactions, the rearrangement of atoms converts one set of substances (the reactants) into another set of substances (the products). The molecular-level rearrangements that occur during a reaction often result in observable macroscopic changes. Typical changes include a transfer of heat, a change in color, or the formation of a solid or gas.

Chemical reactions are represented symbolically by chemical equations, with the formulas of the reactants and the products separated by an arrow. A coefficient precedes each formula to indicate the relative number of molecules or formula units involved in the reaction. A chemical equation is balanced when the number of atoms of each component element is the same in the reactants as in the products.

The products of chemical reactions must be identified experimentally. However, reasonable predictions are possible for five classes of reactions:

1. In *decomposition reactions,* complex substances break down into simpler ones upon heating. Patterns of reactivity predict the products of decomposition reactions.
2. *Combination reactions* make more complex substances from simpler ones. The products of reactions between a metal and a nonmetal are ionic compounds. Knowing the common charges of ions allows us to predict them.
3. In *single-displacement reactions,* an element displaces another element from its compound. An activity series predicts whether a single-displacement reaction will occur.
4. In *double-displacement reactions,* two compounds exchange partners, such as cations and anions. Such reactions occur if the products include an insoluble solid (a precipitate), an insoluble gas, or a stable molecule such as water.
5. *Combustion* is the reaction of a substance with oxygen, often rapidly producing heat and flames. When an excess of oxygen is available, the products of the combustion of hydrocarbons are water vapor and carbon dioxide.

An ionic equation accurately describes reactions that occur in solution, because it represents soluble electrolytes with the formulas of their component ions. Any ions that occur on both sides of the equation are spectator ions. They don't participate in the chemical reaction. If the spectator ions are eliminated from the equation, what remains is a net ionic equation. Such equations focus only on the changes that occur during a reaction.

KEY TERMS

activity series (5.4)

anhydrous (5.4)

balanced equation (5.3)

chemical equation (5.3)

combination reaction (5.4)

combustion reaction (5.4)

corrosion (5.4)

decomposition reaction (5.4)

double-displacement reaction (5.4)

ionic equation (5.5)

molecular equation (5.5)

net ionic equation (5.5)

neutralization reaction (5.4)

precipitate (5.4)

precipitation reaction (5.4)

product (5.1)

reactant (5.1)

single-displacement reaction (5.4)

spectator ion (5.5)

QUESTIONS AND PROBLEMS

The following questions and problems, except for those in the *Additional Questions* section, are paired. Questions in a pair focus on the same concept. Answers to the odd-numbered questions and problems are in Appendix D.

Matching Definitions with Key Terms

5.1   Match the key terms with the descriptions provided.
   (a) a reaction in which a free element displaces another element from a compound to produce a different compound and a different free element
   (b) a compound free of molecular water
   (c) a form of a chemical equation in which substances are represented as if they existed as molecules (using formula units), even though the substances may exist in solution as ions
   (d) a chemical reaction in which a substance breaks down into simpler compounds or elements
   (e) a chemical equation that has coefficients in front of each formula to make the number of atoms of each element the same on both sides of the equation
   (f) a substance that is converted to another substance during a chemical reaction
   (g) an ion that is not involved in a chemical reaction
   (h) a reaction that involves molecular oxygen as a reactant and that rapidly produces heat and flames
   (i) an insoluble solid deposited from a solution

5.2   Match the key terms with the descriptions provided.

(a) a reaction in which two elements, an element and a compound, or two compounds join to form a new compound

(b) a list of metals in order of activity

(c) an abbreviated representation of a chemical reaction consisting of chemical symbols and formulas

(d) a substance formed from another substance during a chemical reaction

(e) a form of a chemical equation in which ionic compounds are represented as the separated ions

(f) a reaction of an acid and a base to form an ionic compound and water

(g) a reaction in which the reactants in solution exchange ions to form a solid

(h) a form of a chemical equation in which ionic compounds are represented as the separated ions, and spectator ions are eliminated

(i) a reaction in which two compounds exchange ions or elements to form new compounds

What Is a Chemical Reaction?

5.3   When sodium metal mixes with chlorine gas, a reaction occurs, giving an intense flash of light and solid sodium chloride. Identify the reactants and products in this reaction.

5.4   Nitrogen gas and solid iodine form in an explosive reaction when nitrogen triiodide is detonated. Identify the reactants and products in this reaction.

5.5   Xenon gas reacts with fluorine gas to form xenon tetrafluoride gas. Identify which image represents reactants and then which image represents products.

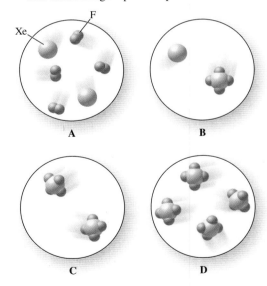

5.6   Ozone gas (O_3) reacts with carbon monoxide gas to form oxygen gas and carbon dioxide gas. Identify which image represents reactants and which image represents products.

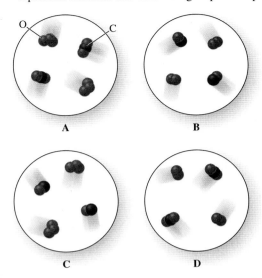

5.7   This molecular-level diagram represents a chemical reaction between nitrogen and hydrogen gases. Is it accurate? If not, what is wrong with it and how would you fix it?

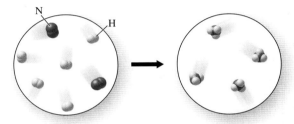

5.8   This molecular-level diagram represents a chemical reaction between hydrogen and oxygen gases. Is it accurate? If not, what is wrong with it and how would you fix it?

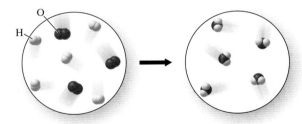

5.9   Examine the following molecular-level diagram of a chemical reaction. The product image is incomplete. How would you modify this to show that the law of conservation of mass is obeyed?

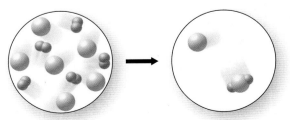

5.10   Examine the following molecular-level diagram of a chemical reaction. The product image is incomplete. How would you modify this to show that the law of conservation of mass is obeyed?

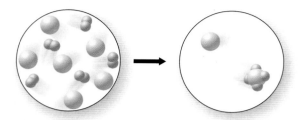

5.11   Sulfur dioxide reacts with oxygen to form sulfur trioxide. Consider the following molecular diagram of a mixture of reactants. Draw a diagram of the products when the reaction is complete.

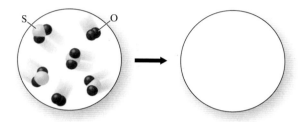

5.12   Nitrogen monoxide reacts with oxygen to form nitrogen dioxide. Consider the following molecular diagram of a mixture of reactants. Draw a diagram of the products when the reaction is complete.

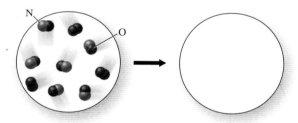

How Do We Know a Chemical Reaction Occurs?

5.13   Examine the following photograph. Identify any evidence that a chemical reaction may be occurring.

5.14   Examine the following photograph. Identify any evidence that a chemical reaction may be occurring.

5.15   A sample of dry ice (solid CO_2) starts to form a white cloud upon standing. Gaseous carbon dioxide is formed. Is this change a chemical reaction? Explain your answer.

5.16   A piece of copper metal is placed in a solution of silver nitrate. The solution turns blue, and a black deposit coats the copper metal. Is this change a chemical reaction? Explain your answer.

5.17   Examine the following molecular-level diagram. Is a chemical reaction occurring? Explain your answer.

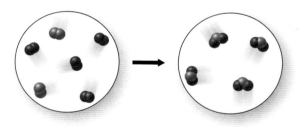

5.18   Examine the following molecular-level diagram. Is a chemical reaction occurring? Explain your answer.

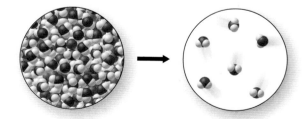

5.19   Examine the following molecular-level diagram. Is a chemical reaction occurring? Explain your answer.

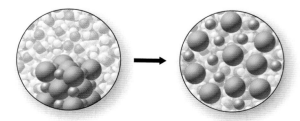

5.20 Examine the following molecular-level diagram. Is a chemical reaction occurring? Explain your answer.

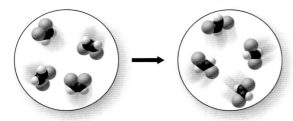

Writing Chemical Equations

5.21 What is a chemical equation?

5.22 What is the difference between a chemical equation and a chemical reaction?

5.23 Consider the following equations. For each, decide whether it represents a chemical reaction or a physical change.
   (a) $CO_2(g) + H_2O(l) \longrightarrow H_2CO_3(aq)$
   (b) $H_2O(s) \longrightarrow H_2O(l)$
   (c) $HOCN(g) \longrightarrow HCNO(g)$

5.24 Consider the following equations. For each, decide whether it represents a chemical reaction or a physical change.
   (a) $SnCl_4(aq) + 4H_2O(l) \longrightarrow Sn(OH)_4(s) + 4HCl(aq)$
   (b) $4K(s) + O_2(g) \longrightarrow 2K_2O(s)$
   (c) $NaCl(s) \longrightarrow Na^+(aq) + Cl^-(aq)$

5.25 Why must a chemical equation be balanced?

5.26 Why can't subscripts be changed to balance a chemical equation?

5.27 Write complete, balanced equations for each of the following reactions.
   (a) When solid sodium hydride (NaH) is added to water, hydrogen gas is released and aqueous sodium hydroxide forms.
   (b) Solid magnesium reacts with solid silicon dioxide to form solid silicon and solid magnesium oxide.

5.28 Write complete, balanced equations for each of the following reactions.
   (a) Solid magnesium reacts with aqueous hydrochloric acid to form gaseous hydrogen and aqueous magnesium chloride.
   (b) Solid copper(II) nitrate is heated to produce solid copper(II) oxide, gaseous nitrogen dioxide, and oxygen gas.

5.29 Consider the following molecular-level diagram of a chemical reaction between $N_2(g)$ and $Cl_2(g)$. Write a balanced equation to represent this reaction.

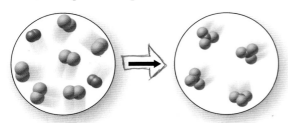

5.30 Consider the following molecular-level diagram of a chemical reaction between $CH_4(g)$ and $O_2(g)$. Write a balanced equation to represent this reaction.

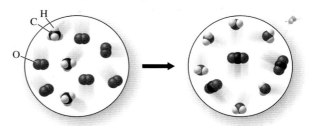

5.31 A piece of magnesium metal is ignited and burns with an intense white flame, leaving behind a white ashlike substance. Consider the following molecular diagrams. Which matches this description of a chemical reaction? Write an equation to represent this reaction.

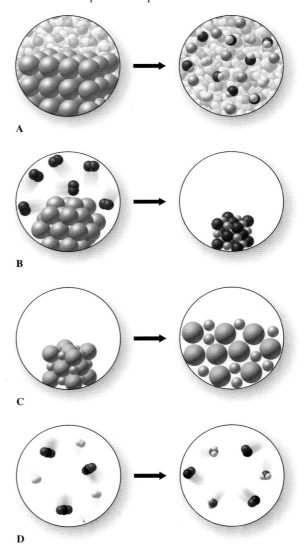

A

B

C

D

5.32 When the silver-colored metal sodium reacts with water, it forms a solution of sodium hydroxide and a molecular gas bubbles out of the solution. Consider the molecular diagrams shown in Question 5.31. Which matches this description of a chemical reaction? Write an equation to represent this reaction.

5.33 Consider the incomplete molecular-level diagram of the following reaction:

$$H_2(g) + I_2(g) \longrightarrow 2HI(g)$$

Complete this diagram to show the products expected for this reaction.

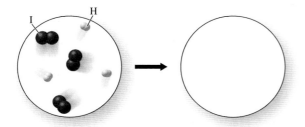

5.34 Consider the incomplete molecular-level diagram of the following reaction:

$$2NO(g) + 2CO(g) \longrightarrow N_2(g) + 2CO_2(g)$$

Complete this diagram to show the products expected for this reaction.

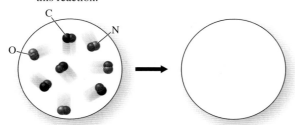

5.35 Consider the following equation:

$$2NO_2(g) \longrightarrow N_2O_4(g)$$

Draw a molecular diagram that shows the reactants and the products for this reaction.

5.36 Consider the following equation:

$$2F_2(g) + 2H_2O(g) \longrightarrow 4HF(g) + O_2(g)$$

Draw a molecular-level diagram that shows the reactants and the products for this reaction.

5.37 Balance each of the following skeletal equations.
(a) $C_2H_6(g) + O_2(g) \longrightarrow CO_2(g) + H_2O(g)$
(b) $NH_3(g) + O_2(g) \longrightarrow NO_2(g) + H_2O(g)$
(c) $NO_2(g) + H_2O(l) + O_2(g) \longrightarrow HNO_3(aq)$

5.38 Balance each of the following skeletal equations.
(a) $C_6H_{10}(l) + O_2(g) \longrightarrow CO_2(g) + H_2O(g)$
(b) $C_6H_5OH(l) + O_3(g) \longrightarrow H_2C_2O_4(s) + O_2(g)$
(c) $Sb_2S_3(s) + O_2(g) \longrightarrow Sb_2O_3(s) + SO_2(g)$

5.39 Write a balanced equation that corresponds to the following word equation for a reaction in aqueous solution:

$$\text{copper} + \text{silver nitrate} \longrightarrow \text{copper(II) nitrate} + \text{silver}$$

5.40 Write a balanced equation that corresponds to the following word equation for a reaction in aqueous solution:

$$\text{barium chloride} + \text{sulfuric acid} \longrightarrow$$
$$\text{barium sulfate} + \text{hydrochloric acid}$$

5.41 Balance each of the following skeletal equations.
(a) $CuCl_2(aq) + AgNO_3(aq) \longrightarrow Cu(NO_3)_2(aq) + AgCl(s)$
(b) $S_8(s) + O_2(g) \longrightarrow SO_2(g)$
(c) $C_3H_8(g) + O_2(g) \longrightarrow CO_2(g) + H_2O(g)$

5.42 Balance each of the following skeletal equations.
(a) $Al_2O_3(s) + HCl(aq) \longrightarrow AlCl_3(aq) + H_2O(l)$
(b) $PCl_3(l) + AgF(s) \longrightarrow PF_3(g) + AgCl(s)$
(c) $NO_2(g) \longrightarrow NO(g) + O_2(g)$

Predicting Chemical Reactions

5.43 Describe the identifying characteristics of the following classes of reactions: decomposition, combination, single-displacement, and double-displacement.

5.44 Give examples of each of the following classes of reactions: decomposition, combination, single-displacement, and double-displacement.

5.45 Determine which classes of reactions would fit into each of the following categories for the reactants and products. Choices are decomposition, combination, single-displacement, double-displacement, and combustion.

Reactants	Products
(a) Two elements	One compound
(b) One element and one compound	One element and one compound
(c) One compound	Two elements

5.46 Determine which classes of reactions would fit into each of the following categories for the reactants and products. Choices are decomposition, combination, single-displacement, double-displacement, and combustion.

Reactants	Products
(a) One compound	One element and one compound
(b) Two compounds	Two compounds
(c) Two compounds	One compound

5.47 A solution of sodium chloride is mixed with a solution of lead(II) nitrate. A precipitate of lead(II) chloride results, leaving a solution of sodium nitrate. Into which class(es) does this reaction fit? Choices are decomposition, combination, single-displacement, double-displacement, and combustion.

5.48 Solid sulfur is ignited and burns in oxygen gas with a blue flame to form sulfur dioxide. Into which class(es) does this reaction fit? Choices are decomposition, combination, single-displacement, double-displacement, and combustion.

5.49 Consider the following molecular-level diagrams of chemical reactions. Identify each as decomposition, combination, single-displacement, or double-displacement.

(a)

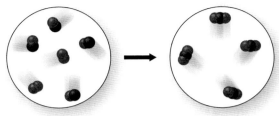

(b)

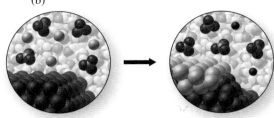

5.50 Consider the following molecular-level diagrams of chemical reactions. Identify each as decomposition, combination, single-displacement, or double-displacement.

(a)

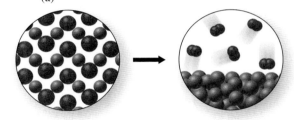

(b)

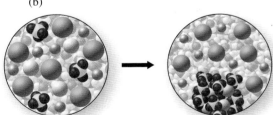

5.51 Balance the following chemical equations. Classify the reactions as decomposition, combination, single-displacement, double-displacement, or combustion.

(a) $CaCl_2(aq) + Na_2SO_4(aq) \longrightarrow CaSO_4(s) + NaCl(aq)$

(b) $Ba(s) + HCl(aq) \longrightarrow BaCl_2(aq) + H_2(g)$

(c) $Al(s) + O_2(g) \xrightarrow{heat} Al_2O_3(s)$

(d) $FeO(s) + CO(g) \xrightarrow{heat} Fe(s) + CO_2(g)$

(e) $CaO(s) + H_2O(l) \longrightarrow Ca(OH)_2(aq)$

(f) $Na_2CrO_4(aq) + Pb(NO_3)_2(aq) \longrightarrow$
$PbCrO_4(s) + NaNO_3(aq)$

(g) $KI(aq) + Cl_2(g) \longrightarrow KCl(aq) + I_2(aq)$

(h) $NaHCO_3(s) \xrightarrow{heat} Na_2CO_3(s) + CO_2(g) + H_2O(g)$

5.52 Balance the following equations and classify the reactions as decomposition, combination, single-displacement, double-displacement, or combustion.

(a) $GaH_3 + N(CH_3)_3 \longrightarrow (CH_3)_3NGaH_3$

(b) $GeCl_2 + Cl_2 \longrightarrow GeCl_4$

(c) $N_2(g) + CaC_2(s) \longrightarrow C(s) + CaNCN(s)$

(d) $N_2(g) + Mg(s) \longrightarrow Mg_3N_2(s)$

(e) $NH_4Cl(s) \longrightarrow NH_3(g) + HCl(g)$

(f) $CaO(s) + SO_3(g) \xrightarrow{heat} CaSO_4(s)$

(g) $PCl_5(g) \xrightarrow{heat} PCl_3(g) + Cl_2(g)$

(h) $Ca_3N_2(s) + H_2O(l) \xrightarrow{heat} Ca(OH)_2(aq) + NH_3(g)$

5.53 When heated, nickel(II) carbonate undergoes a decomposition reaction. Write a balanced equation to describe this reaction.

5.54 When heated, platinum(IV) chloride undergoes a decomposition reaction. Write a balanced equation to describe this reaction.

5.55 Complete and balance the equation for each of the following decomposition reactions.

(a) $CaCO_3(s) \xrightarrow{heat}$

(b) $CuSO_4 \cdot 5H_2O(s) \xrightarrow{heat}$

5.56 Complete and balance the equation for each of the following decomposition reactions.

(a) $Cu(OH)_2(s) \xrightarrow{heat}$

(b) $HgO(s) \xrightarrow{heat}$

5.57 Magnesium metal reacts with oxygen gas in a combination reaction. Write a balanced equation to describe this reaction.

5.58 Sodium metal reacts with chlorine gas in a combination reaction. Write a balanced equation to describe this reaction.

5.59 Complete and balance the equation for each of the following combination reactions.

(a) $Ca(s) + N_2(g) \xrightarrow{heat}$

(b) $SO_3(g) + H_2O(l) \longrightarrow$

(c) $Al(s) + O_2(g) \longrightarrow$

5.60 Complete and balance the equation for each of the following combination reactions.

(a) $Li_2O(s) + H_2O(l) \longrightarrow$

(b) $SO_2(g) + O_2(g) \longrightarrow$

(c) $CaO(s) + H_2O(l) \longrightarrow$

5.61 Zinc metal reacts with a solution of tin(II) chloride in a single-displacement reaction. Write a balanced equation to describe this reaction.

5.62 Iron metal reacts with a solution of copper(II) sulfate in a single-displacement reaction. Write a balanced equation to describe this reaction.

5.63 Consider the following metals. Determine which will react with water and which will react with a solution of hydrochloric acid. Write balanced equations to describe any reactions that occur.

(a) Ca   (b) Fe   (c) Cu

5.64 Consider the following metals. Determine which will react with water and which will react with a solution of hydrochloric acid. Write balanced equations to describe any reactions that occur.
(a) Cr   (b) Bi   (c) K

5.65 Complete and balance the equations for each of the following single-displacement reactions.
(a) $Zn(s) + AgNO_3(aq) \longrightarrow$
(b) $Na(s) + FeCl_2(s) \xrightarrow{\text{heat}}$

5.66 Complete and balance each of the following single-displacement reactions.
(a) $Al(s) + CuSO_4(aq) \longrightarrow$
(b) $Cs(s) + H_2O(l) \longrightarrow$

5.67 Which of the following single-displacement reactions will actually occur if the indicated reactants are mixed?
(a) $3Mg(s) + 2AlCl_3(aq) \longrightarrow 3MgCl_2(aq) + 2Al(s)$
(b) $Zn(s) + MgCl_2(aq) \longrightarrow ZnCl_2(aq) + Mg(s)$
(c) $Cu(s) + Pb(NO_3)_2(aq) \longrightarrow Cu(NO_3)_2(aq) + Pb(s)$
(d) $Ni(s) + 2AgNO_3(aq) \longrightarrow Ni(NO_3)_2(aq) + 2Ag(s)$

5.68 Which of the following single-displacement reactions will actually occur if the indicated reactants are mixed?
(a) $Ni(s) + FeCl_2(aq) \longrightarrow NiCl_2(aq) + Fe(s)$
(b) $2Al(s) + 3NiCl_2(aq) \longrightarrow 2AlCl_3(aq) + 3Ni(s)$
(c) $Fe(s) + CuCl_2(aq) \longrightarrow FeCl_2(aq) + Cu(s)$
(d) $3Sn(s) + 2AlCl_3(aq) \longrightarrow 3SnCl_2(aq) + 2Al(s)$

5.69 Aqueous calcium chloride reacts with aqueous potassium carbonate in a double-displacement reaction. Write a balanced equation to describe this reaction.

5.70 Aqueous ammonium chromate reacts with aqueous lead(II) nitrate in a double-displacement reaction. Write a balanced equation to describe this reaction.

5.71 Complete and balance the equation for each of the following double-displacement reactions.
(a) $CaCO_3(s) + H_2SO_4(aq) \longrightarrow$
(b) $SnCl_2(aq) + AgNO_3(aq) \longrightarrow$

5.72 Complete and balance the equation for each of the following double-displacement reactions.
(a) $NaOH(aq) + CuCl_2(aq) \longrightarrow$
(b) $H_2SO_4(aq) + KOH(aq) \longrightarrow$

5.73 Which of the following ionic compounds would be expected to be soluble in water?
(a) $CuCl_2$ (b) $AgNO_3$ (c) $PbCl_2$ (d) $Cu(OH)_2$

5.74 Which of the following ionic compounds would be expected to be insoluble when mixed with water?
(a) Cr_2S_3 (b) $Ca(OH)_2$ (c) $BaSO_4$ (d) $(NH_4)_2CO_3$

5.75 Aqueous solutions of sodium sulfate and lead(II) nitrate are mixed and a white solid forms. Identify the solid.

5.76 Aqueous solutions of mercury(I) nitrate and calcium iodide are mixed and a yellow solid forms. Identify the solid.

5.77 Write a balanced equation to describe any precipitation reaction that might occur when the following substances are mixed.
(a) $K_2CO_3(aq) + BaCl_2(aq)$
(b) $CaS(aq) + Hg(NO_3)_2(aq)$
(c) $Pb(NO_3)_2(aq) + K_2SO_4(aq)$

5.78 Write a balanced equation to describe any precipitation reaction that might occur when the following substances are mixed.
(a) $MgSO_4(aq) + BaCl_2(aq)$
(b) $K_2SO_4(aq) + MgCl_2(aq)$
(c) $MgCl_2(aq) + Pb(NO_3)_2(aq)$

5.79 Consider the following double-displacement reactions. For each, identify the driving force that causes the reaction to go to completion.
(a) $Hg(NO_3)_2(aq) + H_2S(aq) \longrightarrow HgS(s) + 2HNO_3(aq)$
(b) $MnS(s) + 2HCl(aq) \longrightarrow MnCl_2(aq) + H_2S(g)$
(c) $Ba(OH)_2(aq) + H_2SO_4(aq) \longrightarrow BaSO_4(s) + 2H_2O(l)$

5.80 Consider the following double-displacement reactions. For each, identify the driving force that causes the reaction to go to completion.
(a) $Al_2O_3(s) + 3H_2SO_4(aq) \longrightarrow Al_2(SO_4)_3(aq) + 3H_2O(l)$
(b) $CaCO_3(s) + H_2SO_4(aq) \longrightarrow CaSO_4(s) + H_2O(l) + CO_2(g)$
(c) $BaCl_2(aq) + Na_2SO_4(aq) \longrightarrow BaSO_4(s) + 2NaCl(aq)$

5.81 Write a balanced equation to describe any acid-base neutralization reaction that might occur when the following substances are mixed.
(a) $H_2S(aq) + Cu(OH)_2(s)$
(b) $CH_4(g) + NaOH(aq)$
(c) $KHSO_4(aq) + KOH(aq)$

5.82 Write a balanced equation to describe any acid-base neutralization reaction that might occur when the following substances are mixed.
(a) $H_2(g) + KOH(aq)$
(b) $H_3PO_4(aq) + NaOH(aq)$
(c) $H_2SO_4(aq) + Fe(OH)_2(s)$

5.83 Sulfur dioxide gas reacts with oxygen gas in a combustion reaction. Write a balanced equation to describe the reaction.

5.84 Hydrogen sulfide gas reacts with oxygen gas in a combustion reaction. Write a balanced equation to describe the reaction.

5.85 Write a formula for a product of the combustion reaction of the following elements with molecular oxygen.
(a) Cs   (b) Pb   (c) Al   (d) H_2
(e) C

5.86 Write a formula for the product of the combustion reaction of the following elements with molecular oxygen.
(a) Mg   (b) Fe   (c) Li   (d) S_8
(e) N_2

5.87 Write formulas for the products of the reactions of the following compounds with molecular oxygen.
(a) CH_4 (b) CO   (c) CuS   (d) CH_3OH

5.88 Write formulas for the products of the reactions of the following compounds with molecular oxygen.
(a) Li_3P (b) CS_2 (c) C_2H_6 (d) CH_3CO_2H

Representing Reactions in Aqueous Solution

5.89   Distinguish between an electrolyte and a nonelectrolyte.

5.90   Describe an experimental method for determining whether a solid is an electrolyte or a nonelectrolyte.

5.91   Indicate whether each of the following substances is an electrolyte or a nonelectrolyte.
   (a)   NaOH(*aq*)
   (b)   HCl(*aq*)
   (c) $C_{12}H_{22}O_{11}$(*aq*) (sucrose solution)

5.92   Indicate whether each of the following substances is an electrolyte or a nonelectrolyte.
   (a) CH_3CH_2OH(*aq*)
   (b) H_2O(*l*)
   (c)   NaCl(*aq*)

5.93   Consider the following molecular-level diagram of a solution of a compound in water. Is this substance an electrolyte or a nonelectrolyte?

5.94   Consider the following molecular-level diagram of a solution of a compound in water. Is this substance an electrolyte or a nonelectrolyte?

5.95   Indicate whether aqueous solutions of the following substances contain ions. If so, write formulas for the ions in solution.
   (a) $C_6H_{12}O_6$ (glucose)
   (b) CH_4
   (c)   NaCl

5.96   Indicate whether aqueous solutions of the following substances contain ions. If so, write formulas for the ions in solution.
   (a) I_2
   (b)   KI
   (c) CH_3OH

5.97   Distinguish between a molecular, an ionic, and a net ionic equation.

5.98   Why is it necessary to identify a substance as an electrolyte or a nonelectrolyte when writing a net ionic equation?

5.99   What are spectator ions?

5.100   In which type of equation—molecular, ionic, or net ionic—do we find spectator ions?

5.101   Aqueous calcium chloride reacts with aqueous silver nitrate to form a precipitate of silver chloride and a solution of calcium nitrate. Write a net ionic equation for this reaction.

5.102   Aqueous chromium(III) sulfate reacts with aqueous sodium hydroxide to form a precipitate of chromium(III) hydroxide and a solution of sodium sulfate. Write a net ionic equation for this reaction.

5.103   Consider the following molecular-level diagram for the reaction between Na and H_2O in aqueous solution. Write a net ionic equation for this reaction.

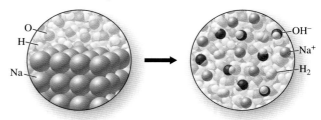

5.104   Consider the following molecular-level diagram for the reaction between Cu and $AgNO_3$ in aqueous solution. Write a net ionic equation for this reaction.

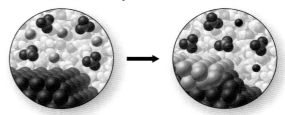

5.105   Write balanced molecular and net ionic equations for the following reactions in aqueous solution.
   (a)   copper + silver nitrate $\longrightarrow$
                                             copper(II) nitrate + silver
   (b)   iron(II) oxide + hydrochloric acid $\longrightarrow$
                                             iron(II) chloride + water

5.106   Write balanced molecular and net ionic equations for the following reactions in aqueous solution.
   (a)   calcium + water $\longrightarrow$
                                             calcium hydroxide + hydrogen
   (b)   barium chloride + sodium sulfate $\longrightarrow$
                                             barium sulfate + sodium chloride

5.107   Add the physical state for each substance in the following aqueous reactions. Then write and balance a net ionic equation for each reaction.
   (a) $NaCl + Ag_2SO_4 \longrightarrow Na_2SO_4 + AgCl$
   (b) $Cu(OH)_2 + HCl \longrightarrow CuCl_2 + H_2O$
   (c) $BaCl_2 + Ag_2SO_4 \longrightarrow BaSO_4 + AgCl$

5.108   Add the physical state for each substance in the following aqueous reactions. Then write and balance a net ionic equation for each reaction.
(a) $Cr_2(SO_4)_3 + KOH \longrightarrow Cr(OH)_3 + K_2SO_4$
(b) $K_2CrO_4 + PbCl_2 \longrightarrow KCl + PbCrO_4$
(c) $Na_2SO_3 + H_2SO_4 \longrightarrow Na_2SO_4 + H_2O + SO_2$

5.109   When aqueous solutions of sodium carbonate and calcium bromide mix, a white solid forms.
(a) Identify the solid.
(b) Write a balanced molecular equation for the reaction.
(c) Write a balanced ionic equation for the reaction.
(d) Identify any spectator ions in the reaction.
(e) Write a balanced net ionic equation for the reaction.

5.110   When aqueous solutions of potassium chromate and barium chloride mix, a yellow solid forms.
(a) Identify the solid.
(b) Write a balanced molecular equation for the reaction.
(c) Write a balanced ionic equation for the reaction.
(d) Identify any spectator ions in the reaction.
(e) Write a balanced net ionic equation for the reaction.

5.111   Predict whether reactions should occur between aqueous solutions of the following compounds. If so, write balanced molecular and net ionic equations for the reactions.
(a) $Sr(NO_3)_2 + H_2SO_4$
(b) $Zn(NO_3)_2 + Na_2SO_4$
(c) $CuSO_4 + BaS$

5.112   Predict whether reactions should occur between aqueous solutions of the following compounds. If so, write balanced molecular and net ionic equations for the reactions.
(a) $(NH_4)_2CO_3 + CrCl_3$
(b) $Ba(OH)_2 + HCl$
(c) $FeCl_3 + NaOH$

Additional Questions

5.113   Bismuth selenide (Bi_2Se_3) is used in semiconductor research. It can be prepared directly from its elements.

$$2Bi + 3Se \longrightarrow Bi_2Se_3$$

Classify the reaction as decomposition, combination, single-displacement, double-displacement, or combustion.

5.114   Heating causes the copper mineral malachite, $CuCO_3 \cdot Cu(OH)_2$ to decompose. Write a balanced equation for the reaction.

5.115   Write a balanced equation to describe any precipitation reaction that might occur when the following substances are mixed.
(a) $ZnSO_4(aq) + Ba(NO_3)_2(aq)$
(b) $Ca(NO_3)_2(aq) + K_3PO_4(aq)$
(c) $ZnSO_4(aq) + BaCl_2(aq)$
(d) $KOH(aq) + MgCl_2(aq)$
(e) $CuSO_4(aq) + BaS(aq)$

5.116   Indicate whether each of the following substances reacts with hydrochloric acid to form water.
(a) $NaOH(aq)$
(b) $Zn(s)$
(c) $CaCO_3(s)$
(d) $Cu(OH)_2(s)$

5.117   Sodium metal reacts with liquid water to form a solution of sodium hydroxide and hydrogen gas. Potassium reacts in a similar manner. Write a balanced equation to represent the reaction of potassium with water.

5.118   The solid compounds NaCl and $Cu(NO_3)_2$ are added to a beaker of water. Describe what happens. Include in your description the identities of any substances that dissolve and any substances that precipitate.

5.119 An aqueous solution contains silver nitrate. Which of the following substances could be added to this solution to precipitate the silver ions as silver chloride?
(a) $Cl_2(g)$
(b) $NaCl(s)$
(c) $PbCl_2(s)$
(d) $CCl_4(l)$

5.120 A solution contains the ionic compounds $BaCl_2$ and $NaCl$. What ionic compound could be added that would remove the barium ions from the solution but not introduce any new kinds of ions?

5.121 Consider the following net ionic equation:

$$H^+(aq) + OH^-(aq) \longrightarrow H_2O(l)$$

Make a list of three different sets of two chemical compounds, each of which would result in this net ionic equation when they are mixed.

5.122 A solution contains cobalt(II) chloride. Identify three metals that could be placed in the solution to cause the cobalt to be plated out as a metal.

5.123 What reaction, if any, occurs if a piece of copper metal is placed in an aqueous solution containing iron(II) sulfate?

5.124 A solution contains both copper(II) nitrate and lead(II) nitrate. What substance could be added to the solution to precipitate the lead(II) ions, but leave the copper(II) ions in solution? After removal of the lead(II) compound, what could then be added to the solution to precipitate the copper(II) ions?

5.125 Draw molecular-level diagrams to show that a solution of KI is a strong electrolyte and that a solution of CO is a nonelectrolyte.

5.126 Write equations for reactions that could be used to release each of the following elements from their compounds.
(a) O_2
(b) H_2
(c) Cl_2

5.127 Identify a substance that could be used to release the following metals from their compounds, using a different substance for each metal.
(a) Cr
(b) Fe
(c) Al
(d) Pb

5.128 Suggest chemical reactions that could be used to identify the contents of four unlabeled bottles containing solutions of the following substances: $CuSO_4$, $Al_2(SO_4)_3$, $Pb(NO_3)_2$, and $FeSO_4$.

5.129 Suggest chemical methods that would separate the following ions or compounds from one another.
(a) Fe^{3+} and Al^{3+}
(b) Ba^{2+} and Mg^{2+}
(c) Cl^- and NO_3^-
(d) $BaSO_4$ and $MgSO_4$

5.130 When copper metal is exposed to the atmosphere over a long period of time, it takes on a green patina (a coating of copper compounds). The green color of copper roofs and the Statue of Liberty are examples. Initially the copper metal reacts with oxygen gas. The product of this reaction reacts in part with water vapor and in part with carbon dioxide in the air. Predict the products and write balanced equations for the three reactions.

6

Quantities in Chemical Reactions

Imagine the car you will drive five years from now. What will power it? Will it be gasoline? ethanol? an electric battery? hydrogen?

Gasoline, a form of petroleum, has been used to power the petroleum-fuel internal-combustion engine (ICE) of automobiles since their invention in the 1860s. Despite steady improvements to the ICE, today's ICE vehicles are only 20% to 25% efficient. This means that 75% to 80% of the energy in gasoline is wasted. This is not the only disadvantage. The combustion of gasoline in the ICE releases pollutants into the atmosphere, and our reliance on gasoline for transportation places a heavy demand on a limited—and rapidly dwindling—natural resource. Although the auto industry has found ways to significantly reduce many toxic gas emissions, the non-toxic carbon dioxide that is released instead is a major factor in the warming of the Earth. It is a contributor to the "greenhouse effect," which is responsible for the recent increase of Earth's average temperature often called "global warming."

Because Lily is concerned about the environmental impacts of burning fuels, she drives a car that runs on an alternative fuel, an ethanol blend called E-85. It is 85% ethanol (ethyl alcohol, an oxygen-containing fuel) and 15% gasoline. She chose the car because ethanol is a cleaner-burning fuel than gasoline. She's happy that ethanol—unlike gasoline—is a renewable fuel. It can be made from the fermentation of corn or any other carbohydrate. She's proud that she bought her car for about the same price as a gasoline-powered car, but she admits that E-85 isn't always easy to find at service stations.

Joel paid more for his battery-powered electric vehicle (EV), another alternative to the ICE. Instead of filling up with gasoline, Joel recharges his car battery. His engine is about 70% efficient, has fewer mechanical parts than an ICE, and gives off no tailpipe emissions. However, the electricity that recharges Joel's car battery comes from a power plant that burns fossil fuel. The combustion process generates carbon dioxide, so there is still the problem of increased CO_2 in the atmosphere, although it doesn't come directly from Joel's car. Also, the overall efficiency drops to less than 40% when the loss of energy between the power plant and the wall outlet in Joel's garage is taken into account. The drain on scarce fossil fuel resources is no less than using gasoline.

Lily and Joel have heard that hydrogen-powered cars may someday be available. How can hydrogen power a vehicle? When hydrogen reacts with oxygen to form water, a large amount of energy is released. Hydrogen is the major fuel that powers the space shuttle throughout takeoff (Figure 6.1). Hydrogen-powered cars do not burn hydrogen directly; instead they use a fuel cell, in which the reaction between hydrogen and oxygen is carried out in a controlled way.

A fuel cell works on a different principle than an ICE (Figure 6.2). In an ICE, gasoline burns with air in the cylinder when it is ignited with a spark plug. The production of hot gases at regular intervals causes the piston to move up and down. A crankshaft converts the up-and-down action into a spinning motion, turning the wheels of the car. In a fuel cell, the hydrogen does not actually burn. Hydrogen gas runs through one side of the cell, while oxygen runs through the other. When hydrogen molecules interact with a catalytic membrane, electrons are stripped from the molecules to give free hydrogen ions (H^+) and electrons. The freed electrons create a current that runs the vehicle's electric motor. They are used on the oxygen side as they combine with oxygen molecules to form oxide ions (O^{2-}). The H^+ ions travel through the membrane to the oxygen side, where they react with the O^{2-} ions to form water molecules. In some ways, the hydrogen fuel cell looks and works like a battery, but there is a major difference. The chemicals that run a battery are stored in it, and they eventually run out. In a fuel cell, the hydrogen and oxygen must be continuously delivered to the cell from external storage tanks that can be refilled.

Figure 6.1
The reaction between hydrogen and oxygen produces the energy to propel the space shuttle into space. The product of this combustion reaction, H_2O, is visible as clouds of steam.

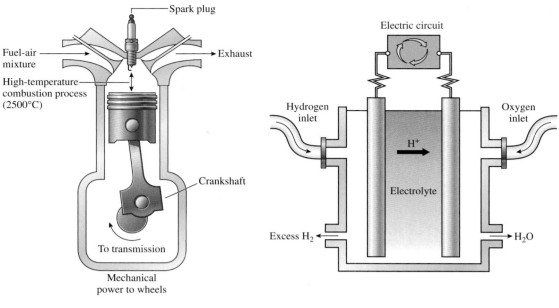

A Single cylinder internal-combustion engine (ICE)

B Hydrogen fuel cell

Figure 6.2
(A) In an internal-combustion engine (ICE), a fuel-air mixture burns inside a combustion chamber, providing the force needed to turn a crankshaft. (B) A hydrogen fuel cell works something like a battery. A controlled reaction creates a current that runs the vehicle's electric motor.

Many people think hydrogen will be the fuel of the future because it is efficient and clean burning. There are, however, problems to overcome. Storing hydrogen gas under high pressure in a vehicle is a safety hazard. Also, hydrogen gas is explosive in air when not controlled. Solutions to these problems have focused on storing the hydrogen as part of a metal hydride compound such as TiH_3, or in methanol (CH_3OH). Hydrogen can then be obtained from chemical reactions of these compounds as needed. Another problem is that hydrogen is not readily available. It is most easily obtained from reactions of methanol or other fuels, which also form carbon monoxide or carbon dioxide. In this case, no reduction of air pollution occurs.

An alternative is obtaining hydrogen from water by electrolysis, a process that requires an electric current. If electricity from a fossil-fuel power plant is used for the purpose, there's no gain in environmental quality or resource conservation. It would be better if the electricity came from wind farms or solar energy. In Iceland, hydrogen fuel cells are used to run buses. Electricity is cheap there because it is produced from geothermal power. Electrolysis at a hydrogen refueling station produces hydrogen for the fuel cells that run the buses. As a result of the success in Iceland, other countries such as Japan are piloting hydrogen-fueled vehicles (Figure 6.3).

As drivers of alternative-fuel vehicles, Lily and Joel are interested in the amount of fuel they need for their vehicles and the energy output provided by the fuel. Whatever the chemical reaction may be, it is often important to know the quantities of the reactants and products and the amount of energy involved in a reaction. For example, we might want to know the amount of oxygen needed to react with the hydrogen that powers a space-shuttle liftoff, or we might want to produce a certain amount of hydrogen from a reaction of methanol. In another situation we might be interested in producing a certain amount of energy in a combustion reaction.

In this chapter we will look at the quantitative aspects of chemical reactions. We will show how to determine the amounts of reactants that are needed and the amounts of products that should be formed in a given reaction. We will investigate the amounts

Figure 6.3
This hydrogen refueling station in Japan is part of a test program.

of energy produced in chemical reactions such as the combustion of fuels, and we will compare the amount of heat released by different fuels. While combustion reactions release energy, some chemical reactions absorb energy. We will look at how to measure the amount of energy released or absorbed during chemical reactions.

Questions for Consideration

6.1 What do the coefficients in balanced equations represent?

6.2 How can we use a balanced equation to relate the number of moles of reactants and products in a chemical reaction?

6.3 How can we use a balanced equation to relate the mass of reactants and products in a chemical reaction?

6.4 How do we determine which reactant limits the amount of product that can form?

6.5 How can we compare the amount of product we actually obtain to the amount we expect to obtain?

6.6 How can we describe and measure energy changes?

Math Tools Used in This Chapter

Units and Conversions (Math Toolbox 1.3)

6.1   THE MEANING OF A BALANCED EQUATION

As you learned in Chapter 5, a balanced equation represents the identity and physical state of the reactants and products in a chemical reaction. In addition the coefficients in the equation show the relative amounts of reactants that combine and the relative amounts of products expected to form. Let's consider the combustion reaction of propane, a chemical reaction that occurs when you light a gas grill. The balanced equation for the combustion of propane (Figure 6.4) shows that when 1 propane (C_3H_8) molecule reacts, 5 O_2 molecules combine with it to form 3 CO_2 molecules and 4 H_2O molecules:

$$C_3H_8(g) + 5O_2(g) \longrightarrow 3CO_2(g) + 4H_2O(g)$$

Since the coefficients in this equation represent the *relative* numbers of reactant and product molecules, the equation reveals the amounts of reactants and products at any scale. For example, if 2 molecules of C_3H_8 react, they must combine with 10 molecules of O_2 producing 6 molecules of CO_2 and 8 molecules of H_2O. The relative numbers of reactants and products remain the same (1:5:3:4).

Propane for barbeque grills is stored under pressure in its liquid state. Air, which is about 20% oxygen, provides the oxygen for the reaction.

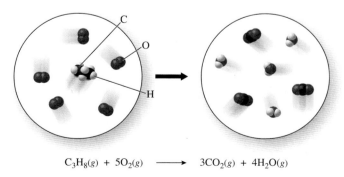

$$C_3H_8(g) + 5O_2(g) \longrightarrow 3CO_2(g) + 4H_2O(g)$$

Figure 6.4
The balanced equation for the combustion of propane shows the relative numbers of propane and oxygen molecules that react and the relative numbers of carbon dioxide and water molecules that are produced.

TABLE 6.1	**Relationships between Reactants and Products**		
$C_3H_8(g)$ +	$5O_2(g)$ $\longrightarrow$	$3CO_2(g)$ +	$4H_2O(g)$
1 molecule	5 molecules	3 molecules	4 molecules
2 molecules	10 molecules	6 molecules	8 molecules
100 molecules	500 molecules	300 molecules	400 molecules
6.022×10^{23} molecules	$5 \times (6.022 \times 10^{23})$ molecules	$3 \times (6.022 \times 10^{23})$ molecules	$4 \times (6.022 \times 10^{23})$ molecules
1.000 mol	5.000 mol	3.000 mol	4.000 mol

Suppose 100 molecules of C_3H_8 react, or suppose 6.022×10^{23} molecules (a mole of molecules) react. How many O_2 molecules will combine with each of these quantities? How many molecules of CO_2 and H_2O should form in each case? Compare your answers to those shown in Table 6.1.

Because 6.022×10^{23} molecules equals 1.000 mol of molecules, we can also say that 1.000 mol of C_3H_8 reacts with 5.000 mol of O_2 to produce 3.000 mol of CO_2 and 4.000 mol of H_2O. A balanced equation, therefore, also describes the *relative number of moles* of reactants and products. Table 6.1 summarizes the relationships between reactants and products for the combustion of propane.

If we know the amount of any reactant or product, we can determine the amount of any other reactant or product in the same reaction by using the balanced equation. The process of determining the amounts of substances in a chemical reaction is called **stoichiometry.** In the next sections, we will use stoichiometry to relate amounts of reactants and products using balanced equations.

6.2   MOLE-MOLE CONVERSIONS

A convenient way to relate the moles of a reactant or a product to other reactants or products is to use *mole ratios.* Mole ratios are obtained from the coefficients in a balanced equation. Let's look again at the balanced equation for the combustion of propane:

$$C_3H_8(g) + 5O_2(g) \longrightarrow 3CO_2(g) + 4H_2O(g)$$

The mole ratio of O_2 to C_3H_8 is 5/1, or $\dfrac{5 \text{ mol } O_2}{1 \text{ mol } C_3H_8}$, because the coefficient in front of O_2 is 5 and the coefficient in front of C_3H_8 is an implied 1. This mole ratio tells us that 5 mol of O_2 are needed for every 1 mol of C_3H_8 that reacts. The same equation also shows that the mole ratio of the product CO_2 to the reactant O_2 is 3/5, or $\dfrac{3 \text{ mol } CO_2}{5 \text{ mol } O_2}$. Mole ratios help us determine the moles of one substance in a reaction when the number of moles of another substance in a reaction is known:

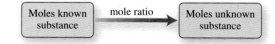

Suppose we want to know the amount of CO_2 that will be produced when 9.21 mol of C_3H_8 reacts? What mole ratio do we use? From the coefficients in the equation, the mole ratio relating CO_2 and C_3H_8 can be written in two ways:

$$\dfrac{3 \text{ mol } CO_2}{1 \text{ mol } C_3H_8} \quad \text{and} \quad \dfrac{1 \text{ mol } C_3H_8}{3 \text{ mol } CO_2}$$

We can use these mole ratios as conversion ratios. (See Math Toolbox 1.3 for a review of unit conversions.) We multiply the known moles of C_3H_8 by the ratio that will cancel the old units (mol C_3H_8) and introduce the new units (mol CO_2):

Moles C_3H_8 — mole ratio → Moles CO_2

$$\text{mol } CO_2 = 9.21 \ \cancel{\text{mol } C_3H_8} \times \frac{3 \text{ mol } CO_2}{1 \ \cancel{\text{mol } C_3H_8}} = 27.6 \text{ mol } CO_2$$

Example 6.1 gives you the opportunity to practice using mole ratios.

EXAMPLE 6.1   Mole-Mole Conversions

If 1.14 mol of CO_2 was formed by the combustion of C_3H_8, how many moles of H_2O were also formed?

$$C_3H_8(g) + 5O_2(g) \longrightarrow 3CO_2(g) + 4H_2O(g)$$

Solution:

We know the number of moles of C_3H_8 and we want to know the moles of CO_2:

Moles C_3H_8 — ? → Moles CO_2

The relationship we use to convert from moles of C_3H_8 to moles of CO_2 is the mole ratio we get from the balanced equation:

Moles C_3H_8 — mole ratio → Moles CO_2

First we must ensure that the equation is balanced. Yes it is, so the coefficients in the equation give mole relationships between CO_2 and H_2O, which can be written in two ways:

$$\frac{3 \text{ mol } CO_2}{4 \text{ mol } H_2O} \quad \text{and} \quad \frac{4 \text{ mol } H_2O}{3 \text{ mol } CO_2}$$

We multiply the known moles of CO_2 by the ratio that will cancel the old units (mol CO_2) and introduce the new units (mol H_2O):

$$\text{mol } H_2O = 1.14 \ \cancel{\text{mol } CO_2} \times \frac{4 \text{ mol } H_2O}{3 \ \cancel{\text{mol } CO_2}} = 1.52 \text{ mol } H_2O$$

Notice that the units cancel properly.

Practice Problem 6.1

Pure methanol is used as a fuel for all race cars in the Indy Racing League and in the Championship Auto Racing Teams. It is used because methanol fires are easier to put out with water than the fires of most other fuels. The balanced equation for the combustion of methanol is

$$2CH_3OH(l) + 3O_2(g) \longrightarrow 2CO_2(g) + 4H_2O(g)$$

Given that 1.00 gal of methanol contains 94.5 mol, how many moles of oxygen gas will react with 1.00 gal of methanol?

Further Practice: 6.11 and 6.12

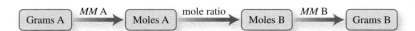

Figure 6.5
The process for converting from the mass of one reactant or product to another involves three basic steps.

6.3 MASS-MASS CONVERSIONS

When we measure the amount of a reactant or product, we do not measure moles because our measuring devices do not count moles. Instead we typically measure the mass of a reactant or product, but a chemical equation does not directly tell us the *mass* relationship between reactants and products. It tells us the *mole* relationship. This is not a problem, however, because we can convert from mass units (grams) to moles using the molar mass (*MM*) of the substance, as you learned in Chapter 4. Once we convert to moles, we can determine the moles of another substance in the equation using the mole ratio as we did in the last section. Then if we want to know the mass of the substance, we can convert to grams using its molar mass. The process always involves the same general conversion steps illustrated in Figure 6.5.

To illustrate a mass-mass conversion, let's consider the combination reaction of sodium metal and chlorine gas to form sodium chloride:

$$2Na(s) + Cl_2(g) \longrightarrow 2NaCl(s)$$

Suppose we have 9.20 g of sodium metal. How many grams of chlorine gas should react with this amount of sodium in this reaction? Let's follow the general procedure for a mass-mass problem shown in Figure 6.5. Because the balanced equation does not give us a relationship between the grams of Na and the grams of Cl_2, we must first convert grams of Na to moles of Na using its molar mass, 22.99 g/mol. The molar mass of Na gives the ratio relationship between grams of Na and moles of Na. This ratio can be written in two ways:

$$\frac{1 \text{ mol Na}}{22.99 \text{ g Na}} \quad \text{and} \quad \frac{22.99 \text{ g Na}}{1 \text{ mol Na}}$$

When we convert from grams to moles, we multiply the known amount in grams by the ratio that will cancel the old units (grams) and introduce the new units (moles):

$$\text{Moles of Na} = 9.20 \text{ g Na} \times \frac{1 \text{ mol Na}}{22.99 \text{ g Na}} = 0.400 \text{ mol Na}$$

Now that we know the moles of the reactant, Na, we can determine the moles of Cl_2 using the mole ratio of Cl_2 to Na obtained from the balanced equation:

$$\boxed{\text{Moles Na}} \xrightarrow{\text{mole ratio}} \boxed{\text{Moles } Cl_2}$$

We multiply the known moles of Na by the ratio that will cancel the old units (mol Na) and introduce the new units (mol Cl_2):

$$\text{Moles of } Cl_2 = 0.400 \text{ mol Na} \times \frac{1 \text{ mol } Cl_2}{2 \text{ mol Na}} = 0.200 \text{ mol } Cl_2$$

Once we know the moles of Cl_2, all we have left to do is convert from moles to grams using the molar mass of Cl_2, 70.90 g/mol.

$$\boxed{\text{Moles } Cl_2} \xrightarrow{MM \ Cl_2} \boxed{\text{Grams } Cl_2}$$

Although chemicals are often measured by mass, some substances are more conveniently measured by volume. If the substance is pure and the density is known, then the mass can be determined from the density and volume using the relationship you learned in Chapter 1:

$$\text{Density} = \frac{\text{mass}}{\text{volume}}$$

$$\text{Mass} = \text{density} \times \text{volume}$$

Sometimes reactants are in solution or are in the gaseous state. Calculations involving these types of substances will be discussed in Chapters 9 and 11.

We multiply 0.200 mol Cl_2 by the mole ratio that cancels the old units (moles) and introduces the new units (grams):

$$\text{Mass of } Cl_2 = 0.200 \ \cancel{\text{mol } Cl_2} \times \frac{70.90 \text{ g } Cl_2}{1 \ \cancel{\text{mol } Cl_2}} = 14.2 \text{ g } Cl_2$$

When we convert from moles to grams, we always *multiply* by the molar mass.

Alternatively we can do all three conversions consecutively in one calculation:

$$\boxed{\text{Grams Na}} \xrightarrow{\textit{MM} \text{ Na}} \boxed{\text{Moles Na}} \xrightarrow{\text{mole ratio}} \boxed{\text{Moles } Cl_2} \xrightarrow{\textit{MM} \ Cl_2} \boxed{\text{Grams } Cl_2}$$

$$\text{Mass of } Cl_2 = 9.20 \ \cancel{\text{g Na}} \times \frac{1 \ \cancel{\text{mol Na}}}{22.99 \ \cancel{\text{g Na}}} \times \frac{1 \ \cancel{\text{mol } Cl_2}}{2 \ \cancel{\text{mol Na}}} \times \frac{70.90 \text{ g } Cl_2}{1 \ \cancel{\text{mol } Cl_2}} = 14.2 \text{ g } Cl_2$$

All mass-mass conversion problems involve the same three-step calculation. Writing the conversion calculation with all steps combined has advantages. It is easier to ensure that the units cancel properly. Also you are less likely to make mistakes.

This calculation shows that 9.20 g of Na should react with 14.2 g of Cl_2. Now let's determine how much NaCl should form. Since we know the masses of all but one substance in the reaction, NaCl, we can use the *law of conservation of mass* to determine the mass of NaCl. The mass of the reactants must equal the mass of the products (Figure 6.6):

$$\text{Mass of reactants} = \text{mass of products}$$

$$\text{Mass Na} + \text{mass } Cl_2 = \text{mass NaCl}$$

$$9.20 \text{ g Na} + 14.2 \text{ g } Cl_2 = 23.4 \text{ g NaCl}$$

In this example, we could calculate the mass of NaCl in this way because we knew the mass of all the other substances in the reaction. Usually this is not the case, and we have to use the mass-mass conversion procedure to determine the mass of a reactant or product from a known mass of another reactant or product. Try calculating the mass of NaCl from the mass of either Na (9.20 g) or Cl_2 (14.2 g) using mass-mass conversions. You should get the same answer (23.4 g NaCl) as you would by summing the masses of the reactants. Example 6.2 gives you practice converting from the mass of one reactant or product to the mass of another.

> The way you round numbers will affect your answer. You'll get one result if you round after each step, but another if you hold the answers to each step in your calculator. Don't be alarmed if your answers are *slightly* different from those given in this textbook.

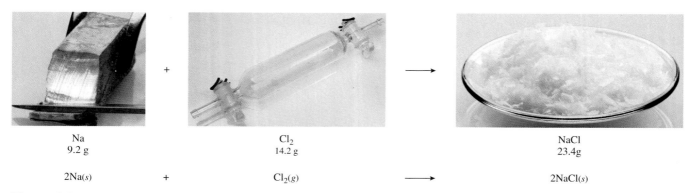

Na	Cl_2	NaCl
9.2 g	14.2 g	23.4g

$$2Na(s) \quad + \quad Cl_2(g) \quad \longrightarrow \quad 2NaCl(s)$$

Figure 6.6
The law of conservation of mass states that the mass of the reactants must equal the mass of the products. Mass is not gained or lost.

EXAMPLE 6.2   Mass-Mass Conversions

The decomposition of nitroglycerin is an explosive reaction that produces four different gases. What mass of water vapor should be produced when 1.0 g of nitroglycerin decomposes by the following reaction?

$$4C_3H_5O_9N_3(l) \longrightarrow 12CO_2(g) + 6N_2(g) + O_2(g) + 10H_2O(g)$$

Solution:

Because we are given grams of nitroglycerin ($C_3H_5O_9N_3$) we must first convert to moles using its molar mass, 227.10 g/mol:

$$\boxed{\text{Grams } C_3H_5O_9N_3} \xrightarrow{MM \; C_3H_5O_9N_3} \boxed{\text{Moles } C_3H_5O_9N_3}$$

$$\text{Moles of } C_3H_5O_9N_3 = 1.0 \; \text{g } C_3H_5O_9N_3 \times \frac{1 \; \text{mol } C_3H_5O_9N_3}{227.10 \; \text{g } C_3H_5O_9N_3}$$

$$= 0.0044 \; \text{mol } C_3H_5O_9N_3$$

Next we use the mole ratio of H_2O to $C_3H_5O_9N_3$ to convert to moles of H_2O:

$$\boxed{\text{Moles } C_3H_5O_9N_3} \xrightarrow{\text{mole ratio}} \boxed{\text{Moles } H_2O}$$

$$\text{Moles of } H_2O = 0.0044 \; \text{mol } C_3H_5O_9N_3 \times \frac{10 \; \text{mol } H_2O}{4 \; \text{mol } C_3H_5O_9N_3}$$

$$= 0.011 \; \text{mol } H_2O$$

Now that we know the moles of H_2O, all we have left to do is convert moles to grams by multiplying by the molar mass of H_2O. Note that the units cancel properly.

$$\boxed{\text{Moles } H_2O} \xrightarrow{MM \; H_2O} \boxed{\text{Grams } H_2O}$$

$$\text{Mass of } H_2O = 0.011 \; \text{mol } H_2O \times \frac{18.02 \; \text{g } H_2O}{1 \; \text{mol } H_2O} = 0.20 \; \text{g } H_2O$$

Alternatively we can do all three conversions consecutively in one calculation:

$$\text{Mass of } H_2O = 1.0 \; \text{g } C_3H_5O_9N_3 \times \frac{1 \; \text{mol } C_3H_5O_9N_3}{227.10 \; \text{g } C_3H_5O_9N_3} \times$$

$$\frac{10 \; \text{mol } H_2O}{4 \; \text{mol } C_3H_5O_9N_3} \times \frac{18.02 \; \text{g } H_2O}{1 \; \text{mol } H_2O} = 0.20 \; \text{g } H_2O$$

When 1.0 g of nitroglycerin decomposes, 0.20 g of water vapor is produced.

Practice Problem 6.2

Hydrogen and oxygen are used as fuel for liftoff during a space-shuttle launch. The hydrogen and oxygen are stored within the shuttle as compressed liquids.

During an average liftoff, 180,000 kg of hydrogen burns as it reacts with oxygen to form water vapor. The balanced equation for this reaction is

$$2H_2(g) + O_2(g) \longrightarrow 2H_2O(g)$$

(a) What mass of oxygen gas is consumed?
(b) What mass of water vapor is produced?
(c) Does the mass of water vapor equal the sum of the masses of the reactants? What law is followed here?

Further Practice: 6.21 and 6.22

6.4   LIMITING REACTANTS

Animation: Limiting Reagent
Online Learning Center, Animations

Chemistry Animations Library, Limiting Reagent, Limiting_Reagent.exe.

When Lily burns ethanol in her car, she relies on oxygen in the air to support the combustion. She doesn't worry about running out of oxygen. A chemist would say that oxygen is in excess for the reaction.

When the space shuttle is in space, however, there is no air to provide oxygen. The shuttle must take the right amounts of hydrogen and oxygen with it. If too much of one reactant were transported, the extra reactant would have nothing to react with. It would simply add extra mass to the shuttle. Mass-mass calculations are important in such situations.

When a space shuttle takes off, it leaves with measured amounts of hydrogen and oxygen so that no extra mass is added. Many reactions, however, occur with an excess of one reactant. Consider a combination reaction between sodium metal and chlorine gas:

$$2Na(s) + Cl_2(g) \longrightarrow 2NaCl(s)$$

How do we know if all of each reactant is used up in the reaction? If one of the reactants does not react completely, there will be some of it left over at the end of the reaction. The leftover amount will be present mixed with the product as shown in Figure 6.7. In which of the pictures of Figure 6.7 is sodium left over? In which is chlorine left over?

A   B

Figure 6.7
When reactants are not mixed in relative amounts as described by the balanced equation, one reactant does not react completely. Here sodium metal reacts with chlorine gas to form sodium chloride, the salt on the wall of the flasks. When the reaction is complete, one reactant remains in excess. What is the excess reactant in flask A? in flask B?

When there is excess sodium, we see shiny sodium metal mixed in with the white salt crystals. When there is excess chlorine, we see greenish-yellow gas remaining with the product. If we do not want any unreacted sodium or chlorine left over, we should mix the reactants in the proper quantities so that their mole ratio is the same as that shown by the coefficients in the balanced equation:

$$\frac{2 \text{ mol Na}}{1 \text{ mol Cl}_2}$$

If the reactants mixed together are *not* present in the exact mole ratio, then we will have too much of one reactant and not enough of another. One reactant reacts completely, and some of the other is left unreacted. The reactant that reacts completely is called the **limiting reactant;** it limits the amount of the other reactant that can react, and it limits the amount of product that can form. A reactant that does not react completely is *in excess.*

EXAMPLE 6.3　　　**Identifying the Limiting Reactant**

Iron metal is added to an aqueous solution of copper(II) sulfate. The solution is blue because copper(II) ions in solution produce a blue color. When the reaction is complete, the solution is colorless. The piece of iron is coated with a granular, brownish-black metal. Identify the limiting reactant and the reactant that is in excess.

Solution:

The loss of all color from the solution indicates that the copper(II) sulfate reacted completely. It is the limiting reactant. The iron metal is in excess because some remains unreacted.

Practice Problem 6.3

In Example 6.3 you identified the limiting reactant. Describe what you would observe if the other reactant was the limiting reactant.

Further Practice: 6.29 and 6.30

To gain a better understanding of limiting reactants, let's look at a system that is not a chemical reaction. Suppose Joel wants to build a model solar-powered

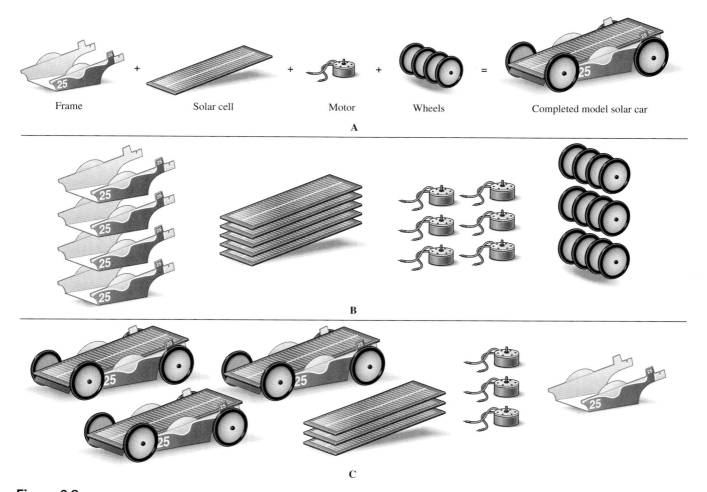

Figure 6.8
To make a model solar car, a specific number of each part is required. To make more than one car, the relative number of parts remains the same. If there is a shortage of a particular part, that part limits the number of solar model cars that can be built. It is the limiting part.

car. He needs 1 frame, 1 solar cell, 1 electric motor, and 4 wheels (Figure 6.8A). We can write an equation to represent the numbers of parts required to make the car:

1 frame + 1 solar cell + 1 electric motor + 4 wheels = 1 model solar car

To make one or more solar cars, the ratio of cells to motors to frames to wheels is 1:1:1:4. If the numbers of parts are present in the same ratio, then all the parts can be used to make cars. For example, if Joel has 10 cells, 10 electric motors, 10 frames, and 40 wheels, he can make 10 cars with no parts left over. If the number of parts is different from this ratio, then the number of cars will be limited by at least one of the parts, and Joel will have leftover parts. For example, suppose he has 5 solar cells, 6 motors, 4 frames, and 12 wheels as shown in Figure 6.8B. How many cars can he make? He has enough frames to make 4 cars. He has more than enough cells and motors, but does he have enough wheels? No, he would need 16 wheels to go with the 4 frames, and he has only 12. So the number of wheels limits the number of cars to 3. The wheels are the *limiting part*. He has too many of the other parts, so they are in excess. Notice that the terms we are using are similar to those we use in discussing a chemical reaction. The difference is that we are describing parts instead of reactants.

EXAMPLE 6.4   Limiting Parts

Suppose Joel received a shipment of 88 wheels, so he now has a total of 100. He still has 5 solar cells, 6 motors, and 4 frames.

(a) Use the parts ratio described in Figure 6.8A to determine the limiting part(s).
(b) How many cars can Joel make?
(c) Identify which parts are in excess and how many of each will be left over.

Solution:

(a) The large shipment of wheels provides excess wheels. They are not the limiting part. The number of cells, motors, and frames needed per car is the same (1:1:1). Since Joel has only 4 frames, and more of each of the other parts, the frame is the limiting part.
(b) The number of limiting parts (frames) determines the number of cars that can be built. Since each car requires 1 frame, a maximum of 4 cars can be built.
(c) Joel has 1 excess cell, 2 excess motors, and 84 excess wheels. They will be left over after 4 complete cars are constructed.

Practice Problem 6.4

Suppose Joel got a shipment of 10 additional frames.

(a) Use the parts ratio described in Figure 6.8A to determine the limiting part(s).
(b) How many cars can Joel make?
(c) Identify which parts are in excess and how many of each will be left over.

Further Practice: 6.31 and 6.32

Now let's look at what occurs at the molecular level in a reaction that has a limiting reactant. Consider the combustion of hydrogen to form water. The molecular-level diagram and balanced equation in Figure 6.9 show that the molecule ratio of H_2 to O_2 is $\dfrac{2\ H_2\ \text{molecules}}{1\ O_2\ \text{molecule}}$. The *molecule ratio* is the same as the *mole ratio* because both are related to the coefficients in the balanced equation.

In this chemical reaction, the reactant molecules are the "parts" that combine in a specific ratio to make the product H_2O molecules. If the number of H_2 molecules mixed is exactly two times the number of O_2 molecules, then all the H_2 and O_2 molecules will react, and no reactants will be left over. If the ratio of H_2 to O_2 molecules mixed together is *not* exactly $\dfrac{2\ H_2\ \text{molecules}}{1\ O_2\ \text{molecule}}$, then one reactant is the limiting

Figure 6.9
For every 2 H_2 molecules that react, 1 O_2 molecule reacts with it to form 2 H_2O molecules.

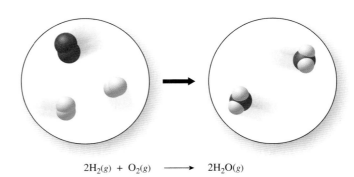

$$2H_2(g) + O_2(g) \longrightarrow 2H_2O(g)$$

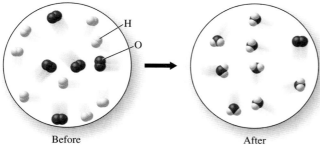

Before   After

$$2H_2(g) + O_2(g) \longrightarrow 2H_2O(g)$$

Figure 6.10
When 8 H_2 molecules and 5 O_2 molecules are mixed, all of the H_2 reacts, and 1 O_2 molecule is left over. The limiting reactant is H_2, and O_2 is in excess.

reactant and the other is in excess. For example, suppose 8 molecules of H_2 are mixed with 5 molecules of O_2 as shown in Figure 6.10. The reactant ratio is not $\dfrac{2\ H_2\ \text{molecules}}{1\ O_2\ \text{molecule}}$, so only one of the reactants is completely consumed. The other is in excess. After the reactants combine in this 2 to 1 ratio, 8 H_2O molecules form. The H_2 reactant is the limiting reactant because it was completely used up. It limited the number of H_2O molecules that could form. The O_2 reactant is left over. It was in excess. When the reaction is complete, the 1 leftover O_2 molecule that did not react will also be present in the reaction mixture along with the 8 H_2O molecules that formed.

Suppose there was no "after" picture in Figure 6.10. Could you still determine the limiting reactant? Yes, it could be done in a way similar to the method used to determine the limiting part for building model cars. The molecule ratio in the balanced equation shows the number of molecules of one reactant that is required to react with the given number of molecules of the other reactant. Let's arbitrarily start by determining the number of H_2 molecules needed to react with the 5 O_2 molecules given in Figure 6.10. We multiply the given number of O_2 molecules by the molecule ratio to get the number of H_2 molecules:

$$\text{Molecules of } H_2 = 5\ \cancel{O_2\ \text{molecules}} \times \frac{2\ H_2\ \text{molecules}}{1\ \cancel{O_2\ \text{molecule}}} = 10\ H_2\ \text{molecules}$$

This calculation tells us that we need 10 H_2 molecules to react with the 5 O_2 molecules. Since we actually have only 8 H_2 molecules, we know that

$$\text{Calculated } H_2 > \text{actual } H_2$$

Since there is not enough H_2 to react with all the O_2, H_2 is the limiting reactant. We come to the same conclusion if we calculate the number of O_2 molecules required to react with the 8 H_2 molecules:

$$\text{Molecules of } O_2 = 8\ \cancel{H_2\ \text{molecules}} \times \frac{1\ O_2\ \text{molecule}}{2\ \cancel{H_2\ \text{molecules}}} = 4\ O_2\ \text{molecules}$$

This calculation tells us that we need 4 O_2 molecules to react with the 8 H_2 molecules. We actually have 5 O_2 molecules:

$$\text{Calculated } O_2 < \text{actual } O_2$$

We have 1 extra O_2 molecule, so O_2 is in excess. The H_2 must be the limiting reactant. All 8 H_2 molecules will react with 4 of the O_2 molecules, and there will be 1 leftover O_2 molecule. Does this match the "after" picture in Figure 6.10? It should.

It makes no difference which reactant you start with to determine the limiting reactant. Your conclusion will be the same. The following table shows what is mixed, what reacts, and the final composition of the mixture.

	$2H_2(g)$	+	$O_2(g)$	$\longrightarrow$	$2H_2O(g)$
Initially mixed	8 molecules		5 molecules		0 molecules
How much reacts	8 molecules		4 molecules		
Composition of final mixture	0 molecules		1 molecule		8 molecules

The following summarizes the method for determining the limiting reactant.

If you calculate the amount of product predicted from each given amount of reactant, the calculation that gives the *least* amount of product identifies the limiting reactant.

Steps for Determining the Limiting Reactant

1. Calculate the amount of one reactant (B) needed to react with the other reactant (A).
2. Compare the calculated amount of B (amount needed) to the actual amount of B that is present initially.
   (a) If calculated B = actual B, there is no limiting reactant. Both A and B will react completely.
   (b) If calculated B > actual B, B is the limiting reactant. Only B will react completely.
   (c) If calculated B < actual B, A is the limiting reactant. Only A will react completely.

Example 6.5 will give you practice identifying limiting reactants in a chemical reaction.

EXAMPLE 6.5   Limiting Reactants—Molecular Level

Consider the combination reaction of nitrogen gas with hydrogen gas to form ammonia gas:

$$N_2(g) + 3H_2(g) \longrightarrow 2NH_3(g)$$

The molecular diagram shows the number of N_2 and H_2 molecules mixed together before the reaction occurs.

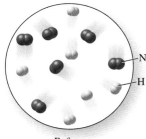

Before

(a) Identify the limiting reactant.
(b) Draw the "after" picture. Include the correct number of product molecules and any leftover reactant molecules.
(c) Determine the reactant that is in excess.

Solution:

(a) The limiting reactant can be identified by determining the number of molecules of one reactant needed to react with the given number of molecules of the other. Here we will calculate the number of N_2 molecules needed to react with the given number of H_2 molecules:

$$\text{Molecules of } N_2 = 6 \ H_2 \ \text{molecules} \times \frac{1 \ N_2 \ \text{molecule}}{3 \ H_2 \ \text{molecules}} = 2 \ N_2 \ \text{molecules}$$

We need 2 N_2 molecules and we actually have 6.

$$\text{Calculated } N_2 < \text{actual } N_2$$

We have 4 extra N_2 molecules, so N_2 is in excess. The H_2 must be the limiting reactant.

(b) Your "after" picture should have 4 NH_3 molecules and 4 N_2 molecules.

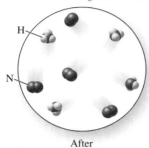

After

(c) Because some N_2 molecules are left over, there was too much to react, so it is in excess. The following table summarizes this reaction.

	$N_2(g)$	+	$3H_2(g)$	$\longrightarrow$	$2NH_3(g)$
Initially mixed	6 molecules		6 molecules		0 molecules
How much reacts	2 molecules		6 molecules		
Composition of final mixture	4 molecules		0 molecules		4 molecules

Practice Problem 6.5

Consider the combination reaction of nitrogen gas with hydrogen gas to form ammonia gas.

$$N_2(g) + 3H_2(g) \longrightarrow 2NH_3(g)$$

The molecular diagram shows the number of N_2 and H_2 molecules mixed together before the reaction occurs.

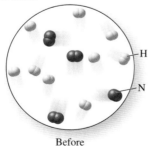

Before

(a) Identify the limiting reactant.

(b) Draw the "after" picture. Include the correct number of product molecules and any leftover reactant molecules.

(c) Determine the reactant that is in excess.

Further Practice: 6.35 and 6.36

Now that we have identified limiting reactants on a molecular scale, let's look at the much larger mole scale. The procedure is similar because the molecule ratio and the mole ratio are the same and are given by the balanced equation. Let's look at the reaction of sodium metal and chlorine gas:

$$2Na(s) + Cl_2(g) \longrightarrow 2NaCl(s)$$

Suppose we mix 0.50 mol Na with 0.20 mol Cl_2. What is the limiting reactant? We'll follow a similar procedure to what we did when we started with the number of molecules of each reactant. Let's begin by finding the moles of Na needed to react with the 0.20 mol Cl_2:

$$\text{Moles of Na} = 0.20 \ \cancel{\text{mol } Cl_2} \times \frac{2 \text{ mol Na}}{1 \ \cancel{\text{mol } Cl_2}} = 0.40 \text{ mol Na}$$

We need 0.40 mol Na, and we actually have 0.50 mol Na:

$$\text{Calculated Na} < \text{actual Na}$$

Na is in excess by 0.10 mol. The Cl_2 must be the limiting reactant. We can expect all the Cl_2 gas to react and some (0.10 mol) of the Na metal to be left over after the reaction.

We can determine how much NaCl should form by doing a mole-mole calculation. Which reactant do we use? We use the limiting reactant Cl_2 because it reacts completely. We multiply the moles of limiting reactant Cl_2 by the mole ratio of NaCl to Cl_2 to determine the moles of NaCl that can form:

$$\text{Moles of NaCl} = 0.20 \ \cancel{\text{mol } Cl_2} \times \frac{2 \text{ mol NaCl}}{1 \ \cancel{\text{mol } Cl_2}} = 0.40 \text{ mol NaCl}$$

The following table summarizes this reaction.

	2Na(s)	**+**	**Cl$_2$(g)**	$\longrightarrow$	**2NaCl(s)**
Initially mixed	0.50 mol		0.20 mol		0.00 mol
How much reacts	0.40 mol		0.20 mol		
Composition of final mixture	0.10 mol		0.00 mol		0.40 mol

When 0.50 mol Na is mixed with 0.20 mol Cl_2, 0.40 mol Na reacts with 0.20 mol Cl_2. When the reaction is complete, 0.40 mol NaCl is produced. Because the Na is in excess by 0.10 mol, that amount, along with the 0.40 mol NaCl that forms, will be present when the reaction is complete.

Example 6.6 will give you practice determining the limiting reactant and the amount of product that can form.

EXAMPLE 6.6   Limiting Reactants—Mole Scale

The ethanol used in Lily's car could be made from the hydrocarbon ethylene (C_2H_4). Ethylene, a gas at room temperature, also burns with oxygen. The balanced equation is

$$C_2H_4(g) + 3O_2(g) \longrightarrow 2CO_2(g) + 2H_2O(g)$$

Suppose we mix 0.25 mol C_2H_4 with 1.0 mol O_2. Identify the limiting reactant and the number of moles of CO_2 that can be formed.

Solution:

To determine the limiting reactant, first determine if there is enough C_2H_4 to react with 1.0 mol O_2:

$$\text{Moles of } C_2H_4 \ = \ 1.0 \ \text{mol } O_2 \ \times \ \frac{1 \ \text{mol } C_2H_4}{3 \ \text{mol } O_2} \ = \ 0.33 \ \text{mol } C_2H_4$$

We need 0.33 mol C_2H_4, and we actually have only 0.25 mol C_2H_4:

$$\text{Calculated } C_2H_4 > \text{actual } C_2H_4$$

Because there is not enough C_2H_4 to react with all the O_2, C_2H_4 is the limiting reactant. (Assure yourself that you would reach the same conclusion if you started the calculation with the given amount of C_2H_4 instead of the given amount of O_2.) We use the amount of limiting reactant to calculate the moles of CO_2 product:

$$\text{Moles of } CO_2 = 0.25 \ \text{mol } C_2H_4 \ \times \ \frac{2 \ \text{mol } CO_2}{1 \ \text{mol } C_2H_4} \ = \ 0.50 \ \text{mol } CO_2$$

We also use the amount of the limiting reactant to determine the amount of O_2 that will react:

$$\text{Moles of } O_2 \text{ that react} = 0.25 \ \text{mol } C_2H_4 \ \times \ \frac{3 \ \text{mol } O_2}{1 \ \text{mol } C_2H_4} \ = \ 0.75 \ \text{mol } O_2$$

This reaction is summarized in the following table.

	$C_2H_4(g)$	+	$3O_2(g)$	$\longrightarrow$	$2CO_2(g)$	+	$2H_2O(g)$
Initially mixed	0.25 mol		1.00 mol		0.00 mol		0.00 mol
How much reacts	0.25 mol		0.75 mol				
Composition of final mixture	0.00 mol		0.25 mol		0.50 mol		0.50 mol

Practice Problem 6.6

In the double-displacement reaction of aqueous silver nitrate and magnesium chloride, silver chloride precipitates and aqueous magnesium nitrate forms:

$$2AgNO_3(aq) + MgCl_2(aq) \longrightarrow 2AgCl(s) + Mg(NO_3)_2(aq)$$

Suppose we mix 1.0 mol of silver nitrate with 1.0 mol of magnesium chloride. Identify the limiting reactant and the number of moles of solid AgCl that can form.

Further Practice: 6.41 and 6.42

Generally we measure amounts of reactants in units of grams. Given grams of reactants, we must first convert from grams to moles for each reactant. Example 6.7 shows how to determine the limiting reactant when grams of reactants are given.

EXAMPLE 6.7

Limiting Reactants When Given the Mass of Reactants

When liquid bromine is mixed with aluminum metal, a combination reaction occurs, forming aluminum bromide:

$$2Al(s) + 3Br_2(l) \longrightarrow 2AlBr_3(s)$$

Determine the limiting reactant and mass of $AlBr_3$ produced when 45.0 g of Br_2 are added to 30.0 g of Al.

Solution:

Because we are given the mass of reactants, we must first convert from grams to moles for each reactant:

$$\text{Moles of Al} = 30.0 \ \cancel{g \ Al} \times \frac{1 \ \text{mol Al}}{26.98 \ \cancel{g \ Al}} = 1.11 \ \text{mol Al}$$

$$\text{Moles of Br}_2 = 45.0 \ \cancel{g \ Br_2} \times \frac{1 \ \text{mol Br}_2}{159.89 \ \cancel{g \ Br_2}} = 0.281 \ \text{mol Br}_2$$

Next we determine the number of moles of one reactant needed to react with the other. Here we will determine the moles of Br_2 needed to react with 1.11 mol Al:

$$\text{Moles of Br}_2 = 1.11 \ \cancel{\text{mol Al}} \times \frac{3 \ \text{mol Br}_2}{2 \ \cancel{\text{mol Al}}} = 1.67 \ \text{mol Br}_2$$

We need 1.67 mol Br_2, but we actually have only 0.281 mol Br_2:

$$\text{Calculated Br}_2 > \text{actual Br}_2$$

Because Br_2 is in short supply, it is the limiting reactant. (Assure yourself that you would come to the same conclusion if you started the calculation with the given amount of Br_2 instead of the given amount of Al.)

You may have been able to see merely by looking at the balanced equation that Br_2 is the limiting reactant. The balanced equation shows that the number of moles of Br_2 must be greater than that of Al, but the number of moles of Br_2 we have is less than that of Al. Also note that although the mass ratio of Br_2 to Al mixed together is (45 g)/(30 g), or 3/2, this is not the mole ratio. When given the mass of reactants, we must always first convert to moles.

To determine the mass of product, we can do a mass-mass conversion starting with the given grams of Br_2. Or we can start with the moles of Br_2 calculated from the grams of Br_2, convert moles Br_2 to moles $AlBr_3$, and then convert moles of $AlBr_3$ to grams of $AlBr_3$:

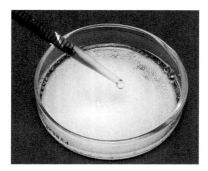

$$\text{Mass of AlBr}_3 \;=\; 0.281 \;\text{mol Br}_2 \;\times\; \frac{2\;\text{mol AlBr}_3}{3\;\text{mol Br}_2}\;\times\;\frac{266.7\;\text{g AlBr}_3}{1\;\text{mol AlBr}_3}$$

$$=\; 50.0\;\text{g AlBr}_3$$

The maximum amount of $AlBr_3$ that can be produced is 50.0 g.

Practice Problem 6.7

Sodium carbonate reacts with hydrochloric acid to produce aqueous sodium chloride, water, and carbon dioxide gas. The balanced equation is

$$Na_2CO_3(s) + 2HCl(aq) \longrightarrow 2NaCl(aq) + H_2O(l) + CO_2(g)$$

Determine the limiting reactant and mass of CO_2 gas produced when 11.0 g of Na_2CO_3 is added to a solution that contains 11.0 g of HCl.

Further Practice: 6.43 and 6.44

6.5 PERCENT YIELD

When the manufacturers of E-85 produce ethanol to power Lily's car, they calculate the amount of ethanol product that *should* form from a reaction. This is the **theoretical yield,** the maximum amount of product that can be obtained from given amounts of reactants. In Example 6.7 for instance, we calculated the theoretical yield of $AlBr_3$ as 50.0 g.

When chemists do experiments and when companies manufacture fuels, they seldom obtain the amount of product predicted by the calculation. They fall short of the theoretical yield for any number of reasons. Reactants and products can be lost due to spillage or during transfer from one container to another (Figure 6.11); and alternative reactions (side reactions) can occur that produce different products. The amount of product we measure in the laboratory is the **actual yield.** The actual yield is usually less than the theoretical yield; however, it can appear greater if a solid is weighed when still wet or if a product is contaminated.

The **percent yield** describes how much of a product is actually formed in comparison to how much should have been formed. It is the actual yield expressed as a percentage of the theoretical yield:

$$\text{Percent yield} \;=\; \frac{\text{actual yield}}{\text{theoretical yield}}\;\times\;100$$

Figure 6.11
The actual yield is usually less than the theoretical yield due to loss during reaction steps, purification, and transfer processes.

The actual yield and theoretical yield can be expressed in any units, as long as those units are the same. Usually yields are reported in mass units, but yields in units of moles or molecules are acceptable.

Suppose the reaction of aluminum metal and liquid bromine described in Example 6.7 produced only 26.8 g $AlBr_3$. What is the percent yield? Since we calculated the amount of $AlBr_3$ that should form as 50.0 g, we know the theoretical yield. The percent yield is the ratio of the actual yield to the theoretical yield multiplied by 100:

$$\text{Percent yield} = \frac{26.8 \text{ g}}{50.0 \text{ g}} \times 100 = 53.6\%$$

Example 6.8 will help you better understand percent yield, theoretical yield, and actual yield.

EXAMPLE 6.8   **Percent Yield, Theoretical Yield, and Actual Yield**

Sodium metal reacts with water in a single-displacement reaction to produce aqueous sodium hydroxide and hydrogen gas. The balanced equation is

$$2Na(s) + 2H_2O(l) \longrightarrow 2NaOH(aq) + H_2(g)$$

When 0.50 mol Na is placed in water, all the sodium metal reacts, and the hydrogen gas produced is isolated. It is determined that 0.21 mol H_2 has been produced. What is the percent yield of H_2?

Solution:

To calculate percent yield, we divide the actual yield by the theoretical yield, and then multiply by 100:

$$\text{Percent yield} = \frac{\text{actual yield}}{\text{theoretical yield}} \times 100$$

The actual yield is the number of moles of H_2 that was isolated once the reaction occurred (0.21 mol). The theoretical yield must be calculated from the amount of limiting reactant that reacted. In this case we can assume the Na is the limiting reactant (not the water) because all the Na reacted. We calculate the theoretical yield by calculating the maximum number of moles of H_2 that should be produced from the moles of limiting reactant Na:

$$\text{Moles of } H_2 = 0.50 \text{ mol Na} \times \frac{1 \text{ mol } H_2}{2 \text{ mol Na}} = 0.25 \text{ mol } H_2 \text{ (theoretical yield)}$$

We substitute the values of actual and theoretical yield into the percent yield equation and solve for percent yield:

$$\text{Percent yield} = \frac{0.21 \text{ mol } H_2}{0.25 \text{ mol } H_2} \times 100 = 84\%$$

Practice Problem 6.8

Toxic carbon monoxide is a by-product of the preparation of hydrogen fuel from methanol for use in hydrogen-powered vehicles. When carbon monoxide reacts with oxygen gas, the greenhouse gas carbon dioxide is produced:

$$2CO(g) + O_2(g) \longrightarrow 2CO_2(g)$$

When 5.0 mol CO is mixed with excess O_2, the reaction occurs to give 3.4 mol CO_2. What is the percent yield of CO_2?

Further Practice: 6.55 and 6.56

If we know the expected percent yield for a reaction, we can predict the amount of product that can realistically be made. Such calculations are important in industry where percent yield is a significant factor in determining profits.

6.6   ENERGY CHANGES

When hydrogen and oxygen react to produce water, energy is released. In a hydrogen fuel cell, most of this energy is converted to electrical energy, which can be used to run a vehicle. Suppose we combine the same amounts of hydrogen and oxygen, but without the fuel cell. Would the same amount of energy be released by the reaction? The answer is yes, but the energy would not be in such a usable form. Some would be lost as *heat* to the surroundings, causing the temperature of the environment to increase. This is what happens when we burn other fuels, such as propane in a gas grill or wood in a bonfire. Whether the energy is converted to useful work or is dissipated into the environment as heat, the amount of energy released is the same. So far in this chapter, we have looked only at the amounts of reactants and products in a chemical reaction. Here we will investigate the amount of energy absorbed or released by a chemical reaction.

Law of Conservation of Energy

As described in Chapter 1, energy can take many forms, and it can be converted from one form to another. Energy can also be transferred as heat. In either case, the amount of energy remains the same. The **law of conservation of energy** expresses this principle: Energy can be converted or transferred, but it cannot be created or destroyed. Thus, when we "use" energy, we are not using it up. We are converting it to a different form of energy. However, we may be converting it into forms that are not useful, such as heat lost to the surrounding environment. When gasoline is burned in an engine, the chemical energy in the gasoline is converted to mechanical energy and heat energy (Figure 6.12). The mechanical energy does useful work, but the heat energy is generally lost to the atmosphere. If gasoline is burned in a large pit, most of the energy would be lost as heat. In order to make the energy possessed by an object most useful to us, we try to minimize the amount of energy lost as heat.

Efficiency is a measure of the amount of useful work that is achieved from an energy conversion. No conversion is totally efficient, as shown in Figure 6.13. The process of conversion always generates some heat that is lost even before useful work can be done. The efficiency of a conversion is the percentage that ends up in the form we want. Consider, for example, the series of conversions from the chemical energy stored in coal to the light energy we see when we turn on a desk lamp (Figure 6.14). When coal is burned in a steam boiler, about 88% of its chemical energy is converted to heat energy, which can be used to heat water and form

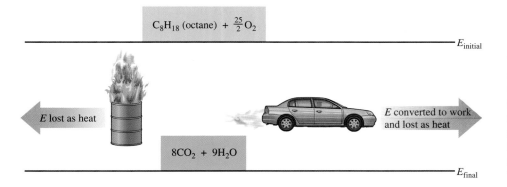

C_8H_{18} (octane) + $\frac{25}{2}O_2$

$E_{initial}$

E lost as heat

E converted to work and lost as heat

$8CO_2 + 9H_2O$

E_{final}

Figure 6.12
When gasoline burns in an ICE, about 32% of the energy released is converted to work and 68% is immediately lost to the surrounding environment as heat. When gasoline is burned in a large pit, 100% of the energy released is immediately lost as heat. The total energy released is the same.

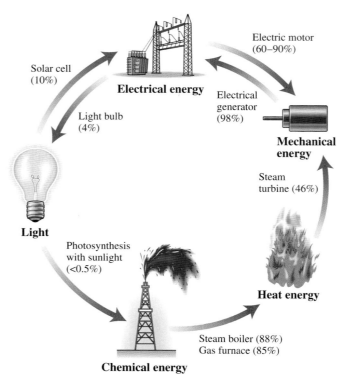

Figure 6.13
The conversion of one kind of energy to another is usually not completely efficient, because some energy is converted into undesired forms. Some typical conversion efficiencies are shown.

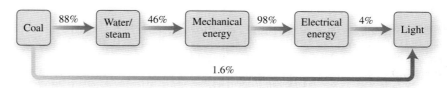

Figure 6.14
From power plant to desk lamp, the overall conversion efficiency is only 1.6%. The other 98.4% of coal's energy is converted to heat energy before ever performing the desired function.

Regular incandescent lightbulbs emit a large amount of energy as heat. Fluorescent lightbulbs generate light by a different process, which does not produce much heat. As a result, fluorescent lightbulbs are four to six times more efficient than incandescent lightbulbs. A 15-W fluorescent lightbulb delivers as much light as a 60-W incandescent lightbulb.

steam. A steam turbine at the power plant can convert about 46% of the heat energy in the steam to mechanical energy, and an electric generator can convert about 98% of the mechanical energy to electrical energy. Finally, when the electrical energy is used to operate a lightbulb, only about 4% is converted to light, while the rest is lost to the surroundings as heat. In the overall process of burning coal to operate a lightbulb, only 1.6% of the chemical energy ends up as light energy. The rest has been lost along the way because of inefficient conversions. Most of the "lost" energy ends up as heat dissipated to the air or into the bodies of water used to cool power plants.

Exothermic Reactions   **Endothermic Reactions**

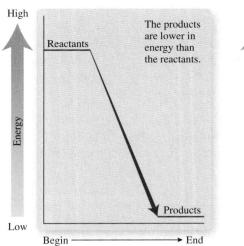

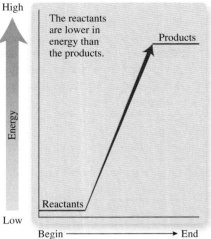

Figure 6.15
An exothermic reaction is thought of as being "downhill" because the products contain less stored energy than the reactants. An endothermic reaction is thought of as being "uphill" because the products contain more potential energy.

Energy Changes That Accompany Chemical Reactions

Lily wonders how the alternative fuel she uses provides the energy that moves her car from one place to another. Where does the energy come from? It comes from chemical energy that is stored as potential energy in the fuel. It is stored in the *bonds* that hold atoms together in molecules. Some substances release some of their potential energy when they react to form substances that store less potential energy. Ethanol, for example, stores a large amount of potential energy in the bonds that hold its atoms together. Some of that energy is released as heat when it reacts with oxygen. A reaction that releases energy is called an **exothermic reaction.**

Not all reactions release energy. Some require an energy input. A reaction that absorbs energy from the surroundings is an **endothermic reaction.** In an endothermic reaction, the products have more potential energy than the reactants. Figure 6.15 shows *reaction profiles,* which describe the overall energy changes that accompany exothermic and endothermic reactions. As shown by the endothermic reaction profile, the energy of the reactants is less than the energy of the products. The difference is the energy absorbed as the reaction takes place. In the exothermic reaction profile, the reactants contain more energy than the products. The difference is the energy released during the reaction. These reaction profiles also show us that if a reaction is exothermic, its reverse reaction is endothermic. And if a reaction is endothermic, its reverse reaction is exothermic.

Physical changes, such as phase changes, can also be endothermic or exothermic. Sweating is a natural process that helps cool our bodies when the water in sweat evaporates. Evaporation is an endothermic process, so it absorbs heat from our skin. Our skin cools, helping to maintain normal body temperature.

EXAMPLE 6.9   Endothermic and Exothermic Reactions

The following is the reaction used in a hydrogen fuel cell:

$$2H_2(g) + O_2(g) \longrightarrow 2H_2O(g)$$

(a) Is this reaction endothermic or exothermic?
(b) The reverse of this reaction, the formation of hydrogen and oxygen from water, can be accomplished by running electricity through water. Is this reaction endothermic or exothermic? Explain.

Solution:

(a) The reaction releases the energy that is converted to useful energy by the fuel cell. Since energy is released by the reaction, it is exothermic.

(b) The reverse of this reaction requires a continuous energy input (electricity) to convert from reactants to products, so it is an endothermic reaction. If a reaction is exothermic, its reverse reaction is endothermic.

Practice Problem 6.9

When a car battery does not work, it can usually be recharged so that it is useful again. Is recharging a battery an endothermic or exothermic process? Explain.

Further Practice: 6.67 and 6.68

Quantities of Heat

For a chemical reaction, the energy change that is easiest to measure is the *heat* absorbed or released. In order to understand how we measure heat changes, we must first learn more about heat and how it is related to temperature changes. **Heat** is energy that is transferred between two objects because of a difference in their temperatures. First think about the heat associated with some physical changes. If we place warm pancakes on a plate, for example, heat is transferred from the pancakes to the plate. The pancakes lose heat and get cooler, while the plate gains heat and gets hotter until the temperature of the pancakes and the temperature of the plate are the same. Heat is always transferred from the hotter to the colder object. When two objects that are in contact become equal in temperature, they have reached a state of *thermal equilibrium.*

In chemistry, quantities of energy and heat are usually given in units of joules (J) or calories (cal). A calorie equals 4.184 J:

$$1 \text{ cal} = 4.184 \text{ J}$$

The Calorie (Cal), used by nutritionists is actually a kilocalorie (kcal), or 1000 calories. For a review of these energy units and conversions between them, see Chapter 1 and Math Toolbox 1.3.

Specific Heat   When heat is added to a substance, the temperature of that substance increases. The amount of heat that must be added to 1 g of a substance to raise its temperature by 1°C is a unique property of that substance, called its **specific heat.** Specific heat values are expressed in units of joules per gram per degree Celsius [J/(g °C)] or calories per gram per degree Celsius [cal/(g °C)]. The specific heat of liquid water is 1.000 cal/(g °C) or 4.184 J/(g °C). This means that if we have a 1.00-g sample of water and we want to increase its temperature by 1.00°C, we must add 1.00 cal or 4.18 J of heat. Similarly, if we have a 2.00-g sample of water and want to increase its temperature by 1.00°C, we must add 2.00 cal or 8.37 J of heat. The more of that substance we have, the more heat we need to add. How much energy will increase the temperature of a 1.00-g sample by 10.0°C? You need 10.0 times as much heat for a 10.0° increase than for a 1.00° increase; you need 10.0 cal or 41.8 J. Some representative values of specific heat for various substances are presented in Table 6.2.

The amount of heat absorbed by a substance when it increases in temperature equals the amount of heat released when it cools back down to the same temperature. The amount of heat transferred to or from a substance is related to the mass, the specific heat, and the change, or difference, in temperature.

When you eat pizza fresh from the oven, why do you burn the top of your mouth and not your tongue? One factor is that the specific heat of cheese is greater than that of the crust. The cheese transfers a greater amount of heat for a given temperature change. Another factor is the greater ability of the cheese to conduct heat. It transfers heat faster.

TABLE 6.2	Specific Heat		Specific Heats of Some Substances	Specific Heat	
Substance	J/(g °C)	cal/(g °C)	Substance	J/(g °C)	cal/(g °C)
Aluminum (s)	0.895	0.214	Water (s)	2.027	0.484
Carbon (diamond)	0.508	0.121	Water (l)	4.184	1.000
Carbon (graphite)	0.708	0.169	Water (g)	2.015	0.482
Calcium (s)	0.656	0.157	Asphalt	0.920	0.220
Chromium (s)	0.450	0.108	Bone	0.440	0.105
Copper (s)	0.377	0.0900	Brick	0.84	0.20
Gold (s)	0.129	0.0310	Cheddar cheese	2.60	0.621
Iodine (s)	0.214	0.0510	Concrete	0.88	0.21
Iron (s)	0.448	0.107	Glass	0.84	0.20
Lead (s)	0.129	0.0310	Granite	0.79	0.19
Mercury (l)	0.140	0.0335	Marble	0.86	0.21
Silver (s)	0.234	0.0560	Olive oil	1.79	0.428
Tin (s)	0.222	0.0530	Sand	0.835	0.200
Uranium (s)	0.117	0.0280	Strawberries	3.89	0.930
			Wax	2.89	0.69

EXAMPLE 6.10   Factors Affecting Heat

Samples of different metals having the same mass are heated to the same temperature and are then allowed to cool in wax. When all have cooled to the same temperature, they have melted through different amounts of the wax. Which of the metals has the greatest specific heat?

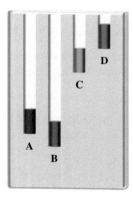

Solution:

From the figure, we see that metal B melted the wax to the lowest level. It did so because it released the greatest amount of heat. Consider the factors that determine the amount of heat transferred. Greater heat could arise from a greater mass, greater specific heat, or greater temperature change. We started with samples of the same mass, and they all underwent the same temperature change, so these are not factors. Therefore the specific heat must be greater for metal B than for the other metals.

Practice Problem 6.10

A drop of boiling water spilled on your hand stings, but a cup of boiling water spilled on your hand burns. What factor explains the difference?

Further Practice: 6.73 and 6.74

If we know the specific heat of an object, its mass, and its change in temperature, we can determine the amount of heat absorbed or released. We can calculate the heat (q) associated with a temperature change by multiplying the mass in grams m of a substance by its specific heat C and by the temperature change ΔT (which is the difference between the final temperature and the initial temperature, $T_f - T_i$):

$$q \;=\; m \;\times\; C \;\times\; \Delta T$$

Heat Mass Specific heat Temperature change

When the temperature of a substance increases (ΔT is positive), q has a positive value; heat is absorbed by the substance. When the temperature decreases (ΔT is negative), q is negative; heat is released by the substance.

We can't measure heat directly. There is no heat-measuring device. Instead, we measure the initial and final temperature of an object and calculate heat from the specific heat, mass, and temperature change. Example 6.11 demonstrates this.

EXAMPLE 6.11   Heat and Temperature Change

A simple solar energy heater is a glass-topped box containing rocks. Sunlight heats the rocks during the day, and air is blown over them at night to heat a house. The box contains 7.5×10^4 g of rocks, which have a specific heat of 0.49 J/(g °C). How much additional heat can the rocks store when they warm from a nighttime temperature of 18°C (65°F) to a daytime temperature of 43°C (110°F)?

Solution:

The temperature of the rocks increases, so they absorb heat, and we expect the calculated value of q to be positive. To calculate heat we use the following equation:

$$q = m \times C \times \Delta T$$

The mass m and specific heat C values are given. The temperature change ΔT is not given, but we can calculate ΔT from the final and initial temperatures:

$$\Delta T = T_f - T_i = 43°C - 18°C = 25°C$$

Substituting values of mass, specific heat, and temperature change into the equation gives the heat change for the rocks:

$$
\begin{aligned}
q &= m \;\times\; C \;\times\; \Delta T \\[4pt]
&= 7.5 \times 10^4 \;\cancel{g} \times 0.49 \;\frac{J}{\cancel{g}\,\cancel{°C}} \times 25\,\cancel{°C} \\[4pt]
&= 7.5 \times 10^4 \times 0.49 \;J \times 25 \\[4pt]
&= 9.2 \times 10^5 \;J \quad \text{or} \quad 9.2 \times 10^2 \;kJ
\end{aligned}
$$

Note that q is positive as we predicted, which means that the rocks have absorbed heat.

Practice Problem 6.11

The specific heat of aluminum is 0.895 J/(g °C). If 156 g of aluminum at 75.0°C is cooled to 25.5°C, how much heat is transferred? What is the sign of q, and what is its significance?

Further Practice: 6.77 and 6.78

We have calculated heat from known values of mass, specific heat, and temperature change. Since the heat equation uses just these four quantities, any one can be determined if the other three are known.

In the examples we have looked at so far, we have focused on the amount of heat absorbed or released by an object such as a rock or a metal. Sometimes, however, we are interested not in an object, but in a process such as a chemical reaction. We call whatever we are interested in—whether an object or a process—the *system.* When a system absorbs heat, where does the heat come from? When a system releases heat, where does the heat go? We know from the law of conservation of energy that energy is not created or destroyed. Heat absorbed by a system comes from its *surroundings,* which might be the air around it—or if in solution—the water around it. Heat released by a system is transferred to its surroundings.

Consider a car engine that heats up due to the combustion reaction in the cylinder. If we describe the system as the combustion reaction in the cylinder, is the value of q for the system positive or negative? We can't take the temperature of the combustion reaction, but we can tell that the surroundings of the combustion reaction (mostly the engine) increases in temperature. The car's engine must be absorbing heat. From the law of conservation of energy, we know that the heat is coming from the combustion reaction. The combustion reaction is releasing heat, so its value of q is negative. Figure 6.16 shows the relationship between a system, its surroundings, and heat transfer.

For some substances such as water, it's easy to calculate the amount of heat absorbed or released. We know the specific heat, and mass is easily measured. The initial and final temperatures of the water are easily measured, too. But the calculation is more difficult for an object such as a metal pipe. We might not know its specific heat, and we can't directly measure the temperature change of the pipe. Instead, we must design a heat transfer process that will let us determine the amount of heat transferred from the pipe to its surroundings, such as a known amount of water. In this case, we call the piece of pipe the system because it's what we are interested in. If we ensure that heat does not escape to anything but the water, then the water is the surroundings. There is no net loss of heat, so the heat change of the system plus the heat change of the surroundings equals zero:

$$q_{system} + q_{surroundings} = 0$$

In this example

$$q_{pipe} + q_{water} = 0$$

Suppose a 92.0-g piece of pipe is heated and then placed in an insulated container filled with 100.0 g of water at 25.00°C (Figure 6.17). The final temperature of the mixture is 29.45°C. If we assume no heat is lost from the water to the container, we

> A *system* can be an object such as a piece of pipe, or a process, such as a physical or a chemical change. The *surroundings* is everything immediately around the system. The surroundings is commonly the air or the liquid around the system.

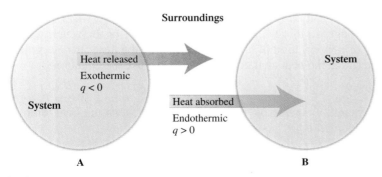

Figure 6.16
(A) When the system releases heat to the surroundings, q is a negative value ($q < 0$).
(B) When the system absorbs heat from the surroundings, q is a positive value ($q > 0$).

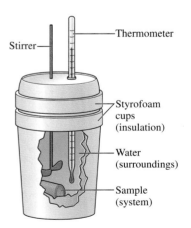

Figure 6.17
The amount of heat released by an object can be determined by measuring the amount of heat absorbed by the water. This experimental process is called *calorimetry.* The insulated container is called a *calorimeter.*

can calculate the amount of heat the pipe loses to the water. It is the same as the amount of heat the water gains, which equals the mass of the water multiplied by the specific heat of water and its temperature change:

$$
\begin{aligned}
q_{water} &= m \times C \times \Delta T \\
&= 100.0\,\cancel{g} \times 4.184\,\frac{J}{\cancel{g}\,\cancel{°C}} \times (29.45\,\cancel{°C} - 25.00\,\cancel{°C}) \\
&= 1860\,J
\end{aligned}
$$

From the law of conservation of energy, the heat change of the pipe plus the heat change of the water equals zero:

$$q_{pipe} + q_{water} = 0$$

The heat change for the pipe then is the same value, but with the opposite sign:

$$q_{pipe} = -\,q_{water} = -1860\,J$$

The process just described is commonly used to determine the heat change for an object for which we do not know the specific heat, or if the initial temperature of the object is difficult to measure. The insulated container is commonly called a *calorimeter*. The measurement of heat transfer using a calorimeter is called *calorimetry*.

EXAMPLE 6.12   Calorimetry

A brick is placed in an insulated calorimeter containing 5.000 kg of water. The temperature of the water decreases from 25.0°C to 19.4°C when thermal equilibrium is reached.

(a) Was the initial temperature of the brick greater than or less than the initial temperature of the water? Explain.
(b) What is the heat change q of the brick?

Solution:

(a) The temperature of the water decreased, so the temperature of the brick must have increased. Since they both have the same temperature at thermal equilibrium, the brick must have started out at a lower temperature than the initial temperature of the water.

(b) We don't know the mass or temperature of the brick, so we can't determine the heat change q for the brick directly. However we do know the mass, specific heat, and temperature change of the water. The water is the brick's surroundings, so if we determine the heat change of the water, we can use the law of conservation of energy to determine the heat change of the brick. To determine the heat change of the water we use the equation

$$q_{water} = m \times C \times \Delta T$$

The temperature change ΔT is the difference between the final temperature and the initial temperature:

$$
\begin{aligned}
\Delta T &= T_f - T_i \\
&= 19.4°C - 25.0°C \\
&= -5.6°C
\end{aligned}
$$

The mass of the water must be in units of grams, so we must convert kilograms to grams:

$$5.000 \ \cancel{kg} \times \frac{1000 \ g}{1 \ \cancel{kg}} = 5.000 \times 10^3 \ g$$

Substituting values for the mass of the water, specific heat of the water, and the temperature change of the water, we get

$$q_{water} = 5.000 \times 10^3 \ \cancel{g} \times 4.184 \ \frac{J}{\cancel{g} \ \cancel{°C}} \times (-5.6 \ \cancel{°C})$$

$$q_{water} = -1.2 \times 10^5 \ J$$

The negative value of q_{water} means that the water released heat. From the law of conservation of energy, we know that the brick absorbed the same amount of heat that the water released. The heat change of the brick plus the heat change of the water must sum to zero. This means that the heat change of the water is equal and opposite in sign to the heat change of the brick:

$$q_{brick} + q_{water} = 0$$

$$q_{brick} = -q_{water} = +1.2 \times 10^5 \ J$$

Practice Problem 6.12

A sample of a metal alloy is heated and then placed in 125.0 g of water held in a calorimeter at 22.5°C. The final temperature of the water is 29.0°C. Assume heat exchange only occurs between the water and the alloy.

(a) Was the initial temperature of the alloy greater than or less than the initial temperature of the water? Explain.

(b) What is the heat change of the alloy?

Further Practice: 6.81 and 6.82

Quantities of Heat Changes That Accompany Chemical Reactions

Lily wonders how the amount of heat released by combustion of a fuel such as ethanol is determined. A common way is to use a special type of calorimeter called a bomb calorimeter (Figure 6.18). It's similar to the calorimeter used for objects like pipes and bricks, but the reaction is kept separate from the water in a compartment that contains an ignition wire. A specific amount of the reactant is placed in the compartment. A small amount of electricity sent through the ignition wire gets the reaction going. Since all combustion reactions are exothermic, the reaction releases heat that the water absorbs.

By measuring the temperature change of the water, we can calculate the heat change of the water, q_{water}, using its specific heat. Assuming no heat is absorbed by the calorimeter, we can determine the heat change for the reaction, $q_{reaction}$, using the law of conservation of energy:

$$q_{reaction} + q_{water} = 0$$

$$q_{reaction} = -q_{water}$$

For example, if a 5.0-g combustible substance burns in a calorimeter, and the surrounding water absorbs 48 kJ of heat ($q_{water} = +48$ kJ), then the combustion reaction must have released 48 kJ of heat:

$$q_{reaction} = -q_{water} = -48 \ kJ$$

A bomb calorimeter can be used to determine the Calorie (energy) content of foods. Foods are fuels. They release their energy slowly when oxidized in the body or rapidly when burned in a calorimeter. Either way, the amount of energy is the same.

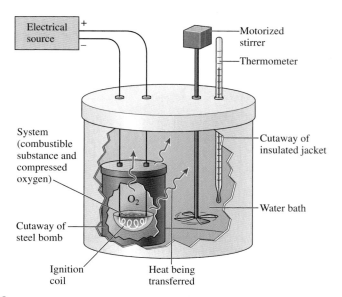

Figure 6.18

A bomb calorimeter is commonly used to measure the amount of heat released in combustion reactions. The water absorbs the heat released by the reaction.

The amount of energy released depends on the amount of substance burned. A few drops of a substance will produce much less heat than a gallon of that substance. In combustion reactions, it's common to report the heat change in terms of the amount of the substance that reacts with oxygen. One common way to report heat changes is in units of *kilojoules per gram* (kJ/g). In the previous example, 5.0 g of compound burned. The heat change *per gram* is one-fifth of the heat change when 5.0 g burns:

$$\frac{-48 \text{ kJ}}{5.0 \text{ g}} = -9.6 \text{ kJ/g}$$

Another way to report heat changes in chemical reactions is in units of *kilojoules per mole* (kJ/mol). Consider the combustion of propane gas (C_3H_8):

$$C_3H_8(g) + 5O_2(g) \longrightarrow 3CO_2(g) + 4H_2O(g)$$

When 1 mol of propane burns, 2220 kJ of heat is released. The heat change for the combustion of propane, then, is –2220 kJ per mol, or

$$q_{reaction} = -2220 \text{ kJ/mol}$$

This ratio allows us to calculate the heat change when any molar amount of propane reacts. For example, if 2.00 mol of propane reacts, twice as much heat is released. This can be calculated by multiplying the 2.00 mol propane by the heat change per mole:

$$\text{Heat change} = 2.00 \text{ mol } C_3H_8 \times \frac{-2220 \text{ kJ}}{1 \text{ mol } C_3H_8} = -4440 \text{ kJ}$$

EXAMPLE 6.13   Heat Changes in Chemical Reactions

The ethanol (CH_3CH_2OH) in Lily's car burns with excess oxygen according to the following equation:

$$CH_3CH_2OH(l) + 3O_2(g) \longrightarrow 2CO_2(g) + 3H_2O(l)$$

The heat change q for this reaction is –1367 kJ per mol of CH_3CH_2OH that reacts.

(a) Is this reaction endothermic or exothermic?
(b) What is the heat change when 0.200 mol of ethanol burns?

Solution:

(a) The value of q is negative, so the reaction is exothermic; energy is released by the reaction.
(b) To calculate the heat change when 0.200 mol CH_3CH_2OH reacts, multiply 0.200 mol CH_3CH_2OH by the heat change per mol of CH_3CH_2OH:

$$\text{Heat change} = 0.200 \ \cancel{\text{mol}} \ CH_3CH_2OH \times \frac{-1367 \text{ kJ}}{1 \ \cancel{\text{mol}} \ CH_3CH_2OH} = -273 \text{ kJ}$$

When 0.200 mol CH_3CH_2OH reacts, 273 kJ of heat is *released* to the surroundings.

Practice Problem 6.13

The balanced equation for the combination reaction of hydrogen gas and solid iodine is

$$H_2(g) + I_2(s) \longrightarrow 2HI(g)$$

The heat change q for this reaction is +53.00 kJ per mole of I_2 that reacts.

(a) Is this reaction endothermic or exothermic?
(b) What is the heat change when 2.50 mol of I_2 reacts?

Further Practice: 6.87 and 6.88

Foods and fuels are usually sold by volume or weight, so the kJ/g ratio is often useful. A mole of liquid octane, for example, has a much greater mass and volume than a mole of liquid propane. Table 6.3 shows the energy changes that correspond to the combustion of 1 g of a variety of fuels. Note that the energy released *per mole* generally increases with the number of carbon atoms in the molecular formula. However the energy released *per gram* is relatively constant for compounds that contain only carbon and hydrogen. The energy released per gram is lower for methanol and ethanol, which also contain oxygen. Hydrogen provides the most energy per gram, which is one reason why it offers promise as the fuel of the future.

Fuels aren't the only substances that can be described in terms of energy content and mass. The energy content of foods is also reported in energy units per gram, and most of the world reports the energy content of foods in kilojoules. In the United States, the preferred unit is Cal/g, which is the same as kcal/g. The energy content of foods comes primarily from proteins, fats, and carbohydrates. The energy values of these components are listed in Table 6.4. Note that fats have a much higher energy value per gram than proteins or carbohydrates. This is one reason why many weight watchers limit the amount of fat in their diet.

Notice in Table 6.3 that a relatively small amount of energy is released when TNT (trinitrotoluene) burns. Other fuels release much more energy per mole or per gram. Is this surprising? The explosive nature of TNT has to do with its detonation *velocity*. The *rate* at which the reaction occurs is greater than for other fuels.

TABLE 6.3	Heat Changes for the Combustion of Fuels $Substance + xO_2(g) \longrightarrow yCO_2(g) + zH_2O(l)$		
		Heat Change	
Fuel	**Formula**	**kJ/mol**	**kJ/g**
Hydrogen	$H_2(g)$	−286	−143
Methanol	$CH_3OH(l)$	−726	−23
Methane	$CH_4(g)$	−891	−56
Ethanol	$CH_3CH_2OH(l)$	−1367	−30
Acetylene	$C_2H_2(g)$	−1301	−50
Ethylene	$C_2H_4(g)$	−1411	−50
Ethane	$C_2H_6(g)$	−1561	−52
Propane	$C_3H_8(g)$	−2219	−50
Butane	$C_4H_{10}(l)$	−2878	−50
Octane	$C_8H_{18}(l)$	−5471	−48
Wood	—	—	−15
TNT	$C_7H_5(NO_2)_3(s)$	−3434	−15

TABLE 6.4	Energy Values of Food Components	
Food Component	**kcal/g (Cal/g)**	**kJ/g**
Proteins	4	17
Fats	9	38
Carbohydrates	4	17

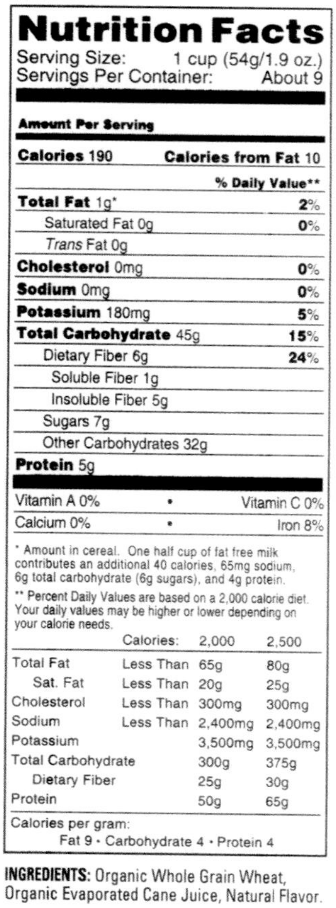

Nutrition Facts

Serving Size: 1 cup (54g/1.9 oz.)
Servings Per Container: About 9

Amount Per Serving

Calories 190	Calories from Fat 10

	% Daily Value**
Total Fat 1g*	**2%**
Saturated Fat 0g	**0%**
Trans Fat 0g	
Cholesterol 0mg	**0%**
Sodium 0mg	**0%**
Potassium 180mg	**5%**
Total Carbohydrate 45g	**15%**
Dietary Fiber 6g	**24%**
Soluble Fiber 1g	
Insoluble Fiber 5g	
Sugars 7g	
Other Carbohydrates 32g	
Protein 5g	

Vitamin A 0%	•	Vitamin C 0%
Calcium 0%	•	Iron 8%

* Amount in cereal. One half cup of fat free milk contributes an additional 40 calories, 65mg sodium, 6g total carbohydrate (6g sugars), and 4g protein.
** Percent Daily Values are based on a 2,000 calorie diet. Your daily values may be higher or lower depending on your calorie needs.

	Calories:	2,000	2,500
Total Fat	Less Than	65g	80g
Sat. Fat	Less Than	20g	25g
Cholesterol	Less Than	300mg	300mg
Sodium	Less Than	2,400mg	2,400mg
Potassium		3,500mg	3,500mg
Total Carbohydrate		300g	375g
Dietary Fiber		25g	30g
Protein		50g	65g

Calories per gram:
Fat 9 • Carbohydrate 4 • Protein 4

INGREDIENTS: Organic Whole Grain Wheat, Organic Evaporated Cane Juice, Natural Flavor.

Figure 6.19
A nutritional label gives the energy content in Cal (or kcal) per serving. Also listed are the grams of fats, carbohydrates, and proteins. This label also lists the Calories from fat.

The data in Table 6.4 are used to determine the energy content of foods as presented in most nutritional labels (Figure 6.19). Instead of burning every food in a calorimeter, food industry scientists determine the grams of protein, fat, and carbohydrate present in one serving. The mass of each is multiplied by their respective energy values. The calories from proteins, fats, and carbohydrates are then added to get the total energy content. A typical food label also lists the number of calories from fat.

SUMMARY

In this chapter we used balanced equations to determine quantities of reactants and products involved in chemical reactions. The coefficients in a balanced equation can be combined to give mole ratios (or molecule ratios), which are relationships between the moles of one reactant or product and another in the same reaction. If the mass of a reactant or product is given, the mass must first be converted to moles using the molar mass of the substance.

When a chemical reaction occurs, the reactants are usually not present in exactly the right amounts for a complete reaction of all reactants. In such cases, only the limiting reactant reacts completely. Once the limiting reactant is identified, it is used to determine the theoretical yield, or the maximum amount of product that can form. If the reaction is done in the laboratory, the amount of product actually obtained can be compared to the theoretical yield by calculating a percent yield.

Energy changes accompany chemical and physical changes. Endothermic reactions absorb energy from the surroundings; exothermic reactions release energy to the surroundings. Generally the energy transferred is in the form of heat. Temperature changes indicate the amount of heat transferred to or from an object. Specific heat is the amount of heat needed to raise the temperature of 1 g of a substance by 1°C. The relationship between the heat change of a substance, q, its mass m, specific heat C, and temperature change ΔT is given by the equation

$$q = m \times C \times \Delta T$$

Sometimes the amount of heat absorbed or released by a system—whether an object or a chemical reaction—cannot be determined directly. Instead, the heat change of the surroundings, usually water, is determined. As described by the law of conservation of energy, the heat change for the system is equal to the heat change of the surrounding water, but opposite in sign. Calorimetry is used to determine the heat change for many chemical reactions. The amount of heat absorbed or released during a chemical reaction depends on the mass of each reactant. The greater the amount of reactants, the greater the heat change. Heat changes for chemical reactions are usually reported in units of kJ/mol or kJ/g of a specific reactant.

KEY RELATIONSHIPS

Relationship	Equation
The percent yield for a reaction is a ratio of the amount of product actually obtained in a chemical reaction to the theoretical yield, the amount calculated from the limiting reactant.	$\text{Percent yield} = \dfrac{\text{actual yield}}{\text{theoretical yield}} \times 100$
The heat change for an object is equal to the product of its mass, its specific heat, and its temperature change. The temperature change ΔT is the difference between the final temperature and the initial temperature.	$q = m \times C \times \Delta T$

KEY TERMS

actual yield (6.5)

endothermic reaction (6.6)

exothermic reaction (6.6)

heat (6.6)

law of conservation of energy (6.6)

limiting reactant (6.4)

percent yield (6.5)

specific heat (6.6)

stoichiometry (6.1)

theoretical yield (6.5)

QUESTIONS AND PROBLEMS

The following questions and problems, except for those in the *Additional Questions* section are paired. Questions in a pair focus on the same concept. Answers to the odd-numbered questions and problems are in Appendix D.

Matching Definitions with Key Terms

6.1   Match the key terms with the descriptions provided.
   (a) the process of determining the amounts of reactants and products in a chemical reaction
   (b) the energy that is transferred between objects due to a difference in their temperatures
   (c) a chemical change that absorbs heat
   (d) the amount of heat required to raise the temperature of 1 g of water by 1°C
   (e) the amount of product actually obtained in the laboratory from a reaction

6.2   Match the key terms with the descriptions provided.
   (a) the reactant that is completely used up in a reaction and determines the amount of product that can form
   (b) a chemical change that releases heat
   (c) the maximum amount of product that can be obtained in a chemical reaction from known amounts of reactants; the amount of product calculated by assuming all the limiting reactant is consumed
   (d) the principle that states that the total amount of energy remains the same no matter how a physical or chemical change occurs
   (e) the amount of a product actually obtained as a proportion of the amount expected, expressed as a percentage

The Meaning of a Balanced Equation

6.3   When one molecule of glucose ($C_6H_{12}O_6$) is metabolized in our bodies, it combines with six O_2 molecules to form six CO_2 molecules and six H_2O molecules.
   (a) Write a balanced equation for this reaction.
   (b) How many CO_2 molecules are formed when 12 glucose molecules react?
   (c) If 30 O_2 molecules react, how many glucose molecules also react?

6.4   When two molecules of methanol (CH_3OH) react with oxygen, they combine with three O_2 molecules to form two CO_2 molecules and four H_2O molecules.
   (a) Write a balanced equation for this reaction.
   (b) How many H_2O molecules are formed when 10 methanol molecules react?
   (c) If 30 O_2 molecules react, how many methanol molecules react?

6.5   What do the coefficients in a balanced equation represent?

6.6   Why is a balanced equation important when determining the amount of product that can be produced from a given amount of reactant?

6.7   The balanced equation for the dissolving of magnesium nitrate in water is

$$Mg(NO_3)_2(s) \xrightarrow{H_2O} Mg^{2+}(aq) + 2NO_3^-(aq)$$

   (a) How many $Mg^{2+}(aq)$ and $NO_3^-(aq)$ ions form for each $Mg(NO_3)_2$ formula unit that dissolves?
   (b) How many moles of Mg^{2+} and NO_3^- form for each mole of $Mg(NO_3)_2$ that dissolves?

6.8   The balanced equation for the dissolving of sodium phosphate in water is

$$Na_3PO_4(s) \xrightarrow{H_2O} 3Na^+(aq) + PO_4^{3-}(aq)$$

   (a) How many Na^+ and PO_4^{3-} ions form for each Na_3PO_4 formula unit that dissolves?
   (b) How many moles of Na^+ and PO_4^{3-} form for each mole of Na_3PO_4 that dissolves?

6.9   The molecular-level diagram shows the reactants (before the reaction) and the products (after the reaction). Which of the following is the best balanced equation to represent this reaction? For the choices you did not select, explain why they are incorrect.

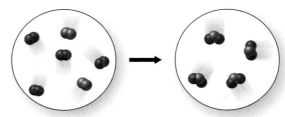

   (a) $2A + 4B \longrightarrow 4AB$
   (b) $A_2 + B_2 \longrightarrow AB_2$
   (c) $2A_2 + 4B_2 \longrightarrow 4AB_2$
   (d) $A_2 + 2B_2 \longrightarrow 2AB_2$

6.10   The molecular-level diagram shows the reactants (before the reaction) and the products (after the reaction). Which of the following is the best balanced equation to represent this reaction on any scale? For the choices you did not select, explain why they are incorrect.

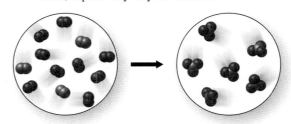

   (a) $A_2 + B_2 \longrightarrow AB_3$
   (b) $A_2 + 2B_2 \longrightarrow 2AB_3$
   (c) $A_2 + 3B_2 \longrightarrow 2AB_3$
   (d) $3A_2 + 9B_2 \longrightarrow 6AB_3$

Mole-Mole Conversions

6.11 Benzene (C_6H_6) burns in air according to the following equation:

$$2C_6H_6(l) + 15O_2(g) \longrightarrow 12CO_2(g) + 6H_2O(g)$$

(a) What is the mole ratio of O_2 to C_6H_6?

(b) How many moles of O_2 are required to react with each mole of C_6H_6?

(c) How many moles of O_2 are required to react with 0.38 mol of C_6H_6?

6.12 Aluminum reacts with chlorine gas according to the following equation:

$$2Al(s) + 3Cl_2(g) \longrightarrow 2AlCl_3(s)$$

(a) What is the mole ratio of Cl_2 to Al?

(b) How many moles of Cl_2 are required to react with each mole of Al?

(c) How many moles of Cl_2 are required to react with 0.085 mol of Al?

6.13 Suppose you have a piece of aluminum that you want to react completely by the following single-displacement reaction:

$$2Al(s) + 6HNO_3(aq) \longrightarrow 2Al(NO_3)_3(aq) + 3H_2(g)$$

(a) What mole ratio would you use in the following equation to determine the number of moles of HNO_3 needed to react with a known amount of Al?

$$\text{mol Al} \times \underline{\hspace{3cm}} = \text{mol } HNO_3$$

(b) If you add more than enough nitric acid so that all the aluminum reacts, what mole ratio would you use in the following equation to determine the moles of H_2 produced?

$$\text{mol Al} \times \underline{\hspace{3cm}} = \text{mol } H_2$$

(c) Suppose you know the number of moles of H_2 product formed and you want to know the number of moles of Al that reacted. What mole ratio would you use in the following equation?

$$\text{mol } H_2 \times \underline{\hspace{3cm}} = \text{mol Al}$$

6.14 Suppose you want to convert iron ore to a specific amount of pure iron using the following reaction:

$$Fe_3O_4(s) + 4CO(g) \longrightarrow 3Fe(s) + 4CO_2(g)$$

(a) What mole ratio would you use in the following equation to determine the number of moles of CO needed to react with a known amount of Fe_3O_4?

$$\text{mol } Fe_3O_4 \times \underline{\hspace{3cm}} = \text{mol CO}$$

(b) If you add more than enough CO so that all the Fe_3O_4 reacts, what mole ratio would you use in the following equation to determine the moles of CO_2 produced?

$$\text{mol } Fe_3O_4 \times \underline{\hspace{3cm}} = \text{mol } CO_2$$

(c) Suppose you know the number of moles of Fe product formed and you want to know the number of moles of CO that reacted. What mole ratio would you use in the following equation?

$$\text{mol Fe} \times \underline{\hspace{3cm}} = \text{mol CO}$$

6.15 Consider the combustion reaction of propane:

$$C_3H_8(g) + 5O_2(g) \longrightarrow 3CO_2(g) + 4H_2O(g)$$

Which of the following is conserved in this reaction?

(a) moles of molecules

(b) moles of atoms

(c) atoms

(d) mass

(e) Which of your answers to (a) through (d) are true for any reaction?

6.16 Consider the combination reaction of hydrogen and iodine:

$$H_2(g) + I_2(s) \longrightarrow 2HI(g)$$

Which of the following is conserved in this reaction?

(a) moles of molecules

(b) moles of atoms

(c) atoms

(d) mass

(e) Which of your answers to (a) through (d) are true for any reaction?

6.17 The balanced equation for the decomposition of TNT, $C_7H_5(NO_2)_3$, is

$$2C_7H_5(NO_2)_3(s) \longrightarrow 7C(s) + 7CO(g) + 3N_2(g) + 5H_2O(g)$$

(a) Write the mole ratios that relate moles of each product to the reactant $C_7H_5(NO_2)_3$?

(b) How many moles of each product would form if 1.00 mol of $C_7H_5(NO_2)_3$ reacts?

(c) How many moles of each product would form if 6.25 mol of $C_7H_5(NO_2)_3$ react?

6.18 The balanced equation for the decomposition of ammonium nitrate is

$$2NH_4NO_3(s) \longrightarrow 2N_2(g) + O_2(g) + 4H_2O(g)$$

(a) Write the mole ratios that relate moles of each product to the reactant NH_4NO_3?

(b) How many moles of each product would form if 1.00 mol of NH_4NO_3 reacts?

(c) How many moles of each product would form if 12.4 mol of NH_4NO_3 react?

Mass-Mass Conversions

6.19 When copper(II) sulfate pentahydrate ($CuSO_4 \cdot 5H_2O$) is heated, it decomposes to the dehydrated form. The waters of hydration are released from the solid crystal and form water vapor. The hydrated form is medium blue, and the dehydrated solid is light blue. The balanced equation is

$$CuSO_4 \cdot 5H_2O(s) \xrightarrow{\text{heat}} CuSO_4(s) + 5H_2O(g)$$

(a) What is the molar mass of $CuSO_4 \cdot 5H_2O$?

(b) What is the molar mass of $CuSO_4$?

(c) If 1.00 g $CuSO_4 \cdot 5H_2O$ is decomposed to $CuSO_4$, predict the mass of the remaining light blue solid.

6.20 When the reddish-brown mercury(II) oxide (HgO) is heated, it decomposes to its elements, liquid mercury metal and oxygen gas:

$$2HgO(s) \xrightarrow{\text{heat}} 2Hg(l) + O_2(g)$$

(a) What is the molar mass of HgO?
(b) What is the molar mass of Hg?
(c) If 2.00 g HgO is decomposed to Hg, predict the mass of the pure Hg metal produced.

6.21 When active metals such as sodium are exposed to air, they quickly form a coating of metal oxide. The balanced equation for the reaction of sodium metal with oxygen gas is

$$4Na(s) + O_2(g) \longrightarrow 2Na_2O(s)$$

Suppose a piece of sodium metal gains 2.05 g of mass after being exposed to air. Assume that this gain can be attributed to its reaction with oxygen.
(a) What mass of O_2 reacted with the Na?
(b) What mass of Na reacted?
(c) What mass of Na_2O formed?

6.22 When aluminum metal is exposed to oxygen gas, a coating of aluminum oxide forms on the surface of the aluminum. The balanced equation for the reaction of aluminum metal with oxygen gas is

$$4Al(s) + 3O_2(g) \longrightarrow 2Al_2O_3(s)$$

Suppose a sheet of pure aluminum gains 0.0900 g of mass when exposed to air. Assume that this gain can be attributed to its reaction with oxygen.
(a) What mass of O_2 reacted with the Al?
(b) What mass of Al reacted?
(c) What mass of Al_2O_3 formed?

6.23 Diiodine pentoxide is used in respirators to remove carbon monoxide from air:

$$I_2O_5(s) + 5CO(g) \longrightarrow I_2(s) + 5CO_2(g)$$

(a) What mass of carbon monoxide could be removed from air by a respirator that contains 50.0 g of diiodine pentoxide?
(b) What mass of I_2 would remain in the respirator?

6.24 Ethanol, used in alcoholic beverages, can be produced by fermentation of sucrose, which is found in sugar cane and other plants. The balanced equation for the fermentation process is

$$C_{12}H_{22}O_{11}(s) + H_2O(l) \longrightarrow 4C_2H_5OH(l) + 4CO_2(g)$$

(a) What mass of ethanol (C_2H_5OH) would be produced when 2.50 g sucrose reacts by this process?
(b) What mass of CO_2 would also be produced?

6.25 For each of the following *unbalanced* equations, balance the equation and then determine how many grams of the second reactant would be required to react completely with 0.600 g of the first reactant.
(a) $Fe(s) + Cl_2(g) \longrightarrow FeCl_3(s)$
(b) $KO_2(s) + H_2O(l) \longrightarrow O_2(g) + KOH(s)$
(c) $C_5H_{12}(g) + O_2(g) \longrightarrow CO_2(g) + H_2O(g)$
(d) $K(s) + Cl_2(g) \longrightarrow KCl(s)$

6.26 For each of the following *unbalanced* equations, balance the equation and then determine how many grams of the first reactant would be required to react completely to give 1.70 g of the product indicated in boldface.
(a) $C_6H_{12}(l) + O_2(g) \longrightarrow \mathbf{CO_2}(g) + H_2O(g)$
(b) $NI_3(s) \longrightarrow N_2(g) + \mathbf{I_2}(s)$
(c) $Na(s) + O_2(g) \longrightarrow \mathbf{Na_2O}(s)$
(d) $Ca(s) + HCl(aq) \longrightarrow CaCl_2(aq) + \mathbf{H_2}(g)$

6.27 Calcium carbonate ($CaCO_3$) is often used in commercial antacids. It acts to reduce the acidity in the stomach by neutralizing stomach acid, which is mostly HCl, by the following reaction:

$$CaCO_3(s) + 2HCl(aq) \longrightarrow CaCl_2(aq) + CO_2(g) + H_2O(l)$$

What mass of $CaCO_3$ is needed to neutralize 0.020 mol HCl?

6.28 Sulfuric acid is commonly used as an electrolyte in car batteries. Suppose you spill some on your garage floor. Before cleaning it up, you wisely decide to neutralize it with sodium bicarbonate (baking soda) from your kitchen. The reaction of sodium bicarbonate and sulfuric acid is

$$2NaHCO_3(s) + H_2SO_4(aq) \longrightarrow Na_2SO_4(aq) + 2CO_2(g) + 2H_2O(l)$$

You estimate that your acid spill contains about 2.0 mol H_2SO_4. What mass of $NaHCO_3$ do you need to neutralize the acid?

Limiting Reactants

6.29 When a fuel is burned in the air, what is typically the limiting reactant?

6.30 An acidic solution is added to a basic solution. If the resulting solution is basic, is the acid or the base the limiting reactant?

6.31 Suppose you are making turkey sandwiches for your friends. You have lots of friends so you want to make as many sandwiches as possible. Because you are on a budget, you only have 24 pieces of bread and 15 pieces of turkey. Each sandwich consists of 2 pieces of bread and 1 piece of turkey.
(a) How many sandwiches can you make?
(b) What is the limiting part?
(c) What part is left over? How many are left over?

6.32 Your little sister is having a party and you are in charge of the party goody bags. You decide that each bag should have 3 candy bars, 1 bottle of nail polish, and 2 packs of bubble gum. You bought a 18-pack of candy bars, 12 bottles of nail polish, and 20 pieces of bubble gum.

   (a) What is the ratio of candy bars to bottles of nail polish to pieces of bubble gum in the goody bags?

   (b) What is the limiting "goody"?

   (c) Assuming you have an unlimited supply of bags, how many goody bags can you make?

   (d) How many of each type of goody will be left over?

6.33 The balanced equation for the reaction of nitrogen and hydrogen gas to form ammonia is

$$N_2(g) + 3H_2(g) \longrightarrow 2NH_3(g)$$

What is the limiting reactant when each of the following sets of quantities of reactants is mixed?

   (a) 9 N_2 molecules and 9 H_2 molecules

   (b) 5 N_2 molecules and 20 H_2 molecules

   (c) 6 N_2 molecules and 18 H_2 molecules

6.34 The balanced equation for the reaction of aluminum metal and chlorine gas is

$$2Al(s) + 3Cl_2(g) \longrightarrow 2AlCl_3(s)$$

What is the limiting reactant when each of the following sets of quantities of reactants is mixed?

   (a) 4 Al atoms and 6 Cl_2 molecules

   (b) 6 Al atoms and 10 Cl_2 molecules

   (c) 12 Al atoms and 20 Cl_2 molecules

6.35 The molecular-level diagram shows a mixture of reactant molecules (three O_2 molecules and eight H_2 molecules) for the following reaction:

$$2H_2(g) + O_2(g) \longrightarrow 2H_2O(g)$$

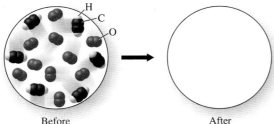

Before   After

   (a) Draw what the mixture should look like when the reaction is complete.

   (b) What is the limiting reactant?

   (c) Which reactant is left over?

6.36 The molecular-level diagram shows a mixture of reactant molecules for the following reaction:

$$2C_2H_2(g) + 5O_2(g) \longrightarrow 4CO_2(g) + 2H_2O(g)$$

Before   After

   (a) Draw what the mixture should look like when the reaction is complete.

   (b) What is the limiting reactant?

   (c) Which reactant is left over?

6.37 Use the balanced equation for the combustion of ethane to complete the table.

$2C_2H_6(g)$ + $7O_2(g)$ $\longrightarrow$ $4CO_2(g)$ + $6H_2O(g)$			
Initially mixed	6 molecules	18 molecules	0 molecules	0 molecules
How much reacts			—	—
Composition of final mixture				

6.38 Use the balanced equation for the combustion of ethane to complete the table.

$2C_2H_6(g)$ + $7O_2(g)$ $\longrightarrow$ $4CO_2(g)$ + $6H_2O(g)$			
Initially mixed	10 molecules	20 molecules	0 molecules	0 molecules
How much reacts			—	—
Composition of final mixture				

6.39 Use the balanced equation for the combustion of butane to complete the table.

$2C_4H_{10}(g)$ + $13O_2(g)$ $\longrightarrow$ $8CO_2(g)$ + $10H_2O(g)$			
Initially mixed	3.10 mol	13.0 mol	0.00 mol	0.00 mol
How much reacts			—	—
Composition of final mixture				

6.40 Use the balanced equation for the combustion of butane to complete the table.

$2C_4H_{10}(g)$ + $13O_2(g)$ $\longrightarrow$ $8CO_2(g)$ + $10H_2O(g)$			
Initially mixed	0.500 mol	3.50 mol	0.00 mol	0.00 mol
How much reacts			—	—
Composition of final mixture				

6.41 The balanced equation for the reaction of phosphorus and oxygen gas to form tetraphosphorus decoxide is

$$P_4(g) + 5O_2(g) \longrightarrow P_4O_{10}(s)$$

What is the limiting reactant when each of the following sets of quantities of reactants is mixed?
(a) 0.50 mol P_4 and 5.0 mol O_2
(b) 0.20 mol P_4 and 1.0 mol O_2
(c) 0.25 mol P_4 and 0.75 mol O_2

6.42 Tetraphosphorus trisulfide (P_4S_3) is used in the heads of wooden matches. This material is manufactured by the heating of a mixture of red phosphorus and sulfur.

$$8P_4(s) + 3S_8(s) \longrightarrow 8P_4S_3(s)$$

What is the limiting reactant when each of the following sets of quantities of reactants is mixed?
(a) 4.4 mol P_4 and 1.1 mol S_8
(b) 0.20 mol P_4 and 0.10 mol S_8
(c) 0.75 mol P_4 and 0.28 mol S_8

6.43 The balanced equation for the reaction of nitrogen and hydrogen to form ammonia is

$$N_2(g) + 3H_2(g) \longrightarrow 2NH_3(g)$$

Assume that 10.0 g of N_2 is mixed with 10.0 g of H_2.
(a) What is the limiting reactant?
(b) What is the maximum amount of NH_3, in grams, that can be produced?

6.44 The balanced equation for the reaction of aluminum metal and chlorine gas is

$$2Al(s) + 3Cl_2(g) \longrightarrow 2AlCl_3(s)$$

Assume that 0.40 g Al is mixed with 0.60 g Cl_2.
(a) What is the limiting reactant?
(b) What is the maximum amount of $AlCl_3$, in grams, that can be produced?

6.45 Sodium bicarbonate reacts with hydrochloric acid in a gas-forming reaction to produce aqueous sodium chloride, water, and carbon dioxide gas:

$$NaHCO_3(s) + HCl(aq) \longrightarrow NaCl(aq) + H_2O(l) + CO_2(g)$$

Determine the limiting reactant and mass of CO_2 gas produced when 8.0 g of $NaHCO_3$ is added to a solution that contains 4.0 g of HCl.

6.46 Sodium metal reacts with water in the following single-displacement reaction:

$$2Na(s) + 2H_2O(l) \longrightarrow 2NaOH(aq) + H_2(g)$$

Determine the limiting reactant and mass of H_2 gas produced when 2.0 g of Na is added to 10.0 g H_2O.

6.47 Use the balanced equation for the combustion of ethane to complete the table.

$2C_2H_6(g)$ +	$7O_2(g)$	$\longrightarrow$ $4CO_2(g)$ +	$6H_2O(g)$
Initially mixed 0.260 g	1.00 g	0.00 g	0.00 g
How much reacts		—	—
Composition of final mixture			

6.48 Use the balanced equation for the combustion of butane to complete the table.

$2C_4H_{10}(g) +$	$14O_2(g)$	$\longrightarrow$ $8CO_2(g) +$	$10H_2O(g)$
Initially mixed 0.130 g	0.580 g	0.00 g	0.00 g
How much reacts		—	—
Composition of final mixture			

6.49 An aqueous solution containing 10.0 g NaOH is added to an aqueous solution containing 10.0 g HCl.
(a) Write a balanced equation for the acid-base reaction that should occur.
(b) Identify the limiting reactant.
(c) Will the solution be acidic or basic when the reaction is complete?

6.50 An aqueous solution containing 10.0 g NaOH is added to an aqueous solution containing 10.0 g H_2SO_4.
(a) Write a balanced equation for the acid-base reaction that should occur.
(b) Identify the limiting reactant.
(c) Will the solution be acidic or basic when the reaction is complete?

Percent Yield

6.51 In a process for obtaining pure copper from copper ore, 55.6 g of copper metal was obtained. Is this the theoretical yield or actual yield?

6.52 After determining the limiting reactant in a reaction, you calculate the amount of product that can form from the limiting reactant. Is this the theoretical yield or the actual yield?

6.53 A student carried out the following precipitation reaction:

$$BaCl_2(aq) + Na_2CO_3(aq) \longrightarrow BaCO_3(s) + 2NaCl(aq)$$

After no more precipitate appeared to form, he filtered the barium carbonate solid and then immediately weighed it. Using this mass, he calculated the percent yield as 105%. Is this possible? Explain.

6.54 What factors may contribute to a percent yield less than 100%?

6.55 A student was synthesizing aspirin in the laboratory. Using the amount of limiting reactant, she calculated the mass of aspirin that should form as 8.95 g. When she weighed her aspirin product on the balance, its mass was 7.44 g.
(a) What is the actual yield of aspirin?
(b) What is the theoretical yield of aspirin?
(c) Calculate the percent yield for this synthesis.

6.56 A student added zinc metal to a copper nitrate solution to obtain copper metal by a single-displacement reaction. From the known moles of copper nitrate in solution, she calculated the expected mass of copper metal as 6.7 g. When she weighed the copper on a balance, the mass was 5.7 g.
(a) What is the actual yield of copper metal?
(b) What is the theoretical yield of copper metal?
(c) Calculate the percent yield.

6.57 The combination reaction of sodium metal and chlorine gas to form sodium chloride is represented by the following balanced equation:

$$2Na(s) + Cl_2(g) \longrightarrow 2NaCl(s)$$

If 5.00 g of sodium metal is reacted with excess chlorine gas, and 11.5 g NaCl is actually obtained, what is the percent yield of sodium chloride?

6.58 Ammonia is synthesized commercially from nitrogen gas and hydrogen gas for the production of fertilizers:

$$N_2(g) + 3H_2(g) \longrightarrow 2NH_3(g)$$

If 100.0 g of nitrogen is reacted with excess hydrogen, and 34.0 g of NH_3 are obtained, what is the percent yield of ammonia?

6.59 When I_2 is mixed with excess H_2, 0.80 mol HI is isolated in the laboratory. If this reaction occurs to give an 85% yield, how many moles of I_2 reacted? (Hint: Write a balanced equation.)

6.60 The reaction of lithium metal and water to form lithium hydroxide and hydrogen gas is represented by the following balanced equation:

$$2Li(s) + 2H_2O(l) \longrightarrow 2LiOH(aq) + H_2(g)$$

When Li is mixed with excess water, 0.30 mol H_2 gas is isolated in the laboratory. If this reaction occurs to give an 85% yield of H_2, how many moles of Li reacted?

6.61 The combination reaction of magnesium metal and bromine to form magnesium bromide is represented by the following balanced equation:

$$Mg(s) + Br_2(l) \longrightarrow MgBr_2(l)$$

If 1.0 mol of Mg is mixed with 2.0 mol of Br_2, and 0.84 mol of $MgBr_2$ is obtained, what is the percent yield for the reaction?

6.62 The combination reaction of sodium metal and nitrogen gas to form sodium nitride is represented by the following balanced equation:

$$6Na(s) + N_2(g) \longrightarrow 2Na_3N(s)$$

If 0.30 mol Na is mixed with 0.60 mol N_2, and 0.092 mol Na_3N is obtained, what is the percent yield for the reaction?

Energy Changes

6.63 If energy cannot be created or destroyed, what happens to the energy of a ball as it rolls down a hill and rests at the bottom?

6.64 If energy cannot be created, what is the source of the energy that is released when gasoline is burned?

6.65 Consider the reaction of hydrogen and oxygen in a hydrogen-fuel vehicle. Describe the transformation of energy in this process in terms of potential and kinetic energy.

6.66 Consider propane fuel being burned in a propane-fueled barbecue grill. Describe the transformation of energy in this process in terms of potential and kinetic energy.

6.67 When the solids ammonium thiocyanate and barium hydroxide are mixed in a beaker, a solution forms, and the temperature of the solution drops to –5°C. Is this an endothermic or exothermic reaction? Explain.

6.68 When baking soda is added to a solution of sulfuric acid, the container becomes warmer to the touch. Is this an endothermic or exothermic reaction? Explain.

6.69 A doctor sprays a liquid on your skin to numb it before injecting a painkiller. You notice that the liquid evaporates quickly. You also notice that your skin feels cold. Explain why your skin feels cold.

6.70 Explain why sweating is a natural way for your body to cool itself.

6.71 What is the relationship between the potential energy of the reactants and products in an exothermic reaction?

6.72 If the reactants have a lower potential energy than the products, is the reaction exothermic or endothermic?

6.73 If the same amount of heat is added to 50.0 g each of the metals, aluminum, copper, and lead, all at the same initial temperature, which metal will have the highest final temperature? Refer to Table 6.2 for specific heat values.

6.74 We add the same amount of heat to three different samples of the same liquid substance. The temperature of the first sample increases in temperature by the greatest amount. How can we explain the different temperature change in each sample?

6.75 How much heat has to be added to 528 g of copper at 22.3°C to raise the temperature of the copper to 49.8°C? (See Table 6.2.)

6.76 How much heat has to be added to 235 g of iron at 25.0°C to raise the temperature of the iron to 250.0°C? (See Table 6.2.)

6.77 What is the heat change when 1.25 g of water vapor (steam) at 185.3°C is cooled to 102.1°C? The specific heat of steam is 2.02 J/(g °C).

6.78 What is the heat change when 55.0 g of water cools from 60.0°C to 25.5°C?

6.79 What are some characteristics of a good calorimeter?

6.80 What information can be obtained from a calorimetry experiment?

6.81 A rock is added to 60.0 g of water. The initial temperature of the water is 25.0°C. The final temperature of the water and the rock is 30.1°C.
(a) Did the rock gain or lose heat?
(b) What is the heat change of the rock?

6.82 A marble is added to 40.0 g of water. The initial temperature of the water is 25.0°C. The final temperature of the water and the marble is 22.2°C.
(a) Did the marble gain or lose heat?
(b) What is the heat change of the marble?

6.83 Why can't you measure the temperature change of a chemical reaction?

6.84 Why do you have to first determine the heat change of the surroundings for a chemical reaction?

6.85 You carry out a chemical reaction in a calorimeter to determine the heat change for the reaction. Identify the system and surroundings.

6.86   When a chemical reaction occurs in a calorimeter containing water, and the temperature of the water decreases, is the reaction endothermic or exothermic?

6.87   A mixture of carbon monoxide, hydrogen, and oxygen gases, known as *water gas,* can be used as an industrial fuel. The reaction of these gases is

$$CO(g) + H_2(g) + O_2(g) \longrightarrow CO_2(g) + H_2O(g)$$

This reaction releases 525 kJ of heat for every mole of carbon monoxide that reacts.
   (a)   Is this reaction endothermic or exothermic?
   (b)   What is the energy change for this reaction in units of kJ/mol of carbon monoxide?

6.88   The reaction that produces the *water gas* mixture, described in Question 6.87, is

$$C(s) + H_2O(g) \longrightarrow CO(g) + H_2(g)$$

This reaction requires an input of 131 kJ of heat for every mole of carbon that reacts.
   (a)   Is this reaction endothermic or exothermic?
   (b)   What is the energy change for this reaction in units of kJ/mol of carbon?

6.89   When a 6.00-g sample of coal is burned, it releases enough heat to raise the temperature of 2010 g of water from 24.0°C to 41.5°C.
   (a)   How much heat did the coal release as it burned?
   (b)   Calculate the heat of combustion of coal in units of kJ/g.

6.90   When a 4.00-g sample of magnesium metal is burned, it produces enough heat to raise the temperature of 2010 g of water from 24.00°C to 33.10°C.
   (a)   How much heat did the magnesium release as it burned?
   (b)   Calculate the heat of combustion of magnesium in units of kJ/g.

6.91   A 2.00-g peanut is burned in a bomb calorimeter containing 1200 g of water. The temperature of the water increases from 25.00°C to 30.25°C.
   (a)   How much heat, in joules, did the peanut release as it burned?
   (b)   Calculate the heat content in units of calories and Calories.
   (c)   Calculate the energy value in units of Cal/g.

6.92   A 5.00-g piece of fat is burned in a bomb calorimeter containing 4050 g of water. The temperature of the water increases by 12.4°C.
   (a)   How much heat, in joules, did the fat release as it burned?
   (b)   Calculate the heat content in units of calories and Calories.
   (c)   Calculate the energy value in units of Cal/g.
   (d)   Calculate the heat content in units of Cal/mol. Assume the fat is all tristearin, a typical fat with the molecular formula $C_{57}H_{110}O_6$.

6.93   The heat change that accompanies the formation of 1.00 mol of carbon dioxide from its elements is −393.7 kJ/mol. What heat change accompanies the formation of 0.650 mol CO_2?

6.94   The decomposition of hydrogen peroxide to form water and oxygen gas releases 196.6 kJ per mol of hydrogen peroxide. How much heat is released when 5.25 mol H_2O_2 decompose?

Additional Questions

6.95   When nitroglycerin explodes, it decomposes to form carbon dioxide gas, nitrogen gas, oxygen gas, and water vapor. The balanced equation is

$$4C_3H_5O_9N_3(l) \longrightarrow 12CO_2(g) + 6N_2(g) + O_2(g) + 10H_2O(g)$$

   (a)   If 1.00 mol $C_3H_5O_9N_3$ decomposes, how many moles of each gaseous product should form?
   (b)   If 2.50 mol $C_3H_5O_9N_3$ decomposes, how many moles of each gaseous product should form?

6.96   Certain drain cleaners are a mixture of sodium hydroxide and powdered aluminum. When dissolved in water, the sodium hydroxide reacts with the aluminum and the water to produce hydrogen gas.

$$2Al(s) + 2NaOH(aq) + 6H_2O(l) \longrightarrow 2NaAl(OH)_4(aq) + 3H_2(g)$$

The sodium hydroxide helps dissolve grease, and the hydrogen gas provides a mixing and scrubbing action. What mass of hydrogen gas would be formed from a reaction of 2.48 g Al and 4.76 g NaOH in water?

6.97   When silver nitrate is added to an aqueous solution of magnesium chloride, a precipitation reaction occurs that removes the chloride ions from solution.

$$2AgNO_3(s) + MgCl_2(aq) \longrightarrow 2AgCl(s) + Mg(NO_3)_2(aq)$$

   (a)   If a solution contains 10.0 g $MgCl_2$, what mass of $AgNO_3$ should be added to remove all the chloride ions from solution?
   (b)   When enough $AgNO_3$ is added so that all 10.0 g $MgCl_2$ react, what mass of the AgCl precipitate should form?

6.98   Sometimes it is more convenient to measure the amount of a liquid reactant in volume units instead of mass units. Given that 1.00 gal is 3.79 L and the density of methanol is 0.793 g/mL, what mass of oxygen is needed to react with 1.00 gal of methanol? The balanced equation is

$$2CH_3OH(l) + 3O_2(g) \longrightarrow 2CO_2(g) + 4H_2O(g)$$

6.99   The balanced equation for the combustion of octane is

$$2C_8H_{18}(l) + 25O_2(g) \longrightarrow 16CO_2(g) + 18H_2O(g)$$

What mass of oxygen is needed to react with 1.00 gal of octane? (A gallon is 3.79 L, and the density of octane is 0.703 g/mL.)

6.100   When lead(II) nitrate is added to an aqueous solution of potassium iodide, a precipitation reaction occurs that removes the iodide ion from solution.

$$Pb(NO_3)_2(s) + 2KI(aq) \longrightarrow 2PbI_2(s) + 2KNO_3(aq)$$

   (a)   If a solution contains 7.00 g KI, what mass of $Pb(NO_3)_2$ should be added to remove all the iodide ion from solution?
   (b)   When enough $Pb(NO_3)_2$ is added so that all 7.00 g KI reacts, what mass of PbI_2 should precipitate?

6.101 A 5.0-g sample of greenish-yellow Cl_2 gas is added to a 10.0-g sample of gray potassium metal to form a white solid in a sealed container.
  (a) Write a balanced equation for the combination reaction that should occur.
  (b) Identify the limiting reactant.
  (c) Predict the appearance of the substances left in the reaction container once the reaction is complete.
  (d) Calculate the mass of product that should form.
  (e) What mass of excess reactant should be mixed with the product after the reaction is complete?

6.102 A 5.00-g piece of copper metal is added to a colorless aqueous solution that contains 20.0 g of $Zn(NO_3)_2$. The products that form are a water-soluble blue ionic compound and a gray metal.
  (a) Write a balanced equation for the single-displacement reaction that should occur.
  (b) Identify the limiting reactant.
  (c) Predict the appearance of the substances left in the reaction container once the reaction is complete.

6.103 The equation for the combustion of hydrogen is

$$2H_2(g) + O_2(g) \longrightarrow 2H_2O(g)$$

  (a) What is the mole ratio of O_2 to H_2?
  (b) If 1.0 g of H_2 reacts, what mass of O_2 will react with it, and what mass of H_2O should form?
  (c) When 1.0 g H_2 is mixed with 4.0 g O_2, what is the theoretical yield of H_2O?

6.104 The decomposition reaction of calcium carbonate is represented by the following balanced equation:

$$CaCO_3(s) \xrightarrow{\text{heat}} CaO(s) + CO_2(g)$$

After a 15.8-g sample of calcium carbonate was heated in an open container to cause decomposition, the mass of the remaining solid was determined to be 9.10 g. The student is unsure if the reaction is complete, so the solid could contain unreacted $CaCO_3$.
  (a) Determine the theoretical yield of CaO and the theoretical yield of CO_2.
  (b) Determine the percent yield of CaO and the percent yield of CO_2.
  (c) Why are the percent yields for CaO and CO_2 different?

6.105 What is the temperature of a 125-g sample of tin, initially at 26.2°C if 155 J of heat is added to the tin? See Table 6.2.

6.106 Calculate the temperature change that results when 1.05 kJ of heat is supplied to 30.0 g of aluminum. See Table 6.2.

6.107 If 20.0 g of an unknown metal cools from 50.0°C to 25.5°C and loses 220.6 J of heat, what is the specific heat of the metal in joules per gram per degree Celsius? Refer to Table 6.2 to determine the possible identity of the metal.

6.108 What is the specific heat of an unknown metal if 658 J of heat increases the temperature of 1.00 kg of the metal by 5.10°C? Refer to Table 6.2 to determine the possible identity of the metal.

6.109 When 0.250 mol of glucose is broken down in your body, 701 kJ of energy is released. What is the energy change in units of kJ/mol and Cal/mol?

6.110 Sodium cyclamate ($NaC_6H_{12}NSO_3$) was a popular nonsugar sweetener until it was banned by the Food and Drug Administration. Its sweetness is about 30 times that of sugar (sucrose, $C_{12}H_{22}O_{11}$). The energy that can be supplied by ingestion of each compound is 16.03 kJ/g for sodium cyclamate and 16.49 kJ/g for sucrose. How many Calories are saved when 1.00 g of sodium cyclamate is used in place of 30.0 g of sucrose?

6.111 In some areas, gasoline is formulated to contain 15% ethanol during some seasons. Gasoline releases 11.4 kcal/g upon combustion, but ethanol releases only 7.12 kcal/g. The mixture releases 10.7 kcal/g. How much less energy in units of kJ/g is provided by the mixture than by pure gasoline?

6.112 A cold object at 5°C is placed in an insulated cup of water at 25°C. Determine whether or not each of the following will occur.
  (a) The increase in temperature of the object will be equal to the decrease in temperature of the water.
  (b) Only the object will change temperature.
  (c) Only the water will change temperature.
  (d) The heat gain by the object will equal the heat lost by the water.
  (e) The final temperature will be somewhere between 5°C and 25°C.

6.113 If 20.0 g iron at 24.5°C is added to 105 g of water at 55.0°C, what will be the final temperature of the mixture? See Table 6.2.

6.114 Methylhydrazine (CH_6N_2) is commonly used as a liquid rocket fuel. The heat of combustion of methylhydrazine is -1.30×10^3 kJ/mol. How much heat is released when 100.0 g of methylhydrazine is burned?

6.115 A 175-g pipe is heated to 78.24°C. The pipe is then placed in 100.0 g of water held in a calorimeter at 25.00°C. The final temperature of the mixture is 33.43°C. The specific heat of water is 4.184 J/g °C. Assuming no heat is transferred out of the water, what is the specific heat of the pipe? Using Table 6.2, determine what the pipe could be made of.

6.116 A 150.0-g sample of copper is heated to 89.3°C. The copper is then placed in 125.0 g of water held in a calorimeter at 22.5°C. The final temperature of the mixture is 29.0°C. Assuming no heat is lost from the water, what is the specific heat of copper?

Electron Structure of the Atom

Andrea, Ben, and Drew are three college students who sit together in chemistry class. They often meet outside of class to go over homework problems and to study for quizzes and exams. After many weeks of hard work, they agree they need to get away from chemistry for a while, so Andrea suggests a weekend trip by car to Las Vegas. They don't suspect that they will encounter chemistry at every turn.

As they approach Las Vegas, the bright lights of the city's signs and advertising displays tell them that their destination is near. Drew, the most studious of the three, remarks that—while the bright lights of Las Vegas are dazzling in their array of colors—good chemistry students should be just as appreciative of the white light that is all around. Andrea and Ben remind Drew that they are supposed to be on vacation, but a discussion about the Sun begins anyway. Drew tells the others that the Sun is the most important source of white light. It provides the light that allows us to see. The light from the Sun provides the energy that plants use to convert carbon dioxide and water into the sugar glucose. As sunlight hits our atmosphere and our planet's surface, some of its energy converts to heat energy that warms the planet.

As they near the city, the friends drive through a brief rain shower and spot a rainbow. Sunlight is composed of all the possible colors of light. The mixture results in a continuum of colors that we see as white light. A rainbow is a separation of the colors in sunlight. It appears when sunlight passes through water droplets in the atmosphere. The water acts as a prism to separate the colors that compose sunlight. Light from a regular (incandescent) lightbulb is also white light. Shine it through a prism and a rainbowlike separation of colors appears (Figure 7.1).

During their first evening in Las Vegas, the friends watch a spectacular fireworks show. Ben begins to wonder how all the different colors and designs are created. Although fireworks date back to ancient China, they were very simple by our standards and produced mostly yellow and orange light. Much work has gone into developing modern fireworks that display a greater variety of colors, including reds, greens, blues, and purples. Fireworks generally consist of two components: the compounds that produce colored light and the explosives. Makers of fireworks use different chemical compounds to get different colors (Figure 7.2). For example, they use strontium nitrate or strontium carbonate to produce red light, and barium nitrate or barium chlorate to make green light. Copper compounds produce bluish-green light. Magnesium and aluminum metals are generally used to produce white fireworks. A potassium nitrate and sulfur mixture produces the white smoke that often accompanies a grand finale.

The explosives in fireworks provide the energy that propels the compounds into the air. Explosions are exothermic reactions that release energy to the surroundings. Some of the energy released changes into kinetic energy, causing the

Animation: Flame Tests
Online Learning Center, Animations

Chemistry Animations Library, Flame Tests, Flame_Test.rm.

Figure 7.1
White light is composed of different colors that can be separated by a prism. A rainbow results from the separation of white sunlight by raindrops.

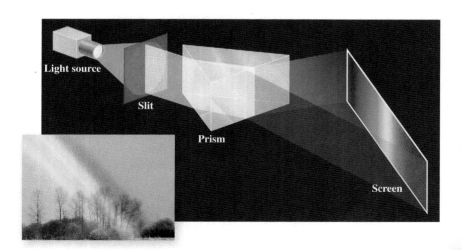

Figure 7.2

Commercial fireworks used by private companies or for sale to the public are composed of two main components: explosive compounds and metals or compounds that will emit a specific type of light. Shown here are finely ground ionic compounds and metals that are used to produce light and special effects. Pure metals, such as magnesium, are often used to produce the white sparkles seen in firework shows.

fireworks' components to move quickly through the air. More important, the explosion causes the compounds to release light. Some of the energy is absorbed by ionic compounds or metal ions, which then release energy in the form of light. It doesn't take an explosion to get such substances to release light. The heat from a simple flame will do. For example, if we hold a small sample of strontium nitrate in a flame, the flame changes to a red color (Figure 7.3). The red color of the flame is the same as we see in strontium nitrate fireworks.

Ben, Andrea, and Drew walk past the casinos and hotels of Las Vegas, admiring the brilliant displays of neon lights and talking about how they work. The neon lights of Las Vegas duplicate all the colors of the rainbow, but the source is not white light. If the light from a neon light passes through a prism, only a few of all possible colors appear. Neon light, then, is different from white light; it does not contain a continuum of colors. Ben says the colored light from neon signs is produced in much the same way as the colors of fireworks, but the energy absorbed by the atoms of gas inside the tube comes from electricity, not from a flame or an explosion. A neon light consists of a tube containing only neon atoms under very low pressure. When an electric current moves through the gas, the neon atoms absorb energy and then emit energy as light. In other words, electric energy is converted to light energy. The light from a neon sign can be separated into its component colors just as can the light from fireworks or colored flames. Neon is used in these types of signs because its light is very bright. When other gases are placed in the tube and electricity is applied, each emits a unique color. The colors we see are usually combinations of more than one color.

Animation: Discharge Tubes
Online Learning Center, Animations

Chemistry Animations Library, Discharge Tubes, Discharge_Tube.avi.

At one time, neon lights contained only the noble gas neon. They glowed a red-orange. Today, we use *neon light* as a general term to describe tubes filled with other noble gases or a mixture of them. The color we see depends on the gases inside, typically pale orange from helium, lavender from argon, gray-green from krypton, and gray or blue-gray from xenon.

Figure 7.3

When a sample of $Sr(NO_3)_2$ is placed in a flame, the flame turns red. This compound is used to make red fireworks.

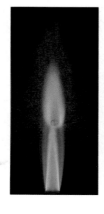

The three students decide to attend a concert in Las Vegas. As they enter the auditorium, a laser light show is in progress. Dozens of beams of colored light move to music in patterns across the stage. Andrea notices that the lines are very narrow and intense, much different from the light given off by the fireworks show and the neon lights. Laser light, she remembers from chemistry class, consists of only one color of light that travels in a narrow and intensely colored beam. Different lasers produce different colors of light (Figure 7.4). Because visible laser light is fun to look at, lasers are great for entertainment, but they have practical uses, too. In surgery the narrow beam can be used to make tiny incisions in a very small area, such as in the cornea of the eye. Recently lasers have been found effective in collapsing varicose veins without the pain and swelling that accompanies traditional surgical methods.

Whether we're looking at neon lights, fireworks, or laser lights, we see light that is constantly being released by atoms. Learning about light and how it interacts with atoms will help you better understand the arrangement and behavior of electrons in atoms. It will also help you relate the microscopic and macroscopic properties of the elements to their positions in the periodic table.

Figure 7.4
Laser lights can be entertaining. They can also be used for practical purposes such as surgery.

Questions for Consideration

7.1 What is light and how can we describe it?

7.2 How can we describe the behavior of the electron in a hydrogen atom?

7.3 How can we describe electrons in all atoms?

7.4 How does the arrangement of electrons in atoms relate to the arrangement of elements in the periodic table?

7.5 Which electrons are chemically important in an atom?

7.6 How does the electron arrangement in ions differ from that in atoms?

7.7 What properties of atoms are related to their electron arrangements?

Math Tools Used in This Chapter

Units and Conversions (Math Toolbox 1.3)

7.1   ELECTROMAGNETIC RADIATION AND ENERGY

Light is a form of energy called **electromagnetic radiation,** or *radiant energy.* All types of electromagnetic radiation travel through space as oscillating waves, and—as their name suggests—they are electric and magnetic fields. They all move in a vacuum at the same speed, the *speed of light,* 3.0×10^8 m/s. The light we see with our eyes, visible light, is only a small fraction of the wide range of electromagnetic radiation. Other forms include X-rays, ultraviolet light, infrared light, microwaves, and radio waves.

Properties of Electromagnetic Radiation

What makes different kinds of electromagnetic radiation different? One variable property is their wavelengths. As shown in Figure 7.5, the **wavelength** (symbolized by the Greek letter lambda, λ) is the distance between two corresponding points on a wave. They can be two crests, two troughs, or any other two corresponding points along the wave. The wavelengths are characteristics of different kinds of electromagnetic radiation. The shorter wavelengths are the gamma rays and X-rays. The longer wavelengths are microwaves and radio waves.

Although we describe frequency as wave cycles per second, the units of frequency are usually written as s⁻¹ or 1/s, pronounced "inverse seconds" or "per second." Wave cycles are not considered units, so there is no numerator unit; the denominator unit is seconds.

For more information on direct and inverse relationships, see Math Toolbox 9.2.

Although we cannot see infrared radiation, we can sometimes feel its effects as heat. All objects emit infrared radiation. The wavelength at which an object radiates most intensely depends on its temperature. The cooler the object, the longer the wavelength. We can detect infrared radiation using special cameras and film that detect differences in temperature. In the picture, the yellow areas are the warmest and the blue and black areas are the coldest.

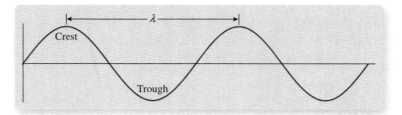

Figure 7.5

Wavelength (λ) is the distance between two corresponding points on a wave. Here it is measured from crest to crest. The same value would be obtained if the measurement were made between two troughs or between any other two corresponding points along the wave.

Another property is frequency. To understand it, imagine that you are standing on a boat dock on a calm day. The waves hit the dock at regular, slow intervals. Now imagine that you are standing on that same dock on a windy day. The waves hit the dock much more frequently. These waves have a higher *frequency* than the waves on a calm day. For light, the **frequency** (symbolized by the Greek letter nu, ν) is a measure of the number of wave cycles that move through a point in space in 1 second. It is described in units of 1/s or **hertz (Hz)**.

Both wavelength and frequency can be used to characterize electromagnetic radiation. The *electromagnetic spectrum* in Figure 7.6 shows the different types of electromagnetic radiation, listed from shortest wavelength (gamma rays) to longest wavelength (TV and radio waves). Notice that as wavelength increases, frequency decreases. As one goes up, the other one goes down. That is, *wavelength and frequency are inversely proportional.* Compare the two waves shown in Figure 7.7. Which has the higher frequency? The one with the shorter wavelength.

Our eyes can detect light only in the visible region of the electromagnetic spectrum. This region is composed of the different colors of light, with red having the longest wavelength and violet the shortest. Atoms absorb light. This is why grass

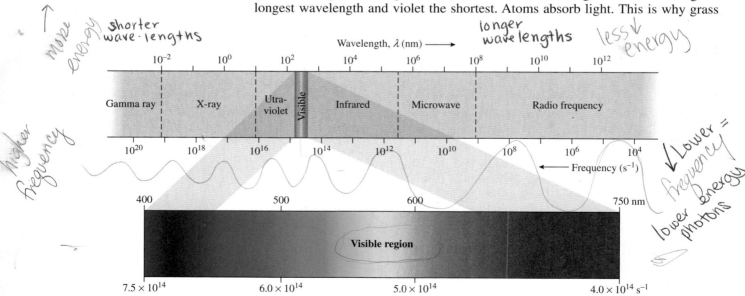

Figure 7.6

The electromagnetic spectrum includes all the types of electromagnetic radiation, shown here in order from shortest wavelength to longest wavelength. As wavelength increases, frequency decreases. The visible region of the electromagnetic spectrum is the region between the ultraviolet and the infrared. It represents a small portion of the electromagnetic spectrum.

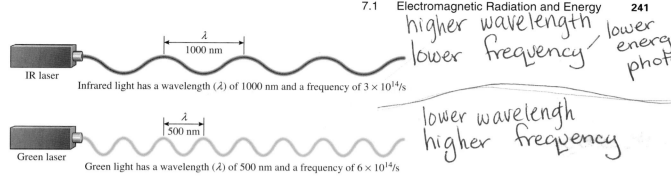

higher wavelength, lower
lower frequency, energy
photons

lower wavelength
higher frequency

Infrared light has a wavelength (λ) of 1000 nm and a frequency of 3×10^{14}/s

Green light has a wavelength (λ) of 500 nm and a frequency of 6×10^{14}/s

IR laser

Green laser

Figure 7.7
The wavelength of the infrared light wave is twice that of the green light. The frequency of the infrared light wave is one-half that of the green. Wavelength and frequency are inversely proportional.

appears green to us. When light from the sun shines on grass, the atoms in the grass absorb *all* the colors of light *except* green. The green wavelengths not absorbed by the grass reflect back to our eyes, and we see green. A white object absorbs little or no visible light. A black object absorbs most or all visible light. These properties of atoms and light can fool our eyes sometimes. Suppose you park your red car at night in a parking lot lit by sodium vapor lamps. They emit only yellow light. Since no red light shines on the car, no red light can reflect back. Your car will look a different color!

Light can be characterized by its wavelength or its frequency, and also by its energy. A unit of light energy can be thought of as a "packet" of energy called a **photon.** The energy E of a photon depends on the wavelength or frequency. Electromagnetic radiation of long wavelength and low frequency, such as radio waves, is composed of low-energy photons. Short-wavelength radiation with high frequencies is composed of high-energy photons. *Photon energy is directly proportional to frequency.* As one gets larger, so does the other. Photon energy is *inversely proportional* to wavelength. As one increases, the other decreases.

A laser such as that used to create a light show produces *monochromatic light,* which is light of only one wavelength and composed of photons of identical energy. One reason that a beam of laser light does not spread out is that the waves are synchronized. They move together so that the peaks are perfectly lined up. Common lasers emit light in the visible or the infrared region of the electromagnetic spectrum. In Example 7.1, we will compare the wavelength, frequency, and photon energy of different lasers and of different types of electromagnetic radiation.

Use a mnemonic to remember wavelengths and colors of visible light: red, orange, yellow, green, blue, indigo, and violet. From longest to shortest wavelength, the first letters of the colors spell a man's name: ROY G. BIV.

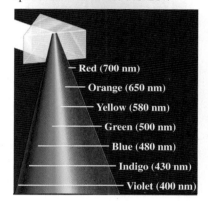

Red (700 nm)
Orange (650 nm)
Yellow (580 nm)
Green (500 nm)
Blue (480 nm)
Indigo (430 nm)
Violet (400 nm)

EXAMPLE 7.1   Relative Values of Wavelength, Frequency, and Photon Energy

Andrea, Ben, and Drew notice that their chemistry teacher uses two different laser pointers. One emits red light and the other emits green light.

(a) Which laser light has the longer wavelength?   *red*
(b) Which has the higher frequency?   *green*
(c) Which is composed of photons of higher energy?   *green*

Solution:

(a) Relative wavelengths are shown in the electromagnetic spectrum in Figure 7.6. Red light is farther to the right with the longer wavelength. Green light has the shorter wavelength. If you do not have an electromagnetic spectrum available, the mnemonic ROYGBIV provides the order, starting with red (R), the color of light with the longest wavelength. Green (G) light falls in the middle of the visible region, with a shorter wavelength. The shortest is violet (V).

(b) Frequency and wavelength are inversely proportional. When frequency is relatively large, wavelength is relatively small. Since green light has the shorter wavelength, it has the higher frequency.

(c) Photon energy is proportional to frequency. Light with a high frequency has a relatively greater photon energy. Since green light has the higher frequency, it also has the greater photon energy. You can also infer relative photon energies by comparing wavelengths. Photon energy is inversely related to wavelength; so green light—with the shorter wavelength—has the higher photon energy.

Practice Problem 7.1

The Sun emits a wide range of electromagnetic radiation. The light that gets to the Earth is mostly visible light, infrared light, and ultraviolet light. We can see the visible light with our eyes. We can feel the infrared light as heat. It warms the earth and the atmosphere. We use sunscreen to protect our skin from the harmful effects of ultraviolet light.

(a) Which type of light has the longest wavelength: visible, infrared, or ultraviolet?
(b) Which has the highest frequency?
(c) Which is composed of the highest-energy photons?

Further Practice 7.13 and 7.14

Mathematical relationships allow us to calculate the wavelength λ, frequency ν, and photon energy E_{photon} of a specific type of electromagnetic radiation. The product of frequency ν and wavelength λ is a constant, c, which is the speed of light, 3.00×10^8 m/s:

$$\nu\lambda = c$$

Note that this equation shows the inverse relationship between wavelength and frequency. Since the speed of light c is a constant, when frequency is large, wavelength must be small; when wavelength is large, frequency must be small. If we know the value of either wavelength or frequency, we can calculate the unknown value by rearranging the equation:

$$\nu = \frac{c}{\lambda} \quad \text{or} \quad \lambda = \frac{c}{\nu}$$

When calculating frequency from wavelength, wavelength must be in units of meters, so it will cancel properly with the units of the speed of light. Because wavelengths are commonly expressed in units of nanometers (1 nm = 10^{-9} m), conversion to meters is required. When calculating wavelength from frequency, note that units cancel to give units of meters.

The mathematical relationship between photon energy and frequency is described by the following equation where h, Planck's constant, is 6.626×10^{-34} J·s:

$$E_{\text{photon}} = h\nu$$

Note that the equation shows the directly proportional relationship between photon energy and frequency. The inverse relationship between photon energy and wavelength is given by the following equation where h is Planck's constant and c is the speed of light:

$$E_{\text{photon}} = \frac{hc}{\lambda}$$

EXAMPLE 7.2   **Calculating Wavelength, Frequency, and Photon Energy**

Infrared light emitted from a TV remote control or from a wireless computer mouse has a wavelength of 805 nm. What are the frequency of the light and the energy of its photons?

Solution:

Recall the relationship between wavelength λ and frequency ν:

$$\nu\lambda = c$$

where c is the speed of light, 3.00×10^8 m/s. We know wavelength, so we can rearrange the equation to solve for frequency:

$$\nu = \frac{c}{\lambda}$$

Wavelength is given in units of nanometers, so we must convert to meters to cancel units with the units of speed of light (m/s). (For a review of unit conversions, see Math Toolbox 1.3.) The relationship between nanometers and meters is

$$1 \text{ m} = 10^9 \text{ nm}$$

This relationship can be written in the following two ways:

$$\frac{10^9 \text{ nm}}{1 \text{ m}} \quad \text{and} \quad \frac{1 \text{ m}}{10^9 \text{ nm}}$$

When we convert from nanometers to meters, we multiply the known value in nanometers by the conversion factor that has the nanometer units in the denominator:

$$805 \text{ nm} \times \frac{1 \text{ m}}{10^9 \text{ nm}} = 8.05 \times 10^{-7} \text{ m}$$

$$\lambda = 8.05 \times 10^{-7} \text{ m}$$

Now that we have the appropriate units for wavelength, we can divide the speed of light by its value to get frequency:

$$\nu = \frac{c}{\lambda}$$

$$= \frac{3.00 \times 10^8 \text{ m/s}}{8.05 \times 10^{-7} \text{ m}} = 3.73 \times 10^{14}\text{/s} \quad \text{or} \quad 3.73 \times 10^{14} \text{ Hz}$$

The energy of a photon can be calculated from frequency using the following equation:

$$E_{\text{photon}} = h\nu$$

where h is Planck's constant, 6.626×10^{-34} J · s. Since we have just calculated frequency, we can substitute values of h and ν into the equation:

$$E_{\text{photon}} = 6.626 \times 10^{-34} \text{ J} \cdot \text{s} \times 3.73 \times 10^{14}\, \frac{1}{\text{s}} = 2.47 \times 10^{-19} \text{ J}$$

The energy of a single photon of this light is 2.47×10^{-19} J.

Practice Problem 7.2

Microwaves are to the right of infrared radiation in the electromagnetic spectrum shown in Figure 7.6, so their wavelengths are longer. The microwaves used to cook food have wavelengths that are about 11 to 12 cm long! If the wavelength of light of a microwave beam is 11.5 cm, what are the frequency of the radiation and the energy of its photons?

Further Practice: 7.17 and 7.18

Atomic Spectra

When visible light passes through a prism, its components separate into a sequence of colors called a *spectrum.* Sunlight or light from an incandescent bulb produces a **continuous spectrum,** which contains all the wavelengths of light in the visible region (Figure 7.8A). A light source that produces a continuous spectrum is called *white light.*

What happens when we pass colored light, such as that from a neon sign, through a prism? Unlike white light, colored light is not composed of all the wavelengths. Its spectrum is not continuous. If we pass the light from a neon sign through a slit and then through a prism, the result is a set of distinct colored lines, each representing a single wavelength of light. The pattern of these lines is a **line spectrum.** The line spectra for neon and some other elements are shown in Figure 7.8B.

When an element is heated or given an electric charge, the energy absorbed is transformed into light energy. The light an element produces in this way contains only specific colors, as shown by its line spectrum. The line spectrum for each element is unique, so it can be used to identify the element. It is like the element's "fingerprint." In Section 7.2 we see how the line spectrum for hydrogen helps to explain the electronic structure of the hydrogen atom.

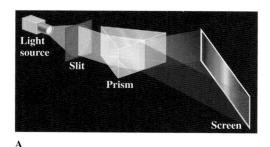

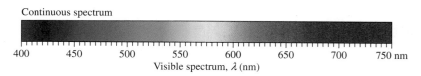

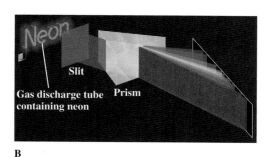

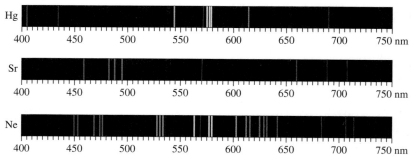

Figure 7.8

(A) When white light from sunlight or an incandescent lightbulb passes through a prism, a continuous spectrum is produced. (B) The colored light emitted by a substance produces a line spectrum that is unique to that substance. The line spectra for mercury, strontium, and neon are shown here.

7.2   THE BOHR MODEL OF THE HYDROGEN ATOM

How can we explain the different colors of light that Andrea, Ben, and Drew saw in Las Vegas? How do neon atoms produce the distinctive red-orange color of a true neon light? In Chapter 2 you learned about the basic structure of the atom. An atom consists of a relatively small, positively charged nucleus made of protons and neutrons. Although the nucleus occupies little space in an atom, it accounts for the majority of its mass. The electrons occupy a relatively large volume of space in the atom surrounding the nucleus.

Recall from Chapter 2 that the nuclear model of the atom was proposed by Rutherford as a result of his gold foil experiment. When Rutherford proposed his theory in 1911, many scientists, especially J. J. Thomson, found it difficult to accept. Although the evidence was compelling, Rutherford's model didn't follow the basic laws of physics known at the time. How could negative and positive charges—which normally attract each other—remain separate? What prevents electrons from tumbling toward the nucleus?

A year after Rutherford published his nuclear model of the atom, a young man named Niels Bohr came to work with Rutherford. Bohr had just completed his doctoral thesis in physics. One of his first assignments with Rutherford was to explain the stability of the arrangement of the electrons in the nuclear model of the atom. One possible explanation that Bohr proposed is described as a planetary model, in which electrons orbit the nucleus similar to the way planets orbit the Sun. Bohr based his ideas on work done by a German physicist, Max Planck, who thought that energy produced by atoms is **quantized;** that is, it can have only certain values. Bohr suggested that the energies of electrons in an atom are quantized. They can have only certain values.

In his model of a hydrogen atom, Bohr proposed that hydrogen's single electron revolves about the nucleus in one of several possible, fixed orbits, each with a specific radius (Figure 7.9). He labeled each orbit with an integer value of n, with n = 1 being the orbit closest to the nucleus. He explained that when the electron is in an orbit close to the nucleus, it has a relatively low energy. The electron can absorb energy and move to a higher-energy orbit that is farther from the nucleus. When the electron drops down to a lower orbit, energy is released. The possible energies of the electron are restricted to those of allowed orbits. The amount of energy that can be absorbed or released is restricted to the possible transitions between the energy levels defined by the orbits. Since the energy of an electron in an atom is quantized, its energy changes are quantized, and the energy absorbed or released is quantized, Bohr said.

The scientific community accepted Bohr's model because it explained the hydrogen line spectrum. As shown in Figure 7.10, the four lines in the line spectrum correlate with four of the electron transitions that Bohr proposed. Each line in the spectrum has a specific wavelength, frequency, and photon energy. Bohr thought that the transition or movement of an electron from a higher-energy orbit to a lower-energy orbit releases energy. A transition to the $n = 2$ orbit produces an emission of visible light. The possible transitions correspond to lines in the hydrogen line spectrum: $n = 6$ to $n = 2$ (violet); $n = 5$ to $n = 2$ (blue); $n = 4$ to $n = 2$ (green); and $n = 3$ to $n = 2$ (red). In Bohr's model, transitions to the $n = 1$ orbit are higher-energy transitions that result in the emission of ultraviolet light. Transitions to the $n = 3$ orbit are lower-energy transitions that result in the emission of infrared light.

Bohr said that when an electron moves from a higher-energy orbit to one that is lower in energy, an electron releases an amount of energy ΔE equal to the *difference* in energy between the two orbits:

$$\Delta E = E_{\text{final}} - E_{\text{initial}}$$

The quantized behavior of an electron in an atom is similar to the notes played on a keyboard. You can play any of the notes that correspond to keys, but you can't play a note between two adjacent keys. The notes you play are quantized because only certain notes are possible. The difference between notes is also quantized.

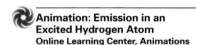

Animation: Emission in an Excited Hydrogen Atom
Online Learning Center, Animations

Chemistry Animations Library, Emission in an Excited Hydrogen Atom, 1Emission_Spectrum.rm.

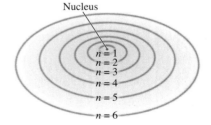

Figure 7.9

In the Bohr model of the hydrogen atom, the nucleus is at the center of the atom, and the electron is located in one of the circular orbits (n =1 through n = 6), each with a specific energy and radius.

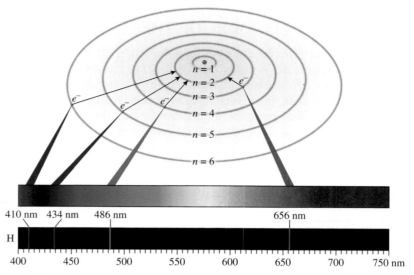

Figure 7.10
Four distinct lines appear in the line spectrum of hydrogen. In the Bohr model, the difference in energy between certain orbits corresponds to the photon energies of the lines that appear in the hydrogen line spectrum. As the electron moves from a higher-energy orbit to a lower-energy orbit, energy is released as light.

The value of ΔE is negative for a downward transition because energy is released as a photon. The energy of the photon, E_{photon}, equals the absolute value of the energy change of the electron. We express this relationship using lines around ΔE to show that we take the absolute value of ΔE, which is its positive value:

$$E_{photon} = |\Delta E|$$

EXAMPLE 7.3   The Bohr Model and the Hydrogen Line Spectrum

Use the Bohr model of the hydrogen atom and the hydrogen line spectrum in Figure 7.10 to answer the following questions.

(a) What is the photon energy of the light that produces the red line in the hydrogen spectrum?
(b) Of the four lines in the hydrogen line spectrum, which has the highest photon energy?

Solution:

(a) In the hydrogen line spectrum in Figure 7.10, the wavelength of red light is given as 656 nm. Knowing the wavelength λ, we can find E_{photon} using the equation that relates wavelength to photon energy:

$$E_{photon} = \frac{hc}{\lambda}$$

We must convert the wavelength units from nanometers to meters so units cancel, just as we did in Example 7.2:

$$656 \; \text{nm} \times \frac{1 \; \text{m}}{10^9 \; \text{nm}} = 6.56 \times 10^{-7} \, \text{m}$$

Now we substitute Planck's constant, the speed of light, and the wavelength into the equation for photon energy:

$$E_{photon} = \frac{6.626 \times 10^{-34} \text{ J} \cdot s \times 3.00 \times 10^{8} \frac{m}{s}}{6.56 \times 10^{-7} m} = 3.03 \times 10^{-19} \text{ J}$$

(b) From the equation that relates photon energy to wavelength, we see that photon energy is *inversely* proportional to wavelength—the shorter the wavelength, the higher the photon energy. In Figure 7.10 we see that the violet line in the hydrogen spectrum has the shortest wavelength, so it has the highest photon energy. The highest photon energy E_{photon} corresponds to the largest electron transition ΔE, which is the difference in energy between the orbits. The $n = 6$ orbit is further in energy from the $n = 2$ orbit than are the other orbits, so the transition from $n = 6$ to $n = 2$ emits light of highest energy.

Practice Problem 7.3

Use the Bohr model of the hydrogen atom and the hydrogen line spectrum in Figure 7.10 to answer the following questions.

(a) What is the photon energy of the light that produces the violet line in the hydrogen spectrum?
(b) Of the four lines in the hydrogen line spectrum, which has the lowest photon energy?

Further Practice: 7.35 and 7.36

While the Bohr model contributed to a better understanding of the atom, it fell short in the long run because it explained only the hydrogen atom. It did not work for atoms that contained more than one electron. In Section 7.3 we will modify the Bohr model to describe our modern concept of the atom.

7.3   THE MODERN MODEL OF THE ATOM

By the 1920s scientists started to work on a new model of the atom that could account for the line spectra of other elements. Erwin Schrödinger, from Austria, developed a mathematical model that seemed to work for all atoms. Schrödinger's model of the atom was similar to Bohr's model in that the energies of electrons were quantized. They could have only certain energies. The new model, however, describes electrons as occupying *orbitals,* not orbits. An **orbital** is a three-dimensional region in space where the electron is likely to be found, not a circular pathway. The mathematical model was based on the probability of finding an electron in a region outside the nucleus. From it chemists created a probability map to show where an electron is likely to be found. The diagram in Figure 7.11A is the probability map for a hydrogen electron in its lowest-energy state. The darker regions represent areas where the electron is most likely to be found. Because the likelihood of finding an electron in the outer regions of the probability map is small, it is customary to show an enclosed region in space where the electron is likely to be found 95% of the time (Figure 7.11B). We will use orbital representations without dots for the rest of this chapter.

In the modern model of the atom, orbitals of similar size are considered to be in the same **principal energy level.** The first six principal energy levels for the electron in a hydrogen atom are shown in Figure 7.12. In the hydrogen atom, the energies of the orbitals are restricted to the energies of the principal energy levels.

A

B

Figure 7.11
(A) This probability map is for the hydrogen electron in its lowest-energy state. The electron is more likely to be found in the darker regions. (B) The size of this orbital is defined by an enclosed region where the likelihood of finding an electron is 95%.

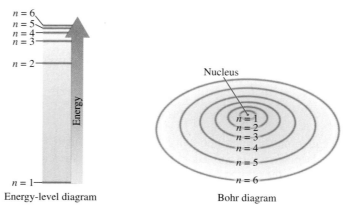

Energy-level diagram   Bohr diagram

Figure 7.12
In the modern model of the atom, principal energy levels describe the allowed energies of the electrons in orbitals. In the hydrogen atom, the allowed energies for the electron are the same as the energies of the Bohr orbits.

Orbitals come in different shapes and sizes. There are four types, which we label with the letters *s, p, d,* and *f.* The general shapes of these orbitals are shown in Figure 7.13. These orbitals are distributed into the principal energy levels.

Animation: Atomic Line Spectra
Online Learning Center, Animations

Chemistry Animations Library, Atomic Line
Spectra, 3Emission_Spectrum.rm.

Orbitals at lower-energy levels are smaller. Orbitals at the higher-energy levels are larger and extend farther from the nucleus. The first principal energy level (*n* = 1) consists of a single *s* orbital. We call this *s* orbital a 1*s* orbital because it is in the *n* = 1 principal energy level and it has the *s* orbital shape.

Principal energy level   Orbital shape

The second principal energy level (*n* = 2) consists of *s* and *p* orbitals called the 2*s* and 2*p* orbitals. The 2*s* orbital is like the 1*s* orbital, but it is larger (Figure 7.14). A single *p* orbital, as depicted in Figure 7.13, has a figure-eight or dumbbell shape. In their respective energy levels, *p* orbitals always come in sets of three (Figure 7.15). The three 2*p* orbitals lie with their centers overlapping at the nucleus and perpendicular to one another. We can think of each two-lobed orbital as directed along one of the *x, y,* or *z* axes that are each perpendicular to one another. We label these orbitals accordingly as p_x, p_y, and p_z orbitals. Because there are two types of orbitals in the second principal energy level, we say that there are two sublevels: the 2*s* sublevel and the 2*p* sublevel. A **sublevel** consists of just one type of orbital at a specific principal energy level. The 2*s* sublevel consists of a single 2*s* orbital, and the 2*p* sublevel consists of three 2*p* orbitals.

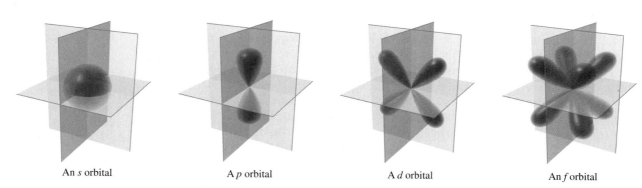

An *s* orbital   A *p* orbital   A *d* orbital   An *f* orbital

Figure 7.13
The *s, p, d,* and *f* orbitals differ in shape and number of lobes.

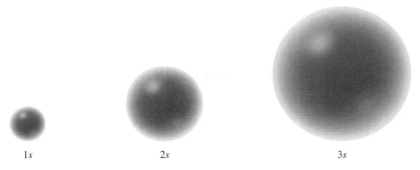

Figure 7.14
The size of the orbital increases as the principal energy level increases.

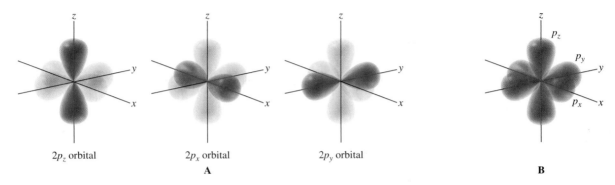

Figure 7.15
(A) Three *p* orbitals compose a *p* sublevel. They have the same shape, but they point in different directions in space. They lie perpendicular to one another, so we can imagine them lying along perpendicular lines that we call *x, y,* and *z* axes. The p_z orbital lies along the *z* axis, the p_x orbital lies along the *x* axis, and the p_y orbital lies along the *y* axis. (B) The three *p* orbitals coexist with their centers at the nucleus.

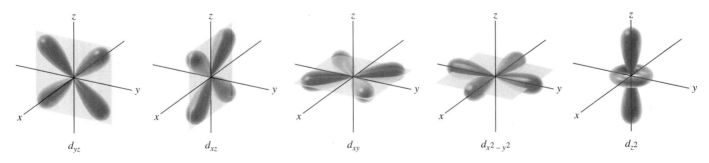

Figure 7.16
There are five *d* orbitals in a *d* sublevel. Most have four lobes that lie along a specified plane in space. One looks like a *p* orbital with a donut shape around its center.

At the third principal energy level ($n = 3$), three sublevels are available: the 3*s*, 3*p*, and the 3*d* sublevels. The 3*s* sublevel consists of a single 3*s* orbital, and the 3*p* sublevel consists of a set of three 3*p* orbitals. The 3*d* sublevel consists of five *d* orbitals (Figure 7.16). The fourth principal energy level ($n = 4$) consists of four sublevels: the 4*s,* the 4*p,* the 4*d,* and the 4*f.* The 4*f* sublevel consists of seven 4*f* orbitals.

A convenient way to show the distribution of sublevels and orbitals in an atom is with an orbital diagram. The one shown in Figure 7.17 is specifically for the hydrogen atom. The boxes represent orbitals and the groups of boxes represent sublevels.

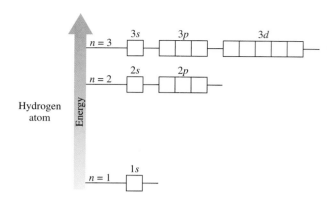

Figure 7.17

The orbital diagram for the hydrogen atom shows the sublevels and orbitals that can exist at each principal energy level. Each box represents an orbital, and groups of boxes represent sublevels. For the hydrogen atom only, the sublevels within a principal energy level all have the same energy. Here we show only the first three principal energy levels for hydrogen.

In the modern model of the hydrogen atom, the energy levels have the same energies as the orbits described in the Bohr model. The hydrogen line spectrum is produced by the movement of electrons in an atom from orbitals in higher principal energy levels to orbitals in lower principal energy levels. The hydrogen atom line spectrum is simple to explain because the orbitals in the same principal energy level have the same energy. This is not the case for atoms with more than one electron.

Orbital Diagrams for Multielectron Atoms

Sublevels in the same principal energy level have the same energy only in atoms with one electron. In atoms with more than one electron, the interaction among electrons causes the sublevels within the same principal energy level to have different energies. The result is an orbital diagram like that shown in Figure 7.18. Notice that, within a given principal energy level, the *p* orbitals are higher in energy than the *s* orbitals, and the *d* orbitals are higher in energy than the *p* orbitals. The 3*d* orbitals are so high that they are slightly higher in energy than the 4*s* orbitals.

Figure 7.18

The orbital diagram for a multielectron atom shows the same sublevels and orbitals as for the hydrogen atom. However, in atoms that contain more than one electron, the sublevels within a principal energy level have different energies. Note that the energy of the 3*d* sublevel is higher than the energy of the 4*s* sublevel.

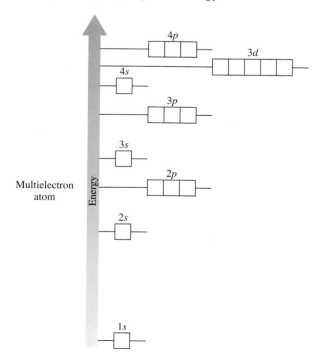

How are electrons in atoms arranged into these principal energy levels, sublevels, and orbitals? Can we predict the electron arrangement for a particular atom? We can determine the arrangement of electrons for an atom in its **ground state,** the lowest-energy state of the atom. Consider the orbital diagram that represents how electrons are arranged in a carbon atom in its ground state:

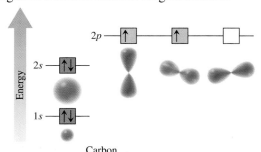

Carbon

The boxes represent orbitals, and the arrows represent electrons. Can you come up with some rules for the arrangement of electrons? What is the maximum number of electrons per orbital? Why do some orbitals contain only one electron?

When we create an orbital diagram for a particular atom, we fill in electrons starting with a bare nucleus and add one electron at a time. (This is not how atoms are created in nature.) You may have noticed that the lowest-energy orbitals have two electrons in them, and some of the higher-energy orbitals have only one or none. This is a result of the **aufbau principle** (from the German *aufbauen,* "to build up"), which states that *electrons fill orbitals starting with the lowest-energy orbitals.* For the hydrogen atom in its ground state, its electron is in the 1s orbital. You may have also noted that no more than two electrons occupy each orbital. This is the basis for the **Pauli exclusion principle,** which states that *a maximum of two electrons can occupy each orbital, and they must have opposite spins.* Electrons in the same orbital are represented by arrows that point in opposite directions. An electron can spin in one of two directions. We represent these spins with an up arrow ($\uparrow$) and a down arrow ($\downarrow$). For example, helium's ground state orbital diagram consists of two electrons in the 1s orbital:

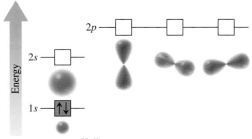

Helium

For lithium, with three electrons, two electrons go in the 1s orbital. The first principal energy level is filled. We can now proceed to the second principal energy level. Because the 2s sublevel is lower in energy than the 2p sublevel, the third electron goes into the 2s orbital:

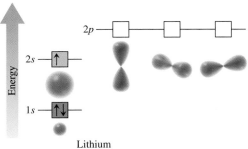

Lithium

Boron, with five electrons, has an orbital diagram with two electrons in the 1s orbital, two electrons in the 2s orbital, and one electron in one of the 2p orbitals. It makes no difference which p orbital we place the electron in because they are equal in energy. The orbital diagram for boron is

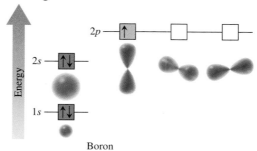

Boron

What happens when there is more than one electron in a sublevel that contains more than one orbital, such as a p sublevel? Are electrons paired up right away, or are they kept in separate orbitals until pairing is required? Did you notice in carbon's orbital diagram that the electrons in the p sublevel are unpaired in separate p orbitals? This brings us to the third rule, **Hund's rule:** *Electrons are distributed into orbitals of identical energy (same sublevel) in such a way as to give the maximum number of unpaired electrons.* Hund's rule is a consequence of the fact that negatively charged electrons repel each other. It follows that electrons should be found in separate orbitals if no energy is required to get there. What would the orbital diagrams for nitrogen and oxygen look like? How many unpaired electrons would these atoms have?

It is common to show a simpler orbital diagram, where all the orbitals are represented on one line, with sublevels labeled. We must remember that the energy of the sublevels increases from left to right. For example, we can represent the orbital diagram for carbon as

EXAMPLE 7.4   Orbital Filling Rules for Ground-State Electron Diagrams

Identify the orbital diagrams that are incorrect. For those that are incorrect, identify the filling rule that is violated.

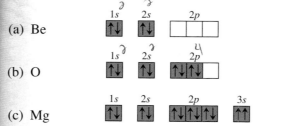

Solution:

(a) The orbital diagram is correct for beryllium.

(b) The orbital diagram for oxygen is not correct. Hund's rule is not obeyed. The orbitals in the 2p sublevel are the same energy. We must place electrons into individual 2p orbitals singly until an electron occupies all three orbitals before starting to pair electrons. Since there are four electrons in the 2*p* sub-level, two should be paired and two should be unpaired:

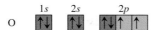

(c) The orbital diagram for magnesium is not correct. The two electrons in the 3*s* orbital have the same spin, so the Pauli exclusion principle is not followed. The arrows in the 3*s* orbital should be pointing in opposite directions to represent electrons with opposite spin:

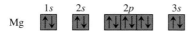

Practice Problem 7.4

Identify the orbital diagrams that are incorrect. For those that are incorrect, identify the filling rule that is violated.

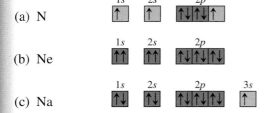

Further Practice: 7.43 and 7.44

Electron Configurations

Although orbital diagrams are useful in describing the arrangement of electrons in an atom, they take up a lot of space. Is there a shorter way to convey the same information? Yes, we can use a shorthand notation, called an **electron configuration,** which shows the distribution of electrons among sublevels. When writing electron configurations, we write the number of the principal energy level followed by the symbol for the sublevel. We then add a superscript to each sublevel symbol to indicate the number of electrons in that sublevel. Consider the arrangement of electrons in a carbon atom, as shown by its orbital diagram:

The lowest-energy sublevel, 1*s*, contains two electrons. This is represented as

$$1s^2$$

The next higher-energy sublevel, 2*s*, also contains two electrons:

$$2s^2$$

The last sublevel that contains electrons is the 2*p* sublevel. We do not have to show the number of electrons in each *p* orbital. We write the electron configuration to show the total number of electrons in the 2*p* sublevel:

$$2p^2$$

Combining the sublevels, we get the electron configuration for carbon:

$$C \qquad 1s^2 2s^2 2p^2$$

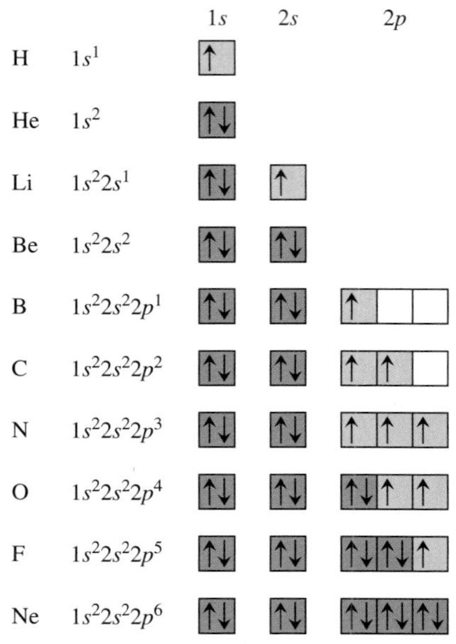

The orbital diagram showing the first 10 elements:

		1s	2s	2p
H	$1s^1$	↑		
He	$1s^2$	↑↓		
Li	$1s^2 2s^1$	↑↓	↑	
Be	$1s^2 2s^2$	↑↓	↑↓	
B	$1s^2 2s^2 2p^1$	↑↓	↑↓	↑
C	$1s^2 2s^2 2p^2$	↑↓	↑↓	↑ ↑
N	$1s^2 2s^2 2p^3$	↑↓	↑↓	↑ ↑ ↑
O	$1s^2 2s^2 2p^4$	↑↓	↑↓	↑↓ ↑ ↑
F	$1s^2 2s^2 2p^5$	↑↓	↑↓	↑↓ ↑↓ ↑
Ne	$1s^2 2s^2 2p^6$	↑↓	↑↓	↑↓ ↑↓ ↑↓

Figure 7.19
Orbital diagrams give information about how electrons are arranged within a sublevel. The orbital diagrams and corresponding electron configurations are shown for the first 10 elements.

Electron configurations are convenient, but they do not show how electrons are arranged in orbitals within the same sublevel. The electron configurations for the first 10 elements are shown in Figure 7.19, along with their orbital diagrams.

EXAMPLE 7.5   Writing Electron Configurations

Use the orbital diagram in Figure 7.18 to write the electron configurations for the following:

(a) Na
(b) Ar
(c) Br

Solution:

(a) The atomic number of sodium is 11, so it has 11 protons and 11 electrons. We distribute electrons into the orbitals in Figure 7.18, starting with the lowest-energy orbital, 1s. By putting a maximum of two electrons in each orbital, we get the electron configuration

$$1s^2 2s^2 2p^6 3s^1$$

(b) The atomic number of argon is 18. It has 18 electrons, which is enough to fill sublevels through 3p. The electron configuration is

$$1s^2 2s^2 2p^6 3s^2 3p^6$$

(c) From the periodic table we see that bromine has 35 electrons. Filling these electrons into the orbital diagram from lowest- to highest-energy sublevels, we get the following electron configuration:

$$1s^2 2s^2 2p^6 3s^2 3p^6 4s^2 3d^{10} 4p^5$$

Notice that electrons fill the 4s orbital before filling the 3d orbitals because the 4s sublevel is lower in energy than 3d.

Practice Problem 7.5

Use the orbital diagram in Figure 7.18 to write the electron configurations for the following:

(a) Al
(b) Sc
(c) K

Further Practice: 7.45 and 7.46

Writing electron configurations for the first 36 elements is easy if you have an orbital diagram like Figure 7.18. In Section 7.4 we will see how the arrangement of elements in the periodic table relates to the electron configurations of the elements. Once you understand the patterns, you won't need an orbital diagram to write electron configurations. You will only need a periodic table.

7.4 PERIODICITY OF ELECTRON CONFIGURATIONS

The neon in the signs that Andrea, Ben, and Drew saw in Las Vegas emits light when charged with electricity, but it is *inert*. That is, it does not react chemically with other elements. The inert character of neon and the other noble gases relates to their similar electron configurations. In Chapter 2 you saw that elements in the same group of the periodic table have similar properties. Here you will see that elements in the same group also have similar electron configurations.

Consider the alkali metals, the group IA (1) elements. The electron configurations for Li, Na, K, and Rb are

$$\begin{array}{ll} \text{Li} & 1s^2 2s^1 \\ \text{Na} & 1s^2 2s^2 2p^6 3s^1 \\ \text{K} & 1s^2 2s^2 2p^6 3s^2 3p^6 4s^1 \\ \text{Rb} & 1s^2 2s^2 2p^6 3s^2 3p^6 4s^2 3d^{10} 4p^6 5s^1 \end{array}$$

What patterns do you see? Notice that these group IA (1) elements all have one electron in the last-filled sublevel, the s sublevel. Each ends in ns^1, where n is the principal energy level.

The group IIA (2) elements have a similar pattern. Each ends in ns^2:

$$\begin{array}{ll} \text{Be} & 1s^2 2s^2 \\ \text{Mg} & 1s^2 2s^2 2p^6 3s^2 \\ \text{Ca} & 1s^2 2s^2 2p^6 3s^2 3p^6 4s^2 \\ \text{Sr} & 1s^2 2s^2 2p^6 3s^2 3p^6 4s^2 3d^{10} 4p^6 5s^2 \end{array}$$

The group IA (1) and IIA (2) elements are in the *s block* of the periodic table because their highest-energy electron is in an s sublevel (Figure 7.20).

Next let's consider a few of the halogens, group VIIA (17). What pattern do you see in the electron configurations of F, Cl, and Br?

$$\begin{array}{ll} \text{F} & 1s^2 2s^2 2p^5 \\ \text{Cl} & 1s^2 2s^2 2p^6 3s^2 3p^5 \\ \text{Br} & 1s^2 2s^2 2p^6 3s^2 3p^6 4s^2 3d^{10} 4p^5 \end{array}$$

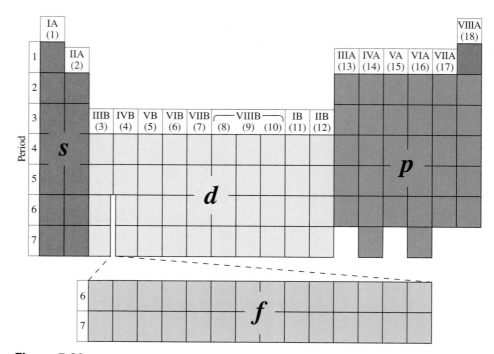

Figure 7.20
The periodic table can be subdivided into *s, p, d,* and *f blocks* that correspond to the last-occupied sublevel for the element in that block.

For the halogens, five electrons occupy the highest-energy sublevel, the *p* sublevel. Each ends in np^5. The noble gases, group VIIIA (18)—each with one more electron than the halogen it follows—have six electrons in the last-filled *p* sublevel. Each ends in np^6:

Ne	$1s^2 2s^2 2p^6$
Ar	$1s^2 2s^2 2p^6 3s^2 3p^6$
Kr	$1s^2 2s^2 2p^6 3s^2 3p^6 4s^2 3d^{10} 4p^6$

The halogens and noble gases, along with the other elements in groups IIIA (13) through VIIIA (18), are in the *p block* of the periodic table because their highest-energy electron is in a *p* sublevel (Figure 7.20). Look at the elements in group VIA (16). How do their electron configurations end? Do they all end with the same number of electrons in the last-occupied sublevel?

Next let's consider the transition metals. As shown in Figure 7.20, the transition metals are in the *d block.* Their highest-energy electrons are in *d* orbitals. Most of the time, the transition metals follow a similar pattern as the main-group elements in the *s* and *p* blocks. For example, the group IIIB (3) transition metals (Sc, Y) have one electron in the last-occupied *d* sublevel (nd^1). The group IIB (12) transition metals (Zn, Cd) have 10 electrons (nd^{10}).

Let's use these periodic trends to write electron configurations for elements. As atomic numbers increase, the number of electrons increases. So if we follow the atomic symbols of the elements in the periodic table, we should be able to write an electron configuration for any element as we go. As we enter a new *s, p,* or *d* block, we know to change to a new sublevel. The principal energy level number *n* written before the sublevel letter is obtained from the period number. For the *s* and *p* blocks, *n* is the period number (row number). For the *d* block, *n* is one less than the period number. Figure 7.21 helps to show the relationship between sublevels and the period numbers.

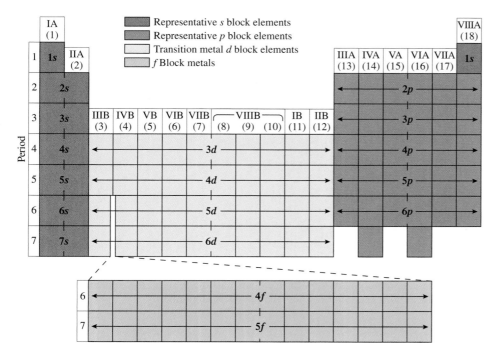

Figure 7.21
The periodic table helps us determine the electron configuration for an element. The principal energy level number n written before the sublevel letter is obtained from the period number. For the s and p blocks, n is the period number (row number). For the d block, n is one less than the period number. For the f block, n is two less than the period number.

Let's write the electron configuration for phosphorus using the periodic table. Phosphorus is a main-group element in the p block, so we expect its highest-energy electrons to be p electrons. Following the arrows in the first periodic table in Figure 7.22, we start in period 1. There is only an s block in period 1, so only the $1s$ orbital is filled. We write

$$1s^2 \qquad \text{(period 1)}$$

In the next row, period 2, there are two columns in the s block and six columns in the p block giving

$$2s^2 2p^6 \qquad \text{(period 2)}$$

In the third row, period 3, there are two columns in the s block. In the p block, we fill in only three electrons because phosphorus is in the third column of the p block:

$$3s^2 3p^3 \qquad \text{(period 3)}$$

Combining the configurations from each period, we get the electron configuration for phosphorus:

$$\text{P} \qquad 1s^2 2s^2 2p^6 3s^2 3p^3$$

The superscripts in the electron configuration for phosphorus should add to 15, the number of electrons in a phosphorus atom.

Next let's write the electron configuration for manganese (Mn). We follow the arrows in the second periodic table in Figure 7.22 from hydrogen to manganese. Manganese is a transition metal in the d block, so we expect its highest-energy electrons to be d electrons. Manganese is in period 4, so the last-occupied d orbitals are the $(4-1)d$ or $3d$ orbitals. To write the electron configuration, we follow the same general procedure for manganese as we did for phosphorus. In period 1 we fill the first principal energy level:

$$1s^2 \qquad \text{(period 1)}$$

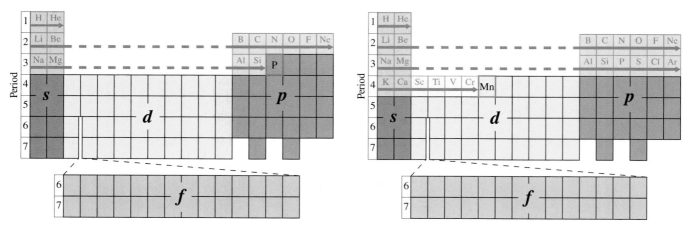

Figure 7.22
To determine the electron configuration for an element, follow the arrows to obtain its electron configuration, filling in one electron for each box (element symbol). When writing the electron configuration for elements past the third period, remember to label the d sublevel with a number one less than the period number.

In period 2 we fill the s and p sublevels in the second principal energy level:

$$2s^2 2p^6 \quad \text{(period 2)}$$

In period 3 we fill the s and p sublevels in the third principal energy level:

$$3s^2 3p^6 \quad \text{(period 3)}$$

In period 4 we first fill the $4s$ sublevel, then we continue to the d block which is the $3d$ sublevel (the period number -1). Manganese is in the fifth column of the d block, so there are five $3d$ electrons. The electron configuration for the s and d block electrons from period 4 is

$$4s^2 3d^5 \quad \text{(period 4)}$$

Combining the configurations from each period, we get the electron configuration for manganese:

$$\text{Mn} \qquad 1s^2 2s^2 2p^6 3s^2 3p^6 4s^2 3d^5$$

Do the superscripts total to the number of electrons in a manganese atom?

Example 7.6 will give you practice at writing electron configurations using the periodic table.

EXAMPLE 7.6 — Writing Electron Configurations Using the Periodic Table

Use a periodic table to write the electron configurations for the following elements.

(a) Si
(b) Zr
(c) Br

Solution:

(a) The atomic symbol for silicon, Si, is in the second column of the p block, and it is in period 3, so its electron configuration will end in $3p^2$. The rest of its electron configuration is obtained by following the periodic table in order of increasing atomic number starting with H and He in period 1. We get $1s^2$ for period 1, $2s^22p^6$ for period 2, and $3s^23p^2$ for period 3. Combining these we get the electron configuration for Si:

$$\text{Si} \qquad 1s^22s^22p^63s^23p^2$$

The superscripts add to 14, the number of electrons in a silicon atom.

(b) The atomic symbol for zirconium, Zr, is in the second column of the *d* block, and it is in period 5. Its electron configuration will end in $4d^2$ (not $5d^2$) because the principal energy level number for the *d*-block elements is always the period number minus one ($5 - 1 = 4$). The electron configuration is obtained in the same way as for Si. For periods 1, 2, and 3, we get the first part of the electron configuration: $1s^22s^22p^63s^23p^6$. In period 4, we add the *s*-block electrons, $4s^2$; the *d*-block electrons, $3d^{10}$; and the *p*-block electrons, $4p^6$. In period 5, we add the *s*-block electrons, $5s^2$; and the two *d*-block electrons, $4d^2$. Combining the sublevels from all five periods, we get the electron configuration for Zr:

$$\text{Zr} \qquad 1s^22s^22p^63s^23p^64s^23d^{10}4p^65s^24d^2$$

The superscripts add to 40, the number of electrons in a zirconium atom.

(c) The atomic symbol for bromine, Br, is in the fifth column of the *p* block, and it is in period 4. Its electron configuration will end in $4p^5$. For periods 1, 2, and 3, we get the first part of the electron configuration: $1s^22s^22p^63s^23p^5$. In period 4, we add the *s*-block electrons, $4s^2$; the *d*-block electrons, $3d^{10}$; and the *p*-block electrons, $4p^5$. Combining the sublevels from all four periods, we get the electron configuration for bromine:

$$\text{Br} \qquad 1s^22s^22p^63s^23p^64s^23d^{10}4p^5$$

Practice Problem 7.6

Use a periodic table to write the electron configurations for the following elements.

(a) S

(b) Rb

(c) Cd

Further Practice: 7.59 and 7.60

For elements with lengthy electron configurations, chemists often use *abbreviated electron configurations*. Notice that argon and calcium differ only by the $4s^2$ electrons at the end of the electron configuration for Ca:

$$\text{Ar} \qquad 1s^22s^22p^63s^23p^6$$
$$\text{Ca} \qquad 1s^22s^22p^63s^23p^64s^2$$

Instead of writing the complete electron configuration for Ca, we will use the symbol [Ar] to represent the part of calcium's electron configuration that corresponds to argon. The abbreviated electron configuration for Ca substitutes [Ar] for the first part of its electron configuration:

$$\text{Ca} \qquad [\text{Ar}]4s^2$$

By convention, electron configurations are only abbreviated using noble-gas configurations. To write an abbreviated electron configuration for an element, locate its position on the periodic table and find the closest noble gas in the period above the element. Use this noble gas in the abbreviation. Once you have written the noble-gas symbol in brackets, add the additional electron configuration needed to reach the element on the periodic table.

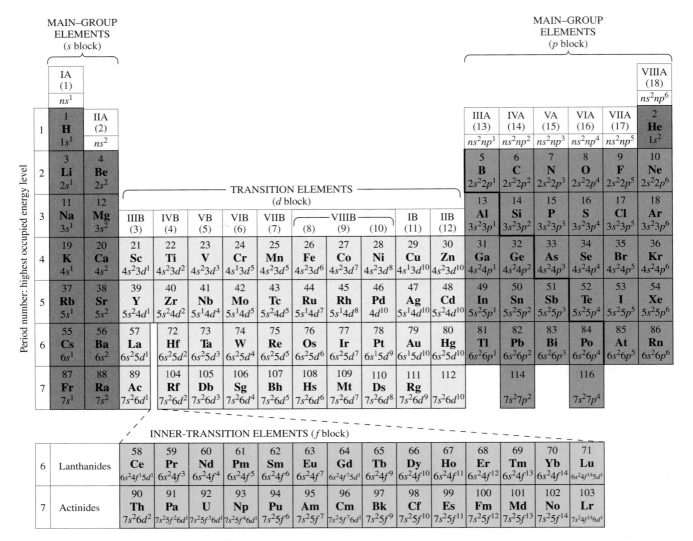

Figure 7.23
This periodic table shows the partial electron configurations (electrons beyond the previous noble gas) for all the elements. Note that the last-filled sublevels correspond to the *s*, *p*, *d* or *f* blocks designated in the periodic table.

There are a few minor exceptions to the normal filling order for electron configurations. Notably, the lanthanides and actinides vary from a predictable pattern. Also some of the transition metal elements in group VIB (6) and group IB (11) have electron configurations different from what the filling rules predict. The deviations result from the extra stability of half-filled and completely filled *d* sublevels. Can you find any exceptions for period 4 in Figure 7.23?

What is the abbreviated electron configuration for bromine (Br)? The closest noble gas with a lower atomic number is argon. We enclose the symbol for argon in brackets to represent its electron configuration. Argon is at the end of period 3, so we add the electrons that correspond to the blocks we must pass through in period 4 to get to bromine, just as we do when we write the complete electron configuration. The abbreviated electron configuration is

$$\text{Br} \qquad [\text{Ar}]4s^2 3d^{10} 4p^5$$

We have not yet mentioned electron configurations that include the *f* orbitals. Cerium (58), the first element of the lanthanide series, is the first element in the *f*-block series (Figure 7.21). It is enough to note here that the lanthanide and actinide elements are in the *f* block because their highest-energy electron is in an *f* sublevel. There are 14 lanthanides, this number corresponding to the number of electrons that can fill the seven 4*f* orbitals. Similarly, the actinide series (elements 90 to 103) corresponds to the filling of the 5*f* orbitals. The lanthanides and actinides are rarely encountered, and many irregularities occur in their electron configurations. The periodic table in Figure 7.23 shows abbreviated electron configurations for all the elements.

7.5   VALENCE ELECTRONS FOR THE MAIN-GROUP ELEMENTS

Elements in the same group have the same number of electrons in their last-filled sublevel and principal energy level. The last-filled principal energy level is called the **valence level,** or *valence shell.* It is highest in energy and contains orbitals that are larger than orbitals in lower principal energy levels. An electron that occupies the valence level is a **valence electron.** An electron in a principal energy level below the valence level is called a **core electron,** or inner electron. The similar properties of same-group elements, discussed in Chapter 2 and in Section 7.7, relate to the number of valence electrons in the atoms. Let's look closely at the valence electrons of the main-group elements.

One way to determine the number of valence electrons for an element is to look at its electron configuration or its abbreviated electron configuration. For example, the electron configuration for silicon is

$$\text{Si} \qquad 1s^2 2s^2 2p^6 3s^2 3p^2 \quad \text{or} \quad [\text{Ne}]3s^2 3p^2 \quad \longleftarrow \text{\small 4}$$

The valence level is the $n = 3$ principal energy level, which contains the $3s$ and $3p$ sublevels. From the electron configuration, we can see that there are a total of four electrons in the $n = 3$ principal energy level. Thus silicon has four valence electrons. Note that the valence electrons in silicon are the electrons that are written after the electron configuration for neon. Electrons that comprise a noble-gas electron configuration are core electrons held very strongly by the nucleus. For some elements, however, some of the core electrons are not part of a noble-gas electron configuration. For example, the abbreviated electron configuration for bromine is

$$\text{Br} \qquad [\text{Ar}]4s^2 3d^{10} 4p^5$$

The valence level is the fourth principal energy level ($n = 4$), which contains seven valence electrons. The $3d$ electrons are not valence electrons since they are in the third principal energy level ($n = 3$). *For any main-group element, the number of valence electrons is equal to the number of electrons in the highest-energy s and p sublevels.*

A more convenient way to determine the valence level and the number of valence electrons is to use the periodic table. In Section 7.4 we saw how the periodic table predicts an element's electron configuration. The principal energy level number is the same as the period number for the s and p sublevels. Now we know that the valence electrons occupy the highest-energy s and p sublevels. This tells us that the period number of the element equals the valence level number. The number of valence electrons can be determined from the Roman numeral group number of the element. Except for helium, the Roman numeral group number corresponds to the number of valence electrons because the group number is a sum of the s and p valence electrons. Silicon, for example, is in group IVA (14); the Roman numeral IV corresponds to four valence electrons, two s electrons and two p electrons. Bromine, in group VIIA (17), has seven valence electrons, as do all the other elements, the halogens, in that group. The periodic table shows that *elements in the same group have the same number of valence electrons.*

Example 7.7 will give you practice determining the number of valence electrons for main-group elements.

EXAMPLE 7.7   Number of Valence Electrons

For each of the following, determine the number of valence electrons.

(a) nitrogen   5
(b) potassium   5
(c) oxygen   6

Solution:

(a) The electron configuration for nitrogen (N) is

$$1s^2 2s^2 2p^3$$

The valence level has the highest principal energy level number, $n = 2$. There are five electrons in the valence level (2 electrons in the $2s$ sublevel and 3 electrons in the $2p$ sublevel), so there are five valence electrons. We can also use the periodic table to determine the number of valence electrons. Nitrogen is in group VA (15). The Roman numeral group number (V) corresponds to a nitrogen atom's five valence electrons. Note that using the periodic table to determine valence electrons is a faster method. We will use it for parts (b) and (c).

(b) Potassium (K) is in group IA, so it has just one valence electron, the $4s^1$ valence electron.

(c) Oxygen (O) is in group VIA (16), so it has six valence electrons, the $2s^2 2p^4$ valence electrons.

Practice Problem 7.7

For each of the following, determine the number of valence electrons.

(a) magnesium
(b) carbon
(c) lead

Further Practice: 7.75 and 7.76

7.6 ELECTRON CONFIGURATIONS FOR IONS

The electron configurations we have written so far are for atoms. Because atoms have no charge, the number of electrons equals the number of protons, which is the atomic number. Ionic substances, such as those that create the colors in fireworks, consist of charged ions, not neutral atoms. Ions are charged because their number of electrons is greater than or less than the number of protons. To write electron configurations for ions, we adjust the number of valence electrons to account for the charge.

Cations have fewer electrons than protons. To write their electron configurations, we remove valence electrons from the atom's electron configuration. Consider calcium as an example. A calcium atom has the following electron configuration:

$$\text{Ca} \qquad 1s^2 2s^2 2p^6 3s^2 3p^6 4s^2$$

When a Ca^{2+} ion forms, two electrons are lost from the valence $4s$ sublevel. We write the electron configuration for a Ca^{2+} ion as

$$Ca^{2+} \qquad 1s^2 2s^2 2p^6 3s^2 3p^6$$

Anions have more electrons than protons. To write their electron configurations, we add electrons to the valence level of the atom's electron configuration. Consider chlorine as an example. A chlorine atom has the following electron configuration:

$$\text{Cl} \qquad 1s^2 2s^2 2p^6 3s^2 3p^5$$

When a Cl^- ion forms from a chlorine atom, one electron is gained. It is added to the valence $3p$ sublevel. The electron configuration for a Cl^- ion is

$$Cl^- \qquad 1s^2 2s^2 2p^6 3s^2 3p^6$$

Note that the electron configurations for Ca^{2+} and Cl^- are identical, and they are the same as for an argon atom. The Ca^{2+} cation, the Cl^- anion, and the Ar atom are **isoelectronic** because they have the same number of electrons. Even so, the three are different because they have different numbers of protons. Can you think of other common ions that are isoelectronic with these three?

EXAMPLE 7.8   Electron Configurations for Ions

Write the abbreviated electron configuration for each of the following ions. For each ion identify another ion that is isoelectronic with it.

(a) S^{2-}
(b) Sr^{2+}
(c) Al^{3+}

Solution:

(a) To write the abbreviated electron configuration for the sulfide ion (S^{2-}), start with the abbreviated electron configuration for the sulfur atom:

$$S \qquad [Ne]3s^23p^4$$

The S^{2-} anion has a 2– charge, so we add two electrons to the highest-energy valence sublevel (the $3p$ sublevel). The electron configuration for S^{2-} is

$$S^{2-} \qquad [Ne]3s^23p^6 \text{ or } [Ar]$$

Another ion that is isoelectronic with S^{2-} must have the same number of electrons and the same electron configuration. Note that the electron configuration for S^{2-} is the same as that of the noble gas argon. Ions that are isoelectronic with Ar are also isoelectronic with S^{2-}. Examples include P^{3-}, Cl^-, K^+, and Ca^{2+}.

(b) To write the abbreviated electron configuration for the strontium ion (Sr^{2+}), start with the abbreviated electron configuration for the strontium atom:

$$Sr \qquad [Kr]5s^2$$

The Sr^{2+} cation has a 2+ charge, so we remove two electrons from the highest-energy valence sublevel (the $5s$ sublevel). The electron configuration for Sr^{2+} is

$$Sr^{2+} \qquad [Kr]$$

Ions that are isoelectronic with Sr^{2+} have the same electron configuration as the noble gas krypton. Ions that are isoelectronic with Kr are also isoelectronic with Sr^{2+}. Examples are Se^{2-}, Br^-, and Rb^+. Some transition metals are also isoelectronic with Sr^{2+}. An example is Zr^{4+}.

(c) To write the abbreviated electron configuration for the Al^{3+} ion, start with the abbreviated electron configuration for the aluminum atom:

$$Al \qquad [Ne]3s^23p^1$$

The Al^{3+} cation has a 3+ charge, so we must remove three electrons from the $n = 3$ valence level. The electron configuration for Al^{3+} is the same as that for neon.

$$Al^{3+} \qquad [Ne]$$

The Al^{3+} ion has 10 electrons like neon. Other ions with 10 electrons include O^{2-}, F^-, Na^+, and Mg^{2+}.

Practice Problem 7.8

Write the abbreviated electron configuration for each of the following ions. For each ion, identify another ion that is isoelectronic with it.

(a) Br^-
(b) N^{3-}
(c) K^+

Further Practice: 7.79 and 7.80

7.7 PERIODIC PROPERTIES OF ATOMS

The electronic structure of atoms explains what happens when light interacts with matter. When atoms absorb energy, electrons move into orbitals at higher energy levels. When the electrons drop to lower energy positions, energy is released in the form of electromagnetic radiation that is sometimes in the visible region of the spectrum. These transitions of electrons account for the colors of light Andrea, Ben, and Drew saw in the fireworks, neon lights, and lasers in Las Vegas.

In this section we'll discuss how electron arrangements relate to other properties of elements, including reactivity, tendency to lose electrons, and atomic size.

Chemical Reactivity and Electron Configurations

Recall from Chapter 5 the activity series that predicts the relative reactivity of metals. Notice the elements that are at the top of the activity series, as shown in Figure 7.24. Can you see the pattern? Most are alkali metals, group IA (1), and alkaline earth metals, group IIA (2). They are the most reactive metals. These metals are so reactive that they are not found in nature as pure elements. They are usually found as ions combined with nonmetals in compounds.

The alkali metals, in the first group in the periodic table, are commonly found in nature as oxides. Their stable compounds with oxygen have the general formula M_2O, where M represents an alkali metal:

$$Li_2O \qquad Na_2O \qquad K_2O \qquad Rb_2O \qquad Cs_2O$$

When the alkali metals combine with oxygen or with other elements in compounds, they are 1+ ions. The electron configurations for the alkali metals explain this similarity:

Li	$1s^22s^1 = [He]2s^1$
Na	$1s^22s^22p^63s^1 = [Ne]3s^1$
K	$1s^22s^22p^63s^23p^64s^1 = [Ar]4s^1$
Rb	$1s^22s^22p^62s^22p^63s^23p^64s^23d^{10}4p^65s^1 = [Kr]5s^1$
Cs	$1s^22s^22p^62s^22p^63s^23p^64s^23d^{10}4p^65s^24d^{10}5p^66s^1 = [Xe]6s^1$

What do these elements have in common? All have a single valence electron in an s orbital. For each, the loss of the valence s electron gives the electron configuration of the noble gas that precedes it in the periodic table.

When the alkali metals form 1+ ions, they lose their only valence electron to get an electron configuration of a noble gas:

Li^+	$1s^2 = [He]$
Na^+	$1s^22s^22p^6 = [Ne]$
K^+	$1s^22s^22p^63s^23p^6 = [Ar]$

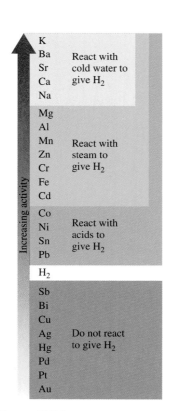

Increasing activity

K Ba Sr Ca Na	React with cold water to give H_2
Mg Al Mn Zn Cr	React with steam to give H_2
Fe Cd	
Co Ni Sn Pb	React with acids to give H_2
H_2	
Sb Bi Cu Ag Hg Pd Pt Au	Do not react to give H_2

Figure 7.24

In this chemical activity series, the most active elements appear at the top.

Rb$^+$ $1s^2 2s^2 2p^6 2s^2 2p^6 3s^2 3p^6 4s^2 3d^{10} 4p^6$ = [Kr]

Cs$^+$ $1s^2 2s^2 2p^6 2s^2 2p^6 3s^2 3p^6 4s^2 3d^{10} 4p^6 5s^2 4d^{10} 5p^6$ = [Xe]

Because valence electrons are farthest from the nucleus, they're the ones that participate in chemical reactions. Chemical reactions of the elements involve changes in the electron configurations—that is, addition, removal, or sharing of electrons—to give more stable configurations. Usually, a change in the number of electrons results in a noble-gas electron configuration:

He $1s^2$

Ne   [He]$2s^2 2p^6$

Ar   [Ne]$3s^2 3p^6$

Kr   [Ar]$3d^{10} 4s^2 4p^6$

Xe   [Kr]$4d^{10} 5s^2 5p^6$

What do the electron configurations of the noble gases have in common? Their valence *s* and *p* sublevels are full. Because very few compounds of the noble gases are known (and must be synthesized under very special conditions), we can conclude that the filled *s* and *p* sublevels give exceptional stability to any atom or ion.

The alkaline earth metals are also quite reactive, although not as reactive as the alkali metals. The abbreviated electron configurations for the alkaline earth metals are

Be   [He]$2s^2$

Mg   [Ne]$3s^2$

Ca   [Ar]$4s^2$

Sr   [Kr]$5s^2$

Ba   [Xe]$6s^2$

These elements are commonly found in nature as 2+ ions combined with oxygen or with halogens. The ions of the alkaline earth metals have noble-gas electron configurations:

Be^{2+} [He]

Mg^{2+} [Ne]

Ca^{2+} [Ar]

Sr^{2+} [Kr]

Ba^{2+} [Xe]

We've considered the metals on the far left of the periodic table. Let's now consider the halogens on the right, next to the noble gases. The halogens are highly reactive nonmetals, so they are rarely found in nature as pure elements. They are found as 1– ions combined with metal ions, such as in NaCl or CaF$_2$. The chemical behavior of the halogen elements is based on *gaining* one electron. How are their electron configurations similar?

F   [He]$2s^2 2p^5$

Cl   [Ne]$3s^2 3p^5$

Br   [Ar]$3d^{10} 4s^2 4p^5$

I   [Kr]$4d^{10} 5s^2 5p^5$

What ions do these elements form? How can we explain their ion formation in terms of forming the more stable electron configuration of a noble gas?

The modern model of the atom, with its description of electronic structure, provides an explanation for the similar properties of elements within the same group of the periodic table. However, elements in the same group do not have exactly the

Animation: Properties of Alkali and Alkaline Earth Materials
Online Learning Center, Animations

Chemistry Animations Library, Properties of Alkali and Alkaline Earth Materials, GroupI_And_II_rxn.rm.

Because francium is very rare and highly radioactive, we won't discuss it here.

Lithium

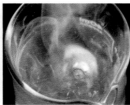
Sodium

Potassium

Rubidium

Cesium

Figure 7.25
All the alkali metals react with water. Their reactivity increases going down the group on the periodic table.

same properties. Many properties vary in a regular fashion moving down a group. As an example, let's look more closely at the chemical reactivity and electron configurations of the alkali metals.

If our students on their Las Vegas trip threw chunks of the alkali metals into one of the spectacular water fountains in Las Vegas, they'd observe some evidence of chemical reactions—some of them quite spectacular. *Of course, the students would not actually do this because the reactions are very dangerous.* The reactions of the alkali metals with water are shown in Figure 7.25.

Beginning with lithium, Ben, Andrea, and Drew might notice the chunk fizzle a little, a sign that a gas is being produced. The gas is hydrogen, which is produced by a single-replacement reaction. Sodium, the next alkali metal down the group in the periodic table, reacts more vigorously. Things get even more interesting when potassium meets water. It reacts violently, and the hydrogen usually ignites to give a flame. Rubidium reacts even more explosively. Cesium's reaction is so violent that it blows out the side of the container. Rubidium and cesium are both so reactive that they will react quickly with the oxygen in the air. Scientists must work with them in an unreactive argon environment.

The reactivity of the alkali metals increases as we go down the group in the periodic table. This general trend is also observed with the alkaline earth metals, group IIA (2), which are less reactive than the alkali metals. How can we explain the variation in reactivity within a group? Next, we will learn about atomic properties that help explain these trends. We will also see why properties change gradually as we move across the periodic table from left to right.

Ionization Energy

When metal elements react, they lose valence electrons and form cations. *The more easily metal atoms give up their valence electrons, the more reactive they are.* One important factor is **ionization energy** (*IE*), which is a measure of the energy required to remove a valence electron from a gaseous atom to form a gaseous ion. It is usually expressed as the energy required to remove an electron from each atom in exactly 1 mol of atoms. For example, consider the ionization process for lithium, which can be represented with the following equation:

$$\text{Li}(g) \longrightarrow \text{Li}^+(g) + \text{e}^- \qquad IE = 520 \text{ kJ/mol}$$

In other words, 520 kJ of energy would be required to remove the outermost electron from each atom in 1 mol of lithium. Ionization energy is related to electron arrangement and helps explain differences in the reactivities of metals. *In general, atoms with low ionization energies do not bind their electrons very tightly, so they are very reactive.*

Trends in Ionization Energies
The ionization energies of the main-group elements are given in Figure 7.26. What patterns can you find? Notice that the ionization energies of the alkali and alkaline earth metals are relatively low. If we look specifically at elements in the same group, we can see that their ionization energies

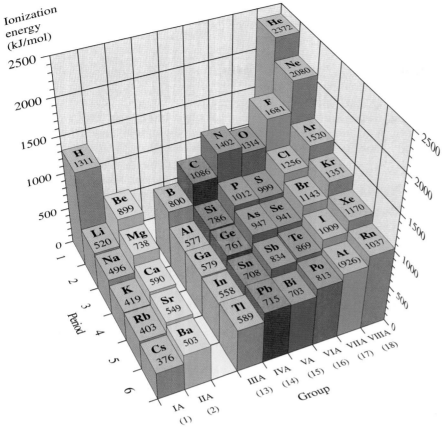

Figure 7.26
Ionization energies for the main-group elements are given in units of kJ/mol. They are the amount of energy required to remove one electron from each atom in 1 mol of atoms of the element when in the gaseous state.

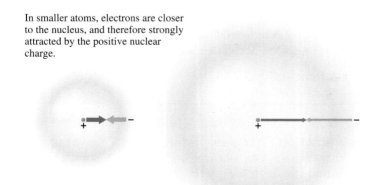

In smaller atoms, electrons are closer to the nucleus, and therefore strongly attracted by the positive nuclear charge.

Figure 7.27
A valence electron in a lower principal energy level is closer to the nucleus than a valence electron in a higher principal energy level. The attraction between the nucleus and the valence electron is greater for smaller atoms, resulting in a greater ionization energy.

decrease as we move down a group. This happens because the valence electron that's removed upon ionization is farther from the nucleus as we move down a group. Electrons farther from the nucleus are easier to remove. They are not as strongly attracted to the positive charge of the nucleus (Figure 7.27). This explains why elements such as cesium are so reactive.

Figure 7.28
Ionization energy tends to increase going up the periodic table and from left to right.

You may have observed another pattern in ionization energies from Figure 7.26. They increase from left to right on the periodic table. What is different about atoms in the same period that makes ionization energy increase? The increase in the number of electrons does not explain this trend. It is the increase in the number of protons. As the atomic number increases across a given period, the number of protons in the nucleus increases, increasing the positive charge in the nucleus. This increase in nuclear charge pulls the electrons in the valence level closer to the nucleus and makes removal of a valence electron more difficult. The general tendency is for ionization energy to increase from bottom to top and from left to right across the periodic table (Figure 7.28).

EXAMPLE 7.9 — Trends in Ionization Energy

Without looking at Figure 7.26, use periodic trends in ionization energy to answer the following questions. Explain your reasoning.

(a) Which element, carbon or fluorine, should have the greater ionization energy? Why?
(b) Which element, strontium or magnesium, should have the greater ionization energy? Why?

Solution:

(a) Carbon and fluorine are in the same period in the periodic table, with fluorine farther to the right. Ionization energy increases from left to right across a period, with exceptions for the first and fourth electrons in the *p* orbitals. Since neither of these elements is one of the exceptions, fluorine should have the greater ionization energy.
(b) Strontium and magnesium are in the same group in the periodic table, with magnesium closer to the top. Ionization energy increases up a group, so magnesium should have the greater ionization energy.

Practice Problem 7.9

Without looking at Figure 7.26, use periodic trends in ionization energy to answer the following questions. Explain your reasoning.

(a) Which element, potassium or lithium, should have the greater ionization energy?
(b) Which element, silicon or chlorine, should have the greater ionization energy?

Further Practice: 7.87 and 7.88

Successive Ionization Energies In this section we have focused on the ionization energy that removes one electron from a neutral atom. This is sometimes referred to as the first ionization energy IE_1. Suppose we remove a second electron. Will its removal require the same amount of energy as the first? For example, consider a lithium atom, which has three electrons. The first ionization energy of lithium corresponds to the removal of the first electron. To remove an electron from 1 mol of lithium atoms requires 520 kJ of energy. We represent the process involved for the first ionization energy IE_1, with the following equation:

$$\text{Li}(g) \longrightarrow \text{Li}^+(g) + e^- \qquad IE_1 = 520 \text{ kJ/mol}$$

To remove an electron from 1 mol of Li^+ ions requires 7298 kJ. This is the second ionization energy IE_2, and it is represented with the following equation:

$$\text{Li}^+(g) \longrightarrow \text{Li}^{2+}(g) + e^- \qquad IE_2 = 7298 \text{ kJ/mol}$$

TABLE 7.1	Successive Ionization Energies for Period 2 Elements									
Element	IE_1	IE_2	IE_3	IE_4	IE_5	IE_6	IE_7	IE_8	IE_9	IE_{10}
Li	520	7298	11,815							
Be	899	1757	14,849	21,006			Core electrons			
B	800	2427	3660	25,026	32,827					
C	1086	2353	4620	6222	37,830	47,277				
N	1402	2856	4582	7475	9445	53,266	64,360			
O	1314	3388	5300	7469	10,989	13,326	71,334	84,078		
F	1681	3374	6050	8408	11,023	15,164	17,868	92,038	106,434	
Ne	2080	3952	6122	9370	12,178	15,238	19,999	23,069	115,379	131,431

Note: Ionization energies given in kJ/mol.

Removal of an electron from a mole of Li^{2+} ions requires 11,815 kJ, which is the third ionization energy IE_3.

$$Li^{2+}(g) \longrightarrow Li^{3+}(g) + e^- \qquad IE_3 = 11,815 \text{ kJ/mol}$$

Notice that ionization energy increases with each successive removal of an electron. How can we explain this trend? Remember that the second electron is removed from a 1+ charged ion. It is harder to remove because there are fewer electrons, so the ones that remain feel the attraction of the positively charged nucleus more strongly. The same is true for further removal of electrons: ($IE_3 > IE_2 > IE_1$). The electrons are held more strongly as the positive charge on the ion increases.

Notice that the increase in ionization energy with successive removal of electrons is not gradual. For lithium, the second ionization energy is more than 10 times greater than the first! How does the electron configuration of lithium explain this?

$$Li \qquad 1s^2 2s^1$$

The first electron to be removed is a $2s$ valence electron. The second electron to be removed is a $1s$ core electron. The nucleus holds core electrons much more tightly than valence electrons. The large second and third ionization energies for lithium show why lithium does not form Li^{2+} or Li^{3+} ions except under extreme conditions.

Table 7.1 shows the successive ionization energies for the elements in period 2. Note that removal of core electrons (to the right of the stair-step line) requires much more energy than removal of valence electrons (to the left of the line).

Atomic Size

Another trend that shows a pattern throughout the periodic table is the size of atoms. The size of an atom is often described in terms of its **atomic radius,** the distance from the center of the nucleus to the outer edge of the atom. The radius of an atom is often reported in units of picometers (1 pm = 10^{-12} m). Because the outer edge of an atom is not sharply defined, scientists measure the distance between the centers of two identical bonded atoms to determine atomic radii. The radius is one-half this distance (Figure 7.29).

Atomic radii for the main-group elements are shown in Figure 7.30. Do you observe any patterns? Notice that the atomic radius generally increases as we move down a group. How might you explain this trend? The electron configurations for the elements in a group provide the answer. For example, recall the electron configurations for the alkali metals, group IA (1):

Li	[He]$2s^1$	Rb	[Kr]$5s^1$
Na	[Ne]$3s^1$	Cs	[Xe]$6s^1$
K	[Ar]$4s^1$		

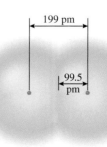

Cl—Cl

Figure 7.29
To determine the size of an atom, scientists measure the distance between the centers of two bonded atoms. For identical atoms, the atomic radius is one-half this distance.

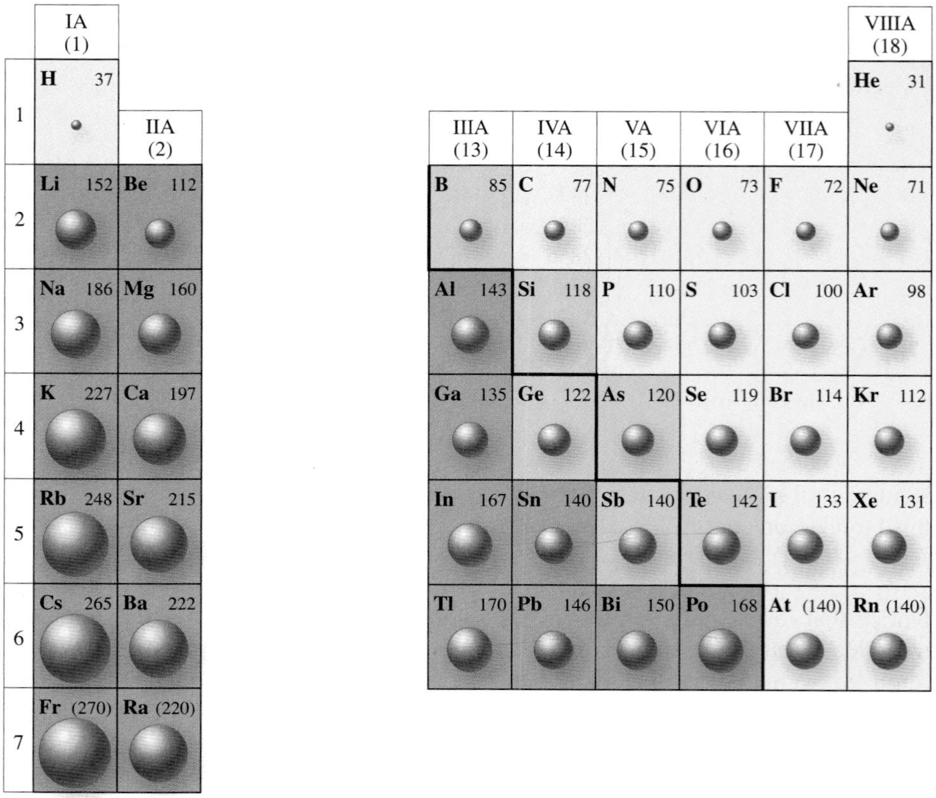

IA (1)										VIIIA (18)

Figure 7.30
This periodic table shows the relative atomic size of the main-group elements, along with their atomic radii in units of picometers.

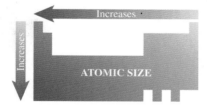

Figure 7.31
Atomic size increases down a group and from right to left in a period.

Lithium's valence electron is in the $2s$ orbital; the valence level is $n = 2$, very close to the nucleus. At the bottom of the group is cesium. Its valence level is $n = 6$, much farther from the nucleus. As we go from top to bottom in a group, the atomic radius increases because the valence electrons are in larger orbitals.

As we move from left to right across a period, atomic size generally decreases. You might think that size should increase because the number of electrons increases, but the electrons are added to the same principal energy level while the positive charge in the nucleus increases. As we move from left to right across a period, the increased positive charge in the nucleus becomes more effective at drawing the valence electrons closer to the center of the atom, causing atoms to the right on the periodic table to be smaller in size. Trends in atomic size are shown in Figure 7.31.

EXAMPLE 7.10   Relative Sizes of Atoms

(a) Which atom is larger, carbon or fluorine? Explain.
(b) Which atom is larger, strontium or magnesium? Explain.

Solution:

(a) As we go from left to right in a period, size decreases because the increase in positive nuclear charge pulls the valence electrons closer to the nucleus. Carbon atoms are larger than fluorine atoms.

(b) When we move down a group on the periodic table, size increases because the valence electrons are in principal energy levels farther from the nucleus. Strontium atoms are bigger than magnesium atoms.

Practice Problem 7.10

(a) Which atom is larger, potassium or lithium? Explain.
(b) Which atom is larger, silicon or chlorine? Explain.

Further Practice: 7.99 and 7.100

Sizes of Ions

When ions form from neutral atoms, the radii of the atoms change. Consider what happens when lithium ionizes (Figure 7.32A). A neutral atom of lithium has 3 electrons and 3 protons. The ion has 2 electrons and 3 protons. The nuclear charge remains the same, while the number of electrons decreases. In the ion, the protons draw the remaining electrons closer to the nucleus. Consequently, the Li^+ ion is smaller than the neutral atom.

The opposite happens with anions. For example, consider the formation of the fluoride ion (Figure 7.32B). The neutral atom has 9 electrons and 9 protons. The F^- ion has 10 electrons and 9 protons. Because there are more electrons, the protons cannot hold the electrons as close to the nucleus. Thus the ion is larger than the neutral atom. A summary of the sizes of many main-group ions is given in Figure 7.33.

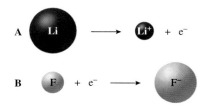

Figure 7.32
(A) A cation is smaller than the neutral atom from which it forms. The nucleus holds the cation's fewer electrons more tightly. (B) An anion is larger than the neutral atom from which it forms. The nucleus holds the anion's greater number of electrons less tightly.

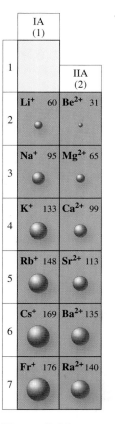
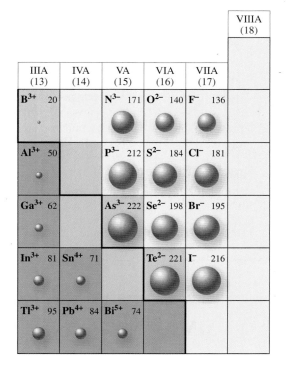

Figure 7.33
Shown here are the ionic radii for ions of the main-group elements. Note the trends down a group and for ions that are isoelectronic.

There are so many factors involved in determining the sizes of ions that no single trend spans the entire periodic table. However, for a series of ions with the same number of electrons—*an isoelectronic series*—consider the following ions:

Ion		Radius (pm)	Electron Configuration	Protons
S^{2-}		184	$1s^2 2s^2 2p^6 3s^2 3p^6$	16
Cl^-		181	$1s^2 2s^2 2p^6 3s^2 3p^6$	17
K^+		133	$1s^2 2s^2 2p^6 3s^2 3p^6$	19
Ca^{2+}		99	$1s^2 2s^2 2p^6 3s^2 3p^6$	20
Sc^{3+}		81	$1s^2 2s^2 2p^6 3s^2 3p^6$	21

Note that all the ions in this series have the same number of electrons and the same electron configuration. However they have different numbers of protons (atomic numbers). *For any isoelectronic series, as the number of protons increases, the ion size decreases.*

EXAMPLE 7.11   Sizes of Ions in an Isoelectronic Series

Arrange the following isoelectronic ions in order of increasing radius: Al^{3+}, F^-, Mg^{2+}, N^{3-}, Na^+, O^{2-}.

Solution:

Each of these ions has 10 electrons. The size of isoelectronic ions increases with decreasing nuclear charge. Thus, N^{3-}, with a nuclear charge of 7+, should be the largest ion, and Al^{3+}, with a nuclear charge of 13+, should be the smallest. The order is $Al^{3+} < Mg^{2+} < Na^+ < F^- < O^{2-} < N^{3-}$.

Practice Problem 7.11

Arrange the following isoelectronic ions in order of increasing radius: Br^-, Se^{2-}, Rb^+, Sr^{2+}, Y^{3+}.

Further Practice: 7.105 and 7.106

SUMMARY

Scientists interested in light led the way to our current understanding of atoms—specifically how electrons are arranged in atoms. Light, a form of electromagnetic radiation, can be described by its wavelength, frequency, or photon energy. The light we see is a small portion of the electromagnetic spectrum, the visible region. When an element is heated or given an electric charge, the energy absorbed is transformed into light energy. The light an element produces in this way contains only specific colors, as shown by its line spectrum. A hydrogen atom produces four lines in its visible line spectrum.

In Bohr's model of the hydrogen atom, electrons are described as circling the nucleus in orbits at specific distances from the nucleus. An electron in an orbit is said to have a specific energy. An electron moves to a higher orbit by absorbing energy, and it emits light energy when it falls back to a lower orbit. Bohr's model of the atom explains only the hydrogen line spectrum and not the spectra for multielectron atoms. However, Bohr's model is important because it describes electrons as having specific allowed energies. Their energies are quantized.

The idea that the energies of electrons are quantized is retained in the modern model of the atom, where electrons are described as being in specific principal energy levels, sublevels, and orbitals. Orbitals are three-dimensional regions in space where an electron is likely to be found. Sublevels contain sets of orbitals with the same letter designations: s, p, d, or f. In multielectron atoms, the sublevels in the same principal energy level have different energies: $s < p < d < f$. We describe the relative energies of the orbitals using an orbital diagram. To write ground-state electron configurations, we follow the three orbital filling rules: the aufbau principle, the Pauli exclusion principle, and Hund's rule. The periodic table is a useful tool for writing electron configurations. It can be divided into s, p, d, and f blocks, which helps us fill electrons into the correct sublevels.

Valence electrons are the electrons in the highest principal energy level. They are held less tightly by the nucleus than core electrons. For main-group elements, the number of valence electrons is the number of electrons in the last-filled s and p sublevels. Elements in the same group have the same number of valence electrons. Electron configurations for ions are written by adjusting the number of valence electrons. Many ions have the same electron configuration as the nearest noble gas, because noble-gas configurations are very stable.

The periodic properties of elements relate to the periodic properties of their atoms. Similar electron configurations and the same number of valence electrons explain the similar properties of elements in the same group. Atomic properties such as ionization energy and atomic size vary regularly in the periodic table, and they help explain the variation in properties within a group or period. Ionization energy (IE) is the amount of energy required to remove the highest-energy valence electron from an atom. Its value increases going up a group and from left to right within a period. Successive electrons are more difficult to remove, so their ionization energies are greater than the first ionization energy for a specific element ($IE_3 > IE_2 > IE_1$).

On the periodic table, atomic size increases when going down a group and from right to left within a period. Cations are smaller than their neutral atoms, and anions are larger. In an isoelectronic series, the ion with the greatest nuclear charge is the smallest.

KEY RELATIONSHIPS

Relationship	Equation
The frequency times the wavelength is equal to a constant, the speed of light. Frequency and wavelength are inversely proportional.	$v\lambda = c$
The energy of a photon is equal to Planck's constant times the light frequency. Photon energy is proportional to frequency.	$E_{photon} = hv$
The energy of a photon is equal to Planck's constant times the speed of light divided by wavelength. The energy of a photon is inversely proportional to wavelength.	$E_{photon} = \dfrac{hc}{\lambda}$
The change in energy of an electron is equal to the difference between the final energy level and the initial energy level occupied by the electron.	$\Delta E = E_{final} - E_{initial}$
The energy of a photon released due to an electron transition is equal to the absolute value of the change in energy of the electron, or the difference in energy between two energy levels.	$E_{photon} = |\Delta E|$

KEY TERMS

atomic radius (7.7)

aufbau principle (7.3)

continuous spectrum (7.1)

core electron (7.5)

electromagnetic radiation (7.1)

electron configuration (7.3)

frequency (7.1)

ground state (7.3)

hertz (Hz) (7.1)

Hund's rule (7.3)

ionization energy (7.7)

isoelectronic (7.6)

line spectrum (7.1)

orbital (7.3)

Pauli exclusion principle (7.3)

photon (7.1)

principal energy level (7.3)

quantized (7.2)

sublevel (7.3)

valence electron (7.5)

valence level (7.5)

wavelength (7.1)

QUESTIONS AND PROBLEMS

The following questions and problems, except for those in the *Additional Questions* section, are paired. Questions in a pair focus on the same concept. Answers to the odd-numbered questions and problems are in Appendix D.

Matching Definitions with Key Terms

7.1 Match the key terms with the descriptions provided.
  (a) energy that travels through space at the speed of light as oscillating waves
  (b) the number of wave cycles passing a stationary point in 1 second
  (c) the amount of energy required to remove a valence electron from an atom or ion in the gaseous state
  (d) a rule stating that electrons are distributed into orbitals of the same energy in such a way as to give the maximum number of unpaired electrons
  (e) a description of the distribution of electrons in an atom's sublevels
  (f) an inner electron that is not a valence electron
  (g) a region in space where an electron is likely to be found
  (h) a spectrum that consists of all the wavelengths of visible light
  (i) having the same number of electrons

7.2 Match the key terms with the descriptions provided.
  (a) the distance between two corresponding points on a wave
  (b) the smallest packet of energy of a specific type of electromagnetic radiation
  (c) a measure of the size of an atom, measured from the nucleus to the outer edge of the atom
  (d) a rule stating that a maximum of two electrons can occupy an orbital and that the electrons must have opposite spin
  (e) an electron that occupies the highest principal energy level
  (f) the lowest-energy electron configuration
  (g) an energy level consisting of orbitals of the same type and energy
  (h) a spectrum that consists of only certain wavelengths of light; a "fingerprint" for an element
  (i) having only specific allowed values

Electromagnetic Radiation and Energy

7.3 List three types of electromagnetic radiation that have longer wavelengths than visible light.

7.4 List three types of electromagnetic radiation that have higher frequencies than visible light.

7.5 Draw a picture of two waves, one with twice the frequency of the other. Label the wave with the higher frequency.

7.6 Draw a picture of two waves, one with three times the wavelength of the other. Label the wave with the longer wavelength.

7.7 List the following colors of visible light from shortest wavelength to longest wavelength: blue, orange, yellow, red.

7.8 List the following colors of visible light from lowest frequency to highest frequency: orange, green, violet, yellow.

7.9 What does it mean when we say that wavelength is inversely proportional to frequency?

7.10 What does it mean when we say that photon energy is proportional to frequency?

7.11 What type of radiation has wavelengths just slightly longer than red light?

7.12 What type of radiation has wavelengths just slightly shorter than violet light?

7.13 Which type of electromagnetic radiation is composed of the highest-energy photons? Has the longest wavelength? Has the highest frequency?

7.14 Which color of light in the visible region is composed of the lowest-energy photons? Has the shortest wavelength? Has the lowest frequency?

7.15 Calculate the frequency of light that has a wavelength of 75.0 nm. What type of radiation is this?

7.16 Calculate the wavelength of light that has a frequency of 5.00×10^{14} s^{-1}. What type of radiation is this?

7.17 What is the photon energy of light with a wavelength of 465 nm? What type of radiation is this?

7.18 What is the frequency of light that is composed of photons that each has an energy of 1.99×10^{-25} J? What type of radiation is this?

7.19 What type of light gives a continuous spectrum?

7.20 Should colored light from a heated ionic compound produce a continuous spectrum or a line spectrum?

7.21 Do any two elements produce the same line spectrum?

7.22 The line spectrum for neon is shown in Figure 7.8B. Are there any conditions when pure neon would not produce the same set of lines?

The Bohr Model of the Hydrogen Atom

7.23 Bohr proposed that the electron in a hydrogen atom circles the nucleus in orbits. In his model, could electrons exist between orbits? Explain.

7.24 In the Bohr model, could the electron ever fall below the $n = 1$ orbit? Explain.

7.25 In the Bohr model, how does the electron move to a higher-energy orbit or to a lower-energy orbit?

7.26 In the Bohr model, is light emitted or absorbed when an electron moves from a higher-energy orbit to a lower-energy orbit?

7.27 In the Bohr model, how many photons are emitted when an electron moves directly from the $n = 4$ to the $n = 2$ orbit?

7.28 In the Bohr model, how many photons are emitted when an electron moves from the $n = 4$ to the $n = 3$ orbit, and then from the $n = 3$ to the $n = 2$ orbit?

7.29 In the Bohr model, which of the following electron transitions in a hydrogen atom results in the emission of the highest-energy photon?
$n = 6$ to $n = 3$
$n = 5$ to $n = 3$

7.30 In the Bohr model, which of the following electron transitions in a hydrogen atom results in the emission of the highest-energy photon?
$n = 3$ to $n = 2$
$n = 4$ to $n = 3$

7.31 In the Bohr model, which of the following electron transitions in a hydrogen atom results in emission of the longest wavelength of light?
$n = 6$ to $n = 3$
$n = 5$ to $n = 3$

7.32 In the Bohr model, which of the following electron transitions in a hydrogen atom results in emission of the longest wavelength of light?
$n = 3$ to $n = 2$
$n = 4$ to $n = 3$

7.33 Explain how the Bohr model accounts for the four colored lines in the hydrogen line spectrum.

7.34 Which electron transition in the Bohr hydrogen atom corresponds to each of the colored lines in the hydrogen line spectrum (red, green, blue, violet)?

7.35 The wavelength of the green light in the hydrogen line spectrum is 434.1 nm. What is the photon energy of the green light emitted by a hydrogen atom?

7.36 The wavelength of the blue light emitted from a hydrogen atom is 410.1 nm. This light is a result of electron transitions from the $n = 5$ to the $n = 2$ energy levels. How much higher in energy is the $n = 5$ energy level than the $n = 2$ energy level?

The Modern Model of the Atom

7.37 How do the *orbitals* described in the modern model of the atom differ from the *orbits* described in the Bohr model?

7.38 Draw a picture of an $n = 1$ Bohr orbit and a $1s$ orbital. Why is it more difficult to draw a $1s$ orbital?

7.39 Draw a picture of a $2p$ orbital and a $3p$ orbital. How do they differ?

7.40 Draw a picture of a typical $3d$ orbital and a typical $4d$ orbital. What is the difference between the $3d$ orbital and the $4d$ orbital?

7.41 How many orbitals are in each of the following sublevels?
(a) $4s$
(b) $2p$
(c) $3d$
(d) $4f$
(e) $2s$
(f) $4p$

7.42 Which sublevels are in each of the following principal energy levels?
(a) $n = 1$
(b) $n = 2$
(c) $n = 3$
(d) $n = 4$

7.43 Complete the following orbital diagrams for the ground-state electron arrangement for silicon, lithium, and phosphorus. Be sure to follow the three orbital filling rules: the aufbau principle, the Pauli exclusion principle, and Hund's rule.

7.44 Complete the following orbital diagrams for the ground-state electron arrangement for aluminum, fluorine, and sulfur. Be sure to follow the three orbital filling rules: the aufbau principle, the Pauli exclusion principle, and Hund's rule.

7.45 Write the electron configuration for each of the following elements.
(a) silicon
(b) lithium
(c) magnesium

7.46 Write the electron configuration for each of the following elements.
(a) sodium
(b) neon
(c) sulfur

7.47   What is the maximum number of electrons that can fill a $2p$ orbital?

7.48   What is the maximum number of electrons that can fill a $2p$ sublevel?

7.49   What is the maximum number of electrons that can fill the second principal energy level?

7.50   What is the maximum number of electrons that can fill the first principal energy level?

7.51   How many unpaired electrons are in a phosphorus atom?

7.52   How many unpaired electrons are in a chlorine atom?

Periodicity of Electron Configurations

7.53   Which elements have five electrons in the highest-energy p sublevel in their ground state?

7.54   Which elements have only one electron in their highest-energy s orbital in their ground state?

7.55   Which elements have a partially filled d sublevel?

7.56   Which elements have a partially filled f sublevel?

7.57   Which element has one electron in the $3p$ sublevel?

7.58   Which element has two electrons in its $3d$ sublevel?

7.59   Write the complete electron configurations for atoms of the following elements.
(a) Na   (b) Mn   (c) Se

7.60   Write the complete electron configurations for atoms of the following elements.
(a) Sc   (b) As   (c) Ba

7.61   What is wrong with the following electron configuration for bromine (Br)?
$$1s^2 2s^2 2p^6 3s^2 3p^6 4s^2 4d^{10} 4p^6$$

7.62   What is wrong with the following electron configuration for rubidium (Rb)?
$$1s^2 2s^2 2p^6 3s^2 3p^6 4s^2 3d^{10} 3p^6 4s^2$$

7.63   Identify the elements that have the following abbreviated electron configurations.
(a) [Ne] $3s^2 3p^4$
(b) [Ar] $4s^2 3d^8$
(c) [Xe] $6s^2$

7.64   Identify the elements that have the following abbreviated electron configurations.
(a) [Ar] $4s^1$
(b) [Xe] $6s^2 4f^{14} 5d^{10}$
(c) [Kr] $5s^2 4d^{10} 5p^5$

7.65   Write the abbreviated electron configuration for each of the following elements.
(a) Na
(b) Mn
(c) Se

7.66   Write the abbreviated electron configuration for each of the following elements.
(a) Sc   (b) As   (c) Ba

7.67   Which element has two filled p sublevels and five electrons in one of its p sublevels?

7.68   Which element has four filled s sublevels and no d electrons?

Valence Electrons for the Main-Group Elements

7.69   How do you identify the valence electrons from an electron configuration?

7.70   How do you identify valence electrons from the periodic table?

7.71   Are the valence electrons always all the electrons that follow the noble-gas electron configuration in an abbreviated electron configuration? Explain.

7.72   For the element gallium (Ga), are the $3d$ electrons classified as valence electrons or core electrons? Explain.

7.73   Explain why the number of valence electrons for a main-group element is equal to its Roman numeral group number.

7.74   Which elements have the same number of valence electrons as selenium?

7.75   Determine the valence level and the number of valence electrons for each of the following elements.
(a) Ca   (b) F   (c) Bi

7.76   Determine the valence level and the number of valence electrons for each of the following elements.
(a) Sr   (b) Ga   (c) I

Electron Configurations for Ions

7.77   How does the electron configuration for a cation differ from that of its neutral atom? How is it similar?

7.78   How does the electron configuration for an anion differ from that of its neutral atom? How is it similar?

7.79   Write the complete and abbreviated electron configuration for each of the following ions. For each ion, identify another ion that is isoelectronic with it.
(a) Mg^{2+}
(b) O^{2-}
(c) Ga^{3+}

7.80   Write the complete and abbreviated electron configuration for each of the following ions. For each ion, identify another ion that is isoelectronic with it.
(a) Na^+
(b) P^{3-}
(c) H^-

7.81   Identify at least three ions that are isoelectronic with neon.

7.82   Identify at least three ions that are isoelectronic with krypton.

Periodic Properties of Atoms

7.83   Explain, in terms of ionization energies, why potassium is more reactive than sodium.

7.84   Explain, in terms of ionization energies, why strontium is more reactive than calcium.

7.85   Explain, in terms of ionization energies, why potassium is more reactive than calcium.

7.86   Explain, in terms of ionization energies, why the alkali and alkaline earth metals are not used to make coins or jewelry.

7.87   Rank the following elements in order of increasing ionization energy: P, O, S.

7.88 Rank the following elements in order of increasing ionization energy: Na, Cl, F.

7.89 Use electron configurations to explain why more energy is required to remove an electron from a lithium atom than from a sodium atom.

7.90 Use electron configurations to explain why more energy is required to remove an electron from a bromine atom than from an iodine atom.

7.91 Use electron configurations to explain why the ionization energy for fluorine is greater than that for oxygen.

7.92 Use electron configurations to explain why the ionization energy for calcium is greater than that for potassium.

7.93 Write balanced equations that represent the processes that correspond to the first and second ionization energies for magnesium.

7.94 Which ionization energy (IE_1, IE_2, IE_3) corresponds to the following process?

$$Al^{2+}(g) \longrightarrow Al^{3+}(g) + e^-$$

7.95 Which do you expect to have the higher third ionization energy (IE_3), magnesium or aluminum? Explain.

7.96 Which do you expect to have the higher second ionization energy (IE_2), potassium or calcium? Explain.

7.97 Explain, in terms of ionization energies, why the following ions do not form under normal conditions.
(a) F^+
(b) Na^{2+}
(c) Ne^+

7.98 Explain, in terms of ionization energies, why the following ions do not form under normal conditions.
(a) Ba^{3+}
(b) O^{2+}
(c) Cl^+

7.99 Rank the following elements in order of increasing atomic size: P, O, S. $O < S < P$

7.100 Rank the following elements in order of increasing atomic size: Na, Cl, F.

7.101 Why are chlorine atoms smaller than sulfur atoms?

7.102 Why are krypton atoms larger than argon atoms?

7.103 For each pair, identify the larger atom or ion.
(a) Al or Al^{3+}
(b) S or S^{2-} ?

7.104 For each pair, identify the larger atom or ion.
(a) F or F^-
(b) Sr or Sr^{2+}

7.105 Which is the larger ion, K^+ or Ca^{2+}? Explain.

7.106 Which is the larger ion, S^{2-} or Cl^-? Explain.

Additional Questions

7.107 In the modern model of the atom, we use orbital diagrams to describe the electron structure of the atom. How is the orbital diagram for the hydrogen atom different from the orbital diagram for all other atoms?

7.108 Consider the electron configurations of the noble gases. Which valence sublevels are completely filled for each noble gas?

7.109 Write the abbreviated electron configuration for each of the following elements.
(a) Bi
(b) Rn
(c) Ra

7.110 When writing electron configurations for elements in period 4, why are electrons placed in the $4s$ sublevel before the $3d$ sublevel?

7.111 How many p orbitals are filled with electrons in a xenon atom?

7.112 How many s orbitals are filled with electrons in a zirconium atom?

7.113 How many d orbitals are filled with electrons in a cadmium atom?

7.114 Is it better to describe the noble gases as having eight valence electrons or zero valence electrons?

7.115 Why is the periodic trend in atomic size opposite to the trend in ionization energy?

7.116 Describe the changes in the sizes of the copper and silver atoms during the following single-displacement reaction.

$$Cu(s) + 2AgNO_3(aq) \longrightarrow Cu(NO_3)_2(aq) + 2Ag(s)$$

7.117 Describe the changes in the sizes of the sodium and chlorine atoms during the following combination reaction.

$$2Na(s) + Cl_2(g) \longrightarrow 2NaCl(s)$$

7.118 How would you explain light *absorption* by an atom in terms of electron transitions?

Chemical Bonding

Michael stops by the snack bar and picks up a hamburger, fries with extra salt, and a soft drink. He takes his food to a park near campus, where he meets Ashley and Amanda for lunch. Ashley brought a salad from home, made of a variety of vegetables, including morel mushrooms that her family had picked in the woods the previous weekend. Amanda has a tuna sandwich on whole wheat bread.

As Michael slathers ketchup on his quarter-pound beef patty, his eating habits draw a little good-natured chiding from Ashley, a serious vegetarian. Michael defends his choices with a chuckle, saying he needs the protein in meat if he is going to lift weights at the gym. Ashley thinks for a moment and then counters that her mushrooms also contain protein. Amanda interrupts, pointing out that it's not the protein itself that is necessary for good nutrition, but the amino acids that proteins are made from. The human body contains enzymes that can break down proteins and other enzymes that reassemble the resulting amino acids into human proteins. Other enzymes process carbohydrates and fats, also needed in a balanced diet.

The three students start to wonder what makes some foods more desirable than others. They decide that appearance, taste, and odor attract us to food, but there is more to food and nutrition than what we observe. Humans must eat food that contains carbon, along with a number of other elements, in order to live, grow, develop, and stay healthy. But why do we have to consume carbon-based food? Carbon atoms form the backbone of most of the molecules that are in our bodies, as well as in the plants and animals that we eat. Carbon is also found in some nonliving things—including rocks, ocean water, and the atmosphere—as well as in coal, oil, and natural gas, which are the remains of plants and animals that lived millions of years ago. Carbon atoms are used over and over again. They move between and among organisms and the environment in a continuous cycle (Figure 8.1).

To see how the carbon cycle works, let's trace the possible history of a carbon atom in a mushroom in Ashley's salad. This carbon atom has been around for a long time, but not always as a part of a molecule in the mushroom. Eons ago it was part of a carbon dioxide molecule in the air. It was taken up by a leaf of a tree in a swampy tropical forest. The tree, through the process of photosynthesis, incorporated the carbon atom along with hydrogen from water into a glucose molecule, releasing oxygen from the water into the atmosphere. After a few centuries, the tree died and decomposed. It sank into the swamp and formed part of a layer of peat, partially carbonized vegetable matter often used as a fuel. Over time the swampy

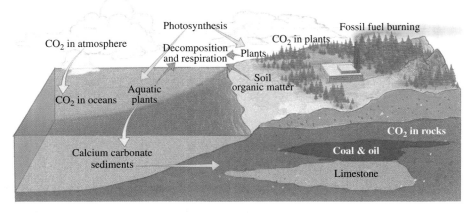

Figure 8.1
The movement of carbon around our planet is summarized by the carbon cycle. Some of the carbon transfer processes are rapid, while others take millions of years.

area dried and a river deposited layers of sediment on top of it, burying the peat and subjecting it to heat and pressure over millions of years. The carbon atom thus became a part of a layer of coal.

The coal was mined during the nineteenth century and burned to produce heat during the cold winters of Civil War times. The combustion caused the carbon atom to combine with oxygen, and it returned to the atmosphere as yet another carbon dioxide molecule. It remained there for over 100 years before being taken up by another tree, which subsequently died and began decomposing. In the rotting log, a mushroom spore began to grow. It absorbed the carbon atom and incorporated it into an amino acid. Inside the cells of the mushroom, many such amino acids were joined to form a protein molecule. The protein ended up in Ashley's salad. When she ate her lunch, she digested the protein molecule into amino acids. They entered her cells and became part of Ashley—whether muscle, hair, skin, or bone.

Amanda suggests that there is more than one possible pathway of a carbon atom in the cycle as she eyes her sandwich with new respect. Most of the carbon atoms in her bread are part of carbohydrates—compounds of carbon, hydrogen, and oxygen. In the body they "burn" to release energy and the waste product carbon dioxide. The CO_2 moves from cells into the bloodstream and then to the lungs, where it is exhaled into the atmosphere. Once there, it is available once more for plants to take up and use in the process of photosynthesis.

Michael says that there is more to the story. Carbon is all around us. Large amounts of carbon dioxide dissolve into the oceans, where the gas is converted to calcium carbonate in coral and in the shells of other marine organisms. Carbon dioxide is also converted to calcium carbonate in rocks such as limestone and marble. Interesting formations of limestone build up in caves, where the mineral dissolves under acidic conditions and then redeposits as stalactites and stalagmites when the water evaporates. Michael recognizes that calcium carbonate has many important uses as he pops an antacid tablet to relieve his upset stomach after lunch.

In this chapter we will examine the ways in which atoms such as carbon can combine to form compounds. What are the ways and why are they different? Much of the answer has to do with the forces that hold atoms together in molecular or ionic compounds. These forces are *chemical bonds*. Understanding chemical bonding helps us explain how compounds form and behave. In this chapter we will see how chemical bonds and the shapes of molecules explain the physical and chemical properties of many compounds, including those that contain carbon and those that don't. Before we explore these topics, consider the following questions.

Questions for Consideration

8.1 How can we classify the types of bonding of different compounds?
8.2 What is the nature of the bonding in ionic compounds?
8.3 What is the nature of the bonding in molecular compounds?
8.4 How do we predict and explain the shapes of molecular compounds and polyatomic ions?

8.1   TYPES OF BONDS

In our review of the carbon cycle, we mentioned two carbon compounds, $CaCO_3$ in limestone and CO_2 in the air. Recall from Chapter 3 that calcium carbonate ($CaCO_3$) is classified as an ionic compound because it is composed of a metal cation and a polyatomic anion. Carbon dioxide (CO_2) is classified as a molecular compound because it is composed of nonmetals. Consider the significantly different physical and chemical properties of these two compounds, as summarized in Table 8.1.

TABLE 8.1	Properties of Two Carbon Compounds	
Compound	**Calcium Carbonate**	**Carbon Dioxide**
Formula	$CaCO_3$	CO_2
Physical state	White solid	Colorless gas
Molar mass (g/mol)	100.1	44.01
Density (g/mL)	2.71	0.00198
Melting point (°C)	1339 (at high pressure)	−56.6 (at 5.11 atm)
Boiling point (°C)	Decomposes	Sublimes at 78.6
Electrical conductivity as solid	Very low	Very low
Electrical conductivity as liquid	High	Very low
Dissolves in	Acids	H_2O, CCl_4

TABLE 8.2	General Properties of Ionic and Covalent Substances
Ionic	**Covalent**
Crystalline solids	Gases, liquids, or solids
Hard and brittle solids	Brittle and weak solids, or soft and waxy solids
Very high melting point	Low melting point
Very high boiling point	Low boiling point
Good electrical conductor when molten or in solution	Poor conductor of electricity and heat
Often soluble in water but not in carbon tetrachloride	Often soluble in carbon tetrachloride but not in water

Why do these compounds have such different properties? To a considerable extent, the differences can be attributed to chemical bonds. A **chemical bond** is a force that holds atoms together in a molecule or compound. Calcium carbonate is an ionic compound made of Ca^{2+} ions and CO_3^{2-} ions. Ionic compounds are held together by *ionic bonds*. Carbon dioxide is a molecular compound made of CO_2 molecules. In each molecule, *covalent bonds* connect the carbon and oxygen atoms. Molecular substances are also called covalent substances.

The nature of these bonds results in the differences in properties observed for the two classes of compounds. General properties of ionic and covalent compounds are summarized in Table 8.2.

Ionic and Covalent Bonding

Electron configurations explain how and why atoms form different types of bonds in compounds that exhibit such different properties. In bonding, atoms tend to lose, gain, or share *valence* electrons to achieve a more stable configuration similar to that of a noble gas. In the simplest type of **ionic bonding,** one or more electrons are transferred from a metal to a nonmetal, forming a positively charged metal cation and a negatively charged nonmetal anion, each with a more stable electron configuration. The ions are held together by *electrostatic forces,* the attractions between the opposite charges of ions. The attractions are offset in part by repulsions between ions of like charge. Because each ion is surrounded by and attracted to several ions of opposite charge (Figure 8.2A), a lot of energy is required to separate the ions from one another. The strong ionic bonds explain the high melting points of ionic solids and the hard, crystalline form that many ionic compounds take.

In **covalent bonding,** electrons are not transferred from one atom to another. Instead they are shared in pairs, forming molecules in which each atom has at least a part of a noble-gas electron configuration. The molecules are only weakly

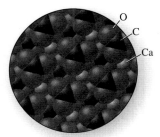

A $CaCO_3$

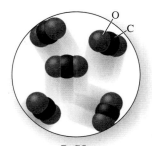

B CO_2

Figure 8.2
(A) Calcium carbonate consists of arrays of calcium cations and carbonate anions held together by attractions between their opposite charges. In the carbonate ions, each carbon is bonded to three oxygens with covalent bonds. (B) Molecules of carbon dioxide interact only weakly.

attracted to other molecules as in CO_2 (Figure 8.2B). This explains why molecular compounds generally have lower melting and boiling points than ionic compounds and usually occur naturally as gases, liquids, or soft solids. Covalent bonding generally occurs between nonmetals.

Notice in Figure 8.2A that calcium carbonate contains calcium ions and carbonate ions. The carbonate ions consist of one carbon atom bonded to three oxygen atoms. These bonds result from the sharing of electrons, so they are covalent. Thus calcium carbonate contains both ionic and covalent bonds.

EXAMPLE 8.1   Types of Bonding

Identify the type of bonding in each of the following substances.

alkalin earth metal   (a) $CaCl_2$ *halogen/non metal*
(b) NO_2
(c) $NaNO_3$

Solution:

(a) $CaCl_2$ contains a metal and a nonmetal, so the bonding is ionic.
(b) NO_2 contains two nonmetals, so the bonding is covalent.
(c) $NaNO_3$ has a metal and two nonmetals. Because the nonmetals form a polyatomic anion, NO_3^-, the bonding between nitrogen and oxygen is covalent. The bonding between the cation and the polyatomic anion is ionic.

Practice Problem 8.1

Identify the type of bonding in each of the following substances.

(a) NaF
(b) ClO_2
(c) $FeSO_4$

Further Practice: 8.9 and 8.10

Consider the following groups of chemical substances organized by type of bonding:

$$CaCl_2, K_2O, CsF \qquad H_2, N_2, O_3 \qquad CO, NO, H_2O$$

How would you classify the bonding in these groups? The first group consists of compounds made of a metal and a nonmetal. The second contains nonmetal-nonmetal molecules of identical elements. The third contains nonmetal-nonmetal molecules of different elements. Based on what you've read so far, you might classify the bonding in the first group as ionic and in the second and third groups as covalent. You would be right, but there is more to the story. Ionic and covalent bonding represent two extremes of a continuous range of bonding possibilities. Ionic bonding assumes a complete transfer of electrons from one atom to another. It best applies to compounds formed between the most metallic and the most nonmetallic elements, such as CsF. Covalent bonding best applies to compounds of two nonmetals. A pair of electrons may be shared equally by the two bonded atoms, as in Cl_2 and H_2. However, because nonmetal elements differ in their tendencies to lose or gain an electron, a covalent bond between two different nonmetals usually does *not* result in an equal sharing of electrons. This is the case in compounds like CO or H_2O. Instead they bond through a partial transfer of electrons from one atom to the other, with unequal sharing of electrons. The resulting bond has some characteristics of an ionic bond and some characteristics of a covalent bond.

A whole range of bonding types has this mixed ionic and covalent character. For convenience the bonds in a compound are usually classified as the dominant type. For example, the bonding in NaCl is considered ionic, and the bonding in CO_2 or HF is called covalent, although each has some of the other's character.

Polar and Nonpolar Covalent Bonds

Let's examine the unequal sharing of electrons in a covalent bond such as that in hydrogen fluoride. On the average, the electron pair is closer to the fluorine atom than to the hydrogen atom. Although HF is classified as covalent, it does have some ionic character because the electron from the hydrogen atom has been partially transferred to the fluorine atom. The amount of ionic character in a principally covalent bond can be judged from the polarity of the bond. **Polarity** is the degree of transfer of the electrons. Bonding patterns fall on a continuum from complete transfer and no sharing of electrons at one extreme to no transfer and equal sharing of electrons at the other extreme. The continuum is shown in Figure 8.3.

As Figure 8.3 shows, covalent bonds can be subdivided into two general types. In a **nonpolar covalent bond,** electrons are equally shared and charge is evenly distributed over the two bonded atoms. In a **polar covalent bond,** the electrons tend to be found closer, on the average, to one of the atoms than the other. This unequal sharing leads to the development of partial charges on the two atoms, as in the compound HF (Figure 8.4). The bonding electrons are closer to the fluorine atom than the hydrogen atom, so fluorine has a partial negative charge ($\delta-$). The hydrogen atom has the bonding electrons further away, so it develops a partial positive charge ($\delta+$).

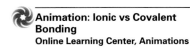

Animation: Ionic vs Covalent Bonding
Online Learning Center, Animations

Chemistry Animations Library, Ionic vs Covalent Bondings, Ionic_vs_Covalent Bonding.swf.

Electronegativity

How do we know whether a covalent bond is polar or nonpolar? In the early 1930s, Linus Pauling developed the idea that charges could be separated in a bond. He examined the difference in properties between the covalent bonds formed between identical elements (such as in H_2 and F_2) and those formed between different elements (such as in HF). The bonds between different elements are stronger. Pauling proposed that a partial transfer of electrons, resulting in partial ionic character, would explain the extra stability. The partial transfer occurs because one atom attracts the shared electrons more strongly than the other does. Pauling called the ability of an atom to attract bonding electrons **electronegativity.** If two atoms have the same electronegativity, the

Animation: Electronegativity
Online Learning Center, Animations

Chemistry Animations Library, Electronegativity, Electronegativity.rm.

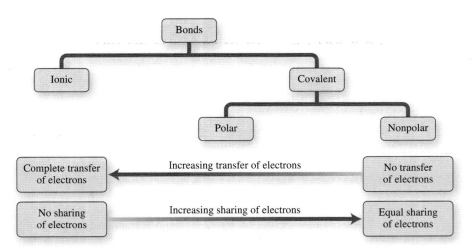

Figure 8.3
Bonds can be classified by the extent of transfer or sharing of electrons.

Figure 8.4
In HF, the shared electrons are closer, on the average, to the fluorine atom than to the hydrogen atom. Consequently fluorine develops a partial negative charge ($\delta-$) and hydrogen a partial positive charge ($\delta+$). The Greek letter delta (δ) is used to stand for partial charge, and is followed by the sign of the charge.

Figure 8.5
Electronegativity values of the
elements.

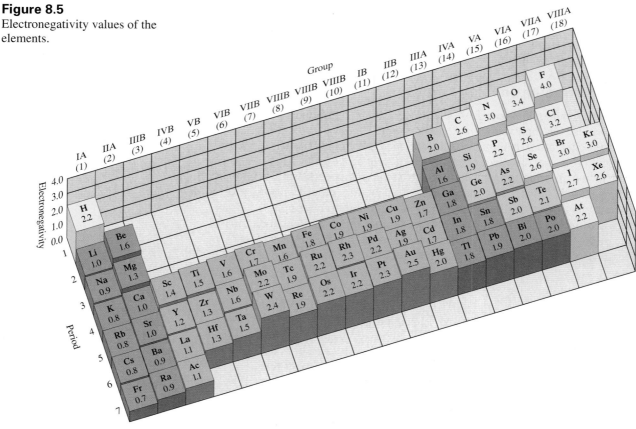

Linus Pauling was a chemistry
professor at the California Institute of
Technology. He received two Nobel
Prizes. The first, the Chemistry Prize
in 1954, was for his work on the
chemical bond. His second, the Peace
Prize in 1963, acknowledged his
efforts to stop nuclear weapons testing.

electrons are shared equally, and the bond is nonpolar covalent. The greater the difference in electronegativity, the more ionic character the bond has, and the greater the partial ionic charges on the atoms and the more polar the bond.

By examining the stability of various bonds, Pauling devised a scale of electronegativity values, with fluorine at the highest value (4.0) and francium at the lowest (0.7). Electronegativity values for most of the elements are given in Figure 8.5. The higher the electronegativity value of an element, the greater its attraction for electrons.

Electronegativity values show periodic trends. The nonmetals tend to gain electrons, so their electronegativity values are higher. The metals tend to lose electrons, so their electronegativity values are lower. Fluorine, with the highest electronegativity value, is the most reactive of the nonmetals. Francium, with the lowest electronegativity value, is the most reactive of the metals. Electronegativity values tend to increase from bottom to top of a group and from left to right across a period, as illustrated in Figure 8.6. We can use the difference in electronegativity to help us decide which bonds are more polar. *The greater the difference in electronegativity values between two elements, the more polar the bond that joins them.*

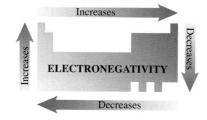

Figure 8.6
Trends in the values of
electronegativity of the elements.

EXAMPLE 8.2   Polarity of Bonds

Which of the following molecules contain polar bonds? If a bond is polar, which atom has the partial negative charge?

(a) CO_2 (b) O_3 (c) NCl_3 (d) CH_4

Which bond is the most polar? (In the formulas of compounds, the first element is the central atom.)

Solution:

To determine whether the bonds in a molecule are polar, we first find the electronegativity values of the bonded atoms. If the values are identical, the bond is nonpolar. If the values are different, the bond is polar, and the atom with the greater electronegativity has the partial negative charge.

(a) The electronegativity of carbon is 2.6. The value for oxygen is 3.4. The bond is polar, with the partial negative charge on the oxygen atom.
(b) The bonds are between three oxygen atoms, so there is no electronegativity difference and the bonds are nonpolar.
(c) Nitrogen and chlorine atoms have electronegativities of 3.0 and 3.2, respectively. The bond is polar, with the partial negative charge on the chlorine atoms.
(d) The bonds are between carbon and hydrogen. The electronegativity of carbon is 2.5, and the value for hydrogen is 2.1. The bond is polar, with the partial negative charge on the carbon atom.

The most polar bond will be the one with the greatest difference in electronegativities. The difference for $C-O$ is $3.4 - 2.6 = 0.8$. The difference for $C-H$ is $2.6 - 2.2 = 0.4$. Thus, the $C-O$ bond is more polar than the $C-H$ bond. The difference for $N-Cl$ is $3.2 - 3.0 = 0.2$, so this bond is less polar than $C-O$ or $C-H$. These results are summarized as follows in order of increasing polarity.

Bond	Difference in Electronegativity Values		Bond Type
$O-O$	$3.5 - 3.5 = 0$		Nonpolar covalent
$N-Cl$	$3.2 - 3.0 = 0.2$		Polar covalent
$C-H$	$2.6 - 2.2 = 0.4$		Polar covalent
$C-O$	$3.4 - 2.6 = 0.8$	(most polar)	Polar covalent

Practice Problem 8.2

Which of the following molecules contain polar bonds? If a bond is polar, which atom has the partial negative charge?

(a) SO_2 (b) N_2 (c) PH_3 (d) CCl_4

Which bond is the most polar? (In the formulas of compounds, the first element is the central atom.)

Further Practice: 8.23 and 8.24

8.2   IONIC BONDING

Nonpolar covalent bonds lie at one end of the continuum shown in Figure 8.3. Ionic bonds lie at the other. They are familiar to you already in compounds such as sodium chloride, the table salt that Michael likes on his fries.

The formation of ions and ionic bonds relates to the electron configurations of the elements. We can see the relationship in the periodic table. Each element immediately *following* a noble gas is a metal, which has a strong tendency to *lose* electrons and thereby achieve a stable noble-gas electron configuration. When giving up electrons to gain a noble-gas configuration, the metal develops a *positive* electrical charge, forming a *cation*. Each element immediately *preceding* a noble gas in the periodic table is a nonmetal, which has a strong tendency to *add* electrons to achieve a noble-gas configuration. In so doing, it develops a *negative* electrical charge, forming an *anion*.

| TABLE 8.3 | Lewis Symbols of the Period 2 Elements |

	Group							
	IA (1)	IIA (2)	IIIA (13)	IVA (14)	VA (15)	VIA (16)	VIIA (17)	VIIIA (18)
Electron configuration	$1s^22s^1$	$1s^22s^2$	$1s^22s^22p^1$	$1s^22s^22p^2$	$1s^22s^22p^3$	$1s^22s^22p^4$	$1s^22s^22p^5$	$1s^22s^22p^6$
Valence electrons	1	2	3	4	5	6	7	8
Lewis symbol	Li·	·Be·	Ḃ·	·Ċ·	:Ṅ·	:Ö·	:F̈·	:N̈e:

Gilbert Newton Lewis was a chemistry professor at the University of California, Berkeley. He proposed the concept of covalent bonding in 1916.

For the main-group elements, the valence electrons are located in ns and np orbitals, where n designates the principal energy level.

Transition metals are not usually represented with Lewis symbols since they may have as many as 10 d electrons, which would make the symbols cumbersome.

Lewis Symbols

To determine the charge of any monatomic ion, we need only examine the atom's valence electrons, because they are the electrons involved in ion formation. A convenient way to show the valence electrons is with a **Lewis symbol** or an *electron dot symbol,* in which dots placed around an element's symbol represent valence electrons. The dots are placed *singly* on the four sides of the elemental symbol, in any order, and then paired as necessary. An element with a noble-gas configuration is surrounded by four pairs of dots representing eight electrons, an *octet.*

Table 8.3 shows the Lewis symbols for the period 2 elements. They follow a progression starting with one valence electron for Li and ending with eight for Ne. Try drawing the Lewis symbols for the period 3 elements, Na through Ar. What pattern develops? It should look the same, starting with one valence electron for the group IA (1) elements and ending with eight in group VIIIA (18). All the noble gases come at the ends of periods. All but He have eight valence electrons occurring in four pairs.

EXAMPLE 8.3   Lewis Symbols of Atoms

The electron configuration of sulfur is $1s^22s^22p^63s^23p^4$. Write the Lewis symbol for the sulfur atom.

Solution:

The valence electrons are those designated $3s^23p^4$. Sulfur has six valence electrons, so it has two electron pairs and two unpaired electrons: :S̈·

Practice Problem 8.3

Write the Lewis symbol for the silicon atom.

Further Practice: 8.29 and 8.30

The Lewis symbols for hydrogen and helium in period 1 do not fit the pattern. Since their valence levels can hold only two electrons, only one side of the elemental symbol is used. Hydrogen is represented as H· , which corresponds to the electron configuration $1s^1$. Helium is He: with two valence electrons, representing the filled level $1s^2$. The hydride ion (H^-) formed by adding one electron to the hydrogen atom, is represented as H:⁻ . The hydrogen ion (H^+) is formed by removing one electron from the hydrogen atom. Since it has no remaining electrons, its Lewis symbol H^+ has no dots.

The loss of electrons from a metal and the gain of electrons by a nonmetal create the simplest kind of ionic bond. The transfer must happen at the same time. A metal can lose electrons to form a cation only when a nonmetal is available to gain them. The nonmetal then forms an anion with a stable octet of valence electrons. The following equations use Lewis symbols to illustrate this process in the formation of sodium chloride:

$$\text{Na·} \longrightarrow \text{Na}^+ + \text{e}^-$$

$$:\ddot{\text{Cl}}· + \text{e}^- \longrightarrow :\ddot{\text{Cl}}:^-$$

$$\text{Na}^+ + :\ddot{\text{Cl}}:^- \longrightarrow (\text{Na}^+)(:\ddot{\text{Cl}}:^-) \quad \text{or} \quad \text{NaCl}$$

The sodium atom loses one electron to form a sodium ion, which has a noble-gas configuration. This electron is transferred to the chlorine atom, which becomes the chloride ion; this ion also has a noble-gas configuration. The ions associate to form sodium chloride, a compound held together by attractions between the opposite charges of the ions.

Because sodium needs to lose one electron and chlorine needs to gain one, the atoms combine in a 1:1 ratio, which leads to electrical neutrality. Other ions may combine in different ratios. Consider, for example, the formation of aluminum chloride:

$$·\dot{\text{Al}}· \longrightarrow \text{Al}^{3+} + 3\text{e}^-$$

$$3:\ddot{\text{Cl}}· + 3\text{e}^- \longrightarrow 3:\ddot{\text{Cl}}:^-$$

$$\text{Al}^{3+} + 3:\ddot{\text{Cl}}:^- \longrightarrow (\text{Al}^{3+})(:\ddot{\text{Cl}}:^-)_3 \quad \text{or} \quad \text{AlCl}_3$$

Aluminum loses three electrons to achieve a noble-gas electron configuration. Since it takes only one electron to complete the chlorine atom's octet, three chlorine atoms are required to use the three electrons given up by the aluminum atom. We now have an aluminum ion and three chloride ions, which associate in an electrically neutral 1:3 ratio to form ionic aluminum chloride (AlCl_3). This ratio is maintained by an array of aluminum cations and chloride anions arranged in a regular pattern.

EXAMPLE 8.4   Lewis Symbols of Ions

Write the Lewis symbols for the magnesium and sulfide ions. Then write a formula for a compound that would form between them, using their Lewis symbols.

Solution:

Magnesium can lose two valence electrons to achieve a noble-gas configuration, giving the Mg^{2+} cation with no remaining valence electrons:

$$·\text{Mg}· \longrightarrow \text{Mg}^{2+} + 2\text{e}^-$$

Sulfur has six valence electrons—two electron pairs and two unpaired electrons. It gains two electrons to form the S^{2-} anion:

$$:\dot{\ddot{\text{S}}}· + 2\text{e}^- \longrightarrow :\ddot{\text{S}}:^{2-}$$

These ions associate in a 1:1 ratio to form MgS:

$$\text{Mg}^{2+} + :\ddot{\text{S}}:^{2-} \longrightarrow (\text{Mg}^{2+})(:\ddot{\text{S}}:^{2-}) \quad \text{or} \quad \text{MgS}$$

Practice Problem 8.4

Write the Lewis symbols for the beryllium and nitrogen ions. Then write a formula for a compound that would form between them, using their Lewis symbols.

Further Practice: 8.31 and 8.32

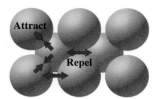

Figure 8.7
In an ionic crystal, ions of like charge repel one another, while ions of opposite charge attract. The regular pattern of their arrangement assures that attractions are stronger than repulsions.

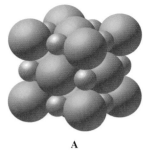

A

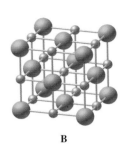

B

C

Figure 8.8
The sodium chloride crystal is represented at the molecular level in two ways. In each, look for the arrangement of one ion surrounded by six ions of opposite charge. (A) In a space-filling model, the smaller spheres are the sodium ions and the larger spheres are the chloride ions. The model shows the relative sizes of the ions accurately, but masks the details of the crystal structure. (B) A ball-and-stick model reveals the inside of the crystal, but does not show relative sizes correctly. (C) A photograph of some NaCl crystals shows how the shape of the crystals corresponds with the arrangement of ions within the crystals.

Structures of Ionic Crystals

The bonds that hold ions together in ionic compounds result from attractions between oppositely charged ions. If like-charged ions are close to one another, their repulsions partially offset these attractive forces, as shown in Figure 8.7. How do the ions fit together to make a stable ionic compound? An ion, which may be represented as a charged sphere, exerts a force equally in all directions, so ions of opposite charge surround it. The pattern is a **crystal lattice**, and the result is an **ionic crystal**, in which the ions are arranged in a regular geometric pattern that maximizes the attractive forces and minimizes the repulsive forces.

The charges and sizes of the ions largely determine the characteristic patterns of ionic crystals. For example, many ionic salts with a 1:1 ratio of ions have the same structure as sodium chloride, with each ion surrounded by six ions of opposite charge (Figure 8.8).

For a crystal lattice to form, cations and anions must come into contact. If the cations are too small or too large, a different structure develops. For example, if a salt contains a large cation, such as a cesium ion, each is surrounded by eight similarly sized ions of opposite charge. The crystal lattice for cesium chloride is shown in Figure 8.9. For other compositions, such as salts with 1:2 or 2:1 ratios of ions, different patterns form. A typical structure shown in Figure 8.10 is that of CaF_2, found in the mineral fluorite.

Ionic bonds are very strong because of the large number of interactions between oppositely charged ions. It takes a lot of energy to separate an ion from the

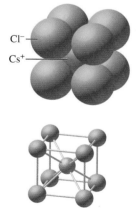

Cl⁻
Cs⁺

Figure 8.9
The cesium chloride structure is adopted when cations and anions are about the same size.

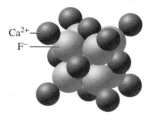

Ca²⁺
F⁻

Figure 8.10
Ionic compounds that have twice as many of one ion as the other often adopt the same structure as CaF_2, found in the mineral fluorite.

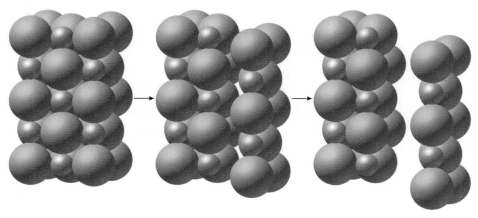

Figure 8.11
When a crystal is struck with sufficient force, layers of ions shift. The attractive forces between ions of opposite charge move out of alignment. Layers of like charge align, and they are pushed apart by repulsive forces. The crystal splits along these lines.

solid structure. Because melting and boiling require the separation of ions, ionic compounds have high melting and boiling points.

Strong attractive forces also make ionic crystals hard and brittle. If struck forcefully, ionic crystals shatter. They break because the force shifts the alignment of ions from the stable interaction of opposite charges to an unstable alignment of like charges (Figure 8.11). Along a shifted layer, strong repulsive forces replace strong attractive forces, and the crystal cracks.

The crystalline structure of ionic compounds also explains why solid ionic salts are poor conductors of electricity. The ions are not mobile but are held in place in the crystal lattice. However, melting greatly increases the electrical conductivity of ionic salts, because the ions in the liquid are free to move around.

8.3   COVALENT BONDING

When two nonmetals form a compound, as in the carbon dioxide that Michael, Ashley, and Amanda exhale, the bond is covalent. Since the atoms do not give up electrons, they cannot achieve a noble-gas configuration, but they can come close by sharing electrons. This compromise allows them to approximate a noble-gas configuration, with the shared electrons being a part of the electron configuration of both atoms.

While ionic bonds consist of electrostatic forces spread out over the crystal, covalent bonds involve a more localized attractive force between atoms. The atoms are held together tightly in molecules. A covalent bond is strong because each shared electron interacts simultaneously with two nuclei. Electrons and nuclei attract one another because of their opposite charges. At the same time, electrons repel one another and nuclei repel each other because of their like charges. When the electrons are located primarily between the two nuclei, the attractions are maximized and the repulsions are minimized, resulting in a covalent bond, as shown in Figure 8.12.

Although the attractions (bonds) between atoms within the molecule are strong, the molecules are not bonded together. Consequently covalent compounds have low melting and boiling points, and they often exist as gases or liquids. As solids, they are brittle or soft, because their molecules are not held together strongly. For example, in solid carbon dioxide (dry ice), the individual molecules remain intact and are identifiable in the molecular model (Figure 8.13).

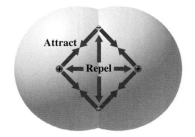

Figure 8.12
A covalent bond results when attractive forces between opposite charges are greater than repulsive forces between like charges.

Figure 8.13
In solid carbon dioxide (CO_2) molecules pack together in a regular array, but they retain their identity and can be removed from the crystal easily.

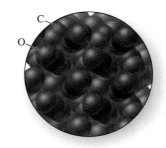
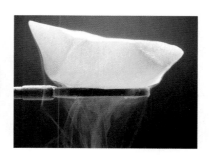

The Octet Rule

The number of bonds an element can form depends on its tendency to achieve a noble-gas configuration with an octet of eight valence electrons. Review the Lewis symbols of the elements carbon through fluorine in Table 8.3. How many bonds do you think each atom will make? Do your predictions result in an electron configuration containing an octet of valence electrons?

The tendency of an atom to achieve an electron configuration having eight valence electrons is known as the **octet rule.** Thus fluorine, with seven valence electrons ($:\ddot{F}\cdot$), normally forms one covalent bond, since it needs one electron to reach a noble-gas configuration. Carbon, with four valence electrons ($\cdot\dot{C}\cdot$), must gain four electrons to reach a noble-gas configuration, so it usually forms four bonds. When these two elements combine in a 1:4 ratio, carbon forms four bonds and each fluorine forms one bond, giving the covalent compound CF_4. As shown in Figure 8.14, if we count shared electrons for both atoms, we find eight electrons around the carbon atom and around each fluorine atom, the same number as in the noble gas neon. Fluorine also forms a covalent compound by sharing electrons with hydrogen. Because they each need one electron to reach a noble-gas configuration, they combine in a 1:1 ratio to form HF (Figure 8.14). While carbon and fluorine form bonds to reach the same electron configuration as neon, hydrogen forms one bond to reach the same configuration as helium. Hydrogen, with only a $1s$ orbital in its valence level, has two electrons when it forms a covalent bond.

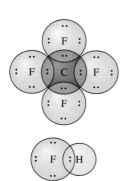

Figure 8.14
If shared electrons are counted for both atoms, each atom in CF_4 has a noble-gas configuration. In HF, the fluorine atom has an octet, but the hydrogen atom has only 2 electrons.

The Lewis symbols for the nonmetals in period 2 show the number of bonds each typically forms:

$$\cdot\dot{C}\cdot \qquad :\dot{N}\cdot \qquad :\ddot{O}\cdot \qquad :\ddot{F}\cdot \qquad :\ddot{Ne}:$$
$$4 \qquad\quad 3 \qquad\quad 2 \qquad\quad 1 \qquad\quad 0$$

To reach an octet, carbon needs four electrons and generally forms four bonds. Nitrogen needs three electrons, so it typically forms three bonds. Oxygen forms two bonds, and fluorine forms one bond. Since neon has a valence level filled with electrons, it forms no bonds.

Lewis Formulas for the Diatomic Elements

In some cases the only way to achieve an octet of electrons around each atom in a diatomic molecule is to share more than two electrons. Depending on the number of electrons the atoms share, covalent bonds are classified as single, double, or triple.

Single Covalent Bonds   A covalent bond that consists of a pair of electrons shared by two atoms is a **single bond.** Each atom has a half-filled orbital in the valence level, and the orbitals overlap to allow the electron pair to be a part of both atoms. For example, in the formation of the diatomic hydrogen molecule, each hydrogen atom has an electron in the $1s$ orbital. When the atoms come together, as shown in Figure 8.15, the orbitals overlap and the atoms share an electron pair:

Figure 8.15
Hydrogen atoms do not combine unless they get close enough together. Hydrogen molecules form when orbitals overlap, creating a region in space where both of their electrons can be found.

$$H\cdot \; + \; \cdot H \;\longrightarrow\; H:H$$

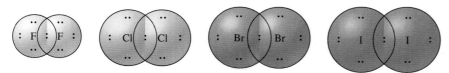

Figure 8.16
All the halogen elements exist as diatomic molecules. Each atom has an octet of electrons, as shown by the circles, because they share one electron pair between the bonded atoms. The shared electron pair is a single covalent bond, shown here in the overlapping circles.

This representation of a hydrogen molecule is a **Lewis formula** (or *electron dot formula*), in which the atoms are shown separately and the valence electrons are represented by dots. Lewis formulas are sometimes simplified by showing each bond not as a pair of dots, but as a line. For H_2, the Lewis formula could be either H:H or H−H.

The halogens also share one pair of electrons in diatomic molecules. Lewis formulas that show the distribution of electrons in the halogen molecules are given in Figure 8.16. Each atom in these structures has an octet, provided that the shared electrons are counted for each atom. Each of the covalent bonds in these molecules is a single bond, since it arises from the sharing of a single electron pair.

Double and Triple Bonds Consider the O_2 and N_2 molecules. Can a single covalent bond be formed between two oxygen atoms or between two nitrogen atoms and satisfy the octet rule? What is the problem? Can you think of some way to resolve this problem?

Some combinations of atoms do not have enough electrons to satisfy the octet rule with a single bond. To achieve an octet, they must share more than one pair of electrons. In O_2, for example, oxygen atoms must share two pairs of electrons to achieve an octet, while nitrogen atoms in N_2 must share three pairs (Figure 8.17). This bonding pattern reflects the number of bonds normally formed by these elements in order to reach an octet. Oxygen usually forms two bonds, and nitrogen normally forms three.

The sharing of two pairs of electrons is a **double bond.** The sharing of three pairs of electrons is a **triple bond.** In Lewis formulas, two pairs of dots or two parallel lines represent a double bond. Three pairs of dots or three parallel lines represent a triple bond.

Figure 8.17
The diatomic oxygen molecule has a double bond, formed by the sharing of two electron pairs between the oxygen atoms. In the diatomic nitrogen molecule, each atom shares three electron pairs with the other atom, forming a triple bond. Note the numbers of electrons shown in the areas where the circles overlap.

Valence Electrons and Number of Bonds

Elements combine in different ways to achieve octets. Oxygen, for example, typically forms two bonds, but it can do so in different ways. When oxygen atoms combine in molecular oxygen (O_2) they form a double bond:

$$:\ddot{O}\cdot + \cdot\ddot{O}: \longrightarrow :\ddot{O}::\ddot{O}:$$

But in combining with hydrogen atoms to form water (H_2O), oxygen may form two single bonds—one with each of two hydrogen atoms:

$$H\cdot + \cdot\ddot{O}\cdot + \cdot H \longrightarrow H:\ddot{O}:H$$

In a different compound, two oxygen atoms join with a single bond, with two hydrogen atoms supplying the additional electrons, to form hydrogen peroxide (H_2O_2):

$$H\cdot + \cdot\ddot{O}\cdot + \cdot\ddot{O}\cdot + \cdot H \longrightarrow H:\ddot{O}:\ddot{O}:H$$

Similarly a nitrogen atom, with five valence electrons, shares three electron pairs with another nitrogen atom to form molecular nitrogen (N_2):

$$:\dot{\ddot{N}}\cdot + \cdot\dot{N}: \longrightarrow :N:::N:$$

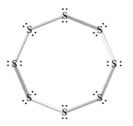

Figure 8.18
Phosphorus forms a molecule containing four atoms. Molecular sulfur contains eight atoms. Single bonds connect the atoms in each molecule.

Nitrogen forms single bonds with three hydrogen atoms in ammonia (NH_3):

$$H\cdot + H\cdot + H\cdot + :\overset{\cdot}{N}\cdot \longrightarrow \begin{matrix} H \\ :\overset{\cdot\cdot}{N}:H \\ H \end{matrix}$$

Like oxygen, nitrogen can bond to itself with a single bond, if another element is also present, as in hydrazine (N_2H_4):

$$H\cdot + H\cdot + \cdot\overset{\cdot}{N}\cdot + \cdot\overset{\cdot}{N}\cdot + H\cdot + H\cdot \longrightarrow \begin{matrix} H\ \ H \\ H:\overset{\cdot\cdot}{N}:\overset{\cdot\cdot}{N}:H \end{matrix}$$

Instead of forming double or triple bonds, some elements form molecules containing more than two atoms, each connected to others with single bonds. Phosphorus, which needs to form three bonds to complete its octet, can bond to three other phosphorus atoms, resulting in a P_4 molecule with each phosphorus atom at the corner of a tetrahedron. In contrast, sulfur needs to form two bonds to complete its octet. It can do this in several ways, but the most common is to form a ring of eight sulfur atoms, giving S_8. The common forms of the elements phosphorus and sulfur are shown in Figure 8.18.

EXAMPLE 8.5   Lewis Formula of a Compound

Determine the chemical formula of a simple compound that follows the octet rule and is formed from sulfur and chlorine atoms. Use Lewis formulas to describe the bonding in this compound.

Solution:

Sulfur has six valence electrons. It can fill its octet by sharing two electrons. Thus a sulfur atom tends to form two single bonds or one double bond. Chlorine has seven valence electrons and needs to share one by forming a single bond. Thus we would expect that two chlorine atoms could combine with one sulfur atom to form SCl_2, with single bonds between the atoms:

$$:\overset{\cdot\cdot}{\underset{\cdot\cdot}{Cl}}:\overset{\cdot\cdot}{\underset{\cdot\cdot}{S}}:\overset{\cdot\cdot}{\underset{\cdot\cdot}{Cl}}: \quad \text{or} \quad :\overset{\cdot\cdot}{\underset{\cdot\cdot}{Cl}}-\overset{\cdot\cdot}{\underset{\cdot\cdot}{S}}-\overset{\cdot\cdot}{\underset{\cdot\cdot}{Cl}}:$$

Notice that each atom has a share of an octet of electrons and that the total number of electrons equals the sum of the valence electrons of the component atoms.

Practice Problem 8.5

Determine the formula of a simple compound that follows the octet rule and is formed from nitrogen and fluorine atoms. Use electron dot structures to describe the bonding in this compound.

Further Practice: 8.53 and 8.54

Structures of Covalent Molecules

Writing Lewis Formulas   Let's look more closely at how to develop Lewis formulas for common molecules and ions. Consider the following small molecules containing carbon, each of which can be found in nature as part of the carbon cycle:

$$\begin{matrix} H & & :\overset{\cdot\cdot}{Cl}: & H \\ H:\overset{\cdot}{\underset{\cdot}{C}}:H & H:\overset{\cdot\cdot}{\underset{\cdot\cdot}{C}}:H & H:\overset{\cdot\cdot}{C}::\overset{\cdot\cdot}{O}: & :\overset{\cdot\cdot}{O}::C::\overset{\cdot\cdot}{O}: \quad :C:::O: \quad H:C:::N: \\ H & :\overset{\cdot\cdot}{Cl}: \end{matrix}$$

Does each atom in these molecules follow the octet rule? You already know that hydrogen does not. It gets a share of only two electrons. What about the other atoms? Using these examples, try to work out a procedure for writing Lewis formulas.

Let's use methanol (CH$_3$OH)—also called wood alcohol because it was originally obtained by distilling wood—as an example to see one way that Lewis formulas can be written.

Step 1. Write an atomic skeleton. Place the symbols of all the elements in the compound in the correct locations with respect to one another. To draw correct formulas, you must know which atoms are connected by bonds.

- The arrangement of atoms is usually symmetrical.
- In a molecule of two different elements, the one with the greater number of atoms usually surrounds the one with the lesser number of atoms.
- The central atom, the one surrounded by the other atoms, tends to be the one that is less electronegative and is present in the least quantity. This atom usually forms the greater number of bonds and is found further toward the bottom left side of the periodic table.
- Hydrogen atoms are generally on the outside of the molecule.
- The chemical formula may give clues about the arrangement of atoms.

Example: The atomic skeleton of methanol has the oxygen bonded to the carbon. It has hydrogen atoms bonded to each of them:

$$
\begin{array}{c}
\text{H} \\
\text{H C O H} \\
\text{H}
\end{array}
$$

Step 2. Sum the valence electrons from each atom to get the total number of valence electrons. If the covalent species is a polyatomic ion, be sure to take into account any ionic charge, subtracting electrons for positive charge and adding electrons for negative charge.

Example: The number of valence electrons in methanol is the sum for each atom: 4 from carbon, 6 from oxygen, and 1 from each hydrogen—for a total of 14.

Step 3. Place 2 electrons, a single bond, between each pair of bonded atoms.

Example: In methanol, place a bond between the carbon atom and the oxygen atom. Place single bonds between each of these atoms and the attached hydrogen atoms:

$$
\begin{array}{c}
\text{H} \\
\text{H:}\overset{..}{\text{C}}\text{:O:H} \\
\text{H}
\end{array}
$$

Step 4. If you have not placed all the valence electrons in the formula, add any remaining electrons as unshared electron pairs, consistent with the octet rule.

- Add pairs of electrons first to complete the octet of atoms surrounding the central atom. Then add any remaining electrons in pairs to the central atom.
- When placing electrons around the atoms, recall that hydrogen has two electrons, and period 2 elements have eight. Elements from periods 3 to 7 usually have eight electrons, but 10 or 12 are also possible.

Example: Methanol has 14 valence electrons, and we used 10 of them to form bonds in step 3. We place the remaining four electrons around the oxygen atom to satisfy its octet:

$$
\begin{array}{c}
\text{H} \\
\text{H:}\overset{..}{\text{C}}\text{:}\overset{..}{\underset{..}{\text{O}}}\text{:H} \\
\text{H}
\end{array}
$$

This uses all 14 valence electrons. Is this a satisfactory Lewis formula? Yes, each hydrogen has a share of two electrons, and the carbon and oxygen atoms have shares of eight electrons:

Elements in the third and higher periods, such as sulfur, phosphorus, and the heavier halogens and noble gases, can form compounds with 10 or even 12 electrons surrounding the central atom. An example is SF$_4$, with 10 electrons around the sulfur:

$$
\begin{array}{c}
:\overset{..}{\text{F}}\quad\overset{..}{\text{F}}: \\
\diagdown\diagup \\
\text{S} \\
\diagup\diagdown \\
:\overset{..}{\underset{..}{\text{F}}}\quad\overset{..}{\underset{..}{\text{F}}}:
\end{array}
$$

Step 5. If necessary to satisfy the octet rule, shift unshared electrons from non-bonded positions on atoms with completed octets to positions between atoms to make double or triple bonds.

Example: Since the octet rule is satisfied for methanol, we do not need to shift electrons to make double bonds. As we will see in Example 8.6, CO_2 is a molecule that requires double bonds to complete the octets of each atom.

EXAMPLE 8.6   Lewis Formulas of Molecules

Write a Lewis formula for carbon dioxide (CO_2), the primary carbon compound found in the atmosphere.

Solution:

Step 1. The atomic skeleton of carbon dioxide has oxygen atoms surrounding the carbon atom:

$$O \quad C \quad O$$

Step 2. The sum of valence electrons in carbon dioxide is 16, 4 from carbon and 6 from each oxygen.

Step 3. Place a bond between the carbon atom and each oxygen atom, but not between the oxygen atoms:

$$O:C:O$$

Step 4. Place six electrons around each oxygen to satisfy the octets of the oxygen atoms:

$$:\ddot{O}:C:\ddot{O}:$$

Step 5. This uses all 16 valence electrons. Is this a satisfactory Lewis formula? No, because the carbon atom does not have an octet of electrons:

$$\text{:O:C:O:}$$

We have to shift a pair of electrons from one of the oxygen atoms to the bond, making a double bond:

$$:\ddot{O}::C:\ddot{O}:$$

How many electrons surround the carbon atom now? Since this still does not satisfy the octet rule for carbon, we have to shift an electron pair from the other oxygen atom to its bond, making another double bond:

$$:\ddot{O}::C::\ddot{O}: \quad \text{or} \quad :\ddot{O}=C=\ddot{O}:$$

Count the electrons now to assure yourself that this structure satisfies the octet rule for each atom.

Practice Problem 8.6

Formaldehyde (CH_2O) is present in smoke from burning wood and coal, such as from the grill used to cook Michael's hamburger. It is toxic and often used as a tissue preservative. A formaldehyde molecule has the two hydrogen atoms and the oxygen atom attached to the carbon atom. Write a Lewis formula for the formaldehyde molecule.

Further Practice: 8.55 and 8.56

After step 4 of the procedure, if the central atom lacks two electrons, it must form a double bond with another atom. If it lacks four electrons, two double bonds may form, as in CO_2. Another possibility is the formation of a triple bond. For example, the Lewis formula for hydrogen cyanide, $H:C:::N:$ or $H-C\equiv N:$, has a triple bond between carbon and nitrogen.

In oxoacids, such as HNO_3 or H_2SO_4, the acidic hydrogen atoms are always attached to the oxygen atoms, which are, in turn, attached to the central atom. Since an oxygen with an attached hydrogen already has two bonds, any necessary double bond will not involve the $O-H$ unit. For example, the Lewis formula for nitric acid has two acceptable forms:

$$H:\overset{..}{\underset{..}{O}}:\overset{..}{\underset{}{N}}::\overset{..}{\underset{..}{O}}: \quad \text{or} \quad H-\overset{..}{\underset{..}{O}}-N\overset{\overset{\displaystyle :\overset{..}{O}:}{|}}{=}\overset{..}{O}:$$

and

$$H:\overset{..}{\underset{..}{O}}:\overset{:O}{\underset{..}{N}}:\overset{..}{\underset{..}{O}}: \quad \text{or} \quad H-\overset{..}{\underset{..}{O}}-\overset{\overset{\displaystyle :\overset{..}{O}}{||}}{N}-\overset{..}{\underset{..}{O}}:$$

The double bond can form between either of the two single oxygens (shown here toward the right and toward the top), but not with the bonded oxygen and nitrogen shown at the left.

Polyatomic ions have Lewis formulas similar to those of polyatomic molecules having the same number of electrons. Removal of a hydrogen ion (H^+) from a molecule leaves an anion with the same number of electrons. For example, removal of H^+ from HCN leaves CN^-, which has the same arrangement of electrons as does HCN:

$$\left[:C:::N:\right]^- \quad \text{or} \quad \left[:C\equiv N:\right]^-$$

(Note that we use square brackets around the polyatomic ion to indicate that the ionic charge is for the entire ion, not just one atom.) The nitrite ion (NO_2^-) has the same number of electrons as SO_2, so it has a similar Lewis formula:

$$\left[:\overset{..}{O}::\overset{..}{N}:\overset{..}{\underset{..}{O}}:\right]^- \quad \text{or} \quad \left[:\overset{..}{O}=\overset{..}{N}-\overset{..}{\underset{..}{O}}:\right]^-$$

$$:\overset{..}{O}::\overset{..}{S}:\overset{..}{\underset{..}{O}}: \quad \text{or} \quad :\overset{..}{O}=\overset{..}{S}-\overset{..}{\underset{..}{O}}:$$

Two other examples are the carbonate ion (CO_3^{2-}) and sulfur trioxide (SO_3) which have similar Lewis formulas:

$$\left[:\overset{..}{O}::\overset{\overset{\displaystyle :\overset{..}{O}:}{}}{C}:\overset{..}{\underset{..}{O}}:\right]^{2-} \quad \text{or} \quad \left[:\overset{..}{O}=\overset{\overset{\displaystyle :\overset{..}{O}:}{|}}{C}-\overset{..}{\underset{..}{O}}:\right]^{2-}$$

$$:\overset{..}{O}::\overset{\overset{\displaystyle :\overset{..}{O}:}{}}{S}:\overset{..}{\underset{..}{O}}: \quad \text{or} \quad :\overset{..}{O}=\overset{\overset{\displaystyle :\overset{..}{O}:}{|}}{S}-\overset{..}{\underset{..}{O}}:$$

Resonance   Sometimes Lewis formulas do not give accurate pictures of the bonding as it is known to exist in nature. Consider the sulfur dioxide molecule, for example. The Lewis formula of this molecule can be drawn in two ways, each having one double bond and one single bond:

$$:\overset{..}{O}::\overset{..}{S}:\overset{..}{\underset{..}{O}}: \qquad :\overset{..}{O}:\overset{..}{S}::\overset{..}{\underset{..}{O}}:$$

Do these formulas satisfy the octet rule for each atom? Yes, but neither of them accurately describes the bonds. Measurements on other molecules show that the normal length of a sulfur-oxygen single bond is 175 pm. The usual length of a sulfur-oxygen double bond is 115 pm. But in sulfur dioxide, both bonds are the

Figure 8.19
Two photographs of a seashell, neither of which accurately represents the true structure of the shell. The two views together are a better representation.

When chemists say electrons are "localized," they mean that the electrons tend to be located in a small region of space primarily between two nuclei, although it is not possible to specify the location of an electron with any certainty. In contrast, "delocalized" electrons are associated with three or more nuclei.

same length, 145 pm, a measurement that falls between those of single and double bonds. How can we explain this kind of bond?

The concept of *resonance* allows us to represent this kind of bonding while retaining the utility and simplicity of Lewis formulas. According to this concept, the electron arrangement in molecules like sulfur dioxide is represented not by a single Lewis formula, but by two or more, each illustrating a different aspect of the true arrangement of electrons. The actual molecule is a composite of the formulas drawn and is called a **resonance hybrid.**

The term resonance may suggest a resonating, or oscillating, movement between the different structures, but this is not the case. The true structure is not one of the Lewis formulas some of the time and the other the rest of the time. The real structure always has some of the characteristics of each contributing formula. It is an "average" of the contributions.

Using Lewis formulas to represent the bonding in molecules like sulfur dioxide is similar to using a photograph to represent the structure of a seashell made of calcium carbonate. The photograph of a seashell is not the seashell but only a representation of it, in the same way that a Lewis formula is only a representation of a molecule. Figure 8.19 shows two aspects of a seashell. The shell is not represented sometimes by one photograph and sometimes by the other, but is better represented as a composite of the two.

To represent a resonance hybrid, we draw the contributing structures in the usual way and connect them with double-headed arrows. The two structures for sulfur dioxide, for instance, are drawn as follows:

$$\ddot{\text{O}}::\ddot{\text{S}}:\ddot{\text{O}}: \longleftrightarrow :\ddot{\text{O}}:\ddot{\text{S}}::\ddot{\text{O}}$$

The composite structure is sometimes depicted as a single formula with a dashed or dotted line to show the sharing of the double-bond character over two sets of bonded atoms:

$$\text{O} \doteq \text{S} \doteq \text{O}$$

Such a structure emphasizes the *delocalized* nature of two electrons in the double bond, with resonance forms differing only in the locations of the electrons. As shown, both resonance forms of sulfur dioxide have an atomic arrangement in which the sulfur atom is between the oxygen atoms. It is not legitimate to change the relative positions of the atoms. If we moved one, we might represent an entirely different molecule, or we might show the molecule rotated in space, not a different resonance form.

As mentioned earlier, oxoacids do not form a double bond with any oxygen that is bonded to a hydrogen atom. Although we might expect nitric acid to have three resonance forms—with a double bond between the nitrogen and each of the oxygen atoms—it only has two:

$$\text{H}:\ddot{\text{O}}:\ddot{\text{N}}::\ddot{\text{O}}: \longleftrightarrow \text{H}:\ddot{\text{O}}:\ddot{\text{N}}:\ddot{\text{O}}:$$

The most common resonance forms have a central atom bonded to two (or more) other identical atoms. One bond is double and the other (or others) is single. The double bond can be exchanged with any single bond to give a valid resonance form. Similarly, single and triple bonds can be changed to two double bonds. Let's examine the resonance forms of the carbonate ion, the primary geological form of carbon, to illustrate these points.

EXAMPLE 8.7   Resonance Structures

Carbon occurs in nature in the form of carbonate ion in limestone and seashells. Draw the resonance forms of the Lewis formula for the carbonate ion (CO_3^{2-}).

Solution:

Following the steps in the procedure outlined earlier, we obtain the first Lewis formula:

$$\left[\begin{array}{c} :\ddot{O}: \\ :\ddot{O}::C:\ddot{O}: \end{array}\right]^{2-}$$

It is possible to show the double bond with either of the other 2 oxygen atoms:

$$\left[\begin{array}{c} :\ddot{O}: \\ :\ddot{O}:C::\ddot{O}: \end{array}\right]^{2-} \quad \text{or} \quad \left[\begin{array}{c} :\ddot{O} \\ :\ddot{O}:C:\ddot{O}: \end{array}\right]^{2-}$$

The actual structure is a resonance hybrid of these three structures:

$$\left[\begin{array}{c} :\ddot{O}: \\ :\ddot{O}::C:\ddot{O}: \end{array}\right]^{2-} \longleftrightarrow \left[\begin{array}{c} :\ddot{O}: \\ :\ddot{O}:C::\ddot{O}: \end{array}\right]^{2-} \longleftrightarrow \left[\begin{array}{c} :\ddot{O} \\ :\ddot{O}:C:\ddot{O}: \end{array}\right]^{2-}$$

Practice Problem 8.7

Before 1842 the most common anesthetics were whiskey and a blow on the head. After that, nitrous oxide, also known as laughing gas, came into use. Along with carbon dioxide, nitrous oxide has been implicated in global warming. It is N_2O, with its atoms arranged as NNO. Draw the resonance forms of the Lewis formula for nitrous oxide.

Further Practice: 8.63 and 8.64

Exceptions to the Octet Rule

Although most molecules and polyatomic ions satisfy the octet rule, many do not. Exceptions generally fall into three categories: odd-electron molecules, incomplete octets, and expanded valence levels. We will consider only the first category here.

In a molecule that contains an odd number of valence electrons, one electron must remain unpaired, so one of the atoms cannot have an octet. Generally the atom with an incomplete octet is the one with the lower electronegativity. An example is nitrogen monoxide (NO) which is an electron-deficient molecule with the following Lewis formula:

$$\cdot\ddot{N}::\ddot{O}:$$

The unpaired electron makes this a reactive molecule, because it will react with any other substance that can readily supply an electron. It may, for example, react with chlorine to form nitrosyl chloride:

$$:\ddot{Cl}:\ddot{N}::\ddot{O}:$$

Color is characteristic of substances composed of odd-electron molecules. Nitrogen monoxide is deep blue in the liquid and solid states, although most other nitrogen oxides are colorless. Another nitrogen oxide with an odd number of electrons is the reddish-brown compound nitrogen dioxide.

Some atoms, notably boron, participate in covalent bonding but do not have enough valence electrons to form an octet. Boron has only three valence electrons, which form three covalent bonds, as in the very reactive BH_3 and BF_3 molecules:

$$\begin{array}{cc} H & :\ddot{F}: \\ H:\ddot{B}:H & :\ddot{F}:B:\ddot{F}: \end{array}$$

In Chapter 7 we saw that color in elements arises from the transition of an electron from one orbital to another having a different energy. The same can happen in molecules when there is an orbital with only one electron. The transfer of electrons from another orbital of different energy to the half-filled orbital results in the absorption or emission of light energy. This causes the substance to be colored. Color also appears in some substances that contain double bonds.

EXAMPLE 8.8 | Lewis Formulas of Odd-Electron Molecules

Write a Lewis formula for the reddish-brown compound NO_2.

Solution:

We will follow the procedure outlined earlier.

Step 1. The oxygen atoms are attached to the nitrogen atom but not to one another:

O N O

Step 2. Each oxygen supplies six valence electrons and the nitrogen has five, giving a total of 17 electrons.

Step 3. We use four electrons to make bonds between the nitrogen and the two oxygen atoms:

O:N:O

Step 4. This leaves 13 electrons. We place six around each oxygen atom to complete their octets, leaving one for the nitrogen atom:

:Ö:N:Ö:

Step 5. The nitrogen atom has only five electrons, so we share a pair of electrons from one of the oxygen atoms:

:Ö::N:Ö:

This gives nitrogen a share of only seven electrons, but it cannot have more than eight, so we cannot satisfy the octet rule by sharing another pair of electrons. Because the double bond could involve the other oxygen atom, we draw a resonance hybrid:

:Ö::N:Ö: ⟷ :Ö:N::Ö:

Practice Problem 8.8

Write a Lewis formula for the yellowish-red compound ClO_2.

Further Practice: 8.67 and 8.68

Bonding in Carbon Compounds

Carbon is part of more different chemical compounds than any other element except hydrogen. Carbon forms the backbone of nearly every molecule that is involved in life processes in plants and animals on Earth, including the bodies of Michael, Ashley, and Amanda. Why does carbon have this role, rather than some other element such as boron, nitrogen, or oxygen? In part the answer lies in carbon's four valence electrons and its ability to form four covalent bonds. Boron, with three valence electrons, and nitrogen, with five, can form only three covalent bonds. Oxygen forms only two. Carbon is versatile in its bonding. It can form single, double, or triple bonds to other carbon atoms or to atoms of other elements:

—C—   C=   =C=   —C≡

In addition, carbon-carbon bonds are very strong, so molecules formed by carbon are very stable. Long chains of carbon atoms are found in nature and in synthetic polymers, which contain very large molecules. Despite their size, all can be derived from the simplest of the carbon compounds, the hydrocarbons.

Hydrocarbons   A compound of hydrogen and carbon is a **hydrocarbon**. Hydrocarbons can be categorized into several classes and subclasses (Figure 8.20). Methane (CH_4) the simplest hydrocarbon, has one carbon atom bonded to four hydrogen atoms (Figure 8.21). Notice the three-dimensional shape around each carbon atom is a tetrahedron, which we will explore in more depth in Section 8.4. The next larger hydrocarbon is ethane (C_2H_6) in which each carbon atom is surrounded by four other atoms arranged at the corners of tetrahedra. This tetrahedral arrangement is continued even

The hydrocarbons and their derivatives are so numerous that they form the subject of an entire branch of chemistry, organic chemistry, which we will explore in more detail in Chapter 16.

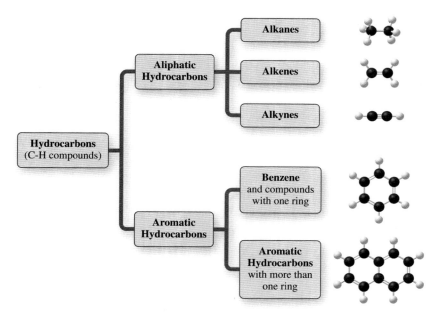

Figure 8.20
Classes of hydrocarbons.

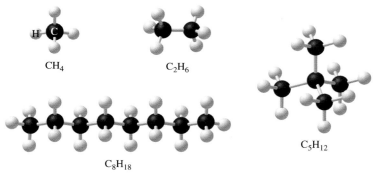

Figure 8.21
Alkanes contain only single bonds. Additional structures could be drawn for C_8H_{18} and C_5H_{12}.

in long chains of carbon atoms, such as C_8H_{18}, called octane, present in gasoline. Hydrocarbons can also be found in branched arrangements, where some carbon atoms attach to three or four other carbon atoms. Any one of these hydrocarbons, which contain only carbon-carbon single bonds, is called an **alkane** (Figure 8.21).

A hydrocarbon that contains a carbon-carbon double bond is an **alkene**. The simplest is ethene (or ethylene) (C_2H_4), which has a double bond between the carbon atoms:

$$\underset{H}{\overset{H}{}}\!\!\diagdown C = C \diagup \overset{H}{\underset{H}{}} \quad \text{or} \quad H_2C{=}CH_2$$

An **alkyne** is a hydrocarbon that contains a carbon-carbon triple bond. The simplest example is ethyne, also known as acetylene, $H{-}C{\equiv}C{-}H$, used in the oxyacetylene welding torch.

Any of these types is an **aliphatic hydrocarbon**, a class in which the bonds are all localized single, double, or triple bonds. Another class of hydrocarbons has carbon atoms arranged in a six-atom ring with alternating single and double bonds—or, more accurately, delocalized bonds. Members of this class are represented by resonance structures. Such a compound is an **aromatic hydrocarbon**. Examples are benzene (C_6H_6) and naphthalene ($C_{10}H_8$), shown in Figure 8.22.

Formulas such as C_8H_{18} don't give much information about how the atoms are connected. In all hydrocarbons, the hydrogen atoms are attached to carbon atoms, while the carbon atoms attach to one another in a variety of ways. Molecules with the same composition but different arrangements of atoms are called *isomers*. Octane (C_8H_{18}) has 15 isomers. The number of isomers increases rapidly as the number of carbons increases. It can be calculated that 75 isomers of $C_{10}H_{22}$ and 366,319 isomers of $C_{20}H_{42}$ are possible.

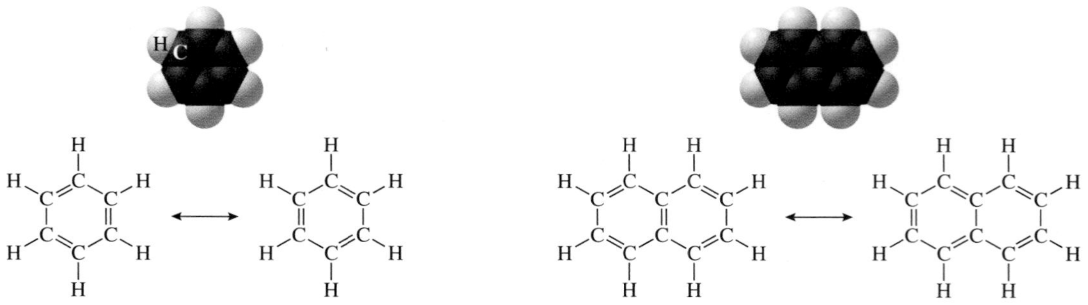

benzene   naphthalene

Figure 8.22
Benzene and naphthalene are aromatic hydrocarbons. Some of their electrons are delocalized in six-membered rings of carbon atoms. Each Lewis formula representation has alternating single and double bonds.

Functional Groups   Many other carbon compounds—including those that are important to the maintenance of life, such as proteins, enzymes, and genetic material—are derivatives of the hydrocarbons; that is, their structures are based on the structures of the hydrocarbons. In such compounds, other groups of atoms are substituted for one or more hydrogen atoms on the hydrocarbon framework. The properties of these derivatives depend on the atoms involved in the substitution. The group that is introduced is called a **functional group**. It is this part of the molecule that gives a class of compounds its characteristic properties.

For example, if a hydroxyl group ($-OH$) replaces a hydrogen atom in the formula of an alkane, the new formula represents an **alcohol**. Methyl alcohol (or methanol) is CH_3OH, derived from the formula for methane. Substituting a hydroxyl group for a hydrogen in the formula for ethane gives the formula for ethyl alcohol (or ethanol), CH_3CH_2OH. The most common functional groups and examples of organic molecules in each class are given in Table 8.4.

TABLE 8.4	Functional Groups in Hydrocarbons		
Class	**Functional Group**	**Example**	**Formula**
Alcohol	$-OH$	Ethyl alcohol	C_2H_5-OH
Ether	$-O-$	Diethyl ether	$C_2H_5-O-C_2H_5$
Aldehyde	$\overset{\displaystyle O}{\overset{\displaystyle \|}{-C-H}}$	Acetaldehyde	$\overset{\displaystyle O}{\overset{\displaystyle \|}{H_3C-C-H}}$
Ketone	$\overset{\displaystyle O}{\overset{\displaystyle \|}{-C-}}$	Acetone	$\overset{\displaystyle O}{\overset{\displaystyle \|}{H_3C-C-CH_3}}$
Carboxylic acid	$\overset{\displaystyle O}{\overset{\displaystyle \|}{-C-OH}}$	Acetic acid	$\overset{\displaystyle O}{\overset{\displaystyle \|}{H_3C-C-OH}}$
Ester	$\overset{\displaystyle O}{\overset{\displaystyle \|}{-C-O-}}$	Ethyl acetate	$\overset{\displaystyle O}{\overset{\displaystyle \|}{H_3C-C-O-C_2H_5}}$
Amine	$\overset{\displaystyle \|}{-N-}$	Methyl amine	$\overset{\displaystyle H}{\overset{\displaystyle \|}{H_3C-N-H}}$

Molecules can contain more than one functional group. For example, the proteins in the mushrooms in Ashley's salad are chains of amino acids, which have both amine and carboxylic acid functional groups. An example is glycine ($H_2NCH_2CO_2H$):

$$H-\underset{\underset{H}{|}}{\overset{\overset{H}{|}}{N}}-\underset{\underset{H}{|}}{\overset{\overset{H}{|}}{C}}-\overset{\overset{O}{||}}{C}-O-H$$

EXAMPLE 8.9   Functional Groups

Draw the structure of an ether that has the formula C_3H_8O.

Solution:

In an ether, the oxygen atom must be located between two of the carbon atoms:

$$C-C-O-C$$

The hydrogen atoms are then bonded to carbon atoms, since the oxygen atom already has its normal two bonds:

$$H-\underset{\underset{H}{|}}{\overset{\overset{H}{|}}{C}}-\underset{\underset{H}{|}}{\overset{\overset{H}{|}}{C}}-O-\underset{\underset{H}{|}}{\overset{\overset{H}{|}}{C}}-H$$

Placing the oxygen atom after the first carbon atom gives the same structure, simply flipped on the page left to right.

Practice Problem 8.9

Draw the structure of a ketone that has the formula C_3H_6O.

Further Practice: 8.71 and 8.72

8.4   SHAPES OF MOLECULES

Michael's and Ashley's lunch foods supply some of the carbon-based molecules their bodies need to sustain their lives. But why did they choose hamburgers and salad instead of rice cakes? They would probably say they like the flavor and aroma of their favorite foods, but how do the human senses of taste and smell actually work? Although these senses must have a chemical basis, determining how they operate has not been simple. It has, however, become clear that there are receptor sites in the nose and tongue where molecules can fit. The receptors are fashioned from the proteins that make up the membranes of cells. The proteins are interlinked in various ways, creating pockets in the structure of the cell membrane. The pockets act as receptors for other molecules if two conditions are met. First the molecules must be the right size and shape to fit into them. Second the molecules must interact in a certain way with atoms in the membrane protein, causing a change that can trigger a nerve impulse. It's the nerve impulse that the brain interprets as, "Gee, that smells great!" or "Ugh! I can't eat that!"

Recognizable tastes fall into several categories. One is sweet, and it is highly attractive to humans for good reason. Lactose is a sweet substance found in mothers' milk. The sweet taste encourages babies to consume more. Other sweet substances include glucose and sucrose, which occur naturally in plants. While it has been difficult to relate the structures of molecules to their tastes, a common feature of sweet molecules is the size of a particular portion of the molecule. This part of a sweet molecule presumably fits into the receptor site. Sweet molecules commonly have an $-H$ (or $-OH$) separated from an $-O-$ atom by about 300 pm, as illustrated in Figure 8.23.

The 2004 Nobel Prize in Physiology or Medicine was given to Richard Axel of Columbia University and Linda Buck of the Fred Hutchinson Cancer Center in Seattle for their studies of the sense of smell.

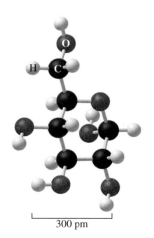

300 pm

Figure 8.23
Glucose is sweet because it contains $-H$ and $-OH$ groups that fit into a taste receptor site on the tongue.

Molecule in Receptor Cavity							
Molecular Characteristics	Disk shape	Spherical shape	Rod shape	Wedge shape	Disk and tail shape	Attraction to negative centers	Attraction to positive centers
Primary Odor	Musky	Camphoric	Ethereal	Pepperminty	Floral	Pungent	Putrid
Chemical and Common Examples	Xylene Musk perfume/ aftershave	Camphor Mothballs	Diethyl ether Pears	Menthol Mint gum/ mouthwash	α-amyl pyridine Roses	Formic acid Vinegar	Hydrogen sulfide Rotten eggs

Figure 8.24

The primary odors arise from interaction between molecules and shaped receptors. However, two of the receptors involve not shape, but interactions with charge separations in molecules.

Although humans can distinguish thousands of odors, one theory proposes that there are only seven primary odors, each associated with a different type of receptor in the nose. When a gaseous molecule enters the nose and interacts with a particular receptor, a nerve impulse is generated and sent to the brain. A molecule that can fit into more than one type of receptor triggers multiple signals, generating a composite odor. The proposed receptor cavities and some molecules that give each of the seven primary odors are shown in Figure 8.24. Which combination of primary odors do you think gives the aroma of Michael's hamburger or Ashley's mushrooms?

The Valence-Shell Electron-Pair Repulsion Theory

As the senses of taste and smell demonstrate, the observable properties of substances derive from the three-dimensional shapes of their molecules. Shape, in turn, is a function of the arrangement of atoms in a molecule. The relative locations of electron pairs around a central atom play a large role in determining a molecule's three-dimensional shape. Negatively charged electrons repel one another, so electron pairs in different orbitals stay as far apart as possible. The tendency of electron pairs to adjust the orientation of their orbitals to maximize the distance between them is the basis of the **valence-shell electron-pair repulsion (VSEPR) theory.** (*Valence shell* is another name for *valence level.*)

To predict the geometric shape of a molecule or ion using VSEPR theory, we first need to know how electrons are arranged in a molecule. In particular we need to know how many unshared electron pairs and atoms surround the central atom. We can get this information by examining the Lewis formula. For example, CH_4 has four atoms and no unshared electron pairs surrounding the carbon atom. In the CO_2 molecule, carbon has only two atoms bonded to it, as shown in the Lewis formula, $:\!\ddot{O}\!:\!:\!C\!:\!:\!\ddot{O}\!:$, and it has no unshared pairs of electrons. The NH_3 molecule has three bonded atoms and one unshared electron pair. The bonded atoms and unshared electron pairs are arranged around the central atom as far apart as possible. The result is a shape characterized by the **bond angle** between the central atom and the atoms bonded to it. What shapes and angles would you predict for molecules having two, three, and four pairs of unshared electrons or atoms around a central atom? How would you make these predictions based on repulsions of electron pairs?

To see what angles to predict, let's examine the shape commonly found in hydrocarbons and many other molecules that arises from four electron pairs or atoms. The Lewis formulas look flat, as in CH_4:

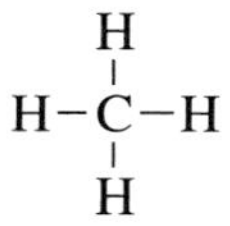

$$H-\underset{\underset{H}{|}}{\overset{\overset{H}{|}}{C}}-H$$

Animation: VSEPR Theory and the Shapes of Molecules
Online Learning Center, Animations

Chemistry Animations Library, VSEPR Theory and the Shapes of Molecules, 01VSEPR.rm.

This structure looks as if it would have bond angles of 90°:

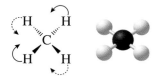

However, this does not accurately represent the structure of the CH₄ molecule, which is not flat. If we move atoms to correct the flatness, we change the bond angles. The larger the angles, the farther away from one another the atoms are. If we move only the carbon atom, the angles actually get smaller. The only way to make the angles larger is to move two hydrogen atoms up and two down:

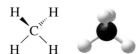

How large can the angles get if we move two up and two down? When the angles all become equal, they have values of 109.5°, giving a tetrahedral shape:

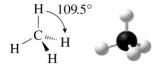

This structure is usually shown in a rotated form:

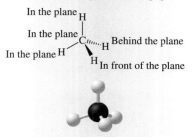

Table 8.5 shows the geometries that achieve the maximum distance between combinations of two, three, or four atoms, or unshared electron pairs. The geometries shown in the table are adopted by any set of objects that are attached at a common point and that stay as far apart as possible. For example, the balloons shown in Figure 8.25 adopt the same geometries as atoms (or unshared pairs of electrons) around a central atom. These structures, called *parent structures,* partly predict a molecule's shape.

Since it is hard to see how atoms are arranged in a two-dimensional line drawing, chemists often use a system of solid lines, dashed lines, and wedges to depict the three-dimensional structure of molecules. Solid lines indicate bonds in the plane of the paper. Dashed lines are bonds that point back from the plane of the paper. Wedges depict bonds that come forward from the plane of the paper:

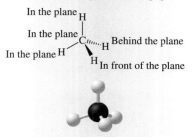

Figure 8.25
Macroscopic objects that are fastened to a common center adopt the same structures as molecules because, like electron pairs, they stay as far apart as possible.

TABLE 8.5	Geometric Structures Arising from Different Numbers of Atoms	
Number of Atoms or Electron Pairs	**Parent Structure**	**Geometric Arrangement of Atoms or Electron Pairs**
2	Linear	180°
3	Trigonal planar	120°
4	Tetrahedral	109.5°

Trigonal planar

$$\begin{bmatrix} \ddot{\,:\ddot{O}:\,} \\ :\ddot{O}:\ddot{N}:\ddot{O}: \end{bmatrix}^{-}$$

$$:\ddot{O}::\ddot{S}:\ddot{O}:$$

$$:\ddot{F}: \\ :\ddot{F}:\ddot{B}:\ddot{F}:$$

Trigonal pyramidal

$$\begin{array}{c} H \\ H:\ddot{N}:H \end{array}$$

$$:\ddot{F}: \\ :\ddot{F}:\ddot{P}:\ddot{F}:$$

$$\begin{bmatrix} :\ddot{O}: \\ :\ddot{O}:\ddot{Cl}:\ddot{O}: \end{bmatrix}^{-}$$

Figure 8.26
Some molecules and ions with the formula AX_3 have a trigonal planar shape, while others have a trigonal pyramidal shape.

Figure 8.27
The four electron pairs in water are arranged around the oxygen atom in orbitals that point approximately to the corners of a tetrahedron. Hydrogen atoms occupy only two of the positions, so the molecular shape is described as bent, not tetrahedral.

EXAMPLE 8.10	**VSEPR Parent Structures**

What is the parent structure for H_2O, in which oxygen is bonded to two other atoms and has two unshared pairs of electrons?

Solution:

The H_2O molecule has four pairs of electrons around the oxygen atom: two bonding pairs and two unshared pairs. Any combination of four unshared electron pairs or atoms has a tetrahedral parent structure.

Practice Problem 8.10

What is the parent structure of the NO_2^- ion, in which nitrogen is bonded to two other atoms and has one unshared pair of electrons?

Further Practice: 8.89 and 8.90

Consider the Lewis formulas for several molecules and ions having the general composition AX_3, shown in Figure 8.26. Some have a trigonal planar shape, and some have a trigonal pyramidal shape. Why do they differ? If the central atom contains any unshared pairs of electrons, the molecular shape is not that of the parent structure, but is derived from it. The shape we assign a molecule includes only the groups of *bonding electrons* within the parent structure, and not the unshared pairs of electrons; the shape is a description of the locations of the atoms in the molecule. Consider, as an example, the water molecule, which has two atoms and two unshared pairs of electrons around the central atom. The four electron pairs in H_2O arrange themselves tetrahedrally around the oxygen atom, as described in Table 8.6 and shown in Figure 8.27. But because we consider only the two bonding pairs when we describe the molecular shape, we do not say that the water molecule is tetrahedral. Rather, we describe it as bent. All possible shapes for molecules that have two to four electron pairs are summarized in Table 8.6. The bond angles for these structures are approximately those of the parent structures.

Notice in Table 8.6 that CO_2 is linear, but SO_2—with the same number of atoms—is bent. We can understand this difference using VSEPR theory. Examine the Lewis formulas of these molecules:

$$:\ddot{O}::C::\ddot{O}:   :\ddot{O}::\ddot{S}:\ddot{O}:$$

What is different about them? The carbon in CO_2 has two bonded atoms and no unshared electron pairs, so it has a linear structure. In contrast sulfur in SO_2 has two bonded atoms and one unshared electron pair. The unshared pair causes the molecule to be bent. Models of these molecules are shown in Figure 8.28.

To predict the shape of a molecule according to VSEPR theory, follow four steps:

1. Draw a Lewis formula.
2. Count the number of atoms bonded to the central atom, and count unshared electron pairs on the central atom.
3. Add the numbers of atoms and the number of unshared electron pairs around the central atom. The total indicates the parent structure.
4. The molecular shape is derived from the parent structure by considering only the positions in the structure occupied by bonded atoms.

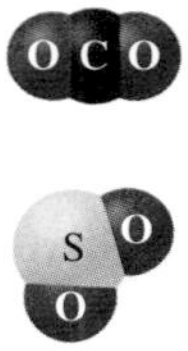

Figure 8.28
Carbon dioxide is linear; sulfur dioxide is bent.

General Formula	Number of Bonded Atoms	Number of Unshared Pairs	Molecular Shape	Examples
TABLE 8.6 — Arrangement of Electron Pairs and Molecular Shapes				
Parent Structure: Linear				
AX_2	2	0	Linear 	$BeCl_2$, CO_2, HCN
Parent Structure: Trigonal Planar				
AX_3	3	0	Trigonal planar 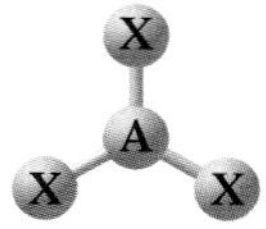	BF_3, BH_3, SO_3, NO_3^-
AX_2	2	1	Bent 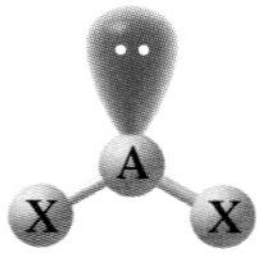	SO_2, NO_2^-
Parent Structure: Tetrahedral				
AX_4	4	0	Tetrahedral 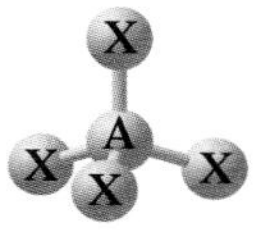	CH_4, CH_2Cl_2, $SiCl_4$, $POCl_3$, BrO_4^-
AX_3	3	1	Trigonal pyramidal 	NH_3, PF_3, NH_2Cl
AX_2	2	2	Bent 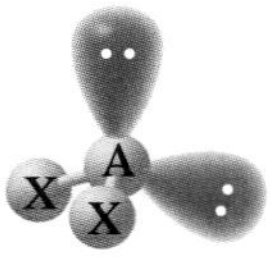	H_2O, F_2O, BrO_2, SO_2, SCl_2

The bond angles in real molecules often differ a little from the values predicted by the VSEPR theory. Differences occur because unshared pairs of electrons occupy more space than bonded pairs. The bonded pairs are forced together somewhat, resulting in slightly smaller bond angles than predicted. For example, ammonia, which has one unshared pair of electrons, has a bond angle of 107° rather than the predicted 109.5°. Water has two unshared pairs of electrons and a bond angle of about 105°. A similar effect is caused by a double bond, which occupies more space than a single bond.

EXAMPLE 8.11   Shapes of Molecules and Ions

The carbonate ion (CO_3^{2-}) is found in limestone and in Michael's antacid tablets. Predict its shape. What are the approximate O—C—O bond angles?

Solution:

1. First, we write a Lewis formula following the procedure from Section 8.3:

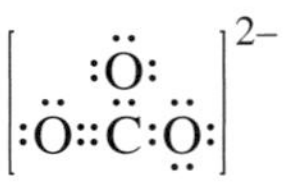

$$\left[\begin{array}{c} :\ddot{O}: \\ :\ddot{O}::C:\ddot{O}: \end{array} \right]^{2-} \quad \text{or} \quad \left[\begin{array}{c} :\ddot{O}: \\ | \\ :\ddot{O}=C-\ddot{O}: \end{array} \right]^{2-}$$

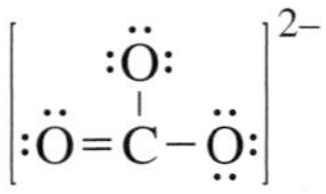

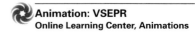
Animation: VSEPR
Online Learning Center, Animations

Chemistry Animations Library, VSEPR, 34VSEPR.swf.

There are two other resonance forms, which have the double bond in other positions, but we need obtain only one Lewis formula to determine the shape.

2. Next, count the bonded atoms and unshared electron pairs on the central atom. The carbon atom has three bonded oxygen atoms and no unshared pairs of electrons around it.

3. The total number of bonded atoms and unshared electron pairs is three, so the parent structure is trigonal planar.

4. Since there are no unshared electron pairs around carbon, the shape of this ion is the same as the parent structure: trigonal planar. The bond angle in a trigonal planar structure is 120°.

Practice Problem 8.11

What is the shape of the nitrite ion (NO_2^-)? What is the O—N—O bond angle?

Further Practice: 8.91 and 8.92

For molecules with more than one central atom, we can repeat the procedure for each central atom, a process that allows us to predict the structures of molecules as large and complex as proteins. Let's consider one amino acid, glycine, which might be found in a protein. It has a sweet taste and is found in gelatin and other animal products; it was probably in Michael's hamburger. Its formula is $H_2NCH_2CO_2H$. How do we predict its structure?

We start, as before, by writing a Lewis formula for the molecule:

$$
\begin{array}{cc}
\text{H} \quad \ddot{\text{O}}: & \text{H} \quad \ddot{\text{O}}: \\
\text{H:N:C:C:O:H} \quad \text{or} & \text{H—N—C—C—O—H} \\
\quad\ \text{H H} & \quad\ \ \text{H H}
\end{array}
$$

This molecule contains four atoms that act as central atoms: the nitrogen atom, both carbon atoms, and one of the oxygen atoms. The nitrogen atom has three atoms bonded to it and one unshared electron pair. Thus the structure around this nitrogen atom is based on a tetrahedral parent structure with one unshared pair of electrons. The shape around the nitrogen atom is trigonal pyramidal, with bond angles of about 109.5°. The first carbon atom is bonded to four other atoms (N, two H, and C) and has no unshared electron pairs, so the shape around it is tetrahedral, with bond angles of about 109.5°. The second carbon atom is bonded to three other atoms (C and two O) and has no unshared electron pairs, so the shape around it is trigonal planar, with bond angles of around 120°. Finally one of the oxygen atoms has two atoms (C and H) bonded to it and has two unshared electron pairs. The structure around it is based on a tetrahedral parent structure, but with two unshared pairs, the shape is bent. The bond angles will be around 109.5°. Verify each of these predictions by examining the molecular model of glycine in Figure 8.29.

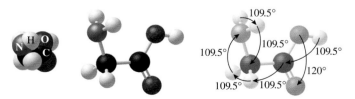

Figure 8.29
VSEPR theory predicts the structure of glycine.

Figure 8.30
(A) The heme group in hemoglobin consists of an iron(II) ion bonded to four nitrogen atoms, each surrounded by a planar arrangement of carbon atoms. (The hydrogen atoms are not shown for clarity.) (B) When oxygen binds to heme, the Fe−O=O arrangement is bent, not linear, since the electrons surrounding each oxygen atom are located at the corners of a trigonal plane. (C) A histidine group in the hemoglobin protein fits into the space next to the oxygen molecule.

Figure 8.31
(A) When carbon monoxide binds to the iron in heme, the Fe−C≡O arrangement is expected to be linear. (B) However, in hemoglobin, a histidine group is close to the heme group, forcing the carbon monoxide to assume a bent arrangement. This makes the iron-carbon bond weaker than if the arrangement remained linear.

Molecular shapes are important in living systems. For example, in humans, red blood cells carry oxygen through the bloodstream to all body cells. They can act as delivery systems because they contain hemoglobin, a large protein molecule that contains four heme groups, one of which is shown in Figure 8.30A. Oxygen molecules bind to the iron ions in heme, as shown in Figure 8.30B. Notice the bent structure, which is consistent with the two unshared pairs of electrons on each oxygen atom:

$$\ddot{O}=\ddot{O}$$

Another part of the hemoglobin protein is an amino acid called histidine. The histidine group projects down near the top of the heme group, just fitting into the space around the oxygen molecule (Figure 8.30C).

This system works well as long as carbon monoxide, a highly poisonous gas, does not get into the bloodstream. If it does, it interferes with the distribution of oxygen molecules to cells. As shown in Figure 8.31A, the arrangement of CO bonded to iron in hemoglobin is linear, since the carbon in carbon monoxide has only one unshared electron pair:

$$:C≡O:$$

In this form, the carbon monoxide binds to the iron 25,000 times more strongly than oxygen does. Under these conditions, any exposure to carbon monoxide would completely remove the ability of heme to bind to oxygen, and there would be no possibility of surviving such exposure.

However, nature has provided some protection. As shown in Figure 8.31B, the histidine group forces the carbon monoxide to assume a bent arrangement. In this form, carbon monoxide binds only 200 times more strongly than oxygen. When hemoglobin binds to carbon monoxide, the red blood cells can no longer transport oxygen to the cells, but continued exposure to oxygen can replace the carbon monoxide, counteracting the toxic effect.

Polarity of Molecules

Animation: Influence of Shape on Polarity
Online Learning Center, Animations

Chemistry Animations Library, Influence of Shape on Polarity, 01Polarity.rm.

In Section 8.1 you learned that bonds are polar if the bonded atoms share electrons unequally, resulting in partial charges on the atoms. Molecules with polar bonds can be polar molecules under certain conditions. Here we will look at how to determine if a molecule is polar.

In a diatomic molecule composed of atoms having different electronegativities, the polarity of the molecule lies along the plane of the bond. Since there are only two atoms, one of them must be partially positive ($\delta+$) and the other partially negative ($\delta-$). This can be represented by an arrow, with the + sign at the position of partial positive charge ($\delta+$) and the arrowhead at the position of partial negative charge ($\delta-$):

$$\overset{\delta+\quad\delta-}{\text{H}-\text{F}}$$
$$\longmapsto\longrightarrow$$

The direction of the polarity is evident in the case of a diatomic molecule with only one bond, but polyatomic molecules with more than one bond are more complex. The polarity of the bonds and the polarity of the molecule may be different.

A polyatomic molecule that has nonpolar bonds cannot be polar. For example, in NBr_3, nitrogen and bromine have the same electronegativity, so each bond is nonpolar and the molecule must be nonpolar. But a polyatomic molecule with polar bonds may be polar or nonpolar, depending on its geometry. We can determine this experimentally by placing a sample of the substance in an electric field, such as between the electrostatically charged plates of a capacitor. The greater the polarity of the substance, the more voltage it takes to charge the plates of the capacitor to the same extent. As shown in Figure 8.32, if the molecules are polar, they align with the electric field. If they are nonpolar, they retain random orientations in the electric field.

Carbon-hydrogen bonds are almost nonpolar, so hydrocarbons are generally considered to be essentially nonpolar. Because of the geometry of hydrocarbons, the slight polarity of the bonds cancels out.

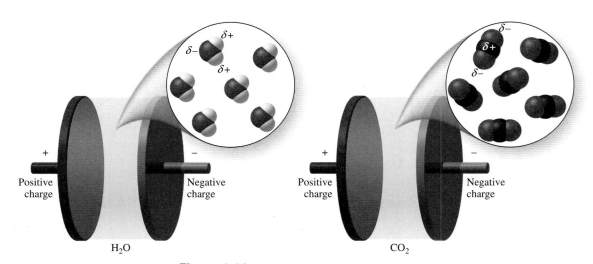

Figure 8.32
The polar molecule H_2O aligns with the electric field in a capacitor. The nonpolar molecule CO_2 does not.

The polarity of a bond can be canceled by the polarity of another bond in the opposite direction. Consider a molecule of gaseous beryllium chloride (linear $BeCl_2$), for example. Although it has two polar Be—Cl bonds, it is nonpolar. The bonds in $BeCl_2$ are 180° apart, and the polarities of the two bonds point in opposite directions. Thus the charge separation places partial negative charges at the two chlorine positions and a partial positive charge at the beryllium position:

$$\overset{\delta-}{Cl}-\overset{\delta+}{Be}-\overset{\delta-}{Cl}$$

No net polarity

The polarities of the two bonds are exactly equal in magnitude and opposite in direction. They cancel each other out, making the molecule nonpolar.

Any molecule in which the relationship of the bonded atoms is completely symmetrical—linear, trigonal planar, or tetrahedral (Table 8.6)—experiences this sort of cancellation and is nonpolar. Thus cancellation of polarity occurs in trigonal planar molecules such as SO_3 and tetrahedral molecules such as CCl_4 (Figure 8.33). In such molecules, the central atom is surrounded by identical atoms. However, if the atoms surrounding the central atom are not identical, the molecule's shape is not completely symmetrical. Such molecules are polar if the bonds are polar. Chloroform ($CHCl_3$) is polar, as are chloromethane (CH_3Cl) and dichloromethane (CH_2Cl_2). In these molecules, the atoms surrounding the carbon are not identical, so the polarities of the bonds do not cancel, resulting in a net polarity for the molecule.

In summary, we must consider two factors when deciding whether a molecule is polar: the bond polarity and the molecular shape. If the bonds are polar *and* the molecular shape is *not* symmetrical, then the molecule is polar. If all the bonds are nonpolar *or* if the molecular shape is symmetrical, the molecule is nonpolar.

Figure 8.33

In symmetrical molecules with identical atoms surrounding a central atom, such as SO_3 and CCl_4, the bond polarities cancel.

EXAMPLE 8.12   **Polarity of Molecules**

Predict whether H_2S is a polar or a nonpolar molecule.

Solution:

The molecule H_2S has polar bonds, since there is a difference of 0.4 in electronegativity between sulfur and hydrogen (from Figure 8.5). To determine the molecular shape, we consider the Lewis formula:

$$H\!:\!\overset{..}{\underset{..}{S}}\!:\!H$$

There are four electron pairs around the sulfur, only two of which are used in bonds, so the molecule is bent. Because the bonds are polar and the molecule is not completely symmetrical, H_2S is polar.

Practice Problem 8.12

Predict whether C_2H_6, NO_2, CO_2, SO_2, and SO_3 are polar or nonpolar molecules.

Further Practice: 8.111 and 8.112

Animation: Polarity of Molecules
Online Learning Center, Animations

Chemistry Animations Library, Polarity of Molecules, Polarity_of_Molecules.swf.

Polarity determines many interactions between molecules. For example, nonpolar molecules exist as gases at lower temperatures than do polar molecules. Polar liquids freeze at higher temperatures than do nonpolar liquids. Ionic salts and polar liquids dissolve better in polar liquids than in nonpolar liquids. On the other hand, nonpolar liquids dissolve better in other nonpolar liquids than in polar liquids. Oil, which is composed of largely nonpolar hydrocarbons, does not dissolve in polar water (Figure 8.34).

Figure 8.34

Nonpolar oil in the top layer does not dissolve in polar water.

SUMMARY

Different chemical bonds give substances different properties. In bonding, atoms tend to lose, gain, or share electrons to achieve a more stable electron configuration, often with eight valence electrons—the octet rule. The tendency is to fill the valence orbitals, so hydrogen usually has two electrons when bonded, and some elements have more than eight.

Ionic bonding involves the complete transfer of one or more electrons from a metal to a nonmetal, often giving both elements a noble-gas configuration. In such compounds, electrostatic forces hold the ions together in crystal lattices. Covalent bonds result from the sharing of a pair of electrons between two atoms. In some molecules, two atoms share two electron pairs to form a double bond or three electron pairs to form a triple bond. Some elements have a stronger tendency to attract electrons (measured by their electronegativity values), and the electrons in a covalent bond are not always shared equally. Such covalent bonds are polar.

Lewis formulas represent the bonding in molecules and ions. In some cases, the actual bonding is a composite of all the resonance structures, and two or more Lewis formulas are needed to show the arrangement of electrons in a covalent molecule or ion.

VSEPR theory, which assumes that electron pairs stay as far apart as possible, predicts the shapes of molecules. The total number of bonded atoms and of unshared electron pairs on the central atom determines the parent structure: linear, trigonal planar, or tetrahedral. The molecular shape is determined by the position of atoms around the central atom, which is affected by the presence of unshared electron pairs.

A nonpolar molecule has either nonpolar bonds or a symmetrical structure. A molecule is polar only if the bonds are polar and the molecule is not symmetrical.

KEY TERMS

alcohol (8.3)

aliphatic hydrocarbon (8.3)

alkane (8.3)

alkene (8.3)

alkyne (8.3)

aromatic hydrocarbon (8.3)

bond angle (8.4)

chemical bond (8.1)

covalent bonding (8.1)

crystal lattice (8.2)

double bond (8.3)

electronegativity (8.1)

functional group (8.3)

hydrocarbon (8.3)

ionic bonding (8.1)

ionic crystal (8.2)

Lewis formula (electron dot formula) (8.3)

Lewis symbol (electron dot symbol) (8.2)

nonpolar covalent bond (8.1)

octet rule (8.3)

polar covalent bond (8.1)

polarity (8.1)

resonance hybrid (8.3)

single bond (8.3)

triple bond (8.3)

valence-shell electron-pair repulsion (VSEPR) theory (8.4)

QUESTIONS AND PROBLEMS

The following questions and problems, except for those in the *Additional Questions* section, are paired. Questions in a pair focus on the same concept. Answers to the odd-numbered questions and problems are in Appendix D.

Matching Definitions with Key Terms

8.1   Match the key terms with the descriptions provided.

    (a) a covalent bond that involves the sharing of one pair of electrons

    (b) a hydrocarbon containing only carbon-carbon single bonds

    (c) bonding between two atoms resulting from the sharing of electrons

    (d) solid structure in which the ions are arranged in a regular repeating pattern

    (e) rule stating that atoms tend to gain, lose, or share electrons to achieve an electron configuration with eight electrons in the valence shell

(f) a bond resulting from the unequal sharing of electron pairs

(g) a hydrocarbon containing carbon-carbon triple bonds

(h) a measure of the ability of an atom to attract electrons within a bond to itself

(i) a covalent bond that involves the sharing of three pairs of electrons

(j) the repeating pattern of atoms or ions in a crystal

(k) representation of a compound consisting of the symbols of the component elements, each surrounded by dots representing shared and unshared electrons

(l) force that holds atoms together in a molecule or compound

8.2 Match the key terms with the descriptions provided.

(a) bonding between cations and anions resulting from electrostatic attractions of opposite charges

(b) a covalent bond that involves the sharing of two pairs of electrons

(c) a group of atoms substituted for hydrogen in the formula of a hydrocarbon that gives the compound its characteristic properties

(d) representation of an atom consisting of the symbol for the element surrounded by a number of dots equal to the number of valence electrons

(e) the angle between the two lines defined by a central atom attached to two surrounding atoms

(f) a bond resulting from the equal sharing of electron pairs

(g) a hydrocarbon containing carbon-carbon double bonds

(h) the separation of electronic charge within a bond (or molecule)

(i) a class of chemical compounds containing only carbon and hydrogen

(j) an average or composite Lewis formula derived from two or more valid Lewis formulas that closely represents the bonding in a molecule

(k) a hydrocarbon derivative containing the −OH group

(l) the shapes of molecules result from the tendency for electron pairs to maximize the distance between them to minimize repulsions

Types of Bonds

8.3 What is a chemical bond?

8.4 Describe the difference between ionic and covalent bonding.

8.5 Which type of elements are most likely to form compounds using ionic bonding?

8.6 Which type of elements are most likely to form compounds using covalent bonding?

8.7 Which of the following compounds are likely to have covalent bonds?
(a) HF (b) NaF (c) NCl_3 (d) $MgBr_2$ (e) CF_4

8.8 Which of the following compounds are likely to have ionic bonds?
(a) CsCl (b) $CaCl_2$ (c) OCl_2 (d) NBr_3 (e) IF_5

8.9 Identify the bonding in each of the following substances as ionic or covalent.
(a) $CuCl_2$ (b) F_2 (c) CO (d) $FeCl_3$

8.10 Identify the bonding in each of the following substances as ionic or covalent.
(a) NaCl (b) H_2 (c) CH_4 (d) SiO_2

8.11 Which of the following images represents an ionic compound?

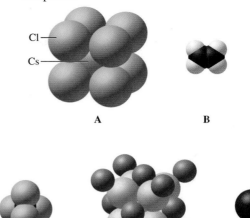

A   B

C   D   E

8.12 Which of the following images represents a covalent compound?

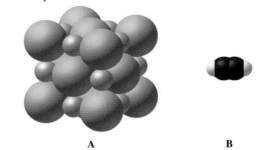

A   B

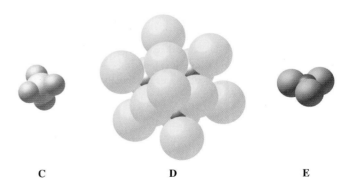

C   D   E

8.13 Which of the following compounds are likely to occur as gases at room temperature?
(a) CH_4 (b) Ca_3N_2 (c) SF_4 (d) KBr (e) HCl

8.14 Which of the following compounds are likely to occur as solids at room temperature?
(a) BrCl (b) $CaCO_3$ (c) SO_2 (d) CaO (e) NO

8.15 Predict whether each of the following substances is likely to have a relatively high or low boiling point.
(a) $TlCl_3$ (b) CsCl (c) CO_2 (d) CaO (e) O_2

8.16 Predict whether each of the following substances is likely to have a relatively high or low melting point.
(a) P_4 (b) $CaCl_2$ (c) CCl_4
(d) AlN   (e) CaC_2

8.17 Describe the experimental basis for the concept of electronegativity.

8.18 Compare the electronegativity of metallic and nonmetallic elements.

8.19 Describe the origin of polarity in bonds.

8.20 Describe how to decide whether a bond is polar.

8.21 Using periodic trends, arrange the following atoms in order of increasing electronegativity.
(a) Br, Cl, F, N, O   (b) C, F, H, N, O

8.22 Using periodic trends, arrange the following atoms in order of increasing electronegativity.
(a) B, C, H, Se, Si   (b) C, Ca, Cl, Cs, Cu

8.23 Decide which molecule in each pair contains a polar bond. Explain why that bond is polar while the other is not. Place $\delta+$ and $\delta-$ on the symbols in the formula to indicate the direction of the polarity.
(a) HF and H_2 (b) ICl and I_2
(c) H_2 and HI

8.24 Decide which molecule in each pair contains a polar bond. Explain why that bond is polar while the other is not. Place $\delta+$ and $\delta-$ on the symbols in the formula to indicate the direction of the polarity.
(a) F_2 and HF   (b) FCl and Cl_2
(c) O_2 and NO

8.25 Arrange the following bonds in order of increasing polarity.
(a) O−H, C−H, H−H, F−H
(b) O−Cl, C−Cl, H−Cl, F−Cl

8.26 Arrange the following bonds in order of increasing polarity.
(a) H−F, F−F, H−H, H−I
(b) B−F, O−F, C−F, H−F

Ionic Bonding

8.27 What information can be determined from a Lewis symbol?

8.28 Why would we represent valence electrons by means of a Lewis symbol instead of designating the valence electron configuration?

8.29 Draw Lewis symbols showing the valence electrons of the following atoms.
(a) As   (b) I   (c) Se
(d) Sr   (e) Cs   (f) Ar

8.30 Draw Lewis symbols showing the valence electrons of the following atoms.
(a) Kr   (b) Sb   (c) F
(d) In   (e) Ba   (f) Si

8.31 Draw Lewis symbols showing the valence electrons of the following ions.
(a) Cl^- (b) Sc^{3+} (c) S^{2-} (d) Ba^{2+} (e) B^{3+}

8.32 Draw Lewis symbols showing the valence electrons of the following ions.
(a) N^{3-} (b) C^{4-} (c) Br^- (d) Mg^{2+} (e) Al^{3+}

8.33 Write a formula for each of the following ionic salts using the Lewis symbols of the ions.
(a) LiCl   (b) $BaCl_2$ (c) BaS

8.34 Write a formula for each of the following ionic salts using the Lewis symbols of the ions.
(a) Na_2O (b) LiOH   (c) CsF

8.35 Explain why K^+ is known in nature, but K^{2+} is not.

8.36 Explain why Cl^- is known in nature, but Cl^{2-} is not.

8.37 Explain why sodium fluoride has the composition represented by the formula NaF.

8.38 Explain why calcium oxide has the composition represented by the formula CaO.

8.39 What is the difference between an ionic crystal and a crystal lattice?

8.40 What holds ions together in a crystal structure?

8.41 Describe the sodium chloride structure shown in Figure 8.8. How many chloride ions surround each sodium ion?

8.42 Describe the cesium chloride structure shown in Figure 8.9. How many chloride ions surround each cesium ion?

8.43 Why does CaF_2 not have the same crystal structure as NaCl?

8.44 Why does $LaCl_3$ not have the same crystal structure as CsCl?

8.45 Would you expect LiCl to have the same crystal structure as NaCl or CsCl? Explain your selection.

8.46 Would you expect FrCl to have the same crystal structure as NaCl or CsCl? Explain your selection.

Covalent Bonding

8.47 Why does hydrogen exist as a diatomic molecule?

8.48 Two chlorine atoms combine to form Cl_2. Why does chlorine form this molecule rather than Cl_3?

8.49 How many single bonds are typically formed by the following atoms?
(a) Be   (b) N   (c) F   (d) Ne ✓ if none then...

8.50 How many single bonds are typically formed by the following atoms?
(a) B   (b) O   (c) H   (d) C

8.51 Distinguish between single, double, and triple bonds.

8.52 How do we decide whether a Lewis formula should have double or triple bonds?

8.53 Identify what main-group element (X) could form each of the following compounds.

8.54 Identify what main-group element (X) could form each of the following compounds.

8.55 Write a Lewis formula for each of the following:
(a) HCN   (b) H_3CCN (c) C_2H_2
(d) C_2H_4 (e) C_2H_6

Br < Cl < N < O < F

8.56 Write a Lewis formula for each of the following:
(a) NH_2OH (b) CCl_4 (c) C_2H_3Cl
(d) C_2Br_2 (e) $HOCl$

8.57 Write a Lewis formula for each of the following:
(a) NO_3^- (b) SO_4^{2-} (c) SO_3^{2-}
(d) NO_2^- (e) NO^+

8.58 Write a Lewis formula for each of the following:
(a) NH_4^+ (b) H_2CO (c) $(CH_3)_2CO$
(d) OH^- (e) CH_3NO_2

8.59 When is it necessary to use the concept of resonance?

8.60 How is the concept of resonance consistent with the octet rule?

8.61 Indicate whether or not each of the following molecules or ions exhibits resonance.
(a) O_2 (b) H_2O (c) SO_2
(d) NO_2 (e) SO_3^{2-}

8.62 Indicate whether or not each of the following molecules or ions exhibits resonance.
(a) N_2 (b) F_2O (c) ClO_2^- (d) CO_2 (e) SO_3

8.63 Write a Lewis formula, including the resonance forms, for each of the following molecules or ions.
(a) SO_2 (b) SO_3 (c) CO_2
(d) CO_3^{2-} (e) NO_3^-

8.64 Write a Lewis formula, including the resonance forms, for each of the following molecules or ions.
(a) CS_2 (b) NCO^- (c) NO_2^-
(d) SO_4^{2-} (e) SO_3^{2-}

8.65 In HF, the hydrogen atom shares two electrons with the fluorine atom, but has no unshared electron pairs. Discuss why this observation is consistent with the principle behind the octet rule.

8.66 Describe the bonding in S_2Cl_2. The atoms are connected in the order, Cl S S Cl. Why does this molecule not exist as SCl?

8.67 Decide whether the indicated atom obeys the octet rule. If not, indicate how the octet rule is broken.
(a) O in H_2O (b) S in SF_4
(c) F in SF_4 (d) S in SF_2

8.68 Decide whether the indicated atom obeys the octet rule. If not, indicate how the octet rule is broken.
(a) B in BCl_3 (b) Cl in ClF_2^- ion
(c) F in F_2 (d) S in SF_6

8.69 An atom in each of the following molecules does not obey the octet rule. Decide which atom violates the rule and explain the nature of the violation.
(a) SF_4 (b) BH_3 (c) XeF_6 (d) ClO_2

8.70 An atom in each of the following molecules does not obey the octet rule. Decide which atom violates the rule and explain the nature of the violation.
(a) NO_2 (b) XeF_2 (c) $BeCl_2$ (d) ICl_3

8.71 What are the different classes of hydrocarbons?

8.72 How many bonds does carbon normally form? What are the different ways it can form this many bonds?

8.73 Draw the Lewis structure of benzene (C_6H_6), a cyclic compound.

8.74 Draw the Lewis structure of ethylene (C_2H_4).

8.75 Identify the class of organic substance for each of the following molecules.
(a) CH_3-OH (b) H_3C-CH_3
(c) $H_3C-O-CH_3$ (d) $H_3C-CH=CH_2$
(e) $H_3C-\overset{\overset{O}{\|}}{C}-CH_3$

8.76 Identify the class of organic substance for each of the following molecules.
(a) H_3C-NH_2 (b) $H_2C=CH_2$
(c) $H_3C-\overset{\overset{O}{\|}}{C}-OH$ (d) $H_3C-\overset{\overset{O}{\|}}{C}-H$
(e) C_6H_6

8.77 Identify the class of organic substance for each of the following molecules.
(a) CH_3CH_2OH (b) CH_3CHCH_2
(c) CH_3CH_2CHO (d) $CH_3CH_2OCH_2CH_3$

8.78 Identify the class of organic substance for each of the following molecules.
(a) $(CH_3)_2NH$ (b) C_2H_2
(c) $CH_3CH_2CO_2H$ (d) $CH_3CH_2COCH_3$

8.79 Identify the class of organic substance for each of the following molecules.

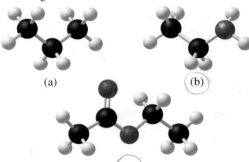

(a)   (b)

(c)

8.80 Identify the class of organic substance for each of the following molecules.

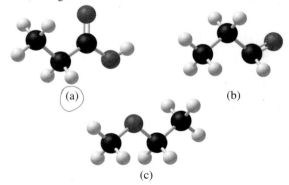

(a)   (b)

(c)

8.81 Draw an aldehyde that has the formula C_4H_8O.

8.82 Draw a ketone that has the formula C_4H_8O.

Shapes of Molecules

8.83 How can VSEPR theory be used to predict molecular shapes?

8.84 Why are unshared pairs of electrons on a central atom not considered to be part of the molecular shape?

8.85 Why is it important to draw Lewis structures before predicting molecular shapes?

8.86 Explain how nonbonding pairs of electrons influence molecular shape.

8.87 Draw each of the following geometric arrangements.
(a) tetrahedral   (b) trigonal planar
(c) bent

8.88 In which of the following molecular shapes would you expect to find one or more unshared pairs of electrons on the central atom?
(a) bent   (b) tetrahedral
(c) trigonal planar   (d) trigonal pyramidal

8.89 Predict the parent structures of the following molecules.
(a) $BeCl_2$ (b) PH_3 (c) SCl_2 (d) SO_2
(e) H_2Te (f) SiH_4 (g) BBr_3 (h) H_2O

8.90 Predict the parent structures of the following molecules.
(a) OCS (b) FNO (c) FCN (d) HN_3
(e) PF_3 (f) SF_2 (g) NO_2

8.91 Predict the shapes and give approximate bond angles for the following molecules.
(a) $BeCl_2$ (b) PH_3 (c) SCl_2 (d) SO_2
(e) H_2Te (f) SiH_4 (g) BBr_3 (h) H_2O

8.92 Predict the shapes and give approximate bond angles for the following molecules.
(a) OCS (b) FNO (c) FCN (d) HN_3
(e) PF_3 (f) SF_2 (g) NO_2

8.93 Predict the bond angles in the following molecules.
(a) NH_3 (b) H_2O (c) HCl (d) HCN
(e) BF_3 (f) H_2CO (g) PCl_3

8.94 Predict the bond angles in the following molecules.
(a) NF_3 (b) SO_2Cl_2 (c) CBr_4
(d) F_2CO (e) PH_3 (f) HOCl

8.95 Give an example of a molecule or polyatomic ion that has the following features.
(a) three bonded atoms, no unshared electrons on the central atom
(b) three bonded atoms, one unshared pair of electrons on the central atom
(c) two bonded atoms, two unshared pairs of electrons on the central atom

8.96 Give an example of a molecule or polyatomic ion that has the following features.
(a) four bonded atoms, no unshared electrons on the central atom
(b) two bonded atoms, no unshared electrons on the central atom
(c) two bonded atoms, one unshared pair of electrons on the central atom

8.97 Identify a molecule or ion that could have the following structures.

(a)   (b)   (c)   (d)

8.98 Identify a molecule or ion that could have the following structures.

(a)   (b)   (c)   (d)

8.99 Is this the structure of NO_3^- or ClO_3^-?

8.100 Is this the structure of SCl_2 or $BeCl_2$?

8.101 Which of the following molecules or ions have unshared electron pairs on the central atom?

A   B   C   D

8.102 Which of the following molecules or ions have unshared electron pairs on the central atom?

A   B   C   D

8.103 Hydrazine (N_2H_4) is a colorless, oily liquid that fumes in air and has an odor much like that of ammonia. It is often used as a rocket fuel. The order of its atoms is H_2NNH_2. How many unshared pairs of electrons are on each nitrogen atom?

8.104 Oxalic acid ($H_2C_2O_4$), a poisonous colorless solid, is found in some vegetables such as spinach and rhubarb. It is present in concentrations well below the toxic limit, so you can't use this as a reason to refuse a helping of spinach. The order of atoms in a molecule of oxalic acid is HO_2CCO_2H. How many unshared pairs of electrons are on the carbon and oxygen atoms?

8.105 Chloropicrin (Cl_3CNO_2) is an insecticide that has been used against insects that attack cereals and grains. It is a liquid with an intense odor. Predict the Cl−C−Cl, Cl−C−N, C−N−O, and O−N−O bond angles in a molecule of chloropicrin.

8.106 Fuel cells are used in many areas, such as the aerospace industry, where energy efficiency is more important than high power output. One fuel cell uses methanol (CH_3OH) as a fuel. Predict the H−C−H, H−C−O, and C−O−H bond angles.

8.107 Distinguish between bond polarity and molecular polarity.

8.108 Why does molecular polarity depend not only on bond polarity but also on the geometry of the molecule?

8.109 Explain how carbon tetrachloride can have polar bonds but still be a nonpolar molecule.

8.110 Explain why hydrocarbons are all essentially nonpolar substances.

8.111 Decide whether each of the following molecules is polar.
(a) HI   (b) CHF_3
(c) SO_2Cl_2 (d) PF_3

8.112 For each pair of molecules decide which molecule is polar and explain why it is polar while the other is not.
(a) SO_2, CO_2 (b) SO_2, SO_3
(c) $SeCl_2$, $BeCl_2$ (d) CH_4, CH_3I

8.113   Explain why the first molecule of each pair is polar and the second is not.
(a) CH_2Cl_2, CCl_4 (b) PF_3, BF_3
(c) BF_2Cl, BF_3 (d) SO_2, CO_2

8.114   For each pair of molecules decide which molecule is most polar and explain your answer.
(a) CCl_4, CH_2Cl_2 (b) CH_3F, CH_3Br
(c) NF_3, NH_3 (d) OF_2, H_2O

8.115   Which of the following molecules would align with an electric field if placed in a capacitor?

A   **B**   **C**   **D**

8.116   Which of the following molecules would align with an electric field if placed in a capacitor?

A   **B**   **C**   **D**

8.117   Which molecule, CF_4 or CCl_2F_2, is most likely to be soluble in water?

8.118   Which substance, SO_2 or CO_2, is likely to be more soluble in water?

8.119   Which of these molecules is polar?

A   **B**

8.120   Which of these molecules is polar?

A   **B**

Additional Questions

8.121   Draw Lewis symbols showing the valence electrons of the following atoms.
(a) Br   (b) Pb   (c) S
(d) Ca   (e) Be   (f) Xe

8.122   Draw Lewis symbols showing the valence electrons of the following ions.
(a) P^{3-} (b) In^{3+} (c) Se^{2-}
(d) Be^{2+} (e) C^{4-}

8.123   Arrange the following atoms in order of decreasing electronegativity: Br, Cl, F, I.

8.124   Which of the following substances is likely to be ionic?
(a) H_2 (b) Li_2O (c) BCl_3
(d) ClBr   (e) SiO_2

8.125   Classify each of the following substances according to the primary type of bonding.
(a) HCN   (b) AgCl   (c) S_8
(d) CH_4 (e) $CoCl_2$

8.126   Write a Lewis formula for each of the following.
(a) $HClO_2$ (b) $HClO_3$ (c) $HClO_4$
(d) BrF_3 (e) ClO_3^-

8.127   Write a Lewis formula for each of the following molecules.
(a) N_2H_2 (b) CS_2 (c) AsF_3
(d) CO_2 (e) CO

8.128   Write a Lewis formula, including the resonance forms, for each of the following molecules or ions.
(a) OCN^- (b) N_3^- (c) ClO_2^-
(d) PO_4^{3-} (e) H_2CO_3

8.129   Draw the best Lewis formula for each of the following. Include any necessary resonance forms.
(a) ClO_4^- (b) NO_2^- (c) NCO^-
(d) HCO_2^- (e) BF_3

8.130   Gaseous aluminum chloride exists as a dimer, Al_2Cl_6, with two chlorines bridging between the two aluminum atoms: $Cl_2AlCl_2AlCl_2$. Draw a Lewis structure for Al_2Cl_6 and for $AlCl_3$. Discuss a reason why $AlCl_3$ forms Al_2Cl_6.

8.131   Describe the molecular shape of the following.
(a) $SiCl_4$ (b) $GaCl_3$ (c) NCl_2^+
(d) IO_3^- (e) PCl_4^+ (f) OF_2
(g) GeH_4 (h) $SOCl_2$ (i) Br_2O (j) ClO_2^-

8.132   Describe the structure and bonding in sulfuric acid (H_2SO_4) and in its two ions, HSO_4^- and SO_4^{2-}.

8.133   Decide which of each pair of gaseous molecules is polar, and explain why that molecule is polar while the other is not.
(a) $BeCl_2$, OCl_2 (b) PH_3, BH_3
(c) BCl_3, $AsCl_3$ (d) SiH_4, NH_3

8.134   Which of the following are nonpolar molecules, even though they have polar bonds?
(a) CH_2Cl_2 (b) $SiCl_4$ (c) SCl_2
(d) ClO_2 (e) PBr_3

Joel is a student in an atmospheric science course. For a class project, he and his lab team are launching a weather balloon. Their balloon is filled with helium and made of a latex material that expands as external pressure decreases. The equipment attached to the balloon includes a number of instruments that Joel and his group will use to monitor atmospheric pressure and temperature as the balloon rises. Joel's project is part of a larger scientific investigation. His launch time is scheduled to synchronize with the release of hundreds of other similar balloons from stations all over the planet (Figure 9.1). After that, the team will work throughout the day to collect and analyze data. Come evening, they will have earned a rest and a small celebration. They plan to get together at Joel's house for burgers cooked on his gas grill.

But first comes the launch. Joel and his team know what to expect. After the release, their balloon will ascend through the lowest layer of Earth's atmosphere. The region from the surface to about 11 km (6.8 mi) above it is the *troposphere* (Figure 9.2). From Earth's surface to the outer edge of the troposphere, the pressure drops by around 80%. There is also a steady decrease in temperature, about 7°C (or 7 K) for every kilometer increase in altitude. Although Joel and the other observers can't see it, the balloon expands as it rises through the atmosphere.

Beyond the troposphere lies the *stratosphere,* a region that extends to about 50 km in altitude above Earth's surface. As the balloon enters it, pressure continues to drop, and the balloon expands even more. Unfortunately, a weather balloon cannot withstand the atmospheric conditions in the upper stratosphere. When it reaches about 30 km (19 mi), the latex stretches beyond its capacity and the balloon bursts. As the data-gathering equipment and transmitter fall, a small parachute opens to ease the instruments' landing. Joel and his group hope they can recover their equipment.

The instruments attached to Joel's weather balloon transmit data about temperature, relative humidity, and pressure. These characteristics of the atmosphere are associated with the behavior of gases. Of the three physical states of matter—gaseous, liquid, and solid—the gaseous state is the simplest, because many properties of gases are the same, regardless of the chemical makeup of the particular gas.

The atmosphere is a natural resource that we too often take for granted. Gases in the atmosphere are necessary to our life and well-being. Both oxygen and carbon dioxide gas, for example, are vital to the maintenance of life. Plants produce oxygen

Figure 9.1
Weather balloons are released simultaneously from hundreds of locations around the planet to forecast and report weather conditions.

List all the ways you can change the size of a balloon containing air. Keep your list handy as we discuss the behavior of gases in this chapter.

↓ temp
↓ pressure

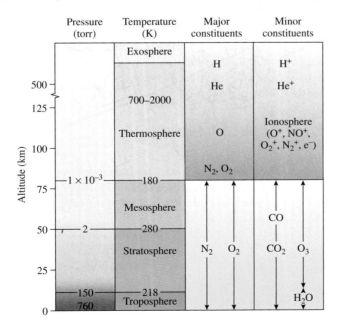

Pressure (torr)	Temperature (K)	Major constituents	Minor constituents
	Exosphere		
		H	H^+
500		He	He^+
125	700–2000		
	Thermosphere	O	Ionosphere (O^+, NO^+, O_2^+, N_2^+, e^-)
100			
		N_2, O_2	
1×10^{-3} — 180			
75	Mesosphere		
			CO
50 — 2 — 280			
	Stratosphere	N_2 O_2	CO_2 O_3
25			
150 — 218			H_2O
760	Troposphere		
0			

Altitude (km)

Figure 9.2
The troposphere is a small portion of the atmosphere surrounding Earth. Notice what happens to temperature and pressure as the distance from the surface increases.

TABLE 9.1		Volume Percent of Gases in the Atmosphere			
Gas	**Volume Percent**	**Gas**	**Volume Percent**	**Gas**	**Volume Percent**
N_2	78.09	CH_4	0.00015	O_3	0.000002
O_2	20.94	Kr	0.0001	NH_3	0.000001
Ar	0.93	H_2	0.00005	NO_2	0.0000001
CO_2	0.032	N_2O	0.000025	SO_2	0.00000002
Ne	0.0018	CO	0.00001	H_2O	Varies
He	0.00052	Xe	0.000008		

Carbon dioxide helps to regulate breathing. A common malfunction of the breathing mechanism is hiccups. One cure involves breathing into a paper bag, which builds up the level of CO_2 in the bloodstream, stimulating normal breathing. The hiccups disappear. High levels of carbon dioxide in the body are not beneficial, however. Breathing air that contains more than 10% CO_2 causes humans to lose consciousness, and continued exposure to high CO_2 levels can lead to respiratory failure and death.

If you live in a metropolitan area, air around your city contains many other gases that make up smog. These gases include NO_2 and SO_2. Small particles like dust also contribute to smog.

The volume percent of water in air is different from relative humidity, which is also a percentage. Relative humidity is a measure of the mass of water vapor present relative to the mass of water needed to saturate the air at a given temperature.

that can be consumed by animals. Carbon dioxide plays a complementary role; it is exhaled by animals and used as a raw material for food production in plants. Although we exhale carbon dioxide as a waste product, it is important to us. Its production during the breakdown of nutrients in the body helps to maintain proper blood acidity, which is essential for good health.

The atmosphere is estimated to have a total mass of about 5.2×10^{18} kg, only about 0.03% of the mass of Earth. Nevertheless the gases in the atmosphere are attracted to Earth by gravity. The levels of the atmosphere differ in their composition (Figure 9.2). The troposphere contains most of the atmosphere's gas molecules, primarily N_2, O_2, CO_2, H_2O, and Ar. The stratosphere contains, in addition to other gases, some ozone (O_3). Ozone is toxic to humans and animals when they breathe it, and it is responsible in part for photochemical smog at low altitudes. In the stratosphere, however, ozone absorbs harmful ultraviolet radiation from the Sun before it reaches Earth. We'll discuss more about ozone in Chapter 12.

Table 9.1 shows the composition of the atmosphere at ground level. The principal components are nitrogen and oxygen, which make up 99% of the substances in air. Air pollution or humidity may cause the composition to differ from that given in the table. The amount of water vapor in the air varies considerably from one location to another, from as high as 5% of the total volume in hot humid areas such as the tropics to as low as 0.01% in cold areas such as polar regions and in dry areas such as deserts.

In this chapter we will examine the physical properties of gases, especially how they behave when subjected to pressure or temperature changes. This behavior is described by several laws, called the *gas laws,* which allow us to predict how gases respond to changes in their environment. To explain the gas laws, we will develop a model for the behavior of gases at an atomic or molecular level. As you read this chapter consider the following questions.

Questions for Consideration

9.1 What are some general properties of gases?

9.2 How does the behavior of gases vary with changes in pressure, temperature, volume, and number of molecules (atoms)?

9.3 What are the mathematical relationships between volume, pressure, temperature, and amount of gas?

9.4 What is the theory that explains the behavior of gases in terms of atomic or molecular motion?

9.5 How can quantities of gases in chemical reactions be calculated?

Math Tools Used in This Chapter

Units and Conversions (Math Toolbox 1.3)

Solving Simple Algebraic Equations (Math Toolbox 9.1)

Graphing (Math Toolbox 9.2)

9.1 THE BEHAVIOR OF GASES

Joel's weather balloon responds to changing conditions because of the unique properties of gases. The gases found in the atmosphere—whether major components like O_2 and N_2, or minor ones like CO and Ar—share some common properties. Among other things, gases can be described in terms of pressure, volume, temperature, and the amount of the gas (number of particles). If one of these properties changes, one or more of the others will change also.

Gases can also be described in terms of density. Recall from Chapter 1 that density is mass per unit volume. Compared to liquids and solids, all gases have relatively low densities. Their densities explain, in part, why they exist above Earth and bubble up through liquids. The density of oxygen, for example, is 0.0013 g/mL under normal conditions of temperature (20°C) and atmospheric pressure (sea level). By comparison, the density of water is 1.0 g/mL. Gases can be compressed into a container, such as the fuel tank of Joel's gas grill, if we exert a force on them. In Joel's grill propane is compressed to the liquid state. When gases are stored under pressure, their densities increase.

Such properties are easy to observe, but what is going on with gas particles that might explain some of the changes we can see and measure? The properties of gas particles—whether atoms or molecules—explain why, among other things, gases have low densities and are compressible.

- *Gases consist of particles that are relatively far apart (as compared to solids and liquids).* The large spaces between particles explain why gases have much lower densities than liquids and solids. Because gases have so much empty space, it is easy to compress them to smaller volumes by applying external pressure.
- *Gas particles move about rapidly.* For example, the average velocity of oxygen gas molecules at 20°C and normal atmospheric pressure is 0.44 km/s (980 mi/h).
- *Gas particles have little effect on one another unless they collide.* Gas particles have essentially no attraction for one another, so they move about freely in all directions. When they collide, gas particles simply bounce off of each other.
- *Gases expand to fill their containers.* Gases take the volume and shape of their containers as a consequence of their random motion.

Temperature and Density

Now let's apply the properties of gas particles to some other observations about gas behavior. All gases expand if heated and contract if cooled. This is one of the properties that cause winds in our atmosphere. Solar heating causes gases in the atmosphere to expand, creating currents that influence weather conditions. Why do gases expand when heated? The additional heat increases the kinetic energy of gas particles, making them move faster. The gas particles move farther away from each other because of the faster motion. Consider the representations for gas particles at different temperatures shown in Figure 9.3. Notice that at the warmer temperature the gas particles are farther apart. In an open system where gases are free to expand or contract, the number of gas particles changes within a given volume if the temperature changes. At warmer temperatures gas particles move farther apart, so there are fewer gas particles in a given volume. Consequently warm gases have lower densities. In the atmosphere, gases at the surface are warmer than those above them. The warmer gas particles with lower densities rise, displacing some of the gas particles at higher altitudes. This movement of warm and cold air creates the currents that cause winds.

What is in the space between gas molecules?

Cool gas

Warm gas

Figure 9.3
Gases expand when heated. At a higher temperature, gas molecules are farther apart. Since density is the ratio of mass to volume, warm gases in an open system have less mass in the same volume, so their density is less than for the same gas at a lower temperature.

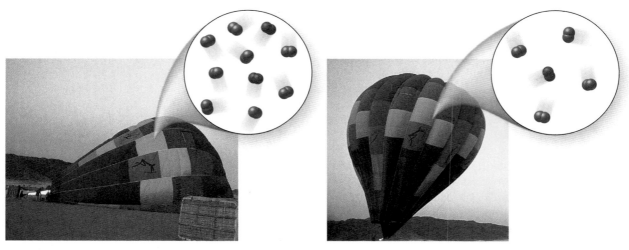

Figure 9.4
When the gas in a hot-air balloon is heated, the gas expands, filling the balloon. The opening in the bottom allows air to move in and out of the balloon. When the balloon is completely filled, the balloon can no longer expand. As the gas in the balloon is heated further, some gas particles escape, decreasing the density of the balloon. Eventually the air in the balloon becomes less dense than the surrounding air. The balloon rises.

You may notice the effect of different air densities if you take a shower in a cool room. The steam rises to the ceiling because the warm water vapor is less dense than the gas that is higher in the room.

Which finger hurts? Use the discussion of pressure to explain your answer.

Figure 9.5
When a flat tire is inflated, gas particles fill the tire and push against its sides. The proper amount of pressure allows the car to be driven safely.

Figure 9.6
In an exercise ball air pressure is great enough to support the weight of an adult human being.

To further explore properties of gases, let's consider a hot-air balloon (Figure 9.4). When the air heats up inside the filled balloon, the gas particles move faster and farther apart. The gas expands, but the balloon's volume is fixed by its inelastic fabric. The only way for the gas to expand is for some of it to escape through the opening at the bottom of the balloon. With fewer gas particles that are now farther apart, the density of the air inside the balloon decreases. Since substances with lower densities float on those of higher densities, the balloon rises.

Pressure

Changes in pressure and amount of gas also affect the behavior of gases. If you've ever put air into a tire that looks low or flat, you've probably noticed that the tire expands as air is forced into it. The increase in volume of the tire results from an increase in the number of gas particles. A pressure gauge indicates when the gas inside the tire exerts enough pressure so the car can be driven safely (Figure 9.5).

As another example, consider an exercise ball used in fitness training (Figure 9.6). The gas particles within the ball exert a pressure that supports the body's weight.

When we blow up a balloon, it inflates because of the increasing number of gas particles colliding against its wall. Gas particles are in constant motion and continuously bounce off the walls of the container. Each time a particle hits the wall, it exerts a force against the container (Figure 9.7). **Pressure** (*P*) is the amount of force applied per unit area:

$$\text{Pressure } (P) = \frac{\text{force}}{\text{area}}$$

We can describe pressure as the force of gas particles hitting the walls of the container divided by the surface area of the container:

$$\text{Pressure } (P) = \frac{\text{force of gas particles}}{\text{area of container}}$$

To distinguish between force and pressure, consider hammering a nail into a piece of wood. If we exert the right amount of force with the hammer on the nail as shown in Figure 9.8A, the nail will penetrate the board. However, if we turn the nail upside down, it will not go into the board, even if we exert the same force (Figure 9.8B). The pressure exerted is significantly reduced when the nail is upside down because the area of the nail head is larger than the area of the tip.

A moderate force can exert a tremendous pressure if the force is applied to a small area. For example, suppose a 115-lb person steps on an upright nail. If the nail point has an area of 0.0100 in^2, the pressure generated is

$$\text{Pressure } (P) = \frac{\text{force}}{\text{area}} = \frac{115 \text{ lb}}{0.0100 \text{ in}^2} = 11{,}500 \text{ lb/in}^2$$

That's enough to push the nail into the person's foot.

Like a hammer pounding a nail, gases exert a force on the walls of a container. Consider the images shown in Figure 9.9. In Figure 9.9A, fewer gas atoms are present in the same size container. Fewer atoms exert less force per unit area on the walls of the container because there are fewer collisions per unit time. The result is a lower pressure. More atoms, as shown in Figure 9.9B, exert more force because there are more collisions per unit time.

Pressure can be expressed in several ways. The one that we encounter most often in our daily lives is pounds per square inch (lb/in^2 or psi). Many automobile tires are inflated to a pressure of 30 psi. Bicycle tires are inflated to pressures between 60 and 100 psi. The pressure exerted by air molecules in the atmosphere is

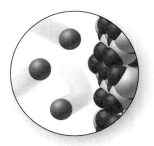

Figure 9.7
The packed structure on the right represents the rigid walls of a container made of glass (SiO$_2$). Gas particles within the container are in constant motion, exerting a force on the glass wall when they hit it.

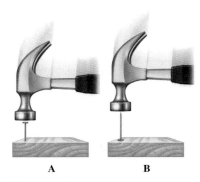

Figure 9.8
In both cases the hammer exerts the same force on the nail. When the nail is upside down, the force is spread over a larger area.

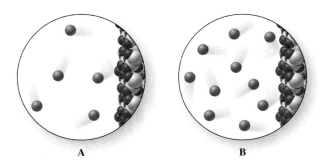

Figure 9.9
(A) Gas particles exert pressure on the wall of the glass container. (B) With a greater number of gas particles, pressure increases because of the increased number of collisions over a given area.

We can relate the pressure that gases exert on the wall of a container to blood pressure. When the heart beats, blood exerts pressure on artery walls in the circulatory system. Blood pressure is typically measured using an inflatable cuff wrapped around the upper arm. As the cuff inflates, gas particles within it exert pressure on the blood vessels in the arm, temporarily cutting off circulation. As air is released from the cuff, blood begins to flow through the arm again, and its pressure can be measured.

Figure 9.10
When the air inside a metal can is evacuated with a vacuum pump, the pressure of the outside air crushes the can.

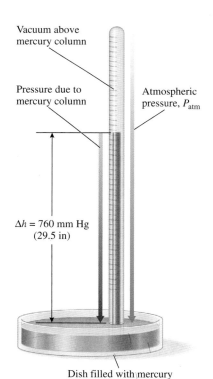

Vacuum above mercury column

Pressure due to mercury column

Atmospheric pressure, P_{atm}

Δh = 760 mm Hg (29.5 in)

Dish filled with mercury

Figure 9.11
Mercury barometers can be used to measure atmospheric pressure. The pressure of the atmosphere on the mercury in the basin forces the column of mercury to a height (Δh) that indicates the atmospheric pressure.

much smaller than what we create inside tires, only about 14.7 psi. However, the pressure of the atmosphere is significant, as Figure 9.10 shows. A vacuum pump attached to the can pumps air out of the can. The pressure of the gases outside the evacuated can is great enough to crush it.

We encounter another measure of pressure in our daily weather reports. For example, a forecaster might say, "Today's barometric pressure is 29.95 inches and falling." How can inches describe pressure? A device called a **barometer** is used to measure atmospheric pressure. The original barometer was a tube closed at one end, filled with mercury, and then inverted into a container of mercury. The height of the mercury column rises or falls so that the pressure caused by the weight of the mercury column equals the pressure exerted by the atmosphere on the mercury pool, as shown in Figure 9.11. Thus changes in atmospheric pressure cause changes in the height of the mercury column. The greater the height of the mercury in the column, the higher the atmospheric pressure. Although modern barometers are designed differently, units based on the height of the mercury column in a barometer, measured in inches, are commonly used to express atmospheric pressures in the United States.

The pressure of the atmosphere at sea level under normal weather conditions corresponds to a column of mercury that is 29.9 in (76.0 cm) tall. This pressure is the basis for a unit, the standard *atmosphere* (atm), which relates to barometric units:

$$1 \text{ atm} = 29.9 \text{ in Hg} = 76 \text{ cm Hg} = 760 \text{ mm Hg}$$

The pressures of gases in chemical systems are often fractions of 1 atm, so they are commonly measured in units of mm Hg, which has been given the name *torr*. Since 1 torr is the same as 1 mm Hg, the torr is related to the unit of atmospheres:

$$1 \text{ atm} = 760 \text{ torr} = 760 \text{ mm Hg}$$

In this book we will generally use units of either atm or torr, since they are most useful for the kinds of measurements chemists make. However, we sometimes need to convert to other units. The official SI unit of pressure is the *pascal* (Pa). This unit is related to the pressure in atmospheres:

$$1 \text{ atm} = 101,325 \text{ Pa}$$

Atmospheric pressure in weather reports broadcast in countries that use the metric system is expressed in hectapascals or kilopascals. Pounds per square inch can also be used to relate to atmospheres:

$$1 \text{ atm} = 14.7 \text{ lb/in}^2$$

Conversions among these units are most easily carried out by using their relationship to a standard atmosphere, as Example 9.1 demonstrates.

EXAMPLE 9.1   Converting Pressure Units

Express the pressure 735 torr in units of atmospheres and pascals.

Solution:
We want to convert pressure units from torr to atmospheres:

The relationship between torr and atm is

$$1 \text{ atm} = 760 \text{ torr}$$

This relationship can be used as a conversion factor as described in Math Toolbox 1.3:

$$P_{atm} = 735 \text{ torr} \times \frac{1 \text{ atm}}{760 \text{ torr}} = 0.967 \text{ atm}$$

We can convert to pascals by using the following conversion factor:

$$1 \text{ atm} = 101{,}325 \text{ Pa}$$

$$P_{Pa} = 0.967 \text{ atm} \times \frac{101{,}325 \text{ Pa}}{1 \text{ atm}} = 9.80 \times 10^4 \text{ Pa}$$

Practice Problem 9.1

The pressure of a gas is 8.25×10^4 Pa. What is this pressure expressed in units of atm and torr?

Further Practice: 9.15 and 9.16

The unit *torr* is used to credit the scientist, Evangelista Torricelli (1608–1647), who discovered the principle associated with the barometer. At the time of his discovery, one of his supporters, Pascal, took a barometer to the top of a mountain in France, measured the pressure at the top, and compared it to the pressure reading taken from a barometer at the bottom. As expected, the pressure on the mountaintop was less than the pressure at the bottom because the density of the atmosphere decreases at higher altitudes.

9.2 FACTORS THAT AFFECT THE PROPERTIES OF GASES

Joel's weather balloon encounters changes in temperature, pressure, and volume as it rises from the surface of Earth to higher altitudes. Each of these factors affects the behavior of a fixed amount of gas. Another factor is the number of gas particles present. In this section we'll describe the relationships among these factors that allow us to make quantitative predictions about gases.

Volume and Pressure

Joel and his team measure changes in atmospheric pressure as their weather balloon rises. The pressure change affects the volume of the balloon. Notice in Figure 9.12 that a balloon expands when atmospheric pressure decreases. Why does this occur? To maintain equal pressures inside and outside the balloon, the gas particles spread out. The only way they can do this in an enclosed balloon is if the volume of the balloon is increased.

As another example, what happens to the size of small gas bubbles released at the bottom of a large fish tank (Figure 9.13)? If you've ever jumped into deep water in a pool or lake, you've felt an increase in pressure below the surface of the water.

Figure 9.12
As the weather balloon in Figure 9.1 ascends, it expands to maintain equal pressures inside and outside the balloon. Because of decreased atmospheric pressure, the pressure of the gas within the balloon decreases as the balloon expands and the volume of the balloon increases. The pressures inside and outside the balloon then become equal. Eventually the balloon expands so much that it bursts.

Figure 9.13
Bubbles in a fish tank increase in size as they move toward the surface of the water. Bubbles on the bottom experience a greater pressure from the surrounding water molecules than those at the top.

Bubbles in a deep fish tank experience the same phenomenon. As a bubble rises, the pressure on it decreases, since there is not as much water pressing down on it. The size of the bubble therefore increases as it rises.

The effect of pressure on gas volume can also be observed in a piston as shown in Figure 9.14. In this system a fixed amount of gas is sealed inside. When the piston is pushed down, external pressure is applied to the gas within the container. Once the piston is still, the pressure exerted by the piston is equal to the pressure on the inside, so the pressure the trapped gas exerts on the inside walls of the container has increased.

If the volume and pressure are measured as the gas is compressed, these quantities can be plotted on a graph as shown in Figure 9.15. From this graph, can you determine a relationship between volume and pressure? What happens to the volume when the pressure increases? What happens to the pressure if the volume increases? What happens to particles inside a container as pressure and volume vary along the curve? Use your interpretation of the graph to answer the questions in Example 9.2.

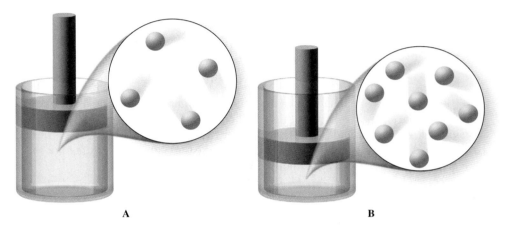

A   B

Figure 9.14
(A) Gas atoms in a cylinder with a movable piston. (B) When the piston moves down, the volume decreases and atoms move closer together, exerting a greater pressure on the walls of the cylinder.

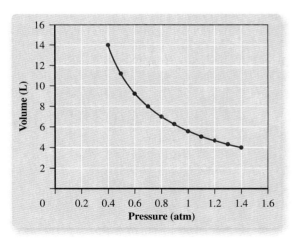

Figure 9.15
This graph shows the relationship of volume and pressure for a gas at constant temperature. What happens to volume when pressure increases?

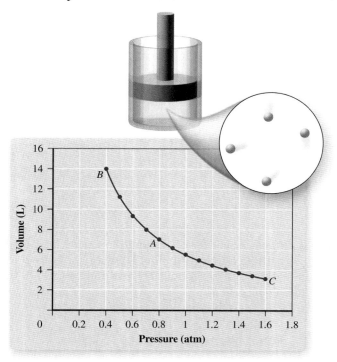

EXAMPLE 9.2

Graphical Relationship of Volume and Pressure for a Gas

The piston shown in the figure represents starting conditions for a helium gas sample. Suppose the volume and pressure correspond to point *A* on the graph. If the pressure increases by a factor of 2, what point along the curve corresponds to the new volume and pressure conditions?

Solution:

Inspection of the graph indicates that when the pressure doubles, the volume is halved. Point *A*, the starting conditions of the gas in the piston, is about 0.8 atm in pressure and approximately 7 L in volume. If the pressure is doubled to 1.6 atm, the volume falls to about 3.5 L. This corresponds to point *C*.

Practice Problem 9.2

The atomic-level image in the figure represents the atoms present in a microscopic volume under the starting conditions defined by point *A*. Draw a picture to represent the atoms in the same microscopic volume at point *C*.

Further Practice: 9.21 and 9.22

The relationship you just examined is known as **Boyle's law**: *For a given mass of gas at constant temperature, volume varies inversely with pressure.* That is, as pressure increases, volume decreases when temperature is constant. Or looking at the relationship another way, as volume increases, pressure decreases. To understand this relationship, consider how gas molecules behave inside a container. They are constantly moving and bouncing off the container walls. The impacts exert force and thus pressure on the inside walls. When an external pressure compresses the gas to a smaller volume, the molecules do not have to travel

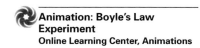

Animation: Boyle's Law Experiment
Online Learning Center, Animations

From Chemical Education Research Group, Department of Chemistry, Iowa State University www.chem.iastate.edu/group/Greenbowe/ sections/projectfolder/flashfiles/gaslaw/boyles_ law_graph.html.

In 1622, Robert Boyle, an English chemist, discovered the relationship between pressure and volume in a gas.

When variables are proportionally related, the symbol ∝ is used to express the mathematical relationship. For example, if x is directly proportional to y, we can express the relationship as $x \propto y$. When two variables are inversely proportional, we can use a similar expression, except we must show the inverse of one of the variables:

$$x \propto \frac{1}{y} \quad \text{or} \quad y \propto \frac{1}{x}$$

For more, see Math Toolbox 9.2.

This picture of a potato chip bag was taken at an elevation of 6000 ft (about 1830 m). Why is it puffier than at lower elevations?

as far to hit the walls, so they collide with it more often. The more frequent the collisions, the more force is exerted on the walls, and the higher the resulting pressure.

If we look at the data in Figure 9.15 more quantitatively, we can see that doubling the pressure halves the volume, tripling the pressure cuts the volume to one-third its original value, and so on. We can express this relationship mathematically, using Boyle's law. The volume occupied by a gas is inversely proportional to its pressure, with V as volume and P as pressure:

$$V \propto \frac{1}{P}$$

We can also use a constant to describe an equivalent relationship:

$$V = \text{constant} \times \frac{1}{P} \quad \text{or} \quad V = \frac{\text{constant}}{P}$$

To illustrate this inverse relationship, the data shown in Figure 9.15 can be graphed in a different way. If we plot the volume versus the inverse of the pressure ($1/P$), we get a straight line as shown in Figure 9.16.

The equation that relates volume and pressure can be rearranged:

$$PV = \text{constant}$$

The constant has the same value for all values of pressure and volume, provided the amount (in terms of mass or moles) of gas and the temperature remain the same. Look again at Figure 9.15. Notice that multiplying volume by pressure at two different points gives the same value. That is, at two different sets of P and V, the values of their products equal a constant:

$$P_1 V_1 = \text{constant} \qquad P_2 V_2 = \text{constant}$$

The subscripts refer to any two points on a plot of volume and pressure. If both $P_1 V_1$ and $P_2 V_2$ equal a constant, they must also equal each other:

$$P_1 V_1 = P_2 V_2$$

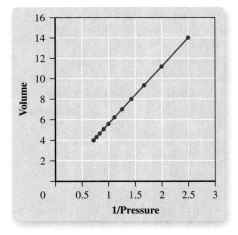

Figure 9.16
Plotting volume against the inverse of pressure ($1/P$) gives a straight line. For more on this relationship, see Math Toolbox 9.2.

As long as the temperature and mass (or moles) of the gas remain constant, Boyle's law holds for many gases under normal atmospheric conditions. A gas that behaves according to predicted linear relationships is called an **ideal gas.**

temp & mass are constant

The equation just given can be used for many pressure-volume calculations. If three of the four values are known, the other can be calculated, as Example 9.3 demonstrates. (See Math Toolbox 9.1 for more information on rearranging simple algebraic expressions.)

EXAMPLE 9.3　Volume-Pressure Relationship

A balloon contains 512 mL of helium when filled at 1.00 atm. What would be the volume of the balloon if it were subjected to 2.50 atm of pressure?

Solution:

Before we use Boyle's law to calculate the final volume, let's consider the problem qualitatively. What do you expect to happen to the balloon if we apply a pressure that's greater than the original pressure? Will the volume increase or decrease? By what factor do you expect the volume to change? We would expect the volume to decrease with an increase in the pressure because pressure and volume are inversely proportional. Since pressure increases by a factor of 2.50, we would expect volume to decrease by a factor of 2.50. Let's now apply Boyle's law to calculate the final pressure. P_1 is 1.00 atm, P_2 is 2.50 atm, V_1 is 512 mL, and V_2 is unknown.

Initial Conditions	Final Conditions
$P_1 = 1.00$ atm	$P_2 = 2.50$ atm
$V_1 = 512$ mL	$V_2 = ?$

We can use the following relationship to determine the new volume:

$$P_1V_1 = P_2V_2$$

This equation can be rearranged to solve for the unknown volume:

$$V_2 = \frac{P_1V_1}{P_2}$$

Now we can substitute P_1, V_1, and P_2 into the expression and solve for V_2:

$$V_2 = \frac{(512 \text{ mL})(1.00 \text{ atm})}{(2.50 \text{ atm})} = 205 \text{ mL}$$

A common error in calculations such as this is switching the values of pressure. Always ask yourself, "Does this answer make sense?" In this case the answer does make sense because the pressure more than doubled and the calculated volume is less than half the original volume.

Practice Problem 9.3

What pressure is needed to compress 455 mL of oxygen gas at 2.50 atm to a volume of 282 mL?

Further Practice: 9.23 and 9.24

Figure 9.17
When an air-filled balloon is cooled to 77 K in liquid nitrogen, the volume of the air inside it decreases drastically. When the balloon warms back up to room temperature, the gas regains its original volume.

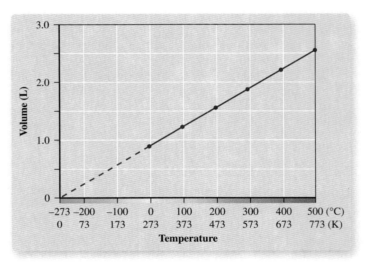

Figure 9.18
What happens to the volume of an ideal gas as temperature increases?

Volume and Temperature

When an inflated balloon is placed in hot sunlight, it gets larger. It may even stretch beyond its limits and burst. When an inflated balloon is placed in a bath of very cold liquid nitrogen, it deflates to a very small volume, but it regains its original volume if allowed to warm to room temperature again, as shown in Figure 9.17. Suppose the volume of a confined gas is carefully measured at constant pressure as the temperature is systematically changed (Figure 9.18). From the graph can you determine a relationship between volume and temperature? What happens to the volume when the temperature increases? Is volume directly proportional or inversely proportional to temperature? What happens to the particles inside the container as volume and temperature vary along the plot? Use the graph to answer the questions in Example 9.4.

EXAMPLE 9.4 **Graphical Relationship of Volume and Temperature for a Gas**

The piston shown in the figure represents starting conditions for a helium gas sample. Pressure is held constant in this container by allowing the piston to move freely as temperature changes. Suppose the initial volume and temperature correspond to point A on the graph. If the kelvin temperature increases by a factor of 2, what point along the line corresponds to the new volume and temperature conditions?

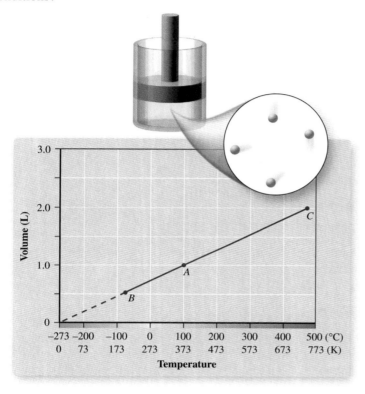

Solution:

The graph shows that volume increases as temperature increases. With the Celsius temperature scale, we have to consider negative values of temperature. The kelvin temperature scale allows us to look at absolute values of temperature, avoiding negative numbers. When the temperature is 373 K, the volume is about 1.0 L. If the temperature doubles to 746 K, the volume increases to about 2.0 L. This corresponds to point C.

Recall that kelvin temperature is equal to the Celsius temperature plus 273.15:

$$T_{kelvin} = T_{Celsius} + 273.15$$

Practice Problem 9.4

The atomic-level image in the figure represents the atoms present in a microscopic volume under the starting conditions defined by point A. Draw a picture to represent the atoms in the same microscopic volume at point C.

Further Practice: 9.33 and 9.34

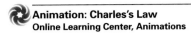

Animation: Charles's Law
Online Learning Center, Animations

Chemistry Animations Library, Charles' Law, Charles_Law.rm.

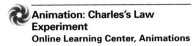

Animation: Charles's Law Experiment
Online Learning Center, Animations

www.chem.iastate.edu/group/Greenbowe/ sections/projectfolder/flashfiles/gaslaw/ charles_law.html. From Chemical Education Research Group, Department of Chemistry, Iowa State University

Jacques Charles and Joseph Gay-Lussac, both French scientists, independently discovered the relationship between temperature and volume. Charles made his discovery in 1787, and Gay-Lussac completed his study in 1802. Both were interested in hot-air balloons. In 1804 Gay-Lussac took a hot-air balloon up to 23,000 ft (about 7000 m) to study the atmosphere.

At absolute zero, 0 K, all motion would cease and the volume of an ideal gas would be zero. This temperature has not been achieved. At normal pressures, gases condense or solidify as absolute zero is approached.

Jacques Charles and Joseph Louis Gay-Lussac investigated the effect of temperature on gas volume. They found that a fixed amount of gas at a fixed pressure increases in volume linearly with temperature, as shown in Figure 9.18. The relationship between temperature and volume is known as **Charles's law**: *For a given mass of gas at constant pressure, volume is directly proportional to temperature on an absolute (kelvin) scale.* In other words, if temperature increases, volume increases. In their calculations, Charles and Gay-Lussac did not see a direct relationship between volume and degrees Celsius. An adjustment of the temperature in degrees Celsius to kelvins avoids negative temperatures and provides a straightforward mathematical relationship between volume and temperature. This can be expressed mathematically with the expression

$$V \propto T$$

We can use a constant to describe an equivalent relationship:

$$V = \text{constant} \times T \qquad \text{or} \qquad \frac{V}{T} = \text{constant}$$

Figure 9.18 shows that dividing volume by temperature at two different points gives the same value. That is, at two different sets of V and T values, their ratios equal a constant (when the pressure is constant):

$$\frac{V_1}{T_1} = \text{constant} \qquad \frac{V_2}{T_2} = \text{constant}$$

The subscripts refer to any two points on a plot of volume and temperature. If both ratios equal a constant, they must also equal each other:

$$\frac{V_1}{T_1} = \frac{V_2}{T_2}$$

This equation can be used to calculate the change in the volume occupied by a fixed amount of gas at constant pressure resulting from a change in the absolute temperature of the gas. It also allows us to calculate the temperature change required to achieve a specific change in volume. If any three values are known, the fourth can be calculated. (See Math Toolbox 9.1 for more information on rearranging simple algebraic expressions.)

EXAMPLE 9.5 **Volume-Temperature Relationship**

If a sample of chlorine gas occupies 50.0 mL at 100.0°C, what is its volume at 25.0°C at constant pressure?

Solution:

Before we use Charles's law to calculate the final volume, let's consider the problem qualitatively. What do you expect to happen to the volume of the balloon when the temperature decreases? Since the gas particles slow down at the lower temperature, their kinetic energy decreases, and the volume of the balloon decreases to maintain a constant pressure. To calculate the final volume we use Charles's law, remembering that we must convert to the kelvin temperature scale. V_1 is 50.0 mL, V_2 is unknown, T_1 is 100.0°C + 273.15 = 373.2 K, and T_2 is 25.0°C + 273.15 = 298.2 K.

Initial Conditions	Final Conditions
$V_1 = 50.0$ mL	$V_2 = ?$
$T_1 = 373.2$ K	$T_2 = 298.2$ K

We can use the following relationship to determine the new volume:

$$\frac{V_1}{T_1} = \frac{V_2}{T_2}$$

This equation can be rearranged algebraically to solve for the unknown volume:

$$V_2 = \frac{V_1 T_2}{T_1}$$

Now we can substitute T_1, V_1, and T_2 into the expression and solve for V_2:

$$V_2 = \frac{(50.0 \text{ mL}) (298.2 \ \cancel{K})}{(373.2 \ \cancel{K})} = 40.0 \text{ mL}$$

As usual, we should ask if the answer makes sense. It does, based on the prediction that V_2 should be less than V_1 when the temperature decreases.

Practice Problem 9.5

A sample of carbon monoxide gas occupies 150.0 mL at 25.0°C. It is then cooled at constant pressure until it occupies 100.0 mL. What is the new temperature in degrees Celsius?

Further Practice: 9.35 and 9.36

Volume, Pressure, and Temperature

The volume of Joel's weather balloon increases as it ascends to higher altitudes. Can the gas laws we've discussed so far explain this change? To answer the question we need to consider how pressure, temperature, and volume relate. Recall that at constant temperature, volume is inversely proportional to pressure:

$$V \propto \frac{1}{P}$$

At constant pressure, volume is directly proportional to temperature:

$$V \propto T$$

The interdependence of these variables can be summarized by the **combined gas law,** which states that for a constant amount of gas, volume is proportional to absolute temperature divided by pressure:

$$V \propto \frac{T}{P}$$

Because of the proportionality relationship, we can use a constant to relate volume to temperature and pressure:

$$V = \text{constant} \times \frac{T}{P} \qquad \text{or} \qquad \frac{PV}{T} = \text{constant}$$

The relationship between P, V, and T holds for two sets of conditions (when the amount of gas is constant):

$$\frac{P_1 V_1}{T_1} = \text{constant} \qquad \frac{P_2 V_2}{T_2} = \text{constant}$$

As usual, for any two sets of conditions, if the quantities equal a constant, they must also equal each other:

$$\frac{P_1 V_1}{T_1} = \frac{P_2 V_2}{T_2}$$

Now that we have a relationship between pressure, temperature, and volume, let's see if we can explain what happens to Joel's weather balloon. Recall that temperature decreases with an increase in altitude. If we assume that the temperature at the surface is about 300 K (approximately room temperature), we see in Figure 9.2 that the temperature decreases to around 218 K at the upper limit of the troposphere. If pressure were held constant, we'd expect the volume of the gas inside the weather balloon to decrease with the drop in temperature. However, pressure is not constant. Figure 9.2 shows that it decreases from 760 to 150 torr as the balloon rises its first 11 km.

We can use the combined gas law to rationalize the observed change in volume for the weather balloon. Rearranging this gas law and solving for V_2 gives:

$$V_2 = V_1 \times \frac{P_1}{P_2} \times \frac{T_2}{T_1}$$

We can substitute these quantities into the combined gas law and calculate how the volume changes with respect to the original volume:

$$V_2 = V_1 \times \frac{760 \text{ torr}}{150 \text{ torr}} \times \frac{218 \text{ K}}{300 \text{ K}} = 3.7 V_1$$

The volume of the balloon should increase by a factor of 3.7.

The combined gas law results from merging Boyle's law and Charles's law. If temperature is held constant ($T_2 = T_1$), the temperature terms cancel, giving Boyle's law:

$$\frac{P_1 V_1}{T_1} = \frac{P_2 V_2}{T_2}$$

$$\frac{P_1 V_1}{T_1} = \frac{P_2 V_2}{T_1}$$

$$P_1 V_1 = P_2 V_2$$

If pressure is held constant ($P_2 = P_1$), the pressure terms cancel, giving Charles's law:

$$\frac{P_1 V_1}{T_1} = \frac{P_1 V_2}{T_2}$$

$$\frac{V_1}{T_1} = \frac{V_2}{T_2}$$

The combined gas law predicts what happens to one of the gas properties—P, V, or T—when one or both of the other properties change. The equation can be rearranged to any form necessary to solve for a given unknown variable, provided the other variables are known.

We can look at this problem in another way. The pressure decreases by a factor of 5 (760/150) and temperature decreases by only a factor of 1.4 (300/218). The change in pressure is greater than the change in temperature, so pressure plays a more important role in explaining the change in volume of the weather balloon.

EXAMPLE 9.6   Combined Gas Law

If a sample of argon gas occupies 2.50 L at 100.0°C and 5.00 atm, what volume will it occupy at 0.0°C and 1.00 atm?

Solution:

Volume, pressure, and temperature change in this process, so we will use the combined gas law. V_1 is 2.50 L, V_2 is unknown, P_1 is 5.00 atm, P_2 is 1.00 atm, T_1 is 100.0°C + 273.15 = 373.2 K, and T_2 is 0.0°C + 273.15 = 273.2 K.

Initial Conditions	Final Conditions
$V_1 = 2.50$ L	$V_2 = ?$
$P_1 = 5.00$ atm	$P_2 = 1.00$ atm
$T_1 = 373.2$ K	$T_2 = 273.2$ K

The combined gas law is

$$\frac{P_1 V_1}{T_1} = \frac{P_2 V_2}{T_2}$$

This equation can be rearranged to solve for the unknown volume:

$$V_2 = \frac{P_1 V_1}{T_1} \times \frac{T_2}{P_2}$$

$$= \frac{(5.00 \text{ atm})(2.50 \text{ L})}{373.2 \text{ K}} \times \frac{273.2 \text{ K}}{1.00 \text{ atm}}$$

$$= 9.15 \text{ L}$$

Practice Problem 9.6

A sample of hydrogen gas occupies 1.25 L at 80.0°C and 2.75 atm. What volume will it occupy at 185°C and 5.00 atm?

Further Practice: 9.49 and 9.50

Gay-Lussac's Law of Combining Volumes

During his investigations of gases and balloons, Gay-Lussac also examined gaseous chemical reactions. In 1808 he measured the volumes occupied by gases before and after reactions. He found that—under conditions of constant pressure and temperature—the volumes of gaseous reactants and gaseous products always relate in a ratio of small whole numbers. For example, as shown in Figure 9.19, two volumes of hydrogen gas react with one volume of oxygen gas to produce two volumes of water vapor. If temperature and pressure conditions remain constant, the volume ratio of reactants and products is the same as the mole ratio, 2:1:2. **Gay-Lussac's law of combining volumes** states that gases combine in simple whole-number volume proportions at constant temperature and pressure. Although Gay-Lussac did not understand the reasons for the ratios, we can now see that the volume proportions of reacting gases are the same as the proportions of reacting molecules.

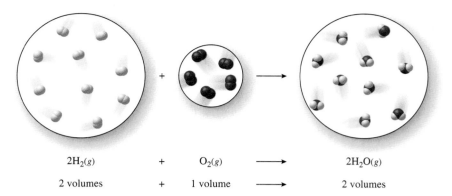

$2H_2(g)$ + $O_2(g)$ ⟶ $2H_2O(g)$

2 volumes   +   1 volume   ⟶   2 volumes

Figure 9.19
When hydrogen gas reacts with oxygen gas to form water vapor at constant temperature and pressure, the volume ratios of reactants and products is the same as the mole ratios given by the chemical equation.

Avogadro's Hypothesis

Consider what happened when Joel filled his weather balloon with helium. When he pumped in the first measured amount, the balloon expanded to a small volume. A second pump of the same amount expanded the balloon to about twice that volume, assuming the number of gas particles was the same each time. The volume, then, depends on the amount of gas blown into the balloon.

Amadeo Avogadro recognized the implications of Gay-Lussac's law of combining volumes. He hypothesized that the volume occupied by a gas at a given temperature and pressure is proportional to the number of gas particles and thus to the moles of the gas. From Figure 9.20 we can see that as the moles of gas particles (n) increases, the volume occupied by the gas also increases:

$$V \propto n$$

We can use a constant to describe an equivalent relationship between volume and amount of gas:

$$V = \text{constant} \times n \qquad \text{or} \qquad \frac{V}{n} = \text{constant}$$

Amadeo Avogadro, an Italian physicist, developed his hypothesis in 1811.

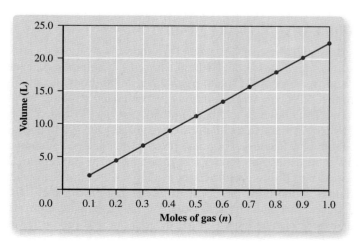

Figure 9.20
What happens to the volume of an ideal gas as the amount of gas increases?

Figure 9.20 reveals that dividing volume by moles of gas at two different points gives the same value. That is, at two different sets of V and n values, their ratios are equal to a constant (when pressure and temperature are constant):

$$\frac{V_1}{n_1} = \text{constant} \qquad \frac{V_2}{n_2} = \text{constant}$$

The subscripts refer to any two points on a plot of volume and moles of gas. If both ratios are equal to a constant, they must also equal each other:

$$\frac{V_1}{n_1} = \frac{V_2}{n_2}$$

This equation can be used to calculate the change in the volume of a gas when the amount of gas changes at constant pressure and temperature.

EXAMPLE 9.7 Volume and Amount of Gas

If a 10.0-L balloon contains 0.80 mol of a gas, what will be the volume of a balloon that contains 0.20 mol of the gas if temperature and pressure remain constant?

Solution:

Before we calculate the final volume, let's consider the problem qualitatively. What do you expect to happen to the volume of the balloon when the number of gas particles decreases? Since there are fewer gas particles, the volume of the balloon must decrease to maintain a constant pressure. To calculate the final volume we can use the following relationship:

$$\frac{V_1}{n_1} = \frac{V_2}{n_2}$$

V_1 is 10.0 L, V_2 is unknown, n_1 is 0.80 mol, and n_2 is 0.20 mol.

Initial Conditions	Final Conditions
$V_1 = 10.0$ L	$V_2 = ?$
$n_1 = 0.80$ mol	$n_2 = 0.20$ mol

Rearranging the equation to solve for the unknown volume, we get

$$V_2 = \frac{V_1 n_2}{n_1}$$

Now we can substitute n_1, V_1, and n_2 into the expression and solve for V_2:

$$V_2 = \frac{(10.00 \text{ L}) (0.20 \text{ mol})}{(0.80 \text{ mol})} = 2.5 \text{ L}$$

Does this answer make sense? It does because we expected the volume to decrease when the amount of gas decreased.

Practice Problem 9.7

If a balloon initially contains 0.35 mol of gas and occupies 8.2 L, what is the volume of the balloon when 1.2 mol of gas is present if temperature and pressure remain constant?

Further Practice: 9.59 and 9.60

Avogadro's hypothesis states that at a given pressure and temperature, equal volumes of all gases contain equal numbers of moles (or particles). In other words, 1 mol of O_2 gas has the same volume as 1 mol of CH_4 gas, if they are both at the same temperature and pressure. At 0°C and 1 atm, 1 mol of any ideal gas occupies a volume of 22.414 L, regardless of the gas. This volume is called the **molar volume.** The temperature and pressure conditions of 0°C and 1 atm are called **standard temperature and pressure (STP).** The volume at STP can be calculated from the number of moles and the molar volume, 22.414 L/mol.

EXAMPLE 9.8   **Avogadro's Hypothesis**

What volume is occupied by 3.00 mol of argon gas at STP?

Solution:

At standard temperature and pressure, 1 mol of gas occupies 22.414 L, the molar volume of a gas. We can use this relationship as a conversion factor to determine the volume occupied by 3.00 mol of gas:

$$V = 3.00 \, \text{mol} \times \frac{22.414 \, \text{L}}{\text{mol}} = 67.2 \, \text{L}$$

Note that the molar volume relationship (22.414 L/mol) holds for a gas at 0°C and 1 atm only.

Practice Problem 9.8

What volume will 7.50 mol of chlorine gas occupy at STP?

Further Practice: 9.61 and 9.62

Animation: Gas Laws
Online Learning Center, Animations

Chemistry Animations Library, Gas Laws, Gas_Laws.exe.

We can calculate the volume occupied by a gas under any other conditions of temperature and pressure from the volume of a sample at STP by using the combined gas law.

EXAMPLE 9.9   **Determining the Volume of a Gas at Non-STP Conditions**

What volume would 1.00 mol of an ideal gas occupy at 25.0°C and 1.00 atm?

Solution:

At standard temperature and pressure, 1 mol of gas occupies 22.414 L, the molar volume of a gas. This gas is at standard pressure (1.00 atm) but not at standard temperature (0°C). However, since we know the molar volume of a gas at STP, we can use the combined gas law to determine the volume at non-STP conditions.

Before we use the combined gas law to calculate the final volume, let's consider the problem qualitatively. What do you expect to happen to the volume with an increase in temperature and a pressure that does not change? Since the gas particles speed up at higher temperatures, the volume should increase. V_1 is the molar volume, 22.414 L; V_2 is unknown; P_1 is the standard pressure, 1.00 atm; P_2 is 1.00 atm; T_1 is the standard temperature, 0.00°C + 273.15 = 273.15 K; and T_2 is 25.0°C + 273.15 = 298.2 K.

Initial Conditions	Final Conditions
$V_1 = 22.414$ L	$V_2 = ?$
$P_1 = 1.00$ atm	$P_2 = 1.00$ atm
$T_1 = 273.15$ K	$T_2 = 298.2$ K

The combined gas law is

$$\frac{P_1 V_1}{T_1} = \frac{P_2 V_2}{T_2}$$

This equation can be rearranged to solve for the unknown volume:

$$V_2 = \frac{P_1 V_1}{T_1} \times \frac{T_2}{P_2}$$

$$= \frac{(1.00 \text{ atm})(22.414 \text{ L})}{273.15 \text{ K}} \times \frac{298.2 \text{ K}}{1.00 \text{ atm}}$$

$$= 24.5 \text{ L}$$

As in previous examples, we ask if the answer makes sense. It does, because an increase in temperature, with pressure staying the same, results in an increase in volume.

Practice Problem 9.9

What volume would 1.00 mol of gas occupy at 50.0°C and 2.50 atm?

Further Practice: 9.65 and 9.66

9.3   THE IDEAL GAS LAW

So far we've used the relationships among pressure (P), volume (V), absolute temperature (T), and number of moles (n) to determine the consequences of changing one of them. Suppose we want to describe a gas under a specific set of static conditions. The relationships among the individual gas laws can be manipulated to give us a general equation that relates volume, pressure, temperature, and amount of gas. Let's review the relationships we've discussed so far.

Boyle's law:	$V \propto \dfrac{1}{P}$	at constant n, T
Charles's law:	$V \propto T$	at constant n, P
Avogadro's hypothesis:	$V \propto n$	at constant P, T

These three equations can be summarized in the form of a single relationship. Since volume is proportional to each of the three factors, it must be simultaneously proportional to all of them:

$$V \propto \frac{nT}{P}$$

Since these variables are proportionally related, we can use a constant to express an equivalent relationship:

$$V = \text{constant} \times \frac{nT}{P}$$

Because it relates all the terms from the individual gas laws, we have a special name for the constant in this expression, the **ideal gas constant,** and we use R to represent it. Substituting the ideal gas constant R into the expression we obtain the following equation:

$$V = \frac{RnT}{P}$$

Rearranging yields an equation known as the **ideal gas law:**

$$PV = nRT$$

An ideal gas is a gas that follows behavior predicted by the ideal gas law. Using the observation that 1 mol of an ideal gas at STP occupies 22.414 L, we can calculate the value of R as follows:

$$
\begin{aligned}
R &= \frac{PV}{nT} \\
&= \frac{(1.000 \text{ atm})(22.414 \text{ L})}{(1.000 \text{ mol})(273.15 \text{ K})} \\
&= 0.08206 \ \frac{\text{L·atm}}{\text{mol·K}}
\end{aligned}
$$

Small gas particles like helium and hydrogen behave most ideally because they have very weak forces of attraction between the atoms or molecules. When there are stronger forces of attraction or repulsion between gas particles, their properties deviate from predictions made by the ideal gas law. This deviation also occurs at very high pressures and low temperatures. When deviations from ideal gas behavior are observed, the gas is called a *real gas.*

This constant is valid only if we measure volume in liters and pressure in atmospheres. Measurements in other units must be converted to use this constant.

The ideal gas law includes all the information summarized by Boyle's law, Charles's law, and Avogadro's hypothesis. Like them, the ideal gas law is valid only for an ideal gas. It indicates that the volume occupied by an ideal gas is directly proportional to the absolute temperature and to the number of moles of gas molecules. It is inversely proportional to pressure. A single equation that relates all the variables is a useful tool, because given any three values and the value of R, we can calculate the unknown quantity.

Calculations with the Ideal Gas Law

To solve a gas law problem, we must first determine what is going on in the gas system and then select the appropriate mathematical relationship. For *changes* in volume, pressure, and temperature, we generally use the combined gas law. In contrast, the ideal gas law allows us to determine any one of the quantities volume, pressure, temperature, or moles of the gas given the other three. If the identity of the gas is known, we can also calculate its mass using its molar mass. The ideal gas law can also be used to determine the density of a gas.

Moles of a Gas   If we are given the P, V, and T conditions for a sample of a gas, we can calculate the number of moles of the gas, using the ideal gas law. We need only rearrange the equation $PV = nRT$ to solve for the number of moles. Remember that all the other quantities must be in units that match the value used for the ideal gas constant.

EXAMPLE 9.10	**Calculating Moles of a Gas**

The volume of a propane cylinder used in Joel's gas grill is 0.960 L. When filled, the cylinder contains liquid propane stored under pressure. When the cylinder is "empty," it contains propane gas molecules at atmospheric pressure and temperature. How many moles of propane gas remain in a cylinder when it is empty if the surrounding atmospheric conditions are 25.0°C and 745 torr?

Solution:

First we must convert to the desired units of T and P:

$$T = 25.0°C + 273.15 = 298.2 \text{ K}$$

$$P = 745 \text{ torr} \times \frac{1 \text{ atm}}{760 \text{ torr}} = 0.980 \text{ atm}$$

Next we rearrange the ideal gas law and substitute the appropriate values to calculate the number of moles:

$$PV = nRT$$

$$n = \frac{PV}{RT}$$

$$= \frac{(0.980 \text{ atm})(0.960 \text{ L})}{\left(0.08206 \dfrac{\text{L} \cdot \text{atm}}{\text{mol} \cdot \text{K}}\right)(298.2 \text{ K})}$$

$$= 0.0384 \text{ mol}$$

As a final step, we should ask if this answer seems reasonable. At standard temperature (273.15 K) and pressure (1.00 atm), 1 mol of an ideal gas occupies 22.414 L, the molar volume. In this case the temperature is 25° higher than standard, the pressure is close to standard pressure, and the volume is significantly less than the molar volume of a gas. We would expect significantly fewer moles of gas present in the cylinder under these conditions, and that's what our calculation shows.

Practice Problem 9.10

The volume of an oxygen cylinder used as a portable breathing supply is 2.025 L. When the cylinder is empty at 29.2°C, it has a pressure of 723 torr. How many moles of oxygen gas remain in the cylinder?

Further Practice: 9.69 and 9.70

Mass of a Gas   Once the number of moles of gas is determined from the ideal gas law, we can calculate the mass of the gas if we know its identity. Knowing the number of moles, we can use the molar mass of the gas to convert moles to grams.

EXAMPLE 9.11   Calculating the Mass of a Gas

What mass of propane (C_3H_8) is left in the cylinder described in Example 9.10?

Solution:

Recall that molar mass provides a relationship between moles of a substance and grams of that substance. The molar mass of propane is 44.10 g/mol. The mass of 0.0384 mol of propane can be determined by multiplying moles by the molar mass:

$$\text{Mass} = 0.0384 \text{ mol } C_3H_8 \times \frac{44.10 \text{ g } C_3H_8}{1 \text{ mol } C_3H_8} = 1.69 \text{ g } C_3H_8$$

Practice Problem 9.11

The volume of an oxygen cylinder used as a portable breathing supply is 1.85 L. What mass of oxygen gas remains in the cylinder when it is empty if the pressure is 755 torr and the temperature is 18.1°C?

Further Practice: 9.79 and 9.80

Density of a Gas   If we know the mass and volume of a gas, we can determine its density. The propane described in Example 9.11 has a mass of 1.69 g. Recall from Example 9.10 that the volume of this gas is 0.960 L. We can now calculate the density:

$$\text{Density} = \frac{\text{mass}}{\text{volume}}$$

$$= \frac{1.69 \text{ g}}{0.960 \text{ L}} = 1.76 \text{ g/L}$$

Dalton's Law of Partial Pressures

If we have a mixture of gases such as those in the Earth's atmosphere, we sometimes want to know the properties of one or more of the individual substances in the mixture. The air we breathe consists of many gases, mostly nitrogen and oxygen. On humid days it contains more water vapor than on dry days. John Dalton studied mixtures of dry air and water vapor. He discovered that the total pressure of the mixture increased incrementally with the amount of water vapor. The consequence of his discovery is **Dalton's law of partial pressures**: *Gases in a mixture behave independently and exert the same pressure they would exert if they were in a container alone.* Thus the total pressure exerted by a mixture of gases is the sum of the *partial pressures* of the component gases:

$$P_{\text{total}} = P_A + P_B + P_C + \ldots$$

where A, B, and C represent the gases in the mixture. For example, the total pressure of the system studied by Dalton is given by the equation:

$$P_{\text{total}} = P_{\text{dry air}} + P_{\text{water}}$$

Mixtures of gases have the same properties as pure gases, provided that all gases in the mixtures are ideal gases and do not react chemically with each other. We often deal with mixtures of gases when we run chemical reactions. Even when a reaction produces only a single gaseous product, the procedure used to collect it may introduce another gas. For example, a technique often used to collect gases from a chemical reaction is the displacement of water from an inverted bottle, as shown in Figure 9.21. This technique involves bubbling the gas through a tube into a bottle filled with water until the gas pushes all the liquid water from the bottle.

For example, heating molten potassium bromate can produce oxygen gas:

$$2KBrO_3(l) \xrightarrow{\text{heat}} 2KBr(s) + 3O_2(g)$$

Since oxygen gas is not very soluble in water and does not react with it, this technique should collect all the gas in a pure state. Unfortunately it does not. The collected gas is a mixture of the oxygen generated by the reaction and some water vapor formed from evaporation. The amount of water vapor contained in the gas is most readily measured by the pressure it exerts at that temperature, called the *vapor pressure*. Vapor pressures for water are given in Table 9.2.

To determine the pressure exerted by a gas collected in this fashion at a specific temperature, we merely subtract the vapor pressure of water from the total pressure

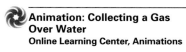
Animation: Collecting a Gas Over Water
Online Learning Center, Animations

Chemistry Animations Library, Collecting a Gas Over Water, 04_Collecting_a_Gas_Over_Water.swf.

When collecting oxygen gas, why do we fill the bottle with water and not air?

Scientists use the term *vapor* to describe the gas phase of a substance that is normally a liquid or solid.

Figure 9.21
Gaseous products can be isolated by displacing water in a closed container. Because water vaporizes to some extent at room temperature, some water vapor will be present with the gaseous product.

TABLE 9.2	Vapor Pressure of Water at Various Temperatures		
Temperature (°C)	Vapor Pressure (torr)	Temperature (°C)	Vapor Pressure (torr)
0	4.6	28	28.3
5	6.5	29	30.0
10	9.2	30	31.8
15	12.8	35	42.2
16	13.6	40	55.3
17	14.5	45	71.9
18	15.5	50	92.5
19	16.5	60	149.4
20	17.5	70	233.7
21	18.6	80	355.1
22	19.8	90	525.8
23	21.1	100	760.0
24	22.4	110	1074.6
25	23.8	150	3570.5
26	25.2	200	11,659.2
27	26.7	300	64,432.8

of the mixture. This calculation is an application of Dalton's law of partial pressures. The total pressure is atmospheric pressure in this case, since the pressure inside the bottle is the same as that outside the bottle. The vapor pressure of water, found in Table 9.2, is subtracted from the total pressure to give the partial pressure of the gas. From the partial pressure of the gas and its volume and temperature, the ideal gas law can be used to calculate the moles of gas collected.

EXAMPLE 9.12 Dalton's Law of Partial Pressures

Suppose we produce gaseous oxygen by the following reaction:

$$2KClO_3(s) \xrightarrow{\text{heat}} 2KCl(s) + 3O_2(g)$$

If we collect 1.50 L of O_2 over water at 27.0°C (300.2 K) and 758 torr using an apparatus like that shown in Figure 9.21, how many moles of O_2 are produced?

Solution:

First, we obtain the partial pressure of water at 27.0°C from Table 9.2. Then we calculate the pressure of the oxygen gas using a rearrangement of Dalton's law:

$$P_{\text{oxygen}} = P_{\text{total}} - P_{\text{water}}$$
$$= 758 \text{ torr} - 26.7 \text{ torr} = 731 \text{ torr}$$

To use the ideal gas law we must convert the pressure in units of torr to atm:

$$P_{\text{oxygen}} = 731 \text{ torr} \times \frac{1 \text{ atm}}{760 \text{ torr}} = 0.962 \text{ atm}$$

Finally we can calculate the number of moles of oxygen using the ideal gas equation:

$$PV = nRT$$

This equation can be rearranged and solved for n:

$$n = \frac{PV}{RT}$$

$$= \frac{(0.962 \text{ atm})(1.50 \text{ L})}{\left(0.08206 \dfrac{\text{L} \cdot \text{atm}}{\text{mol} \cdot \text{K}}\right)(300.2 \text{ K})}$$

$$= 0.0586 \text{ mol}$$

Practice Problem 9.12

Suppose 2.25 L of H_2 gas is collected over water at 18.0°C and 722.8 torr. How many moles of H_2 are produced in this reaction?

Further Practice: 9.83 and 9.84

9.4 KINETIC-MOLECULAR THEORY OF GASES

In Section 9.2 we described an ideal gas as a gas that behaves according to predicted proportional relationships. That is, it obeys the gas laws. The **kinetic-molecular theory of gases** is a model that explains experimental observations about gases under normal temperature and pressure conditions that we encounter in our environment.

Postulates of Kinetic-Molecular Theory

Five postulates of the kinetic-molecular theory of gases allow us to describe and predict ideal gas behavior.

1. *Gases are composed of small and widely separated particles* (molecules or atoms). Because the particles are widely separated, the actual volume of the particles is very small compared to the total volume occupied by the gas, which is mostly empty space. This postulate correctly predicts that the volume occupied by a gas is much larger than that of a liquid or a solid, in which the particles are much closer together. Because gas particles are so widely separated, gases have relatively low densities compared to liquids and solids. They are also highly compressible, since the particles can easily be squeezed much closer together and remain a gas.

2. *Particles of a gas behave independently of one another.* Because gas particles are far apart, they move independently of one another unless they collide. That is, no forces of attraction or repulsion operate between and among gas particles. This postulate explains Dalton's law of partial pressures. If gas particles move independently, the presence of more than one kind is irrelevant to the total pressure exerted by the mixture of gases. Only the total number of particles is important in determining the pressure.

3. *Each particle in a gas is in rapid, straight-line motion, until it collides with another molecule or with its container.* When collisions do occur, the collisions are perfectly elastic. That is, energy may be transferred from one particle to another, but there is no net loss of energy. This postulate also explains why gases fill their containers.

4. *The pressure of a gas arises from the sum of the collisions of the particles with the walls of the container.* This postulate explains Boyle's law, which states that pressure is inversely proportional to volume at constant temperature. For the same amount of gas, the smaller the volume of the container, the more collisions occur per unit area. The average distance traveled by a gas particle before a collision is less in a smaller volume. More collisions in a given area produce greater pressure. This postulate also predicts that pressure should be proportional to the number of moles of gas particles. The more gas particles, the greater the frequency of collisions with the walls, so the greater the pressure.

5. *The average kinetic energy of gas particles depends only on the absolute temperature.* What is average kinetic energy? To answer this question we must examine the velocity, or speed, of gas particles. The relationship between the velocity (v) and kinetic energy (KE) for a given gas particle is given by the equation

$$KE = \tfrac{1}{2}mv^2$$

where m is the mass of a gas particle. In Figure 9.22 look at the line that corresponds to gas particle velocities at low temperature. Notice that it isn't a single

To explain some properties of gases such as why they are easy to compress, some size comparisons are useful. The volume of a helium atom is known to be 1.15×10^{-26} L. The actual volume of 1 mol of helium atoms is 6.92×10^{-3} L. However, the total volume occupied by 1 mol of gaseous helium at STP is 22.4 L. Thus, only 0.03% of the volume occupied by the gas is actually taken up by the atoms.

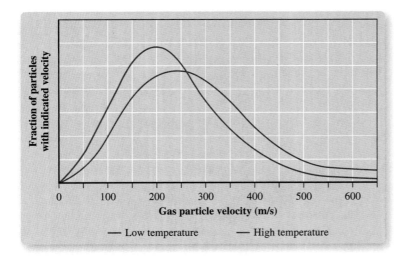

Figure 9.22
Gas particles move at various speeds when the temperature changes. At a given temperature, the speed of molecules varies across a definable range. At the higher temperature, there is a larger fraction of particles moving at greater velocities.

velocity, but a distribution of velocities. To deal with this distribution, it's convenient to describe the average kinetic energy for all the gas particles using the equation

$$KE_{av} = \tfrac{1}{2}m(v_{av})^2$$

where the "av" subscript stands for the average, or mean, of both kinetic energy and velocity. The average velocity is observed for some, but not all, of the gas particles at a given temperature. The rest have velocities either greater than or less than the average value. The distribution of velocities means that the gas particles also have a distribution of kinetic energies. Notice in Figure 9.22 that when the temperature increases, the distribution of velocities shifts. The average velocity is greater for the gas particles at higher temperatures. If the particles move faster, they strike the walls more frequently and with greater energy, so the pressure increases as temperature increases.

Diffusion and Effusion

Kinetic-molecular theory explains much of the behavior of gases. Our everyday experiences tell us that gas particles are in motion. When Joel's friends get close to his house, they can smell the meat he has already begun to cook on his grill. They smell the food because gas particles disperse through the air to their noses. **Diffusion** is the movement of gas particles from regions of high concentration to regions of low concentration (Figure 9.23). Substances can be smelled from long distances because the gaseous molecules diffuse rapidly through the atmosphere.

Although particles of different gases at the same temperature have the same average kinetic energy, these gases have different average velocities. Recall the relationship between velocity and kinetic energy for a collection of gas particles:

$$KE_{av} = \tfrac{1}{2}m(v_{av})^2$$

This equation shows that mass and velocity both influence kinetic energy. For the particles of two gases with different molar masses to have the same average kinetic energy, they must move through space at different average velocities. To compensate for a greater molar mass, heavy gas particles move at lower speeds. Light molecules or atoms move at much higher speeds.

> The average velocity of nitrogen molecules is about 500 m/s at room temperature. If gas molecules move at such high speeds, why does it seem to take a long time for smells to permeate a room?

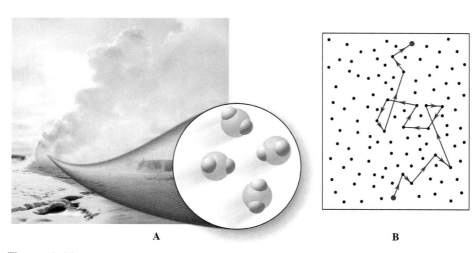

Figure 9.23
(A) The hydrogen sulfide gas that comes out of a "sulfur vent" disperses through the air. Since gas particles are in constant motion, they can diffuse through space. (B) Although gas particles move rapidly, they encounter other gas particles along the way as shown for a typical pathway for a gas particle.

For example, consider the experiment shown in Figure 9.24. In this reaction NH_3 gas reacts with HCl gas to produce solid NH_4Cl. The solid forms closer to the point where HCl is released in the tube because HCl molecules do not diffuse as fast as the lighter NH_3 molecules.

In addition to diffusion, we observe evidence of the motion of gas particles in another way. Consider Joel's weather balloon when it is filled with helium. If he fails to release it and leaves it sitting on the launch pad, it loses its buoyancy and eventually shrinks to a fraction of its inflated size. The escape of helium atoms through the natural holes in the structure of the latex explains this result. The passage of a gas through a small opening is called **effusion** (Figure 9.25). As with diffusion, small particles can effuse faster than large ones.

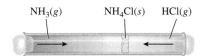

Figure 9.24

Ammonia (NH_3) and hydrogen chloride (HCl) gases react to form solid ammonium chloride (NH_4Cl).

$$NH_3(g) + HCl(g) \longrightarrow NH_4Cl(s)$$

The reactants have to meet before a reaction can occur. Notice that the solid does not form in the middle of the tube. NH_3 is a lighter molecule, so it travels faster than HCl. (Note that HCl and NH_3 are colorless gases.)

9.5   GASES AND CHEMICAL REACTIONS

Chemical reactions produce many of the gases in our atmosphere. For example, the burning of propane in Joel's grill yields carbon dioxide gas and water vapor. Oxygen in the atmosphere is used to burn the propane. Because gases are involved in many chemical reactions, it's important to understand how to calculate the quantities of gaseous reactants and products.

Product Volume from Reactant Volume

Recall that Gay-Lussac found that volumes of gases in reactions occur in small whole-number ratios. The ratios correspond to the coefficients in the balanced equation for the reaction. Consider the reaction of hydrogen with oxygen to form water vapor:

$$2H_2(g) + O_2(g) \longrightarrow 2H_2O(g)$$

If 2.38 L of H_2 at 25.0°C and 720 torr is allowed to react completely with 1.19 L of O_2 under the same conditions of temperature and pressure, we expect to form exactly 2.38 L of gaseous H_2O. This example shows how we can determine the volume of one gaseous substance from the volume of another in a chemical reaction. When temperature and pressure are constant, the volume-to-volume relationship, as described by Gay-Lussac's law of combining volumes, is the same as the mole-to-mole ratios given by the coefficients in the chemical equation. In our example, the ratio of the volumes is 2 H_2 : 1 O_2 : 2 H_2O.

Such simple conversions work only if all the substances are gases at the same pressure and temperature. If temperature and pressure vary, the volume-to-mole relationship is given by the ideal gas law, and the mole-to-mole relationship is provided by the balanced chemical equation:

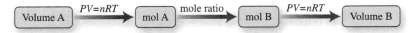

Example 9.13 shows how to perform a calculation of this kind.

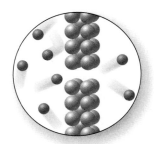

Figure 9.25

Gas particles pass through small openings. This process is called *effusion.*

EXAMPLE 9.13   Product Volume from Reactant Volume

A sample of hydrogen gas has a volume of 8.00 L at a pressure of 5.00 atm and a temperature of 25.0°C. What volume of gaseous water is produced by the following reaction at 150.0°C and 0.947 atm, if all the hydrogen gas reacts with copper(II) oxide?

$$CuO(s) + H_2(g) \longrightarrow Cu(s) + H_2O(g)$$

Solution:

Because the conditions for the products and reactants are different in terms of temperature and pressure, we cannot use Gay-Lussac's law of combining volumes to solve this problem. We'll have to use this sequence of calculations:

$$\boxed{\text{Volume } H_2} \xrightarrow{PV=nRT} \boxed{\text{mol } H_2} \xrightarrow{\text{mole ratio}} \boxed{\text{mol } H_2O} \xrightarrow{PV=nRT} \boxed{\text{Volume } H_2O}$$

We are given gas volume, not moles, so we must first calculate the moles of hydrogen gas from the ideal gas law:

$$\boxed{\text{Volume } H_2} \xrightarrow{PV=nRT} \boxed{\text{mol } H_2}$$

$$PV = n_{H_2} RT$$

$$n_{H_2} = \frac{PV}{RT}$$

$$= \frac{(5.00 \text{ atm})(8.00 \text{ L})}{\left(0.08206 \ \dfrac{\text{L} \cdot \text{atm}}{\text{mol} \cdot \text{K}}\right)(298.2 \text{ K})}$$

$$= 1.63 \text{ mol } H_2$$

Next we use the coefficients from the balanced equation to calculate the moles of gaseous water:

$$\boxed{\text{mol } H_2} \xrightarrow{\text{mole ratio}} \boxed{\text{mol } H_2O}$$

$$n_{H_2O} = 1.63 \text{ mol } H_2 \times \frac{1 \text{ mol } H_2O}{1 \text{ mol } H_2} = 1.63 \text{ mol } H_2O$$

Finally, we calculate the volume of water vapor, using the ideal gas law solved for V:

$$\boxed{\text{mol } H_2O} \xrightarrow{PV=nRT} \boxed{\text{Volume } H_2O}$$

$$V_{H_2O} = \frac{n_{H_2O} RT}{P}$$

$$= \frac{1.63 \text{ mol} \times \left(0.08206 \ \dfrac{\text{L} \cdot \text{atm}}{\text{mol} \cdot \text{K}}\right) \times 423.2 \text{ K}}{0.947 \text{ atm}}$$

$$= 59.8 \text{ L } H_2O$$

Practice Problem 9.13

A sample of hydrogen gas has a volume of 7.49 L at a pressure of 22.0 atm and a temperature of 32.0°C. What volume of gaseous water is produced by the following reaction at 125.0°C and 0.975 atm, if all the hydrogen gas reacts with iron(III) oxide?

$$Fe_2O_3(s) + 3H_2(g) \longrightarrow 2Fe(s) + 3H_2O(g)$$

Further Practice: 9.103 and 9.104

Moles and Mass from Volume

Many chemical reactions have reactants or products in different physical states. For example, solid sodium metal reacts with chlorine gas to produce solid sodium chloride:

$$2Na(s) + Cl_2(g) \longrightarrow 2NaCl(s)$$

If we know the mass of Cl_2, we can calculate the mass of NaCl produced using the procedure described in Section 6.3:

g Cl_2 —molar mass→ mol Cl_2 —mole ratio→ mol NaCl —molar mass→ g NaCl

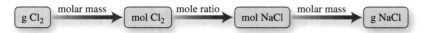

For a review of determining quantities in chemical reactions, see Chapter 6.

Suppose we want to known how much NaCl can be prepared from a known *volume* of Cl_2 under given conditions of temperature and pressure. Knowing the volume of Cl_2 lets us calculate the number of moles of this gaseous reactant using the ideal gas law. This result is used with the coefficients from the balanced chemical equation to calculate the moles of NaCl. The mass of the product can, in turn, be obtained from the number of moles of product and its molar mass. Slight modification of the first step of the previous problem-solving scheme allows us to calculate the mass of NaCl from the volume of Cl_2 gas:

Volume Cl_2 —$PV=nRT$→ mol Cl_2 —mole ratio→ mol NaCl —molar mass→ g NaCl

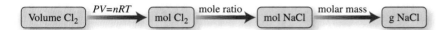

EXAMPLE 9.14 Mass from Volume

How many moles of calcium carbonate form if 3.45 L of CO_2, measured at 45.0°C and 1.37 atm, react with excess CaO? How many grams? The balanced equation is as follows:

$$CaO(s) + CO_2(g) \longrightarrow CaCO_3(s)$$

Solution:

We will have to perform the following series of calculations to solve this problem:

Volume CO_2 —$PV=nRT$→ mol CO_2 —mole ratio→ mol $CaCO_3$ —molar mass→ g $CaCO_3$

First we use the ideal gas law solved for n to calculate the moles of CO_2:

$$n_{CO_2} = \frac{(1.37\,\cancel{atm})\,(3.45\,\cancel{L})}{\left(0.08206\,\frac{\cancel{L}\cdot\cancel{atm}}{mol\cdot\cancel{K}}\right)(318.2\,\cancel{K})}$$

$$= 0.181 \text{ mol } CO_2$$

We calculate the number of moles of calcium carbonate from the number of moles of carbon dioxide by using coefficients in the balanced equation:

mol CO_2 —mole ratio→ mol $CaCO_3$

$$n_{CaCO_3} = 0.181\,\cancel{mol\,CO_2} \times \frac{1 \text{ mol } CaCO_3}{1\,\cancel{mol\,CO_2}} = 0.181 \text{ mol } CaCO_3$$

Then we obtain the mass of calcium carbonate from its molar mass, 100.1 g/mol:

$$mol\ CaCO_3 \xrightarrow{\ molar\ mass\ } Mass\ CaCO_3$$

$$Mass\ of\ CaCO_3 = 0.181\ \cancel{mol\ CaCO_3} \times \frac{100.1\ g\ CaCO_3}{1\ \cancel{mol\ CaCO_3}} = 18.1\ g\ CaCO_3$$

Practice Problem 9.14

How many moles of H_2SO_4 form if 2.38 L of SO_3 gas, measured at 65.0°C and 1.05 atm, react with sufficient water? How many grams? The balanced equation is as follows:

$$SO_3(g) + H_2O(l) \longrightarrow H_2SO_4(l)$$

Further Practice: 9.105 and 9.106

When determining the amount of a gaseous reactant or product in a chemical reaction from the mass of another reactant of product, the reverse of the process described in Example 9.14 can be followed.

SUMMARY

Several properties—including pressure, volume, temperature, and amount—describe the behavior of gases. Pressure is force per unit area. Gas pressure results from collisions of gas particles with the walls of the container. Under conditions of constant temperature, the volume of a gas decreases when pressure increases (Boyle's law). Volume is inversely proportional to pressure. Temperature can also affect gases. Under conditions of constant pressure, volume increases when temperature increases (Charles's law). Thus volume is directly proportional to temperature. When more gas is added to a container that is allowed to expand, the volume increases due to the presence of more gas particles (Avogadro's hypothesis). The combined gas law represents the relationship among temperature, pressure, and volume for a fixed amount (in moles) of gas.

Pressure, volume, temperature, and amount of gas relate mathematically according to the ideal gas law, $PV = nRT$. Given the ideal gas constant and three of the factors, the fourth can be calculated. When a gas contains more than one kind of gas, each acts independently and ideally, and the total pressure of the gas mixture is the sum of each gas pressure (Dalton's law of partial pressures).

The five postulates of the kinetic-molecular theory explain gas behavior at the particulate level. In addition to explaining the gas laws, kinetic-molecular theory accounts for other properties of gases such as diffusion (movement of gas particles through space) and effusion (the movement of gases through a small opening in a container).

We can use the ideal gas law in conjunction with reaction calculations described in previous chapters to calculate quantities of gases involved in chemical reactions.

Math Toolbox 9.1   SOLVING SIMPLE ALGEBRAIC EQUATIONS

Algebraic expressions represent many chemical principles, so understanding the principles requires solving and manipulating such equations. An algebraic equation is a simple statement of equality. For example, the equality $9x + 12x = 63$ is confirmed when $x = 3$:

$$(9 \times 3) + (12 \times 3) = 63$$

We can manipulate an equation in any way that does not destroy the equality. The usual purpose is to obtain a value for an unknown quantity. Operations that will maintain the equality are

- adding the same number to both sides of the equation
- subtracting the same number from both sides of the equation
- multiplying or dividing both sides of the equation by the same number
- raising both sides of the equation to the same power

Consider the equation $16x - 32 = 16$. To solve for x, we first add 32 to both sides of the equation:

$$16x - 32 + 32 = 16 + 32$$
$$16x = 48$$

We then divide both sides of the equation by 16:

$$\frac{16x}{16} = \frac{48}{16}$$
$$x = 3$$

As a second example, consider the equation $\frac{1}{4}x + 4 = 12$. We subtract 4 from both sides of the equation:

$$\frac{1}{4}x + 4 - 4 = 12 - 4$$
$$\frac{1}{4}x = 8$$

Then we multiply each side of the equation by 4:

$$4 \times \frac{1}{4}x = 4 \times 8$$
$$x = 32$$

Now consider $4x = 15 + x$. To solve for x, we begin by moving all the terms that contain x to one side of the equation. Subtracting x from both sides of the equation will accomplish this:

$$4x - x = 15 + x - x$$
$$3x = 15$$

Dividing both sides by 3, we solve for x:

$$\frac{3x}{3} = \frac{15}{3}$$
$$x = 5$$

So far we have considered equations that contain only one variable (unknown). Many chemistry relationships involve more than one variable. For example, to convert temperatures between Celsius (T_C) and Fahrenheit (T_F) scales, we can use the equation

$$T_C = \tfrac{5}{9}(T_F - 32)$$

Using this equation, we can calculate one of the variables if we know the value of the other. For example, if the temperature in a lab is 72.5°F, we can calculate the temperature in degrees Celsius. We begin by assigning the quantity we know to the correct variable.

Given Quantity	Unknown Quantity
$T_F = 72.5°F$	$T_C = ?$

To calculate the temperature in degrees Celsius, substitute the known value for T_F and solve the equation for T_C:

$$T_C = \tfrac{5}{9}(T_F - 32) = \tfrac{5}{9}(72.5°F - 32)$$
$$= \tfrac{5}{9}(40.5°F) = 22.5°C$$

Convince yourself that the problem could be solved for degrees Fahrenheit if you are given a value for degrees Celsius.

It is often useful to rearrange an equation to isolate the variable of interest on one side. For example, the relationship $ab = cde$ contains five variables. Suppose we know a, b, d, and e and want to solve for c. We can rearrange the equation by dividing both sides by d and e. This step removes them from the right side of the equation, leaving an equality statement for c alone:

$$\frac{ab}{de} = \frac{c\,\cancel{d}\,\cancel{e}}{\cancel{d}\,\cancel{e}}$$

$$\frac{ab}{de} = c$$

The equation is usually rearranged so that the unknown variable is on the left of the equality:

$$c = \frac{ab}{de}$$

Using this relationship, calculate c if a is 1.02, b is 0.968, d is 0.08206, and e is 298.15. In problem solving it is often useful to organize the data in some way.

Given Quantities	Unknown Quantity
$a = 1.02$	$c = ?$
$b = 0.968$	
$d = 0.08206$	
$e = 298.15$	

Then substitute the known quantities into the equation and calculate the unknown term:

$$c = \frac{1.02 \times 0.968}{0.08206 \times 298.15} = 0.0404$$

Try the following example: Given that $ab = cde$, calculate e when a is 2.25, b is 12.4, c is 1.43, and d is 0.08206. Did you get 238 as your answer?

Math Toolbox 9.2   GRAPHING

In a controlled scientific experiment involving measured quantities, values for one variable (the dependent variable) are determined as another variable changes (the independent variable). We often find one of two basic relationships between the variables. In a *direct* relationship, both variables increase (or decrease) together. That is, as one increases, so does the other. In an *inverse* relationship, one variable decreases as the other increases.

Large amounts of data often reveal proportional relationships between two variables that can be represented using a single algebraic equation. *Proportional* means the two variables are related by a constant factor, which we can represent in an algebraic equation as k. Proportional relationships are not always obvious from raw data. Determining a relationship—and consequently an equation that relates proportional variables—often requires constructing an appropriate graph. For example, consider the following data of the volume of a gas at different absolute temperatures:

Volume (L)	Temperature (K)
2.20	300
2.37	325
2.59	350
2.95	400
3.30	450
3.71	500
4.37	600
5.17	700

Although the data show that volume increases with absolute temperature (a direct relationship), they do not reveal whether an algebraic proportion exists between the two variables. We have to construct a graph of the data to see if a proportional relationship exists.

If you are creating a graph by hand, the following steps will generally lead to a graph that presents data in a clear and easily interpreted manner.

1. Carefully examine the data to be plotted. Organize the data in a table to facilitate transfer to the graph.
2. Identify and label the axes. The dependent variable is the variable measured experimentally. Plot it on the vertical axis (the y axis). The independent variable is the variable controlled and varied by the experimenter. Plot it on the horizontal axis (the x axis). Clearly label each axis with the name of the variable and the units. Put the units either in parentheses following the name of the variable or separated from the variable name by a comma: volume (L) or volume, L.
3. Decide on the scales and limits of values to be plotted along each axis. The graph should nearly fill the graph paper. Adjust the scale so the data can be entered easily on the graph and subsequently read off the graph with about the same precision as they were measured. The value of each major division on an axis is usually selected so that minor divisions represent workable numbers of units. To avoid using a small area of the graph, it may not always be desirable for the scale of each axis to begin at zero. Increments on the x axis may differ from increments on the y axis.
4. After selecting the scales to be plotted along each axis, clearly mark the value of each major division along each axis.
5. Place dots on the graph at appropriate intersecting y and x values to represent each data point. If necessary to identify the plotted points, draw a small circle around each with a diameter of 1 to 2 mm. If you plot several sets of data on the same graph, use different geometric figures such as squares or triangles to outline the other sets of data points.
6. Draw a smooth line through the data set. Because of experimental error, not all points will fall precisely on the line, but it is generally assumed that the dependent variable actually varies smoothly with the independent variable. For that reason, do *not* connect the points dot-to-dot with separate straight-line segments. If the data appear to follow a straight line, use a straight edge to draw the line. If the data follow a curve, use a graphing aid such as a French curve. Label the completed graph with an appropriate identifying title in a clear space near the top.

Let's use these steps to create a graph of the volume and temperature data given in the earlier table. Assign volume (the dependent variable) to the y axis and temperature (the independent variable) to the x axis. Suppose we have a piece of graph paper that is four major divisions high and six major divisions wide. The y axis then has four major divisions, and we need to cover $5.17\,L - 2.20\,L = 2.97\,L$. To keep even units for each minor division, assign 1.0 unit per major division. On the x axis, we have six major divisions to cover $700\,K - 300\,K = 400\,K$. We assign 100 K per major division. Following the given steps yields this graph:

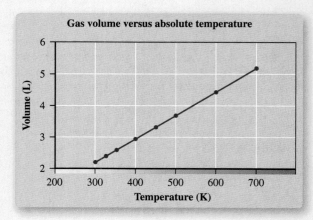

Gas volume versus absolute temperature

In general a straight line through the data points shows that the two variables are directly proportional if the line passes through the origin (0,0). An extended graph would show that volume and temperature are directly proportional. Their relationship can be represented as $y = kx$. (The slope of the line equals k.)

Math Toolbox 9.2 (continued)

Consider another set of data:

Volume (L)	Pressure (atm)
2.20	1.00
2.32	0.95
2.59	0.85
2.93	0.75
3.28	0.67
3.67	0.60
4.40	0.50
5.12	0.43

These data show that volume increases as pressure decreases, although we cannot tell if the relationship is proportional. Following the steps for drawing a graph, we obtain the following plot:

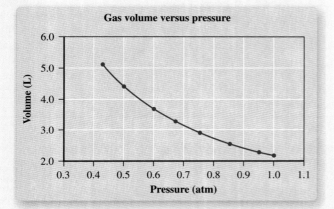

A smooth line through the data points forms a curve. The data reveal an inverse relationship between volume and pressure; that is, volume decreases as pressure increases. However, we don't have a straight line, so it is not easy to see if the relationship is proportional. When scientists encounter such data, they often try to manipulate the values mathematically to see if a straight line can be plotted.

Let's see what happens if we take the reciprocal of pressure. (The reciprocal is 1 divided by the quantity we are interested in—in this case, 1/pressure.)

Volume (L)	1/Pressure
2.20	1.00
2.32	1.05
2.59	1.18
2.93	1.33
3.28	1.49
3.67	1.67
4.40	2.00
5.12	2.33

These data show that as 1/pressure increases, the volume also increases. However, we cannot tell if the data are proportional until we create a graph.

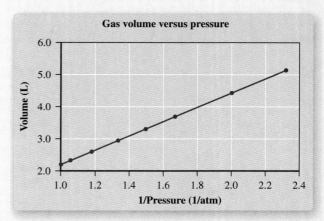

In general, if a graph of one variable versus the reciprocal of another variable yields a straight line that would pass through the origin, then the two variables are inversely (or reciprocally) proportional. That is:

$$x \propto \frac{1}{y} \quad \text{and} \quad y \propto \frac{1}{x}$$

The general equation is $y = k(1/x)$. (Find k by determining the slope of the line.)

We can determine the pressure at a given volume by inspecting the graph. For example, suppose we want to know the pressure when the volume is 3.0 L. At this volume, 1/pressure is about 1.35. Taking the inverse of this quantity, we get a pressure of 0.741 atm. We can also determine the volume at any given pressure. Use the graph to find the volume when the pressure is 0.550 atm. Did you get a value of 4.00 L?

KEY RELATIONSHIPS

Relationship	Equation
Pressure can be expressed in different units.	$1 \text{ atm} = 101{,}325 \text{ Pa} = 760 \text{ mm Hg}$ $= 760 \text{ torr} = 14.7 \text{ lb/in}^2$
Pressure is inversely proportional to volume at constant temperature and moles (Boyle's law).	$P_1 V_1 = P_2 V_2$ (constant T and n)
Temperature is proportional to volume at constant pressure and moles (Charles's law).	$\dfrac{V_1}{T_1} = \dfrac{V_2}{T_2}$ (constant P and n)
Volume is proportional to temperature divided by the pressure if the amount of gas is constant (combined gas law).	$\dfrac{P_1 V_1}{T_1} = \dfrac{P_2 V_2}{T_2}$ (constant n)
Volume is proportional to amount of gas (moles) at constant temperature and pressure (Avogadro's hypothesis).	$\dfrac{V_1}{n_1} = \dfrac{V_2}{n_2}$ (constant T and P)
The amount of gas (moles), and its pressure, volume, and temperature are related by the ideal gas law.	$PV = nRT$
For a mixture of gases, the sum of the individual pressures is equal to the total pressure (Dalton's law of partial pressures).	$P_{\text{Total}} = P_A + P_B + P_C + \dots$
The average kinetic energy of gas particles is related to their mass and average velocity.	$KE_{\text{av}} = \frac{1}{2} m (v_{\text{av}})^2$

KEY TERMS

Avogadro's hypothesis (9.2)

barometer (9.1)

Boyle's law (9.2)

Charles's law (9.2)

combined gas law (9.2)

Dalton's law of partial pressures (9.3)

diffusion (9.4)

effusion (9.4)

Gay-Lussac's law of combining volumes (9.2)

ideal gas (9.2)

ideal gas constant, R (9.3)

ideal gas law (9.3)

kinetic-molecular theory of gases (9.4)

molar volume (9.2)

pressure (9.1)

standard temperature and pressure (STP) (9.2)

QUESTIONS AND PROBLEMS

The following questions and problems, except for those in the *Additional Questions* section, are paired. Questions in a pair focus on the same concept. Answers to the odd-numbered questions and problems are in Appendix D.

Matching Definitions with Key Terms

9.1   Match the key terms with the descriptions provided.
   (a) the movement of gas particles through a small opening
   (b) law stating that at constant temperature, the volume occupied by a fixed amount of a gas is inversely proportional to its pressure
   (c) law that describes the relationship between initial and final conditions of pressure, volume, and temperature for a fixed amount of a gas
   (d) a gas that follows predicted behavior, as described by the ideal gas law
   (e) the amount of force applied per unit area
   (f) law stating that gases in a mixture behave independently and exert the same pressure they would if they were in the container alone
   (g) the volume occupied by 1 mol of a gas, which equals 22.414 L at STP for an ideal gas
   (h) a constant used in the ideal gas law that relates pressure, volume, amount of gas, and temperature

9.2   Match the key terms with the descriptions provided.

(a) the movement of gas particles through the atmosphere from high concentration to low

(b) law stating that at constant pressure, the volume occupied by a fixed amount of gas is directly proportional to its absolute temperature

(c) law stating that in chemical reactions occurring at constant pressure and temperature, the volumes of gaseous reactants and products are in ratios of small whole numbers

(d) law stating the relationship between pressure, volume, temperature, and amount of an ideal gas

(e) a closed tube filled with mercury and inverted into a pool of mercury; instrument used to measure atmospheric pressure

(f) a model of ideal gas behavior; assumes that gas particles occupy no volume and that there are no intermolecular forces between them

(g) 0°C and 1 atm

(h) proposal that equal volumes of all gases contain equal numbers of particles (at constant temperature and pressure)

The Behavior of Gases

9.3   What are some general properties of gases?

9.4   In general, how do the properties of gases differ from the properties of liquids and solids?

9.5   How does the density of warm air differ from the density of cooler air?

9.6   Why does warm air rise?

9.7   The figure shows atoms of a gas at a particular temperature. In the blank circle, show the arrangement of atoms when the temperature decreases and pressure remains constant.

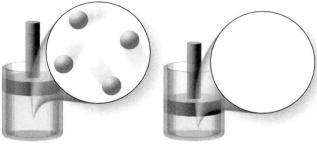

Before   After

9.8   The figure shows atoms of a gas at a particular temperature. Students were asked to select images that show what happens when the temperature increases and pressure remains constant. Many students selected the images shown. What is wrong with each of images (a) to (d)?

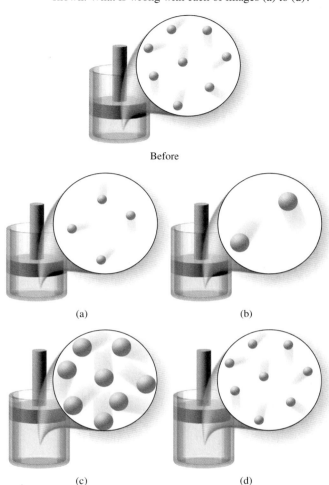

Before

(a)   (b)

(c)   (d)

9.9   What is gas pressure?

9.10   Why do gases exert pressure on the walls of their container?

9.11   How is pressure measured?

9.12   (a) What are the common units of pressure? (b) How are they related?

9.13   The figure shows atoms of a gas at a particular pressure. In the blank circle show the arrangement of atoms when the volume increases and temperature remains constant.

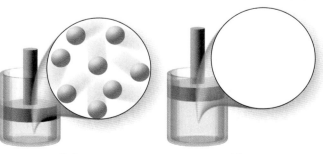

Before   After

9.14 The figure shows atoms of a gas at a particular temperature. Students were asked to select images that show what happens when the volume increases and temperature remains constant. Many students selected the images shown. What is wrong with each of images (a) to (d)?

Before

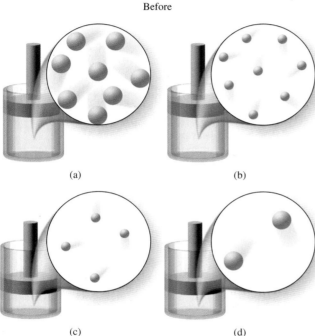

(a)   (b)

(c)   (d)

9.15 Perform the following pressure conversions.
  (a) 745 torr to atm
  (b) 1.23 atm to torr
  (c) 90.1 mm Hg to atm
  (d) 0.643 kPa to Pa
  (e) 1.35×10^5 Pa to mm Hg
  (f) 7.51×10^4 Pa to torr
  (g) 798 torr to Pa
  (h) 29.3 cm Hg to mm Hg

9.16 Perform the following pressure conversions.
  (a) 1.15 atm to torr
  (b) 968 torr to atm
  (c) 2.50×10^5 Pa to atm
  (d) 695 mm Hg to torr
  (e) 0.953 atm to Pa
  (f) 653 torr to mm Hg
  (g) 1545 mm Hg to Pa
  (h) 3.73 kPa to atm

9.17 If the barometric pressure in Denver, Colorado, is 30.24 inches of mercury (in Hg), what is the pressure in units of pascals? (Recall that 2.54 cm = 1 in.)

9.18 If the barometric pressure in Vancouver, Canada, is 101.19 kPa, what is the pressure in units of inches of mercury (in Hg)? (Recall that 2.54 cm = 1 in.)

Factors That Affect the Properties of Gases

9.19 What does Boyle's law tell us about the effect of pressure on the volume of a gas?

9.20 What is Boyle's law in mathematical terms?

9.21 Consider a gas in a container with a movable piston. What happens to the gas particles if the piston is moved to decrease the volume of the container? What happens to the pressure if the piston is moved to give one-half of the original volume of the cylinder? In the blank circle show the arrangement of atoms when the volume decreases by one-half.

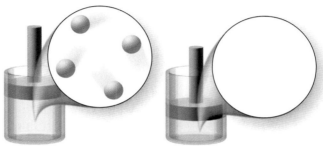

Before   After

9.22 Consider a gas in a container with a movable piston. What happens to the gas particles if the piston is moved to increase the volume of the container? What happens to the pressure if the piston is moved to triple the original volume of the cylinder? In the blank circle show the arrangement of atoms when the volume is tripled.

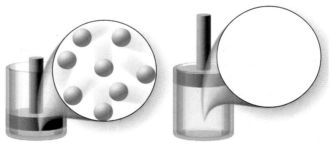

Before   After

9.23 Given a fixed quantity of a gas at constant temperature, calculate the new volume the gas would occupy if the pressure were changed as shown in the following table.

	Initial Volume	Initial Pressure	Final Pressure	Final Volume
(a)	3.00 L	2.00 atm	5.00 atm	?
(b)	30.0 mL	60.0 torr	90.0 torr	?
(c)	2.50 mL	20.0 torr	255 torr	?

9.24   Given a fixed quantity of a gas at constant temperature, calculate the new volume the gas would occupy if the pressure were changed as shown in the following table.

	Initial Volume	Initial Pressure	Final Pressure	Final Volume
(a)	1.50 L	1.50 atm	725 torr	?
(b)	3.25 L	825 torr	456 torr	?
(c)	350 mL	50.0 torr	20.0 torr	?

9.25   Given a fixed quantity of a gas at constant temperature, calculate the new pressure the gas would exert if the volume were changed as shown in the following table.

	Initial Pressure	Initial Volume	Final Volume	Final Pressure
(a)	602 torr	205 mL	1512 mL	?
(b)	0.00100 torr	2.50 L	25.0 mL	?
(c)	0.832 atm	905 L	27.5 L	?

9.26   Given a fixed quantity of a gas at constant temperature, calculate the new pressure the gas would exert if the volume were changed as shown in the following table.

	Initial Pressure	Initial Volume	Final Volume	Final Pressure
(a)	745 torr	155 mL	1.55 L	?
(b)	5.30 atm	1.85 L	4.50 L	?
(c)	755 torr	2.00 L	4500 mL	?

9.27   What pressure is required to compress 925 L of N_2 at 1.25 atm into a container whose volume is 6.35 L?

9.28   What will be the final volume of 186 mL of Cl_2 if the pressure increases from 0.945 atm to 1.76 atm?

9.29   What volume of H_2 must be collected at 725 torr to have 0.450 L of the gas at 1.00 atm?

9.30   A 2.50-L flask is filled with air at 3.00 atm. What size flask is needed to hold this much air at 735 torr?

9.31   What does Charles's law tell us about the effect of temperature on volume of a gas?

9.32   What is Charles's law in mathematical terms?

9.33   Consider a gas in a container that can adjust its volume to maintain constant pressure. Suppose the gas is heated. What happens to the gas particles with the increase in temperature? What happens to the volume of the container?

9.34   Consider a gas in a container that can adjust its volume to maintain constant pressure. Suppose the gas is cooled. What happens to the gas particles with the decrease in temperature? What happens to the volume of the container?

9.35   For a fixed amount of gas held at constant pressure, calculate the new volume the gas would occupy if the temperature were changed as shown in the following table.

	Initial Volume	Initial Temperature	Final Temperature	Final Volume
(a)	5.00 L	30.0°C	0.0°C	?
(b)	212 mL	−60.0°C	501.0°C	?
(c)	37.5 L	212 K	437 K	?

9.36   For a fixed amount of gas held at constant pressure, calculate the new volume the gas would occupy if the temperature were changed as shown in the following table.

	Initial Volume	Initial Temperature	Final Temperature	Final Volume
(a)	224 L	0.0°C	100.0°C	?
(b)	152 mL	45 K	450 K	?
(c)	156 mL	45°C	450°C	?

9.37   For a fixed amount of gas held at constant pressure, calculate the temperature in degrees Celsius to which the gas would have to be changed to achieve the change in volume shown in the following table.

	Initial Temperature	Initial Volume	Final Volume	Final Temperature
(a)	0.0°C	60.0 mL	120.0 mL	?
(b)	−37°C	2.55 L	75 mL	?
(c)	145 K	87.5 L	135 L	?

9.38   For a fixed amount of gas held at constant pressure, calculate the temperature in degrees Celsius to which the gas would have to be changed to achieve the change in volume shown in the following table.

	Initial Temperature	Initial Volume	Final Volume	Final Temperature
(a)	100.0°C	250.0 L	100.0 mL	?
(b)	27.5°C	125 mL	148 mL	?
(c)	300 K	13.7 L	57.2 L	?

9.39   A bubble of air has a volume of 0.350 mL at 24.2°C. If the pressure remains constant, what is the volume of the bubble at 72.5°C?

9.40   If 79.0 L of helium at 27.0°C is compressed to 32.0 L at constant pressure, what is the new temperature?

9.41   Suppose a steel tank is filled with a gas at sea level. This tank keeps the volume constant. If the tank is placed in the bed of a truck and driven up a mountain road to an elevation of 10,000 ft, what happens to the gas particles?

9.42   Suppose a flexible container is filled with a gas at an elevation of 9000 ft. What will happen to this gas if the container is transported to sea level?

9.43 Assume that the volume of a fixed amount of gas in a rigid container does not change. Calculate the pressure the gas would exert if the temperature were changed as shown in the following table.

	Initial Pressure	Initial Temperature	Final Temperature	Final Pressure
(a)	302 torr	0.0°C	105.0°C	?
(b)	735 torr	25.0°C	0.0°C	?
(c)	3.25 atm	273 K	373 K	?

9.44 Assume that the volume of a fixed amount of gas in a rigid container does not change. Calculate the pressure the gas would exert if the temperature were changed as shown in the following table.

	Initial Pressure	Initial Temperature	Final Temperature	Final Pressure
(a)	155 torr	225 K	315 K	?
(b)	795 torr	25°C	206 K	?
(c)	1.74 atm	150°C	23°C	?

9.45 Assume that the volume of a fixed amount of gas in a rigid container does not change. Calculate the temperature in degrees Celsius to which the gas would have to be changed to achieve the change in pressure shown in the following table.

	Initial Temperature	Initial Pressure	Final Pressure	Final Temperature
(a)	30.0°C	1525 torr	1015 torr	?
(b)	250.0°C	0.50 atm	1042 torr	?
(c)	255 K	500.0 torr	1000.0 torr	?

9.46 Assume that the volume of a fixed amount of gas in a rigid container does not change. Calculate the temperature in degrees Celsius to which the gas would have to be changed to achieve the change in pressure shown in the following table.

	Initial Temperature	Initial Pressure	Final Pressure	Final Temperature
(a)	25.0°C	243 torr	735 torr	?
(b)	205 K	2.35 atm	1.20 atm	?
(c)	375 K	875 torr	0.85 atm	?

9.47 A steel tank contains acetylene gas at a pressure of 7.25 atm at 18.5°C. What is the pressure at 37.2°C?

9.48 A steel tank contains gas at a pressure of 5.75 atm at 25.0°C. At what temperature will the pressure decrease to 1.25 atm?

9.49 For a gas under a given initial set of conditions, calculate the final value for the variable indicated if the other two variables change as described in the following table.

	(a)	(b)	(c)
Initial volume	2.50 L	125 L	455 mL
Initial pressure	0.50 atm	0.250 atm	200.0 torr
Initial temperature	20.0°C	25.0°C	300 K
Final volume	?	62.0 L	200.0 mL
Final pressure	760.0 torr	100.0 torr	?
Final temperature	0.0°C	?	327°C

9.50 For a gas under a given initial set of conditions, calculate the final value for the variable indicated if the other two variables change as described in the following table.

	(a)	(b)	(c)
Initial volume	601 mL	237 mL	1.12 L
Initial pressure	900.0 torr	75.0 atm	760.0 torr
Initial temperature	–10.0°C	147 K	0.0°C
Final volume	1200.0 mL	474 mL	?
Final pressure	?	150.0 atm	700.0 torr
Final temperature	0.0°C	?	25.0°C

9.51 What volume of O_2 at STP can be pumped into a 0.500-L tank at 24.5°C to give a pressure of 3.50 atm?

9.52 If 22.0 L of N_2 at 25.0°C and 725 torr are heated to 134°C and compressed to 4.50 L, what is the new pressure?

9.53 What is Gay-Lussac's law?

9.54 What is the relationship between the volume of a gaseous reactant and the volume of a gaseous product when temperature and pressure are constant?

9.55 What is the molar volume of neon gas at standard temperature and pressure (STP)?

9.56 What is the volume occupied by 1.00 mol of H_2 gas at STP?

9.57   Given the following volumes of gases at STP, calculate the number of moles of each gas and the mass of the gas.
(a) 7.62 L CH$_4$
(b) 150.0 mL Xe
(c) 38.1 L CO

9.58   Given the following volumes of gases at STP, calculate the number of moles of each gas and the mass of the gas.
(a) 135 mL H$_2$
(b) 8.96 L N$_2$
(c) 0.75 L He

9.59   The two balloons shown each have a volume of 1.5 L. They are tethered together and are at the same temperature and pressure. Which balloon has the greater number of gas particles? Which balloon has the greater mass? Which balloon has the greater density?

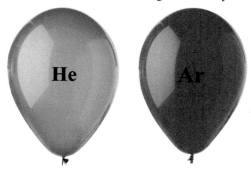

9.60   The two balloons shown each have a volume of 1.5 L. They are tethered together and are at the same temperature and pressure. Which balloon has the greater number of gas particles? Which balloon has the greater mass? Which balloon has the greater density?

9.61   Given the following amounts of gases, calculate the number of moles of each gas. Calculate the volume each amount of gas would occupy at STP.
(a) 6.8 g NH$_3$
(b) 48 g O$_2$
(c) 8.8 g He

9.62   Given the following amounts of gases, calculate the number of moles of each gas. Calculate the volume each amount of gas would occupy at STP.
(a) 2.2 g CO$_2$
(b) 5.6 g N$_2$
(c) 145 g Ar

9.63   Assuming all the gas is removed from the tank when filling balloons, how many 1.0-L balloons can be filled from this tank at STP?

425 g He

9.64   Assuming all the gas is removed from the tank when filling balloons, how many 0.75-L balloons can be filled from this tank at STP?

370 g CO$_2$

9.65   What will be the volume at STP of O$_2$ gas that occupies 12.0 L at 25.0°C assuming constant pressure?

9.66   What will be the volume of 1.50 mol H$_2$ gas at 30.0°C and 0.975 atm?

The Ideal Gas Law

9.67   What is an ideal gas?

9.68   (a) State the ideal gas law? (b) What is it used for?

9.69   Given the following amounts of gases, calculate the number of moles of each gas. Calculate the volume each amount of gas would occupy at 100.0°C and 15.0 atm.
N?
(a) 6.8 g NH$_3$
(b) 48 g O$_2$
(c) 8.8 g He

9.70   Given the following amounts of gases, calculate the number of moles of each gas. Calculate the volume each amount of gas would occupy at 75.0°C and 3.5 atm.
(a) 2.2 g CO$_2$
(b) 5.6 g N$_2$
(c) 7.5 g Ar

9.71   Given the following volumes of gases at 87.5°C and 722 torr, calculate the number of moles of each gas and the mass of the gas.
(a) 7.62 L CH$_4$
(b) 135 mL H$_2$
(c) 8.96 L N$_2$

9.72   Given the following volumes of gases at 54.0°C and 790 torr, calculate the number of moles of each gas and the mass of the gas.
(a) 150.0 mL Xe
(b) 38.1 L CO
(c) 2.5 L O$_2$

9.73   If two balloons at the same temperature and pressure contain equal volumes of CO_2 and He, why does the balloon containing CO_2 stay on the ground and the helium balloon float?

9.74   These two balloons contain the same number of moles of N_2 gas but at different temperatures. Which balloon is at a higher temperature? How do these balloons differ in their densities? Draw representations of gas particles to show how the molecular-level images of N_2 molecules differ in the two balloons.

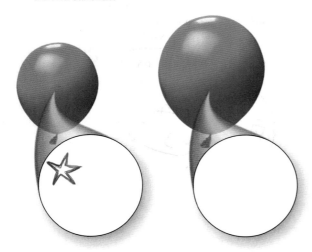

9.75   Calculate the density in grams per liter of the following gases at STP.
(a) NH_3
(b) N_2
(c) N_2O

9.76   Calculate the density in grams per liter of the following gases at STP.
(a) NO
(b) NO_2
(c) O_2

9.77   Calculate the density in grams per liter of the following gases at 25.0°C and 735 torr.
(a) NH_3
(b) N_2
(c) N_2O

9.78   Calculate the density in grams per liter of the following gases at 25.0°C and 735 torr.
(a) NO
(b) NO_2
(c) O_2

9.79   If 5.00-L containers are filled to a pressure of 840 torr at 50.0°C with the following gases, calculate the mass of each gas in its container.
(a) H_2
(b) CH_4
(c) SO_2

9.80   If 2.50-L containers are filled to a pressure of 650 torr at 25.0°C with the following gases, calculate the mass of each gas in its container.
(a) O_2
(b) CO_2
(c) He

9.81   What is Dalton's law of partial pressures?

9.82   How is Dalton's law of partial pressures used to determine the pressure and moles of a sample of gas that is collected over water?

9.83   A sample of oxygen gas saturated with water vapor at 22°C exerts a total pressure of 728 torr. What is the partial pressure of oxygen in the flask? The vapor pressure of water at 22°C is 20.0 torr.

9.84   A total of 0.400 L of hydrogen gas is collected over water at 18°C. The total pressure is 742 torr. If the vapor pressure of water at 18°C is 15.5 torr, what is the partial pressure of hydrogen?

9.85   A tank contains 78.0 g of N_2 and 42.0 g of Ne at a total pressure of 3.75 atm and a temperature of 50.0°C. Calculate the following quantities.
(a) moles of N_2
(b) moles of Ne
(c) partial pressure of N_2
(d) partial pressure of Ne

9.86   A tank contains 150.0 g of CO_2 and 24.0 g of O_2 at a total pressure of 4.25 atm and a temperature of 25.0°C. Calculate the following quantities.
(a) moles of CO_2
(b) moles of O_2
(c) partial pressure of CO_2
(d) partial pressure of O_2

9.87   Given the following densities (in grams per liter) of gases at STP, calculate the molar mass of each gas.
(a) 1.785   (b) 1.340   (c) 2.052
(d) 0.905   (e) 0.714

9.88   Given the following densities (in grams per liter) of gases at 27.0°C and 745 torr, calculate the molar mass of each gas.
(a) 2.436   (b) 0.842   (c) 1.325
(d) 3.450   (e) 1.706

(handwritten top margin: molar mass = g / mole)

Kinetic-Molecular Theory of Gases

(handwritten left margin, rotated: lighter has ↑ velocity)

9.89 State the five postulates of the kinetic-molecular theory.

9.90 How is kinetic energy related to gas pressure?

9.91 Explain in molecular terms why the pressure of a gas increases with increasing temperature when the volume is held constant.

9.92 Joel fills a weather balloon with helium gas in a room at 22°C. He takes the balloon outside where the pressure is the same, but the temperature is 0°C. Draw pictures of particles to show why, at constant pressure, the volume occupied by the helium gas decreases as temperature decreases.

9.93 At constant temperature, the pressure of a gas is inversely proportional to its volume. Explain this relationship in molecular terms.

9.94 Explain why the pressure of a gas sample is directly proportional to the moles of gas particles in the sample.

9.95 At the same temperature, rank the following substances in order of increasing average velocity; CO_2, He, H_2, CH_4.

(handwritten annotations: 44.009, 4.0026, 2.0158, 16.04)

9.96 At the same temperature, rank the following substances in order of increasing average kinetic energy: CO_2, He, H_2, CH_4.

9.97 What is the relationship between rate of diffusion of a gas and its molar mass?

9.98 What is the relationship between rate of effusion of a gas and its molar mass?

9.99 If equal amounts of neon and argon are placed in a porous container, which gas will escape faster?

9.100 If equal amounts of xenon and argon are placed in a porous container, which gas will escape faster?

Gases and Chemical Reactions

9.101 What volume of hydrogen is required to react with 12 L of oxygen under the same conditions?

$$2H_2(g) + O_2(g) \longrightarrow 2H_2O(g)$$

9.102 What volume of nitrogen is required to react with 9 L of hydrogen under the same conditions?

$$3H_2(g) + N_2(g) \longrightarrow 2NH_3(g)$$

9.103 Hexane burns according to the following equation:

$$2C_6H_{14}(g) + 19O_2(g) \longrightarrow 12CO_2(g) + 14H_2O(g)$$

What volume of CO_2 forms when 9.00 L of hexane burn, assuming the two volumes are measured under the same conditions? What volume of oxygen will be needed?

9.104 When nitric acid is synthesized from ammonia, the first step in the process is

$$4NH_3(g) + 5O_2(g) \longrightarrow 4NO(g) + 6H_2O(g)$$

What volume of NO at STP can be formed from 1250 L of NH_3 at 325°C and 4.25 atm, assuming temperature and pressure conditions remain constant?

9.105 Nitrous oxide can be formed by thermal decomposition of ammonium nitrate:

$$NH_4NO_3(s) \xrightarrow{\text{heat}} N_2O(g) + 2H_2O(g)$$

What mass of ammonium nitrate is required to produce 145 L of N_2O at 2850 torr and 42°C?

9.106 Nitric oxide is produced in the reaction between copper metal and nitric acid:

$$3Cu(s) + 8HNO_3(aq) \longrightarrow 3Cu(NO_3)_2(aq) + 4H_2O(l) + 2NO(g)$$

What mass of copper is required to produce 15.0 L of NO at 725 torr and 20.0°C?

Additional Questions

9.107 Why does a weather balloon change in volume when external pressure changes?

9.108 Why do gases expand when they are heated?

9.109 How can a hot-air balloon float in air?

9.110 Why are gas bubbles smaller at the bottom of a fish tank?

9.111 What would have to be the surface temperature to observe no change in volume in a weather balloon once it reached the outer edge of the troposphere? Assume the pressure at the surface is 760 torr and changes to 150 torr. The temperature at the outer edge of the atmosphere is about 218 K.

9.112 If gas molecules move at high speeds, why does it seem to take a long time for smells to permeate a room?

9.113 In macroscopic and molecular-level terms, describe what will happen to this balloon if temperature increases while pressure remains constant. What will happen if temperature decreases at constant pressure? What will happen if the balloon is taken up to an elevation of 10,000 feet?

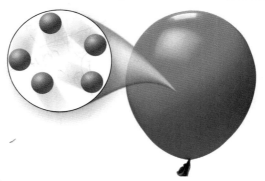

9.114 A propane tank used in gas grills can seem empty at sea level. When a supposedly empty tank is taken to higher elevations, it can be used again to light a grill. Why?

9.115 In air bags used in automobiles, the gas that fills the bags is produced from the reaction of sodium azide (NaN_3):

$$2NaN_3(s) \longrightarrow 2Na(s) + 3N_2(g)$$

What mass of sodium azide is needed to fill a 2.50-L air bag with nitrogen gas at a pressure of 1140 torr and 25°C?

9.116 Should Mount Everest, Death Valley, or Denver have the greatest atmospheric pressure? Why? Which should have the least? Explain.

9.117   Oxygen can be produced by thermal decomposition of mercuric oxide:

$$2HgO(s) \xrightarrow{\text{heat}} 2Hg(l) + O_2(g)$$

What volume of O_2 is produced at 50.0°C and 0.947 atm by decomposition of 27.0 g of HgO?

9.118   What pressure is required to compress 275 L of nitrogen gas at 1.00 atm into a container whose volume is 12.0 L?

9.119   A sample of argon measuring 3.00 L at 27.0°C and 765 torr was bubbled through and collected over water. The pressure of the new system is 778 torr at 15.0°C. What is the volume occupied by the gas? The vapor pressure of water at 15.0°C is 13 torr.

9.120   If 22.0 L of nitrogen gas at STP is heated to 167°C and compressed to a volume of 7.00 L, what will be the final pressure?

9.121   Consider the combustion of butene:

$$C_4H_8(g) + 6O_2(g) \longrightarrow 4CO_2(g) + 4H_2O(g)$$

What volume of butene at 188°C and 2.50 atm can be burned with 12.0 L of O_2 at 745 torr and 25.0°C?

9.122   A 2.00-L tank contains a mixture of 72.0 g N_2 and 66.0 g Ar. Calculate the total pressure of gas in the tank at 45.0°C. Calculate the partial pressure of each gas.

9.123   If 88 L of helium at 27°C is compressed to a volume of 29 L at constant pressure, what will be the new temperature?

9.124   The given images represent a collection of gas molecules at two different pressures. Which image shows molecules at the greatest pressure? What changes in conditions could explain the changes in this gas sample at a molecular level?

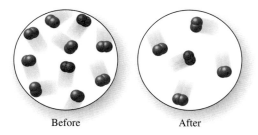

Before   After

9.125   Copper(II) oxide can be reduced to copper metal by heating in a stream of hydrogen gas:

$$CuO(s) + H_2(g) \xrightarrow{\text{heat}} Cu(s) + H_2O(g)$$

What volume of hydrogen at 27.0°C and 722 torr would be required to react with 95.0 g of CuO?

9.126   A cement company uses 1.00×10^5 kg of limestone daily. The limestone decomposes upon heating, forming lime and carbon dioxide:

$$CaCO_3(s) \xrightarrow{\text{heat}} CaO(s) + CO_2(g)$$

What volume of CO_2 at 735 torr and 25.0°C is released into the atmosphere daily by this company?

9.127   If all volumes of gases are measured under the same conditions, what volume of the product will be formed from reaction of 10.0 L of H_2 in each of the following reactions?
    (a) $S(s) + H_2(g) \longrightarrow H_2S(g)$
    (b) $N_2(g) + 3H_2(g) \longrightarrow 2NH_3(g)$
    (c) $C_6H_6(g) + 3H_2(g) \longrightarrow C_6H_{12}(g)$

9.128   Oxygen gas can be generated by heating potassium chlorate:

$$2KClO_3(s) \longrightarrow 2KCl(s) + 3O_2(g)$$

What volume of oxygen gas, collected by displacement of water and measured at 735.0 torr and 70.0°C, will be formed by the decomposition of 13.5 g of potassium chlorate?

9.129   Calcium carbide (CaC_2) is made by heating lime (CaO) with carbon. The lime is made by heating limestone ($CaCO_3$). Acetylene (C_2H_2) can be made from the calcium carbide by reacting it with water:

$$CaC_2(s) + 2H_2O(l) \longrightarrow Ca(OH)_2(aq) + C_2H_2(g)$$

If 1.25 L of acetylene at 0.750 atm can be made from 5.00 g of limestone, what is the temperature of the gas?

9.130   The density of air at 760.0 torr and 25.0°C is 1.186 g/L.
    (a) Calculate the average molar mass of air.
    (b) From this value, and assuming that air contains only molecular nitrogen and molecular oxygen gases, calculate the mass percent of N_2 and of O_2 in air.

9.131   (a) How is the volume occupied by a gas related to the number of moles of that gas?
    (b) What other variable must be held constant to show experimentally that this relationship is valid?

9.132   A sample of an unknown liquid weighing 0.495 g is collected as vapor in a 127-mL flask at 98°C. The pressure of the vapor in the flask is then measured and found to be 691 torr. What is the molar mass of the liquid?

9.133   The explosion of nitroglycerin can be described by the following equation:

$$4C_3H_5(ONO_2)_3(l) \longrightarrow 12CO_2(g) + 10H_2O(g) + 6N_2(g) + O_2(g)$$

What is the total volume of gases produced at 2.00 atm and 275°C from 1.00 g of nitroglycerin?

9.134   Boyle used a U-tube to investigate gas properties. As shown in the figure, a gas was trapped in the closed arm of the U-tube at 29.9 in Hg, and the pressure was varied by adding mercury to the open arm. The total pressure exerted on the gas is the sum of the atmospheric pressure (29.9 in Hg) and the pressure due to the addition of mercury as measured by the difference in mercury height.

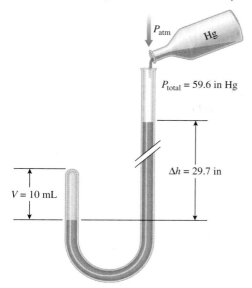

P_{atm} Hg

$P_{total} = 59.6$ in Hg

$\Delta h = 29.7$ in

$V = 10$ mL

Boyle recorded the following data:

Length of Gas Column (in)	Difference between Mercury Levels (in), Δh
48	0.0
44	2.8
40	6.2
36	10.1
32	15.1
28	21.2
24	29.7
22	35.0
20	41.6
18	48.8
16	58.2
14	71.3
12	88.5

Graph these data. What does the graph show about the relationship between volume and pressure?

9.135   For the data in Question 9.134, graph the second column against the inverse of the data in the first column. What does the graph show about the relationship between volume and pressure?

The Liquid
and Solid States

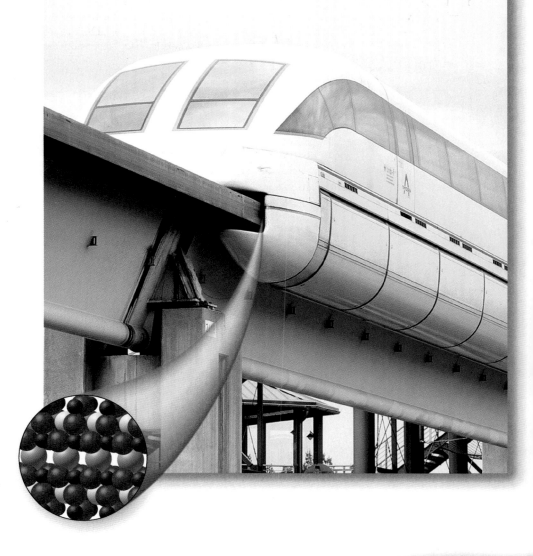

After her chemistry class on Friday, Nicole left for the weekend to attend a family reunion. At the reunion she talked with her mother, grandfather, and great-grandfather about how much she loved chemistry. Her only complaint was that her professor wouldn't allow her to use her new PDA (Personal Digital Assistant, a handheld computer) on exams. She thought she could do better if she didn't have to memorize all those facts and equations. Her mother said that it was similar when she went to college. Students weren't allowed to use their electronic calculators on exams. Her grandfather said that in his day, they weren't allowed to bring slide rules, because their professors feared that math skills would suffer. Her great-grandfather laughed. "I remember going to the blackboard for tests and using chalk to show my work," he said, "and on some tests, even those were forbidden. We had to solve all the problems in our heads!"

The theme of each of the stories was the same. Instructors worried that "modern" technology would interfere with their students' learning. In each case, the innovation came as a result of a new material or a new use for an old one. For example, the first chalkboards of the mid-1800s were made of slate from quarries. They replaced an older technology: wooden boards coated with black, gritty paint to make them easier to write on. Those materials came from natural resources. More recent technologies, such as the LCD (liquid-crystal display) screen used in Nicole's PDA, are made from modern synthetic materials, including polymers and semiconductors.

Materials have been important to humans throughout history, so much so that several historical eras are named for the materials most used at the time. During the Stone Age, for example, tools were made from rocks, minerals, or bone that could be chipped into useful shapes. A popular material was the hard mineral flint, a form of quartz. Around 8000 BC, naturally occurring copper was beaten into shape and used in tools. About 5000 BC, methods for removing copper from its ores were discovered. Around 3500 BC, it was discovered that copper could be made harder by melting it with tin, creating bronze. Bronze tools were much better than copper ones, and we call the era the Bronze Age.

Around 1500 BC iron was recovered from its ores by heating them with charcoal, a discovery probably made by accident. During the Iron Age, tools were made from impure iron by heating, pounding, and reheating the metal, which hardened it. Conversion of a solid to a liquid, modifying the composition of the liquid, and conversion back to a solid are common techniques still used to form new materials.

In modern times, sometimes called the Age of Materials, we have added plastics and many ceramics—along with new metals—to our repertoire of materials that can be used to make tools. For example, superconductors can be used with rapid commuter trains. Materials science is concerned with the synthesis, processing, composition, structure, properties, and performance of new materials.

The properties of a specific material depend on how it is processed. For example, although glass is not a modern material, when it is processed in new ways, it attains new properties. When combined with sodium carbonate, silica (or silicon dioxide) forms window glass. When boron oxide is added, the glass becomes harder and more temperature resistant. It is used for laboratory glassware under brand names such as Pyrex and Kimax. Some baking dishes are also made from this glass. When silver bromide is added, we get glass that darkens when exposed to sunlight and lightens in the dark. This glass is used in some eyeglasses and tinted windows. Glass that has a coating of a metal or metal oxide is reflective glass. It makes windows that reflect sunlight and are hard to see through.

Many modern materials are *alloys* (mixtures of metals) with useful properties. One widely used alloy is Nitinol (Figure 10.1), named for its composition and its discoverers, *Ni*ckel *Ti*tanium *N*aval *O*rdnance *L*aboratory. This "shape-memory"

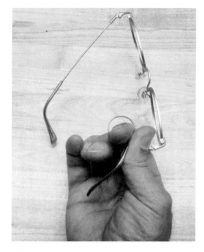

Figure 10.1

Nitinol, a shape-memory metal, is used for eyeglass frames that repair themselves when bent out of shape.

metal is one of several recently developed "smart" materials. If its shape changes, mild heating returns it to its original form. Shape-memory metals are used for self-repairing eyeglass frames as well as life-saving medical devices and a whimsical shirt that rolls up its own sleeves when its environment gets hot.

Besides Nitinol, numerous other materials have been used to make implants for the human body, including heart valves, blood vessels, pacemakers, dental implants, and joint replacements (Figure 10.2). However, many of these implants performed imperfectly because the materials were incompatible with human tissue. The body's immune system rejected them. New research focuses on improving the surface properties of implants to improve biocompatibility. For example, a coating of synthetic hydroxyapatite, $Ca_5(PO_4)_3OH$, encourages tissue to adhere to a stainless-steel or titanium hip replacement. This material is found naturally in bones and teeth and can be synthesized from the calcium carbonate found in coral. When applied to the artificial joint, it behaves like natural bone. Normal proteins build up on its surface, providing lubrication that reduces wear on the implant.

Another new type of material is a class of liquids called ferrofluids. They are used in stereo-speaker drivers as coolants and damping fluids that prevent unwanted vibrations. They work for these purposes because they have the fluid properties of liquids, but they respond to a magnetic field as if they were solids. To prepare ferrofluids, very small particles of a magnetic material like magnetite (Fe_3O_4) are suspended in oil. A long-chain carboxylic acid such as oleic acid $(C_{17}H_{33}CO_2H)$ is added to keep the magnetite particles from precipitating from the mixture. The polar ends of the oleic acid molecules interact with the magnetite particles, while the nonpolar ends interact with the oil, keeping the magnetite particles apart. In the absence of a magnetic field, ferrofluids pour just like the oil they are made from. However, in the presence of a magnetic field, the oil thickens, and the magnetite particles align with the magnetic field. As shown in Figure 10.3, the oil forms spikes in the pattern of the magnetic field. Ferrofluids are used in applications such as magnetic brakes. The oil is suspended between two plates, one of which rotates. When a magnetic field is applied, the oil thickens, and the plate can no longer turn.

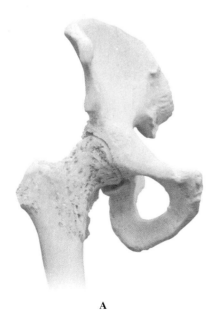

A

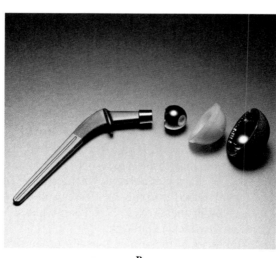

B

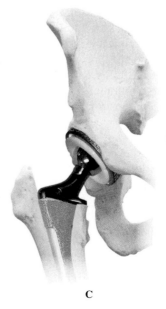

C

Figure 10.2
(A) A hip joint that has been degraded by arthritis can be repaired by replacing the bone in both parts of the joint with (B) an implant. (C) The replacement works better when coated with synthetic hydroxyapatite.

Although many new materials are mixtures of substances, we can learn a lot about their properties by studying the properties of their pure components. These new materials are examples of solids and liquids, so it is important to understand the properties of these states of matter. In this chapter, we will examine the properties and structures of the solid and liquid states. In addition we will investigate the transitions between these states of matter.

Questions for Consideration

10.1 How do the properties of liquids and solids differ, and what happens when substances undergo a change of state?

10.2 Why do atoms or molecules hold together as liquids or solids instead of existing only as gases?

10.3 What are some unique properties of liquids?

10.4 How do atoms or ions fit together to make solids, and how does their arrangement affect their properties?

Math Tools Used in This Chapter

Units and Conversions (Math Toolbox 1.3)

Figure 10.3
In a magnetic field, a ferrofluid thickens and forms spikes in the pattern of the magnetic field much like iron filings form a pattern when sprinkled near a magnet.

10.1 CHANGES OF STATE

In Chapter 1 we discussed the physical states of matter: solid, liquid, and gas. The general properties of these three states are summarized in Table 10.1. To understand their differences, try drawing pictures of atoms or molecules in each of the three states. Then compare your drawings to those shown in Figure 10.4.

Your drawings should show a difference in the distance between the particles in different states of matter. Relatively strong forces hold particles in fixed positions, close together in solids. The particles cannot move readily, so a solid is rigid, retains its shape, and is not easily compressed. A liquid, on the other hand, generally has its molecules farther apart. Attractive forces are not strong enough to hold the particles rigidly in place, so they can slip past one another and move around. Thus a sample of a liquid takes the shape of its container; and because there is some space between particles, a liquid can be compressed slightly. The relative incompressibility of liquids is useful in liquid hydraulic systems such as those used in automobile brakes.

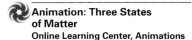
Animation: Three States of Matter
Online Learning Center, Animations
Chemistry Animations Library, Three States of Matter, Matter_States.rm.

Although particles in solids move with difficulty, they are not completely still. They vibrate around an average position. These vibrations increase in frequency as a solid material is heated.

In most substances the particles are closer together in the solid state than in the liquid state, so the solid is denser. Water is an exception. Because of the unique arrangement of its molecules, ice has open spaces in its structure. Thus, on the average, water molecules fill more space in the liquid than in the solid state. How does this fact explain why ice floats on liquid water?

TABLE 10.1	**General Properties of the States of Matter**	
Solid	**Liquid**	**Gas**
Fixed shape	No fixed shape	No fixed shape
Shape not set by container	Takes shape of filled portion of the container	Takes shape of container
Shape remains rigid	Can be poured	Fills container
Particles fixed in place but vibrate around a fixed position	Particles move past one another	Particles move through space
Little or no volume change under moderate pressure	Can be compressed slightly by moderate pressure	Compressed under moderate pressure
Little free space between particles	Some free space between particles	Particles are widely separated with much free space

Solid Cu   Liquid H$_2$O   Gaseous NO$_2$

Figure 10.4
The macroscopic properties of solids, liquids, and gases result from the distance that separates their atoms or molecules and from how freely these particles can move.

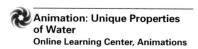
Animation: Unique Properties of Water
Online Learning Center, Animations

Chemistry Animations Library, Unique Properties of Water, Water_Properties.rm.

Contrary to popular science fiction, vaporization does not mean blasting something into oblivion with a ray gun. The term *vapor* refers to the gaseous state of a substance that normally exists as a solid or a liquid. Vaporization is a synonym for evaporation.

All substances can exist in any of the three physical states under certain conditions, which may be extreme. For example, nickel, a component of Nitinol, is a solid unless heated above 1455°C. Then it becomes liquid. Figure 10.5 shows liquid copper being poured at high temperatures. Metals can change to the gaseous state if heated even hotter. Nickel becomes a gas at 2800°C. *Vaporization* of metals is applied in the materials science technique of vapor deposition.

Water is the only common substance that exists in all three physical states under normal atmospheric conditions. Liquid water in underground pools is expelled from a geyser as steam. An iceberg floats in water and gradually melts to produce more liquid water. When an iceberg floats in liquid water, invisible water vapor is also present in the air, so three physical states—gas, liquid, and solid—are present (Figure 10.6). Transitions between these states are *changes of state* or *phase changes*.

Figure 10.5
When heated above 1083°C, copper melts and can be poured just like any other liquid.

Figure 10.6
Water exists in all three physical states under conditions found on the surface of the Earth. Water in the gaseous state, called water vapor, is not visible. Clouds (and steam) are actually tiny droplets of liquid water.

Liquid-Gas Phase Changes

At any given temperature, molecules in a liquid have a range of kinetic energies. Some move faster and some move slower than the average. What happens if a molecule with a high kinetic energy happens to be close to the liquid's surface? If it is moving in the right direction, it may escape into the gas state. This process is called **evaporation** or vaporization (Figure 10.7).

Because the molecules of highest kinetic energy are the ones most likely to escape from the liquid, the average kinetic energy of the remaining liquid molecules decreases as evaporation proceeds. As we saw in Chapter 9, kinetic energy is proportional to absolute temperature, so unless energy is supplied, evaporation decreases the temperature of the liquid. To evaporate at a constant temperature, a liquid must absorb heat. The process is endothermic.

The cooling effect of evaporation is significant in many processes. Some examples include cooling by perspiration (Figure 10.8), evaporative cooling (swamp cooling) in homes, cooling car interiors with liquid sprays, and the use of canvas water bags for desert travel. Have you ever noticed feeling cold when leaving a shower or swimming pool? That's evaporative cooling, and it can be used to treat a sick baby running a high fever. A sponge bath lowers body temperature because the evaporation of water cools the skin. The wind chill factor reported in weather forecasts is another measure of evaporative cooling. Moving air evaporates water from the skin faster than still air, so the air feels colder when the weather is windy. The increased cooling is sensed by the skin as a lower air temperature.

The extent of evaporation depends on how the liquid is contained. In an open container, molecules that are more energetic move away from the liquid's surface into the surrounding atmosphere in the gaseous state. Since the container is open, the process continues until all the liquid has evaporated. Some of the evaporated molecules may return to the liquid state if they happen to move toward the liquid rather than away from it, but most of the evaporated molecules stay in the gaseous state moving farther away. The average energy of the liquid remains constant, because the liquid absorbs heat from its surroundings, maintaining a constant temperature. However, if the container is insulated, heat cannot be absorbed from the surroundings as easily, so the liquid cools as evaporation proceeds. As a result, the process slows, because the average energy of the molecules decreases.

What happens if we place a pure liquid in a closed container (Figure 10.9)? At first only evaporation occurs, since there are no vapor molecules. As evaporation continues, molecules begin to accumulate in the gaseous state. They cannot move

Evaporation

Figure 10.7
At the surface of a liquid, some molecules have sufficient energy to escape into the gas state.

Figure 10.8
Evaporation of a liquid, such as a runner's sweat, is endothermic, so it causes cooling.

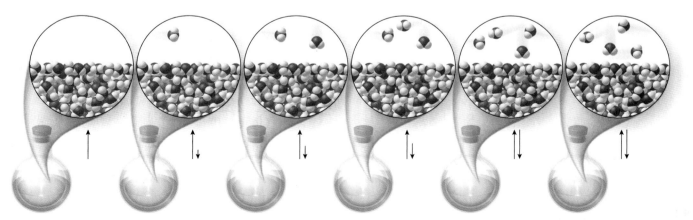

Figure 10.9
In a closed container the number of gas molecules increases until equilibrium is attained. In this state of equilibrium, the rate of liquid molecules evaporating equals the rate of gas molecules condensing. Thus the number of gas molecules remains constant. (Note that original air molecules are not shown.)

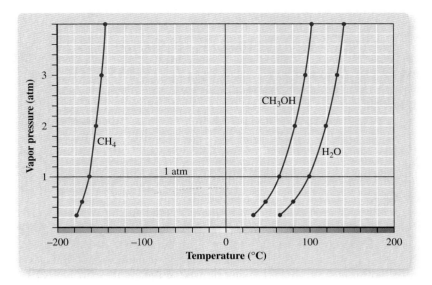

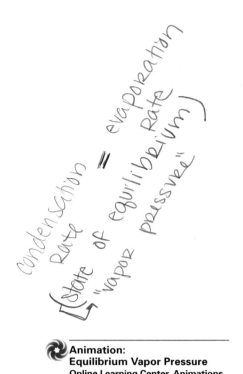

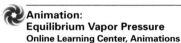

Figure 10.10

The vapor pressure of all substances increases as the temperature increases. The temperature at which the vapor pressure is 1 atm is the normal boiling point. All temperatures corresponding to points on the curves are boiling points for the corresponding pressures.

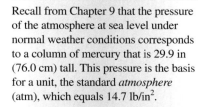

Animation:
Equilibrium Vapor Pressure
Online Learning Center, Animations

Chemistry Animations Library, Equilibrium Vapor Pressure, Vapor_Pressure.exe.

completely away from the liquid, and many move toward it. When they hit the liquid, they reenter the liquid state, a process called **condensation.** Evaporation and condensation continue at rates that depend on two factors: temperature and the concentration of molecules available to undergo each process. Ultimately the rate of condensation equals the rate of evaporation, and the concentration of molecules in the gaseous state becomes constant. This situation, in which opposing processes occur at equal rates, is called a state of **equilibrium.** It can be represented by the equation

$$\text{liquid} \underset{\text{condensation}}{\overset{\text{evaporation}}{\rightleftharpoons}} \text{gas}$$

A double arrow represents a state of equilibrium involving a reversible process.

At equilibrium the rate of evaporation equals the rate of condensation, so the concentration of the substance in the gaseous state is constant, although evaporation and condensation continue to occur. The concentration of gas can be measured by the pressure exerted by the gas molecules above the liquid. At equilibrium this pressure is called the **vapor pressure.** Vapor pressure increases as temperature increases. At higher temperatures, the average kinetic energy of the liquid molecules is greater, so the molecules can escape from the liquid more easily. The concentration of the substance in the gas state at equilibrium is therefore greater at higher temperatures.

When the vapor pressure equals the external atmospheric pressure, the liquid begins to boil. The temperature of a pure, boiling liquid remains constant until all the liquid has vaporized. Adding more heat to the liquid increases the rate of evaporation, not the temperature. The temperature at which boiling occurs—at which the vapor pressure equals the external pressure of the atmosphere—is called the **boiling point.** The boiling point for a given liquid depends on the atmospheric pressure. Figure 10.10 shows that boiling points, which are the temperatures corresponding to points on the curves, increase as the pressure rises. When the atmospheric pressure is exactly 1 atm, the temperature at which boiling occurs is called the **normal boiling point.**

Recall from Chapter 9 that the pressure of the atmosphere at sea level under normal weather conditions corresponds to a column of mercury that is 29.9 in (76.0 cm) tall. This pressure is the basis for a unit, the standard *atmosphere* (atm), which equals 14.7 lb/in^2.

An atmospheric pressure of 1 atm is normal only at sea level, however. At higher altitudes atmospheric pressure is less than 1 atm, and liquids boil at lower temperatures. The normal boiling point of water, for example, is 100.0°C. At about 2750 m (9000 ft) above sea level, such as at the Tuolumne Meadows campground in Yosemite National Park, the atmospheric pressure is about 0.72 atm, and water boils at 91°C. It takes almost twice as long to cook an egg in boiling water at this elevation than at sea level, because it must cook at a lower temperature. Water cannot be heated above its boiling point because the added energy is used to evaporate the water. However, a liquid held under a pressure higher than atmospheric pressure can be heated above its normal boiling point. A pressure cooker has a sealed lid with a pressure valve, so the vapor pressure can rise above atmospheric pressure. Since water boils at a higher temperature under such conditions, the food cooks more rapidly.

100°C - norm boiling pnt

9000 ft above sea level

atm = 0.72 atm

91°C - boiling pnt

EXAMPLE 10.1   Vapor Pressure and Boiling Point

Consider the following vapor pressure curve for acetone, which is used in finger-nail polish removers.

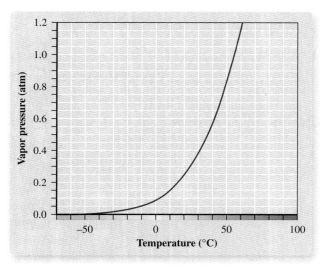

What is the normal boiling point of acetone? What is its boiling point at 0.50 atm?

Solution:

The normal boiling point is the temperature at which the vapor pressure is exactly 1 atm. If we locate the point on the curve at which the pressure is 1 atm and then follow that point down to the temperature axis, we read a temperature of 56°C. The boiling point at 0.50 atm is found in the same way. We locate the point on the curve that has a vapor pressure of 0.50 atm. We follow this point down to the temperature axis and find a value of 36°C.

Practice Problem 10.1

Consider the following vapor pressure curve for propane, which is used as a fuel in barbecue grills.

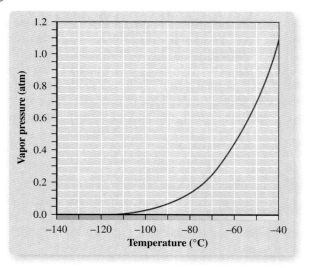

What is the normal boiling point of propane? What is its boiling point at 0.40 atm?

Further Practice: 10.13 and 10.14

Liquid-Solid Phase Changes

The average kinetic energy of liquid molecules decreases as liquids cool. When the kinetic energy of the liquid molecules falls low enough, the molecules become fixed in position in the solid state (although still vibrating), and the liquid freezes. The freezing of a liquid into the solid state occurs at a temperature called the **freezing point.** At this temperature the solid and liquid coexist in a state of equilibrium (Figure 10.11). The temperature is called the **normal freezing point** when this state of equilibrium is reached under a pressure of exactly 1 atm. The temperature remains constant from the time a liquid starts freezing until all the liquid has solidified. After that, release of more heat causes a further decrease in temperature of the solid. Citrus growers use this principle to protect their crops from freezing during cold nights. They spray water on the trees. The freezing of the sprayed water releases enough heat to keep the water inside the plant cells from freezing.

Figure 10.11

At the freezing point (or melting point), the solid and liquid forms of a substance are in a state of equilibrium.

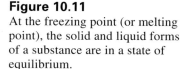

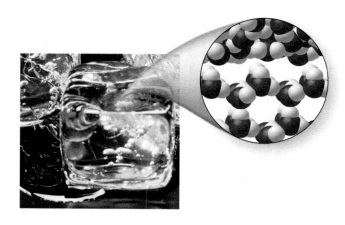

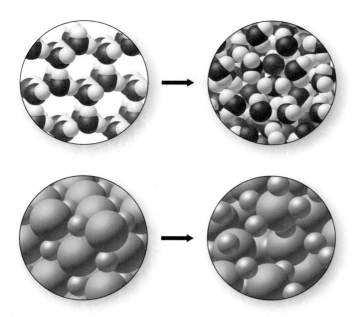

Figure 10.12

When a covalent substance such as water melts, the molecules retain their identity and composition. When an ionic substance such as sodium chloride melts, the ions separate and assume a random arrangement.

Melting is the reverse of freezing. Both processes involve the solid and liquid phases:

$$\text{solid} \underset{\text{freezing}}{\overset{\text{melting}}{\rightleftharpoons}} \text{liquid}$$

Thus the **melting point** of a solid and the freezing point of its liquid state are the same temperature. When a molecular substance melts, the molecules retain their identity. When an ionic substance melts, the ions separate and become randomly arranged (Figure 10.12). More energy is required to melt ionic substances, so they melt at higher temperatures than molecular substances.

Solid-Gas Phase Changes

Solids, like liquids, have a characteristic vapor pressure. The vapor pressures of most solids are rather low, much lower than those of their liquids. However, a few have high vapor pressures and can change directly from the solid to the gaseous state without going through the liquid state. The evaporation of a solid is **sublimation:**

$$\text{solid} \underset{\text{deposition}}{\overset{\text{sublimation}}{\rightleftharpoons}} \text{gas}$$

An example is dry ice (solid CO_2). It sublimes from solid to gas, disappearing into the air. Other common substances that sublime are mothballs and iodine (Figure 10.13). Have you ever noticed how ice cubes get smaller when kept in a freezer for a long time? Solid water sublimes at temperatures below $0°C$. It evaporates because it is in equilibrium with the gaseous state, and no liquid state can exist. Wet clothing hung outside in freezing weather dries slowly for this reason.

The reverse of sublimation is **deposition.** It is a change from gas to solid without going through the liquid state. The formation of snow is a deposition process. The deposition of water vapor in the atmosphere releases heat, so the temperature rises a degree or two when it begins to snow.

Figure 10.13
When solid iodine is heated, it sublimes into the gaseous state. It goes back to the solid state on the cold surface of the evaporating dish on top of the beaker.

EXAMPLE 10.2   Changes in the State of Matter

Identify the process shown in the following diagram.

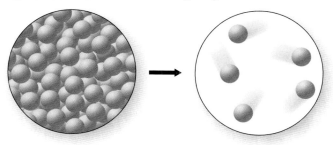

Solution:

We can identify the starting material as a liquid, based on the small amount of space between particles and the random arrangement of particles. In the ending material, there is a large distance between particles, so we can identify it as a gas. A liquid-to-gas transformation is evaporation.

Practice Problem 10.2

Identify the process shown in the following diagram.

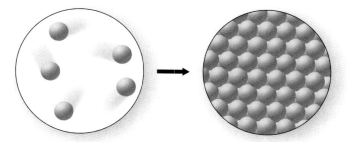

Further Practice: 10.17 and 10.18

Cooling and Heating Curves

Consider a substance in the gaseous state. When cooled at constant pressure, its volume decreases, but it remains as a gas up to a point. When will it convert to a liquid? When the temperature falls enough that the pressure of the gas matches the vapor pressure of the liquid, the gas starts to condense. This happens at the boiling point of the liquid. If we continue to remove heat, more of the gas condenses to the liquid state. When gas and liquid are in equilibrium, removal of more heat does not decrease the temperature. It results only in further condensation. The first plateau in the *cooling curve* shown in Figure 10.14A illustrates this process. Once all the gas condenses, removal of heat again decreases the temperature. The temperature drop halts again at the freezing point, shown as the second plateau in Figure 10.14A. Once all the liquid freezes to solid, further removal of heat lowers the temperature. If we reverse the process and add heat to a solid, we proceed along a heating curve (Figure 10.14B), with plateaus also observed when the substance undergoes changes of state.

In the cooling curves for some substances, we may observe a liquid becoming colder than the freezing point, a phenomenon called *supercooling*. Solidification requires some matrix around which a solid can form—perhaps imperfections in a container wall, pieces of dust, or other solid particles. Thus in a very clean system, solidification may be delayed. Usually, after supercooling a degree or two, the substance begins to crystallize, and the temperature returns to the freezing point. Supercooling is common in substances such as glass, tars, rubbers, and plastics.

EXAMPLE 10.3   Cooling and Heating Curves

The following diagrams represent changes of state.

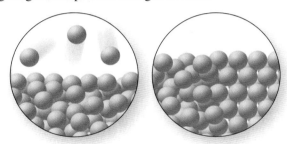

For each change of state, although heat is being added, the temperature does not change. Identify where each change of state would be found on the following heating curve:

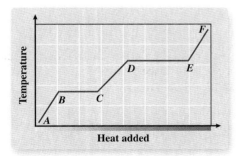

Solution:

The left molecular diagram shows the liquid and gaseous states coexisting. Since the temperature does not change during the process, this corresponds to any point on the *DE* plateau in the heating curve. The right molecular diagram shows solid and liquid together, which corresponds to any point on the *BC* plateau in the heating curve.

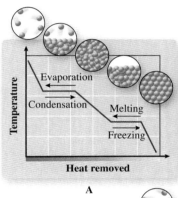

A

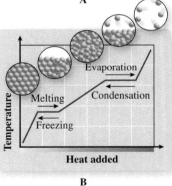

B

Figure 10.14

(A) When heat is removed at a constant rate from a gas, it cools until it reaches the boiling point of the liquid. There the temperature remains constant until the gas completely condenses to liquid. The temperature then drops again until it reaches the melting point of the solid. The temperature remains constant as the liquid converts to the solid state. Then the temperature drops once more. These changes are summarized in a cooling curve. (B) The same changes in reverse order occur when heat is added to a solid, as shown in a heating curve.

Practice Problem 10.3

The following diagrams represent the physical states of a substance.

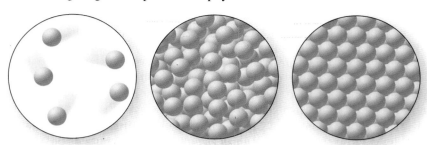

Identify where each state would predominate on the heating curve given in the example.

Further Practice: 10.31 and 10.32

[handwritten margin notes: frozen −40.0°C; 0.0°C melting starts; 100.0°C boiling begins; energy; solid < liquid < gas]

Energy Changes

Each change of state takes place at a constant temperature. Suppose, for example, we have a piece of ice at a temperature of −40.0°C. If we heat the ice at a constant pressure of 1 atm, it will go through the temperature and phase changes shown by the heating curve in Figure 10.14. The temperature will rise until the ice reaches 0.0°C, at which point it will start to melt. The temperature will stay constant until the ice has melted completely, because the added energy is being used to overcome the forces that hold the water molecules together in ice. When the ice has melted, the temperature will start to rise again as determined by the specific heat of water. When the water reaches 100.0°C, it will start to boil; it will stay at 100.0°C until it has completely vaporized. Continued heating will then increase the temperature of the steam to a temperature such as 120.0°C.

The energy required to change the temperature of any physical state of matter is determined by the specific heat of that state. As described in Chapter 6, the heat absorbed or released by a substance when it changes temperature is given by the equation

$$q = m \times C \times \Delta T$$

[handwritten annotations: (the heat of the process); mass; specific heat; temp change]

In this equation, m is the mass, ΔT is the temperature change, and C is the specific heat. The values of the specific heat for the physical states of water are 2.03 J/(g °C) for ice, 4.18 J/(g °C) for liquid water, and 2.02 J/(g °C) for gaseous water.

An energy change accompanies any phase change. In general, the solid form of a substance contains less energy than the liquid form, and the liquid contains less energy than the gas. The change in energy for each process is called the heat of that process (q). For example, the *molar heat of fusion* is the energy required to melt 1 mol of a substance. The molar heat of fusion of ice, $H_2O(s)$, is 6.01 kJ/mol, or 6.01×10^3 J/mol. To melt 5.00 mol of ice, $5.00 \times 6.01 \times 10^3$ J of energy must be added. The *molar heat of vaporization* is the energy required to evaporate 1 mol of a liquid. For liquid water, the molar heat of vaporization is 40.7 kJ/mol, or 4.07×10^4 J/mol.

It is possible to calculate the heat required to carry out the entire process by which ice is transformed to steam, or any other process involving temperature changes and changes in physical state. We can represent the process as follows:

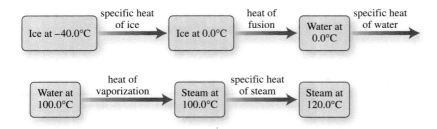

The energy change for the entire process is the sum of the heat used for the individual steps in the process.

$$q = q_{\text{heat ice}} + q_{\text{fusion}} + q_{\text{heat water}} + q_{\text{vaporization}} + q_{\text{heat steam}}$$

Suppose we start with 1.000 mol (18.01 g) of ice at $-40.0°C$. To calculate how much heat is required to raise the temperature of ice to $0.0°C$, the point at which it will start to melt, we use the following equation:

$$q_{\text{heat ice}} = m \times C \times \Delta T$$

The mass is 18.01 g, the specific heat is 2.03 J/(g °C), and the temperature change is 40.0°C:

$$q_{\text{heat ice}} = 18.01 \text{ g} \times 2.03 \text{ J/(g °C)} \times 40.0°C = 1.46 \times 10^3 \text{ J}$$

The molar heat of fusion for the melting of ice is 6010 J/mol. To obtain the heat change, we multiply the molar heat of fusion by the number of moles, n:

$$q_{\text{fusion}} = n \times \text{molar heat of fusion}$$

$$q_{\text{fusion}} = 1.000 \text{ mol} \times 6.01 \times 10^3 \text{ J/mol} = 6.01 \times 10^3 \text{ J}$$

The temperature of the water, now liquid, is then raised by 100.0°C. The heat required for this part of the process is calculated from the specific heat of liquid water:

$$q_{\text{heat water}} = m \times C \times \Delta T$$

$$q_{\text{heat water}} = 18.01 \text{ g} \times 4.18 \text{ J/(g °C)} \times 100.0°C = 7.53 \times 10^3 \text{ J}$$

The evaporation of water, which takes place as the liquid boils, requires 4.07×10^4 J/mol, the molar heat of vaporization. To obtain the heat change, we multiply the molar heat of vaporization by the number of moles, n:

$$q_{\text{vaporization}} = n \times \text{molar heat of vaporization}$$

$$q_{\text{vaporization}} = 1.000 \text{ mol} \times (4.07 \times 10^4 \text{ J/mol}) = 4.07 \times 10^4 \text{ J}$$

The steam is then heated further to 120.0°C, a temperature change of 20.0°C. The change is determined from the specific heat of steam:

$$q_{\text{heat steam}} = m \times C \times \Delta T$$

Steam at 100.0°C $\xrightarrow{\text{specific heat of steam}}$ Steam at 120.0°C

$$q_{\text{heat steam}} = 18.01 \text{ g} \times 2.02 \text{ J/(g °C)} \times 20.0°C = 7.28 \times 10^2 \text{ J}$$

The total heat required is the sum of the changes for each step of the process:

$$q = (1.46 \times 10^3 \text{ J}) + (6.01 \times 10^3 \text{ J}) + (7.53 \times 10^3 \text{ J}) + (4.07 \times 10^4 \text{ J}) + (7.28 \times 10^2 \text{ J})$$
$$= 5.64 \times 10^4 \text{ J} = 56.4 \text{ kJ}$$

EXAMPLE 10.4   Energy for Phase Changes

Calculate the heat required to melt 1.00 g of ice at 0°C. The molar heat of fusion of ice is 6010 J/mol.

Solution:

The heat is calculated from the product of the number of moles and the molar heat of fusion:

$$q_{\text{fusion}} = n \times \text{heat of fusion}$$

Since the amount of ice is given in grams, we must convert to moles. The molar mass of water is used to convert grams to moles, and the molar heat of fusion is used to convert moles to joules:

g H_2O $\xrightarrow{18.01 \text{ g} = 1 \text{ mol}}$ mol H_2O $\xrightarrow{1 \text{ mol} = 6010 \text{ J}}$ J

$$q_{\text{fusion}} = 1.00 \text{ g } H_2O \times \frac{1 \text{ mol } H_2O}{18.01 \text{ g } H_2O} \times \frac{6010 \text{ J}}{1 \text{ mol } H_2O} = 334 \text{ J}$$

Practice Problem 10.4

Calculate the heat required to boil 125 g of water at 100.0°C. The molar heat of vaporization of water is 4.07×10^4 J/mol.

Further Practice: 10.35 and 10.36

10.2 INTERMOLECULAR FORCES

Why do some samples of matter exist under ordinary conditions as gases, while others are liquids or solids? The existence of the liquid and solid states indicates that some attractive force must pull molecules together. An attractive force that operates *between* molecules is called an **intermolecular force.** In contrast, forces *within* molecules, called bonds, hold atoms together in molecules. Intermolecular forces are much weaker than bonding forces. The energies of intermolecular forces are usually in the range of 0.05 to 40 kJ/mol, while those of most bonds are in the range of 200 to 900 kJ/mol.

TABLE 10.2	Intermolecular Forces and Bonds	
Type of Force	**Type of Interaction**	**Occurrence**
London dispersion force	A temporary dipole in one molecule induces the formation of a temporary dipole in a nearby molecule and is attracted to it.	All atoms and molecules
Dipole-dipole force	Polar molecules (permanent dipoles) attract one another.	Polar molecules
Hydrogen-bonding force	Two dipoles, one containing hydrogen bonded to an electronegative element and the other containing an electronegative element, attract one another.	Polar molecules containing unpaired electrons and a hydrogen bonded to nitrogen, oxygen, or fluorine
Covalent bond	Nuclei of two atoms attract the electrons shared between them.	Nonmetal-nonmetal compounds
Ionic bond	Cations and anions attract one another.	Metal-nonmetal compounds

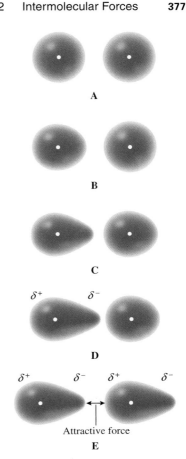

Figure 10.15
London dispersion forces arise from attractions between temporary dipoles. (A) Ordinarily, electrons distribute symmetrically in atoms or nonpolar molecules. (B) An electron cloud may become distorted as electrons travel around the nucleus. (C) If this distortion gets large enough, it may affect adjacent atoms or molecules. (D) The distortion of the electron cloud causes the formation of a temporary or instantaneous dipole. (E) This temporary dipole induces the formation of a dipole in the electron cloud of a nearby atom or molecule, resulting in attractive forces between the instantaneous dipole and the induced dipole. This attraction constitutes the London dispersion force.

What is the origin of these intermolecular forces? They arise from the interaction of positive and negative charges. We will examine three such forces: the *London dispersion force,* the *dipole-dipole force,* and a special case of dipole-dipole attraction, the *hydrogen bond.* Characteristics of these forces as well as chemical bonds are summarized in Table 10.2. We will discuss each intermolecular force in more detail in this section.

London Dispersion Forces

Electrons moving around an atom or nonpolar molecule are not always distributed symmetrically. If the electrons happen to be more on one side at an instant in time, that side will be more negative than normal, and the other side will be more positive. This situation leads to the formation of a temporary dipole, or **instantaneous dipole,** which—like any dipole—contains partial charges (as discussed in Chapter 8). The positive end of the dipole exerts an attractive force on nearby electrons, causing an adjacent atom to develop into another temporary dipole, called an **induced dipole.** This effect passes on to more atoms, resulting in a kind of electron "choreography" in which the movements of the electrons in nearby atoms correlate (Figure 10.15). The attraction between these temporary dipoles is called a **London dispersion force.** It occurs between all atoms and molecules, and it is the only intermolecular force at work in nonpolar substances. Although London dispersion forces are relatively weak, they are strong enough to cause substances normally found as gases, such as neon and methane, to liquefy at high pressures or low temperatures.

London dispersion forces tend to be stronger the larger the atom or molecule. For example, I_2 molecules interact with each other more strongly than Br_2 molecules, which interact more strongly than Cl_2 molecules. This explains why Cl_2 is a gas, Br_2 is a liquid, and I_2 is a solid at room temperature (Figure 10.16). The size of the electron cloud surrounding the atom or molecule, which depends on the number of electrons, causes this trend. The larger the electron cloud, the easier it is to distort it, resulting in stronger London dispersion forces. As the molecules get larger, their intermolecular forces get stronger and this leads to substances existing in liquid or solid states.

London dispersion forces are named after Fritz London, a German-American physicist who first proposed an explanation for them in 1930.

I_2 Cl_2 Br_2 why gas, liquid, solid @ room temp?

Cl$_2$ Br$_2$ I$_2$

Figure 10.16
At room temperature, chlorine is a pale green gas, bromine is a reddish-brown liquid that evaporates easily, and iodine is a purple solid that sublimes easily. The different physical states relate to the strength of the intermolecular forces, which in turn relate to the sizes of the electron clouds in these molecules.

EXAMPLE 10.5   London Dispersion Forces

Which has the stronger London dispersion forces, argon or xenon?

Solution:

Since argon has 18 electrons and xenon has 54 electrons, the electron cloud of xenon is larger than that of argon. The larger electron cloud is easier to distort, so xenon has stronger London dispersion forces.

Practice Problem 10.5

Which has the stronger London dispersion forces, CH_4 or SiH_4?

Further Practice: 10.41 and 10.42

In very large molecules, London dispersion forces can get quite strong. Examples can be found in liquid crystals, such as the display used in Nicole's handheld computer. Liquid crystals can flow like liquids, but have the structural order of solids. Their liquid-crystalline state occurs during melting as a transition between the solid state and the true liquid state. Most liquid-crystal films change color when the molecules change their alignment. The color change can be controlled by temperature, mechanical stress, magnetic fields, or electric fields.

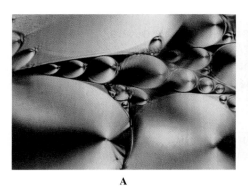

A   B

Figure 10.17
(A) A liquid crystal is magnified 320 times, showing layers of parallel molecules. (B) At a higher magnification, a scanning tunneling microscope image shows individual octadecanol molecules aligned with one another, typical of long organic molecules.

Why do these molecules behave in this way? Liquid-crystalline substances consist of long, narrow, rigid organic molecules. One example is the molecule 4-methoxybenzylidene-4′-n-butylaniline, which behaves as a liquid crystal over the temperature range of 21 to 47°C:

In liquid crystals, molecules align parallel to one another as shown in Figure 10.17. The alignment occurs because strong London dispersion forces operate among large molecules. When an electric field is applied to a film of some liquid crystals, the molecular alignment changes and parts of the film turn black, forming the shapes we see on an LCD screen. Some liquid crystals reflect different colors of light under different conditions, such as changing temperatures.

Dipole-Dipole Forces

In Section 8.4 we discussed the separation of charge within molecules that causes polarity. The attractions among polar molecules result in a second type of intermolecular attraction, the **dipole-dipole force.** With polar molecules, the partially positive end of one molecule attracts the partially negative ends of other molecules (Figure 10.18). In general, attractions between polar molecules are stronger than those between nonpolar molecules of similar size. Consider the boiling points of two molecular substances, O_2 and NO. The boiling points of these substances are −183°C and −152°C, respectively. Why are these differences observed? Both have a double bond between their atoms, and they differ by only one electron. Their electron clouds are similar in size, but the O_2 molecule is nonpolar and the NO molecule is polar. The polar molecule has greater intermolecular forces than the nonpolar molecule, because it has both dipole-dipole and London forces. With stronger intermolecular forces, more energy is required to separate the liquid-state molecules and convert them to gas-state molecules. The increased energy demand causes the boiling point of NO to be higher than that of O_2.

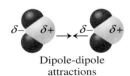

Dipole-dipole
attractions

Figure 10.18
Polar molecules like sulfur dioxide attract each other when the positive end of one molecule aligns with the negative end of another molecule.

EXAMPLE 10.6   Dipole-Dipole Forces

Which of the following molecules experience dipole-dipole forces?

(a) SO_3 (b) CO_2 (c) PCl_3 (d) NO_2

Solution:

Recall from Chapter 8 that two conditions are necessary for a molecule to be polar—which is, in turn, necessary for the molecule to experience dipole-dipole forces. First the molecule cannot be symmetrical. Second there must be an electronegativity difference between the central atom and the atoms surrounding it. If necessary, review Chapter 8 to determine the shapes of these molecules starting with their Lewis formulas.

(a) The SO_3 molecule is trigonal planar, so it is symmetrical.

Despite the difference in electronegativity between sulfur and oxygen, the symmetry of the molecule causes the bond polarities to cancel and make this molecule nonpolar. Consequently it cannot experience dipole-dipole forces.

(b) The CO_2 molecule is linear, so it is symmetrical.

It is nonpolar and does not experience dipole-dipole forces.

(c) The PCl_3 molecule is trigonal pyramidal, so it is not symmetrical.

There is a difference in electronegativity between phosphorus and chlorine, so the bonds are polar. Therefore the molecule is polar and will experience dipole-dipole forces.

(d) The NO_2 molecule is bent, so it is not symmetrical.

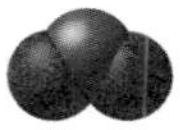

The electronegativities of nitrogen and oxygen are different, so the bonds are polar. The molecule is polar and will experience dipole-dipole forces.

Practice Problem 10.6

Which of the following molecules experience dipole-dipole forces?

(a) SCl_2 (b) CO   (c) NH_3 (d) CCl_4

Further Practice: 10.43 and 10.44

Hydrogen Bonding

The boiling points of different substances vary widely. Consider the boiling points of some elements and nonpolar molecular compounds shown in Figure 10.19. The figure reveals that—in a related series of substances—the boiling point increases as the atoms or molecules get larger. The same is true of polar molecules. The larger the molecule, the higher its boiling point, because the London dispersion forces are greater.

Figure 10.20, however, reveals an irregularity. As expected, the boiling points of the nonpolar hydrides of group IVA elements in the periodic table (beginning

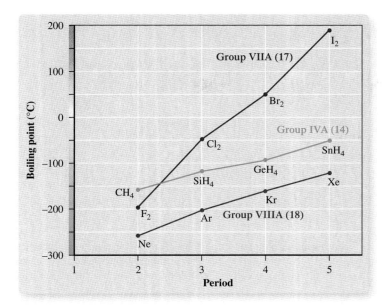

Figure 10.19
Within groups of the periodic table, the boiling points of nonpolar substances increase with the size of the molecule.

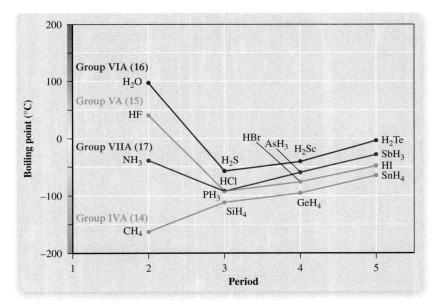

Figure 10.20
Boiling points generally increase with the size of the molecule, but some relatively small molecules have high boiling points, due to hydrogen bonding.

with CH_4) increase as the size of the electron cloud increases. The same occurs with most of the polar hydrides of elements from groups VA (15), VIA (16), and VIIA (17). However, NH_3, H_2O, and HF have unexpectedly high boiling points, and many of their other properties are unusual as well. What is special about these molecules? Their intermolecular forces must be unusually strong.

Especially strong dipole-dipole forces exist between polar molecules that contain hydrogen attached to a small, highly electronegative element (specifically nitrogen, oxygen, or fluorine). Such a force is called a **hydrogen bond,** although there is no bond between molecules in the usual sense. The hydrogen bond is actually a special

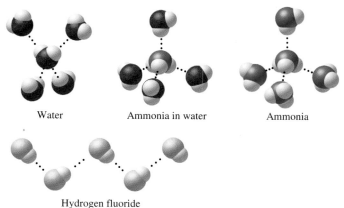

Water   Ammonia in water   Ammonia

Hydrogen fluoride

Figure 10.21
Here three dots represent hydrogen bonding. These forces are experienced between a hydrogen atom on one molecule aligned with an unshared pair of electrons on a highly electronegative atom on another molecule. The two molecules may be alike or different.

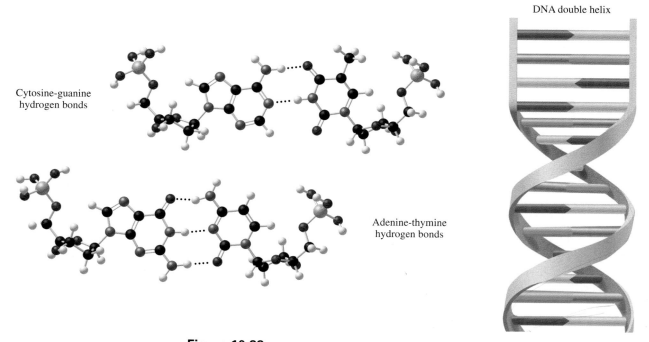

DNA double helix

Cytosine-guanine hydrogen bonds

Adenine-thymine hydrogen bonds

Figure 10.22
In DNA, hydrogen bonds hold together the base pairs adenine-thymine and cytosine-guanine. These bonds form the "rungs" in the spiral ladder that is the DNA double helix.

type of dipole-dipole force. The strength of most hydrogen bonding is in the range of 20 to 40 kJ/mol, although a few examples have energies as high as 150 kJ/mol. This is large for intermolecular forces, but still smaller than the forces in covalent bonds.

Hydrogen bonding occurs in specific directions. The attraction between molecules occurs with a hydrogen atom covalently bonded to a highly electronegative atom in one molecule pointing at an unshared electron pair on a highly electronegative atom in an adjacent molecule. Some examples are shown in Figure 10.21. The hydrogen is closer to the atom to which it is covalently bonded than to the other molecule. Hydrogen bonding is an important force in living systems, where it maintains much of the structure of proteins and nucleic acids, as shown in Figure 10.22. The hydrogen bonding stabilizes molecular shapes, thereby protecting the ability of such molecules to carry out their biological functions.

EXAMPLE 10.7   Hydrogen Bonding

Identify the molecules from the following list that experience hydrogen bonding in the pure liquid state: CH_3F, CH_3CH_2OH, and CH_3-O-CH_3.

Solution:

For a molecule in a pure liquid to experience hydrogen bonding, it must contain a nitrogen, oxygen, or fluorine atom, and a hydrogen atom must be bonded to one of them. The CH_3F molecule contains hydrogen and fluorine, but hydrogen is not bonded to fluorine, so it does not experience hydrogen bonding. The CH_3CH_2OH molecule does contain hydrogen bonded to oxygen, so it will experience hydrogen bonding. The CH_3-O-CH_3 molecule contains hydrogen and oxygen, but hydrogen is not bonded to oxygen. It does not experience hydrogen bonding.

Practice Problem 10.7

Identify the molecules from the following list that experience hydrogen bonding in the pure liquid state: $N(CH_3)_3$, CH_3CO_2H, and HOCl.

Further Practice: 10.49 and 10.50

Trends in Intermolecular Forces

Intermolecular forces vary in strength. Assuming the molecules are similar in size, the magnitudes of intermolecular forces generally compare in the following way:

| molecules experiencing London dispersion forces only | < | molecules also experiencing dipole-dipole forces | < | molecules also experiencing hydrogen bonding |

This relative order may not hold if we compare the London dispersion forces in very large molecules with the dipole-dipole forces in small molecules.

How do we know which intermolecular forces to expect in a given substance? Using the approach developed in Chapter 8, we first decide whether the molecule is polar. If it is, we then look for the occurrence of H−F, H−O, or H−N bonds. If one of those is present, then the strongest force will be hydrogen-bonding forces coupled with London dispersion forces. If one of them is not present, the strongest force will be dipole-dipole forces coupled with London dispersion forces. If the molecule is nonpolar, only London dispersion forces operate. These relationships are summarized in Figure 10.23.

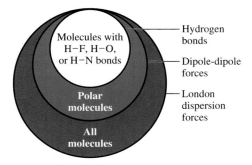

Figure 10.23
All molecules experience London dispersion forces. The subset of polar molecules also exhibits dipole-dipole forces. Only polar molecules that contain H−F, H−O, or H−N bonds are also subject to hydrogen-bonding forces.

EXAMPLE 10.8   Trends in Intermolecular Forces

List the following molecules in order of increasing strength of their intermolecular forces: F_2, HCl, H_2O_2, and Ar.

Solution:

Both F_2 and Ar are nonpolar, so they can involve only London dispersion forces. Since both have the same number of electrons, we would expect their intermolecular forces to be similar. This is confirmed by similar boiling points ($-188°C$ and $-186°C$, respectively). Both HCl and H_2O_2 are polar molecules, each with 18 electrons. Because they are polar, both experience dipole-dipole interactions. Since they are about the same size as F_2 and Ar, the contribution from London dispersion forces should be the same for all four. The H_2O_2 molecule contains hydrogen atoms bonded to oxygen, which also has unshared electron pairs, so it involves hydrogen bonding. Thus, HCl should have intermolecular forces greater than F_2 and Ar, and H_2O_2 should have stronger forces than HCl. The order of increasing strength in intermolecular forces is thus F_2, Ar < HCl < H_2O_2.

Practice Problem 10.8

Indicate which member of each pair of molecules has the stronger intermolecular forces and explain your reasoning.

(a) N_2 and Cl_2 (b) PH_3 and NH_3 (c) SO_2 and CO_2 (d) H_2S and H_2Te

Further Practice: 10.59 and 10.60

Consider the substances methyl alcohol (CH_3OH) and methyl mercaptan (CH_3SH). How should their boiling points compare? Since boiling points increase with increasing intermolecular forces, we must decide which intermolecular forces are present in each substance by first identifying each substance as nonpolar or polar. Each has a methyl group ($-CH_3$) and a hydrogen atom bonded to an atom that contains two unshared electron pairs (O or S), so each is bent and polar. CH_3OH has the hydrogen atom bonded to an oxygen atom, but CH_3SH does not have a hydrogen atom bonded to any F, O, or N. Thus, CH_3OH experiences hydrogen-bonding forces, while CH_3SH experiences only dipole-dipole forces. (Both also experience London forces.) Since hydrogen-bonding forces are stronger than dipole-dipole forces and London forces, we expect CH_3OH to have the higher boiling point. Measurements indicate boiling points of 64.7°C for CH_3OH and 5.95°C for CH_3SH, in agreement with our predictions.

EXAMPLE 10.9   Trends in Boiling Points

Arrange the following liquids in order of increasing boiling points: CH_4, GeH_4, and SiH_4.

Solution:

None of these molecules is polar, so we must consider only the magnitude of the London dispersion forces in each liquid. These forces increase as the molecule becomes larger, due to a more easily distorted electron cloud. As intermolecular forces increase, the energy required to evaporate the liquids increases, and the temperature needed to supply the energy increases. Thus we predict the order of boiling points as $CH_4 < SiH_4 < GeH_4$. This agrees with measured values of the boiling points: CH_4 (–164°C) < SiH_4 (–112°C) < GeH_4 (–89°C)

Practice Problem 10.9

Arrange the following liquids in order of increasing boiling points: $CHBr_3$, $CHCl_3$, and CHI_3.

Further Practice: 10.63 and 10.64

Evaporating a liquid requires overcoming the intermolecular attractions among its molecules. At a constant temperature, vapor pressure decreases with the increasing strength of intermolecular forces. Accordingly we would expect that vapor pressure would be smallest for molecules that have hydrogen bonding. Polar molecules should have the next highest vapor pressures. Nonpolar molecules of similar size should have the highest values. Among nonpolar molecules, those with large electron clouds, such as C_6H_6 and CCl_4, should have lower vapor pressures than those with small electron clouds such as CH_4.

EXAMPLE 10.10   Trends in Vapor Pressure

Explain why the vapor pressure of water is about 40 torr at 35°C, while that of diethyl ether ($CH_3CH_2-O-CH_2CH_3$) is about 760 torr at the same temperature.

Solution:

Both molecules are polar and so undergo dipole-dipole interactions. However, diethyl ether, unlike water, does not have hydrogen bonding. Since hydrogen bonding is stronger than dipole-dipole forces, the intermolecular forces are stronger in water than in diethyl ether. Thus at a given temperature, more diethyl ether evaporates, giving a higher vapor pressure.

Practice Problem 10.10

Which substance, $O_3(l)$ or $O_2(l)$, has the higher vapor pressure at a given temperature? Explain your reasoning.

Further Practice: 10.71 and 10.72

Trends in vapor pressure are opposite those for the melting point and boiling point, but the reasons are the same. Let's consider CH_3OH and CH_3SH again. The boiling points are 64.7°C for CH_3OH and 5.95°C for CH_3SH, consistent with the stronger intermolecular forces in CH_3OH. The stronger the intermolecular forces, the higher the temperature needed to overcome them and evaporate the liquid. At a given temperature then, the substance with the stronger intermolecular forces has the lower concentration of molecules in the gaseous state, and therefore the lower vapor pressure. Observed data are consistent with this conclusion. The values of vapor pressure at 0°C are 0.039 atm for CH_3OH and 0.626 atm for CH_3SH.

| EXAMPLE 10.11 | Trends in Molecular Properties |

Consider the substances CH_4 and NH_3. Which has the stronger intermolecular forces? Which has the higher melting point? boiling point? vapor pressure?

Solution:

The molecule CH_4 is nonpolar, while NH_3 can hydrogen bond. Thus, NH_3 has the stronger intermolecular forces. It takes more energy to overcome the inter-molecular forces in NH_3 than in CH_4, so a higher temperature will be required to melt or boil NH_3. Because of the lower energy required for evaporation, at any given temperature CH_4 will have a higher vapor pressure than NH_3.

Practice Problem 10.11

Consider the substances CO and HF. Which has the stronger intermolecular forces? Which has the higher melting point? boiling point? vapor pressure?

Further Practice: 10.73 and 10.74

A

B

Figure 10.24

(A) When the water in a full bottle freezes, it expands beyond the capacity of the bottle. (B) Freezing water expands in cracks, splitting rocks and paved roads.

10.3 PROPERTIES OF LIQUIDS

The properties of liquids are related to the distance between particles (usually molecules) and to intermolecular forces. Particles in a liquid are much closer together than particles in a gas. However, unlike particles in a solid, liquid particles are not fixed in position. They move about and collide with one another. Because of their intermolecular forces, liquid particles move shorter distances before they collide than do the particles in a gas. They have less independence than gas particles, but more than solid particles.

Density

The densities of the states of matter are related to the distance between particles. Generally liquids are denser than gases by a factor of about 1000 and nearly as dense (90% to 95%) as solids. As you might expect, most substances are denser when solid than when liquid, because their molecules or atoms are closer together. For example, liquid aluminum at 659°C has a density of 2.38 g/mL, while the density of solid aluminum at 25°C is 2.70 g/mL. Water, however, is an exception. Liquid water has its greatest density, 1.0000 g/mL, at 4°C. At 0°C, solid water has a density of 0.9167 g/mL, while liquid water has a density of 0.9998 g/mL. The implication of this density comparison is that water expands when it freezes (Figure 10.24). As a result, ice floats on water, allowing plant and animal life to survive in lakes and streams during the winter. If ice sank, entire lakes would freeze and probably never thaw completely. The expansion of freezing water is responsible for much weathering action. When water freezes in cracks in rock, for example, the rock may be broken up, a process that helps create soil. We see a similar effect in frozen roadbeds, engine blocks, and glass containers, so this action is not always beneficial.

Why does water expand on freezing while most other liquids contract? The answer lies in the arrangement of its molecules. In ice, hydrogen bonding fixes water molecules into a *hexagonal* arrangement. The centers of the hexagons line up to form open channels, leaving a lot of empty space in the solid. In liquid water, the rigid structure is relaxed, and water molecules fill some of the unused space (Figure 10.25).

flurds

gases liquids
↓ viscosity

Figure 10.25
When ice melts, the open spaces in the ice structure fill with water molecules, so the liquid occupies a smaller volume.

Viscosity — solid= ↑ viscosity

Liquids, like gases, are fluids. A **fluid** is any substance that can flow. Most liquids—such as water, alcohol, and gasoline—flow readily, although not as easily as gases. Solids, in contrast, cannot be observed to flow at all over normal time periods, such as days or weeks. The resistance of a substance to flow is its **viscosity.** In general the viscosity of liquids is low, closer to the low viscosity of gases than to the high viscosity of solids.

However, the viscosities of different liquids vary, generally increasing with the magnitude of their intermolecular forces and with molecular size. At the low end of the range is liquid helium-3 below a temperature of 2.1 K. It has such a low viscosity that when it is placed in an open beaker, it actually flows up the sides and out of the beaker. At the other end of the spectrum are such highly viscous liquids as oils (castor oil is about 1000 times more viscous than water), honey, syrup, molasses, and hot tar, a material used to pave roads. The viscosity of a liquid generally decreases as the liquid is heated, because heat energy partially overcomes intermolecular forces. Ferrofluids, discussed in the introduction, normally have the viscosity of oil, but they become more viscous when placed in a magnetic field (Figure 10.3).

The cliché, slow as molasses in January, used to describe high viscosity, may be misleading. On January 15, 1919, a large vat of molasses burst when the temperature reached an unseasonable 43°F, sending over 2 million gallons of molasses down the streets of Boston. The 8-ft wave knocked down buildings and traveled so fast that people in its path couldn't outrun it. The flood killed 21 people and injured many more.

EXAMPLE 10.12   Viscosity

Gasoline contains hydrocarbons such as octane (C_8H_{18}). Motor oils contain much larger molecules, such as $C_{18}H_{38}$. Which is the more viscous liquid?

Solution:

Viscosity is greater for molecules that have stronger intermolecular forces. Both C_8H_{18} and $C_{18}H_{38}$ are nonpolar molecules, so they have only London dispersion forces. Because $C_{18}H_{38}$ is the larger molecule, it should have the stronger intermolecular forces. Thus motor oil is more viscous than gasoline.

Practice Problem 10.12

The Society of Automotive Engineers rates the viscosity of motor oils. Some typical ratings are SAE 40 and SAE 15, where the number is a measure of the relative viscosity. Higher numbers are given to oils that are more viscous. Oils suitable for use during winter have a W placed after the number in the viscosity rating. Which of these oils (SAE 40 or SAE 15) could be used during the winter and thus have a W in its rating?

Further Practice: 10.79 and 10.80

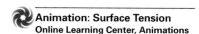

Animation: Surface Tension
Online Learning Center, Animations

Chemistry Animations Library, Surface Tension, Surface_Tension.rm.

Surface Tension

Small droplets of liquids tend to have spherical shapes (Figure 10.26). Water beads up when placed on a waxed surface. Soap bubbles and raindrops are spherical. Why does this happen? As you might expect, it has to do with intermolecular forces. Molecules within a liquid experience forces of attraction equally in all directions, as shown in Figure 10.27. However, there are no forces above the surface of the liquid, but there are normal forces below. Because of this imbalance, forces of attraction tend to pull molecules toward the interior of the liquid. The attraction minimizes the number of molecules at the surface, and thus minimizes the surface area. For a given volume, a sphere is the shape with the smallest surface area.

In order for its surface to be deformed from a spherical shape, the liquid must increase its surface area. To do this, molecules must overcome intermolecular forces in order to move from the interior of the liquid to the surface. The energy required to do this is called **surface tension,** the amount of work necessary to increase the surface area of a liquid by a unit amount. Surface tension causes a liquid surface to behave like a stretched membrane. The greater the intermolecular forces in a liquid, the greater the surface tension. Surface tension decreases somewhat as temperature increases, because intermolecular forces are overcome to some extent by greater kinetic energy at higher temperatures.

Figure 10.26
Water droplets have a spherical shape that minimizes their surface area.

Figure 10.27
Unbalanced intermolecular forces on the surface of a liquid pull molecules back toward the interior. Molecules below the surface experience forces of attraction that are about equal in all directions.

The surface tension of water is great enough to float an insect (Figure 10.28). The force of the insect's mass pushing down is less than the amount of energy needed to overcome intermolecular forces and cause an increase in the water's surface area. The same is true of a steel needle laid carefully on the surface of water. However, if a detergent is added to the water, both the insect and the needle will sink because the surface tension is lowered and the surface is easier to deform. The detergent molecules interfere with the water-water attractive interactions, making it easier to move water molecules apart. This effect is put to use in laundry detergents. They work in part because they reduce the surface tension so the water can wet the clothing better, allowing the detergent to surround the dirt particles.

We observe another result of intermolecular forces when a capillary tube is placed in a liquid. As shown in Figure 10.29, water rises in a glass capillary tube, a phenomenon called *capillary action.* In a tube, water-glass interactions are stronger than the intermolecular forces in water alone, so water rises. The extent of the rise depends on the size of the capillary. Water rises until the water-glass interactions are just offset by the pull of gravity on the column of water. Capillary action is partially responsible for the rise of sap in trees and for a towel's absorption of water.

Intermolecular forces also produce the **meniscus,** the curved surface of a liquid in a container. The meniscus may be either concave or convex, depending on whether the intermolecular forces among liquid molecules are greater than or less than the forces of attraction between the liquid and the container walls. As Figure 10.30 shows, the surface is concave for water because the water-glass interactions are stronger than the water-water interactions. In contrast, the intermolecular forces in mercury are stronger than mercury-glass interactions, so mercury tends to withdraw into itself and repel the glass, forming a convex surface.

Detergents are in a class of compounds called surfactants, which modify surface tension.

Figure 10.28
Water striders can walk across the surface of water because their weight is insufficient to overcome intermolecular forces among water molecules. However, if a detergent is added, surface tension is lessened, and the insect sinks. If it does end up under water, it has to exert considerable force to climb back out. You can examine a similar effect by carefully floating a needle on water, and then adding a drop of dish detergent.

Figure 10.29
(A) Water rises in tubes due to capillary action. A hemocrit (capillary tube) is filled with blood using capillary action. (B) The drying action of a towel is due in part to capillary action.

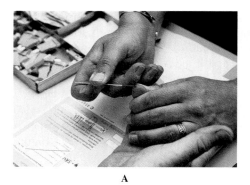

A   B

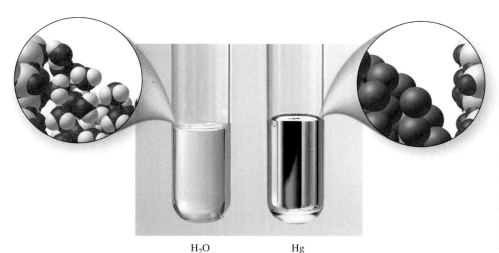

H$_2$O   Hg

Figure 10.30
The meniscus for water (left) is concave because the interactions between the silicon dioxide in the glass and the water molecules are stronger than the interactions among water molecules. The meniscus for mercury (right) is convex because the interactions among mercury atoms are stronger than the interactions between mercury and silicon dioxide.

EXAMPLE 10.13   **Surface Tension**

Water spreads out on a glass surface but beads up on a waxed surface. Explain the difference.

Solution:

The intermolecular forces among water molecules are less than the interactions between water molecules and glass, so the water spreads out on glass. However, on a waxed surface, the opposite is true. The surface water molecules are attracted to the internal water molecules more strongly than to the wax molecules. The water drops form a spherical shape that minimizes their surface area.

Practice Problem 10.13

Scientists want to design a new material that can be coated on a car body so the surface of the car can be wet with water without having the water form beads. What properties of the new material should they be concerned about?

Further Practice: 10.81 and 10.82

10.4 PROPERTIES OF SOLIDS

The most obvious properties of solids are their definite sizes and rigid shapes. Solids are not very compressible, and they usually do not flow. Attractive forces hold the particles (ions, atoms, or molecules) in solids together in a rigid, three-dimensional array. If the temperature of a liquid drops rapidly, the particles may solidify in a partially disordered state, particularly if they are long molecules that can get tangled together. The resulting solid is said to be an **amorphous solid.** Its particles are arranged somewhat randomly, and it lacks a regular form. Glass and most other polymers, such as rubber and plastics, belong in this class. If, in contrast, a liquid solidifies slowly, allowing the array of particles to become well ordered, the result is a **crystalline solid.** Most solids are of this type. Such solids can also be precipitated from solutions containing ionic compounds. The symmetrical arrangement of planar faces typical of a crystalline solid is shown in Figure 10.31. In contrast, amorphous solids rarely show regular planar faces.

Crystals and Crystal Lattices

A **crystal** is an orderly, repeating, three-dimensional assembly of fundamental particles (atoms, molecules, or ions). The pattern formed by these arrays is the **crystal lattice** or *crystal structure*. One layer in the crystal lattice of a typical metal is

Glass panes in old churches in Europe are often thicker at the bottom than at the top. For some time, this was considered evidence that over long periods of time, glass actually flows, a characteristic of liquids. However, investigation of traditional glass-making methods revealed that glass panes were usually not of uniform thickness. It is more likely that the thicker and heavier end of a glass pane was mounted in the bottom of the window frame.

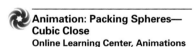
Animation: Packing Spheres—Cubic Close
Online Learning Center, Animations
Chemistry Animations Library, Packing Spheres—Cubic Close, Cubic_Close_Packing.exe.

A   B

Figure 10.31
(A) Quartz, a crystalline solid, has faces that are planes at characteristic angles. (B) Glass, an amorphous solid of nearly the same composition, does not fracture along characteristic planes.

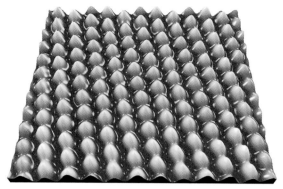

Figure 10.32
A scanning tunneling microscope image of the surface of a gold crystal shows how the atoms pack closely together in an arrangement called close packing or cubic close packing. Count the number of atoms that surround and touch a single atom.

Figure 10.33
Close packing occurs in many places, both naturally and in human-made objects. From the left: pineapple, wasp nest, bubble wrap, and dimples in a golf ball.

Figure 10.34
Grocers stack fruit in the same close-packed arrangement as atoms in metals.

shown in Figure 10.32. Notice the orderly arrangement of atoms. It arises from a tendency of atoms to use space efficiently, that is, to be as close to one another as possible. The arrangement shown is called *close packing* because it is the most efficient way to use space. Six atoms surround and touch each atom in this structure. This arrangement is found at a macroscopic level both in nature and in human endeavors, as shown in Figure 10.33.

In a three-dimensional crystal, more than one layer of atoms must be considered, but each layer has a similar orderly arrangement. The arrangement of atoms in many metals is the same as the arrangement of fruit stacked in grocery stores (Figure 10.34). In these structures, one layer of close-packed atoms is placed on another layer, with the atoms in the second layer fitting into depressions between atoms in the first. Successive layers fit together in a similar fashion.

Use marbles to examine the close-packed structure in three dimensions. Arrange the first layer to place the marbles as close to one another as possible, with six marbles surrounding each individual. Then create a second layer by placing marbles over the depressions in the first layer. Continue for as many layers as you wish to examine.

TABLE 10.3	Classification of Solids			
Type of Solid	**Fundamental Particles**	**Attractive Forces**	**Properties**	**Examples**
Metallic	Atoms	Attractions between nuclei and delocalized valence electrons	Low melting point and soft, or high melting point and hard; good heat and electrical conductors; malleable and ductile	Hg, Na, Fe, Au, Ag, Cu, Cr, alloys such as Nitinol
Ionic	Cations and anions	Ionic bonds	High melting point, hard, brittle; nonconductors when solid; electrical conductors when melted	NaCl, CsCl, CaF_2, $CaCO_3$, KNO_3, $Ca_5(PO_4)_3OH$, $YBa_2Cu_3O_7$
Molecular	Polar molecules	Dipole-dipole forces	Low to moderate melting point, variable hardness, may be brittle; nonconductors	H_2O, NH_3, SO_2, CH_3CO_2H, CH_2Cl_2
	Nonpolar molecules	London dispersion forces	Low melting point, soft; poor heat conductors; electrical insulators	CH_4, O_2, CCl_4, CO_2, C_6H_6, I_2, S_8
Network	Atoms	Covalent bonds	Very high melting point, very hard, somewhat brittle; nonconductors or semiconductors	C (diamond), SiC (carborundum), SiO_2 (silica), Al_2O_3 (alundum), BN (boron nitride)

Types of Crystalline Solids

Solids show a wide range of properties such as melting point, hardness, and malleability. For most solids, the types of forces holding the fundamental particles together explain these properties and correlate with them. The useful classification system presented in Table 10.3 organizes solids into four groups: *metallic, ionic, molecular* (which may be either polar or nonpolar), and *network*. The table describes some of the properties of these classes and gives examples of each.

Consider two solids, the first of which conducts heat, while the second does not. Both solids are soft, but only the first is malleable. The first conducts electricity, but the second is an electrical insulator. How do we classify these solids? Only metallic solids are good heat and electrical conductors, so the first solid must be a metal. We could place the other solid in any of the other classes if we considered only these two properties. However, only molecular solids (along with a few metals, which this solid is not) are soft. So the second solid must be a molecular solid.

We must consider several properties to classify any solid. It is difficult to predict whether a covalent substance will form a molecular solid or a network solid, but we can distinguish between them if we know their properties.

EXAMPLE 10.14   Types of Crystalline Solids

A crystalline solid is hard, breaks when it is struck with a hammer, and is an electrical insulator; but it conducts electricity when it is melted. What type of solid is it?

Solution:

A hard solid could be metallic, ionic, or network. The observation that it does not conduct electricity as a solid is consistent with either an ionic or a network solid. Since it conducts electricity when melted, it must be an ionic solid.

Practice Problem 10.14

A crystalline solid is very hard, does not conduct electricity when solid or when melted, and has a high melting point. What type of solid is it?

Further Practice: 10.97 and 10.98

Metallic Solids   In metals, valence electrons move freely through all parts of the metal. This explains why metals are good conductors of electricity. The bonds between atoms result from attractions between the shared valence electrons and the metal nuclei surrounded by their core electrons, which remain fixed in position. These attractions are spread out over the entire piece of metal. Because the attractions are not localized, it is easy to move atoms by applying force, making metals malleable and ductile. Depending on the strength of the attractions, metals can range from soft to hard and can melt at low to high temperatures. The soft, low-melting-point metals are primarily the group IA (1) elements. The properties of metals can be changed by adding impurities. Remember the bronze that gave its name to an entire age in human history? Mixing copper and tin makes bronze, which has different properties than either of the two pure metals.

Mixing a metal with one or more additional metallic or nonmetallic elements forms an **alloy.** Alloys have properties different from those of their parent elements. For example, as shown in Figure 10.35, zinc added to copper makes the alloy brass. In brass, zinc atoms replace some of the positions in the metal structure formerly occupied by copper atoms. Brass is more resistant to tarnish and takes a better polish than copper, so it is commonly used in plumbing fixtures.

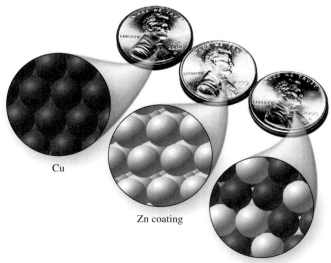

Cu

Zn coating

Brass alloy

Figure 10.35
When copper (left) is chemically treated with zinc metal, it gets a coating of zinc (center). If this mixture is heated, the zinc atoms move into the crystal structure of the copper, giving the alloy brass (right). In brass, zinc atoms occupy some of the positions formerly occupied by copper atoms.

TABLE 10.4	Examples of Alloys		
Alloy	**Metals**	**Properties**	**Uses**
Brass	67% Cu, 33% Zn	Ductile, takes a polish	Hardware
Dental amalgam	70% Ag, 25% Pb, 3% Cu, 2% Hg	Easily worked	Dental fillings
14-karat gold	58% Au, with various amounts of Ag, Cu, and others	Harder than pure gold	Jewelry
Monel metal	69% Ni, 33% Cu, 7% Fe	Resistant to corrosion, bright finish	Kitchen fixtures
Nichrome	60% Ni, 40% Cr	High melting point, low electrical conductivity	Heating coils in stoves, toasters, ovens
Nitinol	49% Ni, 51% Ti	Shape memory	Medical and dental implants, eyeglass frames
Pewter	85% Sn, 6.8% Cu, 6% Bi, 1.7% Sb	Easily cast into shapes	Utensils, cups
Plumber's solder	67% Pb, 33% Sn	Low melting point, soft	Soldering pipe joints
Stainless steel	80.6% Fe, 0.4% C, 18% Cr, 1% Ni	Resistant to corrosion	Tableware
Sterling silver	92.5% Ag, 7.5% Cu	Bright finish	Tableware

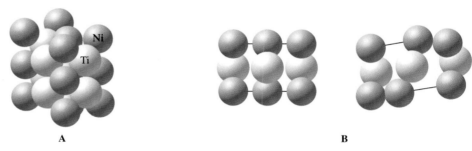

Figure 10.36
(A) Nitinol can change between this and another closely related structure as its temperature changes. (B) An expanded view of a slice through the Nitinol structures shows the small change in position of the atoms that converts one structure to the other.

Pure iron is relatively soft and malleable. In steel, carbon atoms fit into holes in the original iron structure. As more carbon is added, the steel becomes harder and stronger. Other elements, such as chromium, vanadium, nickel, molybdenum, or tungsten, can also be added to iron to make steels with various properties. Some typical alloys, including stainless steel, are described in Table 10.4.

In the introduction, we discussed Nitinol, an alloy with peculiar properties. This alloy containing nickel and titanium can "remember" its original shape. How? This property is related to the structure of the metal (Figure 10.36). Actually the metal has two slightly different structures, one at low temperature and another at high temperature. The two structures change from one to another by relatively minor shifts of the planes of atoms. When the metal is bent at low temperature, stress is placed on the planes of atoms. When the metal is then heated, the stress is relieved by shifts of atoms to the high-temperature structure, giving back the original shape. Upon cooling, the original shape is maintained. If the metal is bent when heated at a high temperature, the new shape will be retained when the metal is cooled.

Ionic Solids Recall from Chapter 8 that ionic solids contain cations and anions arranged in crystalline lattices. The forces holding the particles together are cation-anion attractive electrostatic forces, offset somewhat by cation-cation and anion-anion repulsive electrostatic forces. Because the attractive electrostatic forces, called ionic bonds, are strong, ionic crystals have high melting points. It is difficult to displace ions, so the crystals tend to be hard. They are also brittle. Although displacement is difficult, only a small amount is necessary to break a crystal. Ionic solids are not electrical conductors because the ions are fixed in position. However, melted or dissolved ionic solids, in which the ions are free to move and carry electrical charge, are good conductors.

Many ionic crystal structures resemble the close-packed arrangements we saw for metals. Close-packed structures use only 74% of the available volume, so there are holes left within them. We saw that metals can form alloys by placing other atoms in holes in the metal structure. Ionic crystals are analogous to alloys. In ionic crystals the larger ions (usually the anions) occupy close-packed positions, and the smaller ions (usually cations) fit into holes appropriate to their size. Some ionic crystals formed in this way are shown in Figure 10.37.

Some ionic solids can be classified as *superconducting* materials. They offer no resistance to the conduction of an electric current, and they repel magnetic fields. Early superconducting materials worked only at very low temperatures, such as when cooled by liquid helium to below 4 K, so they had no practical applications. But materials prepared recently are superconductors at higher temperatures. For example, as shown in Figure 10.38, a magnet floats (levitates) above a

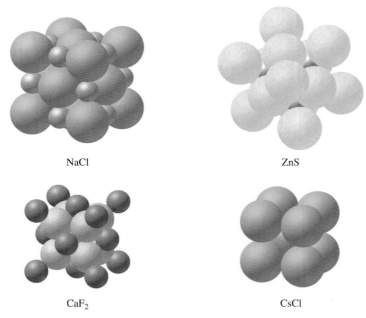

NaCl

ZnS

CaF_2

CsCl

Figure 10.37
Many ionic crystals form by an efficient packing of anions, with cations occupying holes in the resulting structure.

A

B

Figure 10.38
(A) A magnet levitates above a disk of superconducting material. (B) A Japanese train levitates above superconducting tracks.

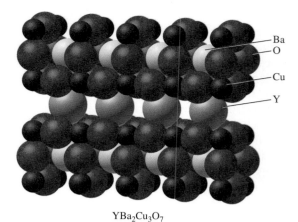

YBa$_2$Cu$_3$O$_7$

Figure 10.39
High-temperature superconductors are ionic solids that form layered structures. In this structure, alternating copper and oxide ions form cubes, barium ions fit in holes in the center of the cubes, and yttrium ions fit into the spaces between layers created by the other ions.

disk of superconducting material cooled to liquid nitrogen temperatures, about 77 K. Such superconductors are used in magnetic resonance imaging (MRI) equipment in hospitals. Extremely fast trains are being tested that can levitate above superconducting tracks (Figure 10.38). Although the nature of superconducting materials is not thoroughly understood, it appears that the crystal structure of the solid material plays a role. The layered structure shown in Figure 10.39 for the superconductor YBa$_2$Cu$_3$O$_7$ seems to be a feature of all superconducting materials discovered so far.

Molecular Solids Intermolecular forces between molecules—not bonds—hold a **molecular solid** together. Such solids are formed mostly from the elements on the right side of the periodic table. Their particles are either atoms, as in the case of the noble gases, or molecules (either nonpolar or polar).

In nonpolar molecular solids, the forces holding molecules together are relatively weak London dispersion forces. As a result, these solids tend to form soft crystals with low melting points. They are electrical insulators and poor conductors of heat. Included in this group are not only the noble gases and the diatomic elements, but also small nonpolar polyatomic molecules, such as carbon dioxide and sulfur hexafluoride. These materials often have structures similar to those of the metals, but with molecules occupying the positions occupied by atoms in metals. Figure 10.40 shows the structure of solid carbon dioxide.

Dipole-dipole forces and London dispersion forces hold polar molecular solids together. Consequently polar molecular solids are often harder than nonpolar molecular solids and have low to moderate melting points. Because they contain no ions, they are electrical insulators even when molten.

Network Solids A **network solid** consists of a giant molecule that forms the entire crystal. Such solids occur mostly among the semiconducting or metalloid elements in the transition region between metals and nonmetals in the periodic table. They also form from elemental carbon in the form of diamond or graphite. Strong covalent bonds connect the atoms in network solids. The crystals of network solids have very stable, three-dimensional structures because their covalent bonds point in specific directions.

Many network solids have the diamond structure (Figure 10.41) or some modification of it, in which each atom is bonded to four other atoms in a tetrahedral arrangement. A typical modification is that in the silicon dioxide crystal, in which

Figure 10.40
In solid carbon dioxide (dry ice), CO$_2$ molecules are arranged in a close-packed pattern similar to that of the atoms in gold metal (Figure 10.32). Molecules of CO$_2$ are arranged at angles relative to one another, providing a better fit into the available space.

Figure 10.41
In diamond, used in jewelry and polishing agents, each carbon atom is bonded to four other carbon atoms. These four carbon atoms are arranged at the corners of a tetrahedron.

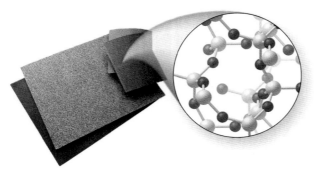

Figure 10.42
In silicon dioxide, used in sandpaper, each silicon atom is bonded to four oxygen atoms in a tetrahedral arrangement. Each oxygen atom is bonded to two silicon atoms. This gives one silicon atom for every two oxygen atoms and an empirical formula of SiO_2.

Figure 10.43
In graphite, each carbon atom is bonded to three other carbon atoms in the same layer. London forces hold layers of carbon atoms together in graphite, which is used in pencil lead. The forces are relatively weak, so layers slide easily over one another. (Note that the distance between layers is exaggerated.)

oxygen atoms form bridges between the silicon atoms (Figure 10.42). In most network solids, the electrons are localized in covalent bonds, making them poor electrical conductors. Some are semiconductors, with a very slight electrical conductivity; they are used to make computer chips. Because their bonds tend to be strong, network solids tend to have high melting points and to be very hard. Among them are the hardest substances known. They are often used in cutting tools.

A different structure is found in the graphite form of carbon. It consists of a two-dimensional network structure of layers of carbon atoms connected by covalent bonds. Relatively weak London dispersion forces hold the layers together (Figure 10.43). Layers of atoms in this structure can slide easily over one another, so graphite is a good lubricant and is used in pencil leads. The electrons in graphite move within the layers readily—in contrast to most network solids—so it is a good electrical conductor in the direction of its layers. For this reason, graphite is often used as an electrode in batteries (see Chapter 14).

EXAMPLE 10.15   Structures of Solids

Identify the type of solid shown by the following molecular-level image. What types of forces hold the particles together in this solid?

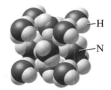

Solution:

It is possible to identify individual molecules, each containing one nitrogen atom and three hydrogen atoms. This composition corresponds to ammonia (NH_3). Because the structure contains molecules, it must be a molecular solid. Covalent bonds hold together the atoms of the ammonia molecule. The molecules are attracted to one another by hydrogen-bonding forces. We know they are at work because hydrogen is attached to nitrogen, one of the elements that participate in hydrogen bonding.

Practice Problem 10.15

Identify the type of solid shown by the following molecular-level image. What types of forces hold the particles together in this solid?

Further Practice: 10.103 and 10.104

SUMMARY

The transitions among the three states of matter are called changes of state. Such changes occur with changes in pressure and temperature. Cooling or heating curves describe the changes in physical state that occur with temperature changes at constant pressure.

Substances exist in the liquid and solid states because of intermolecular forces that hold atoms or molecules together. When comparing molecules of similar size, the weakest intermolecular forces are London dispersion forces, in which an instantaneous dipole causes a nearby atom or molecule to become an induced dipole. These temporary dipoles attract one another. Attractions between permanent dipoles are stronger. The strongest intermolecular forces are those involved in hydrogen bonding. They are dipole-dipole forces between molecules that contain hydrogen bonded to a small-sized, very electronegative atom such as N, O, or F.

The properties of liquids depend on the distance between molecules and the strength of intermolecular forces. Viscosity, or resistance to flow, is one such property. A unique property of liquids is surface tension. Many properties of liquid water and ice can be explained by the existence of strong hydrogen-bonding forces.

Most solids exist in a crystalline form. Crystals contain well-ordered and symmetrical arrangements of planar faces, which are reflected in a similarly ordered arrangement of atoms, molecules, or ions in a crystal lattice. Metals tend to form a structure that gives the most efficient packing of atoms. Alloys form from metals by substitution of atoms of one metal for atoms of another, or by placing other atoms into holes in a metal structure. Many ionic salts adopt structures like those of metals. Crystalline solids can be classified as metallic, ionic, molecular, or network solids. The structures and the types of forces in a solid explain its properties.

KEY RELATIONSHIPS

Relationship	Equation
The heat required to raise the temperature of a substance is directly related to the mass, specific heat, and temperature change.	$q = m \times C \times \Delta T$
The heat required for a change of state of a substance is the product of the number of moles and the molar heat of the change of state.	$q_{fusion} = n \times$ heat of fusion $q_{vaporization} = n \times$ heat of vaporization

KEY TERMS

alloy (10.4)

amorphous solid (10.4)

boiling point (10.1)

condensation (10.1)

crystal (10.4)

crystal lattice (10.4)

crystalline solid (10.4)

deposition (10.1)

dipole-dipole force (10.2)

equilibrium (10.1)

evaporation (10.1)

fluid (10.3)

freezing point (10.1)

hydrogen bond (10.2)

induced dipole (10.2)

instantaneous dipole (10.2)

intermolecular force (10.2)

London dispersion force (10.2)

melting point (10.1)

meniscus (10.3)

molecular solid (10.4)

network solid (10.4)

normal boiling point (10.1)

normal freezing point (10.1)

sublimation (10.1)

surface tension (10.3)

vapor pressure (10.1)

viscosity (10.3)

QUESTIONS AND PROBLEMS

The following questions and problems, except for those in the *Additional Questions* section are paired. Questions in a pair focus on the same concept. Answers to the odd-numbered questions and problems are in Appendix D.

Matching Definitions with Key Terms

10.1   Match the key terms with the descriptions provided.

(a) the partial pressure of gas molecules above a liquid (or solid) when the two states are in equilibrium

(b) temperature at which the liquid and solid states of a substance are in equilibrium (identical to the freezing point)

(c) a state of balance between opposing processes

(d) process by which molecules pass from the liquid state to the gaseous state

(e) a temporary dipole formed by attraction of the electrons in a molecule or atom to a nearby dipole

(f) a mixture of a metal with one or more additional metallic or nonmetallic elements

(g) temperature at which the vapor pressure of a liquid equals the external pressure

(h) vaporization of a solid

(i) a solid made up of discrete molecules held together by intermolecular forces

(j) forces of attraction that occur between molecules

(k) the attractive forces that result from formation of temporary dipoles in molecules

(l) temperature at which the liquid and solid states of a substance are in equilibrium at a pressure of 1 atm

(m) a solid having a shape bounded by planes intersecting at characteristic angles

10.2   Match the key terms with the descriptions provided.
   (a) resistance to flow
   (b) especially strong dipole-dipole forces between polar molecules that contain hydrogen attached to a highly electronegative element
   (c) transition of gas-state molecules to the solid state
   (d) substance that can flow (a gas or a liquid)
   (e) the geometric arrangement of points occupied by particles in a crystalline solid
   (f) curved upper surface of a liquid column
   (g) temperature at which the liquid and solid states of a substance are in equilibrium (identical to the melting point)
   (h) a solid in which all atoms are held together by covalent bonds to make a giant molecule
   (i) a substance in which the particles are arranged in a regular, repeating geometric structure
   (j) temperature at which the vapor pressure of a liquid equals the standard atmospheric pressure of 1 atm
   (k) a temporary dipole formed by the movement of electrons within a molecule
   (l) work necessary to expand the surface of a liquid; the property of a liquid surface that causes it to behave like a stretched membrane
   (m) transition of gas-state molecules to the liquid state

Changes of State

10.3   Compare and contrast liquids and gases at a molecular level.

10.4   Compare and contrast liquids and solids at a molecular level.

10.5   Why does a liquid take the shape of its container, but a solid does not?

10.6   Why does a gas fill its container, but a liquid does not?

10.7   Consider the following picture of a substance. What is its physical state? Explain how you reached your conclusion.

10.8   Consider the following picture of a substance in a container. What is its physical state? Explain how you reached your conclusion.

10.9   Consider the following atomic-level diagram of a substance. What is its physical state? Explain how you reached your conclusion.

10.10   Consider the following atomic-level diagram of a substance. What is its physical state? Explain how you reached your conclusion.

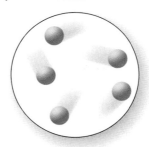

(10.11) Identify each of the phases of water shown in the following image. What terms are used to identify each of the phase changes indicated by the arrows?

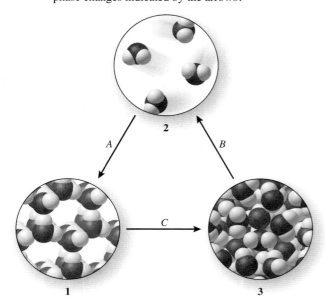

10.12 Identify each of the phases of water shown in the following image. What terms are used to identify each of the phase changes indicated by the arrows?

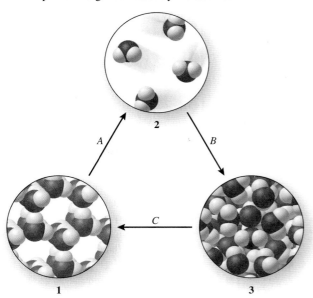

10.13 How is the vapor pressure of a liquid related to its boiling point?

10.14 Why do liquids boil at lower temperatures in the mountains?

10.15 Is evaporation exothermic or endothermic? Describe how energy is involved in this process.

10.16 Is condensation exothermic or endothermic? Describe how energy is involved in this process.

10.17 Draw the phase in the empty circle to illustrate the process of condensation.

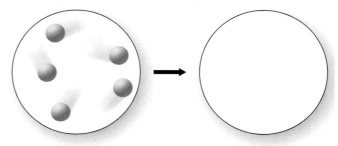

10.18 Draw the phase in the empty circle to illustrate the process of deposition.

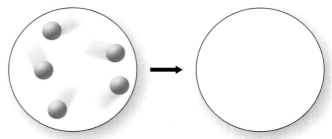

(10.19) Identify the phase change shown in the molecular diagram and indicate whether it is exothermic or endothermic.

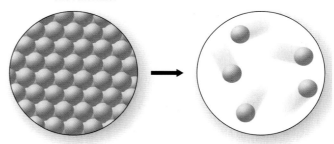

10.20 Identify the phase change shown in the molecular diagram and indicate whether it is exothermic or endothermic.

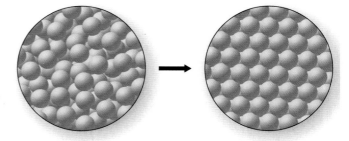

10.21 Explain why a gas converts to a liquid when it is cooled sufficiently.

10.22 Explain why a liquid converts to a solid when it is cooled sufficiently.

10.23 Why can the evaporation of water be used to cool a house?

10.24   Skin swabbed with alcohol is cooled below room temperature, even though the alcohol is at room temperature. Explain this observation.

10.25   Why does pasta take longer to cook in boiling water at a mountaintop campsite than at the beach?

10.26   Would the boiling point of water be above or below 100°C in Death Valley, which is at an altitude of 86 m below sea level?

10.27   Provide a molecular explanation for the increase in vapor pressure with increasing temperature.

10.28   Explain the process of evaporation in terms of molecular behavior in the liquid state.

10.29   Gallium has a melting point of 30°C and a boiling point of 1983°C. What is the physical state of gallium at 100°C? at 15°C?

10.30   Hydrogen has a melting point of −259°C and a boiling point of −253°C. What is the physical state of hydrogen at −263°C? at −255°C?

10.31   The pictured change of state occurs at constant temperature, even though heat is being removed. Where would this change of state occur on the cooling curve?

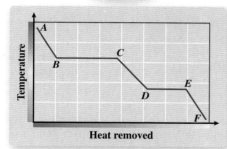

10.32   The pictured change of state occurs at constant temperature, even though heat is being removed. Where would this change of state occur on the cooling curve?

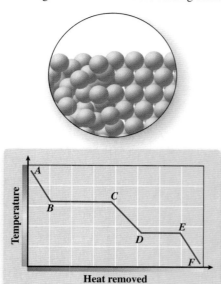

10.33   Calculate the amount of heat required for the evaporation of 105.0 g of water at 100.0°C.

10.34   Calculate the amount of heat required when 1.85 mol of ice is melted to liquid water at 0.0°C.

10.35   Calculate the amount of heat required when 15.0 g of water at 52.5°C is converted to steam at 238.2°C.

10.36   What is the amount of heat required to convert 105 g of ice at −15.2°C to liquid water at 35.6°C?

Intermolecular Forces

10.37   What is meant by the term *intermolecular force*?

10.38   List various types of attractive forces between atoms and molecules.

10.39   What would be the predominant state of matter if there were no intermolecular forces?

10.40   Which physical properties of a substance are influenced by the strength of its intermolecular forces?

10.41   Which of the substances, Ar, Hg, or I_2 is most likely to be a gas at room temperature and atmospheric pressure?

10.42   Which of the substances, Cl_2, Br_2, or I_2 is most likely to be a gas at room temperature and atmospheric pressure?

10.43   Which of the following molecules can experience dipole-dipole forces?
(a) CO_2 (b) NO   (c) NF_3 (d) CH_3Cl

10.44   Which of the following molecules can experience dipole-dipole forces?
(a) SF_2 (b) Cl_2 (c) NO_2 (d) SiH_4

10.45   Should a hydrogen bond be called a "bond"? How is it similar to or different from a covalent bond?

10.46 What molecular properties of ammonia and water cause similarities in their physical properties?

10.47 Indicate whether each of the following molecules has hydrogen bonding in the pure liquid state.

A   **B**   **C**

10.48 Indicate whether each of the following molecules has hydrogen bonding in the pure liquid state.

A   **B**   **C**

10.49 Which of the following substances can participate in hydrogen bonding?
(a) H_2O (b) NH_3 (c) H_2Se

10.50 Which of the following substances can participate in hydrogen bonding?
(a) PH_3 (b) HCO_2H (c) HCl

10.51 Which of the following substances can form hydrogen bonds with water molecules?
(a) SCl_2 (b) HF (c) BeH_2

10.52 Which of the following substances can form hydrogen bonds with water molecules?
(a) CH_4 (b) CH_3OH (c) $CH_3CH_2NH_2$

10.53 Why do different liquids have different vapor pressures?

10.54 Why do different solids have different melting points?

10.55 Why are intermolecular forces greater for polar molecules than for nonpolar molecules that are about the same size?

10.56 Why are intermolecular forces greater for large molecules than for small molecules?

10.57 Under what circumstances might nonpolar molecules have intermolecular forces greater than those of polar molecules?

10.58 The intermolecular forces in water (H_2O) are greater than those in dihydrogen sulfide (H_2S). Explain this observation.

10.59 List all the intermolecular forces in each of the following substances.
(a) C_6H_6 (b) NH_3 (c) CS_2 (d) $CHCl_3$

10.60 List all the intermolecular forces in each of the following substances.
(a) CH_3NH_2 (b) SO_2 (c) CO_2 (d) NO_2

10.61 Describe the interactions that must be overcome to melt or boil the following substances.
(a) krypton   (b) carbon monoxide
(c) methane   (d) ammonia

10.62 Describe the interactions that must be overcome to melt or boil the following substances.
(a) lithium bromide   (b) carbon dioxide
(c) sulfur dioxide   (d) carbon tetrachloride

10.63 Arrange the following substances in order of increasing boiling point: CH_3CH_2OH, CH_3OH, $CH_3CH_2CH_2OH$.

10.64 Arrange the following substances in order of increasing boiling point: AsH_3, NH_3, PH_3.

10.65 Which of the following molecules will have the stronger intermolecular forces?

CO_2 NO_2

10.66 Which of the following molecules will have the stronger intermolecular forces?

C_2H_6 CH_4

10.67 The boiling point of He is 4 K, while that of H_2 is 20 K. Explain why hydrogen boils at a higher temperature than helium.

10.68 The melting point of CF_4 is –184°C, while that of CCl_4 is –23°C. Explain the difference in melting point.

10.69 The boiling point of F_2 is –188°C, while that of HCl is –85°C. Explain the difference in boiling points.

10.70 The boiling point of N_2O is –88°C, while that of O_3 is –112°C. Explain the difference in boiling points.

10.71 Which substance in each pair has the higher equilibrium vapor pressure at a given temperature?
(a) C_2H_5OH or CH_3OH (b) CH_3OH or H_2O
(c) NH_3 or H_2O

10.72 Which substance in each pair has the higher equilibrium vapor pressure at a given temperature?
(a) Cl_2 or Br_2 (b) PH_3 or AsH_3
(c) CH_4 or NH_3

10.73 Arrange the following substances in order of increasing strength of intermolecular forces: H_2O, He, I_2, N_2.

10.74 Arrange the following substances in order of increasing strength of intermolecular forces: NH_3, Ne, O_2, H_2.

10.75 Explain why acetone boils at 56.2°C, while H_2O boils at 100°C. The structure of acetone is

10.76 Explain why the equilibrium vapor pressure of H_2O is 72 torr at 45°C, while that of CS_2 is 760 torr.

Properties of Liquids

10.77 Which properties of liquids are different from those of solids?

10.78 Which properties of liquids are different from those of gases?

10.79 What is viscosity? Discuss properties of liquids that are controlled by viscosity.

10.80 Why does the viscosity of a liquid decrease as the temperature increases?

10.81 What is surface tension? Discuss properties of liquids that are controlled by surface tension.

10.82 Why does the surface tension of water decrease when soap is added to it?

10.83   Why do liquids tend to assume spherical shapes?

10.84   Why do many liquids rise in a tube?

10.85   How would living things be affected if ice were more dense than liquid water?

10.86   In what ways is water different from other liquids? Why is it different?

Properties of Solids

10.87   What is a crystal? Draw a picture of a crystalline substance.

10.88   What is a crystal lattice? Draw a picture of what a crystal lattice might look like.

10.89   Why would we want to know the crystal structure of a solid?

10.90   Why do crystalline solids typically have symmetrical planar faces?

10.91   What is a common structure of metals?

10.92   What tendency exhibited by metal atoms is also found in macroscopic nature?

10.93   Some solids, such as tar, do flow. What is different about these solids?

10.94   Distinguish between amorphous and crystalline solids.

10.95   Why is it useful to classify solids?

10.96   What are the four major types of crystalline solids?

10.97   What type of crystalline solid has a high melting point, conducts electricity when molten but not as a solid, and is brittle?

10.98   What type of crystalline solid conducts electricity as a solid and when molten, is ductile, and has a lustrous appearance?

10.99   Explain why sodium is malleable, but sodium chloride is brittle.

10.100   Explain why solid H_2O is hard, but solid I_2 is soft.

10.101   Compare the structures of metals and the structures of ionic salts.

10.102   How are the structures of ionic crystals related to those of metallic crystals? How are they different?

10.103   Distinguish between the structures of molecular and ionic solids. Give examples of each and contrast their properties.

10.104   Distinguish between the structures of metallic and ionic solids. Give examples of each and contrast their properties.

10.105   What forces hold the following substances together in the solid state?
(a) diamond   (b) CO_2
(c) H_2O (d) SiO_2

10.106   What forces hold the following substances together in the solid state?
(a) graphite   (b) CH_4
(c) NH_3 (d) Cu

10.107   Predict what type of solid (metallic, ionic, molecular, or network) each of the following substances should form.
(a) $CaCl_2$ (b) N_2 (c) Ti
(d) BN   (e) SO_3

10.108   Predict what type of solid (metallic, ionic, molecular, or network) each of the following substances should form.
(a) Cu   (b) P_4 (c) SiO_2
(d) NH_3 (e) CuO

10.109   Indicate whether each of the following solids is a molecular solid.
(a) glass (SiO_2)   (b) ice (H_2O)
(c) $CaCl_2$ (d) Fe
(e) dry ice (CO_2)

10.110   Indicate whether each of the following solids is a network solid.
(a) glass (SiO_2)   (b) ice (H_2O)
(c) $CaCl_2$ (d) Fe
(e) dry ice (CO_2)

10.111   Identify the type of solid shown by the following image.

10.112   Identify the type of solid shown by the following image.

10.113   Consider the melting points of the following solids. Which of these solids is a molecular solid?
SiO_2 1610°C
SiF_4 −90°C
SiC   2700°C

10.114   Consider the melting points of the following solids. Which of these solids is a network solid?
CO_2 −57°C
$GeBr_4$ 26°C
BN   3000°C

Additional Questions

10.115   Explain why wet laundry hung out in freezing weather will eventually dry.

10.116   Arrange the following substances in order of increasing boiling point: Br_2, Cl_2, I_2.

10.117   Match the compounds in the first list with the boiling points in the second list:
CH_4 0°C
C_2H_6 −42°C
C_3H_8 −88°C
C_4H_{10} −162°C

Explain the trend that you have described.

10.118 When a vacuum pump is connected to a container of water, the water begins to boil. Explain this phenomenon.

10.119 Why does liquid water spread out when poured on a flat glass plate, while liquid mercury forms spherical drops?

10.120 Arrange the following substances in order of increasing melting point: CH_4, GeH_4, SiH_4, SnH_4.

10.121 Indicate whether or not each of the following solids is an ionic solid.
   (a) glass (SiO_2)   (b) ice (H_2O)
   (c) $CaCl_2$ (d) Fe
   (e) dry ice (CO_2)

10.122 Explain how an increase in the strength of intermolecular forces affects each of the following properties of a substance.
   (a) vapor pressure of a solid
   (b) hardness of a molecular solid
   (c) boiling point of a liquid
   (d) viscosity of a liquid
   (e) melting point of a molecular solid

10.123 Which substance in each pair has the higher vapor pressure at the same temperature?
   (a) Cl_2 or I_2 (b) CH_3OH or CH_3SH
   (c) PF_3 or NF_3

10.124 Explain why diamond is a good insulator, while graphite is a good conductor of electricity.

10.125 Compare the following substances with respect to electrical conductivity, hardness, melting point, and vapor pressure at room temperature: NaCl, SiC, $SiCl_4$, Fe.

10.126 Which characteristic of molecules changes in magnitude from one state to another and is largely responsible for whether a substance is a gas, a liquid, or a solid?

10.127 List the names for all possible phase changes.

10.128 Indicate which phase changes are exothermic and which are endothermic.

10.129 Draw a picture in the empty circle to illustrate what happens when solid chlorine (Cl_2) melts.

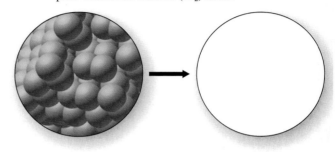

Megan and Derek participate in numerous athletic activities on their college campus. They run or cycle every day, play several team sports, and work out on the free weights at the gym. To maintain their physical fitness they carefully select the foods they consume, and they supplement their diets with protein shakes and vitamins. They also drink sports drinks when they exercise to help replenish lost electrolytes. In their chemistry course they are studying solubility: how well one substance dissolves in another. It doesn't take them long to recognize the importance of this topic to the transport systems that carry nutrients, drugs, and other materials to every cell in the human body.

Anything we ingest is transported through the body in specific ways that depend on its solubility. For example, a drug taken orally must withstand the acidic conditions of the stomach. Some drugs cannot, so they must be administered by injection. A drug taken by injection into fatty tissue must be able to work its way into the circulatory system. If it cannot, it may need to be administered directly into the bloodstream via injection or intravenous (IV) delivery. Because the primary component of IV delivery is water, drugs administered in this way must be water soluble (Figure 11.1).

For a chemistry assignment, Megan and Derek studied the active ingredients in over-the-counter medications used to treat acid indigestion (Figure 11.2). Because antacids neutralize stomach acid, they assumed that all antacids would contain a base. An acid provides H^+ ions in solution and a base often produces OH^- ions. They combine to form water. Megan and Derek found that the bases most commonly used in a chemistry lab, sodium hydroxide and potassium hydroxide, were not present in antacid medications. Instead, they discovered that magnesium hydroxide and calcium hydroxide were commonly used. They wondered why NaOH and KOH were not used, but $Mg(OH)_2$ and $Ca(OH)_2$ were. Derek looked up the solubility rules for hydroxides. He discovered that $Mg(OH)_2$ and $Ca(OH)_2$ are insoluble in water, but NaOH and KOH are soluble. Megan wondered if solubility might explain why some bases are used in antacids and others are not.

Consider the antacid Milk of Magnesia™. Magnesium hydroxide, the active ingredient in Milk of Magnesia, is insoluble in water. It travels down the esophagus and into the stomach without causing damage along the way. Because $Mg(OH)_2$ is insoluble in water, it does not provide very much hydroxide ion until it is in an acidic environment, as in the stomach. Once there, $Mg(OH)_2$ can safely neutralize excess acid. NaOH would not perform in this way. It's very soluble in water and can, therefore, dissolve in saliva in the mouth. It would damage the lining of the mouth and the esophagus before ever getting to the site of action, the stomach. Milk of Magnesia is available as a liquid, but the $Mg(OH)_2$ is not dissolved. It is suspended as tiny particles in the liquid carrier. Many other antacids are available as chewable tablets. When chewed, they form a suspension in the mouth similar in consistency to Milk of Magnesia.

Some drugs can be delivered orally as plain, coated, or encapsulated tablets—the form depending in part on how the drug gets into the circulatory system from the digestive tract and where delivery is required. Most drugs and nutrients are absorbed into the bloodstream through the walls of the small intestine. Encapsulating drugs allows them to get to the small intestine without reacting with the contents of the stomach. Once in the small intestine, capsules can work in several ways. An enzyme present in the intestine can create pores through which drugs simply pass out of the capsule. In other cases enzymes actively lock onto and ferry drugs out of a capsule. A capsule can also work by allowing water to pass into it creating a solution of the drug. This solution can then flow out.

Because our bodies are 70% water, anything we consume interacts with water in some way. As Megan and Derek discovered in their study of antacids, some substances easily dissolve in water and others do not. Those that don't may remain in the body for longer periods of time. For example, vitamin C and the various B vitamins are water soluble. A person must consume them on a regular basis to maintain

Make a list of all the foods and drinks you consume in a week. Which of them are solutions?

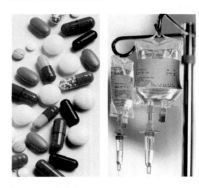

Figure 11.1
Some drugs are delivered orally in the form of tablets or capsules. They are often coated or encapsulated to prevent digestion in the stomach and to allow for time-release of the active ingredient. Other drugs are delivered by syringe or from an IV bag.

Figure 11.2
A variety of antacids is available for people who suffer from indigestion. The active ingredient in some is a base such as $Mg(OH)_2$ that gets to the stomach without injuring the tissues of the mouth and esophagus.

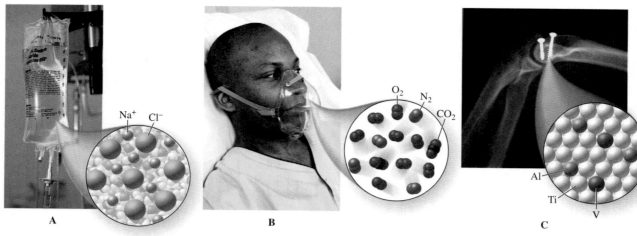

Figure 11.3
(A) A saline IV bag contains a solution of NaCl in water. (B) Gases can also form solutions. Hospital patients can wear masks through which higher oxygen concentrations are provided. Carbon dioxide is also present in the solution because the patient exhales into the mask. (C) Solutions can also be solids. Screws used in orthopedic surgery are often made of an alloy (metal solution) of titanium (90%), aluminum (6%), and vanadium (4%).

good physical health because they are readily excreted in the urine. Other vitamins—including A, D, E, and K—are fat soluble; they are stored in fatty tissue for long periods. These vitamins should not be consumed in large quantities, because they can build up to toxic levels in body tissues.

Although we usually think of solutions as liquids, solids and gases can also form solutions. In the medical field, solid solutions of metals—commonly called alloys—are important in dentistry (braces), optometry (glasses), and surgery (metal implants). The air we breathe is a gaseous solution, mostly of nitrogen and oxygen. The oxygen masks used in hospitals (Figure 11.3) provide higher concentrations of oxygen to patients with respiratory problems. The oxygen tanks that are often seen at sporting events enrich the supply of oxygen that gets to the lungs and help athletes recover more quickly from strenuous activity.

This chapter examines solutions: their composition, how they form, and how dissolved substances affect the properties of solvents. Consider the following questions.

Questions for Consideration

11.1 What is a solution?
11.2 How do substances dissolve to make a solution?
11.3 What determines whether one substance will dissolve in another?
11.4 How can we measure the concentration of a solution?
11.5 In a chemical reaction how can we relate the amount of one reactant or product in solution to the amount of another?
11.6 What properties of solutions depend only on the concentration of dissolved particles, not on the identity of the solute?

Sometimes it is difficult to distinguish between solvent and solute. For example, ethanol (C_2H_5OH) and water can dissolve in each other in all proportions, so the relative amounts do not differentiate the solvent from the solute. In such cases, water is considered the solvent, even when the other substance is present in greater quantity.

11.1   THE COMPOSITION OF SOLUTIONS

Recall from Chapter 1 that a **solution** is a homogeneous mixture with uniform composition throughout. Although solutions can be solids, liquids, or gases, in this chapter we'll focus on solutions made of a substance dissolved in a liquid. The substance being dissolved is the **solute** (usually present in the lesser amount), and the substance doing the dissolving is the **solvent** (usually present in the greater amount). A solution in which the solvent is water is an **aqueous solution.** Figure 11.4 shows the preparation of an aqueous solution of sodium chloride.

Electrolytes

Athletes like Megan and Derek often consume sports drinks to replenish water and electrolytes lost during physical exertion. In the human body, electrolytes serve several purposes. In some cases they participate in chemical reactions. In other cases chemical changes do not occur, but the electrolyte is essential nonetheless. For example, electrolytes help maintain water balance between the interior of cells and the fluid that surrounds them. The electrical properties of electrolytes are also important in nerve function. Imbalances in electrolytes can cause a variety of conditions ranging from muscle cramps to cardiovascular failure.

Electrolyte solutions contain a solute that dissociates or ionizes in a solvent, producing ions. The presence of ions allows electric current to pass through the solution. Examples of electrolytes are soluble ionic compounds such as NaCl, strong acids, and strong bases (Figure 11.5). Acids provide H^+ ions in aqueous solution. Bases neutralize acids and often provide OH^- ions in aqueous solution. Examples of the dissociation or ionization of salts, acids, and bases in water can be described using the following equations:

Figure 11.4
In a sodium chloride solution, the sodium chloride is considered the solute because it dissolves and is present in a lesser amount. Water is the solvent because it does the dissolving and is present in a greater amount.

$$NaCl(s) \xrightarrow{\text{H}_2\text{O}(l)} Na^+(aq) + Cl^-(aq)$$

$$HCl(g) \xrightarrow{\text{H}_2\text{O}(l)} H^+(aq) + Cl^-(aq)$$

$$NaOH(s) \xrightarrow{\text{H}_2\text{O}(l)} Na^+(aq) + OH^-(aq)$$

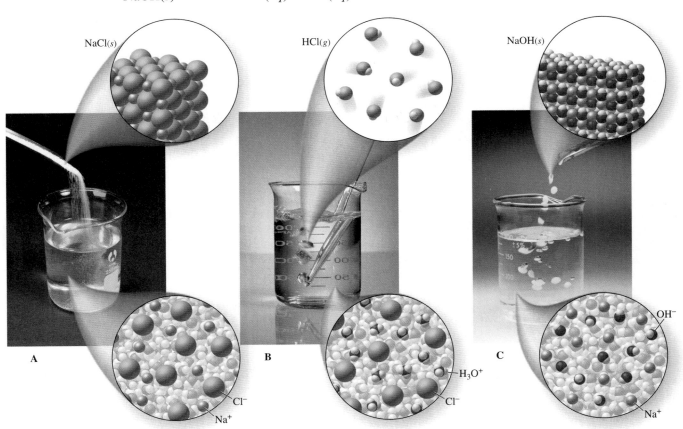

Figure 11.5
Soluble ionic compounds like NaCl (A), strong acids like HCl (B), and strong bases like NaOH (C) dissociate completely in water to produce ions. They are examples of strong electrolytes.

Figure 11.6
Milk of Magnesia™ is a suspension of magnesium hydroxide in water. Because $Mg(OH)_2$ is insoluble, it does not dissociate into ions in water.

$Mg(OH)_2$ crystals

TABLE 11.1	Rules Used to Predict the Solubility of Ionic Salts
Ions	**Rule**
Na^+, K^+, NH_4^+	Most salts of sodium, potassium, and ammonium ions are soluble.
NO_3^-	All nitrates are soluble.
SO_4^{2-}	Most sulfates are soluble. Exceptions: $BaSO_4$, $SrSO_4$, $PbSO_4$, $CaSO_4$, Hg_2SO_4, and Ag_2SO_4.
Cl^-, Br^-, I^-	Most chlorides, bromides, and iodides are soluble. Exceptions: AgX, Hg_2X_2, PbX_2, and HgI_2 ($X = Cl$, Br, or I).
Ag^+	Silver salts, except $AgNO_3$, are insoluble.
O^{2-}, OH^-	Oxides and hydroxides are insoluble. Exceptions: $NaOH$, KOH, $Ba(OH)_2$, and $Ca(OH)_2$ (somewhat soluble).
S^{2-}	Sulfides are insoluble. Exceptions: Salts of Na^+, K^+, NH_4^+, and the alkaline earth metal ions.
CrO_4^{2-}	Most chromates are insoluble. Exceptions: Salts of Na^+, K^+, NH_4^+, Mg^{2+}, Ca^{2+}, Al^{3+}, and Ni^{2+}.
CO_3^{2-}, PO_4^{3-}, SO_3^{2-}, SiO_3^{2-}	Most carbonates, phosphates, sulfites, and silicates are insoluble. Exceptions: Salts of Na^+, K^+, and NH_4^+.

Recall from Chapter 3 that the term *dissociation* describes the process of dissolving for compounds that consist of ions. When a compound that is not ionic provides ions when dissolved in water, the process is called *ionization*.

How can we determine if an ionic compound will dissociate in water to give ions? How readily or completely a solute dissolves in a solvent is its *solubility*. Unfortunately the solubilities for ionic compounds do not follow a predictable pattern. They must be recalled from experimental data. The solubility rules derived from data are summarized in Table 11.1. In general a soluble ionic compound dissociates into ions when dissolved in water. Insoluble compounds remain mostly undissociated in their ionic, crystalline structure. One example is the $Mg(OH)_2$ in Milk of Magnesia (Figure 11.6). Table 11.1 shows that most hydroxides are insoluble, with some exceptions. Since $Mg(OH)_2$ is not one of the exceptions, we can assume it is insoluble—thus it's safe to use as an antacid.

In sports drinks, sodium chloride helps reestablish electrolyte balance. The drinks also contain simple sugars such as fructose or glucose. Solutions of sodium chloride and glucose vary in their ability to conduct electricity because sodium chloride dissociates into ions and glucose does not. Recall from Chapter 3 that we can determine whether a compound has dissociated into ions in solution if the solution conducts electricity. A solute that dissociates completely into ions in aqueous solution is a *strong electrolyte* (Figure 11.7). A *weak electrolyte* dissociates only partially into ions. A *nonelectrolyte* does not dissociate into ions.

Figure 11.8
When soluble nonelectrolytes such as hydrogen peroxide (H_2O_2) dissolve in water, they retain their molecular structure.

Figure 11.7
Soluble ionic compounds, strong acids, and strong bases dissociate completely into ions. They are considered strong electrolytes because their aqueous solutions are good conductors of electricity.

Nonelectrolyte substances retain their molecular structure when dissolved. Glucose and fructose remain as intact molecules when dissolved in water. Both have the same formula, $C_6H_{12}O_6$, but they differ in how their atoms are arranged. They each retain their molecular structure when dissolved in water, in the same way as does the simpler molecular compound, hydrogen peroxide, H_2O_2 (Figure 11.8). A 3% solution of this compound is often used to sterilize cuts and scrapes.

Another nonelectrolyte compound, ethanol (CH_3CH_2OH), behaves the same way. It retains its molecular structure in aqueous solutions. We can describe mixing of ethanol with water using the following equation:

$$CH_3CH_2OH(l) \xrightarrow{\ H_2O(l)\ } CH_3CH_2OH(aq)$$

Ethanol is completely **miscible** with water. That is, water and ethanol can be mixed in any proportion to give a homogenous solution. We'll talk more about miscibility in Section 11.4.

Weak electrolytes are usually weak acids and bases.

EXAMPLE 11.1   Composition of Solutions

What ions, atoms, or molecules are present in solution when the following substances mix with water? Write a balanced equation to represent the process of dissolving each substance in water.

(a) $HNO_3(l)$ (b) $KOH(s)$ (c) $C_{12}H_{22}O_{11}(s)$

Solution:

(a) Nitric acid (HNO_3) is a strong acid that ionizes completely in water to give $H^+(aq)$ and $NO_3^-(aq)$ ions. Water molecules are also present. The equation for dissolving HNO_3 is

$$HNO_3(l) \xrightarrow{H_2O(l)} H^+(aq) + NO_3^-(aq)$$

(b) From Table 11.1 we know that ionic compounds containing K^+ ions are soluble. Potassium hydroxide (KOH), a strong base, gives $K^+(aq)$ and $OH^-(aq)$ ions in water. The equation for dissolving KOH is

$$KOH(s) \xrightarrow{H_2O(l)} K^+(aq) + OH^-(aq)$$

(c) Sucrose ($C_{12}H_{22}O_{11}$) is a molecular compound that does not dissociate into ions when it dissolves. Intact molecules of $C_{12}H_{22}O_{11}$ exist in the solution. The equation for dissolving $C_{12}H_{22}O_{11}$ is

$$C_{12}H_{22}O_{11}(s) \xrightarrow{H_2O(l)} C_{12}H_{22}O_{11}(aq)$$

Practice Problem 11.1

What ions, atoms, or molecules are present in solution when the following substances are mixed with water? Write a balanced equation to represent the process of dissolving each substance in water.

(a) $ZnCl_2(s)$ (b) $CH_3OH(l)$ (c) $KNO_3(s)$

Further Practice: 11.9 and 11.10

Structure and Solubility

Substances differ widely in their solubilities or miscibilities. Oil and water don't mix (Figure 11.9), but ethanol and water mix in any proportion. An organic solvent can remove a grease stain on clothing, but water can't. Why do some substances dissolve in one solvent but not in another? A general rule of thumb is "like dissolves like." That is, when the bonding within a solvent is similar to the bonding in a solute, the solvent will dissolve the solute. Polar covalent solvents dissolve polar covalent or ionic solutes, while nonpolar covalent solvents dissolve nonpolar covalent solutes (Table 11.2).

Consider the vitamin structures shown in Figure 11.10. Except for the −OH group on the end, the vitamin A molecule consists of many carbon-carbon and carbon-hydrogen bonds. It is essentially nonpolar. Fat molecules are also nonpolar because they consist of similar carbon-carbon and carbon-hydrogen bonds (Figure 11.11).

TABLE 11.2

Like Dissolves Like

Solute	Solvent	
	Polar	Nonpolar
Ionic	Soluble	Insoluble
Polar	Soluble	Insoluble
Nonpolar	Insoluble	Soluble

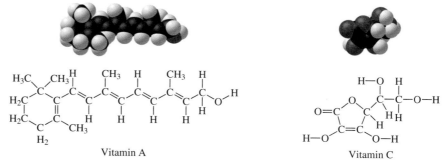

Vitamin A   Vitamin C

Figure 11.10
Vitamin A is a fat-soluble vitamin. It is a ring of carbon atoms attached to a long chain of additional carbon atoms. Although it has a polar −OH group at one end, the molecule is mostly nonpolar. Fat molecules are also mostly nonpolar. Notice that the vitamin C molecule lacks a long nonpolar carbon chain. It is water soluble because it has several −OH groups.

Figure 11.9
Oil and water do not mix.

Figure 11.11
Three long carbon chains make a
fat molecule largely nonpolar.

Since vitamin A and fat molecules are both nonpolar, vitamin A is stored in fatty tissues in the body. In vitamin C several —OH groups make the molecule polar. Since water is also polar, vitamin C is soluble in aqueous body fluids.

11.2   THE SOLUTION PROCESS

A typical sports drink contains carbohydrates, usually sucrose and other sugars; ionic compounds, like sodium chloride and potassium chloride; and fruit juice for flavor. How do these substances form a solution? Let's begin by examining the arrangement of particles in one of the solutes, NaCl, and in the solvent, water. Notice the orderly arrangement of particles in the solute and in the solvent before mixing, as shown in Figure 11.12A. Ionic bonds hold the ions of NaCl together. London dispersion forces and hydrogen bonds hold the molecules of pure water together. Hydrogen bonding is the stronger of the two intermolecular forces in water. After mixing (Figure 11.12C), when a solution has formed, Na^+ and Cl^- ions distribute uniformly through the solvent, water. The formation of the solution must, therefore, involve the destruction of the ionic bonds in NaCl and some of the intermolecular forces in water (Figure 11.12B). Within the solution, new forces of attraction form between the particles that make up the solute and the solvent molecules.

🌀 **Animation: Dissolving Table Salt**
Online Learning Center, Animations

Chemistry Animations Library, Dissolving Table Salt, Dissociation.rm.

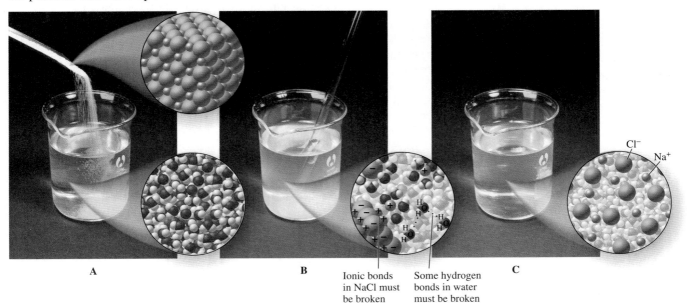

A   B   C

Ionic bonds   Some hydrogen
in NaCl must   bonds in water
be broken   must be broken

Figure 11.12
(A) In solid sodium chloride, ions are held together by ionic bonds. Water molecules are held together by intermolecular forces. (B) The bonds in sodium chloride and some of the hydrogen bonding in water must be broken to form the solution. (C) In solution, Na^+ and Cl^- ions distribute throughout.

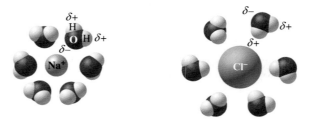

Figure 11.13

When a substance dissolves in water to produce ions, the ions are attracted to water molecules by ion-dipole forces. Notice how the partially negative oxygen atoms on the water molecules arrange themselves around the positive sodium ion. The partially positive hydrogen atoms on water are attracted to the negative chloride ion.

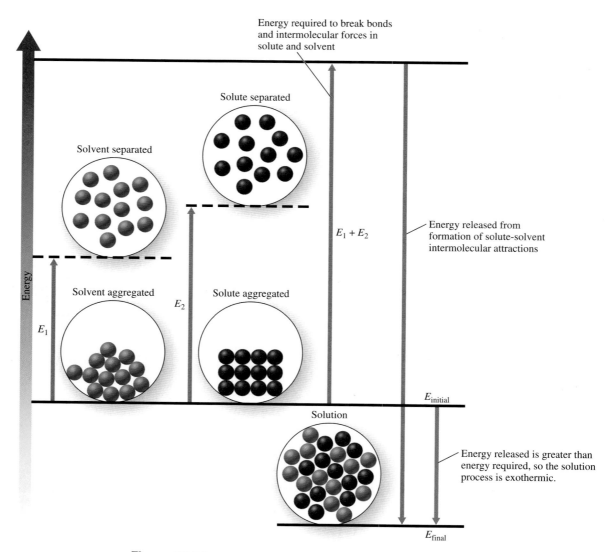

Figure 11.14

(A) The sum of the energies ($E_1 + E_2$) required to separate the solute and to separate the solvent equals the amount that must be absorbed to form a solution. When the new ion-dipole forces of attraction form, energy is released. When this amount is greater than the energy absorbed, the solution process is exothermic.

The force of attraction between an ion and a polar molecule is called an **ion-dipole force** (Figure 11.13). In a sodium chloride solution, several water molecules surround each ion creating multiple ion-dipole forces. The process by which a solute particle is surrounded by solvent water molecules in a solution is called *hydration.* (If the solvent is not water, the process is called *solvation.*) The hydrated ions are surrounded by water molecules oriented so that the partially negative polar ends of the water molecules are closest to the cations and the partially positive polar ends are closest to the anions.

When an ionic compound dissolves in water, the process can be summarized as follows:

• Ionic bonds in the solute break.
• Hydrogen bonds between water molecules break.
• Ion-dipole forces form between ions and water molecules.

As discussed in Chapter 6, changes in energy accompany all physical and chemical processes. In the formation of a solution, energy must be supplied to break the intermolecular forces and/or bonds in both the solute and solvent. It comes, at least in part, from the new intermolecular forces that form between solute and solvent particles. The difference between the energy required for separation of solute and solvent and the energy released upon formation of ion-dipole interactions is the *heat of solution.* The energy changes involved in the solution process are graphically summarized in Figure 11.14, using the formation of two different solutions as examples. When the solvation process provides more energy than is needed to separate the pure solute and solvent particles (Figure 11.14A), the heat of

A solution can contain more than one solute. Aqueous solutions in the human body such as blood, sweat, and tears, contain many solutes.

Animation: The Hydration Process
Online Learning Center, Animations

Chemistry Animation Library, The Hydration Process, 2Hydration.exe.

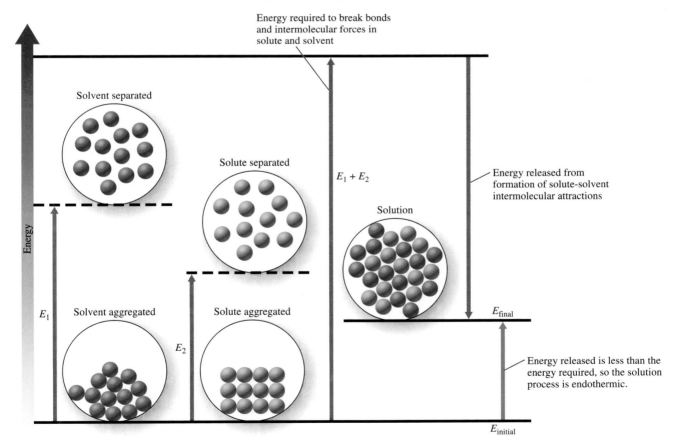

Figure 11.14 cont.
(B) Some solution processes are endothermic. The sum of the energies ($E_1 + E_2$) required to separate solute and to separate solvent particles is greater than the energy released from new solute-solvent interactions.

Athletes often use instant cold packs to reduce swelling and hot packs to relieve pain. Chemical cold packs consist of ammonium nitrate and water in separate compartments in the pouch. When the contents are allowed to mix, the bag feels cold. Is the process endothermic or exothermic? Chemical hot packs usually contain water separated from calcium chloride or magnesium sulfate. These pouches get warm when the seal between the water and the solute is broken and a solution forms. Is the process endothermic or exothermic?

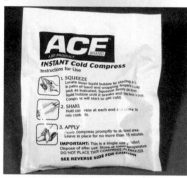

solution is negative, and the overall dissolving process is exothermic. In this case the container in which the solution is made may feel warm because the solution process releases energy.

When the separation processes require more energy than the solvation process releases (Figure 11.14B), the heat of solution is positive, and the overall dissolving process is endothermic. The container in which the solution is made may feel cool to the touch because the solution process absorbs energy from the surroundings.

When a solution forms between a nonpolar solute and a nonpolar solvent, different attractive forces between the molecules must be broken. For example, consider a solution in which iodine (I_2) dissolves in carbon tetrachloride (CCl_4) (Figure 11.15). London dispersion forces are broken between the solute molecules and between the solvent molecules. New London dispersion forces form between the solute and solvent molecules.

In addition to energy, another factor is important in the solution process. This factor, called **entropy**, is a measure of the tendency for matter to become disordered or random in its distribution. Matter spontaneously changes from a state of order to a state of randomness, unless energy changes oppose it. Solutions form because the makeup of a solution naturally tends toward disorganization. For example, consider NaCl and water molecules before they form a solution (Figure 11.16). The ions in NaCl are arranged in an organized pattern. To a lesser extent, so are the water molecules. When combined to make a solution, NaCl and water naturally go to a more random distribution of solute and solvent particles throughout the sample.

EXAMPLE 11.2   The Solution Process

Potassium nitrate (KNO_3) dissolves in water to form an aqueous solution. Describe the forces that must be broken for the solution to form.

Solution:

Potassium nitrate is an ionic compound, so ionic bonds between K^+ and NO_3^- ions must be broken for the solution to form. The solvent, H_2O, is polar covalent, with hydrogen bonding as the primary intermolecular force between water molecules. Some of these hydrogen bonds must be broken for the solution to form.

Figure 11.15
London dispersion forces hold together the molecules of solid iodine (I_2) and the liquid solvent, carbon tetrachloride (CCl_4). When a solution of I_2 and CCl_4 forms, new London dispersion forces form.

NaCl is very ordered and has low entropy

When NaCl dissolves entropy increases →

Ions are in random positions, with high entropy

Figure 11.16
Entropy, the tendency for matter to become more random, contributes to solution formation. Notice that solid NaCl and pure water are more ordered in their arrangements than a solution of NaCl.

Practice Problem 11.2

Iodine (I_2) dissolves in hexane (C_6H_{14}) to form a solution. Describe the forces that must be broken for the solution to form.

Further Practice: 11.21 and 11.22

11.3   FACTORS THAT AFFECT SOLUBILITY

When Megan and Derek studied the active ingredients in antacids, they identified bases such as magnesium hydroxide and calcium hydroxide, both of which are insoluble in water. Based on the rule "like dissolves like," they had expected $Mg(OH)_2$ and $Ca(OH)_2$ to be soluble. If ionic compounds dissolve in polar substances like water, why are $Mg(OH)_2$ and $Ca(OH)_2$ insoluble? To answer this question we have to consider the energy changes associated with the solution process. These bases are insoluble because the ionic bonds between the metal ions and hydroxide ions are very strong. The energy released by formation of ion-dipole attractions does not provide enough energy to break them. Other bases such as sodium hydroxide and potassium hydroxide are soluble because the strength of their ionic bonds is weaker.

The *CRC Handbook of Chemistry and Physics* includes tables of solubility information that are useful in the laboratory.

Structure

As we showed in Section 11.2, when an ionic compound dissolves, the ions become solvated and form ion-dipole interactions with the solvent (Figure 11.13). This occurs only when the solvent molecules are polar. If we reason that an ionic compound is the extreme of polarity, then the rule "like dissolves like" holds for dissolving an ionic compound in a polar solvent.

In order for a solid to be soluble in a given solvent, the energy released by the solvation process must compensate for all (or at least most) of the energy required

Figure 11.17
Carbon tetrachloride (CCl_4) and hexane (C_6H_{14}) are nonpolar liquids, so both have London dispersion forces as their only intermolecular attraction. Because they are alike in their intermolecular forces, they are miscible.

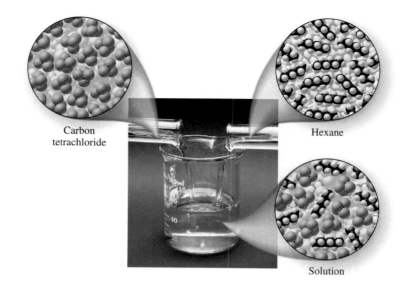

Carbon tetrachloride

Hexane

Solution

Figure 11.18
Oil, a nonpolar substance, does not mix with vinegar, an aqueous solution consisting of polar substances. The intermolecular forces between oil and water are not strong enough to overcome the intermolecular forces between oil molecules and the hydrogen bonding between water molecules. Oil and water are immiscible.

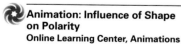
Animation: Influence of Shape on Polarity
Online Learning Center, Animations

Chemistry Animations Library, Influence of Shape on Polarity, 01.Polarity.rm.

For a review of intermolecular forces and their relative strengths, see Chapter 10.

to break up forces at work both within the solute and within the pure solvent. Thus ionic solids are generally soluble in polar solvents, where very strong interactions occur between the dipoles of the solvent and the dissolved ions. A nonpolar solvent cannot interact with dissolved ions strongly enough to overcome the ionic forces within the ionic solid crystal.

In contrast to ionic compounds that are held together by ionic bonds, covalent molecules are held together by intermolecular forces that are relatively weak (Figure 11.17). The solubility (or miscibility) of liquids in other liquids depends largely on the polarity of the solute and solvent molecules. Polar liquids dissolve in other polar liquids because the intermolecular forces formed between solute and solvent molecules are strong enough to overcome the strong solute-solute and solvent-solvent dipole-dipole interactions. Nonpolar liquids do not dissolve in polar liquids—at least not to any great extent—because the intermolecular forces between solvent and solute are not strong enough to overcome the solute-solute and solvent-solvent London dispersion forces (Figure 11.18). Nonpolar liquids dissolve in one another because the forces between the molecules in the pure liquids are weak and also because of the tendency for matter to become more random.

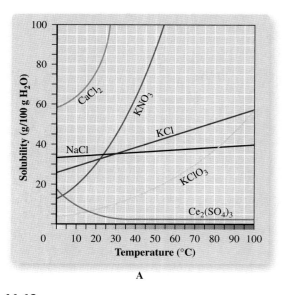

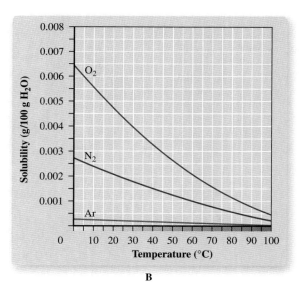

Figure 11.19
(A) In general, the solubility of ionic solids in water increases with increasing temperature.
(B) Gas solubility, in contrast, decreases.

Temperature

Temperature affects the solubility of most substances in water. For example, if you like your tea sweet, you can easily stir more sugar into a cup of hot tea than a frosty glass of iced tea. That's because sugar is more soluble in hot water than in cold. The solubility of most solids increases in water as the temperature of the solution increases. The effect of temperature on solubility in water is shown for some ionic compounds in Figure 11.19. Notice that most ionic compounds are more soluble at higher temperatures, but gases are less soluble as temperature increases.

Gases such as N_2 and O_2 are only slightly soluble in water because they are nonpolar. They become even less soluble as temperature increases. An increase in temperature increases the kinetic energy of the molecules. The increased kinetic energy breaks the intermolecular forces between the solute gas particles and the solvent molecules, allowing dissolved gas to escape more readily from the solution into the gas phase. The flat taste of a carbonated soda that has been left open on the kitchen counter is due in part to the escape of the dissolved CO_2. Sodas do not go flat as quickly if kept in a refrigerator.

An important consequence of reduced gas solubility at higher temperatures occurs in cases of thermal pollution of water. It happens around power plants cooled by a river or stream when warm water is returned to its source after it has been used by the plant for cooling. When the water temperature of the river or stream increases, the solubility of oxygen in the water decreases, threatening the survival of fish and certain other aquatic organisms. Algae blooms are often observed under low-oxygen conditions because algae thrive in this environment.

Pressure

The solubility of gases (but not of liquids or solids) is strongly affected by pressure. The solubility of a gas in a liquid is directly proportional to the pressure of the gas above the liquid. This makes sense in terms of a dynamic equilibrium. An increase in pressure increases the number of gas particles immediately above the solution, increasing the chance that some will dissolve. Figure 11.20 illustrates this principle.

You observe the effect of pressure on solubility every time you open a bottle or can of a carbonated beverage. These beverages are bottled under a pressure of CO_2

Henry's law states that the solubility of a gas in a liquid is directly proportional to the partial pressure of the gas above the liquid.

Figure 11.20
An equilibrium exists between the solute molecules in the gas phase and in the solution. An increase in pressure changes this equilibrium, resulting in increased gas solubility.

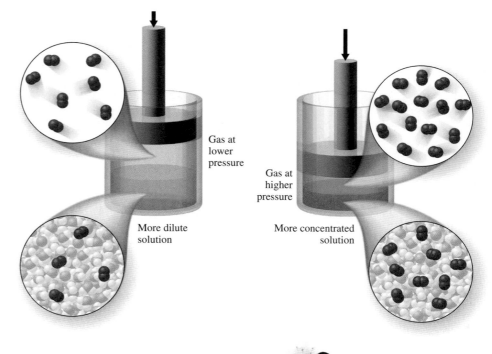

Gas at lower pressure

More dilute solution

Gas at higher pressure

More concentrated solution

Figure 11.21
The effect of pressure on gas solubility can be observed when a soda bottle is opened.

CO_2 under high pressure

Lots of CO_2 dissolved in soda

Pressure released

CO_2 bubbles out of solution

greater than atmospheric pressure. When you remove the bottle cap, the pressure of CO_2 decreases. You can hear the excess gas leave the container. Some of the CO_2 that had been dissolved in the solution escapes and forms gas bubbles (Figure 11.21).

Pressure also affects the solubility of gases in the bloodstream. Hospital patients who suffer from respiratory problems are given higher concentrations of oxygen, often through a mask (Figure 11.3B). Within the mask, the oxygen partial pressure is higher than it is in air. With a higher oxygen pressure, more oxygen dissolves in the patient's bloodstream with each breath, so the patient does not have to breathe as often to maintain an adequate oxygen supply.

Scuba divers must consider the effect of gas solubility at increased pressures under water. Recreational divers typically breathe ordinary air that has been compressed into a tank. It is about 78% nitrogen, which the body can handle under normal conditions of surface atmospheric pressure. As a diver descends, however, the increase in pressure causes more nitrogen to dissolve in the bloodstream and body tissues. If the diver surfaces too rapidly—causing a rapid decrease in pressure—the nitrogen comes out of solution as bubbles in tissues and in the bloodstream. These bubbles cause the neurological and circulatory symptoms of the "bends." Gas mixtures of helium and oxygen (heliox), which are sometimes used by military and professional divers, do not cause the bends, because helium is not as soluble in the blood and tissues as nitrogen.

The symptoms of the bends were first discovered in the 1800s among bridge workers laying underwater foundations at low depths. Their joints were sore, and they walked bent over. These symptoms were later observed in divers who came up too quickly from dives to great depths. The water pressure lessens by 1 atm for every 10 m (33 ft) the diver ascends.

11.4   MEASURING CONCENTRATIONS OF SOLUTIONS

All the properties of solutions we have discussed so far are affected by **concentration,** the relative amounts of the solute and solvent that make up the solution. In Chapter 4 we described solutions as *dilute* and *concentrated.* These terms compare the relative amounts of solute in solutions, as when we describe one solution as more concentrated or more dilute than another. Another set of qualitative terms describes the degree of solubility of solids in a given solvent. A solid may be called *insoluble, slightly soluble, soluble,* or *very soluble.* All these terms are relative. Strictly speaking, there are no insoluble substances; even glass will dissolve in water to a slight extent. However, for practical purposes, solids with solubilities less than about 0.1 g per 100 g of solvent are considered insoluble, while solutes with solubilities greater than about 2 g per 100 g of solvent are considered very soluble. For example, NaCl is considered very soluble because its solubility in water is 35.9 g per 100 g of water at 25°C.

Quantitatively, **solubility** is defined as a ratio that identifies the maximum amount of a solute that will dissolve in a particular solvent to form a stable solution under specified conditions. Solubility is a measure of maximum concentration. If too much solute is present when maximum concentration is reached, the excess remains undissolved. For example, the solubility of lead(II) iodide in water is 0.1 g PbI_2 per 100 g H_2O at 25°C. Thus if 1 g of PbI_2 is added to 100 g of water, only about one-tenth of it dissolves; the rest remains as undissolved solid. An equilibrium is established between the Pb^{2+} and I^- ions in solution and the PbI_2, which stays in the solid phase. Such a solution is a **saturated solution.** The establishment of equilibrium to give a saturated solution is depicted in Figure 11.22. Once the solution is saturated, undissolved solute continues to dissolve, but at the same time, previously dissolved solute is deposited from solution. The two processes occur at exactly the same rate, so the amount of dissolved solute stays the same. This is a dynamic equilibrium, like the evaporation of a liquid in a closed container discussed in Chapter 10.

A solution that contains less than the maximum possible amount of solute is called an **unsaturated solution.** One that contains more than the expected maximum is a **supersaturated solution.** As you might expect, supersaturation is generally a very unstable situation. When a supersaturated solution is disturbed by agitation or when a small "seed" crystal of solute is added, excess solute precipitates (Figure 11.23). A supersaturated solution is usually obtained by preparing a saturated solution at a high temperature. Rapid cooling causes the excess solute to precipitate, so a hot supersaturated solution must be cooled slowly.

When both the solute and the solvent are liquids, other comparative terms are convenient. Liquids such as ethanol and water are called *completely miscible,* which means that they can form solutions in all proportions. *Partially miscible* liquids form

Figure 11.22
In this solution, some of the PbI_2 dissolves, forming Pb^{2+} and I^- ions. Eventually the remaining undissolved PbI_2 settles to the bottom of the beaker.

Figure 11.23
Supersaturated solutions tend to be sensitive. Cooling too quickly or jolting the flask can cause solid to form. Also, a small seed crystal (or a particle of dust) can cause the homogeneous supersaturated solution to revert to a saturated solution in which excess solid is present.

solutions over a limited range of compositions. For example, it is possible to form a solution by adding butanol to water up to a concentration of about 20% butanol by volume. No higher concentration can be achieved. Liquids such as cyclohexane (C_6H_{12}) and water are essentially insoluble in one another and are called *immiscible*.

Although these relative terms and comparisons have their uses, precise quantitative measurements are sometimes essential. For example, to prevent cellular damage in the human body, a saline IV solution must contain an exact concentration of NaCl that precisely matches its concentration in blood. Too much or too little causes red blood cells to shrivel or swell. To make such solutions and assure the proper concentration, we can use one of several concentration units. The particular unit we choose depends on the purpose of the solution and its method of preparation. All the units express concentration as a relative quantity: the amount of solute in a given amount of solvent or in a given amount of solution:

$$\text{Concentration} = \frac{\text{amount of solute}}{\text{amount of solvent or solution}}$$

Some commonly used units are summarized in Table 11.3 and are discussed in this section. Molality is the only unit that describes concentration in terms of the amount of solvent. The others are expressed in terms of the amount of solution.

TABLE 11.3	Concentration Units		
Unit	**Definition**	**Unit**	**Definition**
Percent by mass	$\dfrac{\text{grams of solute}}{\text{grams of solution}} \times 100\%$	Parts per billion	$\dfrac{\text{grams of solute}}{\text{grams of solution}} \times 10^9$
Percent by volume	$\dfrac{\text{volume of solute}}{\text{volume of solution}} \times 100\%$	Molarity (*M*)	$\dfrac{\text{moles of solute}}{\text{liters of solution}}$
Mass/volume percent	$\dfrac{\text{grams of solute}}{\text{volume of solution}} \times 100\%$	Molality (*m*)	$\dfrac{\text{moles of solute}}{\text{kilograms of solvent}}$
Parts per million	$\dfrac{\text{grams of solute}}{\text{grams of solution}} \times 10^6$		

Percent by Mass

Percent by mass is calculated by dividing the solute mass by the mass of solution and multiplying by 100%:

$$\text{Percent by mass} = \frac{\text{grams of solute}}{\text{grams of solution}} \times 100\%$$

We can determine the mass of solution by adding the mass of solute to the mass of solvent as shown in Example 11.3.

EXAMPLE 11.3   Percent by Mass

Determine the percent by mass of NaCl in a solution prepared by adding 25.0 g NaCl to 125.0 g of water. What mass of NaCl is present in 10.0 g of this solution?

Solution:

In this problem we want to go from the masses of solute and solvent to percent by mass:

To go from these quantities to the quantity of interest we must determine the relationship between them. The percent by mass is calculated from the relationship

$$\text{Percent by mass} = \frac{\text{grams of solute}}{\text{grams of solution}} \times 100\%$$

The grams of solute are given in the problem, 25.0 g NaCl. The mass of the solution is the sum of the masses of solute and solvent:

$$25.0 \text{ g NaCl} + 125.0 \text{ g H}_2\text{O} = 150.0 \text{ g solution}$$

$$\text{Percent by mass} = \frac{25.0 \text{ g}}{150.0 \text{ g}} \times 100\% = 16.7\%$$

This quantity tells us that every 100.0 g of this solution contains a mass of 16.7 g NaCl. Since the solution is 16.7% by mass in NaCl, any amount of the solution will contain a proportionate mass of NaCl. Multiplying the mass of the solution and the percent by mass gives the mass of solute present. The mass of NaCl contained in 10.0 g of the solution is

$$\text{Mass NaCl} = 10.0 \text{ g solution} \times \frac{16.7 \text{ g NaCl}}{100.0 \text{ g solution}} = 1.67 \text{ g NaCl}$$

Practice Problem 11.3

What is the percent-by-mass concentration of acetic acid (CH_3CO_2H) in a vinegar solution that contains 2.70 g of acetic acid and 122.8 g of water? What mass of acetic acid is present in 25.0 g of this solution?

Further Practice: 11.49 and 11.50

Percent by Volume

Percent by volume is most often used when liquids are dissolved in liquids, since it is easy to measure amounts of liquids by their volumes. **Percent by volume** is calculated by dividing the solute volume by the solution volume and multiplying by 100%. Calculating percent by volume is similar to calculating percent by mass. We simply substitute volumes for masses in the equation shown earlier for percent by mass:

$$\text{Percent by volume} = \frac{\text{volume of solute}}{\text{volume of solution}} \times 100\%$$

There is an important difference, however. Masses are additive; that is, the mass of solution is equal to the sum of the masses of solute and solvent. Volumes are not always additive. It is possible for the volume of a solution to be different from the sum of the volumes of the solute and solvent. Depending on the attractive forces between solute and solvent, the solution may have a greater or lesser volume than the separated components. So the total volume of a solution must be measured, not calculated from its components.

EXAMPLE 11.4   Percent by Volume

To prepare a solution, we dissolve 12.5 mL of ethanol in sufficient water to give a total volume of 85.4 mL. What is the percent-by-volume concentration of ethanol?

Solution:

In this problem we want to go from the volumes of solute and solution to percent by volume:

The relationship between these quantities is given by the expression for percent by volume:

$$\text{Percent by volume} = \frac{\text{volume of solute}}{\text{volume of solution}} \times 100\%$$

The two required quantities, volume of solute and volume of solution, are given in the problem:

$$\text{Percent by volume} = \frac{12.5 \text{ mL ethanol}}{85.4 \text{ mL solution}} \times 100\% = 14.6\%$$

Practice Problem 11.4

To prepare a solution, we dissolve 205 mL of ethanol in sufficient water to give a total volume of 235 mL. What is the percent-by-volume concentration of ethanol?

Further Practice: 11.57 and 11.58

Mass/Volume Percent

Mass/volume percent is a concentration unit that is similar to the percents of mass and volume. **Mass/volume percent** is calculated by dividing the solute mass by the volume of solution and multiplying by 100%:

$$\text{Mass/volume percent} = \frac{\text{grams of solute}}{\text{volume of solution}} \times 100\%$$

Mass/volume percent is encountered in several applications, including the medical field. For example, the concentration of NaCl in a saline IV bag is 0.9%. For every 100 mL of solution, the mass of NaCl present is 0.9 g.

Parts per Million and Parts per Billion

We often encounter very dilute solutions in which the concentration is too small to express conveniently as a percentage. For example, the Environmental Protection Agency's maximum acceptable level of arsenic in a water supply is 0.000005%, a number that's difficult to read and even more difficult to remember. For this reason the amounts of pollutants in water are often reported in units of parts per million (ppm) or parts per billion (ppb). A solution that is 1 ppm of a solute contains 1 part solute in 1 million parts of solution. **Parts per million** is calculated by dividing the mass of solute by the mass of solution and multiplying by 1 million (10^6):

$$\text{Parts per million} = \frac{\text{grams of solute}}{\text{grams of solution}} \times 10^6$$

Similarly, a solution that is 1 ppb of a solute contains 1 part solute in 1 billion parts of solution. **Parts per billion** is calculated by dividing the mass of solute by the mass of solution and multiplying by 1 billion (10^9):

$$\text{Parts per billion} = \frac{\text{grams of solute}}{\text{grams of solution}} \times 10^9$$

Note that ppm and ppb express the same concept as percent by mass. The only difference is the multiplier—a million, a billion, or 100.

Because parts per million is often used to report the amount of pollutants in water, scientists sometimes look at this concentration unit in another way. For aqueous solutions, parts per million is essentially equal to milligrams of solute per liter of solution. For example, the acceptable level of arsenic in drinking water is reported as 0.05 ppm. This means that 1 L of drinking water cannot contain more than 0.05 mg of arsenic.

Molarity

Molarity (*M*) is one of the most commonly used expressions of concentration in the laboratory. You may remember from Chapter 4 that **molarity** (molar concentration) is the ratio of moles of solute to the volume of solution in liters:

$$\text{Molarity} = \frac{\text{moles of solute}}{\text{liters of solution}}$$

Calculating molarity is straightforward, given the moles or mass of solute and the volume of solution. We will consider some calculations involving molarity later in Section 11.5. For a review of molarity calculations see Chapter 4 and Example 11.5.

EXAMPLE 11.5   Molarity

A solution contains 78.2 g NaCl dissolved in sufficient water to give a total volume of 0.525 L. The molar mass of NaCl is 58.44 g/mol. What is the molarity of this solution?

Solution:

In this problem we want to go from the mass of solute and volume of solution to molarity:

Mass solute and volume solution → ? → Molarity

Note the difference between mass/volume percent and density. Recall from Chapter 1 that density is mass per volume:

$$\text{Density} = \frac{\text{mass}}{\text{volume}}$$

For a solution, density corresponds to a ratio of the mass of solution to the volume of solution:

$$\text{Density} = \frac{\text{mass of solution}}{\text{volume of solution}}$$

Sometimes the density of a solution is needed to convert from one concentration unit to another. Work Problems 11.53 and 11.54 to get some practice at using density in concentration calculations.

Sometimes parts per million and parts per billion are calculated on a volume basis. The concentration of pollutants in air is often reported this way.

Animation: Making a Solution
Online Learning Center, Animations

Chemistry Animations Library, Making a Solution, Making_a_Solution.swf.

In the past the term *normality* was frequently used in chemistry labs to describe the concentration of a solution. The term was useful because it specified the number of reactant molecules, atoms, or ions—known as *equivalents*—available in a reagent bottle. For example, a 1-L solution that contains 1 mol of H_2SO_4 has 2 equivalents of H^+ available, so it is said to be a 2 *N* solution. On the other hand, a 1-L solution containing 1 mol of HCl has only 1 equivalent available, so it is 1 *N*. Although molarity is more often used in chemistry labs now, equivalents and milliequivalents are still commonly used in the medical profession.

Molarity is a relationship between moles of solute and volume of solution:

$$\text{Molarity} = \frac{\text{moles of solute}}{\text{liters of solution}}$$

To calculate the molarity we need to know the moles of solute. The relationship between moles of a substance and its mass is given by the molar mass. We can calculate the number of moles of NaCl from the mass as follows:

$$\text{mol NaCl} = 78.2 \text{ g NaCl} \times \frac{1 \text{ mol NaCl}}{58.44 \text{ g NaCl}} = 1.34 \text{ mol NaCl}$$

Now we have all the information we need to calculate the molarity:

$$\text{Molarity} = \frac{1.34 \text{ mol NaCl}}{0.525 \text{ L solution}} = 2.55 \text{ mol/L}$$

For convenience we can report this concentration as 2.55 *M*.

Practice Problem 11.5

A solution contains 117 g of potassium hydroxide (KOH) dissolved in sufficient water to give a total volume of 2.00 L. The molar mass of KOH is 56.11 g/mol. What is the molarity of the aqueous potassium hydroxide solution?

Further Practice: 11.59 and 11.60

Molality

Concentrations with units that contain a volume term can vary with temperature. Suppose we have a 0.1 *M* NaCl solution. If the temperature increases, will the molarity stay the same? No, because the volume of the solution will change. When liquids are heated, they expand slightly, so the ions that make up a sodium chloride solution will be spread over a larger volume decreasing the molarity. A concentration unit that uses moles of solute and mass of solvent does not change with temperature. **Molality** (*m*) is such a unit. It is the concentration expressed as moles of solute per kilogram of solvent:

$$\text{Molality} = \frac{\text{moles of solute}}{\text{kilograms of solvent}}$$

Because molality does not vary with temperature, it is useful in experiments involving temperature-dependent properties. It is also useful for examining properties that depend on the number of solute particles in solution, as we will see in Section 11.6. To calculate molality, we must determine the number of moles of solute and the mass of solvent. Note that it is the mass of the solvent—not of the solution—that is needed.

EXAMPLE 11.6 Molality

A solution contains 22.5 g of methanol (CH₃OH) dissolved in sufficient water to give a total mass of 105.3 g. The molar mass of CH₃OH is 32.04 g/mol. What is the molality of the aqueous methanol solution?

Solution:

In this problem we want to go from the mass of solute and mass of solution to molality:

Molality is a relationship between the moles of solute and the mass of solution in kilograms:

$$\text{Molality} = \frac{\text{moles of solute}}{\text{kilograms of solvent}}$$

We can calculate the number of moles of solute from the mass of solute and its molar mass:

$$\text{mol CH}_3\text{OH} = 22.5 \ \text{g CH}_3\text{OH} \times \frac{1 \ \text{mol CH}_3\text{OH}}{32.04 \ \text{g CH}_3\text{OH}} = 0.702 \ \text{mol CH}_3\text{OH}$$

Subtracting the mass of solute from the mass of solution yields the mass of solvent:

$$\text{Mass H}_2\text{O} = 105.3 \ \text{g solution} - 22.5 \ \text{g CH}_3\text{OH} = 82.8 \ \text{g H}_2\text{O}$$

This mass must be converted to units of kilograms:

$$\text{Mass H}_2\text{O} = 82.8 \ \text{g} \times \frac{1 \ \text{kg}}{1000 \ \text{g}} = 0.0828 \ \text{kg H}_2\text{O}$$

Now we have all the information we need to calculate the molality:

$$\text{Molality} = \frac{\text{moles of solute}}{\text{kilograms of solvent}}$$
$$= \frac{0.702 \ \text{mol CH}_3\text{OH}}{0.0828 \ \text{kg H}_2\text{O}}$$
$$= 8.48 \ \text{mol/kg}$$

For convenience this concentration is reported as 8.48 *m*.

Practice Problem 11.6

A solution contains 56.7 g of potassium hydroxide (KOH) dissolved in sufficient water to give a total mass of 245.8 g. The molar mass of KOH is 56.11 g/mol. What is the molality of the aqueous potassium hydroxide solution?

Further Practice: 11.61 and 11.62

11.5   QUANTITIES FOR REACTIONS THAT OCCUR IN AQUEOUS SOLUTION

In this section we will reexamine some of the reactions described in Chapter 5 to see how to calculate quantities for reactions that occur in aqueous solutions. We will consider two types of reactions: precipitation reactions and acid-base neutralization reactions. Both types are examples of double-displacement reactions, in which two compounds exchange ions to form new compounds.

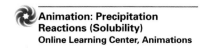

Animation: Precipitation Reactions (Solubility)
Online Learning Center, Animations

Chemistry Animations Library, Precipitation Reactions (Solubility), 3Precipitation_Reactions.swf.

Precipitation Reactions

Patients undergoing X-ray treatment to diagnose digestive tract problems drink a suspension of barium sulfate. Barium sulfate can be prepared from a precipitation reaction in which solutions of barium chloride and potassium sulfate are mixed (Figure 11.24). Recall from Section 5.4 that in a precipitation reaction the cation of one ionic compound combines with the anion of another ionic compound to form an insoluble, solid substance, a *precipitate*. The reaction between barium chloride and potassium sulfate is such a reaction. Both reactants are strong electrolytes, so they dissociate in aqueous solutions. The ions present before any reaction occurs are $Ba^{2+}(aq)$, $Cl^-(aq)$, $K^+(aq)$, and $SO_4^{2-}(aq)$. Possible formulas for the precipitate, then, are $BaCl_2$, K_2SO_4, $BaSO_4$, and KCl. The first two choices can be eliminated because we started with solutions of $BaCl_2$ and K_2SO_4. If they could have precipitated as solids, they would have done so before being mixed. This leaves $BaSO_4$ and KCl. Two solubility rules from Table 11.1 are pertinent to choosing between the possibilities: Most compounds of K^+ are soluble, and most compounds of Cl^- are soluble. From these rules we can conclude that the precipitate is unlikely to be KCl. Therefore it must be $BaSO_4$. The balanced equation for this reaction is

$$BaCl_2(aq) + K_2SO_4(aq) \longrightarrow BaSO_4(s) + 2KCl(aq)$$

Now that we've reviewed characteristics of a precipitation reaction, Example 11.7 shows a calculation involving the formation of $BaSO_4$.

Recall that this is a *molecular equation,* because it represents the substances as if they existed as molecules (or formula units) rather than dissociating into ions. $BaCl_2(aq)$ is a shorthand way of indicating that Ba^{2+} and Cl^- ions are present in the solution. Similarly, for $K_2SO_4(aq)$, K^+ and SO_4^{2-} ions are present. To review reactions of this type see Chapter 5.

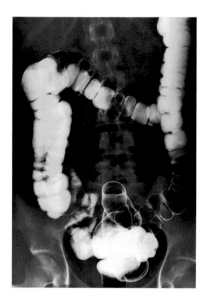

Figure 11.24
Barium sulfate, used in X-ray applications, can be prepared from the reaction of barium chloride with potassium sulfate.

EXAMPLE 11.7   Calculations Involving Precipitation Reactions

Suppose we want to prepare $BaSO_4$ by adding 0.450 M K_2SO_4 to 130.0 mL of 0.250 M $BaCl_2$. What volume of the K_2SO_4 solution is required to react completely with the $BaCl_2$? How many grams of $BaSO_4$ will precipitate?

Solution:

For the first part of this problem we want to go from the volume of one reactant to the volume of another:

$$\boxed{\text{Volume } BaCl_2 \text{ solution}} \xrightarrow{\ ?\ } \boxed{\text{Volume } K_2SO_4 \text{ solution}}$$

There is no direct relationship between these two quantities, but we recognize this as a problem in which we're given a quantity of one substance in a chemical reaction and want to calculate another. We'll have to refer to the balanced equation to determine the mole relationship for the two substances. From the balanced equation we know that 1 mol of $BaCl_2$ reacts with 1 mol of K_2SO_4 to produce 1 mol of $BaSO_4$. Is there a relationship between the volume of $BaCl_2$ and moles of $BaCl_2$?

$$\boxed{\text{Volume } BaCl_2 \text{ solution}} \xrightarrow{\ ?\ } \boxed{\text{mol } BaCl_2}$$

Yes, its molarity is 0.250 M, which tells us that for every liter of solution, 0.250 mol of $BaCl_2$ is present. In Chapter 6 we obtained moles of reactant from grams using the molar mass as a conversion factor. Here we must convert the volume of a reactant in solution to moles using the molar concentration as a conversion factor. In this case we know the volume of 0.250 M $BaCl_2$ is 130.0 mL. We have to first convert milliliters to liters because the volume unit in molarity is liters:

$$\boxed{\text{Volume } BaCl_2 \text{ (in mL)}} \xrightarrow{\text{1000 mL = 1 L}} \boxed{\text{Volume } BaCl_2 \text{ (in L)}} \xrightarrow{\text{molarity}} \boxed{\text{mol } BaCl_2}$$

$$\text{mol } BaCl_2 = 130.0 \text{ mL } BaCl_2 \text{ solution} \times \frac{1 \text{ L}}{1000 \text{ mL}} \times \frac{0.250 \text{ mol } BaCl_2}{\text{L solution}}$$

$$= 0.0325 \text{ mol } BaCl_2$$

We can now use the mole ratio in the chemical equation to calculate the moles of K_2SO_4 required for the reaction:

$$\boxed{\text{mol } BaCl_2} \xrightarrow{\text{mole ratio}} \boxed{\text{mol } K_2SO_4}$$

$$\text{mol } K_2SO_4 = 0.0325 \text{ mol } BaCl_2 \times \frac{1 \text{ mol } K_2SO_4}{1 \text{ mol } BaCl_2} = 0.0325 \text{ mol } K_2SO_4$$

Now that we know the moles of K_2SO_4 required to completely react with the $BaCl_2$, we have to determine the volume of K_2SO_4 solution that will provide the desired number of moles:

$$\boxed{\text{mol } K_2SO_4} \xrightarrow{\text{?}} \boxed{\begin{array}{c}\text{Volume } K_2SO_4 \\ \text{solution (in L)}\end{array}}$$

The molar concentration of K_2SO_4 solution provides a relationship we can use to calculate the volume:

$$\boxed{\text{mol } K_2SO_4} \xrightarrow{\text{molarity}} \boxed{\begin{array}{c}\text{Volume } K_2SO_4 \\ \text{solution (in L)}\end{array}}$$

$$\text{Volume of } K_2SO_4 \text{ solution} = 0.0325 \text{ mol } K_2SO_4 \times \frac{1 \text{ L } K_2SO_4 \text{ solution}}{0.450 \text{ mol } K_2SO_4}$$

$$= 0.0722 \text{ L } K_2SO_4 \text{ solution}$$

We can interpret this volume of K_2SO_4 solution as the amount of solution that has enough moles of K_2SO_4 (0.0325 mol) to react with all the $BaCl_2$. An alternative to this step-by-step approach involves setting up conversions consecutively in one calculation. To determine the volume of K_2SO_4 required, we can set up the following sequence of conversions:

$$\text{Volume of } K_2SO_4 \text{ solution} = 0.1300 \text{ L } BaCl_2 \text{ solution} \times \frac{0.250 \text{ mol } BaCl_2}{1 \text{ L } BaCl_2 \text{ solution}} \times$$

$$\frac{1 \text{ mol } K_2SO_4}{1 \text{ mol } BaCl_2} \times \frac{1 \text{ L } K_2SO_4 \text{ solution}}{0.450 \text{ mol } K_2SO_4}$$

$$= 0.0722 \text{ L } K_2SO_4 \text{ solution}$$

This is a small number more conveniently expressed in milliliters:

$$\text{Volume in mL} = 0.0722 \text{ L } K_2SO_4 \text{ solution} \times \frac{1000 \text{ mL}}{1 \text{ L}}$$

$$= 72.2 \text{ mL } K_2SO_4 \text{ solution}$$

Recall from Chapter 6 that we used the following guide to solve problems involving masses of reactants and products.

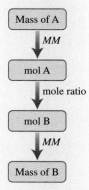

The following sequence summarizes the relationships used to solve this problem:

Now let's calculate the mass of solid $BaSO_4$ that will be produced when adding 72.2 mL of 0.450 M K_2SO_4 solution to 130.0 mL of 0.250 M $BaCl_2$. We know the moles of $BaCl_2$ from a previous calculation. We can use the mole ratio given by the chemical equation to determine the moles of $BaSO_4$ that should form:

$$\text{mol } BaSO_4 = 0.0325 \text{ mol } BaCl_2 \times \frac{1 \text{ mol } BaSO_4}{1 \text{ mol } BaCl_2} = 0.0325 \text{ mol } BaSO_4$$

Do you recall how to calculate grams of a substance from moles of that material? Using the molar mass we can calculate the grams of $BaSO_4$ that will precipitate:

$$\boxed{\text{mol } BaSO_4} \xrightarrow{\text{molar mass}} \boxed{\text{Mass } BaSO_4}$$

$$\text{Mass } BaSO_4 = 0.0325 \text{ mol } BaSO_4 \times \frac{233.4 \text{ g } BaSO_4}{1 \text{ mol } BaSO_4} = 7.59 \text{ g } BaSO_4$$

An alternative to this step-by-step approach involves setting up conversions consecutively in one calculation. To determine the mass of $BaSO_4$ produced we can set up the following sequence of conversions:

$$\text{Mass } BaSO_4 = 0.1300 \text{ L } BaCl_2 \text{ solution} \times \frac{0.250 \text{ mol } BaCl_2}{1 \text{ L } BaCl_2 \text{ solution}} \times$$

$$\frac{1 \text{ mol } BaSO_4}{1 \text{ mol } BaCl_2} \times \frac{233.4 \text{ g } BaSO_4}{1 \text{ mol } BaSO_4}$$

$$= 7.59 \text{ g } BaSO_4$$

$$\boxed{\substack{\text{Volume (in L)} \\ BaCl_2 \text{ solution}}} \xrightarrow{\text{molarity}} \boxed{\substack{\text{mol} \\ BaCl_2}} \xrightarrow{\text{mole ratio}} \boxed{\substack{\text{mol} \\ BaSO_4}} \xrightarrow{\text{molar mass}} \boxed{\substack{\text{Mass} \\ BaSO_4}}$$

The key to solving problems that involve solution concentrations is to realize that concentration is a ratio that can be used as a conversion factor. If volume is known and we want to calculate moles, we simply multiply by the molarity. If moles are given and we want to calculate volume, we divide by molarity.

Practice Problem 11.7

When a solution of lead(II) nitrate is mixed with a solution of potassium chromate, a yellow precipitate forms according to the equation

$$Pb(NO_3)_2(aq) + K_2CrO_4(aq) \longrightarrow PbCrO_4(s) + 2KNO_3(aq)$$

What volume of 0.105 M lead(II) nitrate is required to react with 100.0 mL of 0.120 M potassium chromate? What mass of $PbCrO_4$ solid forms?

Further Practice: 11.69 and 11.70

Acid-Base Titrations

Now let's consider another type of reaction that takes place in solution, an acid-base neutralization reaction. In such a reaction, an acid reacts with a base, normally forming a salt and water. The formation of water molecules from hydrogen ions and hydroxide ions is common to all aqueous neutralization reactions. The acid-base reaction involves the donation of a hydrogen ion by an acid and the acceptance of that hydrogen ion by a base. A common example is the reaction between hydrochloric acid and sodium hydroxide, both of which are strong electrolytes:

hydrochloric acid + sodium hydroxide $\longrightarrow$ sodium chloride + water

The molecular equation for this reaction is

$$HCl(aq) + NaOH(aq) \longrightarrow NaCl(aq) + H_2O(l)$$

The ionic equation for this reaction is

$$H^+(aq) + Cl^-(aq) + Na^+(aq) + OH^-(aq) \longrightarrow Na^+(aq) + Cl^-(aq) + H_2O(l)$$

Note that Na^+ and Cl^- ions appear on both sides of the equation, and they do not participate in the reaction. They are spectator ions, which can be omitted when writing the net ionic equation:

$$H^+(aq) + OH^-(aq) \longrightarrow H_2O(l)$$

This net ionic equation is common to all aqueous strong acid–strong base reactions.

Titration is the process of determining the concentration of one substance in solution by reacting it with a solution of another substance that has a known concentration. Although the technique can be applied to other types of reactions, we will focus on acid-base titrations. To carry out the process, we add one solution to a known volume of the other solution, as shown in Figure 11.25, until just beyond the point at which the reaction between the two substances is complete. We can often tell when we have reached this point, called the *equivalence point,* by a change in the color of an *indicator,* a substance that has been added to the system and is sensitive to changes in acid or base content. Notice the color change in Figure 11.25. The indicator used here, phenolphthalein, changes from colorless to pink when the solution becomes basic. The point at which the indicator changes color is technically called the *end point,* which may be slightly different from the equivalence point but is usually a good approximation.

When running a titration, we use the amount of one substance in a chemical reaction to calculate the amount of another. In titration calculations the volume of the reactant solution of known concentration is used to calculate the moles of that reactant. The moles of the other reactant are obtained from the coefficients in the balanced equation, and this value is used with the volume of the solution of unknown concentration to calculate its molarity. Example 11.8 shows how titration calculations are done.

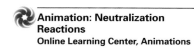 **Animation: Neutralization Reactions**
Online Learning Center, Animations

Chemistry Animations Library, Neutralization Reactions, 09Neutralization_Reactions.swf.

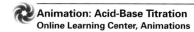 **Animation: Acid-Base Titration**
Online Learning Center, Animations

Chemistry Animations Library, Acid-Base Titration, Acid_Base_Titration.swf.

Indicators will be discussed in more detail in Chapter 13.

Figure 11.25
To determine the concentration of a known volume of acid, a known volume of a base of a known concentration is added until some sign of a reaction can be observed, such as a color change of an indicator. This laboratory technique is called titration.

EXAMPLE 11.8   Acid-Base Titrations

Suppose a titration is run in which 25.05 mL of NaOH solution of unknown concentration reacts with 25.00 mL of 0.1000 M H_2SO_4 solution. The chemical equation that summarizes the reaction is as follows:

$$H_2SO_4(aq) + 2NaOH(aq) \longrightarrow 2H_2O(l) + Na_2SO_4(aq)$$

What is the molarity of the NaOH solution?

Solution:

We want to go from the volume of one reactant to the molarity of another:

$$\boxed{\text{Volume } H_2SO_4 \text{ solution}} \xrightarrow{\quad ? \quad} \boxed{\text{Molarity NaOH solution}}$$

There is no direct relationship between these two quantities. Since the molarity of a solution is a ratio of the moles of a solute to the volume of a solution of that solute, we need to know both of these quantities to determine the molarity of the NaOH solution. The volume of NaOH solution is stated in the problem, but we do not know the moles of NaOH. We'll have to calculate the moles of NaOH that react with the H_2SO_4 from the information we have about H_2SO_4.

From the balanced equation we know that 1 mol of H_2SO_4 reacts with 2 mol of NaOH, so this relationship will be useful in solving the problem. The volume of 0.1000 M H_2SO_4 solution that is required to react completely with the 25.05 mL of NaOH is 25.00 mL. From the information about H_2SO_4 and the mole ratio given in the balanced chemical equation, we can calculate the moles of NaOH that reacted:

$$\boxed{\text{Volume } H_2SO_4 \text{ (in L)}} \xrightarrow{\text{molarity}} \boxed{\text{mol } H_2SO_4} \xrightarrow{\text{mole ratio}} \boxed{\text{mol NaOH}}$$

We begin by converting milliliters to liters because the volume unit in molarity is expressed as liters. Then we substitute relationships that allow us to calculate moles of NaOH that react:

$$\text{mol NaOH} = 25.00 \text{ mL } H_2SO_4 \text{ solution} \times \frac{1 \text{ L}}{1000 \text{ mL}} \times$$

$$\frac{0.1000 \text{ mol } H_2SO_4}{\text{L } H_2SO_4 \text{ solution}} \times \frac{2 \text{ mol NaOH}}{1 \text{ mol } H_2SO_4}$$

$$= 0.005000 \text{ mol NaOH}$$

We now know the number of moles of NaOH present in the original solution. Using this quantity and the volume stated in the problem, we can calculate the molarity. Recall that molarity is a ratio of moles of solute to liters of solution, so we must convert the volume in milliliters to liters. Then we use this quantity to calculate the molarity of the NaOH solution:

$$\text{Volume NaOH solution} = 25.05 \text{ mL NaOH solution} \times \frac{1 \text{ L}}{1000 \text{ mL}}$$

$$= 0.02505 \text{ L NaOH solution}$$

$$\text{Molarity} = \frac{\text{moles of solute}}{\text{liters of solution}}$$

$$= \frac{0.005000 \text{ mol NaOH}}{0.02505 \text{ L solution}} = 0.1996 \text{ } M \text{ NaOH}$$

The relationships used to solve this problem are summarized as follows:

Practice Problem 11.8

We need 28.18 mL of 0.2437 *M* HCl solution to completely titrate 25.00 mL of Ba(OH)$_2$ solution of unknown concentration. What is the molarity of the barium hydroxide solution?

Further Practice: 11.77 and 11.78

Calculations involving quantities in solution are similar to using masses, as described in Chapter 6. In general the following scheme can be used to organize calculations that involve volume and molarity:

11.6   COLLIGATIVE PROPERTIES

When a solute dissolves in a pure solvent, the properties of both may change. We have already examined some reactions that bring about chemical changes, but some changes in physical properties do not require chemical change. They do not depend on the identity of the solute—only on the number of solute particles present. Such a property is called a **colligative property.** Colligative properties vary only with the number of solute particles (molecules or ions) present in a specific quantity of solvent. What the solute is does not matter. Colligative properties include osmotic pressure, vapor pressure lowering, boiling point elevation, and freezing point depression.

Osmotic Pressure

Most processes that occur in the human body involve solutions. The body often works to balance the concentration of solutes inside cells with the concentration outside—in the extracellular fluid. Because the body reacts unfavorably to imbalances in solute concentration, the concentration of particles in IV fluids must be controlled. For example, when the concentration of solutes in the blood is higher than the solute concentration within a red blood cell, the cell shrinks. It expands when the solute concentration in the fluid is lower. Why do these changes occur?

Osmosis is a process in which solvent particles diffuse through a barrier that does not allow the passage of solute particles. Osmosis allows water to pass through plant and animal cell walls and membranes. The movement of water molecules balances differences in solute concentration between cells and extracellular fluid. A membrane that allows the passage of some substances but not others is a **semipermeable membrane.** In crossing semipermeable membranes, water tends to travel preferentially from the side that is more dilute in solute to the side that is more concentrated, as shown in Figure 11.26. This movement of water molecules acts to equalize the solute concentration on both sides of the membrane.

For example, when pure water is placed on one side of a semipermeable membrane and a salt solution on the other side, the volume of the salt solution rises

Animation: Osmosis
Online Learning Center, Animations

Chemistry Animations Library, Osmosis, Osmosis.rm.

Several terms describe the locations of water in the body. Water inside cells is *intra*cellular. Water outside cells is *extra*cellular. It includes both the plasma in blood and *interstitial* water, which lies in the spaces between cells.

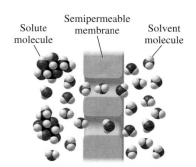

Figure 11.26
When separated by a semipermeable membrane, a solvent will travel from a solution of low solute concentration through the barrier to a solution with higher solute concentration.

Figure 11.27
When a 3% salt solution contained by a semipermeable membrane is placed in pure water, water molecules travel across the membrane into the salt solution, and the solution rises. Can the concentrations on both sides of the membrane ever become equal by osmosis?

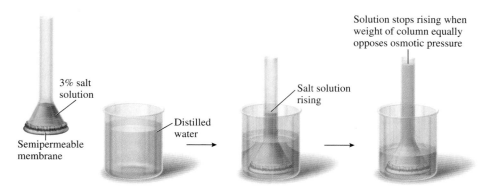

During osmosis the water actually moves in both directions, but it moves more in the direction that leads to equalization of the solute concentration.

The body also controls solute concentrations through dialysis in the kidneys. In this process solute particles selectively pass through semipermeable membranes instead of solvent molecules as in the process of osmosis.

above the level of the water in the beaker (Figure 11.27). The rise happens because water molecules pass through the membrane from the beaker of pure water into the salt solution. Consequently, the salt solution becomes more dilute. One reason this process occurs is entropy, the tendency for matter to become more disordered. When solvent molecules travel through a membrane, the distribution of solute particles in the solution becomes more random.

To prevent osmosis—that is, to prevent water from passing through to the more concentrated side of a membrane—pressure can be exerted on the solution. This pressure, called the *osmotic pressure,* is related to the height a solution will rise up in a column that is separated from a more dilute solution (or pure solvent) by a semipermeable membrane, as shown in Figure 11.28. Like other colligative properties, osmotic pressure is the same regardless of the identity of the solute. It only varies with the concentration of solute particles.

Solvent can be forced to flow in the opposite direction by the application of a pressure greater than the osmotic pressure. This process, called *reverse osmosis,* is the flow of solvent from high solute concentration through a semipermeable membrane to a solution with low solute concentration (Figure 11.29). The process is used in the manufacture of many popular brands of bottled water. It is also important in desalinating seawater. In this application seawater is pumped through a tube containing a semipermeable membrane at a pressure of about 25 atm, much greater than the osmotic pressure of seawater. As a result water molecules flow from the side of the membrane containing a high solute concentration to the side with a lower solute concentration (the reverse of osmosis). This same process, on a smaller scale, is used in the reverse-osmosis units available for use in homes.

Figure 11.28
(A) Solvent moves across the membrane from the pure solvent side to the solution. (B) The solution volume increases, and it becomes less concentrated. The difference in volumes creates an osmotic pressure. (C) To prevent osmosis from occurring, an external pressure must be exerted on the solution side of the membrane to keep the solutions at the same level.

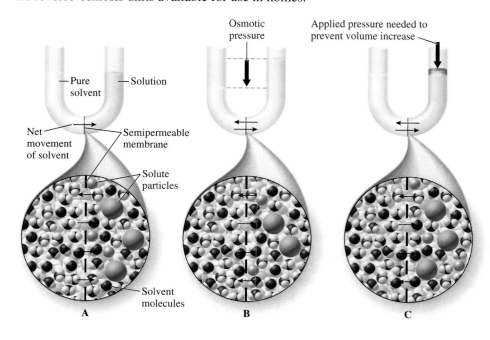

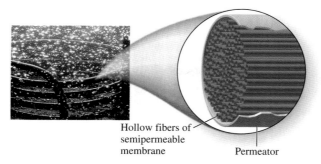

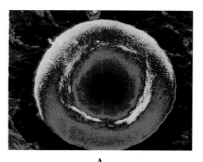

A

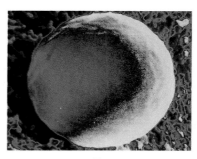

B

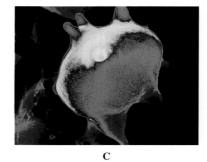

C

Figure 11.29

A typical reverse-osmosis coil contains many hollow fibers made of semipermeable membranes. Seawater is pumped through the coil at high pressure, causing solvent to flow out of the high-concentration solution into the fibers. From them pure water can then be collected.

Now let's return to the question of what happens in the blood of living organisms (Figure 11.30). Blood is a solution of many dissolved substances in which suspended red blood cells are carried throughout the body. If the blood has a higher concentration of dissolved solute than the fluid within red blood cells, water will flow out of the red blood cells and they will shrink. If the blood has a higher concentration of solute, it is *hypertonic* to the cell. If the blood has a lower concentration of solute than the red blood cells, water will flow into the red blood cells causing them to expand. In this case the blood is said to be *hypotonic* to the cell. When the concentration of solute is the same outside and inside the red blood cell, the blood is considered *isotonic* to the cell. Solutions delivered intravenously must be isotonic to blood for this reason.

Vapor Pressure Lowering

Consider a solution of table sugar, or sucrose, in water. If separate beakers of sugar solution and pure water are placed into a sealed container, over time the volume of the sugar solution will increase, and the volume of the pure water will decrease (Figure 11.31). Why? The answer lies in the differences in vapor pressure of the pure solvent and the solution. Since the system is sealed, water that evaporates cannot escape. Instead water evaporates faster from the liquid having the greater vapor pressure and condenses in the liquid with the lower vapor pressure. The presence of solute molecules lowers the vapor pressure of the solvent (Figure 11.32).

Solutes can be described as volatile or nonvolatile. *Volatile* solutes readily form a gas, while *nonvolatile* solutes don't. Ethanol is a volatile solute; sugar is nonvolatile. The effect of a nonvolatile solute in solution on vapor pressure is shown

Figure 11.30

(A) When the surrounding solution has the same solute concentration as the cell contents, it is *isotonic,* and cell size is normal. (B) A *hypotonic* surrounding solution causes water molecules to move into the red blood cell, swelling it. Eventually it will burst. (C) A surrounding solution that is *hypertonic* will cause water molecules to move out of a red blood cell, and it shrinks.

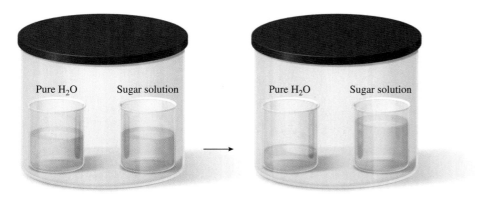

Figure 11.31

Water evaporates faster from the beaker of pure water and condenses into the beaker containing the sugar water.

Figure 11.32
Addition of a solute to a pure solvent reduces the number of solvent molecules in the gas phase, lowering the vapor pressure.

Pure liquid   Solution

graphically in Figure 11.33. Notice that at a given temperature, the vapor pressure of the solution is lower than that of the pure solvent.

The graph in Figure 11.33 is a *phase diagram.* It shows the physical state of a substance at a given temperature and pressure. For example, the curve for the pure solvent that lies between the liquid and gas phases corresponds to temperatures and pressures at which the liquid state is in equilibrium with the gaseous state. As temperature increases, the vapor pressure increases. Notice the position of the liquid-gas curve for a solution. It's lower than the curve for the pure solvent, indicating that at a given temperature, the vapor pressure for the solution is lower than that of the pure solvent.

Boiling Point Elevation

We have already seen that the vapor pressure of a liquid increases with temperature. Notice in Figure 11.33 that when the temperature rises high enough, the vapor pressure of the liquid equals the surrounding atmospheric pressure. When this happens, the liquid has reached its boiling point. The addition of a solute affects the boiling point because it affects the vapor pressure. The presence of solute means that the temperature

Figure 11.33
The addition of a nonvolatile solute lowers the vapor pressure of water. This causes an elevation of the boiling point, ΔT_b, and a depression of the freezing point, ΔT_f.

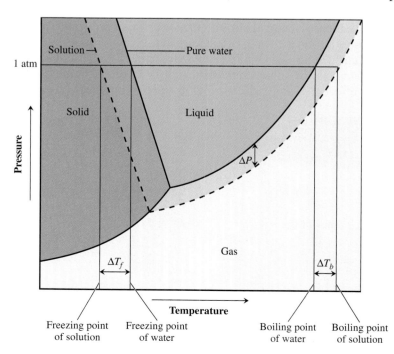

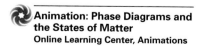

Animation: Phase Diagrams and the States of Matter
Online Learning Center, Animations

Chemistry Animations Library, Phase Diagrams and the States of Matter, 01Phase_Diagram, rm.

must be raised further to raise the vapor pressure to atmospheric pressure, so the boiling point is raised as well. The boiling point elevation is shown in Figure 11.33 as ΔT_b. The lowering of vapor pressure is proportional to the concentration of the solute, so the elevation of the boiling point is also proportional to the concentration of the solute.

Boiling point elevation is related quantitatively to the molality of the solute particles present in the solution. If the solution represented in Figure 11.33 had a greater molality, the difference in boiling point would be correspondingly greater. The change in boiling point is related to molality by the following equation:

$$\Delta T_b = K_b m$$

The term ΔT_b is the difference in boiling point between the solution and pure solvent. The term K_b is a constant characteristic of the solvent, the *boiling point constant.* For water K_b is 0.52°C/m. Recall that m represents the molal concentration of the solution:

$$m = \frac{\text{moles of solute}}{\text{kilograms of solvent}}$$

Freezing Point Depression

The freezing point of a liquid is the temperature at which the liquid and solid states of a substance are at equilibrium. The presence of a solute lowers the freezing point of pure solvent by an amount ΔT_f as shown in Figure 11.33. Like boiling point elevation, freezing point depression is proportional to the molal concentration of solute.

This colligative property finds practical application in the use of ethylene glycol ($HO-CH_2-CH_2-OH$) as an antifreeze in automobile radiators. Ethylene glycol lowers the freezing point of water that is used as a radiator coolant. (It also raises the boiling point, preventing boilover during the summer.) Another widespread application in cold climates is the use of sodium chloride or calcium chloride to melt ice on roads and sidewalks. These compounds work by lowering the freezing point of water.

The relationship of freezing point depression to molality is expressed in an equation similar to that for boiling point elevation:

$$\Delta T_f = K_f m$$

The term ΔT_f is the difference in freezing point between pure solvent and solution. The term K_f is a constant characteristic of the solvent, the *freezing point constant.* For water K_f is −1.86°C/m.

EXAMPLE 11.9   **Boiling Point Elevation and Freezing Point Depression**

What is the change in boiling point for a 1.5 m solution of sucrose in water?

Solution:

We can use the following relationship to determine the change in boiling point for the solution:

$$\Delta T_b = K_b m = \left(\frac{0.52°C}{m} \right) \times 1.5\ m = 0.78°C$$

The boiling point of the solution is 0.78°C higher than the normal boiling point. Since the normal boiling point of water is 100°C, the solution boils at 100.78°C.

Practice Problem 11.9

What is the change in freezing point for a 1.5 m solution of sucrose in water?

Further Practice: 11.91 and 11.92

Colligative Properties and Strong Electrolytes

In Example 11.9, we used the molality to determine the change in boiling point and freezing point for a solution. If we know the change in boiling point (or freezing point), we can reverse the process and calculate the molality of the solution. However, the colligative properties measured for many solutions are larger than we predict from such calculations because many solutes dissociate into ions when they dissolve. Colligative properties are proportional to the number of solute particles present in the solution. When the measured value of a boiling point elevation or freezing point depression is significantly greater than the calculated value, we can infer that the solute is an electrolyte and has dissociated into ions.

The degree of dissociation can be determined by considering how the substance behaves in water. This is often expressed as the average number of particles produced by one formula unit of the solute when it dissolves. For example, when one formula unit of NaCl dissolves, it completely dissociates to form two ions, $Na^+(aq)$ and $Cl^-(aq)$. Every NaCl formula unit that dissolves adds two particles to the solution. In contrast, sugar, a nonelectrolyte, retains its molecular structure when it is mixed with water, so for every molecule that dissolves, only one sugar particle exists in the aqueous solution. Because the concentration of particles present in the solution determines colligative properties, we can predict that a 1 m solution of the strong electrolyte NaCl would have about twice the effect on freezing and boiling points as a 1 m solution of sucrose, a nonelectrolyte.

> Strong electrolytes do not necessarily dissociate 100%. Sometimes oppositely charged ions attract each other in solution and exist loosely as ion pairs. The consequence is slight deviations in the expected colligative properties.

EXAMPLE 11.10 Colligative Properties and Strong Electrolytes

Which of the following aqueous solutions should have the highest boiling point: 1 m sucrose, 1 m NaCl, or 1 m $MgCl_2$?

Solution:

Boiling point elevation depends on the number of particles present in the solution. Sucrose retains its molecular structure when it dissolves in water, so the concentration of particles in a 1 m solution of sucrose is 1 m.

One formula unit of sodium chloride dissociates to give two ions in aqueous solution:

$$NaCl(s) \longrightarrow Na^+(aq) + Cl^-(aq)$$

The concentration of particles in a 1 m NaCl solution is thus about 2 m.

Magnesium chloride also dissociates to give ions in aqueous solution:

$$MgCl_2(s) \longrightarrow Mg^{2+}(aq) + 2Cl^-(aq)$$

Every formula unit of $MgCl_2$ provides three ions. The concentration of particles in a 1 m $MgCl_2$ solution is thus about 3 m. This solution has the greatest concentration of dissolved particles, so it is expected to have the highest boiling point.

Practice Problem 11.10

Which of the following aqueous solutions is expected to have the lowest freezing point: 0.5 m CH_3CH_2OH, 0.5 m $Ca(NO_3)_2$, or 0.5 m KBr?

Further Practice: 11.93 and 11.94

SUMMARY

Solutions are found among any combination of physical states. Metal alloys and gas mixtures are solutions. The most common solutions consist of one or more solids dissolved in a liquid solvent, usually water. Solids that dissolve to form ions in water are electrolytes. In solutions of nonelectrolytes, the solute retains its molecular structure.

Solution formation involves two kinds of forces: those that must be overcome (intermolecular forces and bonds) and those that are made when solute and solvent combine to form a solution. Entropy is another important force that drives the solution process. Polar solvents generally dissolve other polar substances and ionic solids. Nonpolar solvents dissolve other nonpolar substances and covalent solids. The solubility of solids usually increases with temperature. Gases become less soluble in liquids with increasing temperature, but more soluble with increasing pressure.

The concentration of solutions can be expressed in a variety of ways. Commonly used units are percent by mass, percent by volume, parts per million, parts per billion, molarity, and molality. When solutions are involved in chemical reactions, the most common concentration unit is molarity. It can be used to calculate the quantities of solutions used in chemical reactions. Such calculations are often used for precipitation and titration reactions.

Colligative properties of solutions depend only on the concentration of solute particles, not on their identity. These properties include increasing osmotic pressure, vapor pressure lowering, boiling point elevation, and freezing point depression.

KEY RELATIONSHIPS

Relationship	Equation
To calculate percent by mass, divide the mass of solute by the mass of the solution and multiply by 100%.	$\text{Percent by mass} = \dfrac{\text{grams of solute}}{\text{grams of solution}} \times 100\%$
To calculate percent by volume, divide the volume of solute by the volume of the solution and multiply by 100%.	$\text{Percent by volume} = \dfrac{\text{volume of solute}}{\text{volume of solution}} \times 100\%$
To calculate mass/volume percent, divide the mass of solute by the volume of the solution and multiply by 100%.	$\text{Mass/volume percent} = \dfrac{\text{grams of solute}}{\text{volume of solution}} \times 100\%$
To calculate parts per million (ppm), divide the mass of solute by the mass of the solution and multiply by 1 million (10^6).	$\text{ppm} = \dfrac{\text{grams of solute}}{\text{grams of solution}} \times 10^6$
To calculate parts per billion (ppb), divide the mass of solute by the mass of the solution and multiply by 1 billion (10^9).	$\text{ppb} = \dfrac{\text{grams of solute}}{\text{grams of solution}} \times 10^9$
To calculate molarity, divide the moles of solute by the volume of the solution in liters.	$\text{Molarity} = \dfrac{\text{moles of solute}}{\text{liters of solution}}$
To calculate molality, divide the moles of solute by the mass of solvent in kilograms.	$\text{Molality} = \dfrac{\text{moles of solute}}{\text{kilograms of solvent}}$
Boiling point elevation for a solution is proportional to the molality of solute particles in the solution.	$\Delta T_b = K_b m$
Freezing point depression for a solution is proportional to the molality of solute particles in the solution.	$\Delta T_f = K_f m$

KEY TERMS

aqueous solution (11.1)

colligative property (11.6)

concentration (11.4)

entropy (11.2)

ion-dipole force (11.2)

mass/volume percent (11.4)

miscible (11.1)

molality (11.4)

molarity (11.4)

osmosis (11.6)

parts per billion (11.4)

parts per million (11.4)

percent by mass (11.4)

percent by volume (11.4)

saturated solution (11.4)

semipermeable membrane (11.6)

solubility (11.4)

solute (11.1)

solution (11.1)

solvent (11.1)

supersaturated solution (11.4)

titration (11.5)

unsaturated solution (11.4)

QUESTIONS AND PROBLEMS

The following questions and problems, except for those in the *Additional Questions* section, are paired. Questions in a pair focus on the same concept. Answers to the odd-numbered questions and problems are in Appendix D.

Matching Definitions with Key Terms

11.1   Match the key terms with the descriptions provided.
  (a) a solution that is in equilibrium with excess solute
  (b) a homogeneous mixture of two or more substances in which the solvent is water
  (c) moles of solute per liter of solution
  (d) the substance doing the dissolving; usually the component of a solution that is present in the larger amount
  (e) a measure of the tendency for matter to become randomly distributed (disordered)
  (f) a ratio that describes the maximum amount of a solute that dissolves in a particular solvent to form a saturated solution under specified conditions
  (g) mass of solute divided by mass of solution and multiplied by 10^6; in aqueous solutions it is essentially the milligrams of solute per liter of solution
  (h) when liquids can mix in all proportions
  (i) passage of solvent molecules, but not solute particles, through a membrane from a solution that is less concentrated in solute to a solution that is more concentrated
  (j) a property of a solution that depends only on the concentration of solute particles, not on the identity of the solute
  (k) volume of solute divided by volume of solution and multiplied by 100%
  (l) mass of solute divided by volume of solution and multiplied by 100%

11.2   Match the key terms with the descriptions provided.
  (a) an unstable solution that contains more dissolved solute than the maximum dictated by the solubility
  (b) a homogeneous mixture of two or more substances uniformly dispersed at a molecular or ionic level

  (c) moles of solute per kilogram of solvent
  (d) the substance being dissolved; usually the component of a solution that is present in the lesser amount
  (e) an intermolecular force between an ion and a polar molecule
  (f) the relative amounts of solute and solvent in a solution
  (g) mass of solute divided by mass of solution and multiplied by 10^9
  (h) a solution that contains less than the maximum amount of solute possible in a stable system
  (i) barrier that allows the passage of solvent but not of solute particles
  (j) a process of determining the concentration of one substance in solution by reacting it with a solution of another substance of known concentration
  (k) mass of solute divided by mass of solution and multiplied by 100%

The Composition of Solutions

11.3   Identify the solute and solvent in each of the following solutions.
  (a) seawater
  (b) steel, an alloy of iron which contains up to 1.5% carbon
  (c) oxygenated water

11.4   Identify the solute and solvent in each of the following solutions.
  (a) an IV solution that contains a water-soluble drug and water
  (b) nail polish remover, which usually contains mostly ethyl acetate and some ethanol and some water
  (c) carbonated water, which is an aqueous solution of dissolved CO_2 gas

11.5   What types of substances are strong electrolytes? How do they behave when dissolved in water?

11.6   What types of substances are nonelectrolytes? How do they behave when dissolved in water?

11.7 A molecular-level representation of a solution is shown. Is the solute an electrolyte or a nonelectrolyte?

11.8 A molecular-level representation of a solution is shown. Is the solute an electrolyte or a nonelectrolyte?

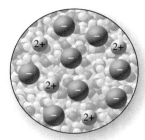

11.9 What ions, atoms, or molecules are present after the following substances mix with water? For ionic compounds, you may want to refer to the solubility rules in Table 11.1.
  (a) HBr
  (b) NH_4Cl
  (c) butanol, $CH_3CH_2CH_2CH_2OH$

11.10 What ions, atoms, or molecules are present after the following substances dissolve in water? For ionic compounds, you may want to refer to the solubility rules in Table 11.1.
  (a) LiOH
  (b) O_2
  (c) $Mg(NO_3)_2$

11.11 Which of the following diagrams best represents an aqueous solution of magnesium chloride ($MgCl_2$)?

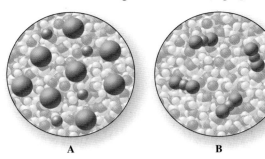

A   B

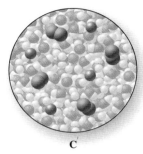

C

11.12 Which of the following diagrams best represents an aqueous solution of sodium sulfate (Na_2SO_4)?

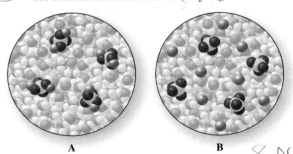

A   B

C

11.13 Write a chemical equation that describes the dissolving of the following substances in water:
  (a) solid calcium hydroxide, $Ca(OH)_2$
  (b) nitrogen gas (N_2)
  (c) liquid methanol (CH_3OH)

11.14   Write a chemical equation that describes the dissolving of the following substances in water:
(a) gaseous hydrogen chloride (HCl)
(b) chlorine gas (Cl_2)
(c) solid magnesium nitrate, $Mg(NO_3)_2$

11.15   Cooking oil, composed mostly of hydrocarbons, is very insoluble in water. Use the rule "like dissolves like" to explain why these two liquids are immiscible.

11.16   Iodine (I_2) is very soluble in hexane (C_6H_{14}) but not in water. Explain this using the rule "like dissolves like."

11.17   Why are ethanol and water so soluble in one another?

11.18   Grease is hard to wash from your hands using only water. What does this tell us about the intermolecular forces in grease?

11.19   Use the rule "like dissolves like" to predict whether each of the following should be soluble in water.
(a) benzene (C_6H_6)
(b) ethylene glycol ($HOCH_2CH_2OH$)
(c) potassium iodide (KI)

11.20   Use the rule "like dissolves like" to predict whether each of the following should be soluble in hexane (C_6H_{14}), a nonpolar hydrocarbon.
(a) sodium nitrate ($NaNO_3$)
(b) methanol (CH_3OH)
(c) bromine (Br_2)

The Solution Process

11.21   What types of forces must be broken when an ionic solid dissolves in water? What new forces are formed?

11.22   What types of forces must be broken when a polar molecular substance dissolves in water? What new forces are formed?

11.23   What types of forces must be broken when a nonpolar molecular substance dissolves in a nonpolar solvent? What new forces are formed?

11.24   Nonpolar substances sometimes dissolve in water to a small extent. Explain why this might occur.

11.25   When ammonium nitrate (NH_4NO_3) dissolves in water, the solution feels cool. What can we say about the relative strengths of the forces between solute and solvent? Between the particles in the pure substances?

11.26   When NaOH dissolves in water, the solution feels warm. What can we say about the relative strengths of the forces between solute and solvent? Between the particles in the pure substances?

11.27   What factors drive the formation of a solution?

11.28   What is entropy? Does it increase or decrease during solution formation?

Factors That Affect Solubility

11.29   In terms of intermolecular forces, explain why ionic compounds are insoluble in nonpolar solvents.

11.30   In terms of intermolecular forces, explain why nonpolar solutes are insoluble in polar solvents.

11.31   In terms of intermolecular forces, explain why NaCl is soluble in water.

11.32   In terms of intermolecular forces, explain why I_2 is soluble in C_6H_{14}.

11.33   Normal tap water contains a small amount of dissolved oxygen. What happens to the solubility of oxygen when the water is heated?

11.34   What is thermal pollution of water, how does it relate to gas solubility, and how does it affect fish?

11.35   Would you expect the solubility of oxygen and nitrogen at high altitudes (lower atmospheric pressure) to be greater or less than that at sea level? Explain.

11.36   How does the concentration of dissolved oxygen and nitrogen in the bloodstream change as a scuba diver swims to lower depths of the ocean? Explain.

11.37   A scuba diver who has the bends is treated by being placed in a chamber that has a high pressure of air. The air pressure in the chamber is lowered gradually. How does this treatment help the diver?

11.38   Why do carbonated beverages go "flat" after being opened for some time?

11.39   The diagram shows a representation of oxygen dissolved in water at about 25°C. Draw a diagram that represents what the solution might look like after the solution temperature increases.

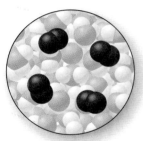

11.40   The diagram in Question 11.39 shows a representation of oxygen dissolved in water. Draw a diagram that represents what the solution might look like after the partial pressure of O_2 above the solution increases.

Measuring Concentrations of Solutions

11.41   The solubility of KCl at 20°C is 34.0 g per 100 g of water. What is the maximum amount (in grams) of KCl that will dissolve in 50.0 g of water at 20°C?

11.42   The solubility of ammonium chloride at 30°C is 41.1 g per 100 g of water. What is the maximum amount (in grams) of ammonium chloride that will dissolve in 346.0 g of water at this temperature?

11.43   What is the difference between a saturated solution and an unsaturated solution?

11.44   What is the difference between a saturated solution and a supersaturated solution?

11.45   How might you prepare a saturated solution of a substance of unknown solubility?

11.46   The solubility of ammonium chloride increases with temperature. How might you prepare a supersaturated solution of ammonium chloride?

11.47   The solubility of $Ca(OH)_2$ at 30°C is 0.15 g per 100 g of water. Describe the solution formed when 1.0 g $Ca(OH)_2$ is added to 100.0 g of water.

11.48   The solubility of $Ca(OH)_2$ is 0.18 g per 100 g of water at 10°C. Describe the solution formed when 0.15 g $Ca(OH)_2$ is added to 100.0 g of water.

11.49   What is the percent-by-mass concentration of NaCl in a solution that is prepared by adding 15.0 g of NaCl to 90.0 g of water?

14.3%

11.50   What is the percent-by-mass concentration of sucrose in a solution that is prepared by adding 25.0 g of sucrose to 100.0 g of water?

11.51   What mass of KI is dissolved in 400.0 g of a solution that is 15.0% KI by mass?

11.52   What mass of sodium phosphate is dissolved in 255 g of a solution that is 20.0% by mass Na_3PO_4?

11.53   What is the percent-by-mass concentration of acetic acid $(HC_2H_3O_2)$ in a vinegar solution that contains 54.50 g acetic acid in a 1.000-L solution? The density of this solution is 1.005 g/mL.

11.54   What is the percent-by-mass concentration of a 100.0-mL H_3PO_4 solution that contains 88.20 g of H_3PO_4? The density of the solution is 1.40 g/mL.

11.55   For what types of solutions are concentrations commonly expressed as percent by volume? Why?

11.56   When dealing with percent-by-volume concentrations of solutions composed of two liquids, why must the total volume be measured instead of using the sum of the volumes of the component liquids?

11.57   A solution of methanol is prepared by dissolving 35.0 mL of methanol in sufficient water to give a total volume of 115.0 mL. What is the percent-by-volume concentration of methanol?

11.58   A solution of bromine is prepared by dissolving 2.0 mL of bromine in sufficient carbon tetrachloride to give a total volume of 40.0 mL. What is the percent-by-volume concentration of bromine?

11.59   How many moles of HCl are contained in 100.0 mL of 0.15 M HCl?

11.60   What volume of 0.25 M NaOH contains 0.050 mol NaOH?

11.61   A solution contains 20.5 g of sodium chloride dissolved in sufficient water to give a total mass of 166.2 g. What is the molality of this solution?

11.62   How many moles of sucrose $(C_{12}H_{22}O_{11})$ are contained in 50.0 g of a 0.85 m sucrose solution?

11.63   What is the molar concentration of ions in a 1.5 M KNO_3 solution? What is the molal concentration of ions in a 1.5 m KNO_3 solution?

11.64   What is the molar concentration of ions in a 2.5 M $MgCl_2$ solution? What is the molal concentration of ions in a 2.5 m $MgCl_2$ solution?

11.65   A 250.0-g sample of pond water contains 2.4 mg arsenic. What is the mass percent of arsenic in the pond? How many parts per million of arsenic are in the pond? How many parts per billion?

11.66   A 375.0-g sample of river water taken near an industrial plant contains 37 mg of chromium. What is the mass percent of chromium in the river? How many parts per million of chromium are in the river? How many parts per billion?

11.67   Drinking water may contain a low concentration of lead ion (Pb^{2+}) due to corrosion of old lead pipes. The EPA has determined that the maximum safe level of lead ion in water is 15 ppb. Suppose a sample of tap water was determined to have a lead ion concentration of 0.0090 ppm. Assume the density of the solution is 1.00 g/mL.

(a) What is the concentration of lead ion in the tap water in units of ppb? Is it safe to drink?

(b) What is the concentration of lead ion in units of mg/mL?

(c) What mass of lead ion is in 100.0 mL of this drinking water?

(d) How many moles of lead ion are in 100.0 mL of the water?

11.68   Excessive levels of nitrates found in drinking water can cause serious illness. When nitrates are converted to nitrites in the body, the nitrites can interfere with the oxygen-carrying capacity of the blood, causing blueness of the skin. The EPA has determined that the maximum safe level of nitrate ion in water is 10 ppm. Suppose a sample of tap water was determined to have a nitrate ion concentration of 95 ppb. Assume the density of the solution is 1.00 g/mL.

(a) What is the concentration of nitrate ion in units of ppm? Is it safe to drink?

(b) What is the concentration of nitrate ion in units of mg/mL?

(c) What mass of nitrate ion is in 100.0 mL of this drinking water?

(d) How many moles of nitrate ion are in 100.0 mL of the water?

Quantities for Reactions That Occur in Aqueous Solution

11.69   One way to remove lead ion from water is to add a source of iodide ion so that lead iodide will precipitate out of solution:

$$Pb^{2+}(aq) + 2I^-(aq) \longrightarrow PbI_2(s)$$

(a) What volume of a 1.0 M KI solution must be added to 100.0 mL of a solution that is 0.15 M in Pb^{2+} ion to precipitate all the lead ion?

(b) What mass of PbI_2 should precipitate?

11.70   If a solution of $AgNO_3$ is added to an HCl solution, insoluble AgCl will precipitate:

$$AgNO_3(aq) + HCl(aq) \longrightarrow AgCl(s) + HNO_3(aq)$$

(a) What volume of a 1.0 M $AgNO_3$ solution must be added to 250.0 mL of a 0.100 M HCl solution to precipitate all the chloride ion?

(b) What mass of AgCl should precipitate?

11.71   How many moles of sodium carbonate (Na_2CO_3) are required to precipitate the calcium ion from 850.0 mL of a 0.35 M $CaCl_2$ solution? (Begin by writing a balanced chemical equation for the reaction.)

11.72   How many moles of calcium chloride $(CaCl_2)$ are required to precipitate the carbonate ion from 1.5 L of a 0.75 M Na_2CO_3 solution? (Begin by writing a balanced chemical equation for the reaction.)

0.015

11.73 What volume of 0.1050 *M* NaOH is required to neutralize 20.00 mL of 0.1500 *M* H_2SO_4? (Begin by writing a balanced chemical equation for the reaction.)

11.74 What volume of 0.850 *M* H_2SO_4 is required to neutralize 20.00 mL of 0.1033 *M* NaOH? (Begin by writing a balanced chemical equation for the reaction.)

11.75 When 20.00 mL of H_2SO_4 is titrated with 0.9854 *M* NaOH, 35.77 mL of NaOH is required to neutralize the H_2SO_4. What is the molarity of the H_2SO_4 solution?

11.76 When 20.00 mL of 1.005 *M* H_2SO_4 is titrated with a solution of NaOH, 44.68 mL of NaOH is required to neutralize the H_2SO_4. What is the molarity of the NaOH solution?

11.77 What volume of 0.1000 *M* NaOH is required to neutralize each of the following solutions?
(a) 10.00 mL of 0.1000 *M* HCl
(b) 15.00 mL of 0.3500 *M* HNO_3
(c) 25.00 mL of 0.0500 *M* H_3PO_4

11.78 What volume of 0.5000 *M* HCl is required to neutralize each of the following solutions?
(a) 35.00 mL of 0.0500 *M* NaOH
(b) 10.00 mL of 0.200 *M* $Ba(OH)_2$
(c) 15.00 mL of 0.2500 *M* NH_3

Colligative Properties

11.79 What is a colligative property?

11.80 List some colligative properties.

11.81 When osmosis occurs between two solutions of different concentrations, what would you observe about each solution? Explain your observations in terms of molecular motion.

11.82 When osmosis occurs between an aqueous solution and pure water, can both sides ever become equal in concentration? Explain.

11.83 Describe the process of reverse osmosis.

11.84 What is desalination?

11.85 What will happen to a blood cell that is placed in an aqueous solution that has a high salt concentration relative to the blood cell concentration?

11.86 What will happen to a blood cell that is placed in pure water?

11.87 What happens to each of the following properties of a liquid when a solute is dissolved in the liquid to make a solution?
(a) vapor pressure
(b) boiling point
(c) freezing point

11.88 What happens to each of the following properties of a solution when the concentration of solute is decreased?
(a) osmotic pressure
(b) boiling point
(c) freezing point

11.89 Sweet tea is often made by dissolving a lot of sugar in water, brewing the tea, and then chilling. Explain what happens to the boiling point of water when sugar is added.

11.90 Some arctic fish produce a substance similar to the ethylene glycol used as antifreeze in cars. In terms of the freezing point of water, explain how the compound might help them survive in their habitat.

11.91 Compare a 2.5 *m* solution of glucose ($C_6H_{12}O_6$) to pure water.
(a) What is the difference in boiling points?
(b) What is the boiling point of the 2.5 *m* glucose solution?

11.92 Compare a 5.0 *m* solution of ethylene glycol ($HOCH_2CH_2OH$) to pure water.
(a) What is the difference in freezing points?
(b) What is the freezing point of a 5.0 *m* ethylene glycol solution?

11.93 Which solution, 1.0 *m* NaCl or 1.0 *m* glucose ($C_6H_{12}O_6$) should have the highest boiling point? Explain.

11.94 Which solution, 1.0 *m* $NaNO_3$ or 1.0 *m* $MgCl_2$, should have the highest boiling point? Explain.

11.95 The following diagrams represent a solution of NaCl, a solution of glucose, and a more dilute solution of glucose. Which represents the aqueous solution with the lowest freezing point? Which solution is which?

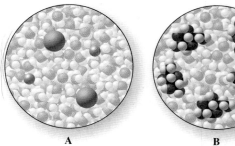

A   B

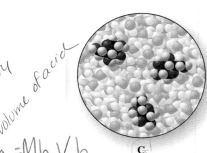

C

[handwritten notes:]

molarity   volume of acid

① $M_a V_a = M_b V_b$ volume base

$[H_2SO_4]$ $(NaOH)$

② count # of H's in the acid ↯ put that # in front of the M
③ then count # of OH's ...

$\dfrac{2 M_a V_a}{2 V_a} = \dfrac{1 M_b V_b}{2 V_a}$

$\dfrac{(0.9854)(35.77)}{2(20.00)} = \boxed{0.8812}$

11.96 The following diagrams represent a solution of $MgCl_2$, a solution of $NaNO_3$, and a more dilute solution of $NaNO_3$. Which represents the aqueous solution with the lowest freezing point? Which solution is which?

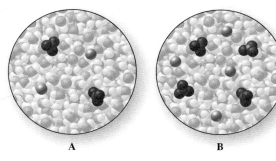

A   B

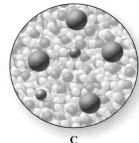

C

11.97 Determine the moles of solute particles dissolved in 1.0 kg of water for each of the following solutions.
(a) 2.0 m CH_3OH
(b) 2.5 m NaCl
(c) 1.0 m $Al(NO_3)_3$

11.98 Determine the moles of solute particles dissolved in 1.0 kg of water for each of the following solutions.
(a) 2.0 m HCl
(b) 0.15 m Na_2SO_4
(c) 1.0 m $C_{12}H_{22}O_{11}$

11.99 Which solution, 1.0 m $MgCl_2$ or 1.0 m $NaNO_3$, should have the greatest osmotic pressure? Explain.

11.100 Which solution, 1.0 m NaCl or 1.0 m CH_3OH, should have the greatest osmotic pressure? Explain.

Additional Questions

11.101 Vitamin D is soluble in fat tissue, which is primarily non-polar. What does this tell us about the polarity of vitamin D molecules?

11.102 Vitamin B is a water-soluble vitamin. What does this tell us about the polarity of vitamin B?

11.103 When an ionic solid dissolves in water, which of the processes involved in the solution process require energy? Which release energy?

11.104 What changes in temperature and pressure increase the solubility of a gas in solution?

11.105 The solubility of NaCl at 30°C is 36.3 g per 100 g of water. What mass of NaCl is dissolved in a saturated solution that contains 500.0 g of water?

11.106 The solubility of KNO_3 increases as the temperature rises. If a saturated solution of KNO_3 is heated to a higher temperature, will it still be a saturated solution? Explain.

11.107 What is the molarity of a concentrated HBr solution if the solution is 48.0% HBr by mass and has a density of 1.50 g/mL?

11.108 A solution of sulfuric acid contains 24.0% by mass of H_2SO_4 and has a density of 1.17 g/mL. What is the molarity of H_2SO_4 in this solution?

11.109 Which of the following concentration units has values that will not change when the temperature of the solution is changed? Explain your reasoning for each.
(a) percent by mass
(b) percent by volume
(c) molarity
(d) molality

11.110 A 1.0 M acetic acid solution has a density of 1.005 g/mL. What is the molality (m) of this solution?

11.111 For dilute aqueous solutions, molality and molarity concentration values are similar. Why?

11.112 For each of the following combinations of solutions, determine the mass of silver carbonate (Ag_2CO_3) that should precipitate. (Begin by writing a chemical equation for the reaction.)
(a) 100.0 mL of 0.200 M $AgNO_3$ and 100.0 mL of 0.200 M Na_2CO_3
(b) 200.0 mL of 0.200 M $AgNO_3$ and 100.0 mL of 0.200 M Na_2CO_3
(c) 100.0 mL of 0.200 M $AgNO_3$ and 200.0 mL of 0.200 M Na_2CO_3

11.113 For each of the following combinations of solutions, determine the mass of barium sulfate ($BaSO_4$) that should precipitate. (Begin by writing a chemical equation for the reaction.)
(a) 500.0 mL of 0.100 M $BaCl_2$ and 90.0 mL of 0.500 M K_2SO_4
(b) 100.0 mL of 0.100 M $BaCl_2$ and 100.0 mL of 0.500 M K_2SO_4
(c) 100.0 mL of 0.100 M $BaCl_2$ and 500.0 mL of 0.500 M K_2SO_4

11.114 What mass of NaOH is required to neutralize 1.8 L of 2.0 M HCl?

11.115 Sodium bicarbonate is sometimes used to neutralize acid spills. If 2.0 L of 6.0 M H_2SO_4 is spilled, what mass of sodium bicarbonate ($NaHCO_3$) is required to neutralize all the H_2SO_4?

11.116 What is the boiling point of a 3.0 m $Ca(NO_3)_2$ solution in water?

11.117 The freezing points of two NaCl solutions were determined to be −1.15°C for solution A and −4.10°C for solution B. Which solution has the highest concentration of NaCl?

11.118 If a 100.0-g sucrose ($C_{12}H_{22}O_{11}$) solution has a boiling point of 101.5°C, how many moles of sucrose are in the solution? How many grams?

Reaction Rates and Chemical Equilibrium

As Ellen and Chad are driving to the mall on a hot summer day, they hear an ozone alert announced on the radio. The newscaster says that the ozone level in the air is around 130 parts per billion (ppb), which is above safe levels. He advises his listeners—especially those with respiratory problems—to stay indoors as much as possible.

"Why is ozone so bad," Ellen asks. "Isn't it just another form of oxygen, and doesn't it shield us from the ultraviolet radiation that causes skin cancer?" As Ellen and Chad discuss these questions, they reach the top of a hill. Looking down on the city in the valley below, Chad notices more smog than usual and wonders if it has anything to do with the ozone warning.

Ozone (O_3) is a form of oxygen that is more reactive than diatomic oxygen (O_2). Breathing ozone damages alveoli, the key oxygen exchange structures in the lungs. Even moderate concentrations of ozone make breathing more difficult, especially during exercise. Higher concentrations can permanently damage lung tissue and increase susceptibility to bacterial pneumonia.

Ozone is only bad for us when its level increases in the air layer closest to the Earth's surface, the troposphere. The concentration of ozone in the troposphere is normally 35 to 50 ppb, which is considered a safe level. A small amount of ozone occurs naturally, mostly as a product of lightning storms. Excess ozone is produced when nitrogen oxides and other pollutant gases from car exhausts and electrical power plants combine with oxygen in the presence of sunlight. When there is little wind to disperse pollutant gases, the ozone level rises in large cities on hot days. Chad was right. The thick smog and the elevated ozone level go together.

Ellen recalls a news report she saw on television about the harmful effects of ozone *depletion* in the stratosphere. The stratosphere is a layer of the atmosphere 18 to 50 km (10 to 30 mi) above the Earth (Figure 12.1). The ozone layer, which contains the highest concentrations of ozone, is in the upper stratosphere. It is beneficial

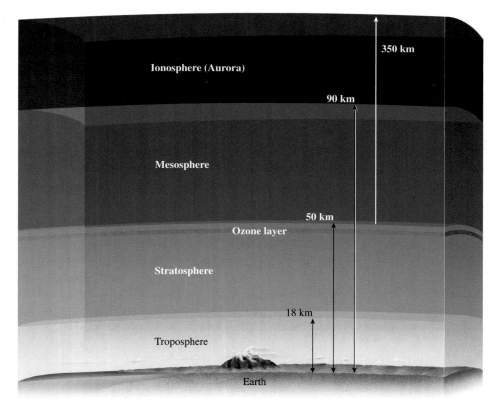

Figure 12.1
The atmosphere is a blanket of air that extends more than 560 km above the Earth's surface. Its four regions differ in chemical composition and density. The troposphere contains the air we breathe. It has the greatest density of gas molecules. The stratosphere, which extends from 18 to 50 km above the Earth's surface, is less dense. The ozone layer is part of the upper stratosphere.

to life on Earth because it acts to filter out harmful ultraviolet (UV) radiation from the Sun. How does it do that? When an ozone molecule absorbs UV radiation, it splits into an oxygen molecule and an oxygen atom. The single oxygen atom is very unstable and reacts quickly with another O_3 molecule to form two O_2 molecules:

$$O_3(g) \xrightarrow{\text{UV light}} O_2(g) + O(g)$$

$$O(g) + O_3(g) \longrightarrow 2O_2(g)$$

The absorption of UV radiation destroys ozone, so for its concentration in the stratosphere to remain constant, ozone must also continually be *created* by chemical reactions. Sunlight helps to create ozone from O_2. Radiation from the Sun can cause O_2 to split into two oxygen atoms. Then each oxygen atom can react with an O_2 molecule to form an O_3 molecule:

$$O_2(g) \xrightarrow{\text{UV light}} 2O(g)$$

$$O_2(g) + O(g) \longrightarrow O_3(g)$$

The amount of ozone in even the most concentrated areas of the stratosphere is about 8 ppm (or 8000 ppb), which is quite small compared to the amount of O_2, which is 20,000 ppm. The *relative amounts* of ozone and oxygen remain constant as the reactions occur repeatedly, breaking down and reforming about 30 million kg of ozone every day.

Many factors influence the delicate balance between the amounts of O_3 and O_2 in the stratosphere. Changes in wind patterns and temperatures cause the balance to fluctuate, so the ozone layer is not the same all over the world and at all times of the year. More worrisome, Chad tells Ellen, is the effect of human-made pollutants such as chlorofluorocarbons (CFCs). Used commercially as refrigerants and propellants, CFCs upset the ozone balance by destroying more ozone molecules than can be replaced by natural processes. An example of a CFC is dichlorodifluoromethane (CF_2Cl_2):

CFCs are very stable and do not react under normal conditions in the troposphere around us. (Before 1973, they were thought to be harmless to the environment because they were so unreactive.) In 1973, Rowland and Molina from the University of California at Irvine began to study the fate of CFCs in the stratosphere. They found that when ultraviolet light interacts with a CFC molecule, it causes a carbon-chlorine bond to break:

$$CF_2Cl_2(g) \xrightarrow{\text{UV light}} CF_2Cl(g) + Cl(g)$$

The chlorine atom reacts readily with ozone to produce oxygen (O_2) and ClO:

$$Cl(g) + O_3(g) \xrightarrow{\text{UV light}} ClO(g) + O_2(g)$$

The ClO is unstable and reacts with another molecule of O_3, forming an O_2 molecule and a Cl atom:

$$ClO(g) + O_3(g) \xrightarrow{\text{UV light}} 2O_2(g) + Cl(g)$$

Two ozone molecules are destroyed, and the Cl atom that started it all is reproduced. It's available to destroy two more ozone molecules. Once a Cl atom forms in the stratosphere, it can destroy thousands of ozone molecules. Because the chlorine atoms speed up the reaction and can be reused, Cl is considered a *catalyst* for the destruction of ozone.

Due to the harmful effects of even small concentrations of CFCs in the stratosphere, CFC production was banned on a global scale in 1995. However, CFCs are still used in older model car and home air conditioners. Because of their continued use and their chemical stability, it will be many years until the CFCs released in the past disappear from the stratosphere. Recent studies have shown that the rate of ozone depletion has slowed, but it will be many decades before the original state of balance can be reestablished.

When, Ellen wonders, will the relative rates of ozone destruction and re-formation balance again? The reaction rate, or how fast a reaction occurs, depends on many factors. One is the presence of a catalyst, such as the Cl atoms in the destruction of ozone. In this chapter, we will look at two important aspects of chemical reactions: reaction rate and chemical equilibrium. We will learn how catalysts, as well as temperature and concentration, influence changes at the molecular level.

Questions for Consideration

12.1 What conditions affect reaction rates?
12.2 How do molecular collisions explain chemical reactions?
12.3 How do concentration, temperature, and catalysts affect molecular collisions and reaction rates?
12.4 What is chemical equilibrium?
12.5 What is the state of equilibrium for a specific reaction?
12.6 What happens when we disturb a system at equilibrium?

12.1   REACTION RATES

Some chemical reactions occur in seconds, while others take days, even years (Figure 12.2). The reactions in the ozone layer are rapid, occurring millions of times daily. Combustion reactions of fuels with oxygen also occur rapidly, but corrosion of metals with oxygen, such as iron in a rusting car, is comparatively slow. Active metals react with water immediately, while a similar reaction with less active metals such as aluminum and zinc takes longer. **Reaction rate,** or *rate of reaction,* is a measure of how fast a reaction occurs.

A reaction can occur at different rates under different conditions. Conditions that affect the rate of reaction include temperature, reactant concentration, surface area, and the presence of a catalyst. Higher temperatures generally cause reactions to occur faster. For example, milk and other foods spoil more quickly when left in a warm place. Can you think of other reactions that occur faster at higher temperatures? Or slower at lower temperatures?

The rate of reaction is analogous to speed. We measure driving speed as distance per unit time, usually in miles per hour (mi/h) or kilometers per hour (km/h). Similarly reaction rates can be described as a change in molarity of a reactant or product per second (M/s).

A light stick starts to glow when the seal on a contained liquid is broken, allowing it to mix with another substance. The chemical reaction gives off light energy. Why does a light stick glow longer if placed in the freezer after the reaction has begun?

Figure 12.2
Reactions occur at a wide range of rates. For example, the sulfur in a safety match reacts with oxygen more quickly than iron metal reacts with oxygen in corrosive reactions.

Figure 12.3

Acetic acid reacts with sodium bicarbonate ($NaHCO_3$) to form CO_2 gas. The photos show the reaction at its beginning and after 10 s, comparing 1 *M* and 3 *M* acetic acid as reactants. Because the reaction with 3 *M* acetic acid produces more gas over the same time period, we can see that its reaction rate is faster.

1 *M*   3 *M*
Before reaction

1 *M*   3 *M*
After 10 s

Which has more surface area, a cube of sugar or granular sugar? Which dissolves more quickly?

A   **B**

Figure 12.4

(A) The heated solid iron nail exposes less surface area and fewer atoms, so it reacts slowly. (B) The greater surface area in steel wool exposes many atoms to the air, so they ignite explosively.

Figure 12.5

Without a catalyst, hydrogen peroxide (H_2O_2) decomposes slowly. The enzyme called catalase catalyzes the decomposition of hydrogen peroxide, so the reaction occurs rapidly. In this photo H_2O_2 breaks down in the presence of the catalase in a piece of liver, releasing bubbles of oxygen gas.

Concentration also influences reaction rate. Consider the reaction that occurs when baking soda is added to vinegar. Vinegar is a solution of acetic acid in water. How does the concentration of acetic acid affect its reaction with baking soda (Figure 12.3)? Increasing the concentration of a reactant generally increases the reaction rate. Can you think of other reactions that occur faster when the concentrations of reactants are increased?

If the reactant is a solid, increasing its surface area also increases reaction rate. Metals in their granular form have greater surface area and react more quickly with air than large pieces of metal. The aluminum used in fireworks is in a granular form so that it will burn quickly. The effect of increased surface area for the reaction of iron with oxygen is shown in Figure 12.4.

Catalysts increase the rate of a reaction. The chemical catalysts in the catalytic converters of automobiles increase the rate at which the polluting gases NO and NO_2 are converted to harmless N_2 and O_2 gases. In living systems, protein enzymes act as catalysts, speeding up the countless chemical reactions that are necessary to sustain life. For example, the harmful compound hydrogen peroxide (H_2O_2) can sometimes form in the human body. A catalyst called *catalase* speeds its decomposition into oxygen and water (Figure 12.5). The effects of catalysts are not always desirable—as, for example, the role of chlorine atoms from CFCs in the depletion of the Earth's ozone layer.

12.2 COLLISION THEORY

Why does the rate of chemical reactions vary so widely? **Collision theory** states that in order for a reaction to occur, reactant molecules must collide in the proper orientation and with sufficient energy. Because only a fraction of the collisions meet these requirements, only a few result in a reaction and yield products. Consider, for example, the reaction between O_3 and NO that occurs in smog:

$$O_3(g) + NO(g) \longrightarrow O_2(g) + NO_2(g)$$

In order for this reaction to occur, an O_3 molecule and an NO molecule must first collide. Of the many such collisions that actually happen, only some of them have the orientation required for the reaction to occur. Convince yourself this is plausible. Draw the structures of the reactants and products. Which bonds must break and which bonds should form during this reaction? How should the reactant molecules be oriented in order for products to form?

Examine this collision:

This shows the oxygen atom from NO (left) heading straight toward an oxygen atom from O_3 (right); but the equation tells us that a new nitrogen-oxygen bond must be formed. This collision will fail. The atoms aren't lined up correctly.

An effective collision must bring together a nitrogen atom and a terminal oxygen atom so that an oxygen atom can be transferred to the nitrogen atom, leaving an O_2 molecule behind. Does the following collision have a proper orientation?

Animation: Trajectory of Reaction
Online Learning Center, Animations

Chemistry Animations Library, Trajectory of Reaction, 1Orientation.avi.

Yes, it does. The nitrogen atom on NO will collide with a terminal oxygen atom on O_3. It is lined up just right so that an oxygen atom from the O_3 molecule can be transferred to the nitrogen atom in the NO molecule.

If an NO molecule and an O_3 molecule collide with the proper orientation, will a reaction automatically occur? No, not unless they collide with sufficient energy. In a given sample, some reactant molecules have more kinetic energy than others. Usually some fraction of the molecules has enough energy to have effective collisions. Reactant molecules that collide with proper orientation and sufficient energy can form a product.

Let's take a closer look at the energy required for a reaction. In a chemical reaction, bonds in the reactants must be broken so that new bonds in the products can form. Different reactions occur at different rates because the amount of energy required to break bonds in the reactant molecules is different. As we discussed in Chapter 6, an energy diagram shows the relative energies of the reactants and products (Figure 12.6). This type of energy diagram, although useful, shows only the *net* energy change for the reaction, not the energy changes that occur *during* it.

> The collision requirements for a reaction are like playing darts. The dart must have sufficient energy (you must throw it fast enough), and it must have the proper orientation (you must aim it correctly).

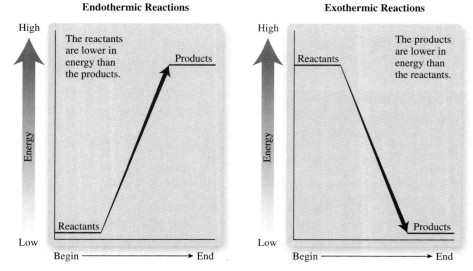

Figure 12.6

In endothermic reactions, the average energy of the products is greater than that of the reactants. Energy is absorbed. In exothermic reactions, the average energy of the products is less than that of the reactants. Energy is released. These energy diagrams show only the difference in energy between the reactants and products, not the energy changes that occur during the reaction.

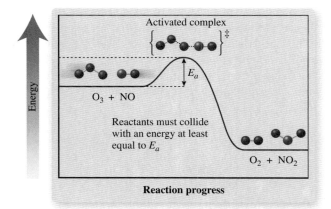

Figure 12.7

This diagram shows the activation energy E_a, the energy barrier that must be overcome by reactants before they can be converted to products. This is the amount of energy required to form the activated complex, which exists briefly as bonds in the reactants break and bonds in the products form.

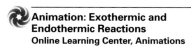

Animation: Exothermic and Endothermic Reactions
Online Learning Center, Animations

Chemistry Animations Library, Exothermic and Endothermic Reactions, 1Energy_Activation.rm.

In both endothermic and exothermic reactions, reactants must overcome an *energy barrier* before they can change to products (Figure 12.7). Energy is needed to break bonds in reactants before the reactants can be converted into products. The minimum amount of energy needed to overcome this energy barrier is called the **activation energy, E_a.** If we compare two similar reactions at the same temperature, the reaction with the lower activation energy is the faster reaction.

Reactant molecules that collide with the proper orientation and sufficient energy form an **activated complex.** It is a short-lived, unstable, high-energy chemical species that must be achieved before products can form. The activated complex is generally designated by a chemical formula in brackets with a superscript double dagger ($\ddagger$), such as $\{O_3NO\}^{\ddagger}$ for the reaction of O_3 and NO in Figure 12.7. In most reactions the activated complex is so short-lived, its actual structure is unknown. However, we can imagine that it is a molecular form that is in the process of forming new bonds while breaking old ones.

Animation: Activation Energy
Online Learning Center, Animations

Chemistry Animations Library, Activation Energy, 2Energy_Activation.exe.

EXAMPLE 12.1 Energy Diagrams with Activated Complexes

The following reaction is exothermic:

$$2HI(g) \longrightarrow H_2(g) + I_2(g)$$

Draw an energy diagram that shows the relative energies of the reactants, products, and the activated complex. Label the diagram with molecular representations of the reactants, products, and a possible structure for the activated complex.

Solution:

The average energy of the products is lower than the average energy of the reactants because the reaction is exothermic. The energy of the activated complex corresponds to the top of the energy barrier. A possible structure for the activated complex shows a species that would result from a collision with the proper orientation, allowing H—H and I—I bonds to form while H—I bonds break.

$$2HI(g) \longrightarrow H_2(g) + I_2(g)$$

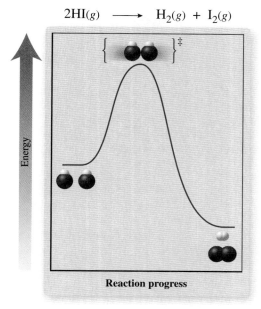

Practice Problem 12.1

The following reaction is an endothermic reaction:

$$2NO_2(g) \longrightarrow 2NO(g) + O_2(g)$$

Draw an energy diagram that shows the relative energies of the reactants, products, and the activated complex. Label the diagram with molecular representations of the reactants, products, and a possible structure for the activated complex.

Further Practice: 12.19 and 12.20

Each reaction has its own specific energy diagram and value for activation energy. Reactions with large activation energies tend to be slow because a relatively small fraction of reactants have sufficient energy for an effective collision. Reactions with small activation energies tend to be fast reactions because a large fraction of reactants have sufficient energy for an effective collision. However, conditions such as concentration and temperature change the rate of a reaction. They change the number of collisions that are effective without changing the activation energy.

12.3   CONDITIONS THAT AFFECT REACTION RATES

The conditions that most often affect reaction rates are changes in concentration, surface area, and temperature. The presence or addition of catalysts is also a significant factor. We will next explore how collision theory explains all these effects.

Figure 12.8

The collision rate increases when the concentration of reactants increases. The fraction of effective collisions, however, remains constant (1:3 in this example), because the kinetic energy of the system is unchanged.

The ratio of effective collisions is 1:3.   The ratio of effective collisions is 2:6, or 1:3.

Concentration and Surface Area

A reaction may go faster when the concentration of one or more of the reactants increases. Such an increase in concentration increases the number of reactants per unit volume. Because molecules are closer together, the number of collisions per unit time increases. As total collisions increase, the number of molecules that have the orientation and energy required for the reaction also increases. However, the *fraction* of effective collisions remains the same, because the temperature and kinetic energy are constant (Figure 12.8).

The effect of increasing the surface area of a solid in a reaction is similar to increasing the concentration. For example, a ground-up piece of iron has more iron atoms exposed to the air than an iron nail, so it will rust more quickly. Increasing the surface area increases the number of atoms that are exposed so that they can collide with other reactants.

Temperature

Why does a reaction proceed more rapidly at higher temperatures? As you learned in Chapter 10, the average kinetic energy of a substance increases when the temperature rises. An increase in kinetic energy causes the reaction rate to increase in two ways: It increases the collision rate, and it increases the fraction of collisions that are effective (Figure 12.9). The collision rate increases because the molecules move faster at higher temperatures, so they collide more frequently. The fraction of effective collisions also increases, because the average kinetic energy of the molecules increases. Consequently more of the reactant molecules attain the required activation energy. Temperature effects on reaction rates are significant. For a typical reaction, the rate approximately doubles for every 10°C increase in temperature.

The chirp rate of the snowy tree cricket is temperature-dependent. It is about 17 chirps per 15 s in cold weather. In very warm weather, it is about 55 chirps per 15 s.

EXAMPLE 12.2	**Collision Rates and the Fraction of Effective Collisions**

Suppose the collision rate between reactant molecules A and B at 25°C is 10,000 collisions per second. The number of effective collisions at this same temperature is 100 per second. How will each of the following affect the *total number of collisions* and the *fraction of effective collisions* between molecules A and B?

(a) The temperature increases.
(b) The concentration of reactant A increases.

Solution:

(a) When the temperature rises, the average kinetic energy and average velocity of the molecules increase. Increased molecular speed causes an increase in the rate of both molecular collisions and the fraction of effective collisions. Thus we expect a total number of collisions greater than 10,000 and a fraction of effective collisions greater than 100 in every 10,000.

(b) When the concentration of A increases, the number of collisions between A and B also increases, because more A molecules can make contact with each B molecule. Thus we expect a total number of collisions greater than 10,000 and a greater number of effective collisions. However, the fraction of effective collisions will stay equal to 100 in every 10,000, because no energy has been added to the system.

Practice Problem 12.2

How will each of the following affect the *total number of collisions* and the *fraction of effective collisions* between molecules A and B as described in the example statement.

(a) The temperature decreases.
(b) The surface area of reactant A increases.

Further Practice: 12.21 and 12.22

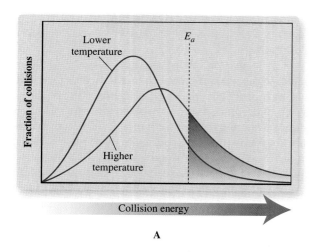

A

Lower temperature
Lower collision rate
Lower rate of effective collisions
Smaller fraction of effective collisions

Higher temperature
Greater collision rate
Greater rate of effective collisions
Larger fraction of effective collisions

B

Figure 12.9
(A) Increasing the temperature of the reactants increases the average kinetic energy. At the higher temperature, a greater fraction of the reactant molecules have sufficient energy (E_a) for a reaction (both colored regions). (B) Increasing the temperature of the reactants also increases their average velocity. Molecules that move faster collide more frequently, so the collision rate also increases.

In Example 12.3, you will use collision theory to help Ellen and Chad solve problems in the laboratory.

EXAMPLE 12.3   **Using Collision Theory to Explain Reaction Rates**

In her chemistry lab Ellen is determining ways to increase the rate of reaction of magnesium metal with hydrochloric acid:

$$Mg(s) + 2HCl(aq) \longrightarrow MgCl_2(aq) + H_2(g)$$

The first time she runs the reaction, she places a 5.0-g strip of magnesium metal in a 0.10 M HCl solution. Suggest ways that Ellen can increase the rate of this reaction. For each suggestion, use collision theory to explain why the reaction rate should increase.

Solution:

One way to increase the reaction rate is to increase the concentration of HCl, possibly to 0.20 M or higher. The increased concentration should increase the collision rate between hydronium ions and magnesium atoms. Although Ellen cannot increase the concentration of magnesium metal, she can increase its surface area. Cutting the strip into tiny pieces would make more magnesium atoms accessible for collisions. The collision rate would increase, causing the reaction rate to increase. A third way is to heat the reactant mixture on a hot plate (hydrogen is flammable). An increase in temperature would increase both the total number of collisions and the fraction of effective collisions, thus increasing the reaction rate.

Practice Problem 12.3

Under normal conditions the iron in steel wool reacts with oxygen in a slow reaction as it corrodes or rusts:

$$4Fe(s) + 3O_2(g) \longrightarrow 2Fe_2O_3(s)$$

In his chemistry lab Chad is working to increase the rate of the rusting process without placing the steel wool in a flame. Suggest ways he can do this. For each suggestion, use collision theory to explain why the reaction rate should increase.

Further Practice: 12.27 and 12.28

Catalysts

A **catalyst** alters the pathway in which a reaction occurs without itself being consumed in the reaction. The new reaction pathway is a lower-energy pathway with lower activation energy. In this way, a catalyst increases the rate of a reaction. For example, consider the energy diagram for the conversion of O_3 to O_2 in Figure 12.10. In the presence of the atomic Cl catalyst, the activation energy is lower. With lower

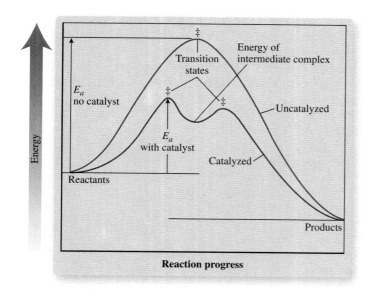

Figure 12.10
The Cl catalyst lowers the activation energy for the reaction of O_3 to form O_2, thereby increasing the rate of the reaction. At a specific temperature, more collisions meet the energy requirement for reaction. Note that the energy diagram for the catalyzed reaction has two transition states. This occurs because the catalyzed reaction occurs in two steps, with the first step forming an intermediate complex with the catalyst which has a lower energy than the two transition states.

activation energy, a greater fraction of reactants can achieve the new minimum energy requirement, and the reaction rate increases.

How does a catalyst lower the activation energy? Let's consider some examples. Most cars with internal combustion engines are equipped with catalytic converters to help reduce the amounts of CO, NO, and NO_2 pollutants released into the air (Figure 12.11A). Reactions of these substances to form nontoxic CO_2 and N_2 are usually very slow. Catalysts within the catalytic converter speed up these reactions so that they occur before the gases leave the car's exhaust system.

Let's look specifically at the conversion of CO to CO_2 catalyzed by platinum and palladium metals. In a chemical equation catalysts are generally written above the arrow because they are not part of the net reaction:

$$2CO(g) + O_2(g) \xrightarrow{\text{Pt/Pd}} 2CO_2(g)$$

The platinum and palladium metal atoms adsorb CO and O_2 molecules onto their surface (Figure 12.11B). Once there, the bonds within the O_2 molecules weaken or break, and the resulting atoms can easily bond with nearby CO molecules. The CO_2 molecules then come off the metal surface and leave through the exhaust. The catalysts lower the activation energy by forming new activated complexes with lower activation energy. The catalyzed conversion of NO to N_2 and O_2 occurs in a similar way.

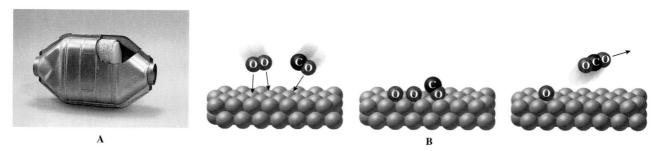

A **B**

Figure 12.11
(A) Catalytic converters in automobiles contain platinum, palladium, and rhodium metals coated on a ceramic surface. The result is a large surface area of catalyst. The rhodium metal catalyst in the converter increases the rate of conversion of NO and NO_2 to N_2 and O_2. (B) Platinum and palladium metals catalyze the conversion of CO to CO_2. The metal atoms at the surface attract the oxygen atoms of both CO and O_2 and form weak metal-oxygen bonds. The interaction between the metal and oxygen weakens the bonds in CO and O_2, so the bond breaking required for the reaction occurs more easily.

An **enzyme** is a molecule that catalyzes specific reactions within living organisms. Without enzymes, life-sustaining chemical processes would be so slow that they would not occur at the relatively low temperatures of plant and animal cells. An enormous number of enzymes—about 25,000—catalyzes all the processes necessary for the proper functioning of the human body. Most enzymes are large protein molecules with molar masses between 12,000 and 40,000 g/mol. The intricate three-dimensional structures of enzymes contain one or more depressions, or holes, which are called *active sites* (Figure 12.12). The shape of an active site is unique to an enzyme and allows it to interact with only one specific kind of reactant molecule, its *substrate.* In order for an enzyme-catalyzed reaction to occur, substrate molecules must fit into the structure of an enzyme at its active site. Once in the site, intermolecular interactions or bonds between the enzyme and the substrate modify the substrate in such a way that conversion to products requires lower activation energy.

To see how an enzyme works, consider sucrose, the ordinary sugar we use in baking or add to coffee and tea. It belongs to a class of sugars called disaccharides because it consists of two monosaccharide components. Sucrose molecules have just the right shape to fit into the active site of the enzyme sucrase. As shown in Figure 12.13, the sucrase enzyme catalyzes the reaction of sucrose with water to form

An enzyme is often named after the substance it catalyzes, using the suffix *-ase.* For example, DNAase is the enzyme that catalyzes the breakdown of a DNA molecule.

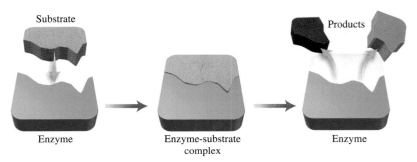

Substrate

Products

Enzyme

Enzyme-substrate complex

Enzyme

Figure 12.12
The unique shape of an enzyme determines its function. It can operate only on a substrate that fits its active site. Once a substrate molecule forms a complex with the enzyme, the reaction proceeds quickly. In this case the substrate decomposes into two smaller molecules.

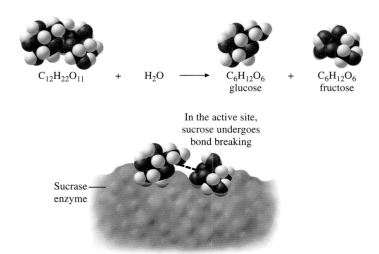

$$C_{12}H_{22}O_{11} \quad + \quad H_2O \longrightarrow \quad C_6H_{12}O_6 \quad + \quad C_6H_{12}O_6$$
glucose   fructose

In the active site, sucrose undergoes bond breaking

Sucrase — enzyme

Figure 12.13
During the digestion of foods, the larger disaccharide sucrose molecule breaks down to the smaller monosaccharides glucose and fructose. The enzyme sucrase catalyzes the reaction. The interaction between the enzyme and the sucrose molecule causes the bond between the monosaccharides to weaken, so it breaks more rapidly.

two monosaccharide molecules, glucose and fructose. Once fitted into the enzyme's active site, the bond between the two components of the sucrose molecule is weakened enough that the breakdown to monosaccharide units is fast.

EXAMPLE 12.4 The Effect of a Catalyst on Activation Energy

Draw an energy diagram for the uncatalyzed oxidation reaction of CO with O_2 to form CO_2. The reaction is exothermic. Label reactants and products. Use a dotted line to show how the energy changes are different when this reaction is catalyzed by a mixture of platinum and palladium metals.

Solution:

The products are lower in energy than the reactants, because the reaction is exothermic. The dotted line represents the energy changes for the catalyzed reaction because catalysts allow the formation of lower-energy activated complexes.

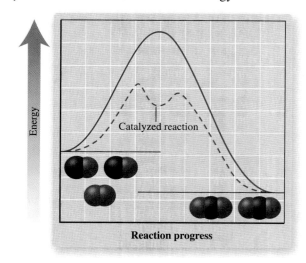

Reaction progress

Practice Problem 12.4

The decomposition of HI is an exothermic reaction:

$$2HI(g) \longrightarrow H_2(g) + I_2(g)$$

Draw an energy diagram for the uncatalyzed reaction. Label reactants and products. Sketch a possible activated complex. Use a dotted line to show how the energy changes are different when platinum metal, a catalyst that increases the rate of the reaction, is added to the system.

Further Practice: 12.29 and 12.30

One important characteristic of a catalyst is that it lowers the activation energy for a reaction. Another is that it remains unchanged after the reaction. In a catalytic converter, for example, once the products leave the metal surface, the metal atoms are free to combine with a new set of reactants. Similarly when products leave the active sites of an enzyme, the active sites are available for new reactants to enter.

Recall the catalytic role of chlorine in the destruction of ozone molecules by CFCs. It comes from the breakdown of CFCs in the presence of ultraviolet light:

$$CF_2Cl_2 \xrightarrow{\text{UV light}} CF_2Cl + Cl$$

The subsequent conversion of O_3 to O_2 in the presence of the Cl catalyst is a two-step reaction:

Step 1: $\quad\quad O_3(g) + Cl(g) \longrightarrow ClO(g) + O_2(g)$

Step 2: $\quad\quad\underline{ClO(g) + O_3(g) \longrightarrow Cl(g) + 2O_2(g)}$

Overall: $2O_3(g) + Cl(g) + ClO(g) \longrightarrow ClO(g) + Cl(g) + 3O_2(g)$

The chlorine atom is not consumed; it is used in the first step and then regenerated in the second. Notice that ClO is formed in the first step and then used in the second. We call ClO an **intermediate** because it forms temporarily during the reaction. When we combine the two steps, we obtain an overall equation in which the Cl catalyst and the ClO intermediate appear on both sides. Both the catalyst and the intermediate cancel out of the equation:

$$2O_3(g) + \cancel{Cl(g)} + \cancel{ClO(g)} \longrightarrow \cancel{ClO(g)} + \cancel{Cl(g)} + 3O_2(g)$$

The net equation shows only the reactants and products, which is the same equation as that for the uncatalyzed reaction:

$$2O_3(g) \longrightarrow 3O_2(g)$$

For any reaction that occurs in more than one step, any intermediates and catalysts are not part of the net equation.

The Cl catalyst lowers the activation energy because the activated complex for the first step is more stable than it would be without the catalyst. In the presence of the Cl catalyst, the activated complex consists of a chemical species in which O—O bonds are breaking and Cl—O bonds are forming:

$$O-O \cdots O \cdots Cl$$

Without the catalyst, the activated complex would consist of a higher-energy species where the O—O bond is breaking:

$$O-O \cdots O$$

EXAMPLE 12.5   **Identifying Intermediates and Catalysts**

Acetaldehyde (CH_3CHO) is a volatile organic compound formed during the combustion of fuels and often found in the emissions from car exhausts. It can decompose to form carbon monoxide and methane by the following two-step process. Identify any intermediates or catalysts.

Step 1: $CH_3CHO + I_2 \longrightarrow CH_3I + CO + HI$

Step 2: $\quad CH_3I + HI \longrightarrow CH_4 + I_2$

Solution:

Iodine, I_2, is used in step 1 and then formed in step 2. Because I_2 is regenerated, it is a catalyst. Both CH_3I and HI are formed in step 1 and then used in step 2. Because CH_3I and HI are formed temporarily during the reaction, they are both intermediates.

Practice Problem 12.5

Ethene ($H_2C{=}CH_2$) can be converted to ethanol (CH_3CH_2OH) by a three-step process. Identify any intermediates or catalysts.

$$H_2C{=}CH_2 + H_3O^+ \longrightarrow H_3C-CH_2^+ + H_2O$$

$$H_3C-CH_2^+ + H_2O \longrightarrow CH_3CH_2OH_2^+$$

$$CH_3CH_2OH_2^+ + H_2O \longrightarrow CH_3CH_2OH + H_3O^+$$

Further Practice: 12.43 and 12.44

12.4 CHEMICAL EQUILIBRIUM

Chemical equilibrium is a state reached by a chemical reaction where there is no change in the concentrations of reactants and products. It is established when a single reaction occurs in which reactants are converted to products, and these products are converted back to reactants by the reverse process at an equal rate.

An example in our atmosphere is the equilibrium between nitrogen dioxide (NO_2) and dinitrogen tetroxide (N_2O_4). Most of the NO_2 in the atmosphere comes from automobiles. It also forms from the breakdown of N_2O_4 in an equilibrium reaction (Figure 12.14). While NO_2 is being formed from N_2O_4, N_2O_4 is being formed from NO_2 by the reverse process:

$$N_2O_4(g) \longrightarrow 2NO_2(g)$$

$$2NO_2(g) \longrightarrow N_2O_4(g)$$

The NO_2 is a brown gas, while the N_2O_4 is colorless. The brown color of smog is partially due to the presence of NO_2. Without the equilibrium that exists between NO_2 and N_2O_4, the smog would appear significantly darker. In this section we'll look at factors that influence both reaction rates and the equilibrium in reactions such as these.

Although it may seem that nothing is happening in a reaction container where a state of equilibrium has been reached, this is not the case. When a reaction reaches equilibrium, the forward reaction continues to occur, but the reverse reaction also occurs at the same rate:

Forward reaction: $N_2O_4(g) \longrightarrow 2NO_2(g)$

Reverse reaction: $2NO_2(g) \longrightarrow N_2O_4(g)$

This state of equilibrium, where the forward and reverse reactions occur at the same rate, is represented using an equilibrium arrow ($\rightleftharpoons$):

$$N_2O_4(g) \rightleftharpoons 2NO_2(g)$$

In the absence of destructive influences such as CFCs, the amounts of O_3 and O_2 in the stratosphere remain constant because both are constantly being destroyed and reformed. A situation of this type is often termed a *dynamic* equilibrium. As important as this balance is, the destruction and formation of ozone are not reverse reactions, so the balance is not called a *chemical* equilibrium.

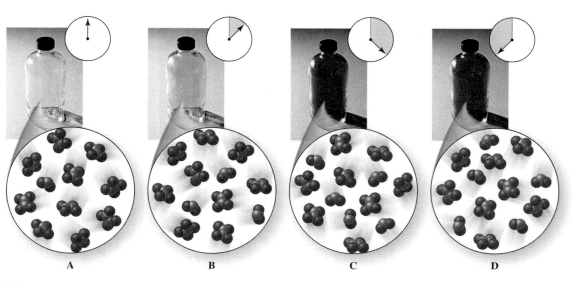

Figure 12.14
The conversion of N_2O_4 to NO_2 is a reversible reaction and reaches a state of chemical equilibrium. (A and B) These show the system before equilibrium. (C and D) These show the system after equilibrium. At equilibrium, the numbers of N_2O_4 and NO_2 molecules do not change. However, N_2O_4 is still being converted to NO_2, while NO_2 is converted back to N_2O_4 at an equal rate.

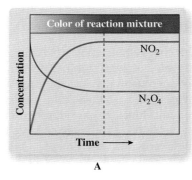
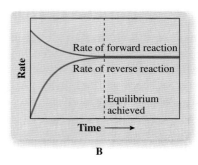

A   B

Figure 12.15

(A) As equilibrium is approached, the concentration of reactants decreases and the concentration of products increases. Equilibrium is established when the concentrations of reactants and products remain constant. (B) As equilibrium is approached, the rate of the forward reaction decreases as the rate of the reverse reaction increases. Equilibrium is established when the rate of the forward reaction is equal to the rate of the reverse reaction.

When any chemical reaction reaches the equilibrium state, the rates of the forward and reverse reactions are equal; there is no net change in the concentrations of reactants and products. Reactions that reach a state of equilibrium are *reversible reactions*. A true equilibrium exists in a closed system, where neither reactants nor products can enter or leave.

To better understand the concept of chemical equilibrium, let's consider an analogy. Recall from the introduction that when Ellen and Chad heard the ozone alert, they were driving to a mall. Once they arrived, they split up, and Ellen went immediately to a two-level store with a single entrance on the first level. She arrived just as the store opened, and—along with a number of other shoppers—she entered on the first floor. At that time, all shoppers were on level 1. None were on level 2. She and several other shoppers headed for the escalator to go up to the second floor. At that time, many people were going up, but no one was going down. After a while, however, as some people finished their shopping on level 2 and decided to go back down to level 1, the number of people on the down escalator grew. At this point, a state of equilibrium was achieved between the numbers of shoppers on level 1 and level 2. People went up and down on the escalators, but no net change occurred in the relative number of shoppers on the two levels (assuming no shoppers entered or left the store). The rate of people going up equaled the rate of people going down. This does not mean that the numbers of people on levels 1 and 2 were equal. In fact, level 2 might always have more shoppers, but the *ratio* of people on the two levels would remain constant.

The example of people riding escalators is like a chemical equilibrium. Let's take a closer look at what occurs as the N_2O_4/NO_2 system *approaches* equilibrium. As the reaction begins, N_2O_4 reacts to form NO_2. As the reaction proceeds, the concentration of N_2O_4 decreases, so the rate of the forward reaction decreases. The concentration of NO_2 increases, so the reverse reaction speeds up (Figure 12.15).

What would happen if we placed pure NO_2 in a closed reaction container? Would the same state of equilibrium be reached as shown in Figure 12.15? Yes, it would be the same as long as the temperature remained the same.

12.5 THE EQUILIBRIUM CONSTANT

When a reaction reaches equilibrium, amounts of reactants and products may be about equal. More often, however, one or the other predominates: either a large amount of reactants and small amount of products, or a small amount of reactants and a large amount of products. When the concentration of reactants at equilibrium is large relative to the concentration of products, we say the equilibrium favors reactants. When the concentration of products at equilibrium is large relative to the concentration of reactants, we say the equilibrium favors products. When describing an equilibrium system in this way, we are describing the *position of equilibrium.*

The Equilibrium Constant Expression

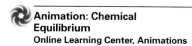
Animation: Chemical Equilibrium
Online Learning Center, Animations
Chemistry Animations Library, Chemical Equilibrium, Chemical_Equilibrium.swf.

Consider the state of equilibrium reached at the two-level store in the mall. Would the same state of equilibrium be reached if the entrance were on level 2 instead of level 1? Yes, the shoppers would distribute themselves in the same way and move up and down the escalators at the same rates. In a chemical reaction, the same state of equilibrium is reached whether we start with all reactants, all products, or a mixture of reactants and products.

How can we describe a state of equilibrium mathematically in terms of the relative concentrations of reactants and products? Let's look at different sets of equilibrium concentrations for a specific reaction to see if we can determine the relationship between them that will give the same value for all sets of equilibrium concentrations. We will consider the following reversible reaction:

$$2HI(g) \rightleftharpoons H_2(g) + I_2(g)$$

Table 12.1 shows the concentrations of reactants and products at equilibrium that result when the reaction is run three times, each time starting with different amounts of reactants and products. The table also shows the results of three different ways to describe the relative amounts of reactants and products. Which expression in Table 12.1 gives relatively constant values for the three experiments?

From Table 12.1 we can see that the expression in the last column gives values that are extremely close in all cases (0.200, 0.201, 0.199). From these three trials we could calculate an average constant of 0.200. The expression that gives a constant in Table 12.1 is the product concentrations multiplied together and divided by the reactant concentration squared. Why is the reactant concentration, [HI], squared? Does it have anything to do with the *coefficient* 2 in front of HI in the balanced equation? Yes, it does. Chemists have studied a variety of reactions under many conditions. They have found that an expression like that in the last column of Table 12.1 gives a constant (the *equilibrium constant, K_{eq}*) for any reaction under conditions of constant temperature. In general, for a reaction with the general form

Square brackets [ ] around the formula for a substance indicate the concentration of that substance in units of mol/L, molarity.

$$aA + bB \rightleftharpoons cC + dD$$

the **equilibrium constant expression** is

$$K_{eq} = \frac{[C]^c [D]^d}{[A]^a [B]^b}$$

where [A], [B], [C], and [D] are the molar concentrations of the reactants and products at equilibrium, and the exponents a, b, c, and d are the values of the coefficients in the balanced equation. The value of the equilibrium constant K_{eq} can be determined when equilibrium concentrations of reactants and products are known.

Note that similar values (5.01, 4.98, 5.02) are also obtained for the expression that is the inverse of the expression in the last column of Table 12.1. This expression also gives a constant.

$$\frac{[H_2][I_2]}{[HI]^2} \xrightarrow{\text{inverse}} \frac{[HI]^2}{[H_2][I_2]}$$

$$\frac{[HI]^2}{[H_2][I_2]} = 5.00$$

By convention, chemists have chosen the expression with the products in the numerator. The inverted expression would be the equilibrium constant expression if the reaction were written in the reverse direction.

$$H_2(g) + I_2(g) \rightleftharpoons 2HI(g)$$

TABLE 12.1	**Equilibrium Concentrations at Constant Temperature: $2HI(g) \rightleftharpoons H_2(g) + I_2(g)$**					
Experiment	**Equilibrium [HI]**	**Equilibrium [H₂]**	**Equilibrium [I₂]**	$\dfrac{[H_2][I_2]}{[HI]}$	$\dfrac{[H_2] + [I_2]}{[HI]}$	$\dfrac{[H_2][I_2]}{[HI]^2}$
1	0.704 M	0.180 M	0.550 M	0.141	1.04	0.200
2	1.44 M	0.757 M	0.550 M	0.186	0.903	0.201
3	0.634 M	0.283 M	0.283 M	0.126	0.893	0.199

The value of the equilibrium constant for a specific reaction is always the same at a specified temperature. If the temperature of the reaction system changes, the value of the equilibrium constant changes.

Let's write the equilibrium constant expression for the formation of ammonia from its elements. The balanced equation is

$$N_2(g) + 3H_2(g) \rightleftharpoons 2NH_3(g)$$

The equilibrium constant expression is

$$K_{eq} = \frac{\left[NH_3\right]^2}{\left[N_2\right]^1\left[H_2\right]^3}$$

Remember that when no coefficient appears in the balanced equation, the coefficient is understood to be 1. We normally leave the exponent 1 out of the equilibrium constant expression:

$$K_{eq} = \frac{\left[NH_3\right]^2}{\left[N_2\right]\left[H_2\right]^3}$$

Suppose we combine N_2 and H_2 gases in a reaction container at 400 K. Figure 12.16 shows the relative concentrations before the reaction occurs and after the reaction has reached equilibrium. From Figure 12.16 we can see that the equilibrium concentrations of reactants and product are as follows:

$$[N_2] = 0.0600 \ M$$

$$[H_2] = 0.180 \ M$$

$$[NH_3] = 0.280 \ M$$

To calculate the value of the equilibrium constant at 400 K, we substitute these equilibrium concentrations into the equilibrium constant expression:

$$K_{eq} = \frac{\left[NH_3\right]^2}{\left[N_2\right]\left[H_2\right]^3}$$

$$= \frac{(0.280 \ M)^2}{(0.0600 \ M)(0.180 \ M)^3} = 224$$

Although you might argue that this calculated value of the equilibrium constant should have units of $1/M^2$, by convention equilibrium constants are written without units.

The value of the equilibrium constant tells us about the position of equilibrium. When its value is much greater than 1, there are more products than reactants at equilibrium, and we say that the position of equilibrium lies to the right. We can see from Figure 12.16 that the equilibrium concentration of the NH_3 product is greater than either one of the reactant equilibrium concentrations. If the value of the equilibrium constant is much less than 1, there are more reactants than products at equilibrium,

Figure 12.16
The graphs show the initial and equilibrium concentrations of reactants and products for the reaction:

$$N_2(g) + 3H_2(g) \rightleftharpoons 2NH_3(g)$$

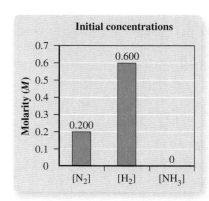
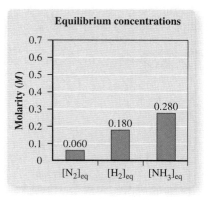

TABLE 12.2	Meaning of the Value of K_{eq}
Value of the Equilibrium Constant K_{eq}	**Position of Equilibrium**
$K_{eq} \gg 1$	Lies to the right. Products favored.
$K_{eq} \ll 1$	Lies to the left. Reactants favored.
$K_{eq} \approx 1$	Lies in the middle. Similar amounts of reactants and products.

and we say that the position of equilibrium lies to the left. If the value of the equilibrium constant is about 1, similar amounts of reactants and products exist at equilibrium. The meaning of the value of the equilibrium constant K_{eq} is summarized in Table 12.2. Given equilibrium concentrations and a balanced equation, you should be able to determine the value of K_{eq} and the position of equilibrium.

EXAMPLE 12.6   Determining K_{eq} from Equilibrium Concentrations

Sulfur trioxide is placed in a reaction container, heated to 130°C, and allowed to reach a state of equilibrium:

$$2SO_3(g) \rightleftharpoons 2SO_2(g) + O_2(g)$$

The equilibrium concentrations are determined to be

$$[SO_2] = 0.026 \ M$$

$$[O_2] = 0.013 \ M$$

$$[SO_3] = 0.12 \ M$$

(a) Write the equilibrium constant expression for this reaction.
(b) Calculate the value of the equilibrium constant at 130°C.
(c) Describe the position of equilibrium.

Solution:

(a) The equilibrium constant expression for this reaction is

$$K_{eq} = \frac{[SO_2]^2 [O_2]}{[SO_3]^2}$$

The concentrations of SO_2 and SO_3 are both squared because the balanced equation shows a coefficient of 2 for each.

(b) We calculate the value of the equilibrium constant by substituting the equilibrium concentrations into the equilibrium constant expression:

$$K_{eq} = \frac{[SO_2]^2 [O_2]}{[SO_3]^2} = \frac{(0.026)^2 (0.013)}{(0.12)^2}$$

$$= 6.1 \times 10^{-4}$$

(c) The value of the equilibrium constant is much less than 1. This tells us that the concentration of the products is small relative to the concentration of the reactant. The position of equilibrium lies to the left. The reaction is reactant-favored at this temperature.

Practice Problem 12.6

At 25°C a pure sample of N_2O_4 is placed in a reaction container and allowed to reach equilibrium:

$$2NO_2(g) \rightleftharpoons N_2O_4(g)$$

The equilibrium concentrations are determined to be

$$[NO_2] = 0.0750 \ M$$

$$[N_2O_4] = 1.25 \ M$$

(a) Write the equilibrium constant expression for this reaction.
(b) Calculate the value of the equilibrium constant at 25°C.
(c) Describe the position of equilibrium.

Further Practice: 12.71 and 12.72

Predicting the Direction of a Reaction

Suppose we want to carry out a reaction that reaches equilibrium, but we start with a mixture of reactants and products. Will the reaction go in the forward or reverse direction? It depends on the relative amounts of reactants and products in the container. If the amounts are equilibrium concentrations, the system will remain at equilibrium and no net reaction will occur. If the amounts are not equilibrium concentrations, a forward or reverse reaction will occur until the concentrations are equilibrium concentrations as described by the equilibrium constant. If we start with more products than there should be at equilibrium, the reaction will proceed in the reverse direction, reducing the amounts of products and increasing the amounts of reactants. If we start with more reactants than there should be at equilibrium, the reaction will proceed in the forward direction, reducing the amounts of reactants and increasing the amounts of products.

How can we predict in which direction a reaction will proceed to reach equilibrium? If we substitute the concentrations of reactants and products in the starting mixture into the equilibrium constant expression, the answer will equal the equilibrium constant value only if the concentrations are equilibrium concentrations. If the value obtained is less than the equilibrium constant value, then there is not enough product and too much reactant. The reaction will proceed in the forward direction to reach equilibrium. If the value obtained is greater than the equilibrium constant value, then there is too much product and not enough reactant. The reaction will proceed in the reverse direction to reach equilibrium.

As an example, consider the following reaction and its equilibrium constant:

$$CO(g) + H_2O(g) \rightleftharpoons CO_2(g) + H_2(g) \qquad K_{eq} = 16$$

The molecular picture in Figure 12.17 shows a mixture of these reactants and products in a given volume. The equilibrium constant expression for this reaction is

$$K_{eq} = \frac{[CO_2][H_2]}{[CO][H_2O]}$$

To determine if this system is at equilibrium, we can substitute the number of molecules of each reactant or product into the equilibrium constant expression:

$$\frac{(9)(9)}{(1)(1)} = 81$$

This value is not equal to the equilibrium constant $K_{eq} = 16$, so the system is not at equilibrium. A reaction not at equilibrium will proceed either in the forward or

$CO(g) + H_2O(g) \rightleftharpoons CO_2(g) + H_2(g)$

$K_{eq} = 16$

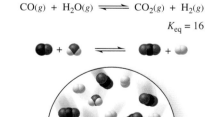

Figure 12.17
Is this system at equilibrium? If not, in which direction will the reaction proceed?

reverse direction. In this case, because the value obtained is larger than the equilibrium constant (81 > 16), we know that there is too much product and not enough reactant. The reaction will proceed in the reverse direction to reach equilibrium. The concentrations of CO_2 and H_2 will decrease and the concentrations of CO and H_2O will increase until the ratio equals the equilibrium constant value of 16.

As another example, consider the formation of ammonia from its elements:

$$N_2(g) + 3H_2(g) \rightleftharpoons 2NH_3(g)$$

Predict in which direction this reaction will proceed when we mix equal amounts of N_2, H_2, and NH_3 at 400 K. The initial concentrations are each 0.20 M.

$$[N_2]_{initial} = 0.20\ M$$

$$[H_2]_{initial} = 0.20\ M$$

$$[NH_3]_{initial} = 0.20\ M$$

We substitute these initial concentrations into the equilibrium constant expression and compare the value we calculate to the known value of the equilibrium constant expression at 400 K, which we calculated earlier to be 224:

$$K_{eq} = \frac{\left[NH_3\right]^2}{\left[N_2\right]\left[H_2\right]^3} = 224 \quad \text{(at equilibrium)}$$

Substituting initial concentrations, we get

$$\frac{(0.20)^2}{(0.20)(0.20)^3} = 25 \quad \text{(not at equilibrium)}$$

Because 25 is less than 224, we can predict that the reaction will proceed in the forward direction until equilibrium is reached.

> In this discussion we substituted numbers of molecules into the equilibrium constant expression instead of molar concentrations. We can do this only when the number of concentration terms in the numerator equals that in the denominator, and volume units from molar concentration values would cancel completely.

EXAMPLE 12.7   Predicting the Direction of a Reaction

Consider this reaction and its equilibrium constant:

$$S_2Cl_2(g) + Cl_2(g) \rightleftharpoons 2SCl_2(g) \qquad K_{eq} = 4$$

The following molecular picture shows a mixture of reactants and products.

(a) Determine whether this system is at equilibrium. If it is not, predict the direction in which the reaction will proceed to reach equilibrium.
(b) What will happen to the concentration of Cl_2?

Solution:

(a) The equilibrium constant expression for this reaction is

$$K_{eq} = \frac{\left[SCl_2\right]^2}{\left[S_2Cl_2\right]\left[Cl_2\right]} = 4$$

In this case we can substitute numbers of molecules instead of concentrations because the number of terms in the numerator is equal to that in the denominator. To determine if the system is at equilibrium, we substitute the numbers of molecules of each reactant and product into the equilibrium constant expression:

$$\frac{(4)^2}{(4)(4)} = 1$$

The value obtained is not equal to the equilibrium constant, $K_{eq} = 4$, so the system is not at equilibrium. Because the obtained value is smaller than the equilibrium constant, there is not enough product and too much reactant. The reaction will proceed in the forward direction to make more product and reach equilibrium.

(b) As the reaction proceeds in the forward direction to reach equilibrium, the number of Cl_2 molecules decreases as Cl_2 reacts with S_2Cl_2. Because the number of Cl_2 molecules decreases, the Cl_2 concentration decreases.

Practice Problem 12.7

Consider again the following reaction and its equilibrium constant:

$$S_2Cl_2(g) + Cl_2(g) \rightleftharpoons 2SCl_2(g) \qquad K_{eq} = 4$$

Suppose we start with the following initial concentrations of the reactants and products:

$$[S_2Cl_2]_{initial} = 0.10 \; M$$

$$[Cl_2]_{initial} = 0.10 \; M$$

$$[SCl_2]_{initial} = 0.30 \; M$$

(a) Determine whether the system is at equilibrium. If it is not, predict the direction in which the reaction will proceed to reach equilibrium.

(b) What will happen to the concentration of Cl_2?

Further Practice: 12.79 and 12.80

Heterogeneous Equilibrium

The equilibria we have considered so far are reactions in which reactants and products are all in the gaseous state. An equilibrium in which reactants and products are in the same physical state is called a **homogeneous equilibrium.** When an equilibrium involves more than one physical state, it is called a **heterogeneous equilibrium.** We will now look at cases of heterogeneous equilibria.

First let's look at an example of a heterogeneous equilibrium that involves a change in physical state: the vaporization of liquid bromine in a closed container. A state of equilibrium is established between bromine's liquid and gaseous states:

$$Br_2(l) \rightleftharpoons Br_2(g)$$

Figure 12.18
The concentration of gaseous bromine in equilibrium with liquid bromine is independent of the amount of liquid bromine, as long as some liquid bromine remains.

This equilibrium is heterogeneous because it involves two different physical states. The position of equilibrium is not dependent on the amount of liquid bromine in the system, as long as some liquid bromine is present. The concentration of gaseous bromine at equilibrium is the same no matter how much liquid bromine you start with (Figure 12.18). If we add more liquid bromine to the container, the concentration of gaseous bromine will not change. As long as some liquid bromine is present at equilibrium, the state of equilibrium does not depend on the amount present.

We can write the equilibrium constant expression for the evaporation of bromine as

$$K_{eq} = \frac{\left[Br_2(g) \right]}{\left[Br_2(l) \right]}$$

Whether at equilibrium or not, the liquid bromine always has the same concentration. That is, the number of moles per liter does not change when some of the liquid bromine vaporizes. Since it is a constant, we can write it as a constant in the equilibrium constant expression:

$$K_{eq} = \frac{\left[Br_2(g) \right]}{constant} \qquad or \qquad K_{eq} = \frac{1}{constant} \times \left[Br_2(g) \right]$$

We are interested only in the concentrations of substances that can change, so by convention we eliminate the term for liquid bromine from the equilibrium constant expression. It is a constant, so it can be incorporated into the equilibrium constant value:

$$K'_{eq} = K_{eq} \times constant = [Br_2(g)]$$

By convention, chemists omit pure liquids and solids from the equilibrium constant expression for all equilibria. Only gases and dissolved substances are included in the equilibrium constant expression.

Figure 12.19
The concentration and pressure of gaseous CO_2 are independent of the amount of $CaCO_3$ and CaO solids as long as both are present.

Now let's consider an example of a heterogeneous equilibrium in which a chemical reaction occurs. The decomposition of calcium carbonate involves two solids and a gas:

$$CaCO_3(s) \rightleftharpoons CaO(s) + CO_2(g)$$

What is the equilibrium constant expression for this heterogeneous equilibrium? Because $CaCO_3$ and CaO are both solids, and CO_2 is a gas, only the concentration of CO_2 is included in the equilibrium constant expression:

$$K_{eq} = \frac{\left[CO_2\right]}{1} = \left[CO_2\right]$$

The equilibrium constant is equal to the concentration of CO_2. Because the equilibrium constant is a constant at a particular temperature, we would not expect the concentration of CO_2 to change when the amounts of the solids $CaCO_3$ and CaO change. As shown in Figure 12.19, the concentration of CO_2 is independent of the amounts of the solids $CaCO_3$ and CaO as long as all three components are present.

So far we have discussed heterogeneous equilibria involving gases, liquids, and solids. Aqueous reactions involve dissolved substances in water. Because dissolved substances can change concentration, they are also included in the equilibrium constant expression. An example is the dissolving of a relatively insoluble ionic compound in water. In Chapter 11 you learned that in a saturated solution of an ionic compound, the solid and its dissolved ions exist in equilibrium. For example, the equilibrium between solid barium fluoride and its dissolved ions is described by the following equation and equilibrium constant:

$$BaF_2(s) \rightleftharpoons Ba^{2+}(aq) + 2F^-(aq) \qquad K_{eq} = 1.7 \times 10^{-6}$$

The equilibrium constant expression includes the ions in the aqueous state, but not the solid BaF_2:

$$K_{eq} = [Ba^{2+}][F^-]^2$$

The equilibrium constant shows that the position of equilibrium is dependent only on the concentrations of dissolved Ba^{2+} and F^- ions.

When writing equilibrium constant expressions, always omit pure liquid and solid reactants and products, but include gaseous and aqueous reactants and products.

| EXAMPLE 12.8 | Equilibrium Constant Expressions for Heterogeneous Equilibria |

Write the equilibrium constant expression for each of the following reactions.

(a) $C(s) + H_2O(g) \rightleftharpoons CO(g) + H_2(g)$
(b) $H_2CO_3(aq) \rightleftharpoons CO_2(g) + H_2O(l)$

Solution:

(a) The reactant carbon is omitted from the equilibrium constant expression because it is a pure solid:

$$K_{eq} = \frac{[CO][H_2]}{[H_2O]}$$

(b) Substances in the aqueous and gas phases are required in the equilibrium constant expression. The H_2O product is in the liquid state, so it is omitted from the equilibrium constant expression:

$$K_{eq} = \frac{[CO_2]}{[H_2CO_3]}$$

Practice Problem 12.8

Write the equilibrium constant expression for each of the following reactions.

(a) $Mg(s) + CO_2(g) \rightleftharpoons MgO(s) + CO(g)$
(b) $PbCl_2(s) \rightleftharpoons Pb^{2+}(aq) + 2Cl^-(aq)$

Further Practice: 12.85 and 12.86

12.6   LE CHATELIER'S PRINCIPLE

Ellen and Chad decide that the only way to support their love of mall shopping is to get jobs. They go to work for a pharmaceutical company that makes a specific drug by a chemical reaction. The equilibrium constant indicates that the equilibrium lies toward the reactants. Chad and Ellen wonder if anything can be done to cause the equilibrium to shift more toward the product side so a higher yield can be obtained. It is possible to increase the amount of product (or reactant) at equilibrium. We'll look at some ways in this section. We'll use **Le Chatelier's principle,** which states that if a system at equilibrium is disrupted, it shifts to establish a new equilibrium. Changes that can disrupt an equilibrium system include changes in the concentration of a reactant or product, changes in the volume of a gas-reaction container, and temperature changes.

In Section 12.4 we used the analogy of the two escalators at the store to describe chemical equilibrium. Recall that a state of equilibrium was reached when the number of people going up the escalator equaled the number of people going down. At that point no net change occurred in the numbers of people on levels 1 and 2. What happens when a busload of people arrives and a hundred new shoppers enter the store? The equilibrium is disrupted because the number of people on level 1 (the entrance level) increases. As some head up the escalator, the

Animation: Le Chatelier's Principle
Online Learning Center, Animations

Chemistry Animations Library, LeChatelier's Principle, 2Equilibrium.exe.

The French chemist Henry Louis Le Chatelier stated his famous principle in 1884.

number of people going up is greater than the number of people going down. After a while, the numbers going up and down become equal, and the numbers on each floor change to new, but constant values. If we assume that the bus contained the same more-or-less balanced mix of shoppers as when the store opened—not a team of soccer players who all make a beeline for the sporting goods department—an equilibrium will be reached with the same ratio of people on level 2 relative to level 1. In a similar fashion, adding more reactant or product to the reaction container can disrupt a chemical equilibrium, but a new equilibrium position will be established in time.

Reactant or Product Concentration

Consider the aqueous reaction of iron(III) ion with thiocyanate ion to form iron thiocyanate ion. It is shown in Figure 12.20 and described by the following net ionic equation:

$$Fe^{3+}(aq) + NCS^-(aq) \rightleftharpoons FeNCS^{2+}(aq)$$

An equilibrium is established after mixing aqueous solutions of iron(III) nitrate, $Fe(NO_3)_3$, and potassium thiocyanate (KNCS). A dilute solution of $Fe(NO_3)_3$ is a pale yellow color, and a solution of KNCS is colorless (Figure 12.20A). The product of the reaction, $FeNCS^{2+}$, forms a solution that is blood red in color. If we mix dilute solutions of the reactants, a state of equilibrium is quickly established between the reactants and products (Figure 12.20B). When more $Fe(NO_3)_3$ is added to the reaction vessel solution, the solution turns a darker red (Figure 12.20C). The darker color indicates that the equilibrium concentration of $FeNCS^{2+}$ has increased.

This example shows that we can increase the equilibrium concentration of a product by increasing the concentration of only one of the reactants. Le Chatelier's principle explains this shift. The disruption of equilibrium in this example is an increase in concentration of one of the reactants, Fe^{3+}. Immediately after the addition of Fe^{3+}, the system is no longer at equilibrium because the concentration of Fe^{3+} is too high. The reaction responds to this disruption by consuming *some* of the added Fe^{3+} by reaction with NCS^-, forming more $FeNCS^{2+}$. The reaction proceeds in the forward direction until equilibrium is reestablished with new equilibrium concentrations. The equilibrium concentration of blood-red-colored $FeNCS^{2+}$ increases due to the shift toward the product side of the equilibrium. The equilibrium concentration of Fe^{3+} also increases because not all the added amount reacts. The equilibrium concentration of NCS^- decreases because some of it reacted with the added Fe^{3+}. Note, however, that the value of the equilibrium constant does not change. Because the reaction responds by reacting in the forward direction, we say the equilibrium shifts to the right. By increasing the concentration of only one of the reactants, we are able to form more product. How would the equilibrium shift if additional KNCS reactant were added to the equilibrium system? What would we observe?

Suppose we want to reduce the amount of $FeNCS^{2+}$ at equilibrium. We can accomplish this by *removing* a reactant. One way to remove Fe^{3+} is to add a new substance that will react only with the Fe^{3+} (such as NaOH), and not with the NCS^-. If we remove some Fe^{3+} from solution, we disrupt the equilibrium because there is less Fe^{3+} than there should be at equilibrium. The reaction responds by reacting in reverse; some of the $FeNCS^{2+}$ dissociates to form more Fe^{3+}, along with more NCS^-. This reverse reaction replaces most of the removed Fe^{3+} and reestablishes a state of equilibrium. Because the reaction responds by reacting in the reverse direction, we say the equilibrium shifts to the left. *For a system at equilibrium, when the concentration of a reactant or product is increased, the equilibrium will shift to consume the added substance. When the concentration of a reactant or product is reduced, the equilibrium will shift to produce more of the removed substance.* Table 12.3 summarizes the effects on equilibrium of adding or removing reactants or products.

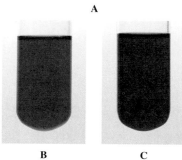

Figure 12.20
(A) An aqueous solution of iron(III) nitrate, $Fe(NO_3)_3$, is pale yellow. An aqueous solution of potassium thiocyanate (KNCS) is colorless. (B) When these solutions mix, the Fe^{3+} ion from the $Fe(NO_3)_3$ reacts with the NCS^- ion from KNCS to form the complex ion iron thiocyanate ($FeNCS^{2+}$), which is blood red in solution. This system is at equilibrium, so Fe^{3+}, NCS^-, and $FeNCS^{2+}$ are all present. (C) The addition of more Fe^{3+} to the solution disrupts the equilibrium. The solution turns darker, showing that more $FeNCS^{2+}$ has formed. The reaction has shifted to the right, consuming some of the added Fe^{3+} ion and forming more red $FeNCS^{2+}$.

TABLE 12.3	Equilibrium Shift Due to Concentration Changes			
General Reaction	**Add Reactant**	**Add Product**	**Remove Reactant**	**Remove Product**
$A(g) + B(g) \rightleftharpoons C(g) + D(g)$	Shift right	Shift left	Shift left	Shift right

EXAMPLE 12.9 Effect of Adding or Removing Reactants or Products

Consider the following system at equilibrium:

$$2CO(g) + O_2(g) \rightleftharpoons 2CO_2(g)$$

Predict the effect of the following changes when the system is initially in a state of equilibrium. Assume that the reaction container volume remains constant.

(a) Carbon dioxide gas is added to the system.
(b) Oxygen gas is added to the system.
(c) Carbon monoxide gas is removed from the system.

Solution:

(a) The addition of CO_2 product increases its concentration. The system is no longer at equilibrium, so it adjusts by shifting to the left. The concentration of CO_2 decreases and the concentrations of CO and O_2 increase until equilibrium is reestablished.
(b) With the addition of O_2 reactant, equilibrium is disrupted. The system will respond by shifting to the right, consuming most of the added O_2 and some CO gas, and producing more CO_2 gas until equilibrium is reestablished.
(c) The removal of CO gas causes the equilibrium to shift left, in effect replacing some of what was removed. In doing so the concentration of O_2 increases and the concentration of CO_2 decreases.

Practice Problem 12.9

Consider the following system at equilibrium:

$$AgI(s) \rightleftharpoons Ag^+(aq) + I^-(aq)$$

Predict the effect of the following changes when the system is initially in a state of equilibrium. Assume that the volume of solution remains constant.

(a) Ag^+ is removed from the system by addition of NaOH.
(b) Solid $AgNO_3$ is added to the system. ($AgNO_3$ is water soluble.)
(c) Solid NaI is added to the system. (NaI is water soluble.)

Further Practice: 12.93 and 12.94

Volume of the Reaction Container

If we increase or decrease the volume of a reaction container of a system at equilibrium, the concentrations of any gases involved in the reaction will change. Because gases fill the reaction container, a change in volume will change the concentrations of all gaseous reactants and products in the container. The reaction will shift so that new equilibrium concentrations are established.

Let's consider the following example of a heterogeneous equilibrium involving a product in the gaseous state:

$$CaCO_3(s) \rightleftharpoons CaO(s) + CO_2(g)$$

The substances in the solid state are left out of the equilibrium constant expression:

$$K_{eq} = [CO_2]$$

If we place a sample of solid $CaCO_3$ in a closed container at an elevated temperature, an equilibrium is established between $CaCO_3$ and the products CaO and CO_2 (Figure 12.21A). The equilibrium constant expression shows that the position of equilibrium is dependent only on the concentration of CO_2. At equilibrium, the concentration of CO_2 is equal to the equilibrium constant. What happens to the concentration of CO_2 when we reduce the volume of the container? The decrease in volume causes the concentration of CO_2 to increase, so the system is not at equilibrium (Figure 12.21B). The reaction will proceed in the reverse direction, decreasing the concentration of CO_2 until it again equals K_{eq} (Figure 12.21C). The opposite shift will occur if we increase the volume of the container. The volume increase will cause the concentration of CO_2 to decrease. The equilibrium will shift to the right to produce more CO_2.

What if a system at equilibrium has both gaseous reactants and gaseous products? Will a change in volume cause an equilibrium shift? The answer depends upon the relative number of gaseous reactants and gaseous products in the balanced equation for the reaction. If there are more gas molecules on one side of the equation, a change in volume will cause a shift. To see how this works, let's consider the equilibrium between dinitrogen tetroxide gas and nitrogen dioxide gas:

$$N_2O_4(g) \rightleftharpoons 2NO_2(g)$$

This equation has 1 gas molecule on the reactant side and 2 gas molecules on the product side. In an N_2O_4/NO_2 system at equilibrium (Figure 12.22A), a reduction in the volume of the container increases the concentrations of both N_2O_4 and NO_2 (Figure 12.22B). The equilibrium is disrupted. Although the system cannot reduce the concentrations of both the reactant and the product, it can decrease the concentration of one of them so the total concentration of gases decreases (Figure 12.22C).

Figure 12.21
(A) At well above room temperature, an equilibrium exists between $CaCO_3$, CaO, and CO_2. The position of equilibrium depends only on the concentration of CO_2, because CO_2 is a gas and the other two compounds in the reaction are solids. (B) Changing the volume of the container disrupts the equilibrium. A reduction in the volume causes the CO_2 concentration to be too high (greater than K_{eq}). (C) To reestablish equilibrium the reaction runs in reverse, reducing the concentration of CO_2 to its original concentration.

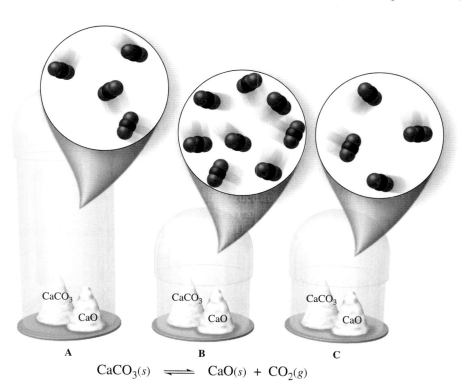

$$CaCO_3(s) \rightleftharpoons CaO(s) + CO_2(g)$$

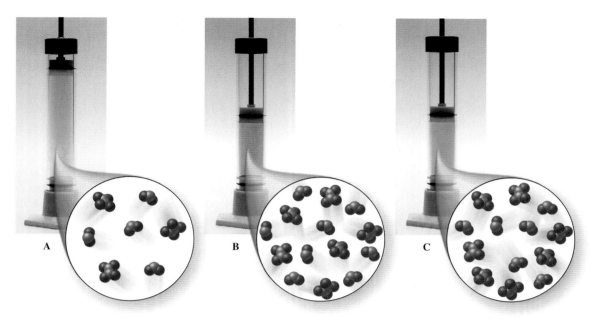

Figure 12.22
(A) A state of equilibrium exists between N_2O_4 and NO_2. The N_2O_4 is colorless and the NO_2 is brown. The position of equilibrium is dependent on both N_2O_4 and NO_2. (B) A reduction in volume initially causes the concentrations of both N_2O_4 and NO_2 gases to increase, disrupting the equilibrium. The mixture turns darker. (C) After equilibrium is reestablished, the color of the mixture is darker than before the equilibrium was disrupted, but lighter than after the initial reduction in volume. The color has lightened because the equilibrium shifted, decreasing the concentration of NO_2. For every two NO_2 gas molecules that reacted, only one N_2O_4 gas molecule formed.

Note that when the reverse reaction occurs, one N_2O_4 molecule is produced for every two NO_2 molecules that react. The reverse reaction, then, causes a net decrease in the total number of molecules in the container, decreasing total gas concentration.

Suppose the volume of the container is increased. How will the N_2O_4/NO_2 reaction respond? The increase in volume causes the gas concentrations to decrease. The reaction shifts to the side of the equation with the greater number of gas molecules—in this case, in the forward direction toward the product side.

Will a change in volume always disrupt a system at equilibrium and cause a shift? Consider the following example of a reaction where there are equal numbers of gas molecules on each side of the equation:

$$H_2(g) + I_2(g) \rightleftharpoons 2HI(g)$$

A decrease in volume would increase the concentrations of both reactants and product, but could the equilibrium shift to decrease the total number of molecules in the container? In this case a change in volume would have no effect on the equilibrium, and the reaction would not shift in either direction. Table 12.4 summarizes how volume changes affect different reactions at equilibrium.

TABLE 12.4	**Equilibrium Shifts Due to Volume Changes**		
Relative Number of Gaseous Molecules in Balanced Equation	**Example**	**Increase Volume**	**Decrease Volume**
reactants < products	$N_2O_4(g) \rightleftharpoons 2NO_2(g)$	Shift right	Shift left
reactants > products	$N_2(g) + 3H_2(g) \rightleftharpoons 2NH_3(g)$	Shift left	Shift right
reactants = products	$2NO(g) \rightleftharpoons N_2(g) + O_2(g)$	No shift	No shift

High temperature

Low temperature

Figure 12.23
The position of equilibrium between N_2O_4 and NO_2 depends on temperature. The mixture at the higher temperature has a higher concentration of brown NO_2 than the mixture at the lower temperature. This indicates that the value of the equilibrium constant for the reaction $N_2O_4(g) \rightleftharpoons 2NO_2(g)$ is greater at higher temperatures.

EXAMPLE 12.10   **Effect of Volume Change**

How will an increase in container volume affect the following reactions at equilibrium? Explain.

(a) $SO_2Cl_2(g) \rightleftharpoons SO_2(g) + Cl_2(g)$
(b) $H_2O(l) + CO_2(g) \rightleftharpoons H_2CO_3(aq)$

Solution:

(a) An increase in volume will decrease the concentration of all the reactants and products, because they are all in the gaseous state. There are two gaseous product molecules and only one gaseous reactant molecule in the balanced equation, so the effect is greater for the products. The reaction will shift right to increase the concentration of the products, reestablishing equilibrium.

(b) An increase in volume will decrease the concentration of only the CO_2, because it is the only molecule in the gaseous state. The equilibrium will shift left to increase the CO_2 concentration and reestablish equilibrium.

Practice Problem 12.10

How will a decrease in container volume affect the following reactions at equilibrium? Explain.

(a) $2NOBr(g) \rightleftharpoons 2NO(g) + Br_2(g)$
(b) $CuO(s) + H_2(g) \rightleftharpoons Cu(s) + H_2O(g)$

Further Practice: 12.97 and 12.98

Temperature

Consider once more the state of equilibrium at the store where Ellen is shopping. What happens when a store employee announces a 1-hour shoe sale on level 1? Suddenly the state of equilibrium is disrupted as more people head down the escalator than up. The number of people on level 2 decreases, and the number of people on level 1 increases. This disruption results in a new ratio of people on the two levels, so a new equilibrium is reached. This change in the position of equilibrium is similar to the response of a chemical equilibrium to a change in temperature. The value of the equilibrium constant changes when the temperature of the reaction system changes.

Consider the now familiar example of the equilibrium between N_2O_4 and NO_2:

$$N_2O_4(g) \rightleftharpoons 2NO_2(g)$$

Figure 12.23 shows the reaction mixture at two different temperatures. At the higher temperature, the reaction mixture is darker in color, indicating a higher concentration of brown NO_2. Because there is more product at the higher temperature, the equilibrium lies farther to the right at higher temperatures.

Not all equilibria shift to make more products when the temperature rises. Consider the formation of ammonia from nitrogen gas and hydrogen gas:

$$N_2(g) + 3H_2(g) \rightleftharpoons 2NH_3(g)$$

At 25°C this reaction reaches an equilibrium consisting mostly of products. However, at 500°C the equilibrium mixture consists of similar amounts of reactants and products. This reaction shifts to make more reactants at higher temperatures.

What is different about these two reactions that causes them to respond to a rise in temperature in opposite ways? The first reaction, the decomposition of N_2O_4, is endothermic:

$$N_2O_4(g) \rightleftharpoons 2NO_2(g) \qquad \text{endothermic}$$

TABLE 12.5	Effect of Temperature Changes on the Position of Equilibrium		
Type of Reaction	**Equation**	**Increase Temperature**	**Decrease Temperature**
Endothermic reaction	heat + A + B $\rightleftharpoons$ C + D	Shift right. K_{eq} increases.	Shift left. K_{eq} decreases.
Exothermic reaction	A + B $\rightleftharpoons$ C + D + heat	Shift left. K_{eq} decreases.	Shift right. K_{eq} increases.

Because endothermic reactions absorb heat when they proceed in the forward direction, we sometimes write the word *heat* on the reactant side of the equation:

$$\text{heat} + N_2O_4(g) \rightleftharpoons 2NO_2(g)$$

When we are dealing with an endothermic reaction, adding heat (by increasing temperature) has the same effect as adding a reactant. The equilibrium shifts to make more products. The opposite occurs when the temperature is lowered. In an endothermic reaction, removal of heat is like removing a reactant. The equilibrium shifts to make more reactants.

In contrast, the formation of ammonia from its elements is an exothermic reaction:

$$N_2(g) + 3H_2(g) \rightleftharpoons 2NH_3(g) \qquad \text{exothermic}$$

Because exothermic reactions release heat when they proceed in the forward direction, we sometimes write the word *heat* on the product side of the equation:

$$N_2(g) + 3H_2(g) \rightleftharpoons 2NH_3(g) + \text{heat}$$

For an exothermic reaction, increasing the temperature of a system at equilibrium has the same effect as adding a product. The equilibrium shifts to the left to make more reactants. The opposite occurs when the temperature is lowered. In an exothermic reaction, removal of heat is like removing a product. The equilibrium shifts to the right to make more product. Table 12.5 summarizes the effect of temperature on endothermic and exothermic reactions.

EXAMPLE 12.11   Effect of Temperature Change

How will a decrease in temperature affect the equilibrium concentration of product in each reaction? How will the value of K_{eq} change in each case?

(a) $2SO_2(g) + O_2(g) \rightleftharpoons 2SO_3(g) \qquad$ exothermic
(b) $3O_2(g) \rightleftharpoons 2O_3(g) \qquad$ endothermic

Solution:

(a) This reaction is exothermic, so we can write the reaction with heat on the product side:

$$2SO_2(g) + O_2(g) \rightleftharpoons 2SO_3(g) + \text{heat}$$

A decrease in temperature results in loss of heat from the equilibrium system. We can think of this as removing a product. The equilibrium will shift to make more product (SO_3), shifting the position of equilibrium to the right. Since there will be more product and less reactant, the value of K_{eq} will be a larger value at the lower temperature.

(b) This reaction is endothermic, so we can write the reaction with heat on the reactant side:

$$\text{heat} + 3O_2(g) \rightleftharpoons 2O_3(g)$$

A decrease in temperature results in loss of heat from the equilibrium system. We can think of this as removing a reactant. The reaction will make more O_2 and less O_3, shifting the equilibrium position to the left. Since there will be more reactant and less product, the value of K_{eq} will be a smaller value at the lower temperature.

Practice Problem 12.11

Consider the following equilibrium:

$$Fe^{3+}(aq) + NCS^-(aq) \rightleftharpoons FeNCS^{2+}(aq)$$

When the temperature is increased, the solution turns darker, indicating a higher concentration of $FeNCS^{2+}$ product. Is this reaction endothermic or exothermic?

Further Practice: 12.101 and 12.102

Catalysts

A catalyst speeds up a reaction by lowering the activation energy. For a reaction that reaches a state of equilibrium, a catalyst increases the rate at which equilibrium is reached. A catalyst does not change the position of equilibrium or affect a system that is in a state of equilibrium. Why? A catalyst is not a reactant or a product in the net reaction. A catalyst acts to lower the activation energy and increase the rate of both the forward and the reverse reaction proportionately.

Increasing Product Yield

Le Chatelier's principle is useful for predicting the effects of changes on a chemical equilibrium. Companies such as the pharmaceutical company that employs Ellen and Chad want to maximize their yield of product. They do this by imposing conditions that shift the equilibrium toward the product side of a reaction. Example 12.12 shows how a variety of factors can be used to maximize product yield.

EXAMPLE 12.12   **Using Le Chatelier's Principle to Increase Product Yield**

Consider the following reaction that has reached a state of equilibrium at 200°C:

$$4HCl(g) + O_2(g) \rightleftharpoons 2Cl_2(g) + 2H_2O(g) \qquad \text{exothermic}$$

Determine whether each of the following will increase the equilibrium concentration of Cl_2 product. Explain your reasoning.

(a) Remove H_2O gas.
(b) Remove HCl gas.
(c) Increase the temperature.
(d) Reduce the volume.
(e) Add a catalyst that increases the reaction rate.

Solution:

(a) The removal of H_2O should increase the concentration of Cl_2 because the equilibrium will shift right, producing more products.

(b) The removal of HCl should not increase the concentration of Cl_2. Instead, the concentration of Cl_2 will decrease because the reaction will shift left, consuming Cl_2 in the process.

(c) Because this reaction is exothermic, we can write the reaction with heat on the product side of the equation:

$$4HCl(g) + O_2(g) \rightleftharpoons 2Cl_2(g) + 2H_2O(g) + \text{heat}$$

An increase in temperature, or addition of heat, has the same effect as addition of product. The reaction will shift left, consuming Cl_2 and H_2O and forming more HCl and O_2. The equilibrium concentration of Cl_2 will fall.

(d) A decrease in volume causes all concentrations to increase followed by a shift of the equilibrium to the side with the fewer number of gaseous reactants or products. The reactant side has 5 gaseous molecules and the product side has only 4 gaseous molecules, so the reaction will shift right, further increasing the concentration of Cl_2.

(e) Addition of a catalyst does not affect the position of equilibrium; it only changes the rate at which the reaction reaches equilibrium.

Practice Problem 12.12

Consider the following reaction that has reached a state of equilibrium at 500°C:

$$PCl_5(g) \rightleftharpoons PCl_3(g) + Cl_2(g) \qquad \text{endothermic}$$

Determine whether each of the following will increase the equilibrium concentration of Cl_2 product. Explain your reasoning.

(a) Add more PCl_3 gas.
(b) Remove PCl_3 gas.
(c) Increase the temperature.
(d) Reduce the volume.
(e) Add a catalyst that increases the reaction rate.

Further Practice: 12.109 and 12.110

SUMMARY

Reaction rate is a measure of how fast a reaction occurs. Although different reactions occur at different rates, reaction conditions can affect the rate of any reaction. Generally, increasing temperature, concentration, or surface area increases the reaction rate, as does adding an appropriate catalyst. Collision theory explains why these changes affect reaction rates. It states that in order for a reaction to occur, reactant molecules must collide in the proper orientation and with sufficient energy. Because of these requirements, only a fraction of the collisions result in a reaction. The energy requirement for a reaction is the activation energy, which is the energy required to form the activated complex that must exist, however briefly, as reactants convert to products. Increasing concentration or surface area increases the collision rate, so the rate of effective collisions also increases. Increasing the temperature increases the average kinetic energy of the molecules, so both the collision rate and the fraction of effective collisions increase. Adding an appropriate catalyst increases the reaction rate by lowering the activation energy, increasing the fraction of molecules with sufficient energy to react.

Although catalysts are changed chemically in a chemical reaction, they are regenerated and can be reused. Intermediates form temporarily in a multistep reaction. Catalysts and intermediates are not part of the overall reaction.

At chemical equilibrium a reaction's forward and reverse processes occur at the same rate, and the concentrations of reactants and products remain constant. The position of equilibrium for a reaction is described by the value of its equilibrium constant K_{eq}. The equilibrium constant expression is obtained from the balanced equation for the reaction. Reactants and products in the liquid or solid state are left out of the equilibrium constant expression because their concentrations do not change; only gases and dissolved substances are included.

Le Chatelier's principle predicts the effects of changes on a system at equilibrium. Changing the concentration of a reactant or product, or changing the volume of a reaction container, can upset a system. The system then must shift to reestablish equilibrium. Changing the temperature of a reaction causes its position of equilibrium to shift farther toward reactants or toward products, changing the value of the equilibrium constant. The addition of a catalyst, however, has no effect on the position of equilibrium. Le Chatelier's principle is often used as a guide for increasing the percent yield of a reaction.

KEY RELATIONSHIPS

Relationship	Equation
The equilibrium constant is the ratio of the equilibrium concentrations of the products to the equilibrium concentrations of the reactants. The expression corresponds to a generic reaction and its equation: $aA + bB \rightleftharpoons cC + dD$	$K_{eq} = \dfrac{[C]^c [D]^d}{[A]^a [B]^b}$

KEY TERMS

activated complex (12.2)

activation energy E_a (12.2)

catalyst (12.3)

chemical equilibrium (12.4)

collision theory (12.2)

enzyme (12.3)

equilibrium constant expression (12.5)

heterogeneous equilibrium (12.5)

homogeneous equilibrium (12.5)

intermediate (12.3)

Le Chatelier's principle (12.6)

reaction rate (12.1)

QUESTIONS AND PROBLEMS

The following questions and problems, except for those in the *Additional Questions* section, are paired. Questions in a pair focus on the same concept. Answers to the odd-numbered questions and problems are in Appendix D.

Matching Definitions with Key Terms

12.1 Match the key terms with the descriptions provided.

(a) the energy barrier that reactants must overcome before they can be converted to products; the difference between the energy of the activated complex and the average energy of the reactants

(b) a chemical species produced in a step during a reaction that is relatively unstable and reacts in a subsequent step

(c) an equilibrium in which all reactants and products are in the same physical state

(d) a principle stating that a system at equilibrium responds to a change in concentration, volume, or temperature in such a way as to counteract the disruption and reestablish equilibrium

(e) a theory that explains the temperature and concentration dependence of the reaction rate on the basis of molecular collisions, molecular orientation, and kinetic energy

(f) the expression obtained from a reaction's balanced equation that relates the relative amount of products to reactants at equilibrium

12.2   Match the key terms with the descriptions provided.
  (a) the highest-energy chemical species that is formed as reactants convert to products
  (b) a protein molecule that acts as a catalyst to a specific reaction or to a narrow range of reactions
  (c) a state where the rates of the forward reaction and the reverse reaction are equal so the net amounts of reactants and products do not change
  (d) an equilibrium in which reactants and products are not all in the same physical state
  (e) a measure of how fast reactants are converted to products, described as the change in concentration of a reactant or product per unit time
  (f) a substance that changes the rate of a reaction, but can be recovered unchanged after the reaction

Reaction Rates

12.3   List three ways to increase the rate of a reaction.

12.4   List three ways to decrease the rate of a reaction.

12.5   What are three ways to increase the rate of the following reaction?

$$NaHCO_3(s) + HCl(aq) \longrightarrow NaCl(aq) + CO_2(g) + H_2O(l)$$

12.6   What are two ways to increase the rate of the following reaction?

$$2Mg(s) + O_2(g) \longrightarrow 2MgO(s)$$

12.7   What type of substance, when added to a reaction, generally increases the rate of the reaction?

12.8   What role do free chlorine atoms play in the following reaction?

$$2O_3(g) \longrightarrow 3O_2(g)$$

Collision Theory

12.9   Are all collisions between reactants effective? Why or why not?

12.10   What are the requirements of an effective collision?

12.11   Consider the following reaction:

$$2HI(g) \longrightarrow H_2(g) + I_2(g)$$

Determine if each of the following collisions has the proper orientation for an effective collision.

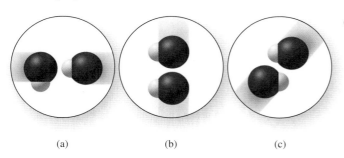

(a)   (b)   (c)

12.12   Draw a picture of 2 NO_2 molecules colliding with the proper orientation to produce 1 N_2O_4 molecule.

$$2NO_2(g) \longrightarrow N_2O_4(g)$$

12.13   What high-energy chemical species is produced when an effective collision occurs?

12.14   How is an activated complex represented?

12.15   What is activation energy?

12.16   Draw a generic energy diagram that shows the energies of reactants, products, and the activated complex. Label the activation energy.

12.17   How does the magnitude of the activation energy affect reaction rate?

12.18   If two similar reactions under the same conditions have different activation energies, which reaction should occur faster: the reaction with the higher activation energy or the reaction with the lower activation energy? Explain.

12.19   The following reaction is exothermic:

$$2O_3(g) \longrightarrow 3O_2(g)$$

Draw an energy diagram that shows the relative energies of the reactants, products, and the activated complex. Label the diagram with molecular representations of reactants, products, and a possible structure for the activated complex.

12.20   The following reaction is endothermic:

$$CF_2Cl_2(g) \longrightarrow CF_2Cl(g) + Cl(g)$$

Draw an energy diagram that shows the relative energies of the reactants, products, and the activated complex. Label the diagram with molecular representations of reactants, products, and a possible structure for the activated complex.

Conditions That Affect Reaction Rates

12.21   Use collision theory to explain why reaction rates generally increase when the temperature increases.

12.22   Use collision theory to explain why reaction rates generally increase when the concentration of one or more of the reactants increases.

12.23   Consider the following factors: increase in temperature, increase in concentration, addition of a catalyst. Which increases the average kinetic energy of the reactants?

12.24   Consider the following factors: increase in temperature, increase in concentration, addition of a catalyst. Which increase the fraction of collisions that are effective collisions?

12.25   The propane used in a gas grill does not usually react with air in a combustion reaction unless first initiated with a spark. Explain using collision theory.

12.26   The fumes from gasoline are more likely than liquid gasoline to catch fire when in contact with static electricity. Explain using collision theory.

12.27   It takes less time to cook steaks than a roast, even when cooked at the same temperature. Explain.

12.28 At altitudes above sea level, the boiling point of water is less than 100°C. Use collision theory to explain why it takes longer to boil an egg at altitudes higher than sea level.

12.29 Explain how a catalyst changes a reaction rate.

12.30 Draw an energy diagram for a reaction with and without the presence of a catalyst that increases the rate of the reaction.

12.31 What kind of catalyst is used in a catalytic converter?

12.32 Write two reactions that are catalyzed in a catalytic converter.

12.33 What part of an enzyme interacts with the reactants when an enzyme acts as a catalyst?

12.34 What reaction does the enzyme sucrase catalyze?

12.35 Why are catalysts not included in a net chemical equation?

12.36 Why would a reaction rate increase with only a small amount of catalyst?

12.37 Why are free chlorine atoms in the stratosphere a serious problem even if their numbers are few?

12.38 If we do not release any more chlorine-atom-producing CFCs into the stratosphere, will the problem of excess ozone destruction go away soon? Explain.

12.39 What is the chemical species that exists temporarily during a multistep reaction?

12.40 What is the difference between an intermediate and an activated complex?

12.41 Consider the following two-step reaction:

$$N_2O_5 \rightleftharpoons NO_3 + NO_2$$
$$NO_3 + NO \longrightarrow 2NO_2$$

(a) Identify any catalysts or intermediates.
(b) Write the net reaction.

12.42 Consider the following two-step reaction:

$$S_2O_8^{2-} + I^- \longrightarrow 2SO_4^{2-} + I^+$$
$$I^+ + I^- \longrightarrow I_2$$

(a) Identify any catalysts or intermediates.
(b) Write the net reaction.

12.43 Consider the following two-step reaction:

$$H_2O_2 + 2Br^- + 2H^+ \longrightarrow 2H_2O + Br_2$$
$$H_2O_2 + Br_2 \longrightarrow 2H^+ + O_2 + 2Br^-$$

(a) Identify any catalysts or intermediates.
(b) Write the net reaction.

12.44 Consider the following two-step reaction:

$$Cu^{2+} + H_2 \longrightarrow CuH^+ + H^+$$
$$CuH^+ + H^+ + H_2C{=}CH_2 \longrightarrow Cu^{2+} + H_3C{-}CH_3$$

(a) Identify any catalysts or intermediates.
(b) Write the net reaction.

Chemical Equilibrium

12.45 Why do some reactions seem to stop when reactants are still present?

12.46 What is a reversible reaction?

12.47 What is chemical equilibrium?

12.48 What happens to the forward and reverse rates as a reaction approaches equilibrium?

12.49 The diagrams show a molecular-level view of a reaction. Which diagram shows the point where equilibrium has been reached?

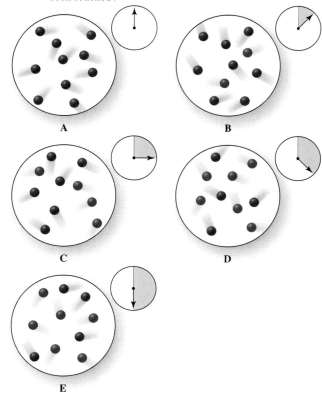

12.50 The graph shows the concentrations of reactant and product as a reaction proceeds. At what point is equilibrium first reached?

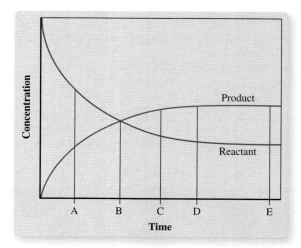

12.51 Consider the following reaction in a sealed container:

$$2SO_3(g) \rightleftharpoons 2SO_2(g) + O_2(g)$$

Can you reach a state of equilibrium if you start with
(a) just SO_3?
(b) just SO_2?
(c) just SO_2 and O_2?
(d) just SO_3 and SO_2?

12.52 Consider the following reaction in a sealed container:

$$N_2(g) + 3H_2(g) \rightleftharpoons 2NH_3(g)$$

Can you reach a state of equilibrium if you start with
(a) just NH_3?
(b) just N_2?
(c) just N_2 and NH_3?
(d) just N_2 and H_2?

12.53 Consider the vaporization of bromine in a closed container:

$$Br_2(l) \rightleftharpoons Br_2(g)$$

What kinds of measurements could you make to determine if this physical change has reached a state of equilibrium?

12.54 Consider the following reaction to form the brown gas NO_2:

$$2NO(g) + O_2(g) \rightleftharpoons 2NO_2(g)$$

What observation would indicate that the reaction has reached a state of equilibrium?

The Equilibrium Constant

12.55 What does the position of equilibrium tell us?

12.56 What can we say about the relative concentrations of reactants and products when the position of equilibrium
(a) lies to the left?
(b) lies to the right?

12.57 Why is it necessary to have a balanced equation before writing an equilibrium constant expression?

12.58 Consider the reaction:

$$H_2(g) + Cl_2(g) \rightleftharpoons 2HCl(g)$$

What is the equilibrium constant expression?

12.59 What does it mean when brackets are placed around the formula for a substance, such as $[HCl]$?

12.60 When we substitute a value for $[HCl]$, what are the units of that quantity?

12.61 Write the equilibrium constant expression for each of the following reactions:
(a) $H_2(g) + F_2(g) \rightleftharpoons 2HF(g)$
(b) $CH_4(g) + 2H_2S(g) \rightleftharpoons CS_2(g) + 4H_2(g)$
(c) $N_2O_4(g) \rightleftharpoons 2NO_2(g)$

12.62 Write the equilibrium constant expression for each of the following reactions:
(a) $2N_2O_5(g) \rightleftharpoons 4NO_2(g) + O_2(g)$
(b) $Br_2(g) + 2HI(g) \rightleftharpoons I_2(g) + 2HBr(g)$
(c) $4NH_3(g) + 3O_2(g) \rightleftharpoons 2N_2(g) + 6H_2O(g)$

12.63 Write the balanced equation that corresponds to the following equilibrium constant expressions. Assume that these are homogeneous equilibria in the gas state.

(a) $K_{eq} = \dfrac{[A][B]}{[C]}$

(b) $K_{eq} = \dfrac{[B]^4[C][D]}{[A]^2}$

12.64 Write the balanced equation that corresponds to the following equilibrium constant expressions. Assume that these are homogeneous equilibria in the gas state.

(a) $K_{eq} = \dfrac{[A]}{[B]^2}$

(b) $K_{eq} = \dfrac{[D]^2}{[A][B][C]^3}$

12.65 (a) What is the mathematical relationship between the following two equilibrium constant expressions?

$$K_{eq} = \frac{[N_2O_4]}{[NO_2]^2} \quad \text{and} \quad K_{eq} = \frac{[NO_2]^2}{[N_2O_4]}$$

(b) Write the balanced equations that correspond to each of these equilibrium constant expressions.

12.66 (a) What is the mathematical relationship between the following two equilibrium constant expressions?

$$K_{eq} = \frac{[NOBr]^2}{[Br_2][NO]^2} \quad \text{and} \quad K_{eq} = \frac{[Br_2][NO]^2}{[NOBr]^2}$$

(b) Write the balanced equations that correspond to each of these equilibrium constant expressions.

12.67 If the equilibrium constant for the reaction $A \rightleftharpoons B$ is 4.0, what is the value of the equilibrium constant for the reaction $B \rightleftharpoons A$?

12.68 If the equilibrium constant for the reaction $C \rightleftharpoons D$ is 0.10, what is the value of the equilibrium constant for the reaction $D \rightleftharpoons C$?

12.69 Under what conditions will the equilibrium constant for a specific reaction change?

12.70 The position of equilibrium for the formation of ammonia is product-favored at 25°C, but reactant-favored at 500°C. Explain why the position of equilibrium changes.

12.71 Consider the following reaction:

$$PCl_5(g) \rightleftharpoons PCl_3(g) + Cl_2(g)$$

At a specific temperature, the equilibrium concentrations were determined to be $[PCl_5] = 0.20\ M$, $[PCl_3] = 0.025\ M$, and $[Cl_2] = 0.025\ M$.
(a) What is the value of the equilibrium constant?
(b) Describe the position of equilibrium.

12.72 Consider the following reaction:

$$CO(g) + NO_2(g) \rightleftharpoons CO_2(g) + NO(g)$$

At a specific temperature, the equilibrium concentrations were determined to be $[CO] = 0.033\ M$, $[NO_2] = 0.021\ M$, $[CO_2] = 0.59\ M$, and $[NO] = 0.59\ M$.
(a) What is the value of the equilibrium constant?
(b) Describe the position of equilibrium.

12.73 Consider the following reaction:

$$CH_4(g) + 2H_2O(g) \rightleftharpoons CO_2(g) + 4H_2(g)$$

At a specific temperature, the equilibrium concentrations were determined to be [CH_4] = 0.098 M, [H_2O] = 0.096 M, [CO_2] = 0.0018 M, and [H_2] = 0.0072 M.
(a) What is the value of the equilibrium constant?
(b) Describe the position of equilibrium.

12.74 Consider the following reaction:

$$2NO(g) + Br_2(g) \rightleftharpoons 2NOBr(g)$$

At a specific temperature, the equilibrium concentrations were determined to be [NO] = 0.0070 M, [Br_2] = 0.018 M, and [NOBr] = 0.088 M.
(a) What is the value of the equilibrium constant?
(b) Describe the position of equilibrium.

12.75 If the initial concentrations of reactants and products are substituted into the equilibrium constant expression, and the value obtained is equal to the equilibrium constant, is the system in a state of equilibrium? If not, in which direction will the reaction shift to reach equilibrium? Explain.

12.76 If the initial concentrations of reactants and products are substituted into the equilibrium constant expression, and the value obtained is greater than the equilibrium constant, is the system in a state of equilibrium? If not, in which direction will the reaction shift to reach equilibrium? Explain.

12.77 Consider the reaction for the formation of ammonia and its equilibrium constant at 400 K:

$$N_2(g) + 3H_2(g) \rightleftharpoons 2NH_3(g) \qquad K_{eq} = 224$$

If 0.050 mol of each reactant and product is mixed into a 1.0-L container, will the reaction proceed in the forward or reverse direction, or is it already at equilibrium?

12.78 Consider the following reaction and its equilibrium constant at 100°C:

$$N_2O_4(g) \rightleftharpoons 2NO_2(g) \qquad K_{eq} = 6.5$$

If 0.050 mol of each reactant and product is mixed into a 1.0-L container, will the reaction proceed in the forward or reverse direction, or is it already at equilibrium?

12.79 Consider the reaction and its equilibrium constant:

$$O_3(g) + NO(g) \rightleftharpoons O_2(g) + NO_2(g) \qquad K_{eq} = 25$$

The molecular-level diagram shows a mixture of reactants and products. Determine whether the system is at equilibrium. If it is not, predict the direction in which the reaction will proceed to reach equilibrium.

12.80 Consider the reaction and its equilibrium constant:

$$O_3(g) + NO(g) \rightleftharpoons O_2(g) + NO_2(g) \qquad K_{eq} = 25$$

The molecular-level diagram shows a mixture of reactants and products. Determine whether or not the system is at equilibrium. If it is not, predict the direction in which the reaction will proceed to reach equilibrium.

12.81 What types of reactions are classified as heterogeneous equilibria?

12.82 What types of reactions are classified as homogeneous equilibria?

12.83 Why are the concentrations of pure liquids and solids not included in the equilibrium constant expression?

12.84 Reactants and products in which physical states are included in the equilibrium constant expression?

12.85 Write the equilibrium constant expression for the following equilibria:
(a) $NH_4Cl(s) \rightleftharpoons NH_3(g) + HCl(g)$
(b) $CaCO_3(s) \rightleftharpoons Ca^{2+}(aq) + CO_3^{2-}(aq)$
(c) $HF(aq) + H_2O(l) \rightleftharpoons H_3O^+(aq) + F^-(aq)$

12.86 Write the equilibrium constant expression for the following equilibria:
(a) $Mg^{2+}(aq) + 2OH^-(aq) \rightleftharpoons Mg(OH)_2(s)$
(b) $2HgO(s) \rightleftharpoons 2Hg(l) + O_2(g)$
(c) $NH_3(g) + H_2O(l) \rightleftharpoons NH_4^+(aq) + OH^-(aq)$

12.87 Consider the following equilibrium and its equilibrium constant:

$$FeO(s) + CO(g) \rightleftharpoons Fe(s) + CO_2(g) \qquad K_{eq} = 0.67$$

If the equilibrium concentration of CO is determined to be 0.20 M, what is the equilibrium concentration of CO_2?

12.88 Consider the following equilibrium and its equilibrium constant:

$$CaCO_3(s) \rightleftharpoons CaO(s) + CO_2(g) \qquad K_{eq} = 0.0033$$

What is the concentration of carbon dioxide at equilibrium?

Le Chatelier's Principle

12.89 What is Le Chatelier's principle?

12.90 How can we apply Le Chatelier's principle to help us form more reactants or more products at equilibrium?

12.91 Consider the following system at equilibrium:

$$CH_4(g) + 2H_2O(g) \rightleftharpoons CO_2(g) + 4H_2(g)$$

Suppose the concentration of CH_4 is increased.
(a) In which direction does the reaction shift to reestablish equilibrium?
(b) What happens to the concentrations of H_2O, CO_2, and H_2 as the reaction shifts to reestablish equilibrium?

12.92 Consider the following system at equilibrium:

$$CH_4(g) + 2H_2O(g) \rightleftharpoons CO_2(g) + 4H_2(g)$$

Suppose the concentration of H_2O is decreased.
(a) In which direction does the reaction shift to reestablish equilibrium?
(b) What happens to the concentrations of CH_4, CO_2, and H_2 as the reaction shifts to reestablish equilibrium?

12.93 Consider the following system at equilibrium:

$$NO(g) + SO_3(g) \rightleftharpoons NO_2(g) + SO_2(g)$$

Indicate the effect of the following changes.
(a) an increase in the concentration of NO
(b) an increase in the concentration of SO_3
(c) a decrease in the concentration of NO_2
(d) an increase in the concentration of NO_2

12.94 Consider the following system at equilibrium:

$$NO(g) + SO_3(g) \rightleftharpoons NO_2(g) + SO_2(g)$$

For each of the following changes, predict whether the equilibrium concentration of SO_2 will increase or decrease.
(a) a decrease in the concentration of NO
(b) an increase in the concentration of SO_3
(c) an increase in the concentration of NO_2
(d) a decrease in the concentration of NO_2

12.95 Consider the following system at equilibrium:

$$PbI_2(s) \rightleftharpoons Pb^{2+}(aq) + 2I^-(aq)$$

Even small quantities of lead ion in drinking water can be toxic. Indicate whether or not the addition of each of the following would reduce the concentration of dissolved lead ion. Explain your reasoning for each.
(a) PbI_2
(b) KI
(c) $Pb(NO_3)_2$

12.96 Consider the following system at equilibrium:

$$2CO_2(aq) \rightleftharpoons 2CO(g) + O_2(g)$$

Carbon monoxide is a toxic gas. Indicate whether or not each of the following would reduce the concentration of CO in the system. Explain your reasoning for each.
(a) Add CO_2.
(b) Add O_2.
(c) Remove CO_2.

12.97 For each of the following systems at equilibrium, predict whether the reaction will shift to the right, left, or not be affected by an increase in the reaction container volume.
(a) $CH_4(g) + 2H_2O(g) \rightleftharpoons CO_2(g) + 4H_2(g)$
(b) $CaCO_3(s) \rightleftharpoons CaO(s) + CO_2(g)$
(c) $NO(g) + SO_3(g) \rightleftharpoons NO_2(g) + SO_2(g)$

12.98 For each of the following systems at equilibrium, predict whether the reaction will shift to the right, left, or not be affected by a decrease in the reaction container volume.
(a) $PCl_5(g) \rightleftharpoons PCl_3(g) + Cl_2(g)$
(b) $N_2(g) + O_2(g) \rightleftharpoons 2NO(g)$
(c) $N_2O_4(g) \rightleftharpoons 2NO_2(g)$

12.99 Consider the following system at equilibrium:

$$PbI_2(s) \rightleftharpoons Pb^{2+}(aq) + 2I^-(aq)$$

(a) Will the concentrations of Pb^{2+} and I^- change if we pour the equilibrium system, including the solid, into a larger container?
(b) Will the number of moles of Pb^{2+} and I^- change if we add water to the equilibrium system?

12.100 Consider the following system at equilibrium:

$$NO(g) + SO_3(g) \rightleftharpoons NO_2(g) + SO_2(g)$$

Will the number of moles of reactants or products change if we increase the volume of the reaction container? Explain your answer.

12.101 For each of the following systems at equilibrium, predict whether the reaction will shift to the right, left, or not be affected by an increase in temperature.
(a) $2H_2O_2(g) \rightleftharpoons 2H_2O(g) + O_2(g)$ exothermic
(b) $N_2(g) + O_2(g) \rightleftharpoons 2NO(g)$ endothermic

12.102 For each of the following systems at equilibrium, predict whether the reaction will shift to the right, left, or not be affected by a decrease in temperature.
(a) $PCl_5(g) \rightleftharpoons PCl_3(g) + Cl_2(g)$ endothermic
(b) $2O_3(g) \rightleftharpoons 3O_2(g)$ exothermic

12.103 If a temperature change causes the equilibrium to shift to the right, does the value of the equilibrium constant K_{eq} increase or decrease? Explain.

12.104   If a temperature change causes the equilibrium constant to become smaller in value, in which direction does the temperature change cause the equilibrium to shift? Explain.

12.105   Does the value of the equilibrium constant increase or decrease when the temperature of each of the following equilibrium systems is increased?
(a) $2H_2O_2(g) \rightleftharpoons 2H_2O(g) + O_2(g)$ exothermic
(b) $N_2(g) + O_2(g) \rightleftharpoons 2NO(g)$ endothermic

12.106   Does the value of the equilibrium constant increase or decrease when the temperature of each of the following equilibrium systems is decreased?
(a) $PCl_5(g) \rightleftharpoons PCl_3(g) + Cl_2(g)$ endothermic
(b) $2O_3(g) \rightleftharpoons 3O_2(g)$ exothermic

12.107   The equilibrium constant K_{eq} is 12.5 at 400°C and 2.4 at 600°C for the reaction

$$CO(g) + H_2O(g) \rightleftharpoons CO_2(g) + H_2(g)$$

Is the reaction endothermic or exothermic?

12.108   The equilibrium constant K_{eq} is 1.0×10^{-6} at 1500 K and 6.2×10^{-4} at 2000 K for the reaction

$$N_2(g) + O_2(g) \rightleftharpoons 2NO(g)$$

Is the reaction endothermic or exothermic?

   Consider the following exothermic reaction:

$$4NH_3(g) + 5O_2(g) \rightleftharpoons 4NO(g) + 6H_2O(l)$$

Which of the following changes will increase the number of moles of NO at equilibrium? Explain why or why not for each change.
(a) Remove H_2O.
(b) Decrease volume.
(c) Decrease temperature.
(d) Add O_2.
(e) Add a catalyst.

12.110   Consider the following endothermic reaction:

$$2Cl_2(g) + 2H_2O(g) \rightleftharpoons 4HCl(g) + O_2(g)$$

Which of the following changes will increase the number of moles of HCl at equilibrium? Explain why or why not for each change.
(a) Remove H_2O.
(b) Increase volume.
(c) Increase temperature.
(d) Add O_2.
(e) Add a catalyst.

Additional Questions

12.111   Why does a light stick glow for a longer period of time when placed in a freezer after the reaction has begun?

12.112   The decomposition of HI into its elements is an exothermic reaction:

$$2HI(g) \longrightarrow H_2(g) + I_2(g)$$

Normally this reaction has an activation energy of 180 kJ/mol. When the reaction is carried out in the presence of platinum metal, the activation energy is lowered to 80 kJ/mol. Draw an energy diagram for the uncatalyzed reaction. Label reactants, products, and a possible activated complex. Use a dotted line to show how the energy changes are different when platinum metal is added to the reaction.

12.113   How does a catalyst that increases reaction rate lower the activation energy?

12.114   Why is it important to give the equation for the reaction when stating the value of its equilibrium constant?

12.115   A sample of hydrogen iodide gas is placed in a reaction container, heated to 450°C, and allowed to reach a state of equilibrium:

$$2HI(g) \rightleftharpoons H_2(g) + I_2(g)$$

The equilibrium concentrations were determined to be

$$[HI] = 0.195\ M$$
$$[H_2] = 0.0275\ M$$
$$[I_2] = 0.0275\ M$$

(a) Write the equilibrium constant expression for this reaction.
(b) Calculate the value of the equilibrium constant at 450°C.
(c) Describe the position of equilibrium.

12.116 Which of the following will cause the value of the equilibrium constant for a specific reaction to change?
    (a) change in the concentration of a reactant or product
    (b) change in volume of the container
    (c) change in temperature
    (d) addition of a catalyst

12.117 What must you know about a reaction before you can predict the effect of a temperature change on the position of equilibrium?

12.118 Use Le Chatelier's principle to explain why the solubility of O_2 in water increases when the partial pressure of O_2 gas above the solution is increased.

12.119 During the Napoleonic wars, the tin buttons used on Russian army uniforms disintegrated in the winter when temperatures fell below 18°C. The culprit was "tin pest," a chemical process described by the equation:

$$Sn_{(white\ metal)} \rightleftharpoons Sn_{(gray\ powder)}$$

Is this process exothermic or endothermic? Explain.

12.120 About 80% of the ammonia produced by industry goes into the production of fertilizers. Ammonia is produced in industry by the reaction of N_2 and H_2 in the following reversible reaction:

$$N_2(g) + 3H_2(g) \rightleftharpoons 2NH_3(g) \qquad exothermic$$

    (a) Which conditions of pressure and temperature favor the formation of the most amounts of ammonia?
    (b) Which conditions of temperature will increase the rate of formation of ammonia?
    (c) Use your answers to parts (a) and (b) to determine whether the reaction should be run at extremely high temperatures, extremely low temperatures, or an intermediate temperature?

 12.121 Oxygen dissolved in human blood binds to hemoglobin (Hb) in the following reversible process:

$$Hb + O_2(aq) \rightleftharpoons HbO_2$$

Carbon monoxide binds strongly to hemoglobin, displacing the oxygen in oxygenated hemoglobin.

$$HbO_2 + CO(aq) \rightleftharpoons HbCO + O_2(aq)$$

What treatment would you suggest for a patient who has been exposed to excessive carbon monoxide? Explain your answer.

Olivia, Jake, and Sara are college students who have part-time jobs during the academic year. Olivia works for a plant nursery where she is in charge of watering and fertilizing perennials. She likes her job because she is a natural at gardening and has taken care of plants since she was a child. She also likes the challenge of figuring out how to help plants thrive. Jake handles customer service and pool maintenance for a swimming pool company. He tests water for several regular customers and adds the right chemicals to maintain safe and attractive pools. Sara cycles to a nearby candy factory on weekends where she helps make and package sour candies. Sara enjoys the free samples, but she also likes dropping in at the plant's quality-control lab where product samples are tested.

Figure 13.1
Acid-base chemistry is important to a healthy swimming pool.

What do Olivia's, Jake's, and Sara's jobs have in common? Each job uses chemistry, especially the chemistry of acids and bases. Olivia's job at the nursery requires her to measure the pH of plant soil solution, the water, and dissolved substances surrounding the plants she tends. Many minerals and organic compounds in soil are acids or bases that can influence pH, and soil pH affects which metal ions dissolve in the soil solution. If the soil is too acidic (pH < 6), Olivia must add lime (CaO), a base that reduces the acidity level of the soil. She's careful to add the correct amount of lime. Too much makes the soil too basic (pH > 8). Under basic conditions, nutrients form insoluble compounds that plants cannot absorb.

Olivia also educates home gardeners about how to maintain healthy soil in their yards. Garden soil can become too acidic for many reasons. One cause is the use of acidifying nitrogen fertilizers. Many popular brands contain the ammonium ion, NH_4^+, which reacts as an acid with soil bacteria. Another cause of acidic soil is *acid rain,* which has a pH below 5.5. Acid rain is produced when sulfur and nitrogen oxides in the air react with water. In acidic soil more nutrients dissolve into the soil solution. Important nutrients may get washed away, or their concentrations may increase to levels toxic to plants.

The gases NO, NO_2, SO_2, and SO_3 react with water in clouds to produce the acids HNO_3, H_2SO_3, and H_2SO_4. The result is acid rain.

When Jake maintains pools, he uses skimmers and filters to remove suspended particles such as leaves, dirt, and hair from the water. He also treats the water to remove—or render harmless—dissolved pollutants such as body wastes, algae, and disease-causing bacteria. He uses chlorine in a variety of forms to kill bacteria. When the concentration of effective chlorine in a pool falls too low, Jake superchlorinates (*shocks*) the pool water by adding calcium hypochlorite, $Ca(OCl)_2$, a source of chlorine. The compound dissolves in water to form $OCl^-(aq)$ ions, which react with water to form hypochlorous acid, $HOCl(aq)$, a strong disinfectant. The HOCl kills bacteria by penetrating their cell walls and destroying cellular proteins.

The pH level of the pool affects the relative amounts of $HOCl(aq)$ and $OCl^-(aq)$ in the water. At a pH of 7.5, the concentration of HOCl is most effective. If the pH falls below 7.2, the pool is too acidic. There is too much HOCl in the pool, and its high levels irritate the eyes of swimmers and can etch plaster walls. Above a pH of 7.8, there is not enough HOCl and scale formation can occur and the pool may become cloudy. Jake measures the pH of the pool regularly as part of his maintenance schedule (Figure 13.1). Usually he uses pH indicators that, when added to a sample of pool water, indicate pH by color. Sometimes he uses a portable pH meter that reads the pH directly. If the pH is too high, the pool water is too basic, so he must add an acid to lower the pH. Hydrochloric acid, HCl, also called muriatic acid, lowers the pH. If the pH is too low (too acidic), the base sodium carbonate, Na_2CO_3, is added to raise it.

The chlorine in swimming pool water is not elemental chlorine, Cl_2. If Cl_2 was added to pool water, it would not stay as Cl_2. It would react with water to form hypochlorous acid, HOCl, and hydrochloric acid, HCl. Because Cl_2 is a gas and very toxic, it is a very dangerous chemical to handle. Other products such as "liquid" chlorine (chlorine dissolved in water) or solid compounds such as NaOCl (bleach) or $Ca(OCl)_2$ are used instead.

Figure 13.2
The taste of sour candy can be attributed to the presence of acids.

Make a list of the acids and bases you encounter in your daily life.

Figure 13.3
Many household products contain acids or bases. Which of these products do you think contain acids and which contain bases?

Sara's job at the candy factory also involves acids (Figure 13.2). Sour candies contain a mixture of sugar and the organic acids malic acid and citric acid. Malic acid gives apples their "bite," and citric acid causes the sour taste of lemons and limes. All acids taste sour, although the degree of sourness varies from acid to acid. In addition to sour candy, many foods such as fruits contain acids. Vinegar, which is found in pickles, salad dressings, and pizza sauce, contains acetic acid. Carbonated sodas contain phosphoric acid, carbonic acid, and sometimes citric acid. The next time you look at a food label, check to see if any acids are listed as ingredients.

While many foods contain acids, bases are rarely found in foods. One reason is that bases taste bitter. Many plants, some poisonous, contain organic bases called alkaloids. Their bitter taste is nature's way of telling us not to eat them. Some bases, however, are useful (Figure 13.3). Antacids, for example, are bases that act to neutralize excess acid in the stomach. While our stomachs can usually tolerate a high concentration of stomach acid (HCl), we sometimes secrete too much and get heartburn. Commonly used antacids contain magnesium hydroxide, $Mg(OH)_2$, aluminum hydroxide, $Al(OH)_3$, or calcium carbonate, $CaCO_3$. Bases are also found in some household cleaners. Ammonia, NH_3, is found in window cleaners, and sodium hydroxide, NaOH, is found in many oven and drain cleaners.

Some of the properties of acids and bases have been known for millennia, but only in the past 100 years have chemists come to understand the behavior of acids and bases at the molecular level. In this chapter we will explore the behavior of acids and bases in detail. We will look at the origin of the pH scale and find out how acid-base buffers work to maintain constant pH levels in solutions such as human blood.

Questions for Consideration

13.1   How do acids and bases differ from other substances?
13.2   How do strong and weak acids differ?
13.3   How can we compare the different strengths of weak acids?
13.4   What causes aqueous solutions to be acidic or basic?
13.5   How is pH related to acidity and basicity?
13.6   What is a buffered solution?

Math Tools Used in This Chapter

Scientific Notation (Math Toolbox 1.1)

Log and Inverse Log Functions (Math Toolbox 13.1)

13.1   WHAT ARE ACIDS AND BASES?

Centuries ago substances were first classified as acids and bases by their observable properties. Like the acids in sour candy or lemons, all acids have a sour taste. They are also corrosive to most metals, so people learned a long time ago not to store fruit juices or vinegar in metal containers. In contrast to acids, bases taste bitter and have a slippery feel. Acids and bases change the color of some dyes. For example, bases turn red litmus dye to blue. Acids change it back to red. When acids and bases react in neutralization reactions, their acidic and basic properties disappear.

Acid and Base Definitions

In the late 1800s a Swedish doctoral student named Svante Arrhenius proposed a definition of acids and bases based on his experiments with electrolytes. Because aqueous solutions of acids and bases conduct electricity, Arrhenius knew that the compounds were forming positive and negative ions in solution. Arrhenius proposed that acids and bases produce specific types of ions that other compounds do not. In the **Arrhenius model of acids and bases,** an acid in aqueous solution produces hydrogen ions, H^+, and a base produces hydroxide ions, OH^-. For example, hydrochloric acid ionizes in water to form $H^+(aq)$ and $Cl^-(aq)$:

$$HCl(g) \xrightarrow{\text{H}_2\text{O}} H^+(aq) + Cl^-(aq)$$

The base sodium hydroxide dissociates when it dissolves in water to form sodium ions and hydroxide ions:

$$NaOH(s) \xrightarrow{\text{H}_2\text{O}} Na^+(aq) + OH^-(aq)$$

The Arrhenius model explains how acids and bases neutralize each other. In a neutralization reaction of HCl and NaOH, for example, the H^+ from the HCl combines with the OH^- from NaOH to produce water:

$$H^+(aq) + OH^-(aq) \longrightarrow H_2O(l)$$

Although his model does not adequately describe acids and bases as we know them today, Arrhenius earned the 1903 Nobel Prize in Chemistry because he insisted that the $H^+(aq)$ and $OH^-(aq)$ ions are important in acid-base behavior. Because of the simplicity of its definition, the model is still used when only a brief explanation of acid-base behavior is needed.

The Arrhenius definition of acids and bases is limited by some fundamental problems. The hydrogen ion, H^+, essentially a proton with an extremely small radius, has a very concentrated positive charge. As a result, free H^+ ions are unlikely to exist in aqueous solution. Instead they are associated with surrounding water molecules. Chemists today normally represent the aqueous H^+ ion as $H_3O^+(aq)$, the **hydronium ion** (Figure 13.4). Another limitation to the Arrhenius model is that it assumes all bases contain OH^- ions, but some do not. Many other ionic compounds, commonly called salts, have basic properties such as the ability to neutralize acids. Examples of salts that are bases are metal oxides, carbonates, and fluorides. Some molecular compounds that contain no –OH group, such as ammonia, NH_3, also have basic properties.

In 1923 J. N. Brønsted, a Danish chemist, and T. M. Lowry, a British chemist, independently proposed a new theory to deal with the limitations of the Arrhenius model. The **Brønsted-Lowry theory** defines an acid as any substance that can donate an H^+ ion to another substance; it defines a base as any substance that can accept an H^+ ion from another substance. This definition includes all the Arrhenius acids and bases plus some ions in salts and some molecular compounds such as ammonia. According to the Brønsted-Lowry theory, when HCl gas dissolves in water, it donates an H^+ to H_2O to form hydronium ions and chloride ions:

$$HCl(g) + H_2O(l) \longrightarrow H_3O^+(aq) + Cl^-(aq)$$

Note that H_2O acts as a Brønsted-Lowry base by accepting the H^+ from HCl.

Ammonia dissolved in water acts as a Brønsted-Lowry base by the following equation:

$$NH_3(aq) + H_2O(l) \rightleftharpoons NH_4^+(aq) + OH^-(aq)$$

In this case water acts as an acid by donating the H^+ to the NH_3 molecule. Note that $OH^-(aq)$ ions form, but the –OH group is not present as part of the base initially. Hydroxide ions form as a result of the Brønsted-Lowry acid-base reaction between NH_3 and H_2O. There are many organic amines (compounds that contain the –NH$_2$ group) that act as bases in water. An example is methylamine, CH_3NH_2:

$$CH_3NH_2(aq) + H_2O(l) \rightleftharpoons CH_3NH_3^+(aq) + OH^-(aq)$$

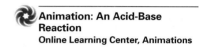

Animation: An Acid-Base Reaction
Online Learning Center, Animations

Chemistry Animations Library, An Acid-Base Reaction, Neturalization.rm.

Figure 13.4
The aqueous H^+ ion is strongly associated with water molecules, most commonly surrounded by four molecules as shown here giving a formula of $H_9O_4^+(aq)$. Chemists, however, generally represent aqueous H^+ as the hydronium ion, $H_3O^+(aq)$.

Recall from Chapter 12 that equilibrium arrows ($\rightleftharpoons$) show a system at equilibrium. The forward and reverse reactions occur at equal rates when the system is at equilibrium.

Some Brønsted-Lowry acid-base reactions take place without water. For example, ammonia gas reacts with hydrogen chloride gas to form ammonium ions and chloride ions.

$$\underset{\text{base}}{NH_3(g)} + \underset{\text{acid}}{HCl(g)} \longrightarrow NH_4^+(g) + Cl^-(g)$$

The ammonium and chloride ions immediately combine to form the solid ionic compound, ammonium chloride, $NH_4Cl(s)$.

Brønsted-Lowry bases also include anions such as the carbonate ion, CO_3^{2-}, which is found in the compound Na_2CO_3 used to alter the pH of swimming pool water. When added to water, Na_2CO_3 first dissociates completely into its ions:

$$Na_2CO_3(s) \xrightarrow{H_2O} 2Na^+(aq) + CO_3^{2-}(aq)$$

The CO_3^{2-} ion acts as a base in water by accepting an H^+ from H_2O:

$$CO_3^{2-}(aq) + H_2O(l) \rightleftharpoons HCO_3^-(aq) + OH^-(aq)$$

Brønsted-Lowry acid-base reactions do not need to include water as the acid or base. Sometimes another substance in the water is better able to give or receive an H^+ ion. In the following reaction, hydrofluoric acid donates an H^+ to the carbonate ion base to form fluoride ion and bicarbonate ion:

$$\underset{\text{acid}}{HF(aq)} + \underset{\text{base}}{CO_3^{2-}(aq)} \rightleftharpoons F^-(aq) + HCO_3^-(aq)$$

Use Example 13.1 to practice identifying the Brønsted-Lowry acid and base in an acid-base reaction.

EXAMPLE 13.1   Brønsted-Lowry Acids and Bases

For each reaction, identify the Brønsted-Lowry acid and base reactants.

(a) $HNO_3(l) + H_2O(l) \longrightarrow NO_3^-(aq) + H_3O^+(aq)$
(b) $SO_4^{2-}(aq) + H_2O(l) \rightleftharpoons HSO_4^-(aq) + OH^-(aq)$
(c) $H_2CO_3(aq) + CH_3NH_2(aq) \rightleftharpoons HCO_3^-(aq) + CH_3NH_3^+(aq)$

Solution:

(a) The name of HNO_3, nitric acid, tells us that it is an acid. However, we can also identify it as an acid because it donates an H^+ to water as shown in the equation. Water accepts H^+ from HNO_3 to form $H_3O^+(aq)$, so H_2O is the base.
(b) The sulfate ion, SO_4^{2-}, does not have any hydrogens to donate, so it cannot be the acid. It is the base because it accepts H^+ to form $HSO_4^-(aq)$. Water donates the H^+, forming $OH^-(aq)$, so H_2O is the acid in this reaction.
(c) Methylamine, CH_3NH_2, acts as a base in water because it accepts an H^+, forming $CH_3NH_3^+(aq)$. Carbonic acid, H_2CO_3, donates the H^+, forming $HCO_3^-(aq)$, so H_2CO_3 is the acid in this reaction.

Practice Problem 13.1

For each reaction, identify the Brønsted-Lowry acid and base reactants.

(a) $OCl^-(aq) + H_2O(l) \rightleftharpoons HOCl(aq) + OH^-(aq)$
(b) $H_2SO_4(aq) + F^-(aq) \longrightarrow HSO_4^-(aq) + HF(aq)$
(c) $NH_4^+(aq) + H_2O(l) \rightleftharpoons NH_3(aq) + H_3O^+(aq)$

Further Practice: 13.11 and 13.12

Conjugate Acid-Base Pairs

When an acid donates an H^+ to a base, the two products differ from the reactants by one H^+ ion. The product that forms as a result of gaining an H^+ ion is called the **conjugate acid** of the base from which it forms. The product that forms as a result of losing an H^+ ion is called the **conjugate base** of the acid from which it forms. Consider the reaction of hydrogen chloride gas with water:

$$\underset{\text{acid}}{HCl(g)} + \underset{\text{base}}{H_2O(l)} \longrightarrow \underset{\text{conjugate acid}}{H_3O^+(aq)} + \underset{\text{conjugate base}}{Cl^-(aq)}$$

In water the acid HCl reacts to form its conjugate base, $Cl^-(aq)$. The base, H_2O, reacts to form its conjugate acid, $H_3O^+(aq)$. The acid reactant and the conjugate base product are a *conjugate acid-base pair*. Likewise the base reactant and the conjugate acid product are a conjugate acid-base pair. Conjugate acid-base pairs always differ by one H^+ ion. The conjugate acid of a substance has one more H^+; the conjugate base of a substance has one less H^+. The conjugate base of an acid is what is left after the acid donates an H^+ to another substance. The conjugate base, then, has one less H atom in its formula and a decrease in charge of 1. Given that a substance reacts as an acid or a base, you should be able to identify its conjugate.

EXAMPLE 13.2   Conjugate Acids and Bases

Identify the conjugate base for each acid and explain its charge.

(a) $HOCl \longrightarrow OCl^-$
(b) $H_2PO_4^-$
(c) H_2O

Solution:

(a) The conjugate base of $HOCl$ is OCl^-. It has one less H atom than the acid and a charge of 1−, one less than the zero charge on $HOCl$.
(b) The conjugate base of $H_2PO_4^-$ is HPO_4^{2-}, which has one less H atom than the acid and a charge of 2−, one less than the 1− charge on $H_2PO_4^-$.
(c) The conjugate base of H_2O is OH^-, which has one less H atom than H_2O and a charge of 1−, one less than the zero charge on H_2O.

Practice Problem 13.2

Identify the conjugate acid for each base.

(a) F^- HF
(b) HCO_3^- H_2CO_3
(c) H_2O H_3O

Further Practice: 13.13 and 13.14

In some of the examples we have looked at, water acts as an acid, while in others it acts as a base. Water can act as an acid by donating an H^+ to another substance, forming hydroxide ion, $OH^-(aq)$. It can also act as a base by accepting an H^+ from another substance, forming hydronium ion, $H_3O^+(aq)$. The conjugate of water, therefore, depends on whether water is acting as an acid or a base in a particular reaction. A substance that can act as either an acid or a base is called an **amphoteric substance.** Water is the most common amphoteric substance. Another is the bicarbonate ion, HCO_3^-. This anion is found in sodium bicarbonate, which is used to neutralize both acids and bases. When mixed with a basic solution containing $OH^-(aq)$, the bicarbonate ion acts as an acid and its conjugate base is CO_3^{2-}:

$$\underset{\text{acid}}{HCO_3^-(aq)} + OH^-(aq) \longrightarrow CO_3^{2-}(aq) + H_2O(l)$$

When mixed with an acidic solution containing $H_3O^+(aq)$, the bicarbonate ion acts as a base and its conjugate acid is H_2CO_3:

$$\underset{\text{base}}{HCO_3^-(aq)} + H_3O^+(aq) \longrightarrow H_2CO_3(aq) + H_2O(l)$$

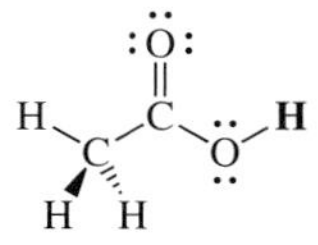

acetic acid

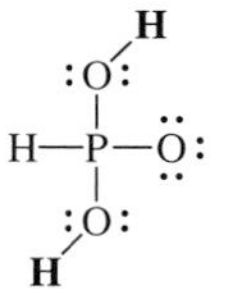

phosphorous acid

Figure 13.5
In oxoacids, the acidic hydrogens are bonded to oxygen atoms. How many acidic hydrogens do these acids have?

Acidic Hydrogen Atoms

In acids that contain more than one hydrogen atom, how do we know which hydrogen atoms are acidic? In other words, which hydrogen atoms ionize in water? An example is acetic acid, a carboxylic acid that is commonly represented with the formula CH_3CO_2H. Its formula shows that each molecule contains four hydrogen atoms. Which are acidic ones? The Lewis structure of acetic acid is shown in Figure 13.5. From its Lewis structure we see that one hydrogen atom is bonded to an oxygen atom, and the other three are bonded to a carbon atom. Only the hydrogen atom bonded to an oxygen atom is acidic; it is the only one that ionizes when acetic acid dissolves in water. The hydrogen atoms bonded to carbon are not acidic. When there is more than one hydrogen atom in an oxoacid, the acidic hydrogen atoms are usually bonded to oxygen. This is the case for phosphorous acid, H_3PO_3. Its Lewis structure (Figure 13.5) shows that two of the hydrogen atoms are bonded to oxygen atoms. They are the acidic hydrogen atoms. The hydrogen atom bonded to the phosphorus atom is not acidic. It is sometimes difficult to determine the number of acidic hydrogen atoms from the molecular formula of an acid, but it's easy when using a Lewis structure or molecular model.

EXAMPLE 13.3   **Identifying Acidic Hydrogens**

Identify the acidic hydrogen atom(s) in hypophosphorous acid, H_3PO_2.

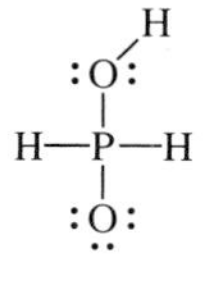

Solution:

Hypophosphorous acid, H_3PO_2, has only one acidic hydrogen. It is the hydrogen bonded to the oxygen atom.

Practice Problem 13.3

Identify the acidic hydrogen atom(s) in the amino acid glycine.

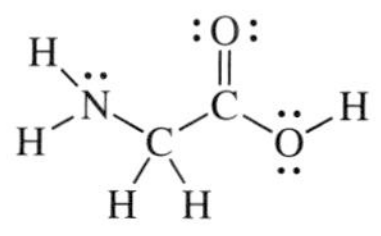

Further Practice: 13.21 and 13.22

Animation: Dissociation of Strong and Weak Acids
Online Learning Center, Animations

Chemistry Animations Library, Dissociation of Strong and Weak Acids, 1Acid_Dissociation.rm.

13.2 STRONG AND WEAK ACIDS AND BASES

Jake uses a solution of HCl to lower the pH of pool water that is too basic. One reason that he uses HCl is that it is a strong electrolyte that ionizes completely in solution. As described in Chapter 11, some acids and bases are strong electrolytes, and others are weak electrolytes. An acid or base that is a strong electrolyte and *completely* ionizes or dissociates in water is a **strong acid** or a **strong base.** An acid or a base that is a weak electrolyte and only *partially* ionizes in water is a **weak acid** or a **weak base.**

$$HCl(g) + H_2O(l) \longrightarrow H_3O^+(aq) + Cl^-(aq)$$

Figure 13.6
Hydrochloric acid is a strong acid; it ionizes completely in water to form $H_3O^+(aq)$ and $Cl^-(aq)$.

TABLE 13.1	Common Strong Acids
Formula	**Name**
HCl	hydrochloric acid
HBr	hydrobromic acid
HI	hydroiodic acid
HNO_3	nitric acid
$HClO_3$	chloric acid
$HClO_4$	perchloric acid
H_2SO_4	sulfuric acid (only one H^+ ionizes completely)

Strong Acids

Strong acids ionize completely when dissolved in water. As shown in Figure 13.6, when the strong acid HCl ionizes in water, the resulting solution consists of only H_3O^+, Cl^-, and H_2O. It contains essentially no un-ionized HCl. Table 13.1 lists the common strong acids.

Some of the strong acids listed in Table 13.1 should be familiar to you. Hydrochloric acid, HCl, is the acid in your stomach that aids in digestion. It is also the acid that pool maintenance people like Jake use to wash the plaster of swimming pools. Sulfuric acid, H_2SO_4, is the acid in most car batteries, and it occurs in acid rain. Nitric acid, HNO_3, also occurs in acid rain.

Strong Bases

Strong bases dissociate completely when dissolved in water. As shown in Figure 13.7, the strong base sodium hydroxide, NaOH, dissociates completely in water to form $Na^+(aq)$ and $OH^-(aq)$. Essentially no undissociated NaOH(aq) remains. Most of the common strong bases are the ionic hydroxides of the group IA (1) and IIA (2) metals. Table 13.2 lists the common strong bases. Commercial oven and drain cleaners contain the strong base sodium hydroxide, NaOH. Sodium hydroxide is also used in the processing of paper and canned olives. Calcium hydroxide,

$$NaOH(s) \xrightarrow{\text{H}_2\text{O}} Na^+(aq) + OH^-(aq)$$

Figure 13.7
The strong base NaOH dissociates completely in water to form $Na^+(aq)$ and $OH^-(aq)$.

TABLE 13.2	Common Strong Bases		
Formula	**Name**	**Formula***	**Name**
LiOH	lithium hydroxide	$Mg(OH)_2$	magnesium hydroxide
NaOH	sodium hydroxide	$Ca(OH)_2$	calcium hydroxide
KOH	potassium hydroxide	$Ba(OH)_2$	barium hydroxide

* Although the group IIA metal hydroxides are not completely water soluble, they are strong bases because the amount that dissolves dissociates almost completely.

Calcium oxide, CaO, has basic properties because when added to water it forms calcium hydroxide, $Ca(OH)_2$. Calcium oxide, commonly called lime, is used to reduce the acidity levels in soil.

$Ca(OH)_2$, sometimes called slaked lime, forms when lime (CaO) mixes with water. Magnesium hydroxide, $Mg(OH)_2$, is a strong base used as an antacid, but as described in Chapter 11, it is relatively insoluble in water.

Weak Acids

How do we know if an acid is a weak acid? *Any acid that is not a strong acid is a weak acid.* A weak acid does not ionize completely when dissolved in water. An equilibrium forms between a weak acid and its conjugate base, so both the reactants and the products are present in solution. Examples of weak acids are citric acid, malic acid, and acetic acid. They are commonly found in fruits and other foods. Table 13.3 is a list of weak acids and where they commonly occur.

When a weak acid dissolves in water, usually fewer than 10% of the molecules transfer an H^+ to the water, leaving the conjugate base. The rest remain in the molecular form. The equilibrium lies far to the left. We use equilibrium arrows ($\rightleftharpoons$) when writing an equation to describe the ionization of a weak acid such as acetic acid, CH_3CO_2H, in water (Figure 13.8):

$$CH_3CO_2H(aq) + H_2O(l) \rightleftharpoons CH_3CO_2^-(aq) + H_3O^+(aq)$$

TABLE 13.3	**Common Weak Acids**	
Name	**Formula**	**Occurrence**
Acetic acid	CH_3CO_2H	Vinegar, sour wine
Carbonic acid	H_2CO_3	Soda, blood
Citric acid	$H_3C_6H_5O_7$	Fruit, soda
Hydrofluoric acid	HF	Used in glass etching and semiconductor manufacturing
Hypochlorous acid	HOCl	Used to sanitize pool and drinking water
Lactic acid	$HC_3H_5O_3$	Milk
Malic acid	$HC_4H_4O_5$	Fruit
Oxalic acid	$H_2C_2O_4$	Nuts, cocoa, parsley, rhubarb
Phosphoric acid	H_3PO_4	Soda, blood
Tartaric acid	$H_2C_4H_4O_6$	Candy, wine, grapes

$$CH_3CO_2H(aq) + H_2O(l) \rightleftharpoons CH_3CO_2^-(aq) + H_3O^+(aq)$$

Figure 13.8
Acetic acid, CH_3CO_2H, is a weak acid because only a small portion of the acid molecules ionize in water. An aqueous solution of acetic acid has a much higher concentration of CH_3CO_2H molecules than of $CH_3CO_2^-(aq)$ and $H_3O^+(aq)$ ions.

Weak Bases

A weak base does not completely ionize when dissolved in water. An equilibrium forms between a weak base and its conjugate acid. Ammonia is a common weak base as shown in Figure 13.9. Most common weak bases are organic compounds that contain the amine group ($-NH_2$), such as methylamine, CH_3NH_2. Methylamine is a weak base, so we show its ionization in water with equilibrium arrows:

$$CH_3NH_2(aq) + H_2O(l) \rightleftharpoons CH_3NH_3^+(aq) + OH^-(aq)$$

Other weak bases that ionize in water are anions that are the conjugate bases of weak acids. Hypochlorite ion, OCl^-, is an example of a weak base. It is found in the compound calcium hypochlorite, $Ca(OCl)_2$, that Jake uses to superchlorinate the water in swimming pools. How does $Ca(OCl)_2$ act as a base? Consider what happens when $Ca(OCl)_2$ dissolves in water. Like any other soluble ionic compound dissolved in water, it dissociates to its aqueous cations and anions:

$$Ca(OCl)_2(s) \xrightarrow{H_2O} Ca^{2+}(aq) + 2OCl^-(aq)$$

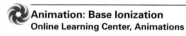

Animation: Base Ionization
Online Learning Center, Animations

Chemistry Animations Library, Base Ionization, Base_Ionization.swf.

Animation: Acid Ionization
Online Learning Center, Animations

Chemistry Animations Library, Acid Ionization, Acid_Ionization.exe.

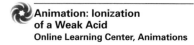

Animation: Ionization of a Weak Acid
Online Learning Center, Animations

Chemistry Animations Library, Ionization of a Weak Acid, Weak_Acid_Ionization.rm.

Figure 13.9
Ammonia is a weak base, so only some of the NH_3 molecules ionize in water to form $NH_4^+(aq)$ and $OH^-(aq)$. An aqueous solution of ammonia has a much higher concentration of NH_3 molecules than of $NH_4^+(aq)$ and $OH^-(aq)$ ions.

$$NH_3(aq) + H_2O(l) \rightleftharpoons NH_4^+(aq) + OH^-(aq)$$

TABLE 13.4	Common Weak Bases	
Name	**Formula**	**Occurrence**
Ammonia	NH_3	Glass cleaners
Calcium carbonate	$CaCO_3$	Antacids, minerals
Calcium hypochlorite	$Ca(OCl)_2$	Chlorine source for swimming pools
Methylamine	CH_3NH_2	Herring brine
Trimethylamine	$(CH_3)_3N$	Rotting fish

The $OCl^-(aq)$ reacts with water as a base and accepts an H^+ ion from water:

$$\underset{\text{weak base}}{OCl^-(aq)} + H_2O(l) \rightleftharpoons HOCl(aq) + OH^-(aq)$$

Table 13.4 lists some common weak bases and where they are found. For those that are ionic compounds, identify the anion that acts as a base when the compound dissolves in water.

EXAMPLE 13.4   Strong and Weak Acids and Bases in Aqueous Solution

Identify each of the following as a strong acid, weak acid, strong base, or weak base. Write an equation to describe its reaction with water.

(a) $CH_3CH_2NH_2(aq)$
(b) $HF(aq)$
(c) $NaF(aq)$
(d) $KOH(aq)$

Solution:

(a) The compound $CH_3CH_2NH_2$ is an organic compound called an amine because it contains the $-NH_2$ group. It is a weak base like NH_3. The equation for its ionization in water is

$$CH_3CH_2NH_2(aq) + H_2O(l) \rightleftharpoons CH_3CH_2NH_3^+(aq) + OH^-(aq)$$

Because $CH_3CH_2NH_2$ is a base, it ionizes in water to form $OH^-(aq)$ ions. Equilibrium arrows are used because $CH_3CH_2NH_2$ is a weak base and ionizes only partially.

(b) Hydrofluoric acid, HF, is an acid. Since it is not on the list of strong acids in Table 13.1, it is a weak acid. The equation for its ionization in water is

$$HF(aq) + H_2O(l) \rightleftharpoons F^-(aq) + H_3O^+(aq)$$

Because HF is an acid, it reacts with water to form $H_3O^+(aq)$ ions. Equilibrium arrows are used because HF is a weak acid and ionizes only partially.

(c) Sodium fluoride is an ionic compound that completely dissociates in water to form $Na^+(aq)$ and $F^-(aq)$. The F^- ion is the conjugate base of the weak acid HF, so it is a weak base. The equation for its reaction with water is

$$F^-(aq) + H_2O(l) \rightleftharpoons HF(aq) + OH^-(aq)$$

Because F^- is a base, it reacts with water to form OH^- ions. Equilibrium arrows are used because F^- is a weak base.

(d) KOH is a strong base because it is a group IA (1) metal hydroxide. The equation for its dissociation in water is

$$KOH(aq) \longrightarrow K^+(aq) + OH^-(aq)$$

Because KOH is a base, it produces hydroxide ions in water. A one-directional arrow is used because KOH is a strong base and dissociates completely.

Practice Problem 13.4

Identify each of the following as a strong acid, weak acid, strong base, or weak base. Write an equation to describe its ionization in water.

(a) $HI(aq)$
(b) $NaCH_3CO_2(aq)$
(c) $NH_4^+(aq)$
(d) $NH_3(aq)$

Further Practice: 13.25 and 13.26

Example 13.5 will help you distinguish between strong and weak acids at a molecular level.

EXAMPLE 13.5   Molecular-level Representations of Strong and Weak Acids

One of the diagrams shown represents a solution of HF, and the other represents an aqueous solution of HCl. Which is which? Explain your reasoning.

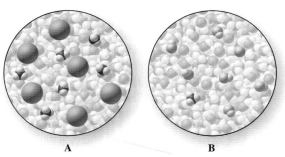

A   B

Solution:

The difference between the two diagrams is the degree of ionization. Diagram A shows no acid molecules, indicating complete ionization of the acid. It represents a strong acid. Diagram B shows many un-ionized acid molecules and a few that are ionized, indicating partial ionization. It represents a weak acid. Of the two acids, HCl is the strong acid and is represented by diagram A. Diagram B represents a solution of HF, the weak acid.

Practice Problem 13.5

One of the diagrams represents a solution of $HClO_4$, and the other represents an aqueous solution of HSO_4^-. Which is which? Explain your reasoning.

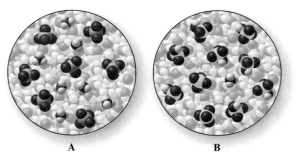

A   B

Further Practice: 13.27 and 13.28

13.3   RELATIVE STRENGTHS OF WEAK ACIDS

Some of the sour candies made at the factory where Sara works are coated with a mixture of sugar and citric acid. Citric acid has a stronger sour flavor than many other edible acids. Why? Is it more acidic than other weak acids? How can we compare the acid strength of different weak acids? Acid strength depends on the relative number of acid molecules that ionize when dissolved in water—that is, the degree of ionization. The percentage of molecules that ionize varies among weak acids. The stronger the acid, the larger its equilibrium constant value for ionization in water, and the greater the percentage of $H_3O^+(aq)$ and conjugate base ions produced. Remember from Chapter 12 that an equilibrium constant K_{eq} describes the relative amounts of reactants and products. The larger the value of K_{eq}, the more products exist at equilibrium.

Consider the ionization in water of two weak acids: acetic acid, CH_3CO_2H, and hydrocyanic acid, HCN:

$$CH_3CO_2H(aq) + H_2O(l) \rightleftharpoons H_3O^+(aq) + CH_3CO_2^-(aq) \quad K_{eq} = 1.8 \times 10^{-5}$$

$$HCN(aq) + H_2O(l) \rightleftharpoons H_3O^+(aq) + CN^-(aq) \quad K_{eq} = 6.2 \times 10^{-10}$$

Which acid, CH_3CO_2H or HCN, is stronger? Their equilibrium constants provide the answer. The stronger acid ionizes to a larger extent, producing more $H_3O^+(aq)$ ions and conjugate base than the weaker acid. The stronger acid has more products at equilibrium and therefore has an equilibrium that lies farther to the right than the weaker acid. Because acetic acid, CH_3CO_2H, has the greater equilibrium constant value for ionization, it is the stronger acid.

Acid Ionization Constants

To distinguish the equilibrium constant for the ionization of acids in water from other reactions acids can undergo, chemists use a special symbol K_a, the **acid**

The equilibrium constant represents the ratio of the concentration of products to reactants at equilibrium, leaving out any substances that are in the liquid or solid phase. For example, the equilibrium constant expression for the ionization of acetic acid, CH_3CO_2H, in water is

$$K_{eq} = \frac{[H_3O^+][CH_3CO_2^-]}{[CH_3CO_2H]}$$

TABLE 13.5	Weak Acids, Their K_a Values at 25°C, and Their Conjugate Bases			

	Acid	K_a Values	Conjugate Base	
Strongest acids	HF	6.3×10^{-4}	F$^-$	**Weakest bases**
	HNO$_2$	5.6×10^{-4}	NO$_2^-$	
	HCO$_2$H	1.8×10^{-4}	HCO$_2^-$	
	CH$_3$CO$_2$H	1.8×10^{-5}	CH$_3$CO$_2^-$	
	HOCl	4.0×10^{-8}	OCl$^-$	
	HCN	6.2×10^{-10}	CN$^-$	
Weakest acids	NH$_4^+$	5.6×10^{-10}	NH$_3$	**Strongest bases**

ionization constant. It describes the equilibrium that forms when an acid reacts with water. Thus the larger the K_a value, the stronger the acid. The K_{eq} values given for the ionization of acetic acid and hydrocyanic acid are also their K_a values:

$$CH_3CO_2H(aq) + H_2O(l) \rightleftharpoons H_3O^+(aq) + CH_3CO_2^-(aq) \quad K_a = 1.8 \times 10^{-5}$$

$$HCN(aq) + H_2O(l) \rightleftharpoons H_3O^+(aq) + CN^-(aq) \quad K_a = 6.2 \times 10^{-10}$$

Given K_a values for weak acids, we can easily compare the relative amounts of $H_3O^+(aq)$ in different weak acid solutions if the solutions are identical in concentration of acid. What can we say about the relative amounts of $H_3O^+(aq)$ in a 0.10 M CH$_3$CO$_2$H solution compared to a 0.10 M HCN solution? Acetic acid, CH$_3$CO$_2$H, with the larger K_a value, ionizes to the larger extent, yielding a greater concentration of H_3O^+. However, the concentration of H_3O^+ in both weak acid solutions is much less than the H_3O^+ concentration in a strong acid solution of the same concentration, such as 0.10 M HCl. We cannot easily compare H_3O^+ concentrations in solutions of weak acids of different concentrations because the concentration of the acid is also a factor. Table 13.5 shows several weak acids and their K_a values, along with the conjugate base of each acid. It also shows that the strength of the conjugate base is inversely related to the strength of the acid. In other words, the stronger the acid, the weaker the conjugate base. The conjugate bases of strong acids (such as Cl$^-$) are so weak that they do not react with water.

How is acid strength different from acid concentration?

EXAMPLE 13.6   Significance of K_a

Which acid ionizes in water to the greater extent, hypochlorous acid, HOCl, or hydrocyanic acid, HCN? Consult Table 13.5 for K_a values.

Solution:

From Table 13.5 we see that the K_a value for HOCl, 4.0×10^{-8}, is larger than the K_a value for HCN, 6.2×10^{-10}. The larger K_a value for HOCl indicates that a greater percentage of HOCl reacts, so it ionizes to H_3O^+ and OCl$^-$ to a greater extent.

Practice Problem 13.6

Which solution has the greater concentration of H_3O^+, a 0.10 M solution of hypochlorous acid, HOCl, or a 0.10 M solution of hydrocyanic acid, HCN?

Further Practice: 13.31 and 13.32

H_3PO_4

H_2CO_3

H_2SO_4

Figure 13.10
Polyprotic acids have more than one acidic hydrogen. How many acidic hydrogen atoms does each of these acids have?

Polyprotic Acids

Some acids contain more than one acidic hydrogen atom. Such an acid, which can donate more than one H^+ ion, is called a **polyprotic acid** (Figure 13.10). For example, a phosphoric acid molecule, H_3PO_4, has three acidic hydrogens, each bonded to an oxygen atom. Carbonic acid, H_2CO_3, and sulfuric acid, H_2SO_4, are also polyprotic acids. Polyprotic acids do not lose all of their acidic hydrogen atoms in water to the same extent. For example, when H_2CO_3 ionizes in water, the resulting solution contains about 99.79% $H_2CO_3(aq)$, about 0.21% $HCO_3^-(aq)$, and about 5000 times less $CO_3^{2-}(aq)$ than $HCO_3^-(aq)$. Because a solution of carbonic acid contains all three species, two acid-ionization equations are required to describe the equilibria that exist in solution, each with its own K_a value:

$$H_2CO_3(aq) + H_2O(l) \rightleftharpoons H_3O^+(aq) + HCO_3^-(aq) \qquad K_{a1} = 4.5 \times 10^{-6}$$

$$HCO_3^-(aq) + H_2O(l) \rightleftharpoons H_3O^+(aq) + CO_3^{2-}(aq) \qquad K_{a2} = 4.7 \times 10^{-11}$$

Sulfuric acid, H_2SO_4, is a polyprotic acid that ionizes completely in water to form $HSO_4^-(aq)$ and $H_3O^+(aq)$:

$$H_2SO_4(aq) + H_2O(l) \longrightarrow H_3O^+(aq) + HSO_4^-(aq) \qquad \text{complete ionization}$$

Once $HSO_4^-(aq)$ forms, it also acts as an acid, but as a weak acid:

$$HSO_4^-(aq) + H_2O(l) \rightleftharpoons H_3O^+(aq) + SO_4^{2-}(aq) \qquad K_a = 1.2 \times 10^{-2}$$

The low K_a value for the second ionization shows that the equilibrium lies toward the reactants; only a small portion of the $HSO_4^-(aq)$ ionizes. A smaller portion of the $HSO_4^-(aq)$ ionizes compared to H_2SO_4 because it is more difficult to remove a positively charged H^+ ion from a negatively charged HSO_4^- ion than from a neutral H_2SO_4 molecule. An aqueous solution of sulfuric acid, then, contains essentially no $H_2SO_4(aq)$, lots of $HSO_4^-(aq)$, and a small amount of $SO_4^{2-}(aq)$ (Figure 13.11).

The K_a values are often labeled to indicate the particular step in the overall ionization process (K_{a1}, K_{a2}, K_{a3}, etc.). Table 13.6 shows several polyprotic acids and their K_a values. Notice that for successive ionizations, the K_a values decrease significantly as H^+ ions are removed.

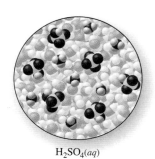

$H_2SO_4(aq)$

Figure 13.11
Sulfuric acid, H_2SO_4, is a strong polyprotic acid. It ionizes completely in water to form $HSO_4^-(aq)$ and $H_3O^+(aq)$. The $HSO_4^-(aq)$ produced also ionizes, but to a much lesser extent. In an aqueous solution of H_2SO_4, there are essentially no H_2SO_4 molecules, lots of $HSO_4^-(aq)$ ions, and a small fraction of $SO_4^{2-}(aq)$ ions. Can you identify the $HSO_4^-(aq)$ and $SO_4^{2-}(aq)$ ions in this figure?

TABLE 13.6	**Acid Ionization Constants for Some Polyprotic Acids (at 25°C)**			
Name	**Formula**	K_{a1}	K_{a2}	K_{a3}
Carbonic acid	H_2CO_3	4.5×10^{-7}	4.7×10^{-11}	
Citric acid	$H_3C_6H_5O_7$	7.4×10^{-4}	1.7×10^{-5}	4.0×10^{-7}
Hydrosulfuric acid	H_2S	8.9×10^{-8}	1.0×10^{-19}	
Oxalic acid	$H_2C_2O_4$	5.6×10^{-2}	1.5×10^{-4}	
Phosphoric acid	H_3PO_4	6.9×10^{-3}	6.2×10^{-8}	4.8×10^{-13}
Sulfuric acid	H_2SO_4	Strong	1.0×10^{-2}	
Tartaric acid	$H_2C_4H_4O_6$	1.0×10^{-3}	4.3×10^{-5}	

| EXAMPLE 13.7 | Behavior of Polyprotic Acids in Water |

The tartaric acid, $H_2C_4H_4O_6$, in grapes promotes a crisp flavor and graceful aging in wine. Its K_a values are given in Table 13.6.

(a) Write equations that show the ionization of tartaric acid in water.
(b) Besides water, which ion or molecule has the highest concentration in solution when tartaric acid is added to water?

Solution:

(a) Tartaric acid has two K_a values, so it has two acidic hydrogen atoms. Two equations are required to describe its behavior in water. The first shows the ionization of the first H^+. The K_a that corresponds to this process is the largest K_a value because the first H^+ always ionizes to the greater extent.

$$H_2C_4H_4O_6(aq) + H_2O(l) \rightleftharpoons HC_4H_4O_6^-(aq) + H_3O^+(aq) \qquad K_{a1} = 1.0 \times 10^{-3}$$

$$HC_4H_4O_6^-(aq) + H_2O(l) \rightleftharpoons C_4H_4O_6^{2-}(aq) + H_3O^+(aq) \qquad K_{a2} = 4.3 \times 10^{-5}$$

(b) The K_{a1} value for $H_2C_4H_4O_6$ shows that it is a weak acid and ionizes only to a small extent. That means that *most* of the $H_2C_4H_4O_6$ does not ionize and is present in water in the highest concentration.

Practice Problem 13.7

Oxalic acid, $H_2C_2O_4$, occurs in plants and foods such as parsley, rhubarb, almonds, and green beans. Its K_a values are listed in Table 13.6.

(a) Write equations that show the ionization of oxalic acid in water.
(b) Besides water, which ion or molecule has the highest concentration in solution when oxalic acid is added to water?

Further Practice: 13.37 and 13.38

13.4 ACIDIC, BASIC, AND NEUTRAL SOLUTIONS

When Olivia tests a soil solution to make sure it is not too acidic, what specifically is she testing for? Remember that acids ionize when dissolved in water to form hydronium ions, $H_3O^+(aq)$. Does the presence of H_3O^+ ions in solution indicate an acidic solution? No. All aqueous solutions, and even pure water, have some H_3O^+ ions (and OH^- ions) in solution. It is the *relative amounts* of these ions that make the difference. In any **acidic solution** the H_3O^+ ion concentration is greater than the OH^- ion concentration. In any **basic solution** the OH^- ion concentration is greater than the H_3O^+ ion concentration. A **neutral solution** is neither acidic nor basic, but has equal concentrations of H_3O^+ and OH^- ions.

The Ion-Product Constant of Water

Pure water, H_2O, is amphoteric. It acts as *both* an acid and a base, but only to a small extent. Water reacts with itself by a process called **self-ionization,** in which an H^+ ion is transferred from one water molecule to another:

$$H_2O(l) + H_2O(l) \rightleftharpoons H_3O^+(aq) + OH^-(aq)$$

Self-ionization is an equilibrium process, so we can write an equilibrium constant expression for it:

$$K_{eq} = [H_3O^+][OH^-]$$

Because the reactant is a liquid, it is not included in the equilibrium constant expression. This equilibrium constant, called the **ion-product constant of water,** is so widely used that it is given a special symbol, $\boldsymbol{K_w}$. At 25°C, K_w has a value of 1.0×10^{-14}:

$$K_w = [H_3O^+][OH^-] = 1.0 \times 10^{-14}$$

The small value of this equilibrium constant indicates that the equilibrium lies very far to the left, so concentrations of H_3O^+ and OH^- are very small. In pure water the concentrations of these ions must be equal, because they are produced in a 1:1 ratio:

$$[H_3O^+] = [OH^-]$$

From the K_w expression, we know that these concentrations must multiply together to give the K_w value of 1.0×10^{-14}:

$$1.0 \times 10^{-14} = [H_3O^+][OH^-]$$

For these concentrations to be equal, each must have a concentration of 1.0×10^{-7} M, because if we substitute this value for the H_3O^+ concentration and the OH^- concentration,

$$1.0 \times 10^{-14} = (1.0 \times 10^{-7}\ M) \times (1.0 \times 10^{-7}\ M)$$

we get

$$[H_3O^+] = [OH^-] = 1.0 \times 10^{-7}\ M$$

> Remember from Chapter 12 that square brackets around a formula represent the molar concentration of that substance. For example, $[H_3O^+]$ represents the molar concentration of hydronium ion.

For our purposes, we will assume the temperature to be 25°C so we will not have to worry about using K_w values at other temperatures.

In a neutral solution such as pure water, the H_3O^+ ion and OH^- ion concentrations are equal. If an acid or base is added to water, the concentrations of both ions change. However, their product continues to be 1.0×10^{-14}, the value of K_w that is constant. How can this be? The reaction of an acid with water increases the H_3O^+ ion concentration to a value greater than 1.0×10^{-7} M. At the same time, the OH^- ion concentration decreases, with the result that the ion-product constant for water remains equal to 1.0×10^{-14}. In a similar fashion, the addition of a base to water increases the OH^- ion concentration and decreases the H_3O^+ ion concentration, but the ion-product constant remains unchanged. Table 13.7 summarizes the definitions of neutral, acidic, and basic solutions in aqueous solution.

Calculating H_3O^+ and OH^- Ion Concentrations

If we know either the H_3O^+ ion concentration or the OH^- ion concentration, we can calculate the other unknown concentration using the ion-product constant of water and its expression:

$$K_w = [H_3O^+][OH^-]$$

TABLE 13.7	**Definitions of Neutral, Acidic, and Basic Solutions in Aqueous Solution**			
Type of Solution	Relative Concentrations	$[H_3O^+]$	$[OH^-]$	K_w
Neutral	$[H_3O^+] = [OH^-]$	$= 1.0 \times 10^{-7}\ M$	$= 1.0 \times 10^{-7}\ M$	1.0×10^{-14}
Acidic	$[H_3O^+] > [OH^-]$	$> 1.0 \times 10^{-7}\ M$	$< 1.0 \times 10^{-7}\ M$	1.0×10^{-14}
Basic	$[OH^-] > [H_3O^+]$	$< 1.0 \times 10^{-7}\ M$	$> 1.0 \times 10^{-7}\ M$	1.0×10^{-14}

If we know the H_3O^+ ion concentration, $[H_3O^+]$, we rearrange the K_w expression to solve for the OH^- ion concentration, $[OH^-]$:

$$[OH^-] = \frac{K_w}{[H_3O^+]} = \frac{1.0 \times 10^{-14}}{[H_3O^+]}$$

If we know the OH^- ion concentration, $[OH^-]$, we rearrange the K_w expression to solve for the H_3O^+ ion concentration, $[H_3O^+]$:

$$[H_3O^+] = \frac{K_w}{[OH^-]} = \frac{1.0 \times 10^{-14}}{[OH^-]}$$

These calculations are shown in Example 13.8.

EXAMPLE 13.8   Calculating H_3O^+ and OH^- Ion Concentrations

Given the concentration of H_3O^+ ion in water, calculate the concentration of OH^- ion in the solution. Identify each solution as acidic, basic, or neutral:

(a) $[H_3O^+] = 1.0 \times 10^{-9}\ M$
(b) $[H_3O^+] = 0.0010\ M$

Solution:

In any aqueous solution at 25°C, the product of the H_3O^+ ion concentration and the OH^- concentration equals 1.0×10^{-14}:

$$K_w = [H_3O^+][OH^-] = 1.0 \times 10^{-14}$$

We rearrange the equation to solve for hydroxide ion concentration:

$$[OH^-] = \frac{1.0 \times 10^{-14}}{[H_3O^+]}$$

(a) We substitute the H_3O^+ ion concentration of $1.0 \times 10^{-9}\ M$ into the equation to solve for the hydroxide ion concentration:

$$[OH^-] = \frac{1.0 \times 10^{-14}}{[H_3O^+]} = \frac{1.0 \times 10^{-14}}{1.0 \times 10^{-9}\ M} = 1.0 \times 10^{-5} M$$

The solution is basic because $[OH^-] > [H_3O^+]$ and because the $[H_3O^+] < 1.0 \times 10^{-7}\ M$.

(b) We substitute the H_3O^+ ion concentration of $0.0010\ M$ into the equation to solve for the OH^- ion concentration:

$$[OH^-] = \frac{1.0 \times 10^{-14}}{[H_3O^+]} = \frac{1.0 \times 10^{-14}}{0.0010\ M} = 1.0 \times 10^{-11} M$$

The solution is acidic because $[H_3O^+] > [OH^-]$ and because the $[H_3O^+] > 1.0 \times 10^{-7}\ M$.

Practice Problem 13.8

Given the concentration of OH^- ion in each solution, calculate the concentration of H_3O^+ in the solution. Identify each solution as acidic, basic, or neutral.

(a) $[OH^-] = 1.0 \times 10^{-8}\ M$
(b) $[OH^-] = 0.010\ M$

Further Practice: 13.47 and 13.48

Suppose we are using a solution in a laboratory that is labeled 0.010 M HCl. Can we determine its concentrations of H_3O^+ ion and OH^- ion? Hydrochloric acid, HCl, is a strong acid, so it ionizes completely to form H_3O^+ ions and Cl^- ions. The concentration of H_3O^+ is actually the concentration of HCl labeled on the bottle:

$$HCl(aq) + H_2O(l) \longrightarrow \underset{0.010\ M}{H_3O^+(aq)} + \underset{0.010\ M}{Cl^-(aq)}$$

From the H_3O^+ concentration, $[H_3O^+]$, and the value of K_w, we can determine the OH^- ion concentration, $[OH^-]$, as we did in Example 13.8. Work the problem and assure yourself that the value is $1.0 \times 10^{-12}\ M$ OH^-.

Suppose we have a 0.10 M solution of sodium hydroxide, NaOH, in the laboratory. What are the H_3O^+ and OH^- ion concentrations in this solution? Sodium hydroxide is a base, so it forms $OH^-(aq)$ ions when dissolved in water. Because it is a strong base with only one OH^- group, $[OH^-]$ is equal to the concentration of NaOH labeled on the bottle:

$$NaOH(s) \xrightarrow{H_2O} \underset{0.10\ M}{Na^+(aq)} + \underset{0.10\ M}{OH^-(aq)}$$

When strong bases such as $Ba(OH)_2$ contain more than one OH^- ion per formula unit, the OH^- concentration is twice the concentration of the $Ba(OH)_2$ solution. Why?

The hydroxide ion concentration, $[OH^-]$, is 0.10 M. Using the K_w expression you should calculate the H_3O^+ ion concentration, $[H_3O^+]$, as $1.0 \times 10^{-13}\ M$.

EXAMPLE 13.9   Calculating $[H_3O^+]$ and $[OH^-]$ in Strong Acid and Strong Base Solutions

Stomach acid is about 1.0 M HCl. What is the H_3O^+ concentration in stomach acid? What is the OH^- concentration?

Solution:

Hydrochloric acid, HCl, is a strong acid, so it ionizes completely in water. For every mole of HCl that ionizes, one mole of H_3O^+ forms. The concentration of H_3O^+ then is 1.0 M. The OH^- concentration is determined from the H_3O^+ concentration and the expression for the ion-product constant for water:

$$[H_3O^+][OH^-] = 1.0 \times 10^{-14}$$

$$[OH^-] = \frac{1.0 \times 10^{-14}}{[H_3O^+]}$$

$$= \frac{1.0 \times 10^{-14}}{1.0\ M}$$

$$= 1.0 \times 10^{-14}\ M$$

Practice Problem 13.9

Sodium hydroxide is sometimes called lye or caustic soda. What is the OH^- concentration in a 0.85 M lye solution? What is the H_3O^+ concentration?

Further Practice: 13.49 and 13.50

13.5   THE pH SCALE

Jake tests the acidity of pool water using pH indicators and a pH meter. Olivia also measures the pH of soil to determine whether to add lime. Perhaps you have heard the term pH, as it is used frequently in advertisements for shampoos and skin-care products. The pH of solutions is easily measured and tells us about the acidity or basicity of a substance. In this section, we will explore how pH is related to the concentration of H_3O^+ (and OH^-) and examine the significance of the pH scale.

Calculating pH

Concentrations of H_3O^+ ions can vary widely among concentrated and dilute acids, so it is convenient to express the acidity of aqueous solutions on a *pH scale*. The **pH** of a solution is defined as the negative logarithm (base 10) of the H_3O^+ concentration:

$$pH = -\log[H_3O^+]$$

For example, the H_3O^+ concentration in pure water is 1.0×10^{-7} *M*. Substituting this value into the equation for pH gives a pH of 7.00:

$$pH = -\log[H_3O^+]$$
$$= -\log(1.0 \times 10^{-7})$$
$$= 7.00$$

Like pure water, all neutral solutions have equal concentrations of H_3O^+ and OH^- (1.0×10^{-7} *M* each), so their pH values are 7. Acidic solutions have pH values less than 7, and basic solutions have pH values greater than 7.

To better understand the meaning of the log function, consider the pH values of solutions that range in H_3O^+ concentration from 1.0×10^{-1} *M* to 1.0×10^{-14} *M*, as shown in Figure 13.12. Notice that each pH value equals the exponent in the H_3O^+ concentration value, but with a positive sign. Use the logarithmic definition of pH to convince yourself that the pH values listed there can be calculated from the H_3O^+ concentration values given.

The relationship between pH and concentration allows us to determine the pH without a calculator only when the H_3O^+ concentration is exactly a power of 10. When it is not—for example, 5.0×10^{-4} *M*—you can use the log function on your calculator to determine the pH:

$$pH = -\log[H_3O^+]$$
$$= -\log(5.0 \times 10^{-4})$$
$$= 3.30$$

The pH of a strong acid solution can be determined if you know the concentration of the acid. For the strong acid HNO_3, for example, the concentration of H_3O^+ equals the HNO_3 concentration because strong acids ionize completely. What is the pH of a 0.10 *M* HNO_3 solution? Because HNO_3 is a strong acid, the concentration of H_3O^+ ion in the solution is 0.10 *M*, or 1.0×10^{-1} *M*. The pH is calculated by substituting the H_3O^+ concentration into the following equation:

$$pH = -\log[H_3O^+]$$
$$= -\log(1.0 \times 10^{-1})$$
$$= 1.00$$

We can also determine the pH of a strong base solution, such as a 0.010 *M* NaOH solution. Strong bases ionize completely, so the concentration of OH^- ion is the

The log function is a button on most calculators. See Math Toolbox 13.1 at the end of the chapter for instructions on using this function on your calculator.

How do we apply rules for significant figures to pH calculations? The rule is that the number of decimal places in a pH value should equal the number of significant figures in the H_3O^+ concentration value.

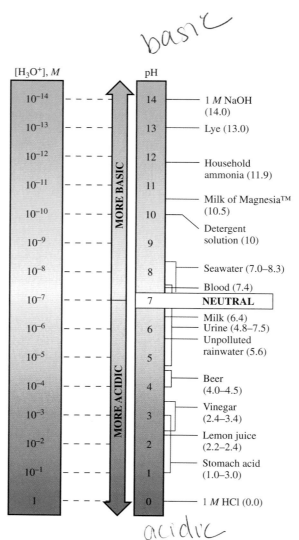

basic

Figure 13.12

acidic

Aqueous solutions we use in our daily lives generally have pH values that range from 0 to 14. However, some acids and bases used in the laboratory are sometimes so concentrated that their pH values are less than 0 or greater than 14. What concentrations of HCl would give a pH less than zero? What concentrations of NaOH would give a pH greater than 14?

same as the concentration of NaOH, $0.010\ M$ or $1.0 \times 10^{-2}\ M$. Because the pH equation requires the H_3O^+ concentration, we first calculate the H_3O^+ concentration from the OH^- concentration using the K_w expression:

$$K_w = [H_3O^+][OH^-] = 1.0 \times 10^{-14}$$

$$[H_3O^+] = \frac{1.0 \times 10^{-14}}{[OH^-]} = \frac{1.0 \times 10^{-14}}{0.010\ M} = 1.0 \times 10^{-12} M$$

Now that we know the H_3O^+ concentration, $1.0 \times 10^{-12}\ M$, we can calculate the pH of the solution:

$$\text{pH} = -\log[H_3O^+]$$

$$= -\log(1.0 \times 10^{-12})$$

$$= 12.00$$

EXAMPLE 13.10   Calculating pH

What is the pH of each of the following solutions? Once you've done the calculation, check your answer to make sure it makes sense.

(a) 0.0010 *M* HBr
(b) 0.035 *M* HNO$_3$
(c) 0.035 *M* KOH

Solution:

(a) Hydrobromic acid is a strong acid, which ionizes completely, so the H$_3$O$^+$ concentration equals the concentration of HBr:

$$[H_3O^+] = 0.0010 \; M$$

The pH can be calculated from the H$_3$O$^+$ concentration:

$$pH = -\log[H_3O^+]$$
$$= -\log(0.0010)$$
$$= 3.00$$

A pH of 3.00 makes sense for an acid because it is less than 7, as we would expect for a solution of any acid.

(b) Nitric acid is also a strong acid, so the H$_3$O$^+$ concentration equals the concentration of HNO$_3$:

$$[H_3O^+] = 0.035 \; M$$

The pH can be calculated from the H$_3$O$^+$ concentration:

$$pH = -\log[H_3O^+]$$
$$= -\log(0.035)$$
$$= 1.46$$

We would expect a lower pH for HNO$_3$ than for HBr because the HNO$_3$ solution was more concentrated.

(c) Potassium hydroxide, KOH, is a strong base, which dissociates completely to form K$^+$(*aq*) and OH$^-$(*aq*) ions. The concentration of KOH, therefore, is also the OH$^-$ concentration:

$$[OH^-] = 0.035 \; M$$

In order to calculate pH, we need to determine the H$_3$O$^+$ concentration from the OH$^-$ concentration and the K_w expression:

$$[H_3O^+][OH^-] = K_w = 1.0 \times 10^{-14}$$

$$[H_3O^+] = \frac{1.0 \times 10^{-14}}{[OH^-]} = \frac{1.0 \times 10^{-14}}{0.035 \; M} = 2.9 \times 10^{-13} M$$

Now that we know the H$_3$O$^+$ concentration, we can calculate the pH of the solution:

$$pH = -\log[H_3O^+]$$
$$= -\log(2.9 \times 10^{-13})$$
$$= 12.54$$

A high pH (greater than 7) is expected because this is a solution of a base.

> ### Practice Problem 13.10
>
> What is the pH of each of the following solutions? Once you've done the calculation, check your answer to make sure it makes sense.
>
> (a) 0.00085 M HCl
> (b) 0.10 M NaOH
> (c) 1.0 M HNO$_3$
>
> **Further Practice:** 13.63 and 13.64

Calculating pOH

Just as pH can be calculated from the H$_3$O$^+$ concentration, a value called pOH can be calculated from the OH$^-$ concentration of the solution. The pOH of a solution is defined as the negative logarithm (base 10) of the hydroxide ion concentration:

$$pOH = -\log[OH^-]$$

Earlier in this section we found that the pH of a 0.010 M NaOH solution is 12.00. The pOH of this strong base can be calculated directly, because the OH$^-$ concentration equals the concentration of NaOH, 0.010 M. This value can be substituted into the pOH equation:

$$
\begin{aligned}
pOH &= -\log[OH^-] \\
&= -\log(0.010) \\
&= -\log(1.0 \times 10^{-2}) \\
&= 2.00
\end{aligned}
$$

Thus the pOH of this 0.010 M NaOH solution is 2.00. Note that the pH and pOH values total 14.00. This is true for any aqueous solution at 25°C.

$$pH + pOH = 14.00$$

This relationship allows us to determine the pH or pOH of a solution if one value or the other is known. For example, if the pOH of a solution is 4.00, its pH is calculated by subtracting the pOH from 14.00:

$$
\begin{aligned}
pH + pOH &= 14.00 \\
pH &= 14.00 - pOH \\
&= 14.00 - 4.00 \\
&= 10.00
\end{aligned}
$$

Why does pH + pOH = 14.00? The relationship can be derived mathematically from the K_w expression of water:

$$[H_3O^+][OH^-] = 1.0 \times 10^{-14}$$

Taking the negative log of both sides of the equation gives

$$-\log([H_3O^+][OH^-]) = -\log(1.0 \times 10^{-14})$$

The log of a product of two values equals the sum of the logs of the terms:

$$-\log[H_3O^+] - \log[OH^-] = -\log(1.0 \times 10^{-14})$$

We know $-\log[H_3O^+]$ and $-\log[OH^-]$ equal pH and pOH, respectively, so we can substitute pH and pOH into this equation:

$$pH + pOH = 14.00$$

Calculating Concentrations from pH or pOH

Olivia measures the pH of a soil sample solution as 6.20, but she wants to know the H$_3$O$^+$ concentration (Figure 13.13). She needs an equation that relates [H$_3$O$^+$] to pH, and the pH equation does just that:

$$pH = -\log[H_3O^+]$$

To get [H$_3$O$^+$] we must rearrange the equation so that [H$_3$O$^+$] is on one side by itself. First we multiply both sides by −1:

$$-pH = \log[H_3O^+]$$

To get rid of the log expression in front of [H$_3$O$^+$], we take the inverse log of both sides. The H$_3$O$^+$ concentration, then, equals the inverse log of the negative pH:

$$\text{Inverse log } (-pH) = [H_3O^+]$$

Figure 13.13
If the pH of a soil sample is 6.20, what is the H$_3$O$^+$ concentration?

Taking the inverse log of a value (x) is the same as 10 raised to the x power. In other words, *the H_3O^+ concentration equals 10 raised to the –pH power:*

$$10^{-pH} = [H_3O^+]$$

The H_3O^+ concentration in a soil sample with a pH of 6.20 can be calculated using this equation:

$$[H_3O^+] = 10^{-pH} = 10^{-6.20} = 6.3 \times 10^{-7}\ M$$

The OH^- concentration can be determined from pOH with a similar equation:

$$[OH^-] = 10^{-pOH}$$

The relationships among pH, pOH, $[H_3O^+]$, and $[OH^-]$ are represented in Figure 13.14. If you know one of these values for a solution, you can calculate any of the other three. Note that there is more than one way to convert from one value to another, as illustrated in Example 13.11.

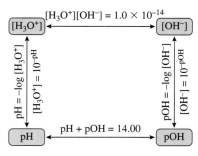

Figure 13.14
If you know pH, pOH, $[H_3O^+]$, or $[OH^-]$ you can calculate any of the other quantities.

EXAMPLE 13.11	Converting Between pH, pOH, $[H_3O^+]$, and $[OH^-]$

Jake measured the pH of water in a swimming pool as 8.10. What is the OH^- concentration in the pool water?

Solution:

Figure 13.14 shows two alternate routes we can take for this calculation. One route is first to convert pH to pOH and then convert pOH to the OH^- concentration:

$$pH \longrightarrow pOH \longrightarrow [OH^-]$$

To convert from pH to pOH, we use the relationship between them:

$$pH + pOH = 14.00$$
$$8.10 + pOH = 14.00$$
$$pOH = 5.90$$

To convert from pOH to $[OH^-]$, we use the relationship that allows us to solve for hydroxide ion concentration:

$$[OH^-] = 10^{-pOH} = 10^{-5.90} = 1.3 \times 10^{-6}\ M$$

The stepwise calculation is summarized as follows:

$$pH \xrightarrow{pH + pOH = 14.00} pOH \xrightarrow{[OH^-] = 10^{-pOH}} [OH^-]$$

An alternative route would be to convert first from pH to $[H_3O^+]$ and then from $[H_3O^+]$ to $[OH^-]$:

$$pH \longrightarrow [H_3O^+] \longrightarrow [OH^-]$$

Try this route. You should get the same answer.

Practice Problem 13.11

When the pH of water in a lake falls below about 4.5, the lake may be considered dead because few organisms can survive in such an acidic environment. What is the pH of a lake that has an OH^- concentration of $1.0 \times 10^{-9}\ M$? Would this lake be considered dead?

Further Practice: 13.65 and 13.66

The inverse log function on your calculator is the 10^x function, which is normally found above the log function as an alternate function for the same key. For more information on how to use the inverse log function, see Math Toolbox 13.1.

Figure 13.15
A pH meter is used to determine the pH of a solution by measuring the voltage that develops when electrodes are dipped into the solution.

Measuring pH

Several methods can be used to measure the pH of a solution. Jake sometimes measures the pH of pool water using a pH meter that is accurate to within hundredths of a pH unit (Figure 13.15). Another less accurate, but often convenient, method is the use of pH *indicators*. They are brightly colored organic dyes that are weak acids or bases. In solution they form an equilibrium with their conjugate bases. The color of the indicator depends on whether the dye is in its acidic or basic form.

Consider, for example, the indicator dye phenolphthalein, which is often used in acid-base titrations. Phenolphthalein has an acidic form (abbreviated HIn) that is colorless and a basic form (In$^-$) that is pink. (The formula HIn is commonly used to represent the acidic form of an indicator. The formula In$^-$ represents the conjugate base of the acid form.)

$$HIn(aq) + H_2O(l) \rightleftharpoons H_3O^+(aq) + In^-(aq)$$
$$\underset{\text{colorless}}{} \underset{\text{phenolphthalein}}{} \qquad\qquad \underset{\text{pink}}{} \underset{\text{phenolphthalein}}{}$$

Phenolphthalein changes from colorless to pink between pH 8.2 and 10. Below pH 8.2 it is primarily in its colorless acidic form because high concentrations of H$_3$O$^+$ cause the equilibrium to lie to the HIn side of the equation. (Recall Le Chatelier's principle from Chapter 12.) Above pH 10 phenolphthalein is primarily in its pink basic form because the low concentration of H$_3$O$^+$ causes the equilibrium to lie to the right side of the equation. The color change ranges for phenolphthalein and other common indicators are shown in Figure 13.16. Note that most of the indicators have acidic and basic forms that are both colored.

An indicator reveals if the pH of a solution is above or below a certain value. Indicators also disclose a specific pH within the indicator's color-change range. For example, the indicator phenol red is sometimes used to find the pH of pool water because its color changes between pH 6.8 and 8.4 (Figure 13.16). Subtle differences in hues are discernible at slightly different pH values.

A mixture of indicators having a variety of colors and color-change ranges can be used to measure the pH of any solution. For example, broad-range pH paper is treated with several indicators. The user reads the pH by comparing the color the

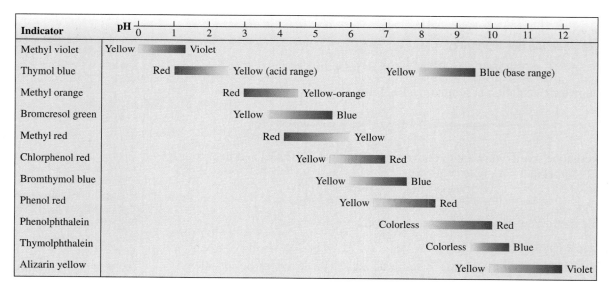

Figure 13.16
Each acid-base indicator has a unique pH color-change range. Outside of its range the indicator exists primarily as the acid form or base form. Within its range, the acid and base forms of the indicator are in equilibrium. The color of the solution within its range can indicate the pH of the solution.

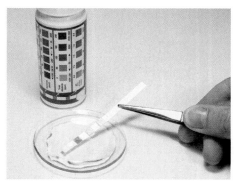

Figure 13.17
Various types of pH paper contain mixtures of different indicators. They are useful for determining pH at a wide range of pH values.

paper turns to a chart of reference colors and pH values printed on the container (Figure 13.17). The liquid universal indicator sometimes used in laboratories contains a mixture of indicators and works on the same principle.

13.6   BUFFERED SOLUTIONS

Olivia's job requires her to find out when there is too much acid or base in garden soil. Because she is considering a career in medicine, she knows what happens when there is too much acid or base in the human bloodstream. Slight changes in blood pH can drastically affect the way our bodies function. For biological reactions to occur normally, blood must stay within the narrow range of pH 7.35 to 7.45. If it rises too high or falls too low, severe health problems, or even death, can occur.

Fortunately our bodies have mechanisms that normally maintain a constant blood pH. One is the action of acid-base buffers in the blood. Blood is a *buffered solution* because it resists changes in pH when limited amounts of acid or base are added to it. A **buffer,** or a *buffer system,* is a combination of a weak acid and its conjugate base (or a weak base and its conjugate acid) in about equal concentrations. The main buffer system in blood is carbonic acid, H_2CO_3, and its conjugate base is the bicarbonate ion, HCO_3^-. The acid and its conjugate base are in equilibrium with each other in solution:

$$H_2CO_3(aq) + H_2O(l) \rightleftharpoons HCO_3^-(aq) + H_3O^+(aq)$$

The equation that describes the H_2CO_3/HCO_3^- buffer system looks just like the equilibrium equation for the ionization of carbonic acid in water. The difference is that in a buffered solution such as blood the concentration of the conjugate base, HCO_3^-, is much greater than the concentration that would result from ionization of the acid alone.

Changes in respiration can affect blood pH by altering the concentration of carbon dioxide in the blood. When people have panic attacks, for example, they breathe very quickly, releasing more CO_2 than normal. This decreases the dissolved CO_2 in the bloodstream. It also causes a decrease of carbonic acid, H_2CO_3, since CO_2 is in equilibrium with carbonic acid:

$$H_2CO_3(aq) \rightleftharpoons CO_2(g) + H_2O(l)$$

Carbonic acid is part of the buffer system in the bloodstream, so changing its concentration affects the pH of the blood. Fortunately our bodies have a way to make a person breathe more slowly. For example, if the panic attack goes on for too long, the person will faint and naturally breathe more slowly. Have you ever wondered

Phenolphthalein, like most acid-base indicators, consists of large organic molecules. The acid form of phenolphthalein is colorless, and the base form is bright pink. Compare the structures of the acid and base forms of phenolphthalein. Can you see that they are a conjugate acid-base pair? Do they differ by one H^+?

Acid, HIn (colorless)

Base, In⁻ (pink)

The desired pH of the water in fish tanks depends on the species of fish that live there. Many fish are most healthy in water with a pH of 7.0. Bromthymol blue indicator is often used in test kits for fish tanks. What other indicators shown in Figure 13.16 might help determine if the pH is 7.0, slightly above 7.0, or slightly below 7.0?

Animation: Buffer Solutions
Online Learning Center, Animations

Chemistry Animations Library, Buffer Solutions, 2Buffer.exe.

The buffers in our bloodstream are effective for only normal fluctuations in acid and base concentrations. If too much acid enters the bloodstream, the pH falls below 7.35 and a condition called *acidosis* occurs. *Alkalosis* occurs when the blood is too basic with a pH above 7.45. Both of these situations are dangerous because they disrupt the body's cellular functions.

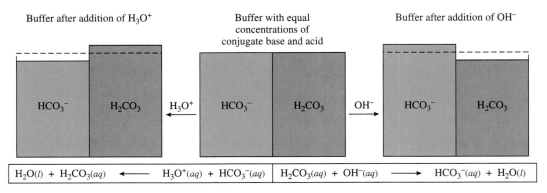

Figure 13.18
When an acid is added to an H_2CO_3/HCO_3^- buffer system, most of the acid is consumed by HCO_3^-. When a base is added, most is consumed by H_2CO_3. The relative amounts of the buffer change, but the pH change is very small.

When the pH of a fish tank is too high or too low, a solution containing an appropriate buffer system is used to bring the pH to its desired level. The $H_2PO_4^-/HPO_4^{2-}$ buffer system is often found in aquarium products that help to maintain a pH 7 environment.

Animation: Using a Buffer
Online Learning Center, Animations

Chemistry Animations Library, Using a Buffer, 1Buffer.rm.

why a remedy for stopping a panic attack is to breathe into a paper bag? Rebreathing the air in the bag increases the concentration of inhaled CO_2, replenishing some that has been lost.

In the laboratory, chemists often prepare buffered solutions if it is important that they maintain a constant pH in a procedure or experiment. To prepare a solution containing the H_2CO_3/HCO_3^- buffer system, for example, both the acid and the base must be added to the solution in similar amounts. In this case the conjugate base is an ion, so it must be added in the form of a soluble salt such as sodium bicarbonate, $NaHCO_3$. Once dissolved in water, the salt dissociates to $Na^+(aq)$ and $HCO_3^-(aq)$:

$$NaHCO_3(s) \xrightarrow{H_2O} Na^+(aq) + HCO_3^-(aq)$$

How does a buffer system prevent large changes in pH when an acid or base is added to the solution? It does so by reacting with most of the H_3O^+ or OH^- from the added acid or base (Figure 13.18). Consider again the H_2CO_3/HCO_3^- buffer system:

$$H_2CO_3(aq) + H_2O(l) \rightleftharpoons HCO_3^-(aq) + H_3O^+(aq)$$

When a base such as NaOH is added, most of the $OH^-(aq)$ from NaOH reacts with the acid, H_2CO_3:

$$H_2CO_3(aq) + OH^-(aq) \longrightarrow HCO_3^-(aq) + H_2O(l)$$

When an acid such as HCl is added, the $H_3O^+(aq)$ ions from HCl react with the base, HCO_3^-:

$$HCO_3^-(aq) + H_3O^+(aq) \longrightarrow H_2CO_3(aq) + H_2O(l)$$

A buffer system works if there is enough acid and conjugate base in the system to react with all the added $H_3O^+(aq)$ or $OH^-(aq)$. The greater the concentrations of the acid and its conjugate base in the buffer mixture, the more effectively the buffer prevents large changes in pH.

EXAMPLE 13.12   **How a Buffer System Works**

Another buffer system in our blood is the $H_2PO_4^-/HPO_4^{2-}$ system. Describe, using a balanced equation, how this system prevents the pH from falling too low when acid enters the bloodstream.

Solution:

The base component of this buffer system, HPO_4^{2-}, reacts with excess H_3O^+ from added acid:

$$HPO_4^{2-}(aq) + H_3O^+(aq) \rightleftharpoons H_2PO_4^-(aq) + H_2O(l)$$

Because most of the excess H_3O^+ is consumed, the pH remains relatively constant.

Practice Problem 13.12

Describe, using a balanced equation, how the $H_2PO_4^-/HPO_4^{2-}$ system in our blood prevents pH levels from rising too high when base enters the bloodstream.

Further Practice: 13.99 and 13.100

Just as a buffer system can moderate pH, the relative amounts of acid and base of a buffer system determine the pH of the solution. In most swimming pool water, for example, an $HOCl/OCl^-$ buffer system is present:

$$HOCl(aq) + H_2O(l) \rightleftharpoons OCl^-(aq) + H_3O^+(aq)$$

Pool water looks best at a pH of 7.5, when the relative amounts of HOCl and OCl^- are equal. At lower pHs, much of the OCl^- is converted to HOCl; at high pHs, much of the HOCl is converted to OCl^-. When the pH is adjusted by adding another kind of acid or base, the equilibrium shifts, so that the relative amounts of HOCl and OCl^- also adjust.

The examples of buffer systems we have discussed are conjugate pairs of weak acids and weak bases. Why wouldn't a strong acid, like HCl, and its conjugate base make an effective buffer? Strong acids ionize completely in water:

$$HCl(aq) + H_2O(l) \longrightarrow H_3O^+(aq) + Cl^-(aq)$$

The Cl^- ion does not react with H_3O^+ to form HCl, because HCl is a strong acid and the acid molecule does not exist in solution. For a buffer system to work, both of the components of the conjugate acid-base pair must exist in equilibrium in solution. This is not the case with a strong acid or base.

EXAMPLE 13.13   Buffer Systems

Which of the following, when added to water, can act as a buffer system? For each buffer system, write a balanced equation.

(a) HF and NaF
(b) HNO_3 and KNO_3
(c) NH_3 and NH_4Cl

Solution:

(a) This is a buffer system because the acid HF is a weak acid and its conjugate base, F^-, forms when the ionic compound NaF dissolves in water. We can ignore the spectator ion Na^+. The following equation shows the conjugate acid-base pair that exists in solution at equilibrium:

$$HF(aq) + H_2O(l) \rightleftharpoons F^-(aq) + H_3O^+(aq)$$

(b) This is not a buffer system, because HNO_3 is a strong acid. Strong acids cannot be components of buffers, because they ionize completely in water and are not in equilibrium with their conjugate bases.

(c) This is a buffer system because NH_3 is a weak base and its conjugate acid, NH_4^+, forms when the ionic compound NH_4Cl dissolves in water. We can ignore the spectator ion Cl^-. The following is a conjugate acid-base pair that will exist at equilibrium in solution:

$$NH_3(aq) + H_2O(l) \rightleftharpoons NH_4^+(aq) + OH^-(aq)$$

Practice Problem 13.13

Which of the following systems, when added to water, can act as a buffer system? For each buffer system, write a balanced equation.

(a) HCl and NaOH
(b) CH_3CO_2H and $NaCH_3CO_2$
(c) HBr and KBr

Further Practice: 13.101 and 13.102

SUMMARY

Acids and bases are important in our everyday lives. They contribute to food tastes, have properties that are helpful for cleaning, and participate in many chemical processes. The Brønsted-Lowry theory is most widely used to describe the behavior of acids and bases. According to this theory, an acid is a substance that donates an H^+ ion to another substance; a base is an H^+ acceptor. Acids and bases can be ranked according to their strength, or how readily they donate or accept H^+ ions. When strong acids dissolve in water, they completely ionize so that essentially no acid molecules remain unreacted. When a weak acid dissolves in water, only a small percentage of the acid molecules ionize. Most remain as molecules. Equilibrium constants called acid ionization constants, K_a, can be used to compare the strengths of different weak acids. The stronger the acid, the larger the K_a value.

The relative concentrations of H_3O^+ and OH^- ions in an aqueous solution at 25°C are determined by the ion-product constant for water, K_w:

$$2H_2O(l) \rightleftharpoons H_3O^+(aq) + OH^-(aq) \qquad K_w = 1.0 \times 10^{-14}$$

$$[H_3O^+][OH^-] = 1.0 \times 10^{-14}$$

Neutral solutions have equal concentrations of H_3O^+ and OH^-. Acidic solutions contain a greater concentration of H_3O^+ than OH^-. Basic solutions contain a greater concentration of OH^- than H_3O^+. The acidity of a solution is commonly expressed in terms of pH, the negative logarithm of the H_3O^+ concentration:

$$pH = -\log[H_3O^+]$$

At 25°C, acidic solutions have a pH less than 7, basic solutions have a pH greater than 7, and neutral solutions have a pH equal to 7.

Buffered solutions contain a weak acid and its conjugate base (or a weak base and its conjugate acid) in similar concentrations. Buffers help to prevent large changes in pH by reacting with small amounts of added acid or base.

Math Toolbox 13.1   LOG AND INVERSE LOG FUNCTIONS

Using Log Functions

The *log* function on your calculator stands for *logarithm to the base 10*. When you take the log of a number, you are asking, "10 to what power will give the value after the word *log*?" For example, the log of 1000 is 3, because 10 must be raised to the third power to get 1000:

$$\log(1000) = 3.0$$

Likewise, the log of 0.001 is –3 because 10 must be raised to the –3 power to get 0.001:

$$\log(0.001) = -3.0$$

To use the log function on most calculators, you first press the log key and then enter the number. (On some calculators, these two steps may be reversed.) Try taking the log of 0.0010 on your calculator. You should get –3.0.

Step 1: Press the log key.
Step 2: Enter 0.001 (or 1×10^{-3}), and then the ENTER or = key.

You should get –3.0. If this does not work, try reordering the key strokes.

Numerical values with large positive or negative exponents, such as concentrations of H_3O^+, are inconvenient to work with. A more convenient way to express such values is to use a logarithmic scale. The pH scale is an example of a logarithmic scale. By convention, the "p" of something is its negative logarithm:

$$pX = -\log X$$

In chemistry the most common examples are the pH scale and the pOH scale:

$$pH = -\log[H^+] = -\log[H_3O^+]$$

$$pOH = -\log[OH^-]$$

To obtain the pH or pOH of a solution, you must first take the log of the concentration and then multiply by –1 (change the sign). Try using your calculator to determine the pH of a solution that has an H_3O^+ concentration of 5.0×10^{-4} M.

$$pH = -\log[H_3O^+]$$

Step 1: Press the +/– (change of sign) key.
Step 2: Press the log key.
Step 3: Enter the H_3O^+ concentration, and then the ENTER or = key.

$$pH = -\log(5.0 \times 10^{-4}) = 3.30$$

If you did not get the right answer, your calculator may require you to reorder the key strokes.

Use your calculator to find the pH of the following solutions.

(a) $[H_3O^+] = 1.0 \times 10^{-8}$ M
(b) $[H_3O^+] = 6.2 \times 10^{-1}$ M

Did you get the following pH values?

(a) pH = 8.00
(b) pH = 0.21

If not, go back and review this section of the Math Toolbox.

Using Inverse Log (10^x) Functions

When you have an equation that involves a log function, you sometimes need to solve for the quantity that follows the log. For example, if you are given pH, you might want to solve for $[H_3O^+]$:

$$pH = -\log[H_3O^+]$$

To solve for $[H_3O^+]$ we must take the inverse log (antilog) of both sides of the equation. To rearrange the equation so that $[H_3O^+]$ is on one side by itself, first multiply both sides by –1:

$$-pH = \log[H_3O^+]$$

To get rid of the log expression in front of $[H_3O^+]$, take the inverse log of both sides:

$$\text{inverse log } (-pH) = \text{inverse log } (\log[H_3O^+])$$

The value of $[H_3O^+]$, then, equals the inverse log of the negative pH:

$$\text{inverse log } (-pH) = [H_3O^+]$$

Taking the inverse log of –pH is the same as 10 raised to the –pH power. The inverse log function on your calculator is the 10^x function, which is often found above the log function as an alternate function for the same key. Another way to describe the relationship between $[H_3O^+]$ and pH is to say $[H_3O^+]$ equals 10 raised to the –pH power:

$$[H_3O^+] = 10^{-pH}$$

To use the inverse log (10^x) function for this calculation on most calculators, you first press the INV, SHIFT, or 2nd button followed by the log key. Then enter the negative of the pH value. The inverse log function, or 10^x function, should appear on your calculator above the log function as the alternative function. By pressing INV, SHIFT, or 2nd button first, you tell the calculator to use the alternative function. (On some calculators, you have to enter the –pH value first.) Try taking the inverse log where pH is 3.00.

$$[H_3O^+] = 10^{-pH}$$

Step 1: Press the INV, SHIFT, or 2nd button.
Step 2: Press the log button (10^x).
Step 3: Enter –3.00, and then the ENTER or = key.

Did you get $[H_3O^+] = 0.0010$ or 1.0×10^{-3}? If not, your calculator may require you to reorder the key strokes. The steps for calculating $[OH^-]$ from pOH are the same as calculating $[H_3O^+]$ from pH:

$$[H_3O^+] = 10^{-pH}$$

$$[OH^-] = 10^{-pOH}$$

The general calculation steps for calculating $[H_3O^+]$ or $[OH^-]$ are

Step 1: Press the INV, SHIFT, or 2nd button.
Step 2: Press the log button.
Step 3: Enter –pH (or –pOH), and then the ENTER or = key.

(On some calculators you may have to reorder the key strokes.)

Math Toolbox 13.1 (continued)

Use your calculator to find the H_3O^+ concentration for the following solutions.

(a) pH = 5.00
(b) pH = 13.60

Did you get the following H_3O^+ concentration values?

(a) $[H_3O^+] = 1.0 \times 10^{-5}$ M or 0.000010 M
(b) $[H_3O^+] = 2.5 \times 10^{-14}$ M

If not, go back and review this section of the Math Toolbox.

KEY RELATIONSHIPS

Relationship	Equation
The self-ionization constant for water equals the product of the hydronium ion concentration and the hydroxide ion concentration. At 25°C, this constant equals 1.0×10^{-14}.	$K_w = [H_3O^+][OH^-] = 1.0 \times 10^{-14}$
The pH of a solution equals the negative log of the hydronium ion concentration. The pOH of a solution equals the negative log of the hydroxide ion concentration.	$pH = -\log[H_3O^+]$ $pOH = -\log[OH^-]$
The hydronium ion concentration equals the inverse log of the negative pH value. The hydroxide ion concentration equals the inverse log of the negative pOH value.	$[H_3O^+] = 10^{-pH}$ $[OH^-] = 10^{-pOH}$
The pH and pOH of a solution total 14.00.	$pH + pOH = 14.00$

KEY TERMS

acid ionization constant, K_a (13.3)

acidic solution (13.4)

amphoteric substance (13.1)

Arrhenius model of acids and bases (13.1)

basic solution (13.4)

Brønsted-Lowry theory (13.1)

buffer (13.6)

conjugate acid (13.1)

conjugate base (13.1)

hydronium ion (13.1)

ion-product constant of water, K_w (13.4)

neutral solution (13.4)

polyprotic acid (13.3)

pH (13.5)

self-ionization (13.4)

strong acid (13.2)

strong base (13.2)

weak acid (13.2)

weak base (13.2)

QUESTIONS AND PROBLEMS

The following questions and problems, except for those in the *Additional Questions* section, are paired. Questions in a pair focus on the same concept. Answers to the odd-numbered questions and problems are in Appendix D.

Matching Definitions with Key Terms

13.1   Match the key terms with the descriptions provided.

(a) a base that ionizes or dissociates completely when dissolved in water

(b) a theory that defines an acid as a substance that donates an H^+ to another substance in solution and a base as a substance that accepts an H^+ in solution

(c) a substance that forms after a base gains an H^+ ion

(d) an acid containing more than one acidic hydrogen

(e) a solution in which the H_3O^+ concentration is greater than the OH^- concentration; a solution with a pH less than 7

(f) a base that does not completely ionize when dissolved in water

(g) a substance that can act as either an acid or a base

(h) the equilibrium constant for the self-ionization of water; the product of the H_3O^+ ion concentration and the OH^- ion concentration in any aqueous solution

(i) a process in which one molecule transfers a proton to another molecule of the same substance; water does this to a very small extent

13.2   Match the key terms with the descriptions provided.
(a) an acid that ionizes completely when dissolved in water
(b) a model that describes acids as substances that generate H^+ ions in solution and bases as substances that generate OH^- ions in solution
(c) an aqueous hydrogen ion; $H_3O^+(aq)$
(d) a solution in which the OH^- concentration is greater than the H_3O^+ concentration; a solution with a pH greater than 7
(e) an acid that does not completely ionize when dissolved in water
(f) a substance that forms after an acid loses an H^+ ion
(g) an equilibrium constant for the ionization of an acid in water; a value that expresses the strength of a weak acid
(h) a measure of the acidity of aqueous solutions; the negative logarithm of the H_3O^+ concentration
(i) a combination of a weak acid and its conjugate base (or a weak base and its conjugate acid) in similar amounts; when in solution, something that helps prevent large changes in pH when small amounts of $H_3O^+(aq)$ or $OH^-(aq)$ are added

What Are Acids and Bases?

13.3   What are some properties of acids?

13.4   What are some properties of bases?

13.5   List some common foods or household products that are bases.

13.6   List some common foods or household products that are acids.

13.7   In terms of the Arrhenius concept of acids and bases, how does an acid behave when dissolved in water? How does an Arrhenius base behave?

13.8   How does the Arrhenius concept of acids and bases emphasize that they are electrolytes?

13.9   How is the Brønsted-Lowry theory of acids and bases different from the Arrhenius model?

13.10   Why is $H_3O^+(aq)$ a more accurate representation of an aqueous H^+ ion than $H^+(aq)$?

13.11   Identify the first reactant in each equation as an acid or a base.
(a) $HF(aq) + H_2O(l) \rightleftharpoons H_3O^+(aq) + F^-(aq)$
(b) $SO_4^{2-}(aq) + H_2O(l) \rightleftharpoons HSO_4^-(aq) + OH^-(aq)$
(c) $C_6H_5OH(aq) + NaOH(aq) \rightleftharpoons$
$H_2O(l) + C_6H_5O^-(aq) + Na^+(aq)$

13.12   Identify the first reactant in each equation as an acid or a base.
(a) $CH_3NH_3^+(aq) + H_2O(l) \rightleftharpoons$
$H_3O^+(aq) + CH_3NH_2(aq)$
(b) $HCN(aq) + H_2O(l) \rightleftharpoons CN^-(aq) + H_3O^+(aq)$
(c) $NaOH(aq) + CH_3CO_2H(aq) \rightleftharpoons$
$H_2O(aq) + CH_3CO_2^-(aq) + Na^+(aq)$

13.13   Write the formula of the conjugate base of each acid.
(a) HNO_2 ~ NO_2^-
(b) HF — F^-
(c) H_3BO_3 ~ $H_2BO_3^-$

13.14   Write the formula of the conjugate base of each acid.
(a) $HOCl$
(b) H_3PO_4
(c) HCO_3^-

13.15   Write the formula of the conjugate acid of each base.
(a) OH^- $-H_2O$
(b) $C_6H_5NH_2$
(c) HCO_3^-

13.16   Write the formula of the conjugate acid of each base.
(a) NH_3
(b) F^-
(c) CO_3^{2-}

13.17   Write an equation to show how the following acids or bases react with water.
(a) $HCl(g)$
(b) $HClO_4(l)$
(c) $CH_3CO_2^-(aq)$

13.18   Write an equation to show how the following acids or bases behave in water.
(a) $NaOH(s)$
(b) $CN^-(aq)$
(c) $HNO_3(l)$

13.19   Which of the following are amphoteric species?
(a) H_2O
(b) HSO_3^-
(c) SO_4^{2-}

13.20   Which of the following are amphoteric species?
(a) NH_4^+
(b) HBO_3^{2-}
(c) $H_2PO_4^-$

13.21   In an H_2CO_3 molecule, both hydrogen atoms are bonded to oxygen atoms. How many acidic hydrogen atoms does carbonic acid have?

13.22   The Lewis structure for the acid phenol is shown. How many acidic hydrogen atoms does phenol have?

Strong and Weak Acids and Bases

13.23   How do strong acids and bases differ from weak acids and bases?

13.24   What do we mean when we say weak acids and bases only partially ionize in water?

13.25   Identify each of the following as a strong acid, weak acid, strong base, or weak base. For each, write an equation that describes its reaction with water.
(a) $H_2SO_4(aq)$ Sa
(b) $Ca(OH)_2(aq)$ sb
(c) $Na_2CO_3(aq)$ wb
(d) $H_3C_6H_5O_7(aq)$ wa
(e) $C_6H_5NH_2(aq)$ wb

13.26 Identify each of the following as a strong acid, weak acid, strong base, or weak base. For each, write an equation that describes its reaction with water.
(a) HCN(*aq*)
(b) NaCN(*aq*)
(c) LiOH(*aq*)
(d) CH_3NH_2(*aq*)
(e) $H_2C_2O_4$(*aq*)

 13.27 Match the molecular-level diagram to each of the following compounds in aqueous solution: HCl, HF, and NH_3.

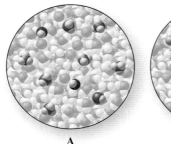 

A **B**

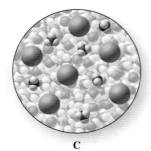

C

13.28 Draw a molecular-level representation of a solution of each of the following acids and bases.
(a) $Ba(OH)_2$
(b) NaOH
(c) HOCl

Relative Strengths of Weak Acids

13.29 If acid A ionizes to a greater extent than acid B, which acid is stronger?

13.30 If acid C has a smaller K_a value than acid D, which acid is stronger?

13.31 Which solution has the highest concentration of H_3O^+, a 0.50 *M* solution of acetic acid, CH_3CO_2H, or a 0.50 *M* solution of formic acid, HCO_2H? Consult Table 13.5 for K_a values.

13.32 Which acid ionizes to the greater extent in water, nitrous acid, HNO_2, or hydrofluoric acid, HF? Consult Table 13.5 for K_a values.

13.33 What is the difference between a concentrated acid solution and a strong acid solution?

13.34 Can a weak acid solution have a higher H_3O^+ concentration than a strong acid solution? Explain.

13.35 What is a polyprotic acid?

13.36 Write the formulas and names of three different polyprotic acids.

13.37 Why does hydrosulfuric acid, H_2S, have two acid ionization constants? Write equations that correspond to each K_a value.

13.38 Which species, H_2S or HS^-, ionizes to the greater extent in water? Explain.

13.39 Which molecules or ions are present in an aqueous solution of oxalic acid, $H_2C_2O_4$, a polyprotic acid? Which is present in the least concentration?

13.40 Which molecules or ions are present in an aqueous solution of citric acid, $H_3C_6H_5O_7$, a triprotic acid? Which is present in the smallest concentration?

Acidic, Basic, and Neutral Solutions

13.41 Under what conditions is the ion-product constant of water, K_w, a constant?

13.42 Does the product of the H_3O^+ ion concentration and the OH^- ion concentration equal 1.0×10^{-14} when substances such as acids and bases are added to water? Explain.

13.43 What are the concentrations of H_3O^+ and OH^- in pure water (at 25°C)?

13.44 What happens to the concentrations of H_3O^+ and OH^- when acid is added to water? When a base is added to water?

 13.45 Identify each of the following conditions as acidic, basic, or neutral.
(a) $[H_3O^+] < [OH^-]$
(b) $[H_3O^+] = 1.0 \times 10^{-5}$ *M*
(c) $[OH^-] = 1.0 \times 10^{-2}$ *M*

13.46 Identify each of the following conditions as acidic, basic, or neutral.
(a) $[OH^-] = 1.0 \times 10^{-12}$ *M*
(b) $[H_3O^+] = 1.0 \times 10^{-8}$ *M*
(c) $[H_3O^+] > [OH^-]$

13.47 Calculate the H_3O^+ ion concentration from the OH^- ion concentration. Then identify the solution as acidic, basic, or neutral.
(a) $[OH^-] = 1.0 \times 10^{-2}$ *M*
(b) $[OH^-] = 1.0 \times 10^{-12}$ *M*
(c) $[OH^-] = 2.5 \times 10^{-8}$ *M*

13.48 Calculate the OH^- ion concentration from the H_3O^+ ion concentration. Then identify the solution as acidic, basic, or neutral.
(a) $[H_3O^+] = 1.0 \times 10^{-2}$ *M*
(b) $[H_3O^+] = 1.0 \times 10^{-6}$ *M*
(c) $[H_3O^+] = 6.4 \times 10^{-10}$ *M*

 13.49 What is the concentration of H_3O^+ in each of the following solutions?
(a) 0.010 *M* HNO_3
(b) 0.075 *M* $HClO_4$
(c) 0.010 *M* NaOH

13.50 What is the concentration of OH^- in each of the following solutions?
  (a) 0.0010 M KOH
  (b) 0.0050 M NaOH
  (c) 0.0010 M HCl

The pH Scale

13.51 Olivia noticed that the pH of her garden soil decreased after it rained. Did her garden soil become more acidic or less acidic?

13.52 Jake noticed that swimming pool water tends to increase in pH more quickly during hot weather. Does pool water tend to become more acidic or more basic during hot weather?

13.53 If one solution has an H_3O^+ concentration that is 10 times that of another solution, what is the difference in their pH values?

13.54 If solution A has a pH that is three pH units greater than that of solution B, how much greater is the H_3O^+ concentration in solution B than in solution A?

13.55 Why does the pH scale typically range from pH 0 to pH 14?

13.56 Can a solution have a pH less than 0 or greater than 14? Explain.

13.57 Can the pH of an acid solution be greater than 7?

13.58 What is the pH range for acidic solutions? For basic solutions?

13.59 What is the pH of solutions having the following H_3O^+ concentrations? Identify each as acidic, basic, or neutral.
  (a) $[H_3O^+] = 1.0 \times 10^{-4}\ M$
  (b) $[H_3O^+] = 1.0 \times 10^{-12}\ M$
  (c) $[H_3O^+] = 6.4 \times 10^{-10}\ M$

13.60 What is the pH of solutions having the following H_3O^+ concentrations? Identify each as acidic, basic, or neutral.
  (a) $[H_3O^+] = 1.0 \times 10^{-8}\ M$
  (b) $[H_3O^+] = 1.0 \times 10^{-14}\ M$
  (c) $[H_3O^+] = 3.8 \times 10^{-3}\ M$

13.61 What is the pH of solutions having the following OH^- concentrations? Identify each as acidic, basic, or neutral.
  (a) $[OH^-] = 1.0 \times 10^{-3}\ M$
  (b) $[OH^-] = 1.0 \times 10^{-7}\ M$
  (c) $[OH^-] = 6.0 \times 10^{-10}\ M$

13.62 What is the pH of solutions having the following OH^- concentrations? Identify each as acidic, basic, or neutral.
  (a) $[OH^-] = 7.0 \times 10^{-4}\ M$
  (b) $[OH^-] = 1.0\ M$
  (c) $[OH^-] = 6.0 \times 10^{-10}\ M$

13.63 What is the pH of the following solutions? Identify each as acidic, basic, or neutral.
  (a) 0.010 M HNO_3
  (b) 0.075 M $HClO_4$
  (c) 0.010 M NaOH

13.64 What is the pH of the following solutions? Identify each as acidic, basic, or neutral.
  (a) 0.0010 M KOH
  (b) 0.0050 M NaOH
  (c) 0.0010 M HCl

13.65 Complete the following table.

	$[H_3O^+]$	$[OH^-]$	pH	pOH	Acidic or Basic?
(a)	1.0×10^{-5}				
(b)		1.0×10^{-4}			
(c)				8.00	
(d)			8.54		
(e)		9.0×10^{-10}			

13.66 Complete the following table.

	$[H_3O^+]$	$[OH^-]$	pH	pOH	Acidic or Basic?
(a)			0.40		
(b)		9.0×10^{-10}			
(c)	1.0×10^{-8}				
(d)				2.00	
(e)	4.5×10^{-2}				

13.67 Determine the pH and pOH of a 0.0050 M HCl solution. What is the relationship between the pH and pOH values?

13.68 Determine the pH and pOH of a 0.050 M NaOH solution. What is the relationship between the pH and pOH values?

13.69 Determine the H_3O^+ concentration in solutions with each of the following pH values. Identify each solution as acidic, basic, or neutral.
  (a) pH = 6.00
  (b) pH = 13.00
  (c) pH = 6.90

13.70 Determine the OH^- concentration given the pH of the following solutions. Identify each solution as acidic, basic, or neutral.
  (a) pH = 5.00
  (b) pH = 12.00
  (c) pH = 11.80

13.71 Determine the H_3O^+ concentration in the following solutions. Identify each solution as acidic, basic, or neutral.
  (a) household ammonia, pH = 11.00
  (b) blood, pH = 7.40
  (c) lime juice, pH = 1.90

13.72 Determine the H_3O^+ concentration in the following solutions. Identify each solution as acidic, basic, or neutral.
(a) orange juice, pH = 3.50
(b) lye, 4.0% sodium hydroxide, pH = 14.00
(c) saliva, pH = 7.00

13.73 What is the OH^- concentration in each of the following solutions?
(a) gastric juice, pH = 1.00
(b) Milk of Magnesia™, pH = 10.50
(c) Seven-Up™, pH = 3.60

13.74 What is the H_3O^+ concentration in each of the following solutions?
(a) lake water, pOH = 6.00
(b) coffee, pOH = 8.90
(c) borax, pOH = 4.50

13.75 Would you expect the pH of 0.010 M acetic acid to be higher or lower than 2.0?

13.76 Would you expect the pH of a 0.010 M NH_3 solution to be higher or lower than 12.0?

13.77 If the pH of an NaOH solution is 13.0, what is the concentration of NaOH?

13.78 If the pH of an HCl solution is 2.0, what is the concentration of HCl?

13.79 If you need to know the pH of a solution to two decimal places, which method of pH measurement discussed in this chapter would be the best choice?

13.80 What are the advantages of using pH paper to measure pH?

13.81 What types of compounds are acid-base indicators?

13.82 Why do we use the symbols HIn and In^- to represent the different forms of an indicator? Which is the acid form and which is the base form of the indicator?

13.83 What is a pH color-change range for an indicator? How is it useful for determining pH between the range of pH values?

13.84 How can an indicator have two color-change ranges? What characteristic must be present?

13.85 Which of the indicators listed in Figure 13.16 would be a good choice for determining the pH of a solution that normally has a pH between 9 and 10?

13.86 Which of the indicators in Figure 13.16 would be a good choice for determining the pH of a solution that normally has a pH between 4 and 5?

13.87 Predict the color of the methyl orange indicator in a solution that has a pH of 3.5.

13.88 Predict the color of the thymol blue indicator in a solution that has a pH of 8.8.

13.89 What can we say about the pH of a solution that is red when the indicator thymol blue is added? When the solution is yellow? When the solution is blue?

13.90 What can we say about the pH of a solution that is colorless when phenolphthalein is added? When the solution is dark pink?

13.91 What is a universal indicator? How are universal indicator and pH paper similar?

13.92 How is litmus paper different from pH paper?

Buffered Solutions

13.93 How is a buffered solution different from a weak acid or a weak base dissolved in water?

13.94 What types of substances make good buffers?

13.95 What is the role of a buffer system in human blood?

13.96 How does the pH change when an acid or base is added to a buffered solution?

13.97 If CH_3CO_2H is added to water, identify an ionic compound that could be added to make a buffered solution.

13.98 If Na_2SO_4 is added to water, what should also be added to make a buffered solution?

13.99 A buffer system used to maintain solution pH values around 12 to 13 is prepared by adding similar concentrations of Na_2HPO_4 and Na_3PO_4 to water.
(a) Write a balanced equation showing the acid and conjugate base in equilibrium. Omit spectator ions.
(b) Describe how this buffer system prevents large pH changes when an acid is added.

13.100 A buffer system used to maintain solution pH values around 5 is prepared by adding similar concentrations of CH_3CO_2H and $NaCH_3CO_2$ to water.
(a) Write a balanced equation showing the acid and conjugate base in equilibrium. Omit spectator ions.
(b) Describe how this buffer system prevents large pH changes when a base is added.

13.101 Which of the following, when added to water, form a buffered solution?
(a) HOCl and NaCl
(b) HNO_2 and KNO_2
(c) CH_3NH_2 and CH_3NH_3Cl

13.102 Which of the following, when added to water, form a buffered solution?
(a) HF and NaF
(b) HI and KI
(c) HOCl and NaOCl

Additional Questions

13.103 Why is NH_3 a base but CH_4 is not? Is it possible for CH_4 to act as a Brønsted-Lowry base?

13.104 Are both acid concentration and acid strength important in determining the concentration of H_3O^+ in an acid solution? Explain.

13.105 Describe what happens when NaF dissolves in water. Describe the acid or base properties of this solution.

13.106 Write an equation that shows what happens when $Ba(OH)_2$ dissolves in water. What is the pH of 0.01 M $Ba(OH)_2$?

13.107 Can the pH of a solution of NaOH ever be less than 7 at 25°C? Explain.

13.108 Can the pH of a solution of HCl ever be greater than 7 at 25°C? Explain.

13.109 Which should have the lower pH, a 0.10 M H_2SO_4 solution or a 0.10 M HNO_3 solution? Explain.

13.110 Consider the reaction classifications we studied in Chapter 5. For each reaction type given, write a chemical reaction in which HCl is a reactant.
  (a) neutralization (double-displacement)
  (b) single-displacement

13.111 Water in a properly maintained swimming pool with a pH of 7.5 has approximately equal concentrations of $HOCl(aq)$ and $OCl^-(aq)$.
  (a) Write a balanced equation showing the acid and conjugate base in equilibrium.
  (b) Describe what happens to this equilibrium when an acid, such as HCl, is added? What happens to the relative concentrations of HOCl and OCl^-? What happens to the pH of the pool water?

13.112 Explain why ions such as Cl^-, NO_3^-, and I^- do not act as bases in water.

14

Oxidation-Reduction Reactions

R emember Anna and Bill from Chapter 1? They had an assignment to wander around campus to identify and classify different forms of matter. They decided that many of the things they observed were either new or rusted metals. For example, at a construction site they noticed shiny, new steel beams, copper pipes, and metallic tubes used for ductwork. They also saw a statue that had corroded over time.

Although we commonly think of metals as shiny, strong, durable, and good electrical conductors, their chemical properties vary considerably. For example, some metals *corrode* (slowly deteriorate due to interaction with the environment) more than others. Compare the rusting of iron on a car to the tarnishing of silver. When iron rusts due to environmental conditions, a compound of iron and oxygen forms and flakes off. The flaking of the rust weakens the body of the car, making it vulnerable to further damage. But when silver tarnishes, a compound containing silver and sulfur forms. Except for some discoloration, the tarnished surface of the silver remains intact.

To some extent nearly all metals are susceptible to reactions with nonmetals in the environment—especially with oxygen—but some are less reactive than others. Aluminum, for example, is structurally very strong, even though it is a low-density metal. Although it is more reactive than iron, it is less susceptible to rusting. The atoms on the surface of an aluminum sheet readily react with oxygen in the atmosphere to form aluminum oxide, Al_2O_3. This compound forms a thin coating on the surface that actually protects the aluminum from further corrosion. Product manufacturers and packagers take advantage of this property. Sometimes they apply a dyed aluminum oxide coating to aluminum objects and give them a bright color.

The corrosion of metals is a serious problem. It's estimated that one-fifth of the steel and iron produced in the United States is used to replace corroded metal. This amounts to billions of wasted dollars every year. What precautions can prevent corrosion? One answer is a protective coating such as paint. We'll discuss other solutions to corrosion later in this chapter.

Why are reactions that corrode metals so prevalent? To answer this question, let's consider the source of the metals we use. Most metals occur in nature as minerals in various ores. That is, instead of finding metals in the form of elements in deposits, we find the metals in the form of compounds—mostly oxides, sulfides, and carbonates. (See Chapter 4.) Through highly involved and costly processes, we extract metals in their elemental form. Over time, however, corrosion returns a metal to its original state as a compound.

Iron is the most widely used of all the metal elements. (Aluminum is second.) One of the primary sources of iron is the mineral hematite, Fe_2O_3 (Figure 14.1). To extract iron for commercial and industrial uses, hematite and other minerals containing iron are loaded into a furnace and allowed to react with carbon:

$$3C(s) + Fe_2O_3(s) \xrightarrow{\text{heat}} 3CO(g) + 2Fe(l)$$

In the extremely high temperature of the furnace the liquid iron flows into another furnace for further purification. Steel, which is mostly iron that contains small amounts of unreacted carbon, results from this purification process. The steel can be used for many purposes, including the manufacture of automobile body panels from thin, flat sheets. Steel car bodies are painted to prevent corrosion, and the surfaces of new cars are protected—for a while. Over time, cars get scratched and dented, and the paint chips away. The iron in the exposed steel is then susceptible to corrosion. Over long periods, the iron reacts with oxygen and water in the environment to form a hydrate of iron(III) oxide. Because this compound does not adhere well to the surface of the steel, it flakes off, exposing more iron metal to environmental conditions that cause further corrosion. Nature provides us with iron in the form of oxides. Through our efforts we obtain iron metal, but, over the years, nature returns the metal to an oxide compound.

Figure 14.1
Hematite is an iron mineral consisting of Fe_2O_3.

For simplicity, rust is sometimes given the formula Fe_2O_3. However, it is a hydrate, so the formula should show the number of water molecules present in a formula unit. This number varies, so it is often written as $Fe_2O_3 \cdot xH_2O$. Since the iron(III) oxide in rust is a hydrate, its properties are different from the properties of the Fe_2O_3 in hematite.

Figure 14.2
The flameless heater used to warm MREs (meals-ready-to-eat) works by adding water to a mixture of magnesium and iron. The reaction generates enough heat to prepare a hot meal of beefsteak with mushrooms and beans or turkey and gravy.

Prior to the invention of an electrolytic process for extraction, aluminum was more expensive than gold, because it was harder to produce in its metallic form. The noblest guests at Napoleon's table used aluminum eating utensils. The lesser guests ate with gold.

There is a fourth category of oxidation-reduction reactions: not spontaneous and undesirable. Why have we skipped such reactions?

Figure 14.3
Different kinds of batteries are used to power numerous, useful devices. Various chemical reactions occur in batteries and generate electricity.

The reaction of carbon with iron(III) oxide to form carbon monoxide and iron metal is a single-displacement reaction (see Chapter 5). Like all single-displacement reactions, it is an oxidation-reduction reaction, which—despite the name—does not always involve oxygen. Instead, an oxidation-reduction reaction involves the transfer of electrons. The corrosion of iron is a good example. In its reaction with oxygen, iron loses electrons and oxygen gains them.

Many oxidation-reduction reactions are exothermic, which allows us to put them to good use. For example, the combustion of methane is an oxidation-reduction reaction that provides heat to homes during the winter. The flameless heaters that soldiers and campers use to prepare hot meals also use an oxidation-reduction reaction (Figure 14.2). The reaction occurs when water mixes with magnesium and iron. It generates enough heat to warm the food.

Oxidation-reduction reactions can also be used to run devices that generate electricity. Such reactions occur in the batteries (Figure 14.3) that power calculators, cell phones, personal digital assistants (PDAs), and other electronic tools. Batteries harness the energy of naturally occurring processes for practical purposes. Like the reactions that occur when materials corrode, the reactions in batteries are *spontaneous*. Once the reaction starts, it proceeds without outside intervention.

Other oxidation-reduction reactions do *not* occur spontaneously. Some external driving force must be applied to make the reaction proceed. For example, until around 1886 there was no practical method for preparing aluminum metal from aluminum oxide, a common compound found in aluminum ores. Passing an electric current through a molten mixture containing aluminum oxide solved the problem. The reaction converts minerals containing aluminum compounds to elemental aluminum metal, but it does not occur spontaneously. It requires a continuous outside energy source that provides electricity. Many pure metals are prepared by a similar process called *electrolysis*.

Oxidation-reduction reactions serve numerous aesthetic purposes, because many compounds of metals are richly colored. For example, the firing processes used by potters to produce colorful ceramics involve oxidation-reduction reactions. Copper(II) oxide and other chemicals form the glaze on the pottery shown in Figure 14.4.

We can look at oxidation-reduction reactions from three different perspectives.

- Some take place spontaneously, but they are undesirable. Metals corrode through such reactions.
- Some occur spontaneously and are desirable, such as reactions in batteries.
- Some do not occur spontaneously, but are desirable. Obtaining aluminum metal from aluminum ores is an example.

In this chapter we will focus on a wide range of oxidation-reduction reactions. As you continue your reading in this chapter, consider the following questions.

Questions for Consideration

14.1   What occurs in an oxidation-reduction reaction?
14.2   How do we account for loss and gain of electrons in an oxidation-reduction reaction?
14.3   How do chemical reactions provide electricity in batteries?
14.4   How are simple oxidation-reduction reactions balanced?
14.5   How are complex oxidation-reduction reactions balanced?
14.6   How do oxidation-reduction reactions generate electricity in spontaneous reactions? How is electricity used to drive oxidation-reduction reactions that do not normally occur?
14.7   How can corrosion be prevented?

14.1   WHAT IS AN OXIDATION-REDUCTION REACTION?

In Chapter 5 we discussed reaction types that include decomposition, combination (or synthesis), single-displacement, double-displacement, and combustion. Our point of reference in Chapter 5 was how *atoms* rearrange to form new substances. Here we want to look at how *electrons* rearrange in chemical processes.

Consider a piece of zinc metal placed in a solution of copper(II) chloride (Figure 14.5). We can represent this reaction with the equation

$$Zn(s) + CuCl_2(aq) \longrightarrow ZnCl_2(aq) + Cu(s)$$

Notice that in this reaction the zinc metal displaces copper from the compound, and we obtain solid copper metal. In Chapter 5 we described this reaction as single-displacement. From a molecular-level perspective, zinc atoms come off the surface of the metal as ions in solution. The copper ions in solution convert to copper atoms and deposit on the surface of the zinc.

Now let's look at this reaction from the point of view of electrons (Figure 14.6). The ionic equation for the reaction makes the process easier to see:

$$Zn(s) + Cu^{2+}(aq) + 2Cl^-(aq) \longrightarrow Zn^{2+}(aq) + 2Cl^-(aq) + Cu(s)$$

Notice that the chloride ions behave as spectators, so we can write a net ionic equation:

$$Zn(s) + Cu^{2+}(aq) \longrightarrow Zn^{2+}(aq) + Cu(s)$$

The two solid metals—$Zn(s)$, a reactant, and $Cu(s)$, a product—both have charges of zero because these elements are in their natural states. As we discussed in Chapter 2, for a zinc atom to form Zn^{2+} ions, the atom has to lose two electrons. For the Cu^{2+} ion to form a copper atom, the ion has to gain two electrons. Electrons are transferred from the zinc metal to the Cu^{2+} ion. This reaction is an oxidation-reduction reaction. An **oxidation-reduction reaction** (or *redox* reaction) is a reaction in which electrons are transferred.

Figure 14.4
The compounds present in different glazes coated on ceramics produce beautiful colors. The green and blue shown here result from different copper compounds. Notice the reddish-orange color near the bottom. It comes from elemental copper produced in the kiln by an oxidation-reduction reaction.

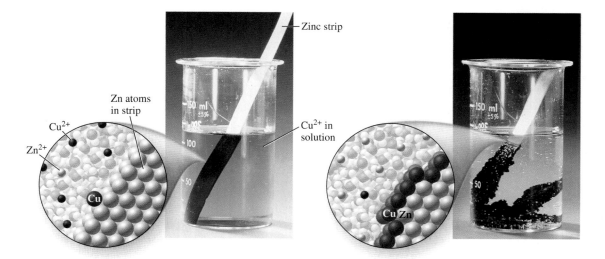

Figure 14.5
When zinc is placed in a solution of $CuCl_2$, copper atoms deposit as a coating on the surface of the zinc. Zinc atoms are converted to Zn^{2+} ions, and copper ions are converted to copper atoms. Chloride ions are spectator ions and they are not shown for simplicity. The net equation for this reaction is

$$Zn(s) + Cu^{2+}(aq) \longrightarrow Zn^{2+}(aq) + Cu(s)$$

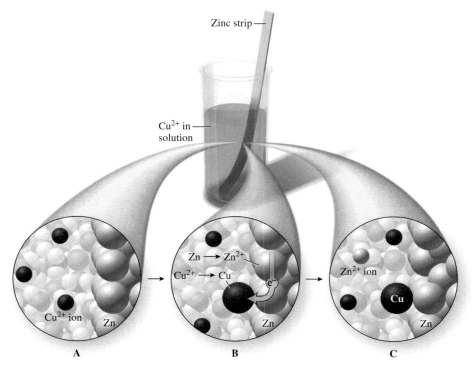

Figure 14.6
(A) For this reaction to occur, Cu^{2+} ions must come into contact with zinc atoms on the metal strip. (B) When a Cu^{2+} ion and a zinc atom touch, two electrons transfer to the copper ion from a zinc atom. (C) Newly formed copper atoms form a coating on the zinc. Zinc ions come off the strip into solution. (Note that spectator ions are not shown.)

Historically the term *oxidation* was used to describe reactions that form oxygen compounds. *Reduction* described reactions in which a compound lost oxygen. Later the definitions were broadened to mean a loss or gain of electrons, but the names stuck. To help keep them straight, use the mnemonic, "LEO says GER." *L*oss of *E*lectrons is *O*xidation. *G*ain of *E*lectrons is *R*eduction.

These reactions are named for the two separate processes that occur together: oxidation and reduction. **Oxidation** is the process of losing one or more electrons. **Reduction** is the process of gaining one or more electrons. Note that oxidation cannot occur without reduction. The two processes go hand in hand.

To probe oxidation-reduction reactions further, let's consider the charges on the zinc, copper, and chlorine in the zinc-copper(II) chloride reaction. The two solid metals each have a charge of zero because they are elements in their natural states. The chloride ion has a charge of 1–, so the copper in $CuCl_2$ must have a charge of 2+. The zinc in zinc chloride also has a charge of 2+. The charges are shown with arrows pointing to each element:

$$\overset{0}{\underset{\downarrow}{Zn}}(s) + \overset{2+}{\underset{\downarrow}{Cu}}\overset{1-}{\underset{\downarrow}{Cl_2}}(aq) \longrightarrow \overset{2+}{\underset{\downarrow}{Zn}}\overset{1-}{\underset{\downarrow}{Cl_2}}(aq) + \overset{0}{\underset{\downarrow}{Cu}}(s)$$

If we look at each species separately as it changes from reactant to product, we can compare charges. Zinc changes from a charge of 0 to 2+. This process involves an increase in charge (a loss of electrons), so it is oxidation. Copper changes from 2+ to 0, which requires a gain of two electrons. The copper ion in copper(II) chloride undergoes reduction. The charge on the chloride ion doesn't change, so it is neither oxidized nor reduced.

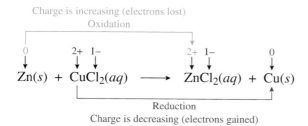

It's convenient to separate the equation into the individual oxidation and reduction processes. For example, let's consider the oxidation of zinc:

$$Zn(s) \longrightarrow Zn^{2+}(aq)$$

Notice that this equation isn't balanced. Do you see why? The charges are unequal: 0 on the left and 2+ on the right. In balancing equations in Chapter 5 we focused strictly on balancing the number of atoms. Here we see that the total charge on each side of the equation must also be equal. To change from an uncharged zinc atom to a Zn^{2+} ion, the metal atom must lose two electrons. We can represent this process with the equation

$$Zn(s) \longrightarrow Zn^{2+}(aq) + 2e^-$$

The equation is now balanced in both atoms and charge. Notice that in this reaction, as in all oxidation processes, the charge on the atom (or ion) losing electrons increases. Since the zinc loses electrons, we include the electrons as a product in the equation.

Now let's consider the reduction of copper ion:

$$Cu^{2+}(aq) \longrightarrow Cu(s)$$

This equation represents the process in which a Cu^{2+} ion forms a copper atom. Again notice that the equation is not balanced because the charges are not equal. For this process to occur, Cu^{2+} must gain two electrons which we write as a reactant:

$$Cu^{2+}(aq) + 2e^- \longrightarrow Cu(s)$$

This equation represents the reduction half of the overall reaction. Here, as in all reduction processes, gaining electrons reduces the charge on the atom or ion.

Chemists commonly describe reactants as agents of oxidation or reduction. A reactant that gives up electrons is oxidized and acts as a **reducing agent.** The reactant that is oxidized is the reducing agent because it provides the electrons to the reactant that gets reduced. The reactant that gains electrons is reduced and acts as an **oxidizing agent.** The reactant that is reduced is the oxidizing agent because it accepts the electrons that are lost by the reactant that is oxidized.

To sort out these definitions let's consider the reaction between zinc and copper(II) chloride again (Figure 14.6). For a reaction to occur, a copper ion must come in contact with a zinc atom. When it does, two electrons are transferred from the zinc atom to the copper ion. In losing two electrons, zinc causes the Cu^{2+} ion to be reduced, so it is the reducing agent. The Cu^{2+} ion is the oxidizing agent because it accepts the two electrons from zinc. In the end, copper atoms deposit on the surface of the zinc metal, and zinc atoms come off the surface of the metal as zinc ions.

Let's return to the net ionic equation for the reaction to summarize the processes:

$$\underset{\substack{\text{reducing} \\ \text{agent}}}{\underset{\text{oxidized}}{Zn(s)}} + \underset{\substack{\text{oxidizing} \\ \text{agent}}}{\underset{\text{reduced}}{Cu^{2+}(aq)}} \longrightarrow Zn^{2+}(aq) + Cu(s)$$

Test your understanding of oxidation-reduction processes by completing Example 14.1.

EXAMPLE 14.1   **Oxidation-Reduction Reactions**

When copper metal is placed in a silver nitrate solution, a single-displacement reaction occurs. Silver deposits on the surface of the copper, and the solution turns blue because Cu^{2+} ions form in the solution.

(a) Write a balanced chemical equation for this reaction.
(b) Identify the element oxidized and the element reduced.
(c) Identify the oxidizing and reducing agents.
(d) On a molecular level, describe what's happening at the surface of the copper metal.

Before   After

Solution:

(a) As we did in Chapter 5, we begin by writing correct formulas for the reactants and products in this single-displacement reaction. Copper metal is one reactant; it is represented as $Cu(s)$. The other reactant, silver nitrate, is an ionic compound composed of silver ions, each with a charge of 1+, and nitrate ions, each with a charge of 1−. Because this compound is soluble in water, we can represent it with the formula $AgNO_3(aq)$. One of the products is a soluble ionic compound composed of copper ions, each with a charge of 2+, and nitrate ions. This product can be represented with the formula $Cu(NO_3)_2(aq)$. The other product is silver metal that has the formula $Ag(s)$. The skeletal equation for this reaction is

$$Cu(s) + AgNO_3(aq) \longrightarrow Cu(NO_3)_2(aq) + Ag(s)$$

We can follow the rules described in Section 5.3 to balance the equation.

$$Cu(s) + 2AgNO_3(aq) \longrightarrow Cu(NO_3)_2(aq) + 2Ag(s)$$

(b) Copper atoms in the copper metal change from 0 to a charge of 2+, so $Cu(s)$ is the reactant that is oxidized. The silver ion from silver nitrate changes from a charge of 1+ to 0, so the Ag^+ ion is reduced.

(c) Because copper metal is the reactant that gives up electrons to cause the reduction of silver ion, it is the reducing agent. Silver ion is the oxidizing agent because it removes electrons from the copper metal, causing its oxidation.

(d) The figure shows that the solution becomes a pale blue as the reaction proceeds indicating that Cu atoms are converted to Cu^{2+} ions. A coating of silver appears on the surface of the copper metal indicating that silver ions, Ag^+, are converted to silver atoms which coat the surface of the copper metal.

Practice Problem 14.1

As shown in the figure, when iron metal is placed in a copper(II) nitrate solution, a single-displacement reaction occurs in which copper deposits on the surface of the iron and aqueous iron(III) nitrate forms.

(a) Write a balanced chemical equation for this reaction.
(b) Identify the element oxidized and the element reduced.
(c) Identify the oxidizing and reducing agents.
(d) On a molecular level, what's happening at the surface of the iron metal?

Before   After

Further Practice: 14.9 and 14.10

14.2   OXIDATION NUMBERS

So far we have described oxidation-reduction reactions of ionic compounds in aqueous solution. We have used electron counts and ionic charges to determine how electrons are lost and gained. Other oxidation-reduction reactions do not take place in solution and, therefore, are not as easy to recognize. For example, consider the oxidation-reduction reaction of carbon with oxygen:

$$C(s) + O_2(g) \longrightarrow CO_2(g)$$

The transfer of electrons is not as obvious here as in a reaction of ionic substances, but this reaction can be identified as oxidation-reduction using a convenient "accounting tool," an oxidation number. An **oxidation number** (sometimes called *oxidation state*) is a charge assigned to the atoms in any compound. For an ionic compound the oxidation numbers are simply the charges on the ions in the compound. Assigning oxidation numbers to atoms in a covalent compound or polyatomic ion is a bookkeeping procedure in which we treat the atoms in the compound as if they were ions with charges that are established by a set of rules. Table 14.1 lists the rules for assigning oxidation numbers. The following points will help you apply the rules.

* The rules are a hierarchy. The first rule that applies takes precedence over any subsequent rules that may apply.
* For an isolated atom or a molecule that contains only one element, only rule 1 need be applied. An uncombined element, whether occurring as an atom such as He or a molecule such as H_2, has an oxidation number of 0. Thus, the oxidation number of sulfur in S, S_2, and S_8 is 0.

- A monatomic ion has an oxidation number equal to its ionic charge, according to rule 2. For example, the oxidation number of lithium in Li^+ is 1+. Similarly, the oxidation number is 1– for chlorine in Cl^-.
- Rule 2 can be applied whenever oxidation numbers have been assigned to all but one element.
- Rule 5 uses position in the periodic table to cover situations that are not handled by one of the other rules.
- To assign oxidation numbers to binary compounds, handle one of the elements with an appropriate rule and the other with rule 2.

Let's apply these rules to the carbon combustion reaction described earlier:

$$C(s) + O_2(g) \longrightarrow CO_2(g)$$

For both reactants, $C(s)$ and $O_2(g)$, the first rule applies, so carbon and oxygen atoms have oxidation numbers of 0. Now let's determine the oxidation numbers of the atoms in carbon dioxide, CO_2. As we proceed through the rules hierarchically, rule 2 applies, but we don't know the oxidation numbers of the atoms in the compound yet. Rule 6 tells us that we should assign an oxidation number of 2– to the oxygen. We can then apply rule 2 to determine the oxidation number of carbon. In Chapter 3 we determined the formulas for ionic compounds using the following relationship:

$$\text{Total positive charge from cations} + \text{total negative charge from anions} = \text{zero net charge}$$

We can modify this relationship to apply to all compounds or ions in the following way:

$$\text{Total positive oxidation numbers} + \text{total negative oxidation numbers} = \text{net charge}$$

This relationship is another way of describing rule 2. For carbon dioxide the net charge is 0. The total of the negative oxidation numbers is 4–, which comes from the two oxygen atoms, each assigned an oxidation number of 2–:

$$\text{Total positive oxidation numbers} + [2 \times (-2)] = 0$$

$$\text{Total positive oxidation numbers} + (-4) = 0$$

$$\text{Total positive oxidation numbers} = 0 + 4$$

$$\text{Total positive oxidation numbers} = 4$$

TABLE 14.1 **Rules for Assigning Oxidation Numbers**

1. The oxidation number of the atoms of an uncombined element is 0.

2. The sum of the oxidation numbers of all atoms in a substance must equal the total charge: 0 for molecules, but the ionic charge for ions (including polyatomic ions).

3. Fluorine has an oxidation number of 1– in its compounds.

4. Hydrogen has an oxidation number of 1+ unless it is combined with metals, where it has an oxidation number of 1–.

5. The position of the element in the periodic table may be useful.
   a. Group IA (1) elements have oxidation numbers of 1+ in their compounds.
   b. Group IIA (2) elements have oxidation numbers of 2+ in their compounds.
   c. Group VIIA (17) elements have oxidation numbers of 1– unless combined with a more electronegative nonmetal.
   d. In binary compounds, group VIA (16) elements have oxidation numbers of 2– unless combined with a more electronegative nonmetal.
   e. In binary compounds, group VA (15) elements have oxidation numbers of 3– unless combined with a more electronegative nonmetal.

6. Oxygen has an oxidation number of 2– except in peroxides (compounds containing the O_2^{2-} ion) in which its oxidation number is 1–.

The oxidation number of carbon must then be 4+.

$$\overset{\overset{0}{\downarrow}}{C(s)} + \overset{\overset{0}{\downarrow}}{O_2(g)} \longrightarrow \overset{\overset{4+\;2-}{\downarrow\;\downarrow}}{CO_2(g)}$$

How would you apply the oxidation number rules to assign charges to the atoms in an ion such as the sulfate ion, SO_4^{2-}? Rule 2 applies, but we must first know the oxidation number of one of the elements in this binary ion. Rule 5 says that group VIA (16) elements are usually assigned an oxidation number of 2–. An exception is when the element is combined with a more electronegative atom such as oxygen. Both cannot have an oxidation number of 2–, so we cannot assign an oxidation number to the sulfur based on this rule. Rule 6 tells us to assign oxygen an oxidation number of 2–. Knowing the oxidation number of oxygen allows us to apply rule 2 to determine the oxidation number of sulfur:

$$\begin{array}{c}\text{Total positive} \\ \text{oxidation numbers}\end{array} + \begin{array}{c}\text{total negative} \\ \text{oxidation numbers}\end{array} = \text{net charge}$$

There are four oxygen atoms in the formula, so we multiply oxygen's oxidation number, 2–, by 4. The net charge on the ion is 2–:

$$\text{Total positive oxidation numbers} + [4 \times (-2)] = -2$$

$$\text{Total positive oxidation numbers} + (-8) = -2$$

$$\text{Total positive oxidation numbers} = -2 + 8$$

$$\text{Total positive oxidation numbers} = 6$$

There is one sulfur atom represented in the formula, so its oxidation number must be 6+.

> To use oxidation numbers in numerical calculations, we'll follow the convention used in math and place the negative sign in front of the number.

EXAMPLE 14.2   Assigning Oxidation Numbers

Assign oxidation numbers to each element in the following formulas.

(a) F_2 (b) PO_4^{3-} (c) $Mg(NO_3)_2$ (d) Cr_2O_3

Solution:

(a) Fluorine, F_2, molecules contain atoms of a single element, fluorine. From rule 1, the oxidation number of F in F_2 is 0.

(b) To assign oxidation numbers for PO_4^{3-}, rule 5e might apply. However, since the phosphorus is a group V (15) element attached to oxygen, which is more electronegative, the rule will not work. We have to use rule 6 to assign an oxidation number of 2– to oxygen. Then using rule 2 we can calculate the oxidation number of phosphorus:

$$\begin{array}{c}\text{Total positive} \\ \text{oxidation numbers}\end{array} + \begin{array}{c}\text{total negative} \\ \text{oxidation numbers}\end{array} = \text{net charge}$$

There are four oxygen atoms in the formula, so we multiply oxygen's oxidation number, 2–, by 4. The net charge on the ion is 3–:

$$\text{Total positive oxidation numbers} + [4 \times (-2)] = -3$$

$$\text{Total positive oxidation numbers} + (-8) = -3$$

$$\text{Total positive oxidation numbers} = -3 + 8$$

$$\text{Total positive oxidation numbers} = 5$$

The oxidation number of phosphorus is 5+.

(c) $Mg(NO_3)_2$ is an ionic compound composed of a metal ion and a polyatomic anion. The polyatomic anion in the compound is nitrate, NO_3^-. Since there are two nitrate ions in a formula unit of the compound, the oxidation number of magnesium must be 2+. (We also know this from magnesium's position in the periodic table, rule 5b.) We can use rule 6 to assign an oxidation number of 2– to oxygen. Then, using rule 2 we can calculate the oxidation number of nitrogen:

$$\text{Total positive oxidation numbers} + \text{total negative oxidation numbers} = \text{net charge}$$

There are three oxygen atoms in the formula for nitrate, so we multiply oxygen's oxidation number, 2–, by 3. The net charge on the ion is 1–:

$$\text{Total positive oxidation numbers} + [3 \times (-2)] = -1$$

$$\text{Total positive oxidation numbers} + (-6) = -1$$

$$\text{Total positive oxidation numbers} = -1 + 6$$

$$\text{Total positive oxidation numbers} = 5$$

The oxidation number for nitrogen is 5+.

(d) Cr_2O_3 is an ionic compound containing a transition metal. Since transition metal charges cannot be predicted from the periodic table, we must begin with rule 6, assigning an oxidation number of 2– to oxygen. Then using rule 2 we can calculate the oxidation number of chromium from the following relationship:

$$\text{Total positive oxidation numbers} + \text{total negative oxidation numbers} = \text{net charge}$$

There are three oxygen atoms in the formula, so we multiply oxygen's oxidation number, 2–, by 3. The net charge on the compound is 0:

$$\text{Total positive oxidation numbers} + [3 \times (-2)] = 0$$

$$\text{Total positive oxidation numbers} + (-6) = 0$$

$$\text{Total positive oxidation numbers} = 0 + 6$$

$$\text{Total positive oxidation numbers} = 6$$

Since there are two chromium atoms in the formula, each one has an oxidation number of 3+.

Practice Problem 14.2

Assign oxidation numbers to each element in the following formulas.

(a) Mg
(b) Mn_2O_3
(c) Na_2SO_4
(d) $Cr_2O_7^{2-}$

Further Practice: 14.19 and 14.20

Let's return to the combustion reaction of carbon:

$$C(s) + O_2(g) \longrightarrow CO_2(g)$$

Recall that we assigned oxidation numbers of 0 to both $C(s)$ and $O_2(g)$. The oxidation number of carbon in CO_2 is 4+. The oxidation number of oxygen is 2–. We can summarize these oxidation numbers as follows:

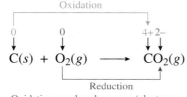

Earlier we used changes in charge to identify oxidation and reduction processes. We can also identify them from changes in oxidation number. *An oxidation-reduction reaction occurs if one or more elements changes its oxidation number.* When an oxidation number increases, electrons have been lost and oxidation has occurred. When an oxidation number decreases, electrons have been gained and reduction has occurred. Notice in this combustion reaction that carbon's oxidation number changes from 0 as a reactant to 4+ as a product. Carbon lost four electrons, so carbon is oxidized. Oxygen's oxidation number changes from 0 as a reactant to 2– as a product. Each oxygen atom gained two electrons, so oxygen is reduced.

Oxidation number increases (electrons lost)

Oxidation

$$C(s) \;+\; O_2(g) \;\longrightarrow\; CO_2(g)$$

Reduction

Oxidation number decreases (electrons gained)

What are the oxidizing and reducing agents in this reaction? Since $C(s)$ is giving up electrons so that $O_2(g)$ can be reduced, $C(s)$ is the reducing agent. Because $O_2(g)$ is taking the electrons from $C(s)$, $O_2(g)$ is the oxidizing agent.

Now that you can identify oxidation-reduction reactions in both ionic compounds and molecular compounds, determine if the reactions in Example 14.3 involve oxidation and reduction.

EXAMPLE 14.3   Identifying Oxidation-Reduction Reactions

Determine which of the following represent oxidation-reduction reactions. For reactions that involve oxidation-reduction, identify the oxidizing and reducing agents.

(a) $AgNO_3(aq) + NaCl(aq) \longrightarrow AgCl(s) + NaNO_3(aq)$
(b) $2Fe(s) + 3Cu(NO_3)_2(aq) \longrightarrow 2Fe(NO_3)_3(aq) + 3Cu(s)$
(c) $CH_4(g) + 2O_2(g) \longrightarrow CO_2(g) + 2H_2O(g)$
(d) $4Fe(s) + 3O_2(g) \longrightarrow 2Fe_2O_3(s)$

Solution:

If an element undergoes a change in oxidation number, an oxidation-reduction reaction has occurred. First assign oxidation numbers to all the elements in the equations using the rules in Table 14.1.

(a) The oxidation numbers of the elements in this reaction are as follows:

$$AgNO_3(aq) + NaCl(aq) \longrightarrow AgCl(s) + NaNO_3(aq)$$

Since none of these elements undergoes a change in oxidation number, this is not an oxidation-reduction reaction.

(b) The oxidation numbers of the elements in this reaction are as follows:

$$\overset{0}{2Fe(s)} + \overset{2+\ 5+\ 2-}{3Cu(NO_3)_2(aq)} \longrightarrow \overset{3+\ 5+\ 2-}{2Fe(NO_3)_3(aq)} + \overset{0}{3Cu(s)}$$

In this reaction the oxidation number of iron changes from 0 to 3+, so it is oxidized. The oxidation number of copper changes from 2+ to 0, so it is reduced. This is an oxidation-reduction reaction. Fe(s) is the reducing agent, and Cu^{2+} is the oxidizing agent.

(c) The oxidation numbers of the elements in this reaction are as follows:

$$\overset{4-\ 1+}{CH_4(g)} + \overset{0}{2O_2(g)} \longrightarrow \overset{4+\ 2-}{CO_2(g)} + \overset{1+\ 2-}{2H_2O(g)}$$

The oxidation number of carbon changes from 4– to 4+, so it is oxidized. The oxidation number of oxygen changes from 0 to 2–, so it is reduced. This is an oxidation-reduction reaction. CH_4 is the reducing agent, and O_2 is the oxidizing agent.

(d) The oxidation numbers of the elements in this reaction are as follows:

$$\overset{0}{4Fe(s)} + \overset{0}{3O_2(g)} \longrightarrow \overset{3+\ 2-}{2Fe_2O_3(s)}$$

The oxidation number of iron changes from 0 to 3+, so it is oxidized. The oxidation number of oxygen changes from 0 to 2–, so it is reduced. This is an oxidation-reduction reaction. Fe is the reducing agent, and O_2 is the oxidizing agent.

Practice Problem 14.3

Determine which of the following represent oxidation-reduction reactions. For reactions that involve oxidation-reduction, identify the oxidizing and reducing agents.

(a) $2H_2O(l) \longrightarrow 2H_2(g) + O_2(g)$
(b) $HCl(aq) + NaOH(aq) \longrightarrow NaCl(aq) + H_2O(l)$
(c) $4Al(s) + 3O_2(g) \longrightarrow 2Al_2O_3(s)$
(d) $Mg(s) + Zn(NO_3)_2(aq) \longrightarrow Mg(NO_3)_2(aq) + Zn(s)$

Further Practice: 14.29 and 14.30

Figure 14.7
Zinc-mercury batteries are used in watches and calculators.

14.3 BATTERIES

All oxidation-reduction reactions that occur without outside intervention release energy. For example, the combustion of methane, CH_4, occurs in some home furnaces. The energy from the combustion reaction provides heat. Heat is useful for heating homes, but not much help when we want to run electronic devices. That's when batteries come in handy. Electricity from a battery is a flow of electrons released from an oxidation-reduction reaction. The correct construction of a battery is essential for harnessing this energy in a useful way.

Consider the zinc-mercury battery used in watches and calculators (Figure 14.7). The reaction inside it is

$$Zn(s) + HgO(s) \longrightarrow ZnO(s) + Hg(l)$$

Although it's not apparent from the equation, the reaction requires the presence of hydroxide ions (OH⁻), supplied by potassium hydroxide. Suppose we simply mixed zinc with mercury(II) oxide in a solution of potassium hydroxide. The energy released by this reaction when electrons are transferred would be lost as heat, which could not be used to power a watch. Instead of allowing all the reactants to mix freely, batteries are constructed so that a barrier separates the oxidation reaction from the reduction reaction. A wire connects them. Rather than being lost as heat, electrical energy is generated as electrons are transferred through a wire from the oxidation reaction to the reduction. If the wire runs through a useful device, such as a watch, the flow of electrons provides the power the device needs to run.

Let's consider another battery constructed from zinc and copper. Recall that earlier we discussed the reaction of zinc metal with copper(II) chloride:

$$Zn(s) + CuCl_2(aq) \longrightarrow ZnCl_2(aq) + Cu(s)$$

If we place a piece of zinc in a solution of copper(II) chloride, copper metal will plate onto the surface of the zinc (Figure 14.5). The blue color of the Cu^{2+} ions in solution disappears, and the temperature rises. The energy is lost to the surroundings as heat. Now consider the same reaction, but with the oxidation and reduction reactions separated (Figure 14.8). The loss of electrons from zinc and the gain of electrons by copper ions lights the bulb, because electrons flow through the wire that connects the separate compartments. A device that sets up a spontaneous chemical reaction to produce electricity in this way is a **voltaic** (or **galvanic**) **cell.** Voltaic cells are the operational parts of a battery.

The tendency for electrons to flow in a cell is measured as *voltage,* and we can use an analogy to understand it. Think of the tendency of electrons to flow as analogous to the flow of water in a river. River water flows sluggishly on a nearly flat landscape, but it flows rapidly and with greater force when the slope is steeply downhill (Figure 14.9). Similarly, chemical reactions vary in their potentials to cause electrons to flow. The potential of a voltaic cell, measured in units of *volts,* is greater in cells containing the strongest reducing and oxidizing agents. A voltmeter can measure the voltage that a voltaic cell produces.

To understand how a voltaic cell produces electricity, let's consider what occurs in each compartment as zinc and copper react:

$$Zn(s) + Cu^{2+}(aq) \longrightarrow Zn^{2+}(aq) + Cu(s)$$

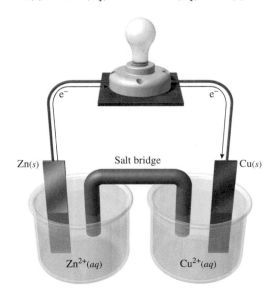

Figure 14.8
When the reactants are separated into half-cells and connected with a wire, electrons flow and the bulb lights.

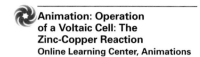

High potential

Low potential

Figure 14.9
The potential for the flow of electrons in a voltaic cell is analogous to the flow of water in a river. The steeper the river bed, the greater the potential for water flow. In a chemical reaction the potential varies depending on the reactants.

The terms *voltage* and *volt* honor Alessandro Volta, who invented the forerunner of the battery around 1800. He is remembered with Luigi Galvani for the work that demonstrated the electrical nature of nerve impulses in animals.

Assigning oxidation numbers we get

$$\overset{0}{\downarrow}\text{Zn}(s) + \overset{2+}{\downarrow}\text{Cu}^{2+}(aq) \longrightarrow \overset{2+}{\downarrow}\text{Zn}^{2+}(aq) + \overset{0}{\downarrow}\text{Cu}(s)$$

The oxidation number of zinc changes from 0 to 2+, so it is the reactant that is oxidized. The oxidation number of copper changes from 2+ to 0, so it is the reactant that is reduced. From this information we can write equations to represent both processes:

Oxidation $\text{Zn}(s) \longrightarrow \text{Zn}^{2+}(aq) + 2e^-$

Reduction $\text{Cu}^{2+}(aq) + 2e^- \longrightarrow \text{Cu}(s)$

Each of these equations is a **half-reaction,** representing either the oxidation or the reduction that occurs in the separate compartments of a voltaic cell. Thus each compartment is called a **half-cell.** Oxidation occurs in one half-cell; reduction occurs in the other. The separation of the oxidation and reduction processes into half-cells harnesses the energy released from a chemical reaction to generate electricity. The potential to give up electrons is greater for zinc than it is for copper. Therefore, electrons flow through the wire from the zinc metal to the copper metal where Cu^{2+} ions in the solution are reduced to copper atoms.

Inside a voltaic cell (Figure 14.10) a solid material that conducts electricity must be present to provide a site for each half-reaction. This material is an **electrode.** The electrode is commonly a metal immersed in an electrolyte solution containing a salt of the same metal. The electrode at which oxidation occurs is the **anode.** The electrode where reduction occurs is the **cathode.** In a cell composed of zinc and copper, the anode is a piece of zinc metal. The electrolyte in the anode half-cell is a compound of zinc such as $\text{Zn(NO}_3)_2$. The cathode is a piece of copper metal. The electrolyte in the cathode half-cell contains a compound of copper such as $\text{Cu(NO}_3)_2$.

Sometimes an inert (unreactive) metal is used as an electrode to serve merely as a site for electron transfer.

Electricity will not flow unless there is a complete circuit between the anode and the cathode. In a voltaic cell like the one shown in Figure 14.10, a salt bridge completes the circuit. A **salt bridge** allows ions to flow so charge balance is maintained. A typical salt bridge is a glass tube in the shape of an inverted U containing an ionic compound such as Na_2SO_4 in a gel. Ions freely flow in and out of the gel and keep charge balanced within the half-cells.

To see how atoms, ions, and electrons flow in a voltaic cell, let's look at what's happening from a molecular-level perspective (Figure 14.11). In the anode half-cell in which oxidation occurs, each atom of zinc metal that reacts loses two electrons. Now an ion, the zinc leaves the surface of the electrode and goes into the solution. The electrons lost by a zinc atom flow toward the cathode through the wire. The electrons come out at the cathode, where they'll come into contact with a Cu^{2+} ion in solution. This ion accepts the electrons, and copper is deposited on the surface of the copper electrode as a neutral atom.

Animation: Galvanic Cell
Online Learning Center, Animations

Chemistry Animations Library, Galvanic Cell, 3Galvanic_Cell.exe.

Figure 14.10
A voltaic cell consists of two electrodes immersed in electrolyte solutions. One electrode is the anode, the other is the cathode. Oxidation occurs at the anode, and reduction occurs at the cathode. Electrons flow through a wire. A salt bridge completes the circuit.

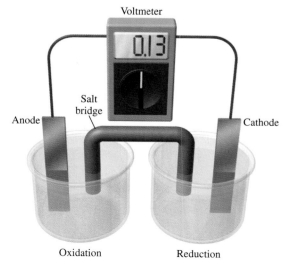

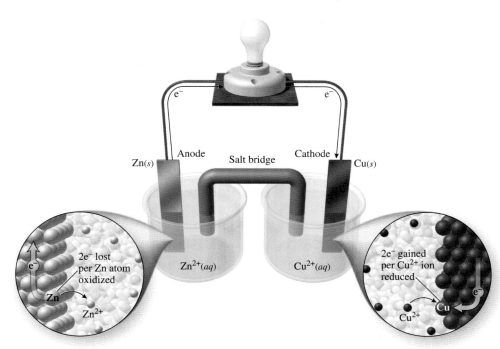

Figure 14.11
The electrons lost by zinc at the anode travel toward the cathode. This flow of electrons causes the bulb to light. At the cathode, a copper ion gains two electrons and gets plated onto its surface. (Note that spectator ions are not shown.)

Now that we can see what's happening at each electrode, can you explain the purpose of the salt bridge? To answer this question, consider what's happening to the charge in each half-cell as the reaction proceeds. At the anode Zn^{2+} ions form from neutral zinc atoms, and the anode half-cell gains some positive charge. Negatively charged ions flow into the anode half-cell from the salt bridge to maintain charge balance. At the cathode Cu^{2+} ions change to neutral copper atoms, and the cathode half-cell gains some negative charge. Positive ions from the salt bridge flow into this half-cell, replacing the positive charge (Figure 14.12). The ions that flow from the salt bridge merely work to balance the charge but do not participate in the chemical reaction.

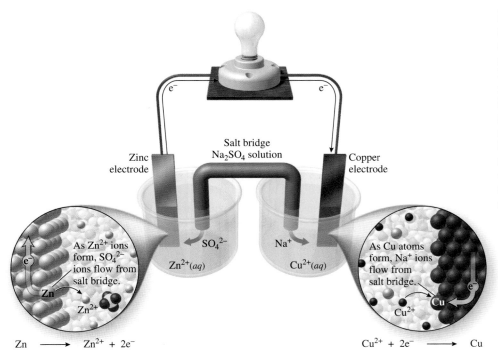

Figure 14.12
Salt bridges commonly contain a solution of an ionic compound such as sodium sulfate, Na_2SO_4. Negatively charged sulfate ions flow from the salt bridge into the anode half-cell where a positive charge is generated. Positively charged sodium ions flow into the cathode half-cell, preventing the buildup of negative charge. A salt bridge completes the circuit. Without it electrons will not flow. (Note that not all spectator ions are shown.)

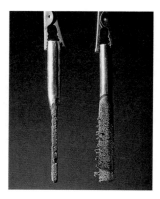

Figure 14.13
As a zinc-copper voltaic cell runs, the zinc electrode deteriorates. The copper electrode increases in mass, because copper atoms plate onto its surface as the reaction proceeds.

What happens when a voltaic cell runs for a long time? What happens to the zinc electrode? Does its mass increase or decrease? What about the copper electrode? Although solutions containing Zn^{2+} ions are colorless, Cu^{2+} ions produce a blue color in water. Will the intensity of the color change as the reaction runs? To answer these questions, look at Figure 14.13. The atoms on the zinc electrode oxidize to aqueous zinc ions, so the anode gets smaller. Copper ions from the aqueous solution deposit on the copper electrode, so the cathode has more copper atoms. Many of the copper ions are no longer present in solution, so the blue color fades.

EXAMPLE 14.4 **Identifying Components of a Voltaic Cell**

Magnesium reacts with copper(II) sulfate according to the following equation:

$$Mg(s) + CuSO_4(aq) \longrightarrow MgSO_4(aq) + Cu(s)$$

(a) Write equations to represent the oxidation and reduction half-reactions.
(b) Determine the oxidizing and reducing agents.
(c) The figure represents components of a voltaic cell in which this reaction takes place. Answer the following questions with the correct component of a voltaic cell, using the letters A to E as shown. Assume that the oxidation half-reaction will occur on the left.
  • In which half-cell will you put the oxidation components?
  • In which half-cell will you put the reduction components?
  • What will you use for the anode and cathode?
  • What electrolyte solutions will you put in each compartment?
  • Which direction will ions flow from the salt bridge?
  • Where do electrons flow?
  • What happens to the mass of the electrodes as the cell runs?

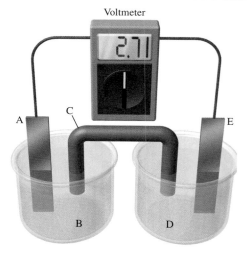

Solution:

(a) Because the sulfate ion, SO_4^{2-}, is a spectator ion, we can simplify the reaction by writing a net ionic equation:

$$Mg(s) + Cu^{2+}(aq) \longrightarrow Mg^{2+}(aq) + Cu(s)$$

Since these are neutral or ionic species, assigning oxidation numbers is straightforward:

$$\overset{0}{Mg}(s) + \overset{2+}{Cu^{2+}}(aq) \longrightarrow \overset{2+}{Mg^{2+}}(aq) + \overset{0}{Cu}(s)$$

Magnesium's oxidation number changes from 0 to 2+, so it is the reactant that is oxidized. The oxidation reaction is a loss of two electrons for each magnesium atom:

$$Mg(s) \longrightarrow Mg^{2+}(aq) + 2e^-$$

Copper's oxidation number changes from 2+ to 0, so it is the reactant that's reduced. The reduction reaction is a gain of two electrons for each Cu^{2+} ion:

$$Cu^{2+}(aq) + 2e^- \longrightarrow Cu(s)$$

From this information we can write half-reaction equations that represent the oxidation and reduction processes:

Oxidation $Mg(s) \longrightarrow Mg^{2+}(aq) + 2e^-$

Reduction $Cu^{2+}(aq) + 2e^- \longrightarrow Cu(s)$

(b) Magnesium is the reactant that's oxidized. The reactant that accepts the electrons from magnesium is Cu^{2+}. The oxidizing agent is Cu^{2+}, which is reduced in the reaction. The reactant that provides the electrons for this reduction is magnesium. Magnesium is the reducing agent.

(c) If we arbitrarily assign the oxidation half-reaction to the compartment on the left, then the answers are as follows:

- The oxidation components will be A and B on the left.
- The right compartment will contain the reduction components D and E.
- Component A will be the anode of magnesium metal. Component E will be the cathode of copper metal.
- The solution in the anode compartment, B, should be an electrolyte of magnesium, such as $Mg(NO_3)_2$. The electrolyte solution in the cathode compartment, D, can be $Cu(NO_3)_2$ or another salt containing Cu^{2+} ions.
- To complete the electric circuit, component C needs to be a salt bridge that allows ions to flow into the two half-cells. Sodium sulfate, Na_2SO_4, is a common ingredient in salt bridges. Sulfate ions, SO_4^{2-}, will flow into the left compartment to balance the positive charge that builds up from the production of Mg^{2+} ions. Na^+ ions will flow into the right half-cell to replace the Cu^{2+} ions that react to form neutral Cu atoms.
- Electrons will flow from electrode A to electrode E through, in this case, a voltmeter.
- The mass of electrode A will decrease because magnesium atoms react to form magnesium ions that flow into the solution. The mass of electrode E will increase because copper ions from the solution react to form copper atoms on the surface of the electrode.

Practice Problem 14.4

Cadmium reacts with nickel(II) nitrate according to the following equation:

$$Cd(s) + Ni(NO_3)_2(aq) \longrightarrow Cd(NO_3)_2(aq) + Ni(s)$$

(a) Write equations to represent the oxidation and reduction half-reactions.
(b) Determine the oxidizing and reducing agents.
(c) Draw a picture like the one given in the example to show the components of this voltaic cell. Label the components A, B, C, D, and E and answer the same questions as in the example.

Further Practice: 14.37 and 14.38

A single cell is often called a battery. Technically, however, a battery is a series of connected voltaic cells.

Figure 14.14
The battery jar was used to provide power for telephones and other devices.

Zinc-copper cells are not what we usually think of when we visualize batteries, but for many years they actually were used to power telephones and other electronic devices. Notice all the plates in the old-fashioned "battery jar" in Figure 14.14. They are a series of anodes and cathodes that provide more electricity than a single pair would. A series of connected voltaic cells that provides a portable source of electrical power is a **battery.**

Zinc-copper batteries have practical limitations because they are "wet cells." Imagine carrying a battery jar around with a cell phone! In spite of its limitations for small electronic devices, a wet cell battery is still in common use today for a much larger application: starting an automobile. The battery used in cars is a lead-acid, or lead storage, battery (Figure 14.15). The reaction that occurs inside it is

$$Pb(s) + PbO_2(s) + 2H_2SO_4(aq) \longrightarrow 2PbSO_4(aq) + 2H_2O(l)$$

In this reaction one form of lead, $Pb(s)$, is oxidized, while another form, $PbO_2(s)$, is reduced:

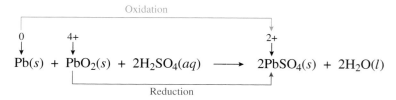

In the lead-acid battery, the anode is constructed of lead metal, with grids that contain a spongy form of lead. The cathode is also made of lead metal, but its grids contain lead(IV) oxide, PbO_2. As in all voltaic cells, we can separate the overall reaction into half-reactions. The anode and cathode half-reactions for the lead-acid battery are not obvious. They are

Anode $\quad Pb(s) + H_2SO_4(aq) \longrightarrow PbSO_4(s) + 2H^+(aq) + 2e^-$

Cathode $\quad PbO_2(s) + H_2SO_4(aq) + 2H^+(aq) + 2e^- \longrightarrow PbSO_4(s) + 2H_2O(l)$

Which of these half-reactions is oxidation and which is reduction?

A single cell for a lead-acid battery has an electrical potential of 2 volts (V). The six cells in a car battery generate a voltage of 12 V. To power small electronic devices, less voltage is needed and "dry cells" are used. These batteries vary in composition and produce different voltages. The 9-V battery (Figure 14.16), for example, is a series of six cells constructed of Zn, MnO_2, and an electrolyte paste of various compounds. The individual cells produce a voltage of 1.5 V. The reaction that occurs in these cells is

$$Zn(s) + MnO_2(s) + H_2O(l) \longrightarrow ZnO(s) + Mn(OH)_2(s)$$

Can you identify the oxidizing and reducing agents in this reaction?

Figure 14.15
The lead-acid battery is a series of six voltaic cells that have alternating anodes and cathodes. It is the most common wet cell battery in use today.

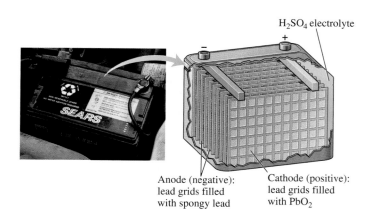

Anode (negative): lead grids filled with spongy lead

Cathode (positive): lead grids filled with PbO_2

Earlier in this chapter we mentioned a battery made of zinc and mercury(II) oxide:

$$Zn(s) + HgO(s) \longrightarrow ZnO(s) + Hg(l)$$

This battery produces a voltage of 1.3 V and is used in watches and calculators. A similar battery is composed of zinc and silver oxide:

$$Zn(s) + Ag_2O(s) \longrightarrow ZnO(s) + 2Ag(s)$$

It produces a voltage of 1.6 V and is commonly used in cameras. Given the overall reaction that occurs in a battery, you should be able to identify the oxidizing and reducing agents.

A

EXAMPLE 14.5   Oxidation and Reduction in Batteries

The reaction that occurs in most watch batteries is

$$Zn(s) + HgO(s) \longrightarrow ZnO(s) + Hg(l)$$

What is the oxidizing agent in this battery? What is the reducing agent?

Solution:

As usual we first assign oxidation numbers:

$$\overset{0}{Zn}(s) + \overset{2+}{Hg}\overset{2-}{O}(s) \longrightarrow \overset{2+}{Zn}\overset{2-}{O}(s) + \overset{0}{Hg}(l)$$

Zinc changes from 0 to 2+, so it is oxidized. Mercury changes from 2+ to 0, so it is reduced. $HgO(s)$ is the oxidizing agent, and $Zn(s)$ is the reducing agent.

Practice Problem 14.5

The reaction that occurs in most camera batteries is

$$Zn(s) + Ag_2O(s) \longrightarrow ZnO(s) + 2Ag(s)$$

What is the oxidizing agent? What is the reducing agent?

Further Practice: 14.41 and 14.42

B

Figure 14.16
(A) A 9-V battery contains (B) a series of six 1.5-V cells.

14.4   BALANCING SIMPLE OXIDATION-REDUCTION EQUATIONS

To construct a battery or to describe an oxidation-reduction reaction, we need to predict the quantities of reactants and products. A balanced equation provides this information. As shown in Chapter 5, the atoms in a chemical equation must balance. In addition, charge must balance, and electrons lost must equal electrons gained.

To see what's involved in balancing a simple oxidation-reduction equation, consider the reaction between copper metal and silver nitrate shown in Figure 14.17. Copper atoms change to Cu^{2+} ions, and Ag^+ ions change to silver atoms. We can represent the reaction using the equation

$$Cu(s) + Ag^+(aq) \longrightarrow Cu^{2+}(aq) + Ag(s)$$

Do you notice anything wrong with this equation? Do the atoms balance? Does the charge balance? Although the atoms balance, the charge does not. The total of the

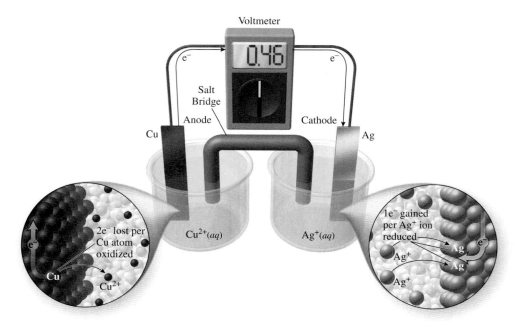

Figure 14.17
The electrons lost by copper at the anode travel toward the cathode. At the cathode, each Ag^+ ion gains an electron and gets plated on the surface of the electrode. Two Ag^+ ions are reduced for every copper atom that's oxidized.

charges of the reactants is 1+. The total of the charges of the products is 2+. In addition, electrons lost do not equal electrons gained. As the equation is written, each copper atom loses two electrons to become Cu^{2+}, and each Ag^+ ion gains one electron to become a silver atom. The equation is not acceptable because it isn't fully balanced.

Let's track the flow of electrons as the reaction proceeds to understand what balancing the equation requires. A copper atom on the surface of the anode loses two electrons to form a Cu^{2+} ion. These extra electrons on the anode flow through the wire toward the cathode. An ion of silver, Ag^+, comes into contact with the surface of the cathode, where an electron reduces the ion to a neutral silver atom. However, the copper atom lost two electrons. If a silver ion takes only one electron, what happens to the other one that was lost by copper? Another silver ion must be reduced to account for both electrons from the copper atom. So, for every copper atom that loses two electrons, two Ag^+ ions must each gain one electron. The balanced equation for the reaction then is

$$Cu(s) + 2Ag^+(aq) \longrightarrow Cu^{2+}(aq) + 2Ag(s)$$

To balance an oxidation-reduction reaction, we first separate the overall reaction into its half-reactions. This allows us to account for electrons gained and lost. It also reveals what happens in the anode and cathode half-cells. Like all other chemical equations, half-reactions must balance. Let's look again at the skeletal equation for the reaction between copper and silver ions:

$$Cu(s) + Ag^+(aq) \longrightarrow Cu^{2+}(aq) + Ag(s)$$

We can begin with copper, the reactant that is oxidized, to write the oxidation half-reaction. Copper metal reacts to form Cu^{2+}:

$$Cu(s) \longrightarrow Cu^{2+}(aq)$$

Is this equation balanced? The atoms balance, but the charge does not. The total charge on the reactant side of the equation is 0, while the total charge on the product side is 2+. We must add electrons to one side of the equation to account for the change in oxidation number. Adding two electrons to the product side gives a total charge of 0 on the right side of the equation:

$$Cu(s) \longrightarrow Cu^{2+}(aq) + 2e^-$$

The equation is balanced because the atoms and charges balance. This is the balanced half-reaction for the oxidation process.

Now let's look at the reduction half-reaction. Silver ion, Ag^+, reacts to form solid silver:

$$Ag^+(aq) \longrightarrow Ag(s)$$

Again the atoms balance, but the charge does not. The charge is 1+ on the reactant side and 0 on the product side. Adding one electron to the reactant side of the equation corrects the imbalance:

$$Ag^+(aq) + e^- \longrightarrow Ag(s)$$

The equation is now balanced because atoms and charges balance.

The oxidation and reduction half-reactions must be considered together to account for equivalence of electrons gained and lost:

$$\text{Oxidation} \qquad Cu(s) \longrightarrow Cu^{2+}(aq) + 2e^-$$

$$\text{Reduction} \qquad Ag^+(aq) + e^- \longrightarrow Ag(s)$$

All the electrons lost by the reactant that's oxidized must be gained by the reactant that's reduced. Multiplying the reduction half-reaction by 2 equalizes electrons gained and lost for the Cu/Ag^+ reaction. When we do this, we multiply each reactant and product by 2:

$$2[Ag^+(aq) + e^- \longrightarrow Ag(s)]$$

$$2Ag^+(aq) + 2e^- \longrightarrow 2Ag(s)$$

The electrons lost in the oxidation half-reaction now equal the electrons gained in the reduction half-reaction. We can add these two half-reactions:

$$Cu(s) \longrightarrow Cu^{2+}(aq) + 2e^-$$
$$\underline{2Ag^+(aq) + 2e^- \longrightarrow 2Ag(s)}$$
$$Cu(s) + 2Ag^+(aq) + 2e^- \longrightarrow Cu^{2+}(aq) + 2Ag(s) + 2e^-$$

Notice that $2e^-$ appears as a reactant and as a product. They cancel:

$$Cu(s) + 2Ag^+(aq) + \cancel{2e^-} \longrightarrow Cu^{2+}(aq) + 2Ag(s) + \cancel{2e^-}$$

$$\text{Overall reaction} \qquad Cu(s) + 2Ag^+(aq) \longrightarrow Cu^{2+}(aq) + 2Ag(s)$$

This equation is now balanced because atoms and charges balance. It is also balanced because electrons gained equal electrons lost. To double-check, count the atoms and determine the total charge of the reactants and the products:

	Reactants	Products
Cu	1	1
Ag	2	2
Charge	2+	2+

You can use the following series of questions to guide you through the balancing process:

1. Which substance is oxidized and which is reduced? To answer this question, assign oxidation numbers to all the elements in the reaction.

2. What are the half-reactions for the oxidation and reduction processes? For each half-reaction,
   a. which element changes in oxidation number?
   b. can any spectator ions be ignored until the final balancing step?
   c. how many electrons must be added to the appropriate side of the equation to account for the change in oxidation number?
3. What factor can be used to multiply each coefficient in the balanced half-reactions to equalize the number of electrons gained or lost?
4. When adding the two half-reactions, can any substances that are present in equal amounts on both sides of the equation be canceled out?

After answering these questions and acting accordingly, you should have a balanced overall equation. Give it a final check by counting the atoms of each element on both sides of the equation. Check to make sure that the sums of the charges are equal on the reactant side and on the product side. Use the steps to work through Example 14.6, balancing an oxidation-reduction equation.

EXAMPLE 14.6 Balancing Oxidation-Reduction Equations

Manganese metal reacts with aqueous iron(III) chloride to produce aqueous manganese(II) chloride and iron metal. Write a balanced chemical equation for this reaction.

Solution:

We begin by writing formulas for the reactants and products and writing a skeletal equation, just as we did in Chapter 5:

$$Mn(s) + FeCl_3(aq) \longrightarrow MnCl_2(aq) + Fe(s)$$

To determine what substances are oxidized and reduced, assign oxidation numbers to all the elements in the skeletal equation:

$$\overset{0}{Mn}(s) + \overset{3+}{Fe}\overset{1-}{Cl_3}(aq) \longrightarrow \overset{2+}{Mn}\overset{1-}{Cl_2}(aq) + \overset{0}{Fe}(s)$$

Notice that the chloride ion does not change in oxidation number, so we assume it is a spectator ion. We can write the skeletal ionic equation as follows:

$$Mn(s) + Fe^{3+}(aq) \longrightarrow Mn^{2+}(aq) + Fe(s)$$

The oxidation number for manganese changes from 0 to 2+, so it is oxidized. We can write an oxidation half-reaction as follows:

$$Mn(s) \longrightarrow Mn^{2+}(aq)$$

We add two electrons to the right side of the equation to account for the change in oxidation number. For a simple process like this, we can look at the addition of electrons in another way. Since the total charge on the left is 0 and the total charge on the right is 2+, we must add two electrons to the product side of the equation to balance the charge:

$$Mn(s) \longrightarrow Mn^{2+}(aq) + 2e^-$$

The atoms and charges are now balanced for this half-reaction.

Now let's consider the other half-reaction. Iron changes from 3+ to 0, so it is the element that is reduced. We begin by writing the reduction half-reaction as follows:

$$Fe^{3+}(aq) \longrightarrow Fe(s)$$

We must add three electrons to the reactant side to balance the charge:

$$Fe^{3+}(aq) + 3e^- \longrightarrow Fe(s)$$

The atoms and charges are now balanced for this half-reaction.

The next step is to consider the oxidation and reduction reactions together:

Oxidation $Mn(s) \longrightarrow Mn^{2+}(aq) + 2e^-$

Reduction $Fe^{3+}(aq) + 3e^- \longrightarrow Fe(s)$

Notice that each manganese atom loses two electrons, but each iron ion gains three. Electrons gained must equal electrons lost, so we multiply the oxidation half-reaction by 3 and the reduction half-reaction by 2:

$$3[Mn(s) \longrightarrow Mn^{2+}(aq) + 2e^-]$$

$$2[Fe^{3+}(aq) + 3e^- \longrightarrow Fe(s)]$$

$$3Mn(s) \longrightarrow 3Mn^{2+}(aq) + 6e^-$$

$$2Fe^{3+}(aq) + 6e^- \longrightarrow 2Fe(s)$$

We can now add the two half-reactions together and cancel out the six electrons that appear on both sides of the equation:

$$3Mn(s) \longrightarrow 3Mn^{2+}(aq) + 6e^-$$
$$2Fe^{3+}(aq) + 6e^- \longrightarrow 2Fe(s)$$
$$\overline{3Mn(s) + 2Fe^{3+}(aq) + \cancel{6e^-} \longrightarrow 3Mn^{2+}(aq) + 2Fe(s) + \cancel{6e^-}}$$

The balanced ionic equation is

$$3Mn(s) + 2Fe^{3+}(aq) \longrightarrow 3Mn^{2+}(aq) + 2Fe(s)$$

The overall equation can be written by returning the chlorides into the formulas for the compounds in aqueous solution:

$$3Mn(s) + 2FeCl_3(aq) \longrightarrow 3MnCl_2(aq) + 2Fe(s)$$

We'll double-check the balancing by counting atoms and charges on both sides of the equation.

	Reactants	Products
Mn	3	3
Fe	2	2
Cl	6	6
Charge	0	0

Practice Problem 14.6

Magnesium metal reacts with aqueous chromium(III) nitrate to produce magnesium nitrate, $Mg(NO_3)_2$, and solid chromium. Write a balanced equation for this reaction.

Further Practice: 14.47 and 14.48

14.5 BALANCING COMPLEX OXIDATION-REDUCTION EQUATIONS

The oxidation-reduction reactions we considered in Section 14.4 are relatively simple processes involving only the oxidizing and reducing agents. Many oxidation-reduction reactions are more complex, and they may involve water, an acid, or a base. Accounting for atoms and electron transfer in such reactions requires a slightly different approach.

Consider a reaction that occurs in a dry cell (Figure 14.18). This battery, like many others, requires the presence of a base for the chemical reaction to occur. The term *alkaline* describes chemical systems that involve bases, so such dry cells are called alkaline batteries. In a typical alkaline battery, the base is a hydroxide paste. The reaction that occurs in a zinc-manganese alkaline battery is

$$Zn(s) + MnO_2(s) + H_2O(l) \longrightarrow ZnO(s) + Mn(OH)_2(s)$$

The base is supplied by a KOH paste that contains the MnO_2. Notice in this reaction that water is a reactant, and a metal hydroxide is one of the products. Also notice that no reactants or products are in aqueous solution and that there are no spectator ions to remove. Is this equation balanced? Yes, the atoms and charges balance. However, the equation doesn't tell us anything about what occurs in the half-cells. The half-cells for this reaction may require some base (OH^-) and water to balance each half-reaction.

Many oxidation-reduction reactions occur in acidic solution, so their equations can be balanced with H^+ ions. For example, the Breathalyzer used by law enforcement officials to test drivers for drunkenness uses an acidic solution of the dichromate ion, $Cr_2O_7^{2}$ (Figure 14.19). In the presence of ethanol, C_2H_5OH, the orange dichromate ion is reduced to the chromium(III) ion, which appears green in solution. The ethanol is oxidized to carbon dioxide. The skeletal equation for this reaction is

$$C_2H_5OH(aq) + Cr_2O_7^{2-}(aq) \longrightarrow Cr^{3+}(aq) + CO_2(g)$$

To balance the half-reactions for this reaction, acid (H^+) and water may have to be added.

Let's separate this reaction into half-reactions to balance it. First assign oxidation numbers to all the atoms in the equation:

$$\overset{2-\ 1+\ 2-\ 1+}{C_2H_5OH}(aq) + \overset{6+\ 2-}{Cr_2O_7^{2-}}(aq) \longrightarrow \overset{3+}{Cr^{3+}}(aq) + \overset{4+\ 2-}{CO_2}(g)$$

> To simplify the balancing process for reactions that occur in acid, we will use H^+ instead of H_3O^+.

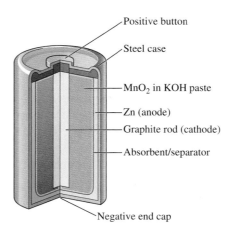

Figure 14.18

A common dry cell battery consists of a zinc anode and an inert graphite rod that serves as the cathode. The reaction requires the presence of a base, so the MnO_2 is in a KOH paste.

Each carbon atom changes from an oxidation number of 2– to 4+, so carbon is oxidized in the half-reaction:

$$C_2H_5OH(l) \longrightarrow CO_2(g)$$

Notice that there are two carbon atoms on the left side of the equation and only one on the right. We must make sure that the atoms of carbon balance before we account for the loss of electrons (Figure 14.20). Each ethanol molecule contains two carbon atoms, so we can place a 2 as a coefficient in front of CO_2 to balance the carbon atoms:

$$\overset{2- \ 1+ \ 2- \ 1+}{C_2H_5OH}(aq) \longrightarrow \overset{4+ \ 2-}{2CO_2}(g)$$

Although this reaction at first appears charge-balanced, we must account for the change in oxidation numbers. Each carbon on the left side of the equation has an oxidation number of 2–, and each carbon on the right has an oxidation number of 4+. Accounting for both carbons on the left gives a combined oxidation number of 4–. Similarly each carbon on the right has an oxidation number of 4+, but because there are two of them, the combined oxidation number is 8+. Thus 12 electrons are lost for every ethanol molecule that reacts:

$$C_2H_5OH(aq) \longrightarrow 2CO_2(g) + 12e^-$$

Now the charge is no longer balanced. The total charge on the left is 0, and the total charge on the right is 12–. Because we've already used electrons to account for the change in oxidation number, we cannot use them to balance the charge. We'll have to add a step to the procedure for balancing oxidation-reduction equations that we used in Section 14.4. Recall that the reaction occurs in the presence of acid, which means H^+ ions are present. To balance charge in an acidic solution, add H^+ ions to the side with the excess negative charge. If we add 12 H^+ ions to the right side of this equation, we can balance the charge:

$$C_2H_5OH(aq) \longrightarrow 2CO_2(g) + 12e^- + 12H^+(aq)$$

The total charges on the left and right are now equal at 0. The only task remaining is to balance oxygen and hydrogen atoms with water molecules. There are 6 hydrogen atoms on the left and 12 on the right. There is one oxygen on the left and four on the right. If we put three H_2O molecules on the left, the atoms of oxygen and hydrogen balance:

$$C_2H_5OH(aq) + 3H_2O(l) \longrightarrow 2CO_2(g) + 12e^- + 12H^+(aq)$$

Now let's consider the reduction half-reaction:

$$Cr_2O_7{}^{2-}(aq) \longrightarrow Cr^{3+}(aq)$$

Figure 14.19
In the Breathalyzers used by law enforcement officials, an oxidation-reduction reaction causes a color change that indicates the amount of alcohol present in a driver's system.

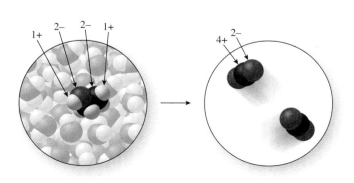

Figure 14.20
Each carbon in the reactant has an oxidation number of 2–. The carbon in the product has an oxidation number of 4+. For every molecule of C_2H_5OH that reacts, two molecules of CO_2 are produced.

Figure 14.21
Each chromium in the reactant has an oxidation number of 6+. The chromium in the product has an oxidation number of 3+. For every ion of $Cr_2O_7^{2-}$ that reacts, two ions of Cr^{3+} must be produced.

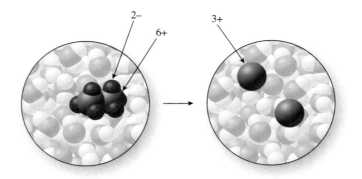

If the reaction occurs in base, OH^- ions are available to balance the charge. If the half-reaction

$$C_2H_5OH(l) \longrightarrow 2CO_2(g) + 12e^-$$

occurs in base, how would you use OH^- and water to balance the equation? Addition of 12 OH^- ions to the reactant side would balance the charge, and 9 H_2O molecules to the product side would balance the oxygen and hydrogen atoms. Convince yourself that this works by writing a balanced equation for the half-reaction and counting the atoms and charges on each side of the equation.

Balance the half-reaction

$$Cr_2O_7^{2-}(aq) + 6e^- \longrightarrow 2Cr^{3+}(aq)$$

if it were to occur in base by adding OH^- and H_2O to appropriate sides of the equation.

Each chromium atom changes from 6+ to 3+. But before we can account for the electron gain, we must make sure that the atoms of chromium balance (Figure 14.21). Because two chromium atoms are present in the dichromate ion, $Cr_2O_7^{2-}$, two Cr^{3+} ions must form for each $Cr_2O_7^{2-}$ that reacts:

$$\overset{6+}{\downarrow}\overset{2-}{\downarrow}\ Cr_2O_7^{2-}(aq) \longrightarrow \overset{3+}{\downarrow}\ 2Cr^{3+}(aq)$$

When we account for both of the chromium atoms on the left, we have a combined oxidation number of 12+. Each chromium atom on the right has an oxidation number of 3+, but because there are two of them, the combined oxidation number is 6+. If we add six electrons to the left we can account for the change in oxidation number:

$$Cr_2O_7^{2-}(aq) + 6e^- \longrightarrow 2Cr^{3+}(aq)$$

Now we have to balance the charge. The total charge on the left is 8–, and the total charge on the right is 6+. Since the reaction occurs in the presence of hydrogen ions, we can use H^+ to balance the charge:

$$Cr_2O_7^{2-}(aq) + 6e^- + 14H^+(aq) \longrightarrow 2Cr^{3+}(aq)$$

The total charges on the left and right are now equal at 6+. The last step is to add water to balance hydrogen and oxygen atoms. There are seven oxygen atoms on the left and none on the right. There are 14 hydrogen atoms on the left and none on the right. If we put 7 H_2O molecules on the product side, oxygen and hydrogen balance:

$$Cr_2O_7^{2-}(aq) + 6e^- + 14H^+(aq) \longrightarrow 2Cr^{3+}(aq) + 7H_2O(l)$$

The reduction half-reaction is balanced.

We'll now look at the oxidation and reduction processes together:

Oxidation $C_2H_5OH(l) + 3H_2O(l) \longrightarrow 2CO_2(g) + 12e^- + 12H^+(aq)$

Reduction $Cr_2O_7^{2-}(aq) + 6e^- + 14H^+(aq) \longrightarrow 2Cr^{3+}(aq) + 7H_2O(l)$

Do electrons lost equal electrons gained? No. Notice that 12 electrons are lost in the oxidation half-reaction. In the reduction half-reaction six electrons are gained. Multiplying the reduction half-reaction by a factor of 2 equalizes the electrons gained and lost:

$$2[Cr_2O_7^{2-}(aq) + 6e^- + 14H^+(aq) \longrightarrow 2Cr^{3+}(aq) + 7H_2O(l)]$$

$$2Cr_2O_7^{2-}(aq) + 12e^- + 28H^+(aq) \longrightarrow 4Cr^{3+}(aq) + 14H_2O(l)$$

We can now combine our oxidation and reduction half-reactions:

$$C_2H_5OH(l) + 3H_2O(l) \longrightarrow 2CO_2(g) + 12e^- + 12H^+(aq)$$
$$2Cr_2O_7^{2-}(aq) + 12e^- + 28H^+(aq) \longrightarrow 4Cr^{3+}(aq) + 14H_2O(l)$$

$$C_2H_5OH(l) + 2Cr_2O_7^{2-}(aq) + 12\cancel{e^-} + 28H^+(aq) + 3H_2O(l) \longrightarrow$$
$$2CO_2(g) + 12\cancel{e^-} + 12H^+(aq) + 4Cr^{3+}(aq) + 14H_2O(l)$$

The 12 electrons on the left and right cancel to give:

$$C_2H_5OH(l) + 2Cr_2O_7^{2-}(aq) + 28H^+(aq) + 3H_2O(l) \longrightarrow$$
$$2CO_2(g) + 12H^+(aq) + 4Cr^{3+}(aq) + 14H_2O(l)$$

There are 28 H^+ ions on the left and 12 on the right, leaving a net of 16 on the reactant side. There are 3 H_2O molecules on the left and 14 on the right, leaving a net of 11 H_2O molecules on the product side. Canceling equal numbers of H^+ ions and H_2O molecules results in an overall balanced equation:

$$C_2H_5OH(l) + 2Cr_2O_7^{2-}(aq) + 16H^+(aq) \longrightarrow 2CO_2(g) + 4Cr^{3+}(aq) + 11H_2O(l)$$

We'll count atoms and charges on both sides of the equation to double-check the balancing.

	Reactants	Products
C	2	2
H	22	22
O	15	15
Cr	4	4
Charge	12+	12+

You can use the following series of questions to guide you through the balancing process if you encounter a reaction that occurs in an acid or base:

1. What reactant is oxidized and what reactant is reduced? Assign oxidation numbers to all the elements in the reaction to make the determination.
2. What are the half-reactions for the oxidation and reduction processes? For each half-reaction, determine each of the following:
   a. Is the element that changes in oxidation number balanced?
   b. Can any spectator ion be ignored until the final balancing step?
   c. How many electrons must be added to the appropriate side of the equation to account for the change in oxidation number?
   d. Is the charge balanced after adding electrons? If not, what ions are present to balance the charge? If it occurs in acid, add H^+ ions. If the reaction occurs in base, add OH^- ions.
   e. Do the hydrogen and oxygen atoms balance? If not, add water molecules to the appropriate side of the equation.
3. What factor can be used to multiply each coefficient to equalize the number of electrons gained and lost?
4. When adding the two half-reactions, are any substances present on both sides of the equation? If so, cancel out an appropriate number. If substances are present on both sides of the equation in unequal amounts, cancel out the appropriate amount of substances on each side.

At this point you should have an overall balanced equation. Make sure atoms and charges balance on both sides of the equation. Apply these steps to Example 14.7.

EXAMPLE 14.7   Balancing Complex Oxidation-Reduction Reactions

Balance the following equation, which occurs in acidic solution:

$$Cu(s) + NO_3^-(aq) \longrightarrow Cu^{2+}(aq) + NO(g)$$

Solution:

We first assign oxidation numbers to identify what is oxidized and what is reduced:

$$\overset{0}{Cu}(s) + \overset{5+\ 2-}{NO_3^-}(aq) \longrightarrow \overset{2+}{Cu^{2+}}(aq) + \overset{2+\ 2-}{NO}(g)$$

Copper changes from 0 to 2+, so it is the reactant that is oxidized:

$$Cu(s) \longrightarrow Cu^{2+}(aq)$$

We add two electrons to the right side of the equation to account for this change in oxidation number:

$$Cu(s) \longrightarrow Cu^{2+}(aq) + 2e^-$$

This half-reaction is now balanced.

Now consider the reduction half-reaction:

$$NO_3^-(aq) \longrightarrow NO(g)$$

Nitrogen changes from 5+ to 2+. Adding three electrons to the left accounts for this change in oxidation number:

$$NO_3^-(aq) + 3e^- \longrightarrow NO(g)$$

The total charge on the left is 4−. On the right it's 0. The reaction occurs in acidic solution, so we add four H^+ ions to the left to balance the charge:

$$4H^+(aq) + NO_3^-(aq) + 3e^- \longrightarrow NO(g)$$

(If the reaction occurs in basic solution, the addition of four OH^- ions to the right balances the charge.) Adding water molecules completes the balancing process for this half-reaction. There are four hydrogen atoms on the left and none on the right. There are three oxygen atoms on the left and one on the right. Putting two H_2O molecules on the product side balances oxygen and hydrogen:

$$4H^+(aq) + NO_3^-(aq) + 3e^- \longrightarrow NO(g) + 2H_2O(l)$$

Now compare the oxidation and reduction half-reactions to see if the electrons lost equal the electrons gained:

Oxidation $\quad Cu(s) \longrightarrow Cu^{2+}(aq) + 2e^-$

Reduction $\quad 4H^+(aq) + NO_3^-(aq) + 3e^- \longrightarrow NO(g) + 2H_2O(l)$

Two electrons appear on the right in the oxidation half-reaction, and three appear on the left. If we multiply the oxidation half-reaction by 3 and the reduction half-reaction by 2, electrons lost will equal electrons gained:

$$3[Cu(s) \longrightarrow Cu^{2+}(aq) + 2e^-]$$
$$2[4H^+(aq) + NO_3^-(aq) + 3e^- \longrightarrow NO(g) + 2H_2O(l)]$$

Now add the two equations:

$$3Cu(s) \longrightarrow 3Cu^{2+}(aq) + 6e^-$$
$$8H^+(aq) + 2NO_3^-(aq) + 6e^- \longrightarrow 2NO(g) + 4H_2O(l)$$
$$\overline{3Cu(s) + 8H^+(aq) + 2NO_3^-(aq) + \cancel{6e^-} \longrightarrow 2NO(g) + 4H_2O(l) + 3Cu^{2+}(aq) + \cancel{6e^-}}$$

Canceling common terms we get

$$3Cu(s) + 8H^+(aq) + 2NO_3^-(aq) \longrightarrow 2NO(g) + 4H_2O(l) + 3Cu^{2+}(aq)$$

We now have a balanced equation. We'll double-check by counting atoms and charges on both sides of the equation.

	Reactants	Products
Cu	3	3
H	8	8
N	2	2
O	6	6
Charge	6+	6+

Practice Problem 14.7

Balance the following equation that occurs in basic solution:

$$Cl_2(aq) \longrightarrow Cl^-(aq) + ClO_3^-(aq)$$

Further Practice: 14.55 and 14.56

14.6   ELECTROCHEMISTRY

When Anna and Bill walked around campus in Chapter 1, they observed several vehicles that run on alternative fuels. You might recall from Chapter 6 that hydrogen-fueled buses are powered by the reaction of hydrogen with oxygen:

$$2H_2(g) + O_2(g) \longrightarrow 2H_2O(g)$$

This reaction, like reactions that occur in batteries, is spontaneous. A chemical reaction or physical change is **spontaneous** if it occurs by itself without outside intervention. That is, it does not need an outside source of energy to progress. Some spontaneous reactions are fast, like the combustion of methane, while others, like rusting, are slow. Note that some spontaneous reactions are desirable while others are not.

The reactions that occur in batteries are spontaneous and desirable because they provide electricity to start cars, run watches, operate power tools, and drive gadgets such as calculators, cell phones, and PDAs. The study of batteries and other voltaic cells constitutes a field of chemistry known as electrochemistry. **Electrochemistry** is the study of the relationship between chemical reactions and electrical work. The study of voltaic cells is a branch of electrochemistry that deals with spontaneous, desirable chemical processes. This branch of electrochemistry deals with oxidation-reduction reactions in which chemistry drives the flow of electricity.

The other branch of electrochemistry is the study of reactions that are desirable but do not occur on their own. They are not spontaneous. This branch of electrochemistry deals with oxidation-reduction reactions in which electricity drives the chemistry. In this section we'll explain more about voltaic cells and introduce a different kind of cell, the *electrolytic* cell.

Voltaic Cells

In Chapter 5 we discussed how single-displacement reactions can be predicted from an activity series (Figure 14.22). All single-displacement reactions in which a metal (metal A) reacts with a compound containing a different metal (metal B) to give a new compound (containing metal A) and a different pure metal (metal B) are oxidation-reduction reactions. The following equation summarizes them:

$$A + BX \longrightarrow AX + B$$

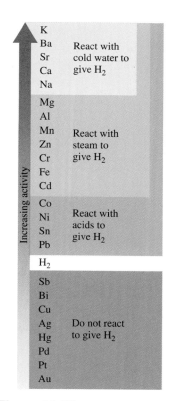

Figure 14.22

A chemical activity series lists metals and H_2 with the most active elements at the top.

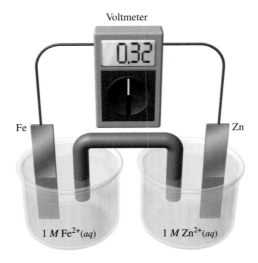

Figure 14.23
In a zinc-iron cell, which reaction will occur?

$$\text{Zn}(s) + \text{FeCl}_2(aq) \longrightarrow \text{ZnCl}_2(aq) + \text{Fe}(s)$$

or

$$\text{Fe}(s) + \text{ZnCl}_2(aq) \longrightarrow \text{FeCl}_2(aq) + \text{Zn}(s)$$

In this equation BX and AX represent ionic compounds containing the metal cations of B and A, and X represents an anion. We can use this equation and the activity series to predict what happens when we construct a voltaic cell. For example, let's set up a cell of zinc and iron (Figure 14.23). We begin with two metals and electrolyte solutions of compounds containing these metals. So in our example, we'll set up one half-cell with zinc as the electrode immersed in a solution of a compound containing zinc, such as ZnCl_2. In the other half-cell we'll use an iron electrode and immerse it in a solution of a compound containing iron, such as FeCl_2. We connect a wire between the two cells and place a salt bridge between the solutions. Can you predict what will occur?

Let's look at the possibilities. From our discussion about single-displacement reactions in Chapter 5, we know that only one of the following reactions occurs spontaneously:

$$\text{Zn}(s) + \text{FeCl}_2(aq) \longrightarrow \text{ZnCl}_2(aq) + \text{Fe}(s)$$
$$\text{Fe}(s) + \text{ZnCl}_2(aq) \longrightarrow \text{FeCl}_2(aq) + \text{Zn}(s)$$

Zinc appears above iron in the activity series, so it is the more reactive metal. It will displace iron from its compounds, so only this reaction will occur:

$$\text{Zn}(s) + \text{FeCl}_2(aq) \longrightarrow \text{ZnCl}_2(aq) + \text{Fe}(s)$$

An activity series predicts the relative strengths of metals as reducing agents and metal compounds as oxidizing agents. As we go up the activity series, the metals become stronger reducing agents. They are more reactive and give up electrons more readily. Metal ions in their compounds increase in oxidizing strength as we move down the activity series. To see how this works, let's return to the zinc-iron example. Zinc acts as the reducing agent because it provides electrons for the reduction of Fe^{2+}. The reaction is spontaneous and can be used to produce iron metal.

EXAMPLE 14.8 Voltaic Cells and the Activity Series

Suppose we want to obtain lead metal from a solution of $Pb(NO_3)_2$. In one half-cell of a voltaic cell we place a lead electrode in a solution of lead(II) nitrate, $Pb(NO_3)_2$. From the activity series, select a metal that we can place in the other half-cell compartment with its corresponding electrolyte solution.

Solution:

To produce lead metal, we should select a metal that can displace lead from its compounds. From the activity series, any metal that appears above lead may be suitable. Examples are Sn, Fe, and Zn. The metals at the top of the series are highly reactive with water, so for safety, and other reasons we should avoid them.

Practice Problem 14.8

Propose a voltaic cell that could be used to deposit chromium metal on the surface of a chromium electrode.

Further Practice: 14.61 and 14.62

Voltaic cells can be made from household materials. For example, a galvanized nail (which is iron coated with zinc) and a copper penny inserted into a lemon can produce a voltage when connected as shown in Figure 14.24.

Electrolytic Cells

Many desirable oxidation-reduction reactions are not spontaneous. An **electrolytic cell** is an electrochemical cell through which electric current is passed to cause a nonspontaneous oxidation-reduction reaction to occur. The process is called **electrolysis.** One of the most important uses of electrolysis is the isolation of elements from their naturally occurring compounds. For example, the primary source of elemental sodium is the electrolysis of molten sodium chloride. In this process sodium chloride is melted, two inert electrodes are inserted into the molten salt, and electricity is passed through it (Figure 14.25). The current causes one of the electrodes to become negatively charged. The other becomes positively charged. Because the electrodes are charged, they attract the ions that move around the molten salt. The negatively charged cathode attracts Na^+ ions; the positively charged anode attracts Cl^- ions. When the proper voltage is applied, Na^+ ions are reduced to sodium atoms at the cathode:

$$Na^+(l) + e^- \longrightarrow Na(l)$$

Animation: Electrolysis of KI
Online Learning Center, Animations

Chemistry Animations Library, Electrolysis, Electrolysis.rm.

  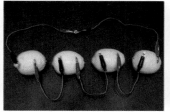 

Figure 14.24

Common household items can be used to produce a battery. In this lemon battery a galvanized nail and a copper penny serve as electrodes. One lemon can generate a voltage of 0.7 V to 0.9 V, which is not enough to light a tiny bulb, but four lemons in series can do the job. Scientists don't fully understand how this battery works.

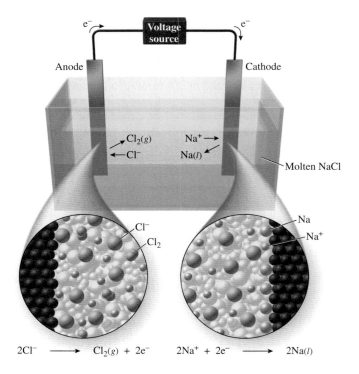

$$2Cl^- \longrightarrow Cl_2(g) + 2e^- \qquad 2Na^+ + 2e^- \longrightarrow 2Na(l)$$

Figure 14.25

When molten sodium chloride is electrolyzed, Cl^- ions are oxidized at the anode to produce Cl_2 molecules. At the cathode, Na^+ ions are reduced to form sodium atoms. Because the reaction is run at a high temperature to keep the NaCl molten, the sodium product is also in the liquid state. The actual cell is designed to remove the products, Na and Cl_2, as they form. A hood placed over the anode prevents the $Cl_2(g)$ from contacting the molten sodium. Why do these products need to be separated as they form?

Electrolysis is commonly used in electroplating jewelry and other items. The item to be plated is immersed in a solution that contains an electrolyte salt of the plate metal. When the power is turned on, the metal ions are reduced to atoms that deposit on the surface of the object.

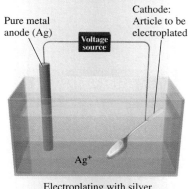

Electroplating with silver

The high temperature required to keep NaCl molten produces liquefied sodium metal at the cathode.

At the anode, chloride ions are oxidized to chlorine atoms which combine to form Cl_2 molecules:

$$2Cl^-(l) \longrightarrow Cl_2(g) + 2e^-$$

The overall chemical reaction for this process is

$$2Na^+(l) + 2e^- \longrightarrow 2Na(l)$$
$$2Cl^-(l) \longrightarrow Cl_2(g) + 2e^-$$
$$\overline{2NaCl(l) \longrightarrow 2Na(l) + Cl_2(g)}$$

Like sodium, calcium and magnesium are produced by electrolysis of molten salts. Other metals are produced using different electrolytic processes.

Electrolysis can also be used to prepare nonmetals. For example, water can be electrolyzed to produce oxygen and hydrogen gas:

$$2H_2O(l) \longrightarrow O_2(g) + 2H_2(g)$$

Electrolysis is also used to recharge batteries. Recall, for example, the reaction that occurs in a lead-acid battery in cars:

$$Pb(s) + PbO_2(s) + 2H_2SO_4(aq) \longrightarrow 2PbSO_4(aq) + 2H_2O(l)$$

An electric current from the car's alternator forces the reaction to go "in reverse." That is, the products are converted back to the reactants, keeping the battery functional:

$$2PbSO_4(aq) + 2H_2O(l) \longrightarrow Pb(s) + PbO_2(s) + 2H_2SO_4(aq)$$

EXAMPLE 14.9   **Electrolytic Cells**

The electrolysis of molten potassium iodide produces liquid potassium and gaseous iodine. Consider the following diagram:

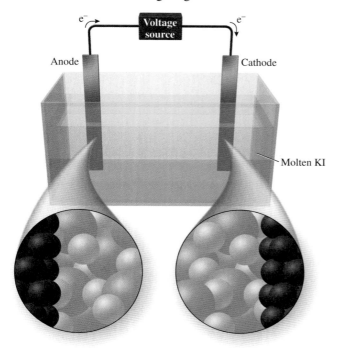

(a) When the power is turned on, what happens at the cathode?
(b) When the power is turned on, what happens at the anode?
(c) Write oxidation and reduction half-reactions.
(d) Write an overall equation for the reaction.

Solution:

(a) Since an oxidation-reduction reaction occurs, one element must be oxidized and the other reduced. Potassium iodide contains K^+ ions and I^- ions. Based on its position in the periodic table, we wouldn't expect potassium to become more oxidized than K^+, so it is reduced. Since reduction always occurs at the cathode, potassium metal is produced there.

(b) Iodide ions form iodine atoms at the anode, because it's the site of oxidation in an electrochemical cell. Two of these atoms combine to form an iodine molecule.

(c) To begin writing half-reactions, first consider the skeletal equation for the overall reaction. The high temperature required to keep the KI molten will make the potassium liquid and the iodine gaseous. Recall that iodine is one of the elements that exists as diatomic molecules:

$$KI(l) \longrightarrow K(l) + I_2(g)$$

Iodine is oxidized in this reaction, so the half-reaction for the oxidation process is

$$2I^-(l) \longrightarrow I_2(g) + 2e^-$$

The half-reaction for the reduction of potassium is

$$K^+(l) + e^- \longrightarrow K(l)$$

(d) To write an overall equation, combine the oxidation and reduction half-reactions:

Oxidation $2I^-(l) \longrightarrow I_2(g) + 2e^-$
Reduction $K^+(l) + e^- \longrightarrow K(l)$

To equalize electrons lost and gained, multiply the reduction half-reaction by 2:

$$2[K^+(l) + e^- \longrightarrow K(l)]$$
$$2K^+(l) + 2e^- \longrightarrow 2K(l)$$

Now add the two half-reactions:

$$2I^-(l) \longrightarrow I_2(g) + 2e^-$$
$$2K^+(l) + 2e^- \longrightarrow 2K(l)$$
———————————————————————————————
$$2K^+(l) + 2I^-(l) + 2e^- \longrightarrow 2K(l) + I_2(g) + 2e^-$$

The first two terms, $K^+(l)$ and $I^-(l)$, can be represented as $KI(l)$. We can also cancel $2e^-$ from both sides of the equation:

$$2KI(l) \longrightarrow 2K(l) + I_2(g)$$

Practice Problem 14.9

The electrolysis of molten calcium chloride produces liquid calcium and chlorine gas. Consider the following diagram:

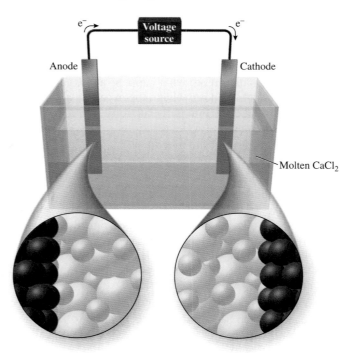

(a) When the power is turned on, what happens at the cathode?
(b) When the power is turned on, what happens at the anode?
(c) Write oxidation and reduction half-reactions.
(d) Write an overall equation for the reaction.

Further Practice: 14.67 and 14.68

14.7 CORROSION PREVENTION

Bill drives a rusty old car. After noticing it, Anna advised him to take steps to prevent further corrosion. The criticism hurt Bill's feelings a little, but he knew Anna was right. We generally define corrosion as a slow deterioration of metals due to interaction with the environment. The rusting of a car is an example of corrosion, because the iron in the steel reacts with oxygen and water in the environment. More specifically, **corrosion** is an oxidation-reduction reaction. Familiar signs of corrosion are the red stains of rust on iron and steel, the green coating on copper and brass, and the black coating on silver. The rusting of the steel in Bill's car is a reaction of iron, oxygen, and water that is spontaneous but undesirable. Dents or scratches and acidic conditions accelerate corrosion. Cars in colder climates rust faster because of the salt that's put on the roads in winter to melt snow and ice (Figure 14.26). Corrosion repair and prevention cost about $275 billion a year in the United States alone.

As Bill and Anna discussed Bill's rusty car, they recalled from their chemistry class that corrosion occurs when an electrochemical cell is set up on the surface of the metal. When iron rusts, a part of the metal acts as an anode, where an oxidation process occurs:

$$Fe(s) \longrightarrow Fe^{2+}(aq) + 2e^-$$

A reduction process occurs simultaneously in another location, which acts as a cathode, with molecular oxygen acting as the oxidizing agent:

$$O_2(g) + 2H_2O(l) + 4e^- \longrightarrow 4OH^-(aq)$$

Electrons released by the oxidation of iron migrate through the metal to the site where the oxygen atoms in molecular oxygen gain electrons (Figure 14.27). In the presence of water, the oxidized metal ions migrate to the site where reduction can occur. The iron(II) ions then combine with hydroxide ions to form insoluble hydroxide salts, which react further with oxygen and water to form the further oxidized hydrated iron(III) oxide, rust. Rust flakes off the surface of the iron metal and exposes fresh metal so iron continues to corrode, ultimately suffering serious damage.

A variety of methods can be used to prevent corrosion. Most often a protective coating such as paint is applied. Other methods, including *cathodic protection* and *plating,* involve less destructive electrochemical cells.

When two metals come in contact in a corrosive environment, the more active of the two metals corrodes, and the other does not. This principle forms the basis for the cathodic protection of a metal. For example, iron pipes or tanks, sometimes buried underground, are often connected to a more active metal, such as magnesium

Figure 14.26
Rust is a serious problem especially in cold climates where salt used to melt ice can accelerate the rusting process.

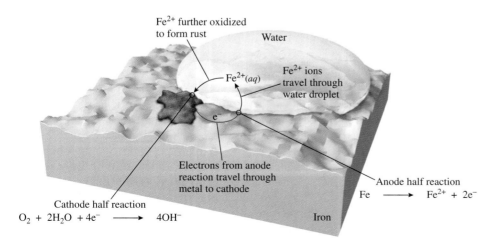

Figure 14.27
When iron corrodes in moist air, the iron oxidizes to form Fe^{2+} ions. These ions migrate through the water droplet to the site of reduction, the cathode. At the cathode, O_2 is reduced. The Fe^{2+} ions oxidize further to form Fe^{3+} ions, which form hydrated iron(III) oxide.

Figure 14.28
Zinc is attached to the trans-Alaskan pipeline to keep the iron from rusting. Because zinc is more reactive, it is oxidized instead of iron. The zinc is attached to the exterior of the pipe, so it can be easily replaced when it's substantially corroded. The iron in the pipeline would be more difficult to replace.

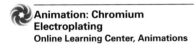
Animation: Chromium Electroplating
Online Learning Center, Animations

Chemistry Animations Library, Chromium Electroplating, Electroplating.rm.

or zinc. The trans-Alaskan pipeline, for example, was constructed with strips of zinc cable for corrosion protection (Figure 14.28). The zinc forms the anode, where oxidation occurs:

$$Zn(s) \longrightarrow Zn^{2+}(aq) + 2e^-$$

Zinc is selectively oxidized instead of iron, so iron metal acts as an inert cathode where reduction of oxygen occurs:

$$O_2(g) + 2H_2O(l) + 4e^- \longrightarrow 4OH^-(aq)$$

The iron acts only as a conductor of electrons and does not take part in either the oxidation or the reduction half-reactions. Thus the iron remains completely uncorroded as long as the zinc remains. It is much simpler to replace the zinc cable with a fresh piece than to replace the pipe. In a similar way magnesium is often used to protect buried gasoline tanks and ships' hulls.

Plating also prevents corrosion. Plating is a variation on cathodic protection. A thin coating of one metal can be applied on top of another by either dipping or electroplating. If the coating metal melts at a reasonably low temperature, dipping is used. For example, iron is galvanized by dipping it in molten zinc. Once it's plated, if a scratch exposes some of the iron to environmental conditions, an oxidation-reduction reaction can occur. However, zinc is a more active metal, so it oxidizes instead of iron. The iron becomes an inert electrode carrying electrons from the zinc for the reduction of molecular oxygen. The iron does not corrode because the zinc is preferentially oxidized. Oxidation of zinc is a very slow process that produces $Zn_2(OH)_2CO_3$. This compound deposits on the iron and acts as a protective coating.

Sometimes we use a less active metal to plate a metal we want to protect. For example, iron is covered with tin to make tin cans. The low activity of tin prevents corrosion. However, the protection of the iron is purely physical. If the tin coating is scratched to expose some iron, the iron corrodes very rapidly at the exposed position. The tin soon flakes off, resulting in even more extensive corrosion. When the iron is exposed, it becomes the anode, since it is the more active metal. The reduction of molecular oxygen occurs on the tin, while the iron is oxidized.

The other type of plating, electroplating, involves covering a more active metal with a less active metal that corrodes slowly. Electroplating uses an electrolytic process in which the metal to be coated acts as an electrode. This metal is placed in a solution that contains a compound of the metal that acts as the coating. When electricity passes through the cell, a thin coating of protective metal forms on the electrode. For example, car bumpers used to be made of steel electroplated with chromium. The steel was immersed in an electrolyte solution that contained a chromium compound. With the passage of electricity, chromium ions were reduced to form chromium metal, which plated onto the surface of the steel. Nickel-plated bathroom fixtures are produced in a similar way.

SUMMARY

Electrons are transferred between reactants in oxidation-reduction reactions. Oxidation is a loss of electrons and an increase in oxidation number. Reduction is a gain of electrons and a decrease in oxidation number. Oxidation numbers are assigned to elements to serve as an accounting tool for determining what is oxidized and what is reduced. Oxidation and reduction always occur together because the element that gains electrons must obtain them from something that gives them up. The reactant containing the element that is oxidized is the reducing agent; it provides electrons to the element that gets reduced. The reactant containing the element that gains electrons is reduced and acts as the oxidizing agent.

Spontaneous oxidation-reduction reactions can generate electrical energy if the reactants are physically separated. If they are not, the energy is lost as heat. The physical arrangement of the separated reactants is described as a cell. The cell is divided into two half-cells that contain the oxidation and reduction half-reactions. Oxidation occurs at the anode, and reduction occurs at the cathode. A cell that produces electricity from a spontaneous chemical reaction is a voltaic cell. Batteries contain one or more voltaic cells that provide electricity to power a device.

Balancing oxidation-reduction reactions begins with the identification of which element is oxidized and which is reduced. Half-reactions account for changes in oxidation number and describe the oxidation and reduction processes that occur at each electrode. To balance them, changes in oxidation number, differences in charge, and numbers of atoms must be balanced. An oxidation-reduction reaction equation is balanced when the number of atoms and charges are the same on both sides of the equation and the number of electrons gained equals the number lost.

Electrochemistry consists of two branches of study. One branch involves the identification and description of reactions in which chemical reactions drive the flow of electrons. These reactions are spontaneous and can be used to generate electricity. The other branch of electrochemistry deals with reactions driven by electricity. These electrolytic reactions are not spontaneous, so they need a continuous energy source to proceed.

Corrosion is generally defined as the slow deterioration of metal in contact with the environment. The rusting of iron, a corrosion process, is an oxidation-reduction reaction. When iron rusts, a part of the metal acts as an anode, where an oxidation process occurs. The reduction of oxygen occurs simultaneously in another location, which acts as a cathode. Corrosion can be prevented in several ways. Most often metal surfaces are painted or covered with a protective coating, either through dipping or electroplating. Another technique, cathodic protection, involves attaching a more reactive metal that corrodes preferentially.

KEY TERMS

anode (14.3)	electrode (14.3)	oxidation (14.1)	reducing agent (14.1)
battery (14.3)	electrolysis (14.6)	oxidation number (14.2)	reduction (14.1)
cathode (14.3)	electrolytic cell (14.6)	oxidation-reduction reaction (14.1)	salt bridge (14.3)
corrosion (14.7)	half-cell (14.3)		spontaneous (14.6)
electrochemistry (14.6)	half-reaction (14.3)	oxidizing agent (14.1)	voltaic (galvanic) cell (14.3)

QUESTIONS AND PROBLEMS

The following questions and problems, except for those in the *Additional Questions* section, are paired. Questions in a pair focus on the same concept. Answers to the odd-numbered questions and problems are in Appendix D.

Matching Definitions with Key Terms

14.1   Match the key terms with the descriptions provided.
    (a) an electrochemical cell through which electric current is passed to cause a nonspontaneous oxidation-reduction reaction to occur
    (b) the electrode at which reduction occurs
    (c) a chemical process in which an atom increases in oxidation number

    (d) an oxidation or reduction process that occurs in one compartment of an electrochemical cell
    (e) the study of the relationship between chemical reactions and electrical work
    (f) a solid material that conducts electrical charge and is used as part of an electrochemical cell
    (g) an atom, ion, or molecule that causes an increase in the oxidation number of another substance
    (h) a reaction in which electrons are transferred
    (i) a slow deterioration of metals due to interaction with the environment
    (j) a charge assigned to the atoms in a compound according to a set of rules used in an electron accounting system

14.2 Match the key terms with the descriptions provided.
(a) an electrochemical cell in which an oxidation-reduction reaction causes electrons to flow through an external circuit
(b) the electrode at which oxidation occurs
(c) a chemical process in which an atom decreases in oxidation number
(d) an individual compartment in which either an oxidation or a reduction reaction occurs
(e) one or more interconnected voltaic cells used as a portable source of electricity
(f) a device used in electrochemical cells that allows migration of ions to maintain electrical neutrality but prevents mixing of the electrolyte solutions
(g) an atom, ion, or molecule that causes a decrease in the oxidation number of another substance
(h) a process in which an electric current forces a non-spontaneous oxidation-reduction reaction to occur
(i) a chemical reaction or physical change that occurs without outside intervention

What Is an Oxidation-Reduction Reaction?

14.3 What is an oxidation-reduction reaction?

14.4 Why is oxidation always coupled with reduction?

14.5 How do you know when something is oxidized?

14.6 How do you know when something is reduced?

14.7 When a strip of magnesium metal is placed in an aqueous solution of copper(II) nitrate, elemental copper coats the surface of the magnesium strip and aqueous magnesium nitrate forms.
(a) Write a balanced equation for this reaction.
(b) What happens to the charge on magnesium as products form? How many electrons are transferred? Does magnesium gain or lose electrons?
(c) What happens to the charge on copper as products form? How many electrons are transferred? Does copper gain or lose electrons?

14.8 When a strip of nickel metal is placed in an aqueous solution of lead(II) nitrate, elemental lead coats the surface of the nickel strip and aqueous nickel(II) nitrate forms.
(a) Write a balanced equation for this reaction.
(b) What happens to the charge on nickel as products form? How many electrons are transferred? Does nickel gain or lose electrons?
(c) What happens to the charge on lead as products form? How many electrons are transferred? Does lead gain or lose electrons?

14.9 Consider the following reaction:

$$Mg(s) + ZnSO_4(aq) \longrightarrow MgSO_4(aq) + Zn(s)$$

(a) Which species is oxidized?
(b) Which species is reduced?
(c) What is the oxidizing agent?
(d) What is the reducing agent?
(e) Draw a molecular-level picture of what happens at the surface of the magnesium metal.

14.10 Consider the following reaction:

$$Mn(s) + CdSO_4(aq) \longrightarrow MnSO_4(aq) + Cd(s)$$

(a) Which species is oxidized?
(b) Which species is reduced?
(c) What is the oxidizing agent?
(d) What is the reducing agent?
(e) Draw a molecular-level picture of what happens at the surface of the manganese metal.

Oxidation Numbers

14.11 What is an oxidation number?

14.12 How do oxidation numbers change during oxidation and during reduction?

14.13 Indicate the oxidation number of each element in the following species.
(a) $Fe(s)$ (b) $Cl_2(g)$ (c) $Fe^{2+}(aq)$ (d) $Cl^-(aq)$

14.14 Indicate the oxidation number of each element in the following species.
(a) $Cu(s)$ (b) $Br_2(l)$ (c) $Cu^{2+}(aq)$ (d) $Br^-(aq)$

14.15 What is the oxidation number of each atom in the compound represented by the following molecule?

14.16 What is the oxidation number of each atom in the compound represented by the following molecule?

14.17 What is the oxidation number of phosphorus in each of the following oxides?
(a) P_4O_{10} (b) P_4O_6 (c) P_4O_8

14.18 Determine the oxidation number of bromine in each of these bromine oxides.
(a) Br_2O (b) BrO_3 (c) Br_2O_3 (d) BrO_2 (e) Br_2O_5

14.19 Indicate the oxidation number of phosphorus in each of the following compounds.
(a) $AlPO_4$ (b) PF_5 (c) H_3PO_4
(d) H_3PO_2 (e) PH_3 (f) H_3PO_3

14.20 Determine the oxidation number of chlorine in each of the following compounds.
(a) $NaCl$ (b) Cl_2 (c) ClO_2
(d) Cl_2O_7 (e) $KClO_4$ (f) $Ca(ClO)_2$

14.21 The ion shown has a charge of 2−. What are the oxidation numbers of all the atoms in this ion?

14.22 The ion shown has a charge of 1−. What are the oxidation numbers of all the atoms in this ion?

14.23 Assign an oxidation number to iodine in each of the following.
(a) IF_7 (b) IO_3^- (c) I^- (d) ICl_3 (e) $IOCl_3$

14.24 Indicate the oxidation number of iron in each of the following compounds.
(a) $FeCl_2$ (b) $FeSO_4$ (c) $Fe(NO_3)_3$ (d) $Fe_2(SO_4)_3$

14.25 Determine the oxidation number of each element in the following compounds.
(a) ClO_2 (b) CaF_2 (c) H_2TeO_3 (d) NaH

14.26 Determine the oxidation number of each element in the following compounds.
(a) Na_2O_2 (b) $Fe(NO_3)_3$ (c) Sc_2O_3 (d) LiH

14.27 Determine the oxidation number of each element in the following ions.
(a) NO_2^- (b) $Cr_2O_7^{2-}$ (c) $AgCl_2^-$ (d) SO_3^{2-} (e) CO_3^{2-}

14.28 Determine the oxidation number of each element in the following ions.
(a) $H_2PO_4^-$ (b) $FeCl_6^{3-}$ (c) ClO_2^- (d) SiF_6^{2-} (e) AsO_4^{3-}

14.29 Determine which of the following represent oxidation-reduction reactions. For reactions that involve oxidation-reduction, determine the oxidizing and reducing agents.
(a) $BaCl_2(aq) + H_2SO_4(aq) \longrightarrow BaSO_4(s) + 2HCl(aq)$
(b) $3H_2(g) + N_2(g) \longrightarrow 2NH_3(g)$
(c) $H_2CO_3(aq) \longrightarrow H_2O(l) + CO_2(g)$
(d) $AgNO_3(aq) + NaCl(aq) \longrightarrow AgCl(s) + NaNO_3(aq)$
(e) $2C_2H_6(g) + 7O_2(g) \longrightarrow 4CO_2(g) + 6H_2O(g)$

14.30 Determine which of the following represent oxidation-reduction reactions. For reactions that involve oxidation-reduction, determine the oxidizing and reducing agents.
(a) $2Na(s) + 2H_2O(l) \longrightarrow 2NaOH(aq) + H_2(g)$
(b) $H_2(g) + F_2(g) \longrightarrow 2HF(g)$
(c) $C(s) + H_2O(s) \longrightarrow CO(g) + H_2(g)$
(d) $Pb(NO_3)_2(aq) + 2NaCl(aq) \longrightarrow$
$PbCl_2(s) + 2NaNO_3(aq)$
(e) $C_2H_5OH(l) + 3O_2(g) \longrightarrow 2CO_2(g) + 3H_2O(l)$

14.31 Under certain conditions, nitrogen gas reacts with oxygen gas to form nitrogen monoxide gas.
(a) Write a balanced equation for this reaction.
(b) What happens to the oxidation number of a nitrogen atom when it changes from a reactant to a product? How many electrons are transferred? Does nitrogen gain them or lose them?
(c) What happens to the oxidation number of an oxygen atom when it changes from a reactant to a product? How many electrons are transferred? Does oxygen gain them or lose them?

14.32 Under certain conditions, solid carbon reacts with oxygen gas to form carbon monoxide gas.
(a) Write a balanced equation for this reaction.
(b) What happens to the oxidation number of a carbon atom when it changes from a reactant to a product? How many electrons are transferred? Does carbon gain them or lose them?
(c) What happens to the oxidation number of an oxygen atom when it changes from a reactant to a product? How many electrons are transferred? Does oxygen gain them or lose them?

14.33 Consider the following reaction:
$$6Fe^{2+}(aq) + Cr_2O_7^{2-}(aq) + 14H^+(aq) \longrightarrow$$
$$6Fe^{3+}(aq) + 2Cr^{3+}(aq) + 7H_2O(l)$$
(a) Which species is oxidized?
(b) Which species is reduced?
(c) What is the oxidizing agent?
(d) What is the reducing agent?

14.34 Consider the following reaction:
$$2MnO_4^-(aq) + 10Cl^-(aq) + 16H^+(aq) \longrightarrow$$
$$2Mn^{2+}(aq) + 5Cl_2(g) + 8H_2O(l)$$
(a) Which species is oxidized?
(b) Which species is reduced?
(c) What is the oxidizing agent?
(d) What is the reducing agent?

14.35 In the following oxidation-reduction reactions, identify the oxidizing agent, the reducing agent, and the number of electrons transferred.
(a) $I_2(aq) + 2OH^-(aq) \longrightarrow I^-(aq) + IO^-(aq) + H_2O(l)$
(b) $Cr(s) + 2H^+(aq) \longrightarrow Cr^{2+}(aq) + H_2(g)$
(c) $2Cr_2O_7^{2-}(aq) + 16H^+(aq) \longrightarrow$
$4Cr^{3+}(aq) + 3O_2(g) + 8H_2O(l)$
(d) $3Fe^{3+}(aq) + Al(s) \longrightarrow 3Fe^{2+}(aq) + Al^{3+}(aq)$

14.36 In the following oxidation-reduction reactions, identify the oxidizing agent, the reducing agent, and the number of electrons transferred.
(a) $6I^-(aq) + BrO_3^-(aq) + 6H^+(aq) \longrightarrow$
$3I_2(aq) + Br^-(aq) + 3H_2O(l)$
(b) $5Br^-(aq) + BrO_3^-(aq) + 6H^+(aq) \longrightarrow$
$3Br_2(aq) + 3H_2O(l)$
(c) $XeF_4(s) + 2H_2O(l) \longrightarrow Xe(g) + 4HF(aq) + O_2(g)$
(d) $Pb(s) + 2H^+(aq) \longrightarrow Pb^{2+}(aq) + H_2(g)$

Batteries

14.37 Draw a diagram of a voltaic cell that corresponds to the following reaction:
$$Fe(s) + Ni^{2+}(aq) \longrightarrow Fe^{2+}(aq) + Ni(s)$$
Label all components of the cell including the anode and the cathode. Identify the electrolyte solutions needed in each half-cell compartment. Write half-reactions to describe the processes that occur at each electrode.

14.38 Draw a diagram of a voltaic cell that corresponds to the following reaction:
$$Mg(s) + Sn^{2+}(aq) \longrightarrow Mg^{2+}(aq) + Sn(s)$$
Label all components of the cell including the anode and the cathode. Identify the electrolyte solutions needed in each half-cell compartment. Write half-reactions to describe the processes that occur at each electrode.

14.39 The figure shows a molecular-level representation for the iron-nickel cell described in Question 14.37 before the reaction begins. Draw a picture of what has happened at each electrode after the cell has run for a long time.

14.40 The figure shows a molecular-level representation for the magnesium-tin cell described in Question 14.38 before the reaction begins. Draw a picture of what has happened at each electrode after the cell has run for a long time.

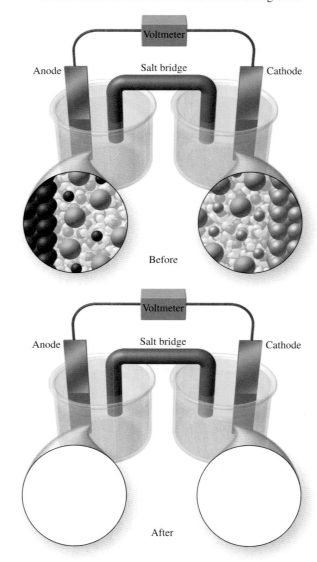

14.41 The reaction that occurs in a lead-acid battery is as follows:

$$Pb(s) + PbO_2(s) + 2H_2SO_4(aq) \longrightarrow 2PbSO_4(aq) + 2H_2O(l)$$

What gets oxidized in this battery? What gets reduced? What is the oxidizing agent? What is the reducing agent?

14.42 The reaction that occurs in a common dry cell battery is as follows:

$$Zn(s) + 2MnO_2(s) + 2NH_4^+(aq) \longrightarrow$$
$$Zn^{2+}(aq) + Mn_2O_3(s) + 2NH_3(aq) + H_2O(l)$$

What gets oxidized in this battery? What gets reduced? What is the oxidizing agent? What is the reducing agent?

14.43 The nickel-cadmium battery is used in portable appliances and power tools. Unbalanced half-reactions that occur in this battery are

$$Cd(s) + 2OH^-(aq) \longrightarrow Cd(OH)_2(s)$$

$$NiO_2(s) + 2H_2O(l) \longrightarrow Ni(OH)_2(s) + 2OH^-(aq)$$

Which equation represents the oxidation half-reaction? Which represents the reduction half-reaction? Add electrons to the appropriate side of each equation to account for the change in oxidation number.

14.44 The zinc-silver oxide battery, although expensive, is used to power satellite systems because of its light weight. Unbalanced half-reactions that occur in this battery are

$$Zn(s) + 2OH^-(aq) \longrightarrow Zn(OH)_2(s)$$

$$2AgO(s) + H_2O(l) \longrightarrow Ag_2O(s) + 2OH^-(aq)$$

Which equation represents the oxidation half-reaction? Which represents the reduction half-reaction? Add electrons to the appropriate side of each equation to account for the change in oxidation number.

Balancing Simple Oxidation-Reduction Equations

14.45 Balance the following half-reactions.
  (a) $Fe^{3+}(aq) \longrightarrow Fe(s)$
  (b) $Zn(s) \longrightarrow Zn^{2+}(aq)$
  (c) $Cl^-(aq) \longrightarrow Cl_2(g)$
  (d) $Fe^{2+}(aq) \longrightarrow Fe^{3+}(aq)$

14.46 Balance the following half-reactions.
  (a) $Ni(s) \longrightarrow Ni^{2+}(aq)$
  (b) $Br_2(g) \longrightarrow Br^-(aq)$
  (c) $Mg^{2+}(aq) \longrightarrow Mg(s)$
  (d) $Cr^{3+}(aq) \longrightarrow Cr^{2+}(aq)$

14.47 For each of the following write balanced half-reaction equations. Also write overall balanced equations.
  (a) $Zn(s) + Fe(NO_3)_3(aq) \longrightarrow Zn(NO_3)_2(aq) + Fe(s)$
  (b) $Mn(s) + HCl(aq) \longrightarrow MnCl_2(aq) + H_2(g)$

14.48 For each of the following write balanced half-reaction equations. Also write overall balanced equations.
  (a) $Al(s) + Fe(NO_3)_3(aq) \longrightarrow$
    $$Al(NO_3)_3(aq) + Fe(NO_3)_2(aq)$$
  (b) $Na(s) + HNO_3(aq) \longrightarrow NaNO_3(aq) + H_2(g)$

14.49 Aqueous iron(III) sulfate reacts with aqueous potassium iodide to form aqueous iron(II) sulfate, aqueous potassium sulfate, and aqueous iodine molecules. Write a balanced equation for this oxidation-reduction reaction.

14.50 When gaseous sulfur dioxide is bubbled into a nitric acid solution, sulfuric acid and gaseous nitrogen dioxide form. Write a balanced equation for this oxidation-reduction reaction.

Balancing Complex Oxidation-Reduction Equations

14.51 Balance the following half-reactions by adding $H^+(aq)$, $H_2O(l)$, and electrons as appropriate.
  (a) $Ba(s) \longrightarrow Ba^{2+}(aq)$
  (b) $HNO_2(aq) \longrightarrow NO(g)$
  (c) $H_2O_2(aq) \longrightarrow H_2O(l)$
  (d) $Cr^{3+}(aq) \longrightarrow Cr_2O_7^{2-}(aq)$

14.52 Balance the following half-reactions by adding $H^+(aq)$, $H_2O(l)$, and electrons as appropriate.
  (a) $I_2(aq) \longrightarrow I^-(aq)$
  (b) $H_2O(l) \longrightarrow O_2(g)$
  (c) $MnO_4^-(aq) \longrightarrow Mn^{2+}(aq)$
  (d) $TiO_2(s) \longrightarrow Ti^{2+}(aq)$

14.53 Balance the following half-reactions by adding $OH^-(aq)$, $H_2O(l)$, and electrons as appropriate.
  (a) $La(s) \longrightarrow La(OH)_3(s)$
  (b) $NO_3^-(aq) \longrightarrow NO_2^-(aq)$
  (c) $H_2O_2(aq) \longrightarrow O_2(g)$
  (d) $Cl_2O_7(aq) \longrightarrow ClO_2^-(aq)$

14.54 Balance the following half-reactions by adding $OH^-(aq)$, $H_2O(l)$, and electrons as appropriate.
  (a) $Cr_2O_7^{2-}(aq) \longrightarrow Cr^{3+}(aq)$
  (b) $ClO_2(aq) \longrightarrow ClO_2^-(aq)$
  (c) $MnO_4^-(aq) \longrightarrow MnO_2(s)$
  (d) $Br^-(aq) \longrightarrow BrO_3^-(aq)$

14.55 Complete and balance the following oxidation-reduction reactions, assuming they occur in acidic solution.
  (a) $H_2S(aq) + Cr_2O_7^{2-}(aq) \longrightarrow S(s) + Cr^{3+}(aq)$
  (b) $V^{2+}(aq) + MnO_4^-(aq) \longrightarrow VO^{2+}(aq) + Mn^{2+}(aq)$
  (c) $Fe^{2+}(aq) + ClO_3^-(aq) \longrightarrow Fe^{3+}(aq) + Cl^-(aq)$

14.56 Complete and balance the following oxidation-reduction reactions, assuming they occur in basic solution.
  (a) $S^{2-}(aq) + CrO_4^{2-}(aq) \longrightarrow S(s) + Cr(OH)_3(s)$
  (b) $MnO_4^-(aq) + CN^-(aq) \longrightarrow MnO_4^{2-}(aq) + CNO^-(aq)$
  (c) $Al(s) + OH^-(aq) \longrightarrow Al(OH)_4^-(aq) + H_2(g)$

14.57 Complete and balance the following oxidation-reduction reactions, assuming they occur in basic solution.
  (a) $NH_3(aq) + ClO^-(aq) \longrightarrow N_2H_4(aq) + Cl^-(aq)$
  (b) $Cr(OH)_4^-(aq) + HO_2^-(aq) \longrightarrow CrO_4^{2-}(aq) + H_2O(l)$
  (c) $Br_2(aq) \longrightarrow Br^-(aq) + BrO^-(aq)$

14.58 Complete and balance the following oxidation-reduction reactions, assuming they occur in acidic solution.
  (a) $C_2O_4^{2-}(aq) + MnO_4^-(aq) \longrightarrow CO_2(g) + Mn^{2+}(aq)$
  (b) $ClO_3^-(aq) + Cl^-(aq) \longrightarrow Cl_2(aq)$
  (c) $S_2O_3^{2-}(aq) + I_2(aq) \longrightarrow SO_4^{2-}(aq) + I^-(aq)$

14.59 Denitrification occurs when nitrogen in the soil is lost to the atmosphere. One way in which denitrification can occur in acidic soils rich in plant matter is described by the following equation:

$$C_6H_{12}O_6(aq) + NO_3^-(aq) \longrightarrow CO_2(g) + N_2(g)$$

Balance this equation, adding $H^+(aq)$ and $H_2O(l)$ as necessary.

14.60 When iodine is oxidized in concentrated HNO_3, it produces the white solid iodic acid (HIO_3):

$$I_2(s) + NO_3^-(aq) \longrightarrow HIO_3(s) + NO_2(g)$$

Balance this equation, adding $H^+(aq)$ and $H_2O(l)$ as necessary.

Electrochemistry

14.61 Consider the partially labeled voltaic cell shown.

Suppose you want to deposit solid iron on the iron electrode. What might you use for the other electrode? What electrolyte would you put in the half-cell with that other electrode?

14.62 Consider the partially labeled voltaic cell shown.

Suppose you want to deposit solid tin on the tin electrode. What might you use for the other electrode? What electrolyte would you put in the half-cell with that other electrode?

14.63 Using the activity series in Figure 14.22, place the following metals in order of increasing reducing strength.

Cu, Ag, Mn, Mg, Na, Au

14.64 Using the activity series in Figure 14.22, place the following metal ions in order of increasing oxidizing strength.

Fe^{2+}, Zn^{2+}, Ag^+, Mg^{2+}, Ni^{2+}, K^+

14.65 What is electrolysis?

14.66 Describe what happens at each electrode during the electrolysis of molten sodium chloride.

14.67 In the electrolysis of molten sodium iodide, liquid sodium and iodine gas are produced. Consider the following diagram:

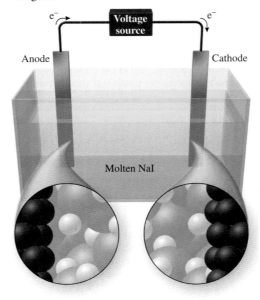

(a) When the power is turned on, what happens at the cathode?
(b) When the power is turned on, what happens at the anode?
(c) Write oxidation and reduction half-reactions.
(d) Write an overall equation for the reaction.

14.68 In the electrolysis of molten aluminum chloride, liquid aluminum and chlorine gas are produced. Consider the following diagram:

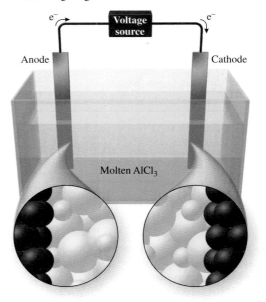

(a) When the power is turned on, what happens at the cathode?
(b) When the power is turned on, what happens at the anode?
(c) Write oxidation and reduction half-reactions.
(d) Write an overall equation for the reaction.

14.69 For Question 14.67 draw molecular-level pictures to represent what happens at the anode and cathode.

14.70 For Question 14.68 draw molecular-level pictures to represent what happens at the anode and cathode.

Corrosion Prevention

14.71 How does attaching a piece of magnesium to the steel hull of a ship prevent the hull from corroding?

14.72 How does zinc attached to an iron pipe prevent it from corroding?

14.73 Iron is often covered with tin, as in a tin can, to protect it from corrosion. If the tin coating is scratched and some iron is exposed, however, the iron corrodes very rapidly. Explain.

14.74 If aluminum siding is fastened to a house with common iron nails, the siding may fall down due to corrosion. Considering the relative activity of aluminum and iron (see Figure 14.22), which metal will corrode? Would corrosion occur if aluminum nails were used? What about nickel nails?

14.75 If the chrome plating on an automobile bumper is damaged, will chromium or iron corrode first? Why?

14.76 The zinc coating of galvanized iron protects the iron from corrosion. If the zinc coating is scratched to expose some iron, will the iron begin to corrode? Why?

Additional Questions

14.77 Indicate the oxidation number of nitrogen in each of the following compounds.
(a) NH_3 (b) N_2H_4 (c) NF_3
(d) NH_2OH (e) $Fe(NO_3)_3$ (f) HNO_2

14.78 The following table gives the results of reactions between five metals and solutions of their nitrate salts. NR means "no reaction" and MD means "metal displaced from solution." From these results, arrange the five metals in order of decreasing chemical activity.

	Ag(s)	**Fe(s)**	**Hg(l)**	**Mn(s)**	**Ni(s)**
$AgNO_3(aq)$	—	MD	NR	MD	MD
$Fe(NO_3)_2(aq)$	NR	—	NR	MD	NR
$Hg(NO_3)_2(aq)$	MD	MD	—	MD	MD
$Mn(NO_3)_2(aq)$	NR	NR	NR	—	NR
$Ni(NO_3)_2(aq)$	NR	MD	NR	MD	—

14.79 Balance the following equations, assuming the reactions occur in basic solution. Also identify the oxidizing agent, the reducing agent, the element oxidized, and the element reduced in each reaction.
(a) $CoCl_2(s) + Na_2O_2(aq) \longrightarrow$
$$Co(OH)_3(s) + Cl^-(aq) + Na^+(aq)$$
(b) $Bi_2O_3(s) + ClO^-(aq) \longrightarrow BiO_3^-(aq) + Cl^-(aq)$

14.80 Aqua regia is a mixture of concentrated HNO_3 and concentrated HCl. It is often used for cleaning laboratory glassware, and it also dissolves platinum. Balance the equation for the dissolution of platinum metal:
$$Pt(s) + NO_3^-(aq) + Cl^-(aq) \longrightarrow PtCl_4^{2-}(aq) + NO_2(g)$$

14.81 Balance the following equation, assuming the reaction occurs in acidic solution:
$$NH_4^+(aq) + NO_3^-(aq) \longrightarrow N_2O(g) + H_2O(l)$$
Also identify the oxidizing agent, the reducing agent, the element oxidized, and the element reduced.

14.82 The reaction that occurs in gas grills involves the burning of propane:
$$C_3H_8(g) + O_2(g) \longrightarrow CO_2(g) + H_2O(g)$$
Assign oxidation numbers to all the reactants and identify the oxidizing and reducing agents.

14.83 In each of the following pairs identify the strongest reducing agent.
(a) Al and Pb
(b) Zn and Ag
(c) Cu and Mn
(d) Cd and Mg

14.84 Why is gold a good metal to use in jewelry?

14.85 Why does brass cause your skin to turn green?

14.86 A voltaic cell is created by using a copper cathode and a magnesium anode. The cathode is immersed in a solution of Cu^{2+} ions, and the anode is immersed in a solution of Mg^{2+} ions. A salt bridge of Na_2SO_4 is also used. What happens to the ions in the salt bridge as the reaction proceeds?

Nuclear Chemistry

D avid was heading to class when he started to feel stomach pains. He decided to go to the student health center, where the results of his physical examination perplexed his doctors so much that they referred him to a specialist for further tests. To locate the source of David's pains, the specialist had a wide range of diagnostic imaging techniques to choose from, including X-rays, computerized tomography (CT) scans, magnetic resonance imaging (MRI), ultrasound studies, and positron emission tomography (PET) scans.

While each technique would provide a different view inside David's abdominal area, all the imaging procedures are alike in one important way: they use some form of *radiation,* either the emission of particles or electromagnetic waves. In Chapter 7 we examined the full spectrum of electromagnetic radiation, increasing in wavelength from gamma rays to radio waves (Figure 7.6). Gamma rays, with the shortest wavelengths, have the highest energies. Radio waves have the longest wavelengths and the lowest energies. The imaging techniques that might be used in David's case involve a variety of types of radiation.

Magnetic resonance imaging, for example, uses strong magnetic fields and radio waves. When the magnetic field is turned on, the patient is irradiated with radio waves. The machine can produce pictures of internal organs because protons in the nuclei of the body's atoms have a spin that aligns with magnetic fields. Radio waves modify the direction of spin, especially in hydrogen atoms. Because the body's soft tissues are made of hydrogen-carbon compounds, the result is an absorption of energy that can be used to map the locations of hydrogen atoms in the body. Energy differences are transmitted as electrical signals that a computer translates into images useful for diagnosis (Figure 15.1).

Other diagnostic options in David's case are X-rays and a CT scan. X-rays have more energy than the radio waves used in an MRI scan. The beam of X-rays is absorbed to varying degrees depending on the density of body tissues. Dense bones show best in an X-ray image. To make the soft tissues of his digestive tract more visible in an X-ray, David drinks a suspension of insoluble barium sulfate before the exam. The compound absorbs X-rays and improves image quality.

The CT scan also uses X-rays. It employs two X-ray beams, allowing the data to be subjected to "computerized tomography," which reconstructs a three-dimensional image of the body. Like MRI, this technique captures images of "slices" through the scanned tissues, as shown in Figure 15.2, providing much more detail than a simple X-ray.

Another imaging method is the PET scan. Like CT and MRI, it employs computer software to create a three-dimensional image, but PET uses the most energetic of electromagnetic radiation, gamma rays. In PET imaging, gamma rays are emitted when a *positron* reacts with an electron. A positron is a relatively rare particle that is like an electron, except that it has a positive charge. The body does not naturally contain many positrons, so they must be introduced before the procedure. This is done with *radioactive isotopes* of certain elements such as fluorine-18, oxygen-15, carbon-11, and nitrogen-13. The isotopes are incorporated in compounds used by cells of the particular organ being examined. If a cancer is present, the compound accumulates in the tumor to a greater or lesser extent than in the surrounding, slower-growing tissue and shows up in the PET scan (Figure 15.3).

As we leave David to talk with his specialist about the diagnostic procedures available to him, we will examine the nature of radiation in more detail. We'll explore nuclear chemistry and the changes that occur in the nuclei of radioactive atoms. We'll look at what happens when isotopes change to other isotopes and give off positrons, gamma rays, or some other type of radiation. We'll also examine the sources of radioactivity, the uses of nuclear reactions as energy sources, and the practical applications of nuclear chemistry, primarily in medicine. Before we explore these topics, consider the following questions.

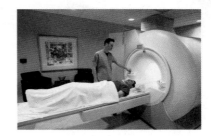

A

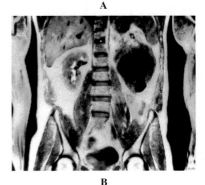

B

Figure 15.1
(A) An MRI instrument. (B) A magnetic resonance image reveals the inner structure of the body.

Tomography reveals only one plane (or "slice") of the internal structure of a body, while eliminating all other planes. A computer assembles the individual planes to produce a three-dimensional image.

For a review of isotopes, see Chapter 2.

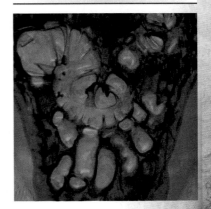

Figure 15.2
A CT scan uses X-rays to produce a two-dimensional "slice" through the body. A series of such scans produces a three-dimensional image, in this case of the intestines.

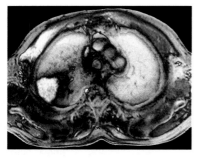

Figure 15.3
Radioactive isotopes and PET imaging reveal a tumor in this patient's abdomen, on the lower left.

When Mendeleev proposed his periodic table, he predicted that a few unknown elements would eventually be found, and he left space for them in the periodic table. Technetium is one of these missing elements.

Questions for Consideration

15.1   Why are some isotopes radioactive and others not?
15.2   What kinds of nuclear reactions do various isotopes undergo?
15.3   Do all radioactive isotopes last the same amount of time?
15.4   How are radioactive isotopes used in medicine?
15.5   Is radiation harmful?
15.6   How can we use radioactive isotopes to generate electricity?

15.1   RADIOACTIVITY

The chemical elements exist in widely varying amounts in the universe and on Earth. Hydrogen is the most abundant element in the universe; nearly 91% of all atoms are hydrogen. In the Earth's crust, the most abundant element is oxygen, which accounts for over 62% of the atoms. At the other extreme, some of the synthetic elements have been prepared in amounts of only a few atoms. These atoms generally exist for only fractions of a second before they destroy themselves by *radioactive decay,* the spontaneous emission of electromagnetic or other types of radiation.

Radiation is defined broadly as energy that comes from a source and travels through matter or space. Radiation may be either of two types: (1) *Electromagnetic* forms include light (visible, infrared, and ultraviolet), as well as gamma rays and X-rays. (2) *Particulate* radiation is mass given off from unstable atoms with the energy of motion (for example, beta particles and alpha particles). Radiation of either type that can produce charged particles in matter is *ionizing radiation,* which includes X-rays and ultraviolet rays. Nonionizing forms of radiation include heat (infrared), radio waves, and microwaves.

Unstable atoms have an excess of matter, energy, or both. If they spontaneously give off this excess as ionizing radiation, they are **radioactive.** Ninety elements occur naturally on Earth. Of these, 81 have at least one stable isotope. They include all the elements from hydrogen through bismuth on the periodic table, with two exceptions. Technetium and promethium have only unstable, radioactive isotopes. The remaining nine natural elements—polonium through uranium—also exist only as radioactive isotopes. The other known elements are all synthetic; they are made with nuclear reactions in a laboratory.

An element is composed of atoms, the smallest particles that retain the properties of that element. An atom can be subdivided, but the unique properties of that element are then lost. Nuclear chemistry is, at least in part, the study of subatomic particles. A nuclear particle is often called a **nucleon,** a general term used to describe either a proton or a neutron. Recall from Chapter 2 that the atomic number, symbolized Z, equals the number of protons in the nucleus. The number of neutrons in the nucleus is the neutron number, symbolized N. The sum of these two quantities is the mass number, symbolized A, which equals the total number of nucleons.

Atoms with the same value of atomic number Z, but different neutron numbers N and mass numbers A, are called *isotopes,* which we can symbolize as follows for element X: $_Z^A X$. An isotope that exists for a measurable length of time and has a defined energy state is a **nuclide.** A nuclide is an atom of a particular atomic number, mass number, and neutron number. In addition its energy state is known and defined. For example, there are three known nuclides of hydrogen, each of which has one proton. The three nuclides have zero, one, and two neutrons, commonly called, respectively, hydrogen, deuterium, and tritium, and symbolized as $_1^1 H$, $_1^2 H$, and $_1^3 H$. Of hydrogen's three nuclides, hydrogen and deuterium are stable, while tritium is radioactive.

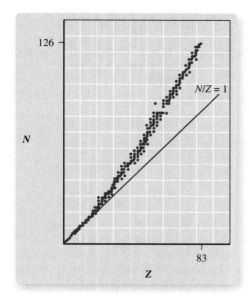

Figure 15.4
A graph of neutron number versus atomic number for the stable isotopes shows that they occur in a narrow band of values, often called the "band of stability."

Nuclear Decay

More than 3000 nuclides are known. Of these, only about 250 are stable. The rest decompose over some period of time, emitting radiation in the process of creating new nuclides. The stable nuclides have a restricted range of combinations of proton and neutron numbers (Z and N), as shown in Figure 15.4. The stable nuclides in this narrow *band of stability* have approximately equal numbers of neutrons and protons in the lighter elements ($Z = 1$ to 20). The ratio of neutrons to protons (N/Z) is about 1 for these elements. The heavier stable nuclides have more neutrons than protons, making the N/Z ratio as high as 1.6 at the upper end of the band of stability. No nuclides heavier than $^{209}_{83}\text{Bi}$ are stable. All nuclides found outside the stability region are radioactive. They undergo nuclear decay reactions. These reactions are different from chemical reactions in that the nuclei undergo a change in composition. Nuclei change into different, more stable nuclei, usually of a different element.

Radiation

When a nucleus undergoes a nuclear reaction, radiation is emitted. Natural radiation associated with radioactive decay can be placed into three classes: alpha particles, beta particles, and gamma rays. These types of radiation behave differently in

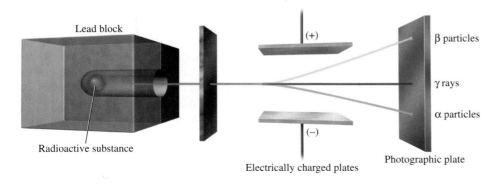

Figure 15.5
The radiation given off by radioactive substances generally falls into one of three categories, depending on how the radiation interacts with an electric field. Alpha particles are attracted to the negative plate. Beta particles are attracted to the positive plate. Gamma rays are unaffected.

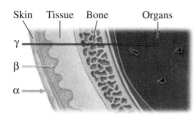

Figure 15.6
Alpha particles barely penetrate the skin, while beta particles can penetrate past the skin into tissue. Gamma rays can penetrate past skin, tissue, and bones to reach internal organs.

TABLE 15.1	**Properties of Types of Radiation**			
Radiation Type	**Notation**	**Mass**	**Charge**	**Penetration into Aluminum**
Alpha	$_2^4\alpha$	4	2+	0.01 mm
Beta (electron)	$_{-1}^0\beta^-$	~0	1−	0.5–1.0 mm
Beta (positron)	$_1^0\beta^+$	~0	1+	(reacts with electrons)
Gamma	γ	0	0	50–110 mm

an electric field, as shown in Figure 15.5. The first two classes actually consist of particles, while the third is pure electromagnetic radiation. Other particles, such as protons ($_1^1p^+$) and neutrons ($_0^1n$), may also be involved in some nuclear reactions.

An **alpha particle,** usually designated as α or $_2^4\alpha$ or $_2^4He^{2+}$, contains two protons and two neutrons. Alpha particles are the nuclei of helium-4 atoms. We often omit the charge of 2+ on the helium nucleus from the symbol for the alpha particle, because it is not necessary to account for electrons when examining nuclear reactions. Of the three types of radiation, alpha particles are the least harmful to animal tissue upon external exposure (Figure 15.6). A 0.01-mm-thick sheet of aluminum foil, a piece of paper, or a layer of skin cells can stop an alpha particle. External alpha particles are not particularly harmful to human tissue, since they do not penetrate the skin deep enough to reach living cells. However, alpha particles inside the body will damage cells.

A **beta particle** is a small, charged particle that can be emitted from unstable atoms at speeds approaching the speed of light. Beta particles are more penetrating than alpha particles, traveling more than 3 m (about 10 ft) through air before being stopped by several sheets of paper or a sheet of aluminum foil about 0.5 to 1 mm thick. They will penetrate through skin into tissue (Figure 15.6). Beta particles are commonly emitted by the picture tubes of television sets and by computer monitors. There are two types of beta particles, electrons and positrons. A **positron** has the same mass as an electron but the opposite charge. Electrons are designated as $_{-1}^0\beta^-$ or $_{-1}^0e^-$, and positrons as $_1^0\beta^+$.

The **gamma ray** is the highest energy and most penetrating type of radiation; a thickness of 50 to 110 mm of aluminum metal is required to stop it. Other types of heavy shielding, such as lead or thick layers of concrete, can also stop gamma rays. They are the most damaging type of natural radiation (Figure 15.6). They are high-energy electromagnetic radiation: energy with neither charge nor mass. The properties of the most common types of radiation are summarized in Table 15.1.

Remember when your parents told you not to sit so close to the TV set? Consider the properties of beta radiation to decide whether their warning was legitimate.

During the Cold War of the 1950s, some people built backyard bomb shelters in hopes of surviving a nuclear attack from the Soviet Union. Why were these shelters made of concrete?

15.2 NUCLEAR REACTIONS

Radioactivity arises from the transformation of one nuclide into another, often resulting in the emission of a particle from the nucleus. As David was having a PET scan to diagnose his stomach complaint, he asked the technician how the radioactive fluorine, ^{18}F, he had been administered would cause a response in the scanning detector. The technician answered that positrons are emitted from the nuclide and react with electrons in David's body, causing a release of gamma rays that the sensor can pick up. David thought about this and reasoned that if the ^{18}F was decaying and giving off positrons, it must be converted to some other nuclide. But which one?

Writing an equation for a nuclear transformation is similar to writing an equation for a chemical reaction, but the procedure is slightly different.

Equations for Nuclear Reactions

To get a feel for nuclear equations, consider the following examples:

$$^{232}_{90}\text{Th} \longrightarrow {}^{228}_{88}\text{Ra} + {}^{4}_{2}\alpha \quad \textit{alpha}$$

$$^{231}_{90}\text{Th} \longrightarrow {}^{231}_{91}\text{Pa} + {}^{0}_{-1}\beta^{-} \quad \textit{beta}$$

$$^{238}_{92}\text{U} + {}^{4}_{2}\alpha \longrightarrow {}^{239}_{94}\text{Pu} + 3{}^{1}_{0}\text{n}$$

Can you see which factors must be considered to balance equations? Can you use this information to identify the unknown product of a nuclear reaction if the type of radiation is known?

Two conditions must be met to balance a nuclear equation: (1) conservation of mass number and (2) conservation of nuclear charge (atomic number). We can identify the particles emitted in a nuclear transformation by checking these two conservation conditions.

Alpha Particle Emission   When a nucleus emits an alpha particle, it loses two protons and two neutrons, so its atomic number decreases by 2 and its mass number decreases by 4, as shown in Figure 15.7. For example, thorium-232 undergoes radioactive decay by alpha particle emission:

$$^{232}_{90}\text{Th} \longrightarrow {}^{228}_{88}\text{Ra} + {}^{4}_{2}\alpha$$

The atomic number decreases by 2 when the nuclide changes from thorium to radium, while the mass number decreases by 4. The emission of an alpha particle conserves the mass number, because the 4 units of mass lost by the nuclide match the mass of the alpha particle that is created. The mass number of the reactant equals the sum of the mass numbers of the products:

$$232 = 228 + 4$$

Similarly the atomic number of the reactant equals the sum of the atomic numbers of the products:

$$90 = 88 + 2$$

With this approach, we can answer questions such as the one that concerned David. Specifically, we can identify the decay product of a radioactive nuclide.

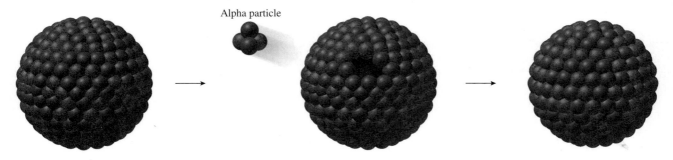

Alpha particle

Figure 15.7
The emission of an alpha particle removes two protons and two neutrons from the decaying nuclide.

EXAMPLE 15.1 — Balancing Equations for Alpha Emission

Radium-228 decays by alpha emission. What new nuclide (X) forms in this reaction?

$$^{228}_{88}\text{Ra} \longrightarrow X + {}^{4}_{2}\alpha$$

Solution:

The mass number and atomic number of the unknown product can be determined using the two conservation conditions. To conserve the mass number, we have

$$228 = A + 4$$
$$A = 228 - 4 = 224$$

To conserve atomic number, we have

$$88 = Z + 2$$
$$Z = 88 - 2 = 86$$

We identify the element from its atomic number. From the periodic table we can see that element 86 is radon. The complete symbol for radon-224 is $^{224}_{86}\text{Rn}$. Thus, the complete reaction is

$$^{228}_{88}\text{Ra} \longrightarrow {}^{224}_{86}\text{Rn} + {}^{4}_{2}\alpha$$

Practice Problem 15.1

What nuclide (X) decays by alpha emission, forming lead-214?

$$X \longrightarrow {}^{214}_{82}\text{Pb} + {}^{4}_{2}\alpha$$

Further Practice: 15.19 and 15.20

Although both electrons and positrons are forms of beta radiation, when no specification is made, the term *beta radiation* will be used in this chapter to mean electrons. To avoid confusion we will use the term *positron* to refer to the other form of beta radiation.

Beta Particle (Electron) Emission   The emission of an electron, a beta particle, from the nucleus results in no change in the mass number but an increase in the atomic number by 1, causing the conversion of a neutron into a proton (Figure 15.8). The number of protons, the atomic number, increases. The total number of nucleons remains the same, and the mass number does not change. For example, thorium-231 decays by beta emission:

$$^{231}_{90}\text{Th} \longrightarrow {}^{231}_{91}\text{Pa} + {}^{0}_{-1}\beta^{-}$$

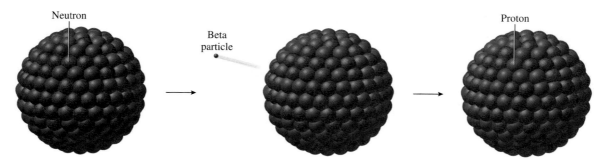

Neutron   Beta particle   Proton

Figure 15.8
Beta emission converts a neutron into a proton.

The mass number stays the same, so it is conserved. The atomic number increases by 1, so the atomic number is also conserved:

$$90 = 91 + (-1)$$

By using the balancing conditions, we can also identify an unknown product or reactant in a beta emission reaction.

EXAMPLE 15.2   **Balancing Equations for Beta Emission**

Carbon-14 undergoes beta emission. Identify the nuclide (X) formed in this reaction and balance the equation:

$$^{14}_{6}C \longrightarrow X + ^{\ 0}_{-1}\beta^{-}$$

Solution:

Because the emitted electron has a mass number of 0, the product X must have a mass number of 14, the same as carbon-14. With a charge of 1– on the emitted electron, the product X must have a nuclear charge greater by 1 than carbon-14, or a value of 7. Atomic number 7 indicates nitrogen, so X must be $^{14}_{7}N$:

$$^{14}_{6}C \longrightarrow {}^{14}_{7}N + ^{\ 0}_{-1}\beta^{-}$$

Practice Problem 15.2

Thorium-234 undergoes beta emission to form a new nuclide (X). Identify the new nuclide and balance the equation for this nuclear reaction:

$$^{234}_{90}Th \longrightarrow X + ^{\ 0}_{-1}\beta^{-}$$

Further Practice: 15.21 and 15.22

Positron Emission   Nuclides that emit positrons are used in generating PET scans for medical diagnosis. A positron has a mass number of 0 and a charge of 1+. Thus when a radioactive nucleus emits a positron, the mass number does not change, but the atomic number decreases by 1. This change converts a proton into a neutron, as shown in Figure 15.9. This change is the reverse of the result of emission of an electron. An example of positron emission is the decay of magnesium-23:

$$^{23}_{12}Mg \longrightarrow {}^{23}_{11}Na + ^{0}_{1}\beta^{+}$$

The mass number is conserved:

$$23 = 23 + 0$$

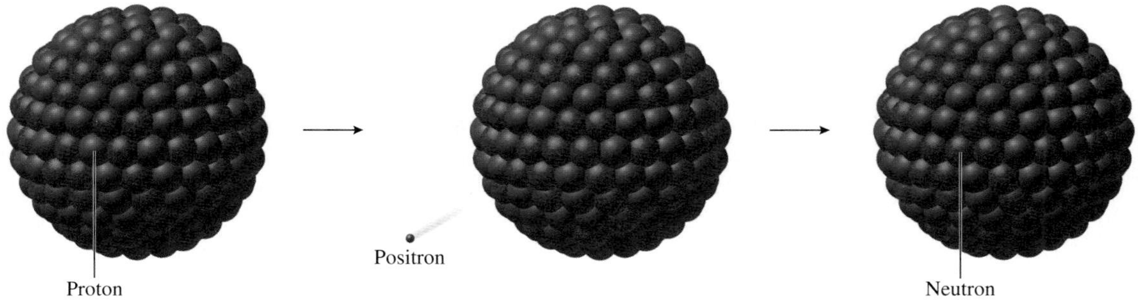

Proton   Positron   Neutron

Figure 15.9
Positron emission converts a proton into a neutron.

The atomic number is also conserved:

$$12 = 11 + 1$$

Thus this nuclear equation is balanced.

EXAMPLE 15.3 | **Balancing Equations for Positron Emission**

David was injected with a compound containing fluorine-18 as a positron emitter for his PET scan. Identify the new nuclide formed by this nuclear decay reaction:

$$^{18}_{9}\text{F} \longrightarrow X + ^{0}_{1}\beta^{+}$$

Solution:

The mass number and atomic number of the unknown product can be determined by the two conservation conditions. To conserve the mass number, we have

$$18 = A + 0$$
$$A = 18 - 0 = 18$$

To conserve atomic number, we have

$$9 = Z + 1$$
$$Z = 9 - 1 = 8$$

The element with atomic number 8 is oxygen, so the unknown product is $^{18}_{8}\text{O}$:

$$^{18}_{9}\text{F} \longrightarrow ^{18}_{8}\text{O} + ^{0}_{1}\beta^{+}$$

Practice Problem 15.3

Determine the identity of the unknown reactant X in the following nuclear reaction:

$$X \longrightarrow ^{11}_{5}\text{B} + ^{0}_{1}\beta^{+}$$

Further Practice: 15.23 and 15.24

Electron Capture   Emission of a positron by a nucleus transforms it into the nucleus of a different element with the same mass number. The new element has an atomic number one less than that of the original element. This same transformation can take place if a nucleus captures a $1s$ electron. As a result of **electron capture**, a proton combines with the electron to form a neutron. This transformation occurs only for a few nuclides. One example is beryllium-7, which undergoes radioactive decay by electron capture:

$$^{7}_{4}\text{Be} + ^{0}_{-1}\text{e}^{-} \longrightarrow ^{7}_{3}\text{Li}$$

The mass number stays the same ($7 + 0 = 7$), but the atomic number decreases by 1 ($4 - 1 = 3$).

Technetium-99*m* is an energetic isomer of technetium-99. The *m* stands for metastable, which means that it decays to a more stable form of the same isotope. Technetium-99*m* decays to technetium-99 with the emission of gamma rays. It is sometimes used in the diagnosis of gastrointestinal disorders.

Gamma Ray Emission   During all of the radioactive decay processes discussed so far, the nucleus changes from a state of higher energy to a state of lower energy. The emission of an alpha or beta particle with a high kinetic energy carries away excess energy. As shown in Figure 15.10, excess energy can also be released in the form of gamma rays, which are pure electromagnetic radiation. The gamma decay process involves no change in mass number or atomic number. One example is the decay of the nuclide ^{99m}Tc:

$$^{99m}_{43}\text{Tc} \longrightarrow ^{99}_{43}\text{Tc} + \gamma$$

Gamma

Energetic nucleus

Stable nucleus

Figure 15.10
The emission of gamma rays converts an energetic nucleus into a lower-energy nucleus.

Different nuclides emit different gamma ray frequencies, from only a few up to as many as 100 different frequencies per nuclide. The frequencies of the gamma rays emitted help to identify the products of nuclear reactions.

Nuclear Bombardment Reactions So far we have mentioned only nuclear reactions that result in the *production* of radiation. In these reactions, one nucleus or isotope spontaneously changes into another. But not all nuclear changes are spontaneous; some are produced intentionally in laboratories by artificial means. In a **nuclear bombardment** reaction, nuclei are hit with a beam of nuclei or nuclear particles to trigger a nuclear reaction. Nuclear physicists and chemists have built particle accelerators (described following Example 15.4) to create a beam of particles and direct it onto a target element. We can represent a successful nuclear bombardment reaction with a nuclear reaction equation that looks much like an equation for a chemical reaction. For example, the bombardment of $^{238}_{92}U$ with neutrons gives $^{239}_{92}U$, which spontaneously decays to $^{239}_{93}Np$:

$$^{238}_{92}U + ^{1}_{0}n \longrightarrow ^{239}_{92}U$$

$$^{239}_{92}U \longrightarrow ^{239}_{93}Np + ^{0}_{-1}\beta^-$$

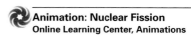

Animation: Nuclear Fission
Online Learning Center, Animations

Chemistry Animations Library, Nuclear Fission, Fission.avi.

All elements with atomic numbers above 92, the atomic number of uranium, are synthetic elements that do not exist naturally on Earth. There are also lighter elements that have never been found in a natural state. For a long time, there were "holes" in the periodic table, shown in Figure 15.11, and no one was able to find elements 43 (Tc), 61 (Pm), or 85 (At). Promethium (Pm) was first identified in 1945 as a decay product of uranium in nuclear reactors.

Technetium (Tc) was first formed in 1937 by the bombardment of molybdenum with deuterium atoms:

$$^{97}_{42}Mo + ^{2}_{1}H \longrightarrow ^{97}_{43}Tc + 2^{1}_{0}n$$

The most useful isotope, technetium-99*m,* is now formed by bombardment of molybdenum-98 with neutrons, giving molybdenum-99, which decays to technetium-99*m:*

$$^{98}_{42}Mo + ^{1}_{0}n \longrightarrow ^{99}_{42}Mo \longrightarrow ^{99m}_{43}Tc + ^{0}_{-1}\beta^-$$

All isotopes of technetium are unstable; most of them undergo radioactive decay in a matter of a few minutes to a few days, but three of its isotopes survive for a few million years. The isotopes that disappear quickly are often used for medical imaging.

Astatine was formed in 1940 by the bombardment of bismuth with alpha particles:

$$^{209}_{83}Bi + ^{4}_{2}\alpha \longrightarrow ^{211}_{85}At + 2^{1}_{0}n$$

All isotopes of astatine are highly radioactive, with the most stable, $^{210}_{85}At$, lasting only a few hours. No more than about 0.05 micrograms of astatine has ever been produced at one time.

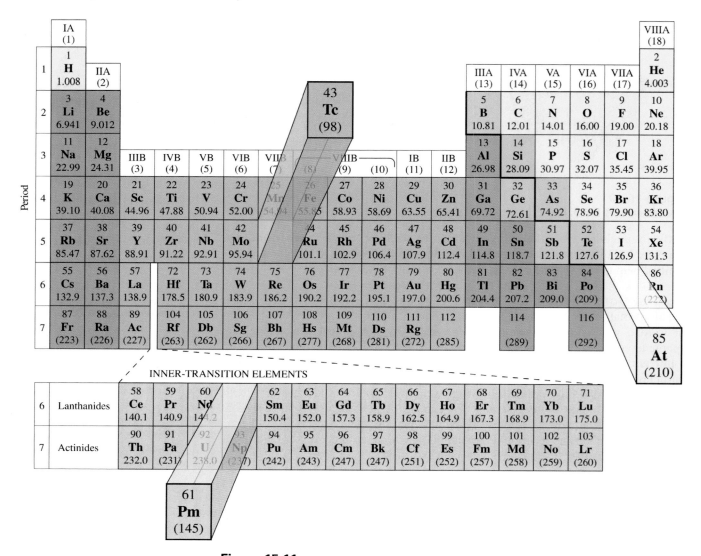

Figure 15.11
Technetium (element 43), promethium (element 61), and astatine (element 85) were missing from the periodic table for a long time, because all their isotopes are radioactive and are not found in nature in significant quantities. Notice that their atomic masses are whole numbers placed in parentheses. This is generally true for elements that do not occur naturally, because the average atomic mass for a sample depends on the specific isotopes in the sample. The number in parentheses is the mass number of the most stable isotope.

Nuclear bombardment reactions have been used to synthesize small amounts of all the *transuranium* elements, those following uranium in the periodic table. The bombardment of uranium with neutrons yields neptunium. Other nuclear particles or other nuclei can be used to bombard heavy nuclei such as uranium, as in the following examples:

$$^{238}_{92}\text{U} + {}^{4}_{2}\alpha \longrightarrow {}^{239}_{94}\text{Pu} + 3{}^{1}_{0}\text{n}$$

$$^{238}_{92}\text{U} + {}^{12}_{6}\text{C} \longrightarrow {}^{246}_{98}\text{Cf} + 4{}^{1}_{0}\text{n}$$

$$^{238}_{92}\text{U} + {}^{14}_{7}\text{N} \longrightarrow {}^{247}_{99}\text{Es} + 5{}^{1}_{0}\text{n}$$

Nuclear scientists continue to use bombardment reactions to prepare even heavier elements. For example, by bombarding bismuth-209 with iron-58 nuclei for one

Meitnerium (Mt) is named in honor of Lise Meitner, a German physicist who fled from Germany during the Second World War because she was Jewish. She is credited along with Otto Hahn and Fritz Strassman for the discovery of nuclear fission.

week, scientists in 1982 finally succeeded in synthesizing a single atom of element 109 with mass number 266, now named meitnerium:

$$^{209}_{83}\text{Bi} + {}^{58}_{26}\text{Fe} \longrightarrow {}^{266}_{109}\text{Mt} + {}^{1}_{0}\text{n}$$

EXAMPLE 15.4 Balancing Equations for Bombardment Reactions

Bombarding a lead-208 target with a beam of another nuclide produces hassium-265 and a neutron. Identify the nuclide used in the bombardment, and write a balanced equation to describe this nuclear reaction.

Solution:

First we find from the periodic table that hassium is element 108. We then write an incomplete equation summarizing the information given:

$$^{208}_{82}\text{Pb} + X \longrightarrow {}^{265}_{108}\text{Hs} + {}^{1}_{0}\text{n}$$

We can apply the conservation of mass condition to the mass numbers:

$$208 + A = 265 + 1$$

$$A = 265 + 1 - 208 = 58$$

We can apply the conservation of charge condition to identify the bombarding element:

$$82 + Z = 108 + 0$$

$$Z = 108 + 0 - 82 = 26$$

Element 26 is iron, so the bombarding particle is iron-58, giving us the balanced equation

$$^{208}_{82}\text{Pb} + {}^{58}_{26}\text{Fe} \longrightarrow {}^{265}_{108}\text{Hs} + {}^{1}_{0}\text{n}$$

Practice Problem 15.4

Bombarding a bismuth-209 target with a beam of another nuclide produces bohrium-262 ($^{262}_{107}\text{Bh}$) and a neutron. Identify the nuclide used in the bombardment, and write a balanced equation to describe this nuclear reaction.

Further Practice: 15.25 and 15.26

Particle Accelerators

We have been discussing bombarding nuclides with subatomic particles or with other nuclides as though this were a simple matter, but how do we get beams of these particles? Generating and manipulating these particles require sophisticated and expensive equipment that greatly increases the velocity of the particles, making them energetic enough to induce a nuclear change. This is done with particle accelerators. Several different types of particle accelerators have been developed. The most common types are linear accelerators and circular accelerators.

In a linear accelerator an ion or electron beam is directed into a series of tubes. Voltage is applied to each tube with an alternating current so that the tube behind the particles repels them and the tube ahead of the particles attracts them. This causes the particles to accelerate. Because the particles go faster as they proceed through the accelerator, the farther the tubes are from the source, the longer the tubes must be (Figure 15.12A). For very high energies, the tubes must be very long (Figure 15.12B).

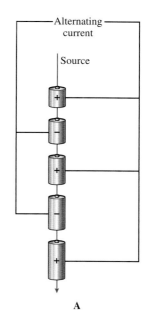

A

B

Figure 15.12
(A) Tubes of increasing length accelerate charged particles through electric fields. To achieve usable energies, such accelerators must be very long. (B) For example, the Stanford Linear Accelerator is built in a 3-km (2-mi) long tunnel covered by laboratory and support facilities.

Figure 15.13
The synchrotron at Fermilab in Batavia, Illinois, has an underground ring that is nearly 6.5 km (4 mi) long. Inside the ring, electromagnets keep accelerating ions moving in a circular path. This laboratory has several circular and linear accelerators. Can you locate them in the photograph?

Most of the accelerators in use today employ a circular design to cut down on the distances needed for the tubes. Perhaps the most successful accelerator—and certainly the largest—is the *synchrotron,* which uses a circular path for the accelerating particles (Figure 15.13). The radius of the larger synchrotrons exceeds 1 km. The synchrotron uses electromagnets to produce a variable magnetic field that maintains the path of the charged particles at a given radius as they accelerate. The electromagnets are enclosed within the hollow ring through which the charged particles pass, so an electromagnet with a radius of 1 km is not necessary.

Spontaneous Nuclear Decay Reactions

We have examined a number of different types of nuclear decay reactions in which one nuclide is transformed into another. But why do some nuclides emit an alpha particle when they decay, while others emit a beta particle or a positron? The tendency for the neutron/proton (N/Z) ratio to move toward the band of stability, shown in Figure 15.14, explains the nuclear reactions of naturally radioactive nuclides. While nuclides with excess mass may emit an alpha particle, those with an N/Z ratio that is too high undergo beta decay. When the N/Z ratio is too low,

Figure 15.14
Each nuclear decay process, except γ emission, changes an unstable nuclide to a new nuclide that is closer to the band of stability. The arrows show the direction in which the nuclide would move on the graph after emission of the indicated radiation. Nuclides above the band of stability undergo beta decay and those below undergo positron emission. Nuclides beyond the band of stability undergo alpha decay.

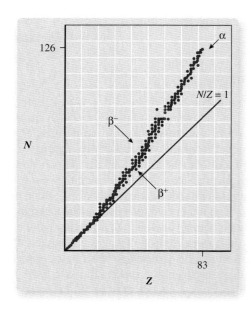

TABLE 15.2	Natural Radioactive Decay Processes			
Reason for Nuclear Instability	**Radioactive Process**	**Emitted Radiation**	**Nuclear Change**	**Change in N/Z Ratio**
Excess mass	Alpha decay	$_2^4\alpha$	Loss of two protons and two neutrons	Slight increase
N/Z too high	Beta decay	$_{-1}^0\beta^-$	Neutron changes to a proton and an electron.	Decrease
N/Z too low	Positron emission	$_1^0\beta^+$	Proton changes to a neutron and a positron.	Increase
N/Z too low	Electron capture	—	Proton combines with an inner-shell electron to become a neutron.	Increase
Energetically excited	γ emission	Gamma ray	Loss of excess energy	None

positron emission or electron capture may occur. High-energy nuclides may lose excess energy through the emission of gamma rays. The characteristics of these nuclear changes are summarized in Table 15.2.

For each process except γ emission, the change that occurs for an unstable nuclide takes it closer to the observed band of stability, as shown by the arrows in Figure 15.14. Radioactive nuclides convert spontaneously, over time, to stable nuclides. This conversion may occur in a single step but more commonly for the heavier isotopes, several successive nuclear reactions are required to reach a stable nuclide.

As shown in Figure 15.14, alpha decay adjusts the nucleus along a line parallel to $N/Z = 1$, while beta decay adjusts the nucleus along a line perpendicular to this. Positron decay accomplishes the opposite of beta decay: protons change to neutrons, so the N/Z ratio increases and the atomic number decreases, as in the decay of fluorine-18:

$$_9^{18}F \longrightarrow {}_8^{18}O + {}_1^0\beta^+$$

Decay of Lighter Elements   The mode of decay is fairly simple to predict for the lighter elements ($Z = 1$ to 20), where stable isotopes have an N/Z ratio of 1. Radioactivity in lighter nuclides also results from an apparently unstable N/Z ratio. The ratios change through beta decay, positron emission, or electron capture, resulting in greater stability.

If the N/Z ratio is too high, a neutron converts to a proton by beta decay (emission of electrons). For example, carbon-14 has a ratio that is too high:

$$\frac{N}{Z} = \frac{14 - 6}{6} = \frac{8}{6} = 1.33$$

A stable state is achieved by beta decay:

$$_6^{14}C \longrightarrow {}_7^{14}N + {}_{-1}^0\beta^-$$

The product, nitrogen-14, has a ratio of 1:

$$\frac{N}{Z} = \frac{14 - 7}{7} = \frac{7}{7} = 1$$

On the other hand, nitrogen-13 has a low ratio:

$$\frac{N}{Z} = \frac{13 - 7}{7} = 0.86$$

Positron decay results in a higher ratio:

$$_7^{13}N \longrightarrow {}_6^{13}C + {}_1^0\beta^+$$

The ratio in the product of this transformation, carbon-13, is not ideal either:

$$\frac{N}{Z} = \frac{13 - 6}{6} = \frac{7}{6} = 1.17$$

Nevertheless, C-13 is the only stable isotope of carbon other than C-12, which has the ideal ratio, $N/Z = 1$. What would happen to the ratio for carbon-13 if it underwent beta emission or electron capture? Would the N/Z ratio get closer to a value of 1?

Electron capture raises the N/Z ratio and occurs preferentially to positron emission in some elements, such as in the decay of beryllium-7, where the ratio changes from $\frac{3}{4}$ to $\frac{4}{3}$:

$$\ce{^{7}_{4}Be} + \ce{^{0}_{-1}e} \longrightarrow \ce{^{7}_{3}Li}$$

The occurrence of electron capture cannot be predicted readily. However, among the first 20 elements, this process apparently occurs only in four cases besides beryllium-7: $\ce{^{22}_{11}Na}$, $\ce{^{37}_{18}Ar}$, $\ce{^{40}_{19}K}$, and $\ce{^{41}_{20}Ca}$.

Decay of Heavier Elements   All isotopes of the elements beyond atomic number 83 (bismuth) are radioactive; they decay to become nuclides of decreased mass and lesser atomic number. They generally decay through both alpha and beta decay processes. The alpha decay gets rid of excess mass with the loss of two neutrons and two protons, as in the decay of uranium-238:

$$\ce{^{238}_{92}U} \longrightarrow \ce{^{234}_{90}Th} + \ce{^{4}_{2}\alpha}$$

An unstable N/Z ratio is not adjusted sufficiently in heavy elements through this initial process, however. Beta decay processes are also necessary to convert a heavy, unstable nuclide to a more stable one. For example, the thorium-234 produced in the decay of uranium-238 is still unstable. Beta decay converts a neutron to a proton, so it decreases the N/Z ratio and increases the atomic number:

$$\ce{^{234}_{90}Th} \longrightarrow \ce{^{234}_{91}Pa} + \ce{^{0}_{-1}\beta^{-}}$$

EXAMPLE 15.5   Predicting the Method of Decay

Predict the method of radioactive decay of the unstable nuclide $\ce{^{24}_{10}Ne}$.

Solution:

The N/Z ratio in this nuclide is

$$\frac{24-10}{10} = \frac{14}{10} = 1.4$$

This ratio is higher than the ideal ratio of 1, so this nuclide undergoes radioactive decay to reduce the value. The emission of an electron (beta decay) converts a neutron to a proton.

$$\ce{^{24}_{10}Ne} \longrightarrow \ce{^{24}_{11}Na} + \ce{^{0}_{-1}\beta^{-}}$$

This sodium nuclide is also radioactive—its N/Z ratio of $\frac{13}{11}$ is greater than 1—so it should undergo further beta decay.

Practice Problem 15.5

Predict the method of radioactive decay of the unstable nuclide $\ce{^{18}_{10}Ne}$.

Further Practice: 15.41 and 15.42

Radioactive Decay Series   Some radioactive nuclides decay to stable isotopes in a single step, but frequently the product of radioactive decay is itself radioactive, especially in the heavier elements. In such cases, a series of alpha and beta decay steps ultimately leads to a stable nuclide. Three such radioactive decay

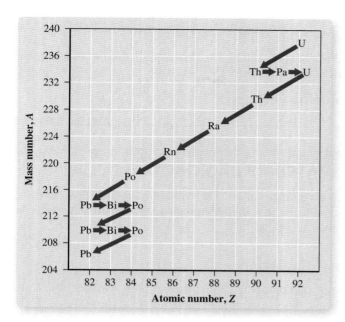

Figure 15.15
Uranium-238 decays to form thorium-234. Through a series of alpha and beta decays, the elements continue to convert to new elements until stable lead-206 forms. All the nuclides in this series except lead-206 are radioactive. This is a very slow process, requiring 4.5 billion years for half of the uranium-238 to decay.

series, which involve only alpha and beta decay, are found among the natural elements. They account for most of the radioactivity among elements 83 through 92. They start with the elements uranium, actinium, and thorium. A fourth series starts with the synthetic element neptunium. The uranium-238 series is shown in Figure 15.15. In each series, when the mass number changes, it differs by a unit of 4, because mass is lost only through loss of alpha particles with mass number 4. The atomic number decreases with alpha decay and increases with beta decay. These decays occur in various combinations until the product is a stable isotope.

EXAMPLE 15.6   Products of Decay Series

Actinium-227 decays and emits five alpha particles and three beta particles in eight steps. What is the final product?

Solution:

As long as we know the total number of particles emitted, we don't have to know the order in which they are emitted. We can write an overall equation for the series of steps:

$$^{227}_{89}\text{Ac} \longrightarrow 5\,^4_2\alpha + 3\,^0_{-1}\beta^- + X$$

We then use the conservation conditions to balance the equation, thereby identifying the unknown product. First we set up the mass balance condition to find the mass number:

$$227 = 5 \times 4 + 3 \times 0 + A$$
$$A = 227 - 20 - 0 = 207$$

Next we use the charge balance condition to find the atomic number of the product:

$$89 = 5 \times 2 + 3 \times (-1) + Z$$
$$Z = 89 - 10 + 3 = 82$$

Element 82 is lead, so the product is lead-207:

$$^{227}_{89}\text{Ac} \longrightarrow 5\,^4_2\alpha + 3\,^0_{-1}\beta^- + \,^{207}_{82}\text{Pb}$$

Practice Problem 15.6

Thorium-232 decays and emits six alpha particles and four beta particles in ten steps. What is the final product?

Further Practice: 15.43 and 15.44

15.3   RATES OF RADIOACTIVE DECAY

Nuclear reactions emit radiation. The emitted radiation has a characteristic kinetic energy and is produced at a characteristic rate. Both can be used to detect nuclides and measure their amounts.

Animation: Radioactive Decay
Online Learning Center, Animations

Chemistry Animations Library, Radioactive Decay, Decay_Radioactive.exe.

Animation: Detecting Radiation
Online Learning Center, Animations

Chemistry Animations Library, Detecting Radiation, Radioactivity.rm.

The exposure of photographic film led to Antoine Henri Becquerel's discovery of spontaneous radioactivity, for which he received a Nobel Prize in 1903.

Detection of Radiation

The exposure of photographic film is the simplest method of detecting radiation. That's why people who work around radiation, including X-ray technicians in hospitals and research scientists in nuclear laboratories, wear film badges. The badges are developed periodically to determine the workers' exposure to radiation. Exposure of the film to radiation darkens the image. The darker the film after development, the greater the radiation exposure. Despite its usefulness, photographic film exposure is a slow and relatively imprecise measure of radiation.

Various instruments have been developed to give speedier and more accurate measures of radiation intensity. One of the earliest was the *Geiger-Müller counter* shown in Figure 15.16. Radiation enters a tube through a thin glass or plastic window and ionizes a small fraction of the argon gas in the tube. A high voltage is applied between the walls of the tube and a wire that passes into its center. When the gas is ionized, it releases electrons that are accelerated by the electrical charge, in turn ionizing other argon atoms. The result is a pulse of electric current, which passes through a digital counter or through a speaker to make a clicking sound. A Geiger-Müller counter can detect alpha, beta, and gamma radiation. It is small and portable, so it can be used in a variety of circumstances outside of laboratories.

Another common and more sensitive detection device is a *scintillation counter.* It contains a tube coated with a material such as zinc sulfide that produces a flash of light when struck by gamma or beta radiation. The flashes cause pulses of electricity to pass through a photoelectric tube, where the electrical signals are amplified to activate a digital counter. The counter measures the intensity of the radiation in units of counts per minute. A scintillation counter can be tuned to detect radiation of

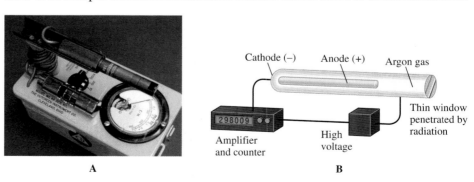

A

B

Figure 15.16
(A) Geiger-Müller counters are small and portable. Units such as this were widely used by uranium prospectors and were kept in civil defense shelters of the 1950s. Modern versions are similar, but much smaller. (B) Radiation ionizes the argon gas, releasing electrons that can be detected.

different energies, so it can measure the energy released from decay of a particular nuclide. Scintillation counters are used extensively in laboratories that study radioactive nuclides or use nuclides to investigate biological processes.

Half-Lives

So far we have been concerned primarily with the types of particles involved in nuclear reactions. Another important characteristic is how fast the reactions proceed. The nuclides used in medical imaging and diagnosis need to be unstable ones that do not remain in the body for too long. That is, they need to have a short **half-life,** which is the time required for half of a sample of a nuclide to decay to a different nuclide. The half-life of the fluorine-18 that David received is 110 min, while that of the C-11 often used for PET scans of the brain is a mere 20 min. Some radioactive elements, in contrast, have half-lives of millions of years.

How does the half-life work? It takes the same time for a fresh sample to decay to one-half the original number of atoms of that nuclide as it does for one-half to decay to one-fourth, and so on (Figure 15.17). As the number of atoms of radioactive nuclide decreases, the amount of radiation, measured by a scintillation counter in counts per minute (cpm), also decreases. If the counts per minute we measure after a time are one-half the original, we know that half the amount of radioactive nuclide has decayed. For the technetium-99*m* used in medical imaging, the half-life is 6.0 h. If we start with a sample that has a radiation intensity of 8000 cpm, after 6.0 h the intensity is 4000 cpm. After an additional 6.0 h, the intensity is 2000 cpm. For each additional 6.0-h period, the intensity of the radiation decreases by one-half of what it was at the beginning of that period. As a general rule, we assume that a nuclide is virtually all gone after 10 half-lives.

The shorter the half-life of a nuclide, the more intense the radiation that it emits, because the radiation is being released at a greater rate. The half-lives for the nuclides involved in the uranium decay series, shown earlier in Figure 15.15, are

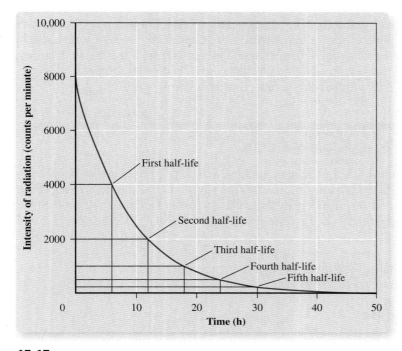

Figure 15.17
In the radioactive decay curve for technetium-99*m*, each successive half-life results in the loss of half of the radioactivity the sample had at the beginning of that period. The time for each half-life is the same, 6.0 h. At the end of the fifth half-life, only $\frac{1}{32}$ (or 3.1%) of the original technetium-99*m* remains. After 10 half-lives, 0.1% of the technetium-99*m* remains.

TABLE 15.3	Half-Lives for the Nuclides in the Uranium Decay Series	
Nuclide	**Method of Decay**	**Half-life**
$^{238}_{92}U$	Alpha	4.46×10^9 years
$^{234}_{90}Th$	Beta	24.1 days
$^{234}_{91}Pa$	Beta	6.75 h
$^{234}_{92}U$	Alpha	2.45×10^5 years
$^{230}_{90}Th$	Alpha	7.54×10^4 years
$^{226}_{88}Ra$	Alpha	1.62×10^3 years
$^{222}_{86}Rn$	Alpha	3.82 days
$^{218}_{84}Po$	Alpha	3.11 min
$^{214}_{82}Pb$	Beta	26.8 min
$^{214}_{83}Bi$	Beta	19.7 min
$^{214}_{84}Po$	Alpha	1.6×10^{-4} s
$^{210}_{82}Pb$	Beta	22.3 years
$^{210}_{83}Bi$	Beta	5.01 days
$^{210}_{84}Po$	Alpha	138.4 days
$^{206}_{82}Pb$	(Stable)	

given in Table 15.3. The table shows that the half-lives of the various nuclides vary over a wide range.

The half-life of a nuclide reveals how long that nuclide will last. Consider, for example, that radioactivity is used not only for diagnosis, but also for treatment of cancer. The patient may be injected with a radioactive nuclide in a compound that is preferentially absorbed by the cancerous cells. For example, radioactive iodide treats thyroid cancer or hyperthyroidism (Grave's disease). In some cases the intensity of the radiation is so high that the patient is a radiation risk to other individuals and must be kept in isolation. In such cases it is desirable to know what the level of radioactivity will be after various time periods.

EXAMPLE 15.7   Half-Life

Strontium-90 is a nuclide found in the fallout from an atomic bomb explosion. It is particularly harmful because it is similar to calcium in size and charge, so it may end up mixed with the calcium phosphate in cows' milk. Strontium-90 undergoes beta decay to yield yttrium-90:

$$^{90}_{38}Sr \longrightarrow \, ^{90}_{39}Y + \, ^{0}_{-1}\beta^-$$

The half-life of this process is 28 years. What percentage of the strontium-90 is left after exactly 56 years?

Solution:

During each half-life, one-half of the strontium-90 present at the beginning of the time period decays. The half-life is 28 years, so 56 years is two half-lives. Let's call the amount present initially 100%. Then at the end of one half-life, the amount will be 50% or half of the starting amount. At the end of two half-lives, one-half of that amount decays, leaving 25%. Thus, 25% of the Sr-90 is left after 56 years.

Practice Problem 15.7

Gold-128 undergoes beta decay to yield mercury-128:

$$^{128}_{80}Au \longrightarrow {}^{128}_{81}Hg + {}^{0}_{-1}\beta^{-}$$

The half-life is 2.7 days. What percentage of the gold-128 is left after 8.1 days?

Further Practice: 15.49 and 15.50

Archeological Dating

The decay of certain radioactive nuclides can be used to measure time on an archeological scale. For example, carbon-containing artifacts from an archeological dig can be dated according to the amount of carbon-14 they contain. The technique is **radiocarbon dating.** The amount of carbon-14 on Earth is small: only 1 ^{14}C atom for every 10^{12} atoms of ^{12}C. The reaction of nitrogen-14 with neutrons from outer space (cosmic rays) continuously produces small amounts of carbon-14 in the upper atmosphere:

$$^{14}_{7}N + {}^{1}_{0}n \longrightarrow {}^{14}_{6}C + {}^{1}_{1}H$$

For purposes of C-14 dating, we assume that the rate of $^{14}_{6}C$ production has remained constant over thousands of years, although it fluctuates slightly during the increased neutron bombardment of solar flares. The C-14 in CO_2 is incorporated into plant tissues during photosynthesis and into animal tissue when plants are consumed. As long as a plant or animal is alive, its C-14 content should match that in the atmosphere. After it dies, because C-14 is no longer being consumed, its C-14 content decreases through beta decay:

$$^{14}_{6}C \longrightarrow {}^{14}_{7}N + {}^{0}_{-1}\beta^{-}$$

Carbon-14 has a half-life of 5730 years. A measurement of the amount of C-14 in the carbon of dead plant or animal material can, therefore, disclose the time elapsed since the organism died. To correct for fluctuations in the intensity of cosmic rays, the method has been calibrated using measurements of the C-14 content of wood in individual tree rings of very old bristlecone pine trees.

Radiocarbon dating can be used to date many different substances, including wood, grains, cloth, bone, shells, peat, charcoal, organic mud, and carbonates. Measurement of the amount of carbon-14 allows dating of objects as old as about 50,000 years. Older objects cannot be dated because they contain too little C-14 to measure.

Animation: Half-Life
Online Learning Center, Animations

Chemistry Animations Library, Half-Life, Half_Life.rm.

Certain bristlecone pine trees were once thought to be the oldest living things on Earth, some of them dating back nearly 5000 years. Recently, however, experts have estimated the age of some desert creosote bushes at around 7000 years.

EXAMPLE 15.8   Radiocarbon Dating

An archaeologist unearths a wooden bowl. It has a carbon-14 content that is 6.25% of the amount in living wood. The half-life of carbon-14 is 5730 years. How old is the bowl?

Solution:

We need to determine the number of half-lives that have elapsed since the wood died. If we assign a value of 100% to the carbon-14 content of the living wood, then the following amounts would be present after successive half-lives, based on the concept that one-half of what is present at the beginning of a half-life period disappears during this time.

Half-lives	0	1	2	3	4	5
^{14}C content	100%	50%	25%	12.5%	6.25%	3.125%

The table shows that four half-lives must have elapsed since the wood died. Assuming that the bowl was made from a freshly cut tree, the age of the bowl must be 4 times the half-life of carbon-14:

$$\text{Age} = 4 \times 5730 \text{ years} = 22{,}920 \text{ years}$$

Practice Problem 15.8

A leather strap found in an archeological dig has a carbon-14 content that is 25% of what can be measured in living tissue. If the half-life of carbon-14 is 5730 years, how old is the leather?

Further Practice: 15.55 and 15.56

Geological Dating

Geologists date their samples in much the same way as archeologists, but using nuclides of longer half-lives than carbon-14 because the time spans are longer. One geological method uses the ratio of potassium-40 to argon-40 in minerals containing potassium. The half-life of potassium-40 is 1.27×10^9 years.

$$^{40}_{19}\text{K} + ^{\ 0}_{-1}\text{e}^- \longrightarrow ^{40}_{18}\text{Ar}$$

Because rocks may have started with different amounts of potassium-40, only the ratio is relevant to determining the date. After one half-life, half of the potassium-40 has been converted to argon-40, giving a ratio of 1:

$$\frac{^{40}\text{K}}{^{40}\text{Ar}} = \frac{50\%}{50\%} = 1$$

After two half-lives, 25% of the K-40 remains and 75% has been converted to Ar-40, giving a ratio of $\frac{1}{3}$:

$$\frac{^{40}\text{K}}{^{40}\text{Ar}} = \frac{25\%}{75\%} = 0.33$$

Similar calculations give ratios of 0.14 after three half-lives and 0.067 after four half-lives. Using this method some terrestrial rocks have been dated to ages as great as two billion (2×10^9) years. The K/Ar method has measured ages as great as 4.5 billion (4.5×10^9) years for meteorites. Techniques that employ other, long-lived radioactive elements yield age estimates of 2.5×10^9 to 3.0×10^9 years for the oldest terrestrial rocks and about 4.5×10^9 years for meteorites. Obtaining the same ages for similar rock samples using different nuclides verifies the accuracy of the methods.

15.4 MEDICAL APPLICATIONS OF ISOTOPES

As David learned when he had stomach pains, many diagnostic procedures employ various types of radiation or isotopes. Radioactive materials and techniques are also used in the treatment of disease and may keep life-saving devices running by providing a reliable power source.

Power Generators

Medical uses aren't the only applications for nuclide power packs. Power packs containing ^{242}Cm, ^{244}Cm, and ^{210}Po have been used to operate instruments in space vehicles and in remote sites, such as the polar regions on Earth.

How small can a power source be? A ^{238}Pu power pack small enough to run a heart pacemaker can be implanted surgically. In these power packs radioactive nuclides generate heat when they decay. This heat is converted to electricity. Such power packs are useful to operate instruments not readily accessible for replacement or maintenance. Such power packs last for a long time, so they don't have to be replaced often. They operate for many years at nearly constant output.

Medical Diagnoses

Many radioactive nuclides are used as tracers to track movements of substances in chemical or biological systems, in much the same way as small radio transmitters track migrating animals. These nuclides can be detected and measured by their characteristic radiation. Tracers find extensive use in medical studies. For example, technetium-99m (^{99m}Tc) emits gamma radiation and decays to ^{99}Tc with a half-life of 6.0 h. When ingested or injected in an appropriate chemical form, this nuclide helps doctors locate tumors in the spleen, liver, brain, skeleton, and thyroid.

The general approach is to use a tracer that concentrates in the tissues to be investigated. Cancerous cells, for example, are often characterized by unusually high rates of metabolism and cell replication, so certain tracers tend to concentrate in them. For example, thyroid cancers can be diagnosed by ingestion of radioactive ^{131}I, which decays to nonradioactive ^{131}Xe by beta and gamma emission with a half-life of 8 days. Scanning the distribution of radiation in the thyroid gland can reveal an abnormal growth (Figure 15.18). The scan for radioactivity creates a picture of the thyroid gland. A tumor shows up as an area of higher or lower radioactivity, depending on the nature of the tumor.

The radioactive labeling of portions of molecules can also help to determine how biological molecules react, either in living systems or in test tubes. For example, iron-55 and iron-59 have been used to determine how fast red blood cells are created in humans and how long they survive. Molecules labeled with phosphorus-32 or phosphorus-33 have been used to study the metabolism of these molecules. Nuclides of oxygen are used in PET studies to reveal which areas of the brain are most active during various emotional states or when performing intellectual tasks.

tumor

Figure 15.18
This scan reveals a tumor of the thyroid gland. In preparation for the scan, the patient ingested a solution of sodium iodide containing iodine-131. After 3 h, the patient was placed between photographic plates. The radiation emitted by the iodine-131 fogged the plates in proportion to the amount of radioactive iodine that had been absorbed. This tumor was diagnosed as benign, or noncancerous, since it did not concentrate the radioactivity and appeared as an area of low intensity in the scan. This conclusion was confirmed by a biopsy after surgery.

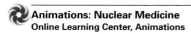

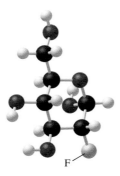

Figure 15.19

Body cells take up glucose and use it as an energy source. To trace the movement of glucose, one of the —OH groups can be replaced by ^{18}F. This molecule is enough like ordinary glucose that cells accept and use it in the normal way.

Positron Emission Tomography

As discussed in the introduction to this chapter, one of the newer medical diagnostic tools is positron emission tomography (PET). A PET scan, like the one David had, detects abnormalities in living tissues without disrupting the tissue. Some radioactive nuclides are incorporated into chemicals that are normally used by the tissues to be investigated. The isotopes used as positron emitters mostly for studies of the brain include ^{11}C (half-life of 20.4 min) and ^{15}O (2.05 min). Studies of blood flow in the heart muscle may employ ^{13}N with a half-life of 9.98 min, while the ^{18}F that David received for his gastrointestinal diagnosis has a half-life of 110 min. Note that all these isotopes have short half-lives, so they don't survive for very long inside the patient.

The radioactive tracer used for a PET scan depends on the diagnostic purpose and the patient's needs. It might be ^{11}C-labeled carbon dioxide or glucose; or it might be glucose in which ^{18}F is substituted for a hydroxyl group (Figure 15.19). Labeled CO_2 is administered to the patient by inhalation; labeled glucose is injected. The patient is then placed into a cylindrical gamma ray detector. A positron emitted within the patient collides with an electron in nearby tissue. Positrons react with electrons to emit two gamma rays, which are sensed by the gamma ray detector:

$$^{0}_{1}\beta^{+} + ^{0}_{-1}e^{-} \longrightarrow 2\,^{0}_{0}\gamma$$

The location of the positron-electron reaction can be determined by tracing back the paths of the two gamma rays to a point in common, just as a detective uses two bullet holes in a wall to determine where a shooter was standing. A computer reconstructs the location of many positron-electron reactions, thereby creating an image of the radioactive nuclide's distribution in the tissues (Figure 15.20).

After David had his PET scan, his doctors determined that he had a stomach ulcer. If you suspect that the stress of taking chemistry gave him an ulcer, you're wrong. The bacterium *Helicobacter pylori* causes ulcers, and David is on his way to a full recovery thanks to a course of antibiotics.

Cancer Therapy

David was fortunate. His PET scan revealed no cancer, but if a tumor had been found, radioactive nuclides—in much higher doses than those used for imaging—might have been used to treat it. Isotopes that emit high-energy gamma rays can destroy young cancer cells. Because most cancer cells have abnormally high growth rates, they preferentially absorb nutrients. If the nutrients contain a gamma-emitting component, the radioactivity becomes concentrated in the cancerous cells, destroying

Figure 15.20

A PET scan of the brain measures brain activity. The red areas are more active than the yellow areas. Black areas indicate no activity. The left image is a "slice" of a patient's brain under normal conditions. The right image is from the same patient during an epileptic seizure.

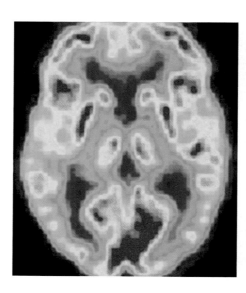

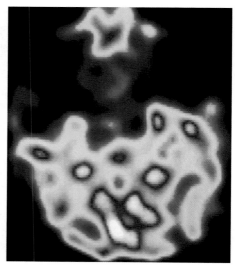

them in greater numbers than normal cells. For example, ^{131}I will destroy thyroid tumors. Gold, as ^{198}Au, has been used to treat lung cancer, and phosphorus as ^{32}P has been used for eye tumors.

15.5 BIOLOGICAL EFFECTS OF RADIATION

Radioactive isotopes can be used to diagnose and treat cancer. Unfortunately they can also cause cancer. Radiation damages living tissue in varying degrees, depending on the intensity and the length of exposure. A certain amount of radiation is natural in the environment. Granite contains radioactive nuclides, as does soil, water, food, and air. Brick and concrete used in construction also contain small amounts of radioactive nuclides. Some exposure to radiation also results from cosmic rays, with the body getting a higher than normal dose during high-altitude airplane flights. Some radioactive nuclides, such as the rare potassium-40, are contained in our bodies. Inhalation of radon gas that can accumulate in homes has received much attention as a cause of cancer.

Radiation Exposure

Radiation can have one of four effects on the functioning of a cell.

1. The radiation can pass through the cell with no damage.
2. The cell can absorb the radiation and be damaged, but it can subsequently repair the damage and resume normal functioning.
3. The cell can be damaged so severely that it cannot repair itself. New cells formed from this cell will be abnormal. This mutant cell can ultimately cause cancer if it continues to proliferate. Such damage can take years to create a tumor.
4. The cell can be so severely damaged that it dies.

The degree of damage depends, in part, on the ionizing ability of the radiation. Of particular interest for health and safety is the ability of radiation to ionize water, which is present in large quantities in plant and animal cells. Ionizing radiation of appropriate energy can strip an electron from a water molecule to form the ion H_2O^+. This ion reacts with another water molecule to form H_3O^+ and the OH (hydroxyl) radical, which is essentially a hydroxide ion that has lost an electron. The hydroxyl radical is very reactive. It readily reacts with many cell components, damaging them so that they do not function normally.

The extent of cell damage in a human depends on the energy of the radiation, its penetrating and ionizing ability, the chemical properties of the source, and the level of exposure. To track the danger, these factors are lumped together in a dose measurement called a *rem* (roentgen equivalent man). A rem is a measure of the danger arising to a human system from the absorption of radiation. Low levels of radiation are often measured in units of 10^{-3} rem, or mrem (millirem).

Normal exposure to radiation in an average area in the United States is about 200 mrem per year (Table 15.4). This amount of radiation has no observable effects on the average person. More important is the effect of dose levels for a single exposure to radiation that lasts for a few minutes to a few hours. A single dose of radiation of 0 to 25 rem causes no observable physical effects. However, radiation levels of 25 to 100 rem cause some temporary blood cell changes. Higher levels of 100 to 300 rem cause symptoms of radiation sickness, including nausea, diarrhea, fever, anemia, and achiness. Such exposure results in an increased risk of cancer. Doses of 400 to 600 rem cause severe radiation sickness, with a 50% chance of death within 30 days. Exposure to 1000 rem and higher levels results in a nearly 100% chance of death within 30 days. The longer someone is exposed to radiation, the greater the chance that cancer will result. It is because of these risks that such great care is taken in the design of nuclear power plants and in medical uses of radioactive nuclides.

The rem unit is named for Wilhelm Conrad Roentgen (1845–1923), the discoverer of X-rays.

TABLE 15.4	Sources of Average Exposure of U.S. Population to Radiation	
Source	**Dose (mrem/year)**
Cosmic rays	50
From the Earth	47
From building materials	3
In human tissue (as ^{40}K)	21
From the air	5
From cigarette smoking (due to ^{210}Po)	1300
Medical (X-rays, radiodiagnosis, radiotherapy)	61
Nuclear power	0.3
Radioactive fallout	4
Consumer products such as TV tubes and watch dials	0.04

Figure 15.21
The strawberries on the left are moldy after 2 weeks of cold storage, while those on the right, which were irradiated with gamma rays, are still fresh.

The destructive effects of radiation on plant and animal cells are being investigated—and, to a limited extent, used—to prevent the spoilage of foods, as shown in Figure 15.21. Upon exposure to appropriate levels of radiation, bacteria, molds, worms, and other parasites are killed without significant changes in the food itself. If food stored in sealed containers is exposed to radiation, the food does not need to be refrigerated or frozen to prevent spoilage. The radiation is usually gamma radiation from cobalt-60. Cobalt-60 is not introduced into the food. Instead the food is exposed only to the gamma radiation from the nuclide, so the food itself does not become radioactive. While the technique is generally considered safe, there is some concern that irradiation of food might alter or destroy nutrients such as amino acids and vitamins.

Radon

Although most concerns about air pollution involve outside air, indoor air pollution is also a health risk. But what could be in the air in our houses that could cause problems such as lung cancer? Second-hand cigarette smoke and carbon monoxide from a faulty gas burner are hazards. But problems arise even for non-smokers who have electric heat. Surprisingly it is a rare noble gas, radon, which has also been implicated as a possible cause of lung cancer. Radon is one of the intermediate products of the radioactive decay of uranium and thorium. It is rare because all its isotopes are radioactive and decay to other elements in relatively short times. The nuclide with the longest half-life, 3.8 days, is ^{222}Rn. It decays further to polonium-218:

$$^{222}_{86}\text{Rn} \longrightarrow {}^{218}_{84}\text{Po} + {}^{4}_{2}\alpha$$

The radon gas that accumulates in houses comes from particular kinds of soil or rock strata. Small amounts of uranium are widely present in soil from granite and shale, and thorium is present in granite and gneiss, a common rock consisting of alternating dark and light bands of material. Breathing radon is no more harmful than breathing any other noble gas, unless a radon atom undergoes radioactive decay while in the lungs. The polonium produced and all its decay products (see Table 15.3) are solids that can attach to lung tissue. When these solids undergo decay, the emitted radiation can cause cell damage.

EXAMPLE 15.9   Radon Decay

If radon-222 decays in the lungs today, what nuclides will still be present tomorrow? Assume that a nuclide is virtually gone after 10 half-lives.

Solution:

Because Rn-222 has a half-life of 3.82 days (Table 15.3), far less than half of it will decay during the first day, assuming it is not exhaled. When Rn-222 decays in the lungs, it forms Po-218, which has a half-life of 3.11 min. After half an hour (about 10 half-lives), most of the Po-218 is gone, forming Pb-214. This nuclide has a half-life of 26.8 min, so it is mostly gone after about 5 h. It decays to form Bi-214, with a half-life of 19.7 min. Most of the Bi-214 is gone after about 3 h, forming Po-214, which has a half-life of 1.6×10^{-4} s. This lasts only a fraction of a second, forming Pb-210, which has a half-life of 22.3 years. This nuclide will stay in the lungs for a considerable period of time, slowly releasing beta particles.

Practice Problem 15.9

If radon-222 decays in the lungs, what types of radiation would be emitted during the first day?

Further Practice: 15.69 and 15.70

The amount of biological damage radon causes depends on how much is in a house and how long its inhabitants are exposed. Several surveys have established a national average of the amount of radon in houses, but values vary over wide ranges. Some houses have as much as 2000 times the average. Even in the average house, radon was found to make a significant contribution to the average radiation dosage experienced annually by residents of the United States. Estimates suggest that people living in homes with radon levels six times the national average experience a 2% increase in lung cancer risk. About 10,000 lung cancer deaths annually can be attributed to radon.

The radon comes from the soil, but how does it get into our houses? No one is certain, but radon has become a much greater problem since we started building airtight houses to save on energy used for heating and air conditioning. Although indoor air is exchanged with outside air every two or three hours in new homes, the exchange is less frequent than in older homes, and radon concentrations are higher as a result.

What can be done about radon in homes? First, residents can determine if the radon level in their home is unusually high. Kits to measure radon levels are commercially available. They consist either of a special piece of plastic or a canister of charcoal. The plastic is damaged by passage of alpha particles released during radon decay, so microscopic examination reveals the number of penetrations that occurred during a set time period. The charcoal adsorbs radon atoms. In both cases, the kits must be returned to a laboratory for evaluation. In some homes unusually high radon levels have been reduced by taking simple steps, such as placing blowers under the house to disperse a buildup of radon.

15.6 NUCLEAR ENERGY

Some heavy radioactive nuclides are so unstable that they decay in two ways: alpha emission and a spontaneous splitting of the nucleus that releases large amounts of energy. **Fission** is the splitting of a heavy nucleus into two or more lighter nuclei and some number of neutrons. Some heavy nuclides will not undergo fission

Figure 15.22
Bombardment with a neutron initiates the fission of a uranium-235 nucleus. The product is unstable uranium-236, which splits apart into lighter nuclides. Some neutrons are also emitted. They can cause other uranium-235 nuclei to undergo fission.

spontaneously, but they will if irradiated or bombarded with appropriate particles, often neutrons. A particularly important fissionable nuclide is uranium-235, which is used extensively in nuclear power plants.

Uranium-235 Fission

The fission of uranium-235 is depicted in Figure 15.22. When a slow neutron—one having a low kinetic energy—strikes a U-235 nucleus, the uranium absorbs the neutron and becomes U-236, which is even less stable than U-235. The U-236 nuclide then splits into two lighter nuclei and emits neutrons and gamma radiation. Many different fission reactions may occur when U-235 is bombarded with neutrons. Over 200 different fission products of this reaction have been identified in varying amounts. Because these large nuclides contain so many neutrons and protons, there are many different ways in which the nuclei can split. As the following examples show, these fission reactions differ only in the manner in which the very unstable U-236 splits or breaks up:

$$^{235}_{92}U + ^{1}_{0}n \longrightarrow ^{92}_{36}Kr + ^{141}_{56}Ba + 3\,^{1}_{0}n$$

$$^{235}_{92}U + ^{1}_{0}n \longrightarrow ^{90}_{38}Sr + ^{143}_{54}Xe + 3\,^{1}_{0}n$$

$$^{235}_{92}U + ^{1}_{0}n \longrightarrow ^{94}_{40}Zr + ^{140}_{58}Ce + 2\,^{1}_{0}n + 6\,^{0}_{-1}\beta^{-}$$

EXAMPLE 15.10   Fission Products

When bombarded by a neutron, uranium-235 undergoes fission, emitting the nuclides antimony-133 and niobium-100. How many neutrons are emitted? Write a balanced nuclear equation to describe this fission process.

Solution:

From the periodic table we can find the atomic numbers of the two products, 51 and 41, respectively:

$$^{235}_{92}\text{U} + ^{1}_{0}\text{n} \longrightarrow ^{133}_{51}\text{Sb} + ^{100}_{41}\text{Nb} + x^{1}_{0}\text{n}$$

We can see that the charges balance, so we can consider the mass balance condition to determine the number of neutrons emitted:

$$235 + 1 = 133 + 100 + x$$

$$x = 235 + 1 - 133 - 100 = 3$$

Thus, three neutrons are emitted, and the balanced equation is

$$^{235}_{92}\text{U} + ^{1}_{0}\text{n} \longrightarrow ^{133}_{51}\text{Sb} + ^{100}_{41}\text{Nb} + 3^{1}_{0}\text{n}$$

Practice Problem 15.10

When bombarded by a neutron, uranium-235 undergoes fission, emitting the nuclides tellurium-137 and zirconium-97. How many neutrons are emitted? Write a balanced nuclear equation to describe this fission process.

Further Practice: 15.73 and 15.74

Chain Reactions

The fission of U-235 can be started by a single slow neutron, but all the resulting nuclear decay reactions produce more neutrons than are consumed. For the overall process, an average of 2.4 neutrons are produced for each neutron used. If all the product neutrons initiate more fission steps, the process continues until all the fissionable material has decayed. Such a process is called a **chain reaction.** It is a reaction in which a product of one step is a reactant in another step.

If we want to use such a reaction as an energy source, the reaction must give off energy continuously. Several conditions must be met for a chain reaction to sustain itself. The neutrons produced by the fission reaction must be of appropriate energy to cause other nuclei to undergo fission. Generally the energy of the neutrons produced by U-235 fission is higher than is optimum for inducing further fission. The neutrons can lose energy in collisions, but if they escape from the uranium sample before they have slowed down sufficiently, they cannot induce fission, because they will not stay in the vicinity of a uranium nucleus long enough to be absorbed. In order for the chain reaction to sustain itself, then, the amount and shape of the sample of fissionable material must be such that the neutrons will not escape. The smallest amount of fissionable material necessary to support a continuing chain reaction is called the **critical mass.**

Another important factor is the rate at which the fission progresses. Ideally, to sustain itself, the process should maintain a constant rate, becoming neither faster nor slower. If other nuclei absorb too few of the emitted neutrons, the chain reaction slows and eventually stops. If other nuclei absorb too many of the emitted neutrons, the process accelerates and culminates in an explosion. This is the principle behind nuclear weapons, which consist of several pieces of uranium. Each is less than the critical mass, so it cannot sustain a chain reaction. But the detonation of a chemical explosion forces the pieces together to form a critical mass, thereby activating an explosion.

Fission Reactors

Fission can be used for peaceful, as well as wartime, purposes. Nuclear power plants use fission to produce electric energy. Such plants are designed to maintain

Animation: Chain Reaction
Online Learning Center, Animations

Chemistry Animations Library, Chain Reaction, Chain_Reaction.avi.

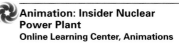

Animation: Insider Nuclear Power Plant
Online Learning Center, Animations

Chemistry Animations Library, Insider Nuclear Power Plant, Nuclear_Power.rm.

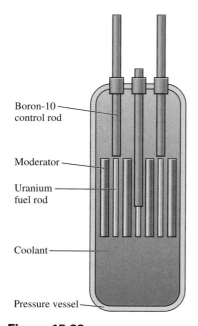

Figure 15.23
Reactor core in a fission reactor.

careful control over the rate of the chain reaction, using neutron absorbers such as cadmium and boron. If the chain reaction is going too quickly, movable *control rods* made of these elements are inserted into a core of uranium fuel in fission reactors (Figure 15.23). The material in the rods reacts with neutrons, preventing them from participating in the chain reaction, as this equation for boron shows:

$$^{10}_{5}B + ^{1}_{0}n \longrightarrow ^{7}_{3}Li + ^{4}_{2}\alpha$$

Although a nuclear weapon needs uranium metal, the uranium fission reactor does not. Normally the uranium fuel is made from the oxide U_3O_8, which is enriched in U-235 from its natural abundance of 0.7% to about 2% or 3%. The neutrons produced by U-235 fission in this fuel are usually too energetic and pass out of the system unchanged, so they do not produce additional fission steps efficiently. To generate power, the neutrons must be slowed down. A *moderator* such as normal water (H_2O), heavy water (D_2O), or graphite is included in the reactor design for this purpose (Figure 15.23).

The fission of U-235 produces about 2×10^{10} kJ/mol of U-235, an amount that is enormous in comparison to the 390 kJ/mol released when burning coal. This energy is released in the form of kinetic energy of the reaction products. This is not usable energy, so it must be converted to another form. A coolant, usually water, circulates around the reactor core (Figure 15.24). The water gets as hot as 310°C, so it is pressurized to prevent boiling. This very hot water, called the primary coolant, then passes through a steam generator. Steam generators are massive structures containing thousands of thin-walled tubes for heat transfer. The primary coolant is pumped through the tubes, and another source of water, called the secondary coolant, is pumped around the tubes and converted to steam. The steam is used to drive a turbine, which operates a generator, producing electricity. The steam must be condensed and cooled so that it can be pumped back through the steam generator. This is often accomplished by circulating a third coolant through a condenser and a cooling tower. Heat is removed from the secondary coolant in the condenser and then released to the environment from massive cooling towers (Figure 15.25). We often recognize a nuclear power plant by the presence of these cooling towers. Because of this design, with three different coolants, water that circulates around the reactor core never leaves the containment building, so there is little danger of the release of radioactive contaminants into the environment.

Figure 15.24
The heat exchange system in a nuclear fission reactor is designed to prevent the loss of any radioactive materials. Heat is transferred to a secondary loop, where it generates steam to operate a turbine and produce electricity.

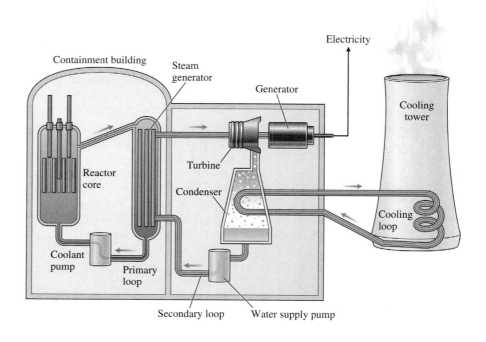

For protection in case of leaks, the reactor core is located in a thick-walled containment building, often built in a dome shape (Figure 15.25). In the event of a reactor leak, the radioactive material would, in principle, be kept within the containment building, and other areas would not become contaminated. The nuclear disaster that occurred in 1986 at the nuclear power plant in Chernobyl, Russia, released a considerable amount of radioactive nuclides into the surrounding environment, some of which spread throughout the world. The Chernobyl reactor differed from U.S. nuclear reactor designs in that it did not include a containment building.

In most cases the immediate products of U-235 fission are themselves radioactive. Once the fission fuel has been used up, a complex mixture of radioactive substances remains. This radioactive waste, as well as other contaminated materials, must be stored so that it can become inactive through decay without causing damage to living things. The best method for disposing of radioactive waste has not yet been determined, but controversy rages and many methods are under investigation. They include reprocessing the fuel and storing it in liquid or solid form in steel tanks or underground.

Figure 15.25
The dome-shaped structure is the containment building. The smoking tower is the cooling tower; the "smoke" is actually condensed water vapor.

Fusion Reactors

Another kind of nuclear reaction that releases large amounts of energy is **fusion,** the combination of light nuclei to form heavier nuclei. A major fusion reaction occurs continuously in the Sun and other stars:

$$4\,^1_1\text{H} \longrightarrow\, ^4_2\text{He} + 2\,^0_1\beta^+$$

This process actually occurs in several steps:

$$^1_1\text{H} + ^1_1\text{H} \longrightarrow\, ^2_1\text{H} + ^0_1\beta^+$$

$$^1_1\text{H} + ^2_1\text{H} \longrightarrow\, ^3_2\text{He}$$

$$^1_1\text{H} + ^3_2\text{He} \longrightarrow\, ^4_2\text{He} + ^0_1\beta^+$$

This fusion process releases about 2.45×10^9 kJ/mol of helium. As the nuclei get close enough together to interact, strong internuclear repulsions occur. To overcome them, considerable energy must be supplied to fuse the nuclei. In the Sun, extremely high temperatures in excess of 10^7 K provide the energy. In a fusion bomb, the initial explosion of a fission bomb provides the high temperatures and pressures needed to initiate fusion. The temperature required to initiate a fusion reaction is called the *ignition temperature.*

Although fusion has never been harnessed for the production of electricity, some researchers are hopeful, and fusion reactors could potentially offer a number of advantages. Fission reactors suffer from a limited supply of fissionable material and the necessity of storing radioactive wastes. Fusion reactors would avoid these shortcomings. They could operate on hydrogen taken from water—a virtually unlimited fuel supply—and create no hazardous by-products.

A variety of fusion reactions could be used in a commercial fusion reactor. The reaction between deuterium and tritium is a likely candidate:

$$^2_1\text{H} + ^3_1\text{H} \longrightarrow\, ^4_2\text{He} + ^1_0\text{n}$$

This reaction produces 1.7×10^9 kJ/mol, over 1700 times as much energy as is needed to get fusion to take place. Enough deuterium is present in ordinary water to last for millions of years. The technology for inexpensively separating deuterium from water has already been developed. Eight gallons of seawater contains about 1 g of deuterium, which can release energy equivalent to burning about 2500 gal of gasoline. Tritium, a radioactive isotope of hydrogen, has a half-life of 12.3 years and emits beta particles. While tritium occurs naturally to some extent, it is not present in nature in sufficient quantity to be used as a fuel. However, tritium can be prepared by neutron bombardment of lithium, which is available in sufficient quantities to provide at least a 1000-year supply.

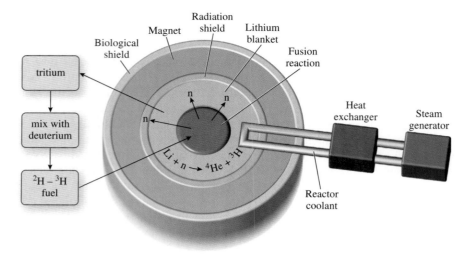

Figure 15.26
Schematic layout of a fusion reactor.

Figure 15.27
A tokamak uses a magnetic field to contain the deuterium-tritium gaseous fuel for a fusion reaction.

In many fusion reactor designs, the reaction chamber would be surrounded by a lithium "blanket," which would transfer the heat from the fusion reaction to a reactor coolant (Figure 15.26). Heat would then be transferred to a steam generator, as in fission reactors. The lithium would also absorb excess neutrons formed in the fusion process, thereby producing more tritium:

$$^{7}_{3}\text{Li} + ^{1}_{0}\text{n} \longrightarrow ^{3}_{1}\text{H} + ^{4}_{2}\text{He} + ^{1}_{0}\text{n}$$

$$^{6}_{3}\text{Li} + ^{1}_{0}\text{n} \longrightarrow ^{3}_{1}\text{H} + ^{4}_{2}\text{He}$$

In fact it is possible to create more tritium than is used, and the excess tritium could be used to fuel other fusion reactors. A reactor that produces fuel that can be used in other reactors is called a *breeder reactor.*

The major problems to be solved before fusion can be used as an energy source are the creation and control of *plasma,* an ionized gas, at temperatures of about 10^8 K. In order to achieve fusion, the gaseous reactants must be condensed to a small volume at high temperatures. This is the purpose of the plasma, which must be kept at a very high temperature for longer than 1 s to initiate the fusion reaction. At this high temperature, however, the plasma will melt any normal container material. Therefore most research has concentrated on containing the plasma with a magnetic field, using doughnut-shaped machines called *tokamaks* (Figure 15.27). A second method confines the fuel in tiny glass pellets that are heated to the required temperatures with laser beams. It is only recently that anyone has been able to get more energy out of fusion than had to be put in, so nuclear fusion is unlikely to prove itself a viable commercial energy source for many years.

EXAMPLE 15.11   Fusion Reactions

What new element would form if two carbon-12 nuclei undergo fusion with no emission of particles?

Solution:

Assuming no particles are emitted by the fused nuclei, adding the mass numbers and atomic numbers provides the answer. The atomic number of carbon is 6, so the fusion of two carbon nuclei should give an atom of magnesium, with an atomic number of 12. The mass number of the carbon nuclide is 12, so when two nuclei fuse, the mass number of the product will be 24. The new nuclide is magnesium-24. The fusion reaction can be represented by the following equation:

$$^{12}_{6}\text{C} + ^{12}_{6}\text{C} \longrightarrow ^{24}_{12}\text{Mg}$$

Practice Problem 15.11

What new element would form if two oxygen-16 nuclei undergo fusion with no emission of particles?

Further Practice: 15.85 and 15.86

SUMMARY

Most nuclides of the elements are radioactive. They decompose to other elements and emit radiation through nuclear reactions. Nuclear reactions are represented by equations similar to chemical equations. They are balanced when both the mass number and the nuclear charge are conserved.

The 250 stable nuclides are all found within a narrow band of stability in which the nuclear composition has nearly equal numbers of protons and neutrons. The heavier radioactive elements decay by combinations of alpha and beta decay processes to form nuclides with lower mass and lower atomic number, eventually reaching the band of stability. The lighter radioactive nuclides decay by several processes: beta (electron) decay, beta (positron) decay, or electron capture. Nuclear bombardment reactions are induced nuclear reactions in which nuclides are bombarded with other particles to create new nuclides.

Nuclear decay reactions have characteristic half-lives, the time required for half of a sample of nuclide to decay to another nuclide. Half-lives and isotope ratios can be used to measure the age of objects of archeological or geological interest.

Radioactive isotopes and their nuclear reactions are used for portable power packs, radioactive tracers, medical imaging, and cancer therapy. However, exposure to too much radiation can cause tissue damage in living things.

Fission is a nuclear reaction in which a nuclide splits to form two other nuclides and, usually, small particles such as neutrons. It is the principle underlying atomic bombs and nuclear power plants. Fusion, the combining of light nuclides to form heavier nuclides, is a potential source of electric energy that has not yet been harnessed.

KEY TERMS

alpha particle (15.1)

beta particle (15.1)

chain reaction (15.6)

critical mass (15.6)

electron capture (15.2)

fission (15.6)

fusion (15.6)

gamma ray (15.1)

half-life (15.3)

nuclear bombardment (15.2)

nucleon (15.1)

nuclide (15.1)

positron 15.1)

radiation (15.1)

radioactive (15.1)

radiocarbon dating (15.3)

QUESTIONS AND PROBLEMS

The following questions and problems, except for those in the *Additional Questions* section, are paired. Questions in a pair focus on the same concept. Answers to the odd-numbered questions and problems are in Appendix D.

Matching Definitions with Key Terms

15.1   Match the key terms with the descriptions provided.
   (a) the smallest amount of fissionable material necessary to support a continuing chain reaction
   (b) the splitting of a heavy atom into two or more lighter atoms and some number of neutrons with the release of energy
   (c) a nuclide containing two protons and two neutrons; the nucleus of a helium-4 atom
   (d) having the ability to emit ionizing radiation spontaneously
   (e) the highest-energy and most-penetrating type of electromagnetic radiation
   (f) a nuclear reaction that is induced by the impact of a nuclide or a subatomic particle on another nuclide
   (g) a nuclear particle (neutron or proton)
   (h) emission of particles or electromagnetic waves

15.2   Match the key terms with the descriptions provided.
   (a) a technique for the establishment of the age of a carbon-containing object by determination of the amount of C-14 present in the object
   (b) a particle having the same mass as an electron but the opposite electrical charge
   (c) incorporation of a 1s electron into the nucleus, resulting in the conversion of a proton into a neutron
   (d) a small, charged particle that can be emitted from atoms at speeds approaching the speed of light; either an electron or a positron
   (e) a reaction in which the product of one step is a reactant in another step
   (f) the combination of light nuclei to form heavier nuclei
   (g) an atom having a specified atomic number, mass number, and nuclear energy
   (h) the time it takes for half of a sample of a nuclide to decay to a different nuclide

Radioactivity

15.3   What is radioactivity?
15.4   Distinguish radioactivity from radiation.
15.5   What are the different types of radiation resulting from radioactivity?
15.6   What happens during nuclear reactions in addition to the emission of radiation?
15.7   Which of the first 90 elements have only radioactive isotopes?
15.8   Which of the elements with atomic numbers greater than 90 are radioactive?
15.9   A sample of a radioactive material emits gamma radiation. What material would you use to enclose the material to prevent the radiation from escaping?

15.10   A sample of a radioactive material emits beta radiation. What material would you use to enclose the material to prevent the radiation from escaping?

Nuclear Reactions

15.11   What features do nuclear decay reactions and nuclear bombardment reactions have in common?
15.12   How do nuclear decay reactions differ from bombardment reactions?
15.13   What are the different types of particles or rays that may result from nuclear decay?
15.14   Which type of radioactivity does not involve the emission of particles?
15.15   Describe the conservation rules that are used to balance nuclear reaction equations.
15.16   What quantity do we examine to determine whether nuclear charge is conserved in a balanced nuclear reaction?
15.17   Describe a nuclear bombardment reaction.
15.18   If we know the starting nuclide and the products of a nuclear bombardment reaction, how do we determine what was used to bombard the starting nuclide?
15.19   What product results from alpha emission by $^{226}_{88}$Ra?
15.20   Write an equation for the alpha decay of $^{228}_{90}$Th.
15.21   What product results from beta (electron) emission by $^{99}_{42}$Mo?
15.22   What is the product of beta (electron) decay of $^{234}_{90}$Th?
15.23   Compounds containing $^{15}_{8}$O are used for PET scans. What nuclide is formed when $^{15}_{8}$O emits a positron?
15.24   A nuclide used in PET scans forms $^{13}_{6}$C when it emits a positron. What is this nuclide?
15.25   Bombardment of $^{250}_{98}$Cf by $^{11}_{5}$B yields four neutrons. What is the other product?
15.26   What isotope is used to bombard $^{97}_{42}$Mo if the products are $^{97}_{43}$Tc and two neutrons?
15.27   Identify the particle emitted by a nucleus that undergoes the following transformation.

Neutron   Proton

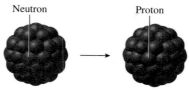

15.28   Identify the particle emitted by a nucleus that undergoes this transformation.

Proton   Neutron

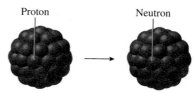

15.29 What nuclide will produce $^7_3$Li by electron capture?

15.30 What product will result from electron capture by $^{44}_{22}$Ti?

15.31 Complete and balance the following equations.

(a) $^{238}_{92}$U $\longrightarrow$ $^{234}_{90}$Th + ?

(b) $^{234}_{90}$Th $\longrightarrow$ $^{234}_{91}$Pa + ?

(c) $^{234}_{92}$U $\longrightarrow$ $^4_2$He + ?

(d) $^{10}_5$B + $^2_1$H $\longrightarrow$? + $^1_0$n

(e) $^{210}_{83}$Bi + $^0_{-1}$e$^-$ $\longrightarrow$?

15.32 Complete and balance the following equations.

(a) $^7_3$Li + ? $\longrightarrow$ 2$^4_2$He

(b) $^{214}_{82}$Pb $\longrightarrow$ $^{214}_{83}$Bi + ?

(c) $^{27}_{13}$Al + $^1_0$n $\longrightarrow$? + $^4_2$He

(d) $^{63}_{29}$Cu + ? $\longrightarrow$ $^{64}_{29}$Cu + $^1_1$H

(e) $^{12}_6$C + ? $\longrightarrow$ $^{13}_7$N

15.33 Write balanced equations for the following nuclear reactions.

(a) alpha decay by $^{227}_{87}$Fr

(b) positron emission by $^{18}_9$F

(c) bombardment of $^{96}_{42}$Mo with protons, resulting in neutron emission

15.34 Write balanced equations for the following nuclear reactions.

(a) beta (electron) emission by $^{114}_{47}$Ag

(b) bombardment of a nuclide with deuterium atoms, producing $^{64}_{30}$Zn and a neutron

(c) electron capture by $^{133}_{56}$Ba

15.35 What is the N/Z ratio for stable isotopes for the first 20 elements?

15.36 What is the N/Z ratio for elements 80 to 90?

15.37 What is the band of stability?

15.38 How is the band of stability used to predict the type of radioactive decay a given nuclide might undergo?

15.39 Which of the following isotopes should be the most stable? Explain your answer.

$$^{16}_8O \quad ^{17}_8O \quad ^{18}_8O$$

15.40 Which of the following isotopes should be the most stable? Explain your answer.

$$^{28}_{14}Si \quad ^{29}_{14}Si \quad ^{30}_{14}Si$$

15.41 Predict the type of nuclear decay expected for the following unstable nuclides.

(a) $^{14}_6$C

(b) $^{234}_{90}$Th

(c) $^{234}_{92}$U

(d) $^{15}_8$O

(e) $^{16}_7$N

15.42 Predict the type of nuclear decay expected for the following unstable nuclides.

(a) $^{19}_8$O

(b) $^{230}_{91}$Pa

(c) $^{10}_6$C

(d) $^{13}_7$N

(e) $^{244}_{94}$Pu

15.43 When $^{234}_{91}$Pa decays, it emits five alpha particles and two beta particles in seven steps. What is the final product of this series of decay steps?

15.44 When $^{226}_{88}$Ra decays, it emits four alpha particles and two beta particles in six steps. What is the final product of this series of decay steps?

Rates of Radioactive Decay

15.45 What is the half-life of a radioactive nuclide?

15.46 What uses are made of the half-life of a radioactive nuclide?

15.47 How is radiation detected?

15.48 How are the half-lives of radioactive nuclides measured?

15.49 The half-life of $^{24}_{11}$Na is 15.0 h. What fraction of the nuclide will remain after 60.0 h?

15.50 The half-life of tritium, $^3_1$H, is 12.3 years. How much of a 100.0-g sample of tritium will be left after a period of 37 years?

15.51 To investigate metabolic pathways, a laboratory rat is injected with a sample containing phosphorus-32, which has a half-life of 14 days. Assuming none of the $^{32}_{15}$P is excreted, what fraction of the $^{32}_{15}$P would be expected to remain in the rat after 4 weeks?

15.52 The half-life of $^{35}_{16}$S is 86.7 days. How long does it take for the radiation intensity to decrease by 75%?

15.53 The half-life of the nuclide $^{211}_{85}$At is 7.5 h. If 0.100 mg of this nuclide is administered for thyroid treatment, how long does it take to reduce the nuclide to 0.0125 mg?

15.54 The half-life of $^{35}_{16}$S is 86.7 days. If 80.0 mg is absorbed by an orange, how long will it take to reduce this radioactive nuclide to 5.0 mg?

15.55 Barley and wheat grains recovered from an ancient Egyptian burial chamber were analyzed for their carbon-14 content. They contained an amount of $^{14}_6$C equal to 50% of that present in living plants. If the half-life of $^{14}_6$C is 5730 years, how old are these grains likely to be?

15.56 A precursor of modern-day bison is *Bison latifrons*. A fossil found in North Dakota was dated with the radiocarbon method. It contained 0.39% as much $^{14}_6$C as modern bison. How old is the fossil likely to be?

15.57   Technetium-99*m* is used for a number of diagnostic purposes in medicine. For example, ^{99m}Tc introduced into the bloodstream is not normally absorbed by brain cells, but it is absorbed by a tumor. A brain scan can locate the tumor. If a sample of ^{99m}Tc is introduced into a body in this way, and 6.25% of its radioactivity remains after 24 h, what is the half-life of ^{99m}Tc?

15.58   A sample of a radioactive nuclide has an activity of 3500 cpm. If the activity has decreased to 1750 cpm 45.0 min later, what is the half-life of the nuclide?

Medical Applications of Isotopes

15.59   Why are pacemakers powered with a $^{238}_{94}$Pu power device rather than a zinc-silver oxide electrochemical cell?

15.60   Explain how radioactive nuclides can be used to study the metabolism of a sugar substitute.

15.61   Why is ^{99m}Tc used for medical diagnosis?

15.62   Why is ^{131}I used for diagnosing thyroid tumors?

15.63   Why is $^{11}_{6}$C, but not $^{14}_{6}$C, used for PET imaging?

15.64   Describe how $^{18}_{9}$F might be used for PET imaging.

Biological Effects of Radiation

15.65   Describe four different effects radiation might have on a living biological cell.

15.66   We are exposed to radiation from many sources. What sources can we control in our daily lives?

15.67   The risk to the general population from radon is about 100 to 1000 times higher than the risk from carbon tetrachloride or benzene, which have been tightly controlled by the U. S. Environmental Protection Agency. Should the EPA generate regulations for radon in houses?

15.68   Radon is estimated to cause 10,000 lung cancer deaths per year, compared to about 110,000 attributed to smoking tobacco. Why is tobacco not more heavily regulated?

15.69   If Po-214 were ingested, what nuclides would be present in appreciable amounts later that day?

15.70   If Po-214 were ingested, what types of radiation would be emitted during the first second after ingestion?

Nuclear Energy

15.71   Distinguish between fission and fusion.

15.72   Write balanced nuclear equations for examples of fission and fusion.

15.73   When bombarded by a neutron, U-235 undergoes fission. If the resulting nuclides are Kr-93 and Ba-140, how many neutrons are emitted?

15.74   When bombarded by a neutron, U-235 undergoes fission. If the resulting nuclides are Sr-91 and Xe-142, how many neutrons are emitted?

15.75   Describe the features of a chain reaction.

15.76   Explain why critical mass is of concern in the design of a fission reactor.

15.77   What are the main components of a fission reactor?

15.78   What is the composition and function of a moderator in a nuclear fission reactor? A control rod?

15.79   Why is the heat from fission reactions not used directly to create steam to operate a turbine?

15.80   What safety features are included in the design of most fission reactors?

15.81   What fuels are most likely to be used as a power source in fusion reactors?

15.82   Discuss the problems that must be overcome before nuclear fusion can become a useful energy source for peaceful purposes.

15.83   Write a balanced nuclear equation for a reaction that might be used in a fusion reactor.

15.84   Write a balanced nuclear equation for a reaction that might occur in a fission reactor.

15.85   What nuclide forms when $^{1}_{1}$H undergoes fusion with $^{2}_{1}$H, assuming only one particle forms?

15.86   What nuclide forms when $^{3}_{1}$H undergoes fusion with $^{4}_{2}$He, assuming only one particle forms?

Additional Questions

15.87   Describe the process of radioactive dating.

15.88   What product will result from neutron capture by $^{238}_{92}$U?

15.89   An old piece of wood has a carbon-14 content of 0.0156 times that of a freshly cut piece of wood. What is the age of the old wood if the half-life of $^{14}_{6}$C is 5730 years?

15.90   The ratio of ^{40}K to ^{40}Ar was used to date moon rocks returned by the Apollo 11 flight. Potassium-40 decays to argon-40 by positron emission with a half-life of 1.3 billion years. Assuming no ^{40}Ar was present initially and that none escaped, how old is a rock that had a K-40 to Ar-40 ratio of 0.143?

15.91   Outline the nuclear reactions that are the primary source of energy in the stars.

15.92   Predict the type of radioactive decay expected for the following unstable nuclides.
(a) $^{17}_{9}$F
(b) $^{21}_{9}$F
(c) $^{216}_{84}$Po
(d) $^{19}_{10}$Ne
(e) $^{15}_{6}$C

15.93   The bombardment of beryllium-9 with alpha particles yields carbon-12. Predict any other products of this reaction and write a balanced equation to describe it.

15.94   What product will result from positron emission by $^{15}_{8}O$?

15.95   Polonium-210 is an alpha-particle emitter with a half-life of 138 days. Static-eliminator brushes for cleaning dust from camera lenses contain a small amount of ^{210}Po at the base of the bristles. Suppose you are in charge of maintaining inventory at a camera store and you are offered a very good buy on a three-year supply of brushes. Should you maintain a stock this size? Consider how much ^{210}Po is left at the end of each year during this period to help you decide.

15.96   Although potassium-40 decays to produce calcium-40 with a half-life of 1400 million years, rocks cannot be dated by measuring the ratio of ^{40}K to ^{40}Ca. Why?

15.97   Neutron activation analysis is a nondestructive method for determining the elements present in materials. The material is bombarded with neutrons, forming radioactive nuclides with decay energies that can be measured. It is often used to verify the authenticity of paintings by matching the composition of paint with that from other paintings by the same artist. Comment on the statement that a painting found to contain chloride could not have been painted in AD 1648, as claimed, since the element chlorine was not discovered until 1774.

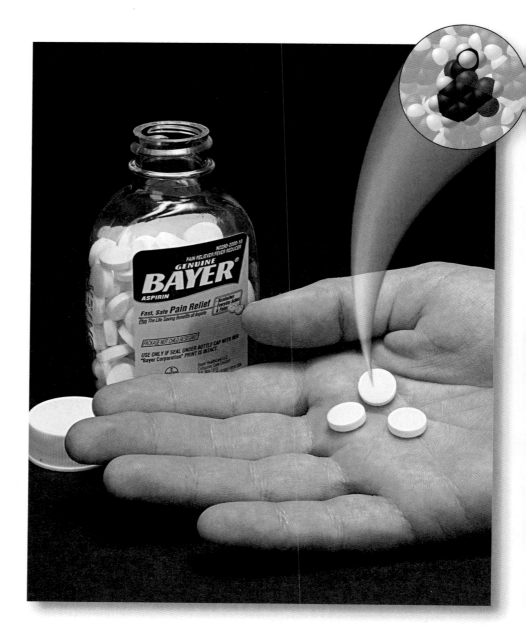

When Scott and Jennifer arrive early for their chemistry class, they find an assignment written on the board: "Identify a molecule containing at least eight atoms that is important to the career you are working toward. Describe how the molecule relates to your career. Draw the molecule's structure and explain how it determines the properties that make this molecule important." Along with other early arrivals, Scott and Jennifer settle into their chairs and begin to discuss their career plans and the molecules important in their chosen fields.

Scott, an environmental science major, is interested in fuel-efficient engines, so he thinks octane should be his choice. He knows that octane is a standard for rating grades of gasoline, but it isn't until he checks his textbook that he learns that octane, C_8H_{18}, has several possible structures, including

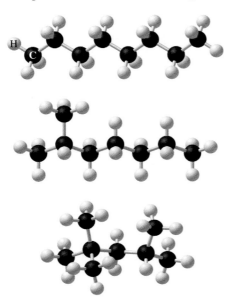

As Scott studies the pictures, he realizes that the octane structures differ only in the way the carbon atoms attach to one another. Because of the large number of atoms, Scott learns that octane molecules have London dispersion forces strong enough to make them liquids at room temperature. (The respective boiling points of the molecules shown are 125.6°C, 117.6°C, and 99.3°C.) Scott also finds that only one of the structures—the last molecule shown—serves as the standard for the octane rating of gasoline. It burns more smoothly than the other molecules.

These octane compounds all have the same formula and the same mass. Why would they have different boiling points?

Jennifer hopes to develop a career in dentistry, so she thinks about mouthwashes and a common ingredient in them: menthol, $C_{10}H_{20}O$. The menthol molecule, she finds, has the following structure:

What group is present in menthol that is not found in octane?

Menthol is also used in local anesthetics such as cough drops, ointments for muscle aches, and anti-itch creams. It is used in aftershave lotions and skin fresheners as well. In all these applications, menthol gives the sensation of cooling. Jennifer guesses that receptors on cell membranes in the mouth, throat, and skin might accept the menthol molecule and generate nerve impulses as a result. If this were true, then the size, shape, and polarity of the menthol molecule would have to be just right for it to interact with a receptor, she reasons. Further reading proves she's right. Although the mechanisms aren't totally understood, research evidence suggests that menthol tricks the temperature receptors into signaling the brain that something cold has come into contact with them.

Omar plans to become a pharmacist. He picks aspirin, $C_9H_8O_4$, as his molecule. He finds the following structure for aspirin, also called acetylsalicylic acid:

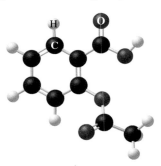

Aspirin blocks the action of an enzyme called cyclooxygenase, Omar learns. The enzyme catalyzes the production and release of prostaglandins. Prostaglandins form whenever tissue is injured, and they send a nerve signal that the brain interprets as pain. Blocking the production of prostaglandins thus decreases the pain felt from an injury, although it does nothing to cure the cause of the pain. Prostaglandins also cause blood platelets to stick together and form clots. Blocking the production of prostaglandins helps to prevent blood clots, so aspirin is often taken to treat or prevent strokes and heart attacks. Omar suspects that aspirin inhibits the functioning of the enzyme by fitting into a cavity in the enzyme structure, blocking substrate molecules from attaching to it.

Would you expect the structure of other painkillers, such as acetaminophen (one brand name is Tylenol), to be similar to that of aspirin?

Maya hopes to become an ophthalmologist. She picks retinal, $C_{20}H_{28}O$, as her molecule since it is important to vision. She finds two different structures for this molecule, called *cis*-retinal (left) and *trans*-retinal (right):

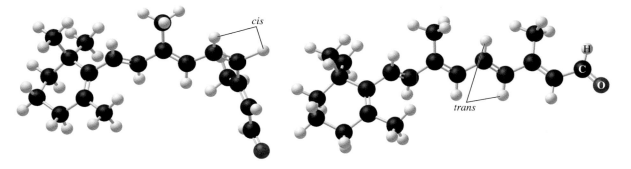

The behavior of retinal is unusual. Hydrocarbons with double bonds do not normally convert from *cis* to *trans* forms upon exposure to light because the conversion requires one of the bonds to break.

The *cis*-retinal molecule has two hydrogen atoms on the same side of one of the carbon-carbon double bonds, while *trans*-retinal has the two hydrogen atoms on opposite sides of this double bond. This difference in structure causes a difference in shape, as seen in the models. The −CHO group on the end of the molecule is quite reactive. It allows a *cis*-retinal molecule to bond to a molecule of opsin, a protein in the rods and cones of the eye. The result is a molecule called rhodopsin (Figure 16.1). The shape of *cis*-retinal allows it to fit into a cavity in the opsin molecule. Visible light converts *cis*-retinal to *trans*-retinal, accompanied by a change from a more bent to a more linear molecular structure. Because it no longer fits well into the opsin molecule cavity, *trans*-retinal detaches from the protein, sending an electrical signal along the optic nerve to the brain. The *trans*-retinal then interacts with an enzyme and changes back to *cis*-retinal, which can reattach to an opsin molecule.

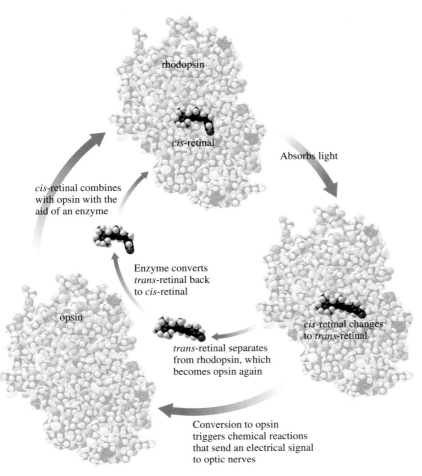

Figure 16.1
Vision depends on the interconversion of *cis*-retinal and *trans*-retinal. The absorption of light causes bond breakage, allowing the interconversion to occur.

Aisha, a criminal justice major, is concerned about law enforcement and drug abuse, so she decides to research the addictive central nervous system stimulant amphetamine, $C_9H_{13}N$. She finds this structure for the molecule:

How are the structures of aspirin and amphetamine alike and different?

The structure and shape of amphetamine are similar to that of norepinephrine, a natural neurotransmitter that carries nerve impulses across the gap between neurons in the brain. Norepinephrine is normally stored inside the ends of neurons before it is released in appropriate amounts. Because of its structure, amphetamine can displace norepinephrine from its storage vesicles. As a result too many norepinephrine molecules are released into the gap, and emotional responses are exaggerated.

What do all these molecules have in common? Did you notice that they all contain carbon? Why is this important? If you had this assignment, what molecule would you select? Would you choose a carbon compound? Try to find a molecule

Some other carbon-containing molecules that are related to specific careers are

lactic acid (in weary muscles)	Sports medicine and physical therapy
linoleic acid (in linseed oil)	Painting
methyl methacrylate (Plexiglas)	Sculpture
glycine (an essential amino acid)	Nutrition
glucose	Emergency medical technician
saturated fats	Food science
diethyl ether	Anesthesiology

The common definition of an organic compound is any compound that contains carbon, but this includes carbonates (CO_3^{2-}), cyanides (CN^-), and other ionic compounds that are usually considered inorganic. Organic chemistry is better defined as the study of carbon compounds in which carbon atoms are bonded to one another and to hydrogen atoms as well, and compounds that are derivatives of these hydrocarbons.

that would meet the criteria outlined by the professor in the first paragraph of this section and carry out the assignment. In this chapter, we'll look at compounds of carbon and hydrogen, and we will explore the field of *organic chemistry.*

Questions for Consideration

16.1   How can we represent organic molecules to illustrate their similarities and differences?
16.2   What are hydrocarbons, and where do they come from?
16.3   What are acyclic hydrocarbons, and how do they react?
16.4   What are cyclic hydrocarbons, and how do they react?
16.5   How do alcohols and ethers differ from the alkanes?
16.6   How are aldehydes and ketones alike and different?
16.7   How do carboxylic acids and esters differ from other organic compounds?
16.8   What is unique about amines?
16.9   What compounds contain multiple functional groups?
16.10 How do we use the IUPAC system to name organic compounds?

16.1   REPRESENTATIONS OF ORGANIC MOLECULES

The compounds that Scott, Jennifer, and the other students chose for the assignment all contain carbon. In combination with a few other elements (primarily H, O, N, and S), carbon forms several million known compounds, far exceeding those of any other element except hydrogen. The most extensive class of carbon compounds consists of *organic compounds,* which contain carbon, hydrogen, and sometimes other elements. The study of organic compounds is **organic chemistry.** The term *organic* relates to the importance of carbon compounds in living systems. Organic compounds need not come from living things, but they are vital to all life processes.

Why is carbon found in so many molecules and in such a variety of structural forms? The answer is found, in part, in the way that carbon bonds to itself and to other elements. A carbon atom can form single bonds, double bonds, or triple bonds in various combinations, for example,

- four single bonds, as in C_2H_6

- two single bonds and one double bond, as in C_2H_4

- two double bonds, as for the central atom in C_3H_4

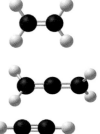

- one single bond and one triple bond, as in C_2H_2

In each case carbon has a total of four bonds. In contrast, hydrogen and fluorine typically form one bond, oxygen forms two bonds, and nitrogen forms three. The greater number of bonds found with carbon gives more possibilities for forming a diverse set of compounds.

A carbon compound that contains only carbon and hydrogen is a **hydrocarbon.** In hydrocarbons, carbon has the ability to bond to other carbon atoms to form long chains or rings of carbon atoms with hydrogen atoms attached. Different branching patterns are possible in many hydrocarbon molecules. Two compounds having the same composition and formula, but with a different arrangement of atoms in their molecular structures, are **isomers** of one another.

Carbon can bond to another nonmetal, such as nitrogen and oxygen, in a few, well-defined ways, giving a characteristic grouping of atoms called a functional group. A **functional group** is a reactive part of a molecule that undergoes characteristic reactions. For example, when a hydroxyl group ($-OH$) is present instead of a hydrogen atom in propane, C_3H_8, we have propyl alcohol or propanol, C_3H_7OH. The functional group in this case is $-OH$, which is characteristic of alcohols.

When the functional group undergoes a reaction, the remainder of the molecule usually remains intact. In reactions, then, the hydrocarbon chain to which the functional group is attached is often unchanged, so for simplicity we often abbreviate the hydrocarbon part of the molecule as $R-$, and show it attached to the functional group. If two hydrocarbon groups are present, the second can be designated $R'-$. The representation of any alcohol, for example, is $R-OH$. The shorthand R allows us to ignore the hydrocarbon chain and concentrate on the part of the molecule that gives it many of its characteristic properties.

A variety of related molecules with different functional groups is shown in Table 16.1. How are these structures related to propane? How can substituting, adding, or removing atoms or groups of atoms create these structures?

Ball-and-stick models are a good way to represent the structure of organic molecules, but other symbols and models have their advantages. The hydrocarbon fuel propane, for example, has the simple *molecular formula* C_3H_8. It has three carbon atoms linked in a chain, with hydrogen atoms bonded to each carbon atom in sufficient numbers to give each carbon atom four bonds: $CH_3CH_2CH_3$. This *structural formula* gives more information about propane than does the molecular formula, because it shows the order in which the atoms are bonded. Butane, another fuel, has the molecular formula C_4H_{10}. It can be represented by a structural formula $CH_3CH_2CH_2CH_3$, or a *condensed structural formula*, $CH_3(CH_2)_2CH_3$, in which identical connected units are grouped. Butane can also exist as an isomer in which one carbon atom is bonded to three $-CH_3$ groups and one hydrogen atom. The condensed structural formula for the isomer is $CH_3CH(CH_3)CH_3$, where the part of the formula in parentheses is understood to be attached to the previous carbon atom.

Another way to represent organic molecules is a *line structure*, which represents connected carbon atoms with a line. The hydrogen atoms bonded to carbon atoms are not shown, because their presence can be deduced from the fact that carbon must form four bonds. Any other atom is specified by its element symbol. In a line structure, the intersection of two lines or any unconnected end of a line represents a carbon atom.

Thus the octane molecules that Scott is interested in can be represented in the following ways:

- Ball-and-stick model:

- Molecular formula: C_8H_{18}
- Structural formula: $CH_3CH_2CH_2CH_2CH_2CH_2CH_2CH_3$
- Condensed structural formula: $CH_3(CH_2)_6CH_3$

- Line structure: ⌃⌄⌃⌄⌃

The ball-and-stick model, condensed structural formula, and line structure for the second isomer of octane that Scott found are

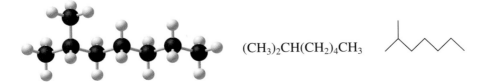

$(CH_3)_2CH(CH_2)_4CH_3$

TABLE 16.1			Functional Groups in Organic Molecules			

Type	Functional Group	Example	Example Structure	Example Name	Example Property
Alkane		$CH_3CH_2CH_3$		propane	Heating fuel for camping stoves
Alkene	$\diagdown C = C \diagup$	$CH_3CH=CH_2$		propene	Used to form polypropylene, a stiff plastic
Alkyne	$-C\equiv C-$	$CH_3C\equiv CH$		propyne	Explosive in air
Alcohol	$-OH$	$CH_3CH_2CH_2OH$		propanol	Alcoholic and slightly stupefying odor
Ether	$-O-$	$CH_3CH_2CH_2OCH_3$		methyl propyl ether	Anesthetic
Ketone	$\overset{O}{\underset{\|\|}{-C-}}$	$CH_3-\overset{O}{\underset{\|\|}{C}}-CH_3$		acetone	Good solvent; fingernail polish remover
Aldehyde	$\overset{O}{\underset{\|\|}{-C-H}}$	CH_3CH_2CHO		propanal	Suffocating odor
Carboxylic acid	$\overset{O}{\underset{\|\|}{-C-OH}}$	$CH_3CH_2CO_2H$		propanoic acid	Odor and flavor of Swiss cheese
Ester	$\overset{O}{\underset{\|\|}{-C-O-}}$	$CH_3CH_2CO_2CH_3$		methyl propanoate	Flavoring agent; fruity rum odor
Amine	$-\overset{\|}{N}-$	$CH_3CH_2CH_2NH_2$		propylamine	Strong ammonia odor

The line structure is an easy way to show the difference in isomers. In these examples, the first isomer is a straight chain of eight carbon atoms, while in the second, one of the carbons is bonded to the second carbon atom in a seven carbon chain.

Line structures also help us focus on the functional group where any chemical reaction is likely to occur. The line structure for Jennifer's menthol molecule is

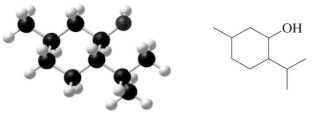

The molecule is a ring, or cycle, of six connected carbon atoms. Attached to this ring are a carbon atom, an −OH group (after skipping one carbon atom in the ring), and then a three-carbon-atom chain connected by its center carbon atom. The menthol molecule has the −OH functional group that is characteristic of alcohols, so it probably behaves and reacts much as other alcohols do.

EXAMPLE 16.1   Drawing Structural Formulas and Line Structures

Aisha studied the amphetamine molecule, using the following ball-and-stick model. Draw both a structural formula and a line structure for this molecule:

Solution:

Examine the ball-and-stick model of this molecule. The structural formula must represent each carbon-carbon bond and include the hydrogen or other atoms attached to each carbon. Since this is not a simple chain structure, we use lines to represent bonds and show how the atoms are connected:

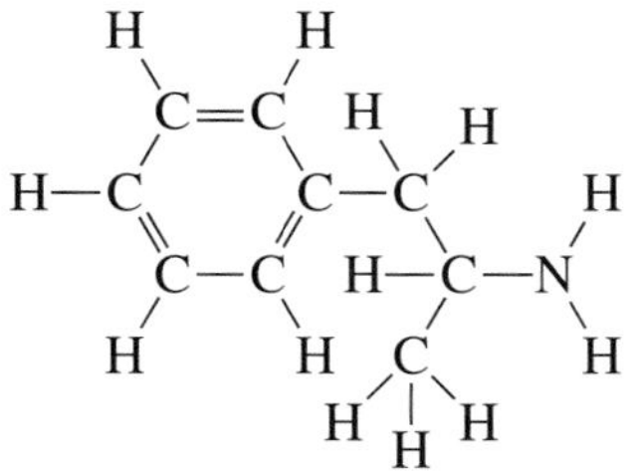

The line structure can then be based on this structural formula, but with the carbon-hydrogen bonds and the symbols for carbon removed:

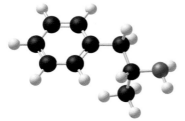

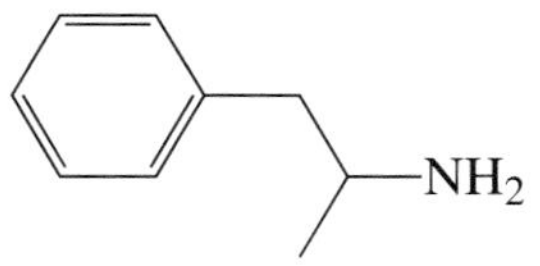

For clarity we leave in the element symbols for the NH_2 group that is characteristic of amines.

Practice Problem 16.1

Omar studied the aspirin molecule with the following ball-and-stick model:

Draw a structural formula and a line structure for this molecule.

Further Practice: 16.7 and 16.8

One additional representation of organic molecules is a *perspective drawing,* which can be used instead of a ball-and-stick model when visualizing the three-dimensional structure of a molecule. In a perspective drawing, solid lines represent bonds that are in the plane of the paper. Dashed wedges represent bonds that extend back behind the plane of the paper. Solid wedges represent bonds that project in front of the plane of the paper. To see how this works, compare the perspective drawing of octane to the ball-and-stick model:

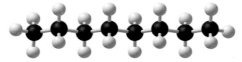

16.2 HYDROCARBONS

Classes of Hydrocarbons

Two major kinds of hydrocarbons, the *aliphatics* and the *aromatics,* were described in Section 8.3 and outlined in Figure 8.20.

Hydrocarbons are a major class of organic compounds, and they form the basic structural components of most other classes. Hydrocarbons differ in the type of carbon-carbon bonds they contain and in whether the carbon atoms form open chains or closed rings. The hydrocarbons can be classified as shown in Figure 16.2.

Acyclic (Open Chain) Hydrocarbons   The first main categories shown in Figure 16.2 are derived from the shape of the molecule. A hydrocarbon molecule that forms an open chain is an **acyclic hydrocarbon.** In such molecules the bonding electrons are *localized;* the electrons are confined to two adjacent atoms. A hydrocarbon with localized bonds is an **aliphatic hydrocarbon.**

A **saturated hydrocarbon** contains only single bonds between the carbon atoms. All the carbon atoms are termed saturated because they are attached to four other atoms—the maximum number possible for carbon. An example is ethane (H_3C-CH_3), which, like all the other compounds in this category, is an alkane. An *alkane* is any straight-chain hydrocarbon containing only single carbon-carbon bonds.

In organic chemistry *saturated* and *unsaturated* have different meanings than when describing the concentration of solutions, but the two uses are related. In each case *unsaturated* means some "holding capacity" remains—either to accept more hydrogen atoms (organic molecules) or to dissolve more solute (in solutions). *Saturated* means that all capacity to hold more has been used.

An **unsaturated hydrocarbon** contains one or more double or triple bonds between carbon atoms. Such molecules are considered unsaturated because two or more of the carbon atoms do not have four other atoms bonded to them. The

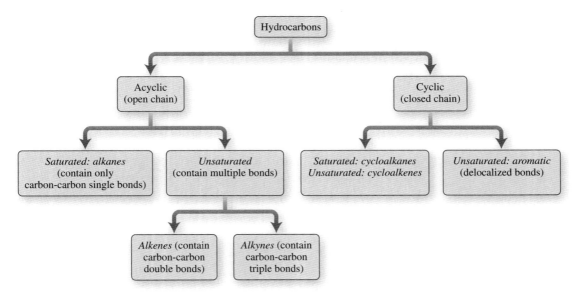

Figure 16.2
Classification of the hydrocarbons.

double-bonded, unsaturated hydrocarbons are the *alkenes* such as $H_2C=CH_2$ (ethene, also called ethylene). The triple-bonded, unsaturated hydrocarbons are the *alkynes* such as $HC\equiv CH$ (ethyne, also called acetylene).

All the acyclic, aliphatic hydrocarbons form open chains of carbon atoms, such as propane, $H_3C-CH_2-CH_3$. The chains may bond to side chains, or branches, such as in 4-ethylheptane:

$$H_2C-CH_3$$
$$|$$
$$H_3C-CH_2-CH_2-CH-CH_2-CH_2-CH_3$$

We'll examine the three categories of acyclic hydrocarbons in more detail in Section 16.3.

Cyclic (Closed Chain) Hydrocarbons   In contrast to the open chain of an aliphatic compound, a **cyclic hydrocarbon** forms a closed chain or ring of carbon atoms, such as in cyclobutane:

$$H_2C-CH_2$$
$$|\qquad|$$
$$H_2C-CH_2$$

Cyclobutane (C_4H_8) is a colorless gas that burns with a luminous flame.

This is an example of a *cycloalkane:* a single-bonded chain that appears to have bent around on itself and attached at its ends. Notice that cycloalkanes, like alkanes, are single-bonded and saturated.

In contrast, an **aromatic hydrocarbon** is a cyclic molecule containing a ring of carbon atoms attached to each other and, at least formally, containing alternating single and double bonds. The double-bond electrons in these compounds are *delocalized,* as explained by resonance (review Section 8.3). For example, a six-carbon ring such as in benzene (C_6H_6) has two possible resonance structures:

However, neither of these drawings represents the actual structure. Instead the double-bond electrons in benzene are delocalized or spread out around the entire ring structure, giving the molecule bonds of equal length and—unlike ordinary double bonds—a tendency toward low reactivity. For these reasons the structure is more accurately represented with a circle inside the ring to indicate the delocalized double bonds:

We'll consider aromatic hydrocarbons in more detail in Section 16.4.

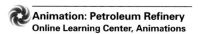

Animation: Petroleum Refinery
Online Learning Center, Animations

Chemistry Animations Library, Petroleum Refinery, Petroleum_Refinery.rm.

Petroleum is a primary source of carbon-containing compounds in plastics and other products. Should we continue to use most of our petroleum as fuel?

Because natural gas has no odor, traces of methyl mercaptan (CH_3SH) are added to commercial supplies so gas leaks in homes and businesses can be detected. It has the odor of rotten eggs.

Petroleum

Along with coal, petroleum is the primary source of hydrocarbons. Their major use is energy production, but around 10% of the coal and petroleum recovered from the Earth is used to make such useful materials and products as plastics, drugs, foods, and synthetic fibers.

Petroleum is usually found naturally in one of two forms, either natural gas or crude oil. Natural gas consists mainly of methane (CH_4), with smaller amounts of ethane (C_2H_6), propane (C_3H_8), and butane (C_4H_{10}). All are colorless, odorless, flammable gases. Natural gas is used primarily for heating and cooking. Crude oil is a dark, thick, odoriferous liquid composed of about 95% hydrocarbons. It must be refined before it can be used. Refining involves a few basic steps. The first is *fractional distillation,* or the separation of the petroleum into fractions having different ranges of boiling points. These fractions are either used directly, purified further, or converted into products that are more desirable. The fractions typically obtained from petroleum are summarized in Table 16.2.

TABLE 16.2	Petroleum Fractions		
Fraction	**Boiling Range (°C)**	**Composition Range**	**Uses**
Natural gas	<20	CH_4 to C_4H_{10}	Fuel, petrochemicals
Petroleum ether	20–60	C_5H_{12}, C_6H_{14}	Solvent
Ligroin, or naphtha	60–100	C_6H_{14}, C_7H_{16}	Solvent, raw material
Gasoline	40–220	C_4H_{10} to $C_{13}H_{28}$, mostly C_6H_{14} to C_8H_{18}	Motor fuel
Kerosene	175–325	C_8H_{16} to $C_{14}H_{30}$	Heating fuel and jet fuel
Gas oil	>275	$C_{12}H_{26}$ to $C_{18}H_{38}$	Diesel fuel and heating fuel
Lubricating oils and greases	High: Viscous liquids	>$C_{18}H_{38}$	Lubrication
Paraffin	High: Melting point 50–60	$C_{23}H_{48}$ to $C_{29}H_{60}$	Wax products
Asphalt or petroleum coke	High: Solid residue		Roofing, paving, fuel, reducing agent

Most petroleum is used to make gasoline, a mixture of hydrocarbons of which only one is octane, the molecule Scott chose for study. Given our reliance on gasoline as a fuel for transportation, the fraction of petroleum that can be directly separated as gasoline is not sufficient to meet demand, because more than half of the hydrocarbons in petroleum have more than 13 carbon atoms, the longest chain found in the gasoline fraction. To obtain larger amounts of gasoline, the process of *catalytic cracking* uses heat and a catalyst to break down the higher-boiling "gas oil" fraction. For example,

$$C_{14}H_{30} \xrightarrow{\text{heat, catalyst}} C_6H_{12} + C_8H_{18}$$

Note the wide range of boiling points in Table 16.2. Why do hydrocarbons boil at such different temperatures? Hydrocarbons differ in their physical characteristics as the number and arrangement of carbon atoms change. The strengths of the intermolecular forces that attract the molecules to one another account for observed differences in properties. For example, as the number of carbon atoms increases, the physical state of the hydrocarbon changes from gas to liquid to solid. What happens to the intermolecular forces? As the molecules get larger, they have greater surface area. Their London dispersion forces increase, so they interact more strongly.

Hydrocarbons containing from 5 to 18 carbon atoms are generally liquid at ordinary environmental temperatures. A number of everyday hydrocarbon products are liquids. Gasoline is made up of the lighter hydrocarbons. Kerosene contains slightly heavier hydrocarbons, and lubricating oil contains even heavier hydrocarbons.

16.3   ACYCLIC HYDROCARBONS

The simplest organic compounds are the open-chain, acyclic hydrocarbons: the alkanes, alkenes, and alkynes. The chains may be straight or branched. The alkanes are single-bonded and saturated. The alkenes and alkynes are unsaturated.

Alkanes

An **alkane** is a saturated hydrocarbon with the general formula C_nH_{2n+2}. The alkanes are obtained either directly from petroleum or indirectly by *cracking*, which breaks larger hydrocarbons into smaller ones. When the value of n is 1 through 4, the alkanes are gases under normal conditions. When n equals 5 through 18, they are liquids; and when n is greater than 18, they are solids. The melting points and boiling points increase as the number of carbon atoms increases. The increase results from greater surface area and stronger intermolecular forces that accompany the larger molecular size. A graph of the change in boiling point as the length of the carbon chain increases is shown in Figure 16.3.

Consider some examples of alkanes that are consistent with the general formula C_nH_{2n+2}. If $n = 1$, $2n + 2 = 4$, and we have CH_4. If $n = 3$, $2n + 2 = 8$, and we have C_3H_8 (propane).

Straight-chain alkanes are not really straight, since each carbon is at the center of a tetrahedron. A straight-chain alkane thus has a zigzag structure. For this reason, some chemists call these alkanes continuous-chain alkanes or unbranched-chain alkanes.

Figure 16.3
The boiling points of straight-chain alkanes increase as the number of carbon atoms in the chain increases.

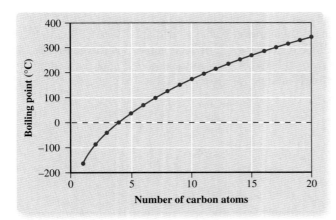

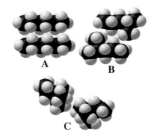

Figure 16.4
Straight-chain alkanes (A) have more surface contact than branched-chain alkanes (B and C) containing the same number of atoms. Thus, straight-chain alkanes experience greater London dispersion forces than do branched-chain alkanes, resulting in higher boiling points. The boiling points of these alkanes, all with the formula C_6H_{14}, are (A) 69.0°C, (B) 60.3°C, and (C) 49.7°C.

As the size of the molecule affects its intermolecular forces, so too does the shape. Straight-chain molecules can have more surface contact with one another than branched-chain molecules (Figure 16.4). The increased surface contact leads to increased London dispersion forces, with temporary dipoles attracting one another. When the molecules are more highly attracted, more energy has to be added to separate the molecules as they escape into the gas phase during boiling.

EXAMPLE 16.2   Intermolecular Forces in Alkanes

Scott found boiling points for two hydrocarbons having eight carbon atoms, represented here by their line structures. Explain the difference in boiling points of these two molecules:

Solution:

Because these molecules contain only carbon and hydrogen, they are nonpolar and they experience only London dispersion forces. They have the same molecular formula, so they have the same number of electrons. The greater surface contact between the linear molecules (left) provides more chances for temporary dipoles to form and interact. Branched-chain molecules (right) have less surface contact than straight-chain molecules, and they don't fit next to one another as well. Thus the branched molecule has the smaller intermolecular forces. It boils at a lower temperature because overcoming its intermolecular forces requires a lesser input of energy.

Practice Problem 16.2

Scott also found a boiling point for another hydrocarbon having eight carbon atoms, represented by this line structure:

Explain why this boiling point is lower than the other two given in the example.

Further Practice: 16.23 and 16.24

Alkane Structure and Nomenclature   The alkanes are named with a prefix denoting the number of carbon atoms in the longest chain followed by the ending *-ane*. The four shorter alkanes—methane (CH_4), ethane (C_2H_6), propane (C_3H_8), and butane (C_4H_{10})—have special names, while the longer alkanes use prefixes derived from the Greek or Latin numbers, such as octane for C_8H_{18}. The names of the first 10 alkanes are listed in Table 16.3.

TABLE 16.3	Names of Simple Alkanes
Alkane Name	**Formula (C_nH_{2n+2})**
methane	CH_4
ethane	C_2H_6
propane	C_3H_8
butane	C_4H_{10}
pentane	C_5H_{12}
hexane	C_6H_{14}
heptane	C_7H_{16}
octane	C_8H_{18}
nonane	C_9H_{20}
decane	$C_{10}H_{22}$

Each successive alkane in the table can be considered to be derived by insertion of one additional $-CH_2-$ group between the first carbon atom and one of its attached hydrogen atoms in the smaller alkane. For example, insertion of a CH_2 group into methane gives ethane, CH_3-CH_2-H or CH_3-CH_3. The third alkane is propane, C_3H_8, which has a second CH_2 group inserted to give $CH_3-CH_2-CH_2-H$ or $CH_3-CH_2-CH_3$. The next molecule in the series is butane, C_4H_{10} or $CH_3-CH_2-CH_2-CH_2-H$ or $CH_3-CH_2-CH_2-CH_3$.

Alkanes in which the carbon atoms form a continuous chain, with no branches, are called *normal alkanes* or *straight-chain alkanes*. The alternate arrangements are called *branched-chain alkanes* and are isomers of the straight-chain compounds. Thus—unlike the first three alkanes—butane has two different structures, or isomers. The first is normal butane (*n*-butane or just butane); it is an unbranched chain of four carbon atoms. The second is isobutane, a chain of three carbon atoms with the added CH_2 group inserted at a center carbon:

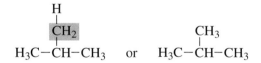

The number of isomers increases rapidly as the number of carbon atoms increases. Pentane (C_5H_{12}) has 3 isomers, $C_{10}H_{22}$ has 75 isomers, and $C_{20}H_{42}$ has 366,319 isomers. Because of the large number of possible isomers, alkanes are named in accordance with a systematic nomenclature established by the International Union of Pure and Applied Chemistry (IUPAC). If common names were applied to all these isomers, chemists would never be able to communicate efficiently. Consider the molecules and their IUPAC names given in Table 16.4. Can you determine the rules used for naming these compounds?

Did you recognize how to name side chains? Did you figure out how to use a numbering system to indicate position? If you need help with the rules, consult Section 16.10. Try to apply your rules to name an alkane with one branch on the chain.

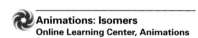

Animations: Isomers
Online Learning Center, Animations

Chemistry Animations Library, Isomers, Hexane_Isomers.rm.

In a molecule containing more than one of a group, prefixes indicate multiples. For example, *di*- means 2 and *tri*- means 3.

TABLE 16.4	Formulas and Names of Some Alkanes
Structural Formula	**Name**

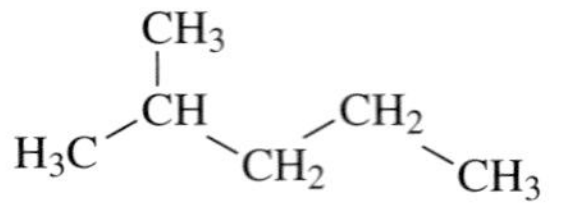

2-methylpentane

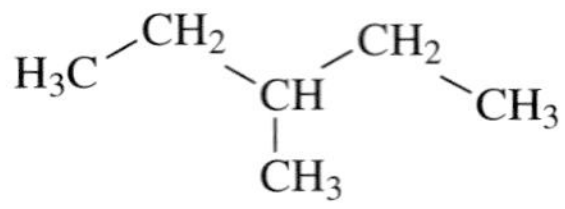

3-methylpentane

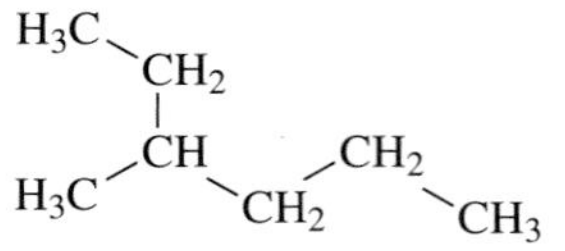

3-methylhexane

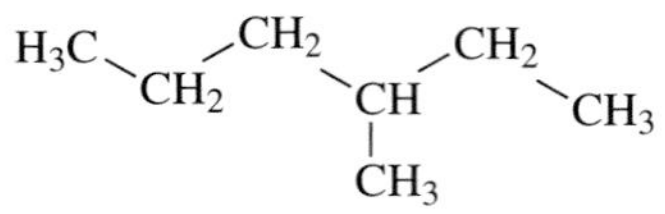

3-methylhexane

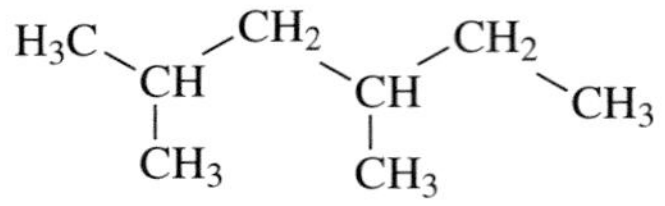

2,4-dimethylhexane

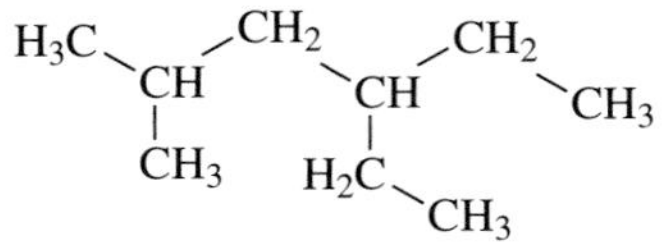

4-ethyl-2-methylhexane

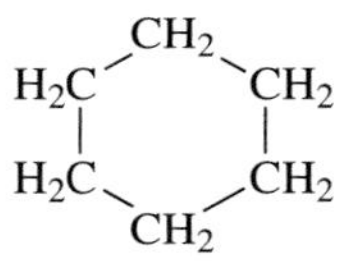

cyclohexane

EXAMPLE 16.3 Names of Alkanes with a Single Branch

Name isopentane according to the IUPAC nomenclature rules:

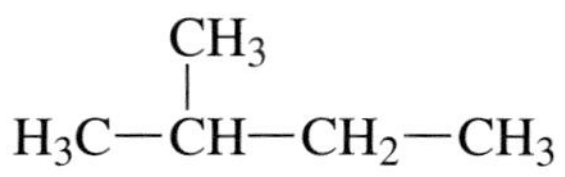

$$\text{H}_3\text{C}-\overset{\overset{\textstyle\text{CH}_3}{|}}{\text{CH}}-\text{CH}_2-\text{CH}_3$$

Solution:

Step 1: The longest continuous carbon chain has four carbons, so the name is based on butane.

Step 2: The carbons in this chain are numbered from the end closer to a branch. The only branch occurs at position 2.

Step 3: The branch occurring at position 2 is CH_3, or methyl, named for the alkane it is derived from, methane.

Step 4: The name is 2-methylbutane.

Practice Problem 16.3

Give the IUPAC name for isobutane.

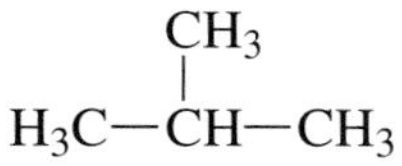

$$CH_3$$
$$|$$
$$H_3C-CH-CH_3$$

Further Practice: 16.31 and 16.32

When naming the branches on alkanes, look for a group derived from an alkane by removal of one hydrogen atom, with a general formula C_nH_{2n+1}. Such a group is an **alkyl group.** The names of the alkyl groups are similar to those of the alkanes (given in Table 16.3), except that the *-ane* ending is replaced by *-yl*. Thus, $-CH_3$ is methyl and $-C_2H_5$ is ethyl.

If an alkane contains more than one branch, each is named separately and in alphabetical order. Consider the following molecule:

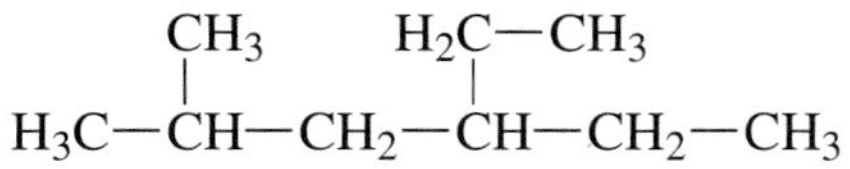

$$CH_3 \qquad H_2C-CH_3$$
$$| \qquad \quad |$$
$$H_3C-CH-CH_2-CH-CH_2-CH_3$$

The longest chain is hexane. A methyl (CH_3) branch occurs at position 2 and an ethyl (C_2H_5) branch occurs at position 4. The numbering starts from the left end of the molecule to give lower numbers. The ethyl group is named first since it occurs first alphabetically. The molecule is named 4-ethyl-2-methylhexane. Now try to name another alkane with more than one branch. If you need help with the rules, consult Section 16.10.

EXAMPLE 16.4   Names of Alkanes with Multiple Branches

Name the isomer of $C_{12}H_{26}$ that has the following structure:

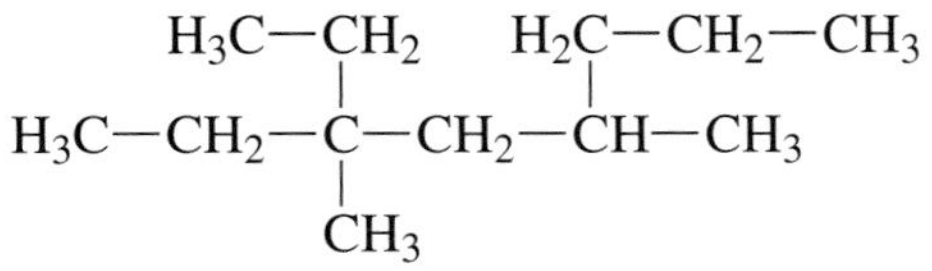

$$H_3C-CH_2 \qquad H_2C-CH_2-CH_3$$
$$| \qquad \qquad |$$
$$H_3C-CH_2-C-CH_2-CH-CH_3$$
$$|$$
$$CH_3$$

Solution:

The longest continuous chain has eight carbons, as shown by the highlighted part of the structure, so the compound is named as a derivative of octane.

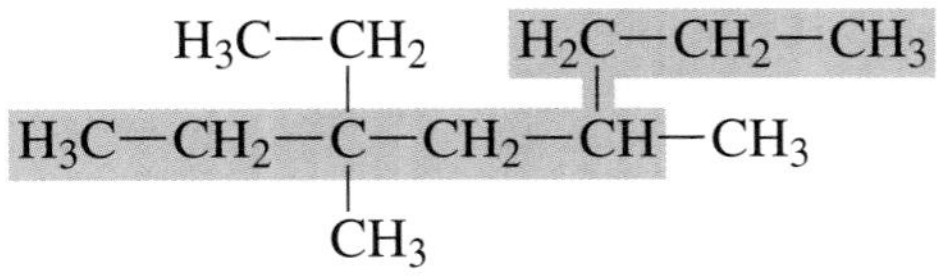

$$H_3C-CH_2 \qquad H_2C-CH_2-CH_3$$
$$| \qquad \qquad |$$
$$H_3C-CH_2-C-CH_2-CH-CH_3$$
$$|$$
$$CH_3$$

There are two methyl groups, one at carbon 3 and one at carbon 5, and one ethyl group at carbon 3. The ethyl group should be named first, because it occurs first alphabetically. The methyl groups can be named together, using the carbon numbers to indicate their positions. The full name is thus 3-ethyl-3,5-dimethyloctane.

Practice Problem 16.4

Name the isomer of $C_{10}H_{22}$ that has this structure:

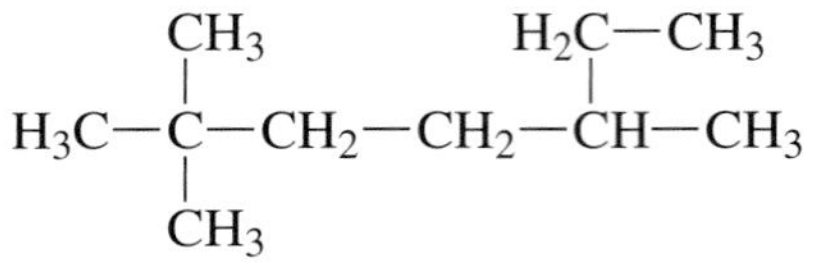

$$\underset{\displaystyle CH_3}{\overset{\displaystyle CH_3}{H_3C-\underset{|}{\overset{|}{C}}-CH_2-CH_2-\underset{\displaystyle \overset{|}{H_2C-CH_3}}{CH}-CH_3}}$$

Further Practice: 16.33 and 16.34

If you are given a name, can you figure out the structure of the compound? You can use the rules in reverse to yield the structural formula from the IUPAC name.

EXAMPLE 16.5   Structural Formulas from Names of Alkanes

Although Scott found three isomers having the chemical formula C_8H_{18}, there are actually more. One is 2,2,3,3-tetramethylbutane. Draw a structural formula for this molecule.

Solution:

The molecule name contains butane, so the longest chain must contain four carbon atoms:

$$C-C-C-C$$

Of the four methyl groups ($-CH_3$), two are attached to carbon 2. The other two are attached to carbon 3.

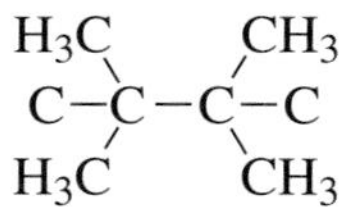

$$\underset{\displaystyle H_3C \qquad CH_3}{\overset{\displaystyle H_3C \qquad CH_3}{C-C-C-C}}$$

Finally we add enough hydrogen atoms to give four bonds to each carbon atom:

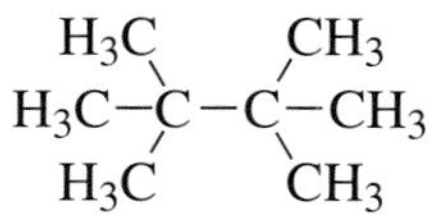

$$\underset{\displaystyle H_3C \qquad CH_3}{\overset{\displaystyle H_3C \qquad CH_3}{H_3C-C-C-CH_3}}$$

Practice Problem 16.5

Another isomer of C_8H_{18} is 2,2,3-trimethylpentane. Draw its structural formula.

Further Practice: 16.35 and 16.36

Reactions of Alkanes Alkanes don't readily undergo chemical reactions, because both carbon-hydrogen and carbon-carbon single bonds are strong. Supplying heat or light provides the energy needed to get alkanes to react. For example, in petroleum refining, heating in the presence of a catalyst cracks or breaks large alkanes into smaller molecules. Such reactions produce shorter-chain alkanes, cycloalkanes, alkenes, and molecular hydrogen.

Alkanes undergo oxidation in exothermic *combustion* reactions, in which they act as fuels and burn when ignited in the presence of oxygen gas to form carbon monoxide or carbon dioxide and water:

$$C_3H_8 + 5O_2 \longrightarrow 3CO_2 + 4H_2O$$

In the presence of sunlight or ultraviolet radiation, alkanes will react with halogens. As shown in Figure 16.5, light provides the energy to break the halogen-halogen bonds and initiate the reaction:

$$C_6H_{14} + Br_2 \xrightarrow{\text{light}} C_6H_{13}Br + HBr$$

A

EXAMPLE 16.6   Reactions of Alkanes

Predict the products of the reaction of propane with bromine in the dark at room temperature.

Solution:

Alkanes do not react with bromine unless subjected to heat, sunlight, or ultraviolet radiation. Thus, there would be no reaction.

Practice Problem 16.6

Predict the products of the reaction of butane with bromine in sunlight.

Further Practice: 16.37 and 16.38

B

Figure 16.5
Hexane does not react with brown bromine unless exposed to bright light. (A) Mixture in the absence of light. (B) Mixture subjected to bright light.

Alkenes and Alkynes

Maya's molecule, retinal, contains two types of functional groups: a carbon-carbon double bond and a −CHO group. Later we'll look at the properties of the −CHO group. Here we will be concerned about the effect that double bonds have on the properties of hydrocarbons. In this section, we consider two such classes of hydrocarbons: alkenes, containing double bonds, and alkynes, containing triple bonds.

Alkene Structure and Nomenclature   **Alkenes** contain at least one C=C double bond, and with just one double bond have the general formula C_nH_{2n}. They are generally obtained from petroleum by the cracking of alkanes, using heat and a catalyst. The simplest such reaction is

$$CH_3-CH_3 \xrightarrow{\text{heat, catalyst}} H_2C=CH_2 + H_2$$

The product is named ethene (IUPAC name), or ethylene (common name).

An alkene molecule may also have more than one double bond. For example, $H_2C=CH-CH=CH_2$, with double bonds at carbons 1 and 3, is 1,3-butadiene. Alkenes can also form *polymers,* molecules made of many repeating units. Polymers are used to make a variety of useful items including rubber tires.

Examine the structural formulas and names of some alkenes given in Table 16.5. Try to determine the rules for naming them.

Did you come up with some rules? How do you name the main chain? How do you designate the positions of double bonds? How do you designate the positions and names of branches? If you need help, consult Section 16.10. Use your rules to name some alkenes.

Ethene seems to play a part in the ripening of fruit, so green fruit is often shipped under an atmosphere of ethene, so that ripening occurs in transit.

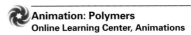

Animation: Polymers
Online Learning Center, Animations

Chemistry Animations Library, Polymers, Polymers.rm.

TABLE 16.5	Formulas and Names of Some Alkenes
Structural Formula	**Name**
$H_2C=CH_2$	ethene
$H_2C=CH-CH_3$	propene
$H_2C=CH-CH_2-CH_3$	1-butene
$H_3C-CH_2-CH=CH_2$	1-butene
$H_3C-CH=CH-CH_3$	2-butene
$H_3C-\overset{\overset{\textstyle CH_3}{\textstyle \|}}{C}=CH-CH_3$	2-methyl-2-butene
$H_2C=CH-\overset{\overset{\textstyle CH_3}{\textstyle \|}}{C}H-CH_3$	3-methyl-1-butene
$H_2C=\overset{\overset{\textstyle H_3C}{\textstyle \|}}{C}-\overset{\overset{\textstyle CH_3}{\textstyle \|}}{C}H-CH_3$	2,3-dimethyl-1-butene

EXAMPLE 16.7   Names of Alkenes

Name this compound:

$$H_3C-\overset{\overset{\textstyle |}{\textstyle CH_3}}{C}H-CH=CH-CH_3$$

Solution:

The longest chain has five carbons, so the compound is a derivative of pentane, making the name pentene (because of the double bond). The carbon atoms are numbered from the end nearest the double bond. The position of the double bond is after the second carbon from the right. Counting from that end, the methyl side chain is attached to the fourth carbon atom. The name is 4-methyl-2-pentene.

Practice Problem 16.7

Name this compound:

$$H_3C-CH_2-\overset{\overset{\textstyle |}{\textstyle CH_3}}{C}=CH_2$$

Further Practice: 16.41 and 16.42

In isomers of alkenes, rotation is restricted around the C=C double bond. Because of the restricted rotation, the relative positions of side chains attached to the carbons in the double bond are fixed. To have such *geometric isomers,* two different groups must attach to each of the carbon atoms in the double bond. The attached groups may be on the same side of the molecule (a *cis* isomer) or on opposite sides (a *trans* isomer). Thus Maya found two different isomers of retinal—a *cis* form and a *trans* form—that differed in the positions of atoms attached to one of the C=C units in the molecule. A simpler example is 2-butene, with the two hydrogen atoms on the same or opposite sides of the double bond:

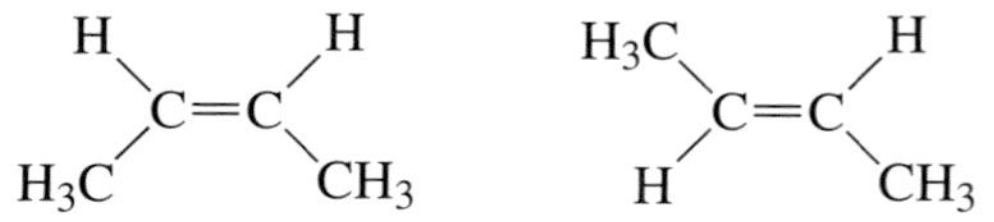

These two molecules have different properties because their structures are different. *Cis*-2-butene has a boiling point of 4°C, but that of *trans*-2-butene is 1°C. The *cis* isomer has stronger intermolecular forces because it is slightly polar, resulting in a higher boiling point.

EXAMPLE 16.8   *Cis* and *Trans* Isomers

Which of these molecules can have *cis* and *trans* isomers? Draw structures for any possible isomers.

(a) $CH_3CH=CHCH_2CH_3$
(b) $(CH_3)_2C=CHCH_2CH_3$

Solution:

(a) Each double-bonded carbon has two groups attached to it, and the two groups in each case are different from each other. Thus, it is possible to have both *cis* and *trans* isomers:

$$\underset{H_3C}{\overset{H}{\diagdown}}C=C\underset{CH_2-CH_3}{\overset{H}{\diagup}} \qquad \underset{H_3C}{\overset{H}{\diagdown}}C=C\underset{H}{\overset{CH_2-CH_3}{\diagup}}$$

(b) Each double-bonded carbon has two groups attached to it, but the first (left) carbon has identical groups, so there will be no *cis* and *trans* isomers.

Practice Problem 16.8

Which of these molecules can have *cis* and *trans* isomers? Draw structures for any possible isomers.

(a) $CH_3CH_2CH=CH_2$
(b) $CH_3CH_2CH=CHCH_2CH_3$

Further Practice: 16.43 and 16.44

Alkyne Structure and Nomenclature   **Alkynes** contain a C≡C triple bond and have the general formula C_nH_{2n-2}. Alkynes are also called acetylenes, from the common name of the simplest alkyne, HC≡CH. Ethyne (IUPAC name) or acetylene (common name) is generally prepared from calcium carbide (Figure 16.6).

$$CaC_2(s) + 2H_2O(l) \longrightarrow HC\equiv CH(g) + Ca(OH)_2(s)$$

Alkynes are named with the same rules as alkenes, but using the ending *-yne* instead of *-ene*. Thus, $CH_3C\equiv CCH_3$ is 2-butyne and $HC\equiv CCH_2CH_3$ is 1-butyne. There are no *cis* and *trans* isomers for alkynes.

Reactions of Alkenes and Alkynes   An *addition reaction* is one in which extra atoms or groups are added to a simpler hydrocarbon. Alkenes and alkynes typically undergo addition reactions across the double or triple bond:

$$CH_2=CH_2 + HBr \longrightarrow CH_3CH_2Br$$

Many of these reactions require a catalyst in order to occur at an appreciable rate. For example, if ethene is placed in an environment of H_2 gas, no reaction occurs. In

Figure 16.6
Before portable electric lamps became available, miners and cave explorers used lamps in which water dripped onto calcium carbide. The acetylene that was produced was burned to give light. Changing the rate of addition of water adjusted the size of the flame.

Figure 16.7
Many consumer products are made of polyethylene.

Figure 16.8
Polystyrene is used in products ranging from clear plastic to Styrofoam.

the presence of a metal catalyst such as platinum, however, ethene converts to ethane:

$$CH_2{=}CH_2 + H_2 \xrightarrow{\text{Pt}} CH_3CH_3$$

Another important reaction is *polymerization,* in which small molecules containing double bonds combine together to form large molecules like plastics. The polymer is made of smaller, repeating units called *monomers.* The formation of polyethylene is an example of an *addition polymerization* reaction, in which monomers add to one another without losing any atoms. The joining of monomer units at the position of the monomer double bond forms the polymer, as shown here for ethylene:

$$H_2C{=}CH_2 + H_2C{=}CH_2 + H_2C{=}CH_2 + ... \longrightarrow$$
$$-H_2C{-}CH_2{-}H_2C{-}CH_2{-}H_2C{-}CH_2{-} ...$$

This equation can be generalized for any number (n) of monomers as follows:

$$nH_2C{=}CH_2 \longrightarrow \left(CH_2{-}CH_2\right)_n$$

Polyethylene is a useful polymer. It can be heated to soften and shape it, then cooled to a plastic substance with its original properties. That's why polyethylene bags can be heat-sealed. Polyethylene films hold in or keep out water vapor but allow the passage of oxygen. Polyethylene is a good electrical insulator. It is commonly used in soda bottles, milk cartons, produce bags, food wrap, wire coatings, and more (Figure 16.7).

Propylene is polymerized similarly to ethylene, but the polymer, polypropylene, is harder and less flexible than polyethylene due to the presence of its methyl side chains.

$$nH_2C{=}\overset{\overset{\displaystyle CH_3}{\displaystyle |}}{CH} \longrightarrow \left(CH_2{-}\overset{\overset{\displaystyle CH_3}{\displaystyle |}}{CH}\right)_n$$

The groups shown in the parentheses are repeated in a chain n times, where n is a large number.

Another related material is the polymer of styrene ($C_6H_5CH{=}CH_2$):

$$nH_2C{=}CH \longrightarrow \left(CH_2{-}CH\right)_n$$

Polystyrene is used in packaging, food and household containers (Figure 16.8), combs, and coatings for television lead-in wire. It is rigid, clear, and nontoxic, and it won't dissolve in most aqueous solutions. It can be prepared in the form of expandable beads, which can be molded and shaped. The result is Styrofoam.

EXAMPLE 16.9   Reactions of Alkenes

Write an equation to describe the reaction between 2-hexene and H_2.

Solution:

This is an example of an addition reaction, also called a hydrogenation reaction when the molecule being added is hydrogen. The hydrogen molecule is split apart with the aid of a catalyst, and the hydrogen atoms are added to the carbon atoms involved in the double bond:

$$CH_3CH{=}CHCH_2CH_2CH_3 + H_2 \xrightarrow{\text{Pt}} CH_3CH_2CH_2CH_2CH_2CH_3$$

Practice Problem 16.9

Write an equation to describe the reaction that would occur between 2-butene and HBr.

Further Practice: 16.47 and 16.48

16.4 CYCLIC HYDROCARBONS

Cycloalkanes and Cycloalkenes

Imagine a hydrocarbon chain of single-bonded carbons with a hydrogen atom removed on each end and the two carbon atoms at the end then bonded together. The result would be the ring structure of a saturated cyclic hydrocarbon, or **cycloalkane,** with the general formula C_nH_{2n}. The simplest cycloalkane is cyclopropane, C_3H_6. It is a colorless, flammable gas that is explosive in air. It was once used as a general anesthetic.

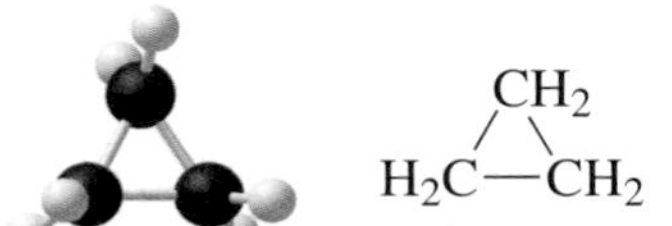

cyclopropane

Many cyclic molecules, especially those in which the rings are small, are strained because they cannot adopt the normal tetrahedral bond angles (109.5°) around carbon. Thus they tend to be somewhat more reactive than the alkanes. The most common cycloalkanes are the low-boiling liquids cyclopentane (C_5H_{10}) and cyclohexane (C_6H_{12}):

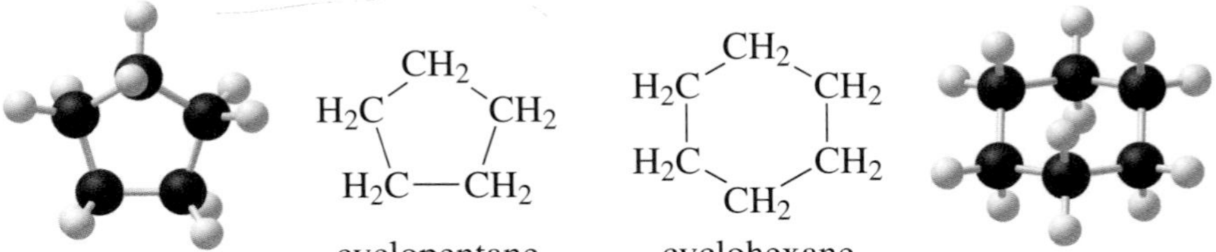

cyclopentane   cyclohexane

Cyclopentane is a flammable, colorless liquid that smells like gasoline. Cyclohexane has similar properties and is used as a solvent for lacquers and resins and in paint removers.

Some alkenes also form ring structures. Molecules containing multiple double bonds and double bonds in ring structures are common in nature. We saw the example of *cis-* and *trans-*11-retinal that Maya studied; it has five carbon-carbon double bonds, one of which is in a cyclohexene ring:

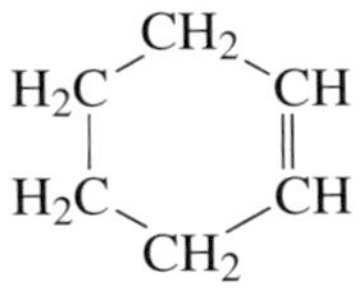

cyclohexene

The official name of retinal has the number 11 in it to designate the position of the double bond that gives rise to the isomers. Some other common examples in nature include vitamin A, required for good vision, and limonene, found in the oils in citrus fruits:

vitamin A   limonene

Aromatic Hydrocarbons

Omar was studying aspirin, which contains a six-carbon ring having alternating single and double bonds. Aisha was interested in amphetamine, which also contains such a six-carbon ring. Although each of these molecules has additional functional groups, both are classified as aromatic. Common aromatic hydrocarbons are cyclic hydrocarbons with alternating single and double carbon-carbon bonds.

The actual structure of aromatic hydrocarbons such as C_6H_6 has the electrons in the double bonds spread out or delocalized over the entire ring system, as discussed in Section 16.2. Double bonds with electrons delocalized over the entire ring are less reactive than ordinary double bonds. The resulting inertness is referred to as *aromatic character*. The delocalized bonding in benzene, C_6H_6, is usually represented by a circle inside the structure, and the hydrogen atoms, one attached to each carbon atom, are not shown:

Aspirin and amphetamine can be represented by the following structural formulas:

aspirin   amphetamine

Nomenclature of Aromatic Hydrocarbons Derivatives of benzene are generally named with benzene as the base name. Many of these molecules have special names as well. The following molecule, for example, is called both methylbenzene and, more commonly, toluene. A methyl group is substituted for one hydrogen atom on the benzene ring:

Examine the names of the aromatic compounds in Table 16.6. Can you come up with some rules for naming them?

TABLE 16.6 **Some Formulas and Names of Aromatic Hydrocarbons**

Structural Formula	Name
CH_3 (benzene ring)	methylbenzene (or toluene)
Cl (benzene ring)	chlorobenzene
NH_2 (benzene ring)	aminobenzene (or aniline)
H_2C-CH_3 (benzene ring)	ethylbenzene
NH_2 (benzene ring) Cl	1-amino-4-chlorobenzene (or 4-chloroaniline)
CH_3, CH_3 (benzene ring)	1,2-dimethylbenzene
CH_3, CH_3 (benzene ring)	1,3-dimethylbenzene
CH_3, CH_3 (benzene ring)	1,4-dimethylbenzene
CH_3, O_2N, NO_2, NO_2 (benzene ring)	2,4,6-trinitrotoluene (TNT)

Did you come up with some rules? How do you name the ring? How do you designate the positions and names of attached groups? If you need help, consult Section 16.10. See if you can apply your rules to naming some aromatic compounds.

EXAMPLE 16.10   Names of Aromatic Compounds

Name this aromatic compound.

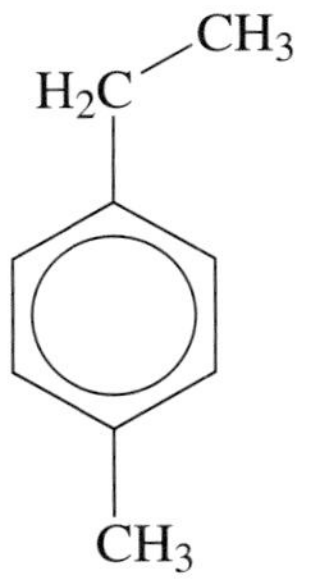

Solution:

The base structure is a six-carbon aromatic ring, so the compound is named as a derivative of benzene. We start numbering the ring at the first attached group (in alphabetical order), ethyl, so the methyl group is at carbon 4. The name is thus 1-ethyl-4-methylbenzene.

Practice Problem 16.10

Name this aromatic compound.

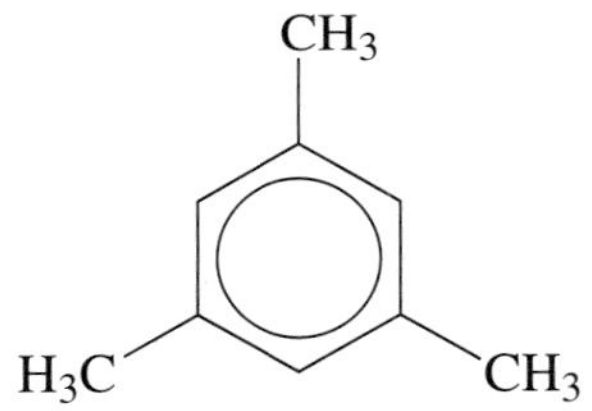

Further Practice: 16.51 and 16.52

Reactions of Aromatic Hydrocarbons   Benzene and other aromatic hydrocarbons lack the reactivity of the double bond in alkenes. Normally, benzene does not react at all, but with the right catalyst, it can undergo a *substitution reaction* in which some functional group replaces one or more hydrogens on the benzene ring. For example, benzene reacts with chlorine in the presence of the catalyst $FeCl_3$ to form chlorobenzene:

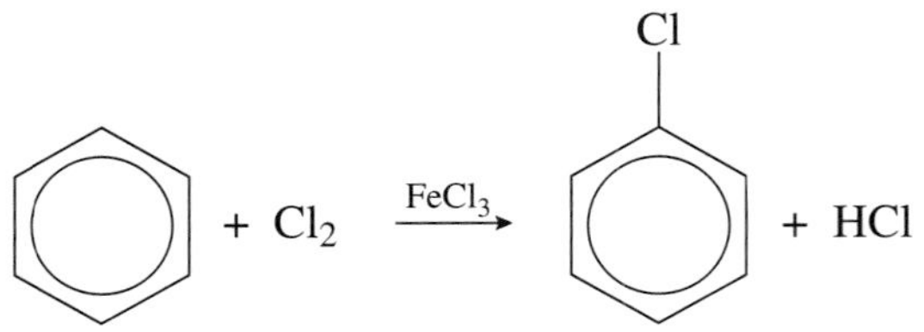

It reacts with chloromethane in the presence of the catalyst $AlCl_3$ to form toluene:

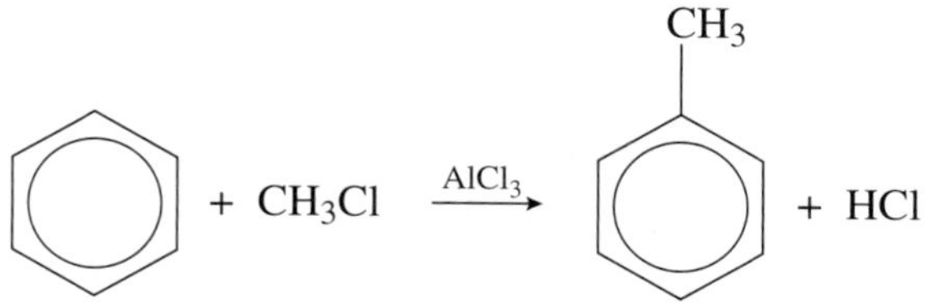

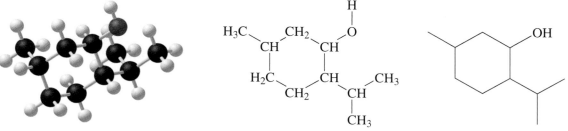

16.5   ALCOHOLS AND ETHERS

Alcohols

Jennifer's molecule, menthol, is an example of an **alcohol,** which is characterized by the functional group, $-OH$:

Alcohols are useful substances, and some of them are among the top 50 chemicals produced in the United States. Alcohols produced in large amounts include methanol (CH_3OH), ethylene glycol ($HOCH_2CH_2OH$), isopropanol or 2-propanol (C_3H_7OH), and ethanol (C_2H_5OH).

The $-OH$ group makes alcohols polar, and it can participate in hydrogen-bonding interactions. As a result even the lower-molar-mass alcohols are liquids, while alkanes of similar size are gases. The short-chain alcohols are water soluble, but solubility decreases as chain length increases. For example, C_4H_9OH is only slightly soluble in water, and alcohols containing five or more carbon atoms are essentially insoluble in water. Alcohols of similar size are soluble in one another. For example, methanol dissolves in ethanol. Methanol is a colorless liquid so toxic that breathing its vapors can blind or kill. It is often added to nonbeverage ethanol, to make it unfit for drinking. Modified ethanol can be sold for scientific use and as an additive for gasoline without the federal taxes levied on beverage alcohol. In this role, methanol is called a denaturant, and the modified ethanol is called *denatured alcohol.*

Because of the polarity and hydrogen bonding of the alcohols, they have significantly higher boiling points than alkanes of comparable mass. For example, the boiling point of CH_3CH_2OH is 78.5°C, but that of $CH_3CH_2CH_3$ is –42.1°C. The alkane molecules are attracted to one another only by London dispersion forces, while alcohol molecules are attracted to one another by stronger dipole forces and hydrogen-bonding forces.

Carbohydrates such as glucose and fructose contain multiple alcohol functional groups:

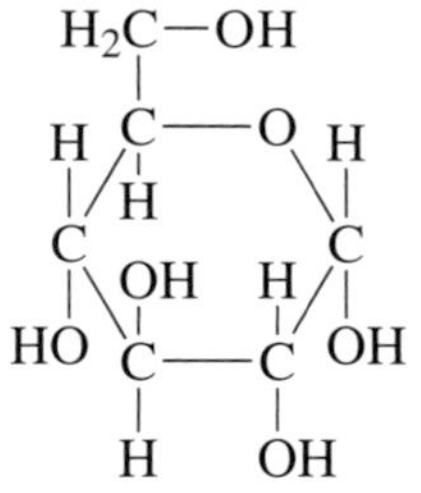

Ethylene glycol, or 1,2-ethanediol, $HO-CH_2-CH_2-OH$, is an alcohol containing two $-OH$ groups. This colorless liquid is commonly used as an antifreeze in automobile radiators.

glucose

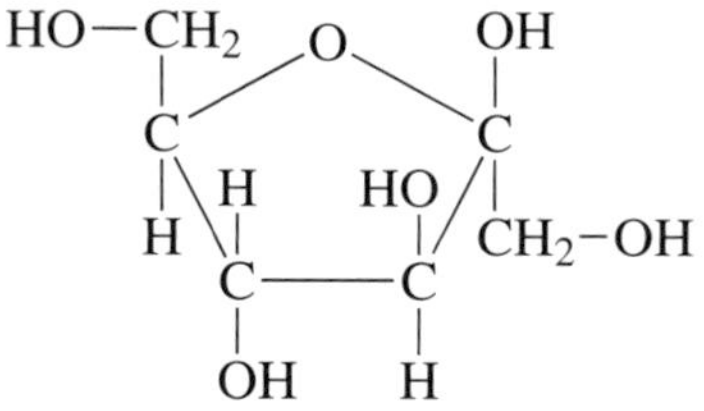

fructose

Also notice the presence of an ether functional group, an $-O-$ atom bonded to two carbon atoms in these simple sugars. Larger sugars such as sucrose and maltose contain two glucose and/or fructose units connected by ether linkages. Starch and cellulose are larger polymers of glucose, also connected by ether linkages. Humans and other vertebrate animals have the enzymes necessary to metabolize starch, converting it back to glucose, which cells use as their primary energy source. However, they cannot digest the cellulose present in grass and most other plant foods. Grazers such as cows rely on bacteria that live in their guts to do the job for them.

Nomenclature of Alcohols   Examine the names and formulas of some alcohols given in Table 16.7. Try to determine the rules for naming these compounds.

Did you come up with some rules? How do you name the chain? How do you designate the position of any branches? How do you designate the position of the $-OH$ group? If you need help, consult Section 16.10. Try to use your rules to name some alcohols.

EXAMPLE 16.11   **Names of Alcohols**

Name this alcohol:

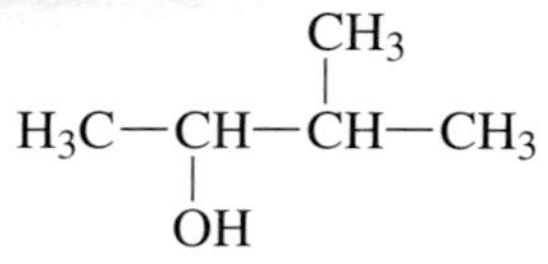

$$H_3C-CH-CH-CH_3$$

with a CH_3 group on the third carbon and an OH group on the second carbon.

Solution:

The longest chain has four carbons, so this is a derivative of butane. The *–ane* ending is changed to *–anol,* making the name butanol. The $-OH$ group is attached to the second carbon, so the chain name is 2-butanol. A methyl group is attached to carbon 3, so the full name is 3-methyl-2-butanol.

Practice Problem 16.11

Name this alcohol:

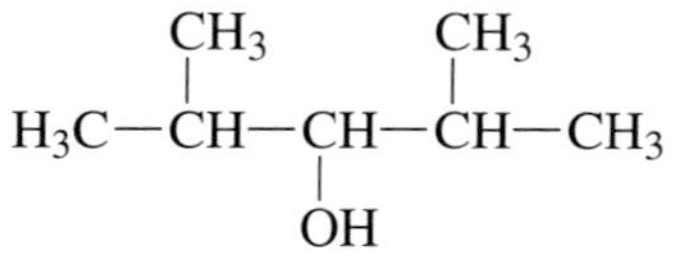

$$H_3C-CH-CH-CH-CH_3$$

with CH_3 groups on the second and fourth carbons and an OH group on the third carbon.

Further Practice: 16.61 and 16.62

An $-OH$ group may be attached directly to an aromatic hydrocarbon group, as in C_6H_5OH:

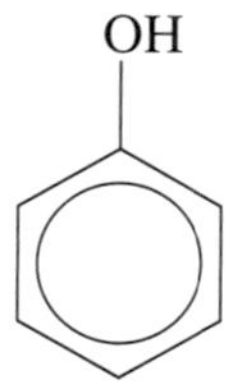

OH

phenol

This molecule is known as phenol. The chemistry of the $-OH$ group changes greatly when it is attached directly to an aromatic ring, and such compounds are classified not as alcohols, but as phenols.

TABLE 16.7	Formulas and Names of Some Alcohols
Structural Formula	**Name**
H_3C-OH	methanol
H_3C-CH_2-OH	ethanol
$H_3C-CH_2-CH_2-OH$	1-propanol
$H_3C-\underset{\underset{OH}{\vert}}{CH}-CH_3$	2-propanol
$H_3C-CH_2-CH_2-CH_2-OH$	1-butanol
$H_3C-CH_2-\underset{\underset{OH}{\vert}}{CH}-CH_3$	2-butanol
$H_3C-\overset{\overset{CH_3}{\vert}}{\underset{\underset{OH}{\vert}}{C}}-CH_3$	2-methyl-2-propanol

Ethers

An **ether** is an organic molecule that contains an $-O-$ atom bonded to two carbon atoms. Ethers are related to the alcohols, but they have two hydrocarbon groups attached to an oxygen: $R-O-R'$. The names of the ethers include the names of the attached alkyl groups, listed alphabetically and followed by the word *ether*, as in diethyl ether ($CH_3CH_2-O-CH_2CH_3$). The ethers tend to be more volatile than the corresponding alcohols. Consider the boiling points of an alkane, an ether, and an alcohol of comparable mass:

Formula	$CH_3CH_2CH_3$	CH_3OCH_3	CH_3CH_2OH
Boiling point (°C)	−42.1	−23.6	78.5

Ethers are polar molecules, so they have stronger intermolecular forces and higher boiling points than alkanes. With no hydrogen atom attached to the oxygen atom in an ether, there is no hydrogen bonding, so it has a lower boiling point than the comparable alcohol.

One of the molecules first used as a general anesthetic was diethyl ether:

$$H_3C \overset{\textstyle CH_2}{\diagdown} \underset{\textstyle O}{} \overset{\textstyle CH_2}{\diagup} CH_3$$

MTBE, or methyl tertiary-butyl ether, is added to some formulations of gasoline to ensure more complete combustion:

$$H_3C \overset{\textstyle O}{\diagdown} \underset{H_3C}{\overset{}{\diagup}} C \overset{-CH_3}{\diagdown}_{CH_3}$$

16.6   ALDEHYDES AND KETONES

Aldehydes

Earlier we discussed Maya's favorite molecules, *cis*- and *trans*-retinal, as examples of alkenes. However, these molecules contain the $-CHO$ functional group, so they are also aldehydes. An **aldehyde** contains a $C=O$ (carbonyl) group in which at least one hydrogen atom is attached to the carbon atom. Aldehydes have the general formula $R-CHO$, in which the hydrogen and the oxygen are both bonded to the carbon, and the oxygen-carbon bond is a double bond:

$$\overset{\textstyle O}{\underset{R-\overset{\Vert}{C}-H}{}}$$

The term resin was originally used to describe the thick, odoriferous sap exuded by cuts in the bark of pine trees. It is now used in a more generic sense to include the many synthetic, amorphous solid materials that are used in plastics, varnishes, inks, and medicines.

Figure 16.9
Urea-formaldehyde foam used for insulation is formed by mixing two liquids. They react to yield a polymer and a gas. The gas causes the polymer to expand and form a foam. In a short time the foam hardens into a rigid solid containing a lot of trapped gas.

The most common aldehydes are formaldehyde (HCHO) and acetaldehyde (CH_3CHO). Formaldehyde (common name), or methanal (IUPAC name), is a colorless gas with an irritating odor. It is highly soluble in water and is commonly sold as a 37% solution known as *formalin*. Formalin causes coagulation of proteins, making it useful as a preservative of tissues. Formaldehyde is also used with phenol to form polymer resins, which are used in plastics and as adhesives in plywood and particleboard. Urea-formaldehyde foam (Figure 16.9) was once widely used for house insulation, until it was banned because of the toxic properties of formaldehyde.

Acetaldehyde (common name), or ethanal (IUPAC name), CH_3CHO, is a volatile, colorless liquid that is soluble in water. It has a characteristic, pungent odor much like that of cut green apples. It is toxic and flammable. Acetaldehyde is used in perfumes, aniline dyes, plastics, and synthetic rubber, and as a reducing agent for silvering mirrors (Figure 16.10).

Ketones

Ketones resemble aldehydes in that both contain a carbon double-bonded to an oxygen, or C=O. In an aldehyde, this group is located on an end carbon. In a **ketone,** however, the organic groups, often alkyl groups—such as methyl or ethyl here designated R and R'—lie on either side of the C=O group:

$$\overset{\displaystyle O}{\underset{\displaystyle R-C-R'}{\|}}$$

The simplest and most important of the ketones is acetone, $(H_3C)_2C{=}O$, which is officially named propanone. Acetone is a colorless, highly flammable liquid with a characteristic pungent odor. It is readily soluble in water. Acetone is widely used as a solvent for fats, oils, waxes, resins, rubber, plastics, lacquers, varnishes, and rubber cements. It is the principal ingredient in some fingernail polish removers.

Figure 16.10
During the Tollens test for aldehydes, silver(I) ion is reduced to metallic silver and the aldehyde is oxidized. Here the reaction with formaldehyde creates a silver mirror on the flask. Ketones are not easily oxidized and do not react in this way.

16.7 CARBOXYLIC ACIDS AND ESTERS

Carboxylic Acids

Omar's favorite molecule, aspirin, contains a carboxylic acid functional group. The characteristic functional group in a **carboxylic acid** molecule is a carbon atom double-bonded to an oxygen atom and also bonded to an −OH group, $-CO_2H$:

$$\overset{\displaystyle O}{\underset{\displaystyle R-C-OH}{\|}}$$

Carboxylic acids are typically weak acids, and their reactions often involve loss of H^+.

The simplest carboxylic acid is formic acid (methanoic acid), HCO_2H, which was first obtained from red ants. It is the chemical agent that causes ant bites, bee stings, and nettle stings to be so painful. Formic acid is a colorless, water-soluble liquid with a pungent odor. It is used in dyeing wool, dehairing and tanning hides, electroplating, coagulating rubber latex, and regenerating old rubber.

Another example is the lactic acid produced in muscles from the metabolism of glucose:

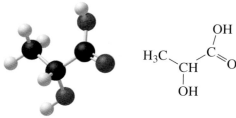

It causes muscle soreness when its concentration builds up. The lactic acid molecule also contains an alcohol functional group.

Another carboxylic acid is linoleic acid. It is a nutritionally essential fatty acid that is also used in manufacturing paints. It also contains two alkene functional groups:

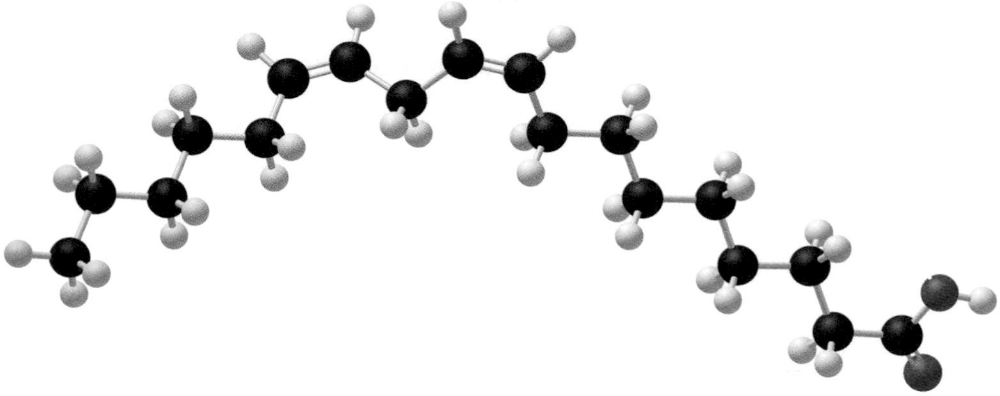

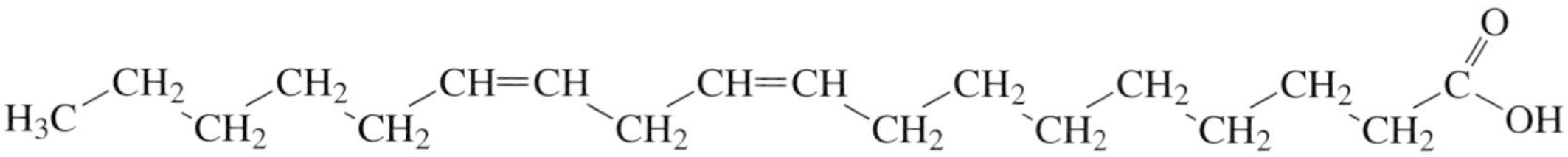

$$H_3C-CH_2-CH_2-CH_2-CH_2-CH=CH-CH_2-CH=CH-CH_2-CH_2-CH_2-CH_2-CH_2-CH_2-CH_2-C(=O)-OH$$

linoleic acid

The most common carboxylic acid is acetic acid, CH_3CO_2H, which is produced as a solution in water, called vinegar, when cider ferments. Fermentation causes the sugar to form ethanol in the presence of catalytic yeast enzymes. In the presence of air, the ethanol oxidizes further to acetic acid:

$$CH_3CH_2OH + O_2 \xrightarrow{\text{yeast}} CH_3CO_2H + H_2O$$

Pure acetic acid is a liquid that has a pungent odor and burns the skin. It is completely soluble in water and is an excellent solvent for many organic compounds, including resins and oils. Acetic acid is used in the synthesis of rubber and various acetate plastics. It is used in printing calico and dyeing silk.

Esters

An **ester** is a compound with the general formula RCO_2R'. In esters, one of the organic groups (often alkyl) is directly attached to the carbon, another is bonded to one of the oxygens, and the other oxygen is double-bonded to the carbon:

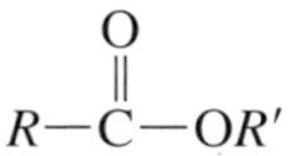

$$R-\overset{\overset{\displaystyle O}{\|}}{C}-OR'$$

The aspirin that interests Omar is both a carboxylic acid and an ester. Another example of an ester is methyl methacrylate, which is used to form the polymer Plexiglas:

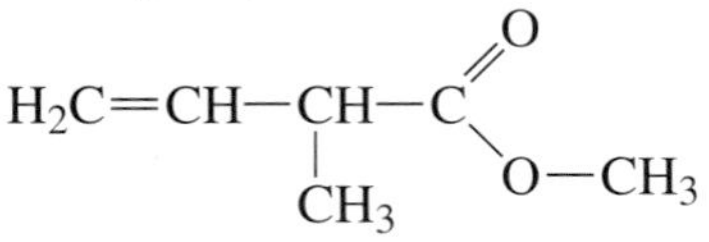

$$H_2C=CH-\underset{\underset{\displaystyle CH_3}{|}}{CH}-C\overset{\displaystyle O}{\underset{\displaystyle O-CH_3}{}}$$

Esters can be formed by a *condensation reaction* of a carboxylic acid with an alcohol. Water is removed in the process, as shown here for the formation of ethyl acetate by the reaction of acetic acid with ethanol:

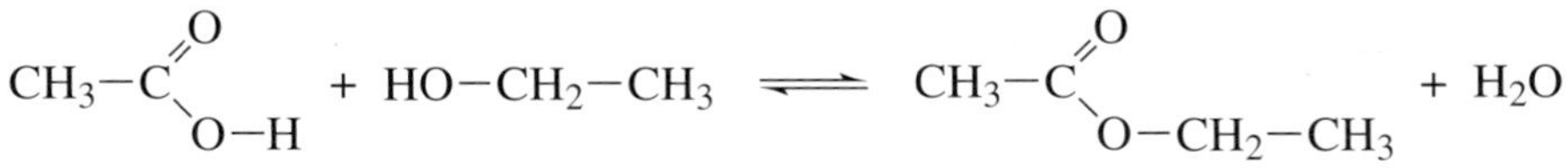

$$CH_3-C\overset{\displaystyle O}{\underset{\displaystyle O-H}{}} + HO-CH_2-CH_3 \rightleftharpoons CH_3-C\overset{\displaystyle O}{\underset{\displaystyle O-CH_2-CH_3}{}} + H_2O$$

Many fruits and flowers contain esters or mixtures of esters that give them their characteristic odors. Some examples are

methyl salicylate   wintergreen

$$\begin{array}{c} O \\ \| \\ C-O-CH_3 \end{array}$$

OH

isoamyl acetate   banana
$(CH_3)_2CHCH_2CH_2O_2CCH_3$

octyl acetate   orange
$CH_3(CH_2)_6CH_2O_2CCH_3$

methyl butyrate   apple
$CH_3O_2CCH_2CH_2CH_3$

Fats and oils are used in the body to synthesize essential tissue constituents. The body also uses them as energy sources. Because they are insoluble, they can be deposited in cells and can remain there unchanged until energy is needed.

Triglycerides are the principal form of fats circulating in the human bloodstream. Most body fat also is in the form of triglycerides. A simple blood test measures the amount of triglycerides in blood. It is one measure of the risk of developing coronary artery disease.

These reactions are reversible and usually require an acid catalyst. Addition of a base neutralizes and removes the carboxylic acid, causing the equilibrium reaction to shift to the left. Most esters are volatile liquids with pleasant odors and flavors. The natural fragrances of flowers and plants often come from esters. Esters are often used as flavoring agents in synthetic and processed foods.

The common animal and vegetable oils and fats are esters, mostly based on an alcohol called glycerol (or 1,2,3-propanetriol). Glycerol reacts with carboxylic acids to form an ester called a *triglyceride:*

$$\begin{array}{ccc}
H_2C-OH & & H_2C-O-\overset{\displaystyle O}{\overset{\|}{C}}-CH_3 \\[2pt]
HC-OH \;+\; 3CH_3CO_2H & \longrightarrow & HC-O-\overset{\displaystyle O}{\overset{\|}{C}}-CH_3 \;+\; 3H_2O \\[2pt]
H_2C-OH & & H_2C-O-\overset{\displaystyle O}{\overset{\|}{C}}-CH_3
\end{array}$$

Esters that form from the reaction of glycerol with high-molar-mass acids are liquid oils or solid fats, depending on the number of double bonds in the alkyl group of the acid. (Such carboxylic acids are called fatty acids.) If the fatty acid is saturated (no double bonds), the triglyceride is a solid, or *fat;* and if the fatty acid is highly unsaturated, with many carbon-carbon double bonds, the triglyceride is a liquid, or *oil.* The greater the unsaturation, the more liquidlike the triglyceride. Common fats include lard, tallow, and milk fat. Common oils are linseed, cottonseed, coconut, soybean, peanut, corn, and olive oils.

A solid ester forms from glycerol and stearic acid, which is a saturated fatty acid, $C_{17}H_{35}CO_2H$:

$$\begin{array}{c}
H_2C-O-\overset{\displaystyle O}{\overset{\|}{C}}-C_{17}H_{35} \\[2pt]
HC-O-\overset{\displaystyle O}{\overset{\|}{C}}-C_{17}H_{35} \\[2pt]
H_2C-O-\overset{\displaystyle O}{\overset{\|}{C}}-C_{17}H_{35}
\end{array}$$

A liquid ester forms from glycerol and linoleic acid, which is an unsaturated fatty acid, $C_{17}H_{31}CO_2H$:

$$\begin{array}{c}
H_2C-O-\overset{\displaystyle O}{\overset{\|}{C}}-C_{17}H_{31} \\[2pt]
HC-O-\overset{\displaystyle O}{\overset{\|}{C}}-C_{17}H_{31} \\[2pt]
H_2C-O-\overset{\displaystyle O}{\overset{\|}{C}}-C_{17}H_{31}
\end{array}$$

Why do these two esters—one a solid and the other a liquid at room temperature—have different melting points? The answer lies in the structure of the carboxylic acids from which the esters were derived:

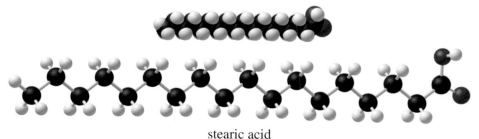

stearic acid

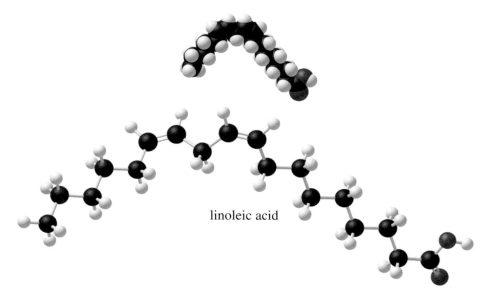

linoleic acid

Note that linoleic acid is bent because of a *cis* arrangement around a carbon-carbon double bond. Which of these acids would have the greater intermolecular forces? Recall the arguments we made for greater intermolecular forces in straight-chain hydrocarbons than in branched-chain hydrocarbons. Since the straight-chain molecules can interact to a greater extent than ones with bent shapes, they have greater intermolecular forces and, as a result, higher melting points. The same logic applies here. Linoleic acid has a melting point of $-12°C$, while stearic acid has a melting point of $70°C$. Thus the esters derived from them have melting points that differ in the same way.

The glyceride of linoleic acid is found in the oils of many seeds and vegetables. Stearic acid comes from animal fats and oils. Do all solid fats come from animals, while liquid fats come from vegetables? Generally, yes, but oils from vegetables can be converted in the laboratory into solids by the process of hydrogenation, in which hydrogen molecules add across the double bonds in the presence of an appropriate catalyst. This is how margarine is made.

Hydrogenation: $C_3H_5(C_{17}H_{31}CO_2)_3 + 6H_2 \longrightarrow C_3H_5(C_{17}H_{35}CO_2)_3$
                                unsaturated   saturated

The sodium compounds of the carboxylic acids obtained from fats and oils are used in soaps. Their anions have polar ends that attract water and nonpolar ends that attract grease and oil, dispersing them into the water. Soap is prepared from animal fats by *saponification,* a reaction with sodium hydroxide (lye) that cleaves the ester linkages and converts the ester to sodium carboxylates and glycerol:

$$C_3H_5(O_2CC_{17}H_{35})_3 + 3NaOH \xrightarrow{\text{heat}} 3NaO_2CC_{17}H_{35} + C_3H_5(OH)_3$$

16.8 AMINES

Aisha was interested in the aromatic molecule amphetamine. Because it contains a nitrogen atom singly bonded to a carbon, amphetamine is also an example of an organic amine. Another aromatic amine is aniline, $C_6H_5NH_2$, which is used in dyes. An **amine** can be considered a derivative of ammonia, NH_3, with one or more hydrogen atoms replaced by organic groups. Like ammonia, amines act as bases in chemical reactions. These covalent substances have a disagreeable fishy smell. Many complex amines, such as caffeine and nicotine, interact with living cells and affect their functioning.

caffeine   nicotine

16.9 MOLECULES WITH MULTIPLE FUNCTIONAL GROUPS

Many important molecules contain more than one functional group, especially those that operate in biological systems. Aspirin, selected by Omar, has both the carboxylic acid and the ester functional groups. Retinal, selected by Maya, has aldehyde and alkene functional groups.

Another important category is the amino acids, which are the building blocks of proteins in plants and animals. Proteins make up most of body tissue. Enzymes, viruses, antibodies, muscle, tendons, hair, and fingernails are all composed at least in part of proteins. Proteins are composed largely of carbon (~52%), hydrogen (~7%), oxygen (~23%), nitrogen (~16%), and lesser amounts of other elements such as sulfur, iodine, phosphorus, iron, and copper. They are very complex substances of high molar mass (about 15,000 to several million g/mol) composed of various combinations of 20 different amino acids.

An **amino acid** is a relatively small organic molecule that contains both the amine group ($-NH_2$) and the carboxylic acid ($-CO_2H$) functional groups. Amino acids are amphoteric. They can consume OH^- by converting $-CO_2H$ to $-CO_2^-$. They can consume H^+ by converting $-NH_2$ to $-NH_3^+$.

The amino acids that are important in living things are the *alpha amino acids*. The structures of some amino acids are shown in Figure 16.11. In each one, the $-NH_2$ and $-CO_2H$ groups are attached to the same carbon. The simplest of the amino acids is glycine:

$$H_2N-\underset{\underset{H}{|}}{\overset{\overset{CO_2H}{|}}{C}}-H$$

All the amino acids except proline have this structure in common:

$$H_2N-\underset{\underset{R}{|}}{\overset{\overset{CO_2H}{|}}{C}}-H$$

The other parts of their structures—the side chains R on this group—are all different. The side chains of the amino acids determine the differences in their properties, such as solubility and acidity.

Proteins are built up from amino acids, as shown in Figure 16.12, by the elimination of a water molecule during a *condensation reaction* of the $-CO_2H$ group from one amino acid with the $-NH_2$ group from another. The result is a peptide linkage or *peptide bond:*

$$-\underset{\overset{|}{H}}{\overset{\overset{H}{|}}{N}}\boxed{-H \quad HO}-\overset{\overset{O}{\|}}{C}- \longrightarrow -\underset{\overset{|}{N}}{\overset{\overset{H}{|}}{N}}-\overset{\overset{O}{\|}}{C}- + H_2O$$

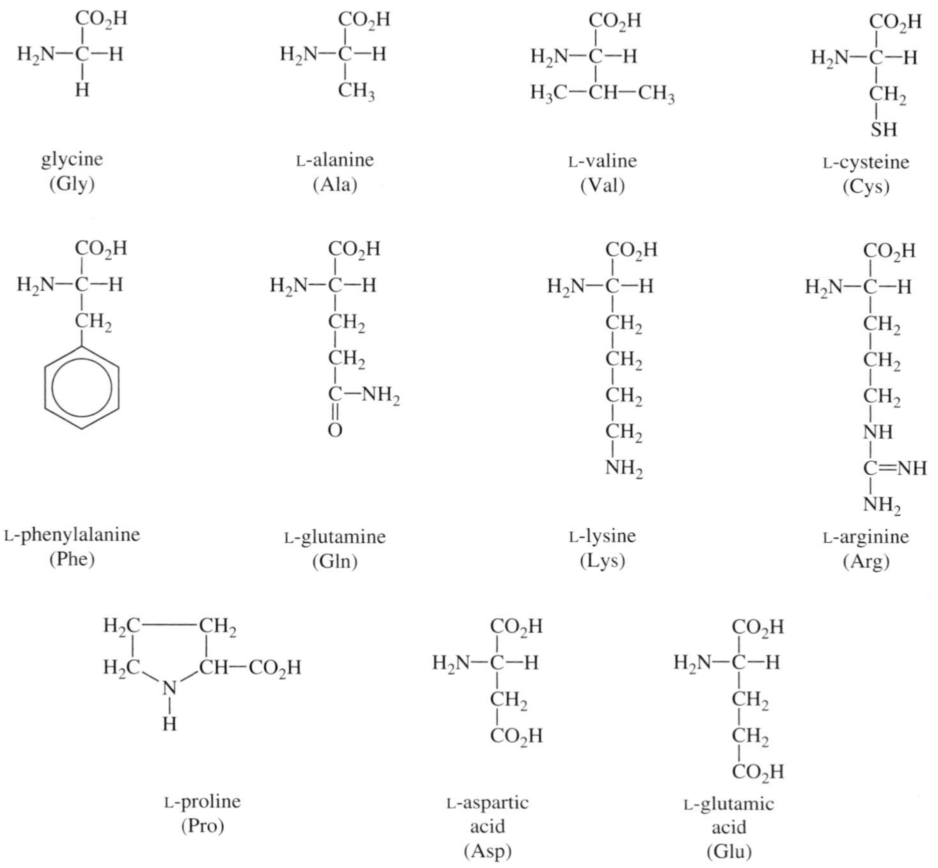

$$CO_2H$$
$$H_2N-\overset{|}{\underset{|}{C}}-H$$
$$H$$

glycine
(Gly)

$$CO_2H$$
$$H_2N-\overset{|}{\underset{|}{C}}-H$$
$$CH_3$$

L-alanine
(Ala)

$$CO_2H$$
$$H_2N-\overset{|}{\underset{|}{C}}-H$$
$$H_3C-CH-CH_3$$

L-valine
(Val)

$$CO_2H$$
$$H_2N-\overset{|}{\underset{|}{C}}-H$$
$$CH_2$$
$$SH$$

L-cysteine
(Cys)

Figure 16.11
Structures of some alpha amino acids.

$$CO_2H$$
$$H_2N-\overset{|}{\underset{|}{C}}-H$$
$$CH_2$$

L-phenylalanine
(Phe)

$$CO_2H$$
$$H_2N-\overset{|}{\underset{|}{C}}-H$$
$$CH_2$$
$$CH_2$$
$$\overset{|}{C}-NH_2$$
$$O$$

L-glutamine
(Gln)

$$CO_2H$$
$$H_2N-\overset{|}{\underset{|}{C}}-H$$
$$CH_2$$
$$CH_2$$
$$CH_2$$
$$CH_2$$
$$NH_2$$

L-lysine
(Lys)

$$CO_2H$$
$$H_2N-\overset{|}{\underset{|}{C}}-H$$
$$CH_2$$
$$CH_2$$
$$CH_2$$
$$NH$$
$$C=NH$$
$$NH_2$$

L-arginine
(Arg)

$$H_2C-CH_2$$
$$H_2C\qquad CH-CO_2H$$
$$N$$
$$H$$

L-proline
(Pro)

$$CO_2H$$
$$H_2N-\overset{|}{\underset{|}{C}}-H$$
$$CH_2$$
$$CO_2H$$

L-aspartic
acid
(Asp)

$$CO_2H$$
$$H_2N-\overset{|}{\underset{|}{C}}-H$$
$$CH_2$$
$$CH_2$$
$$CO_2H$$

L-glutamic
acid
(Glu)

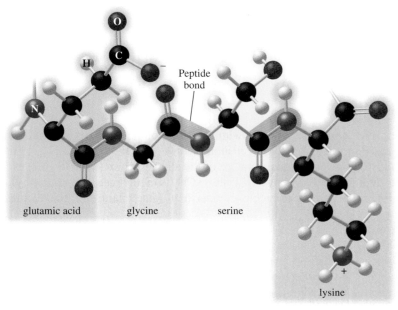

glutamic acid   glycine   serine   lysine

Peptide bond

Figure 16.12
A protein is one or more chains of amino acids linked together by peptide bonds.
Eliminating a water molecule forms a peptide bond.

The formation of a peptide bond is analogous to the formation of esters with alcohols. Both eliminate a water molecule when forming the new linkage. Consider the following reaction between two amino acids:

$$\underset{\substack{\text{HO}-\text{C}-\text{CH}-\text{NH}_2}}{\overset{\substack{\text{O} \;\; \text{CH}_3}}{}} + \underset{\substack{\text{HO}-\text{C}-\text{CH}_2-\text{NH}_2}}{\overset{\text{O}}{}} \longrightarrow \underset{\substack{\text{HO}-\text{C}-\text{CH}-\text{N}-\text{C}-\text{CH}_2-\text{NH}_2}}{\overset{\substack{\text{O} \;\; \text{CH}_3 \;\;\;\; \text{O}}}{}} + \text{H}_2\text{O}$$

Note that a molecule of water forms from an −OH group on the carboxylic acid and a hydrogen on the amine group. The condensation product has a feature in common with the reactant amino acids, since it has a carboxylic acid group at one end and an amine group at the other end. Thus this condensation product can react further with another amino acid:

$$\underset{\substack{\text{HO}-\text{C}-\text{CH}_2-\text{NH}_2}}{\overset{\text{O}}{}} + \underset{\substack{\text{HO}-\text{C}-\text{CH}-\text{N}-\text{C}-\text{CH}_2-\text{NH}_2}}{\overset{\substack{\text{O} \;\; \text{CH}_3 \;\;\;\; \text{O}}}{}} \longrightarrow$$

$$\underset{\substack{\text{HO}-\text{C}-\text{CH}_2-\text{N}-\text{C}-\text{CH}-\text{N}-\text{C}-\text{CH}_2-\text{NH}_2}}{\overset{\substack{\text{O} \;\;\;\;\;\; \text{O} \;\; \text{CH}_3 \;\;\;\; \text{O}}}{}} + \text{H}_2\text{O}$$

This process can continue to form longer and longer chains. When two amino acids react to form a molecule, that molecule is called a *dipeptide*. Three amino acids condense to form a *tripeptide*, and a polymer consisting of many amino acid units is called a *polypeptide*. Although the borderline is arbitrary, a polypeptide having a molar mass greater than about 5000 g/mol is called a *protein*.

16.10 ORGANIC NOMENCLATURE

Organic nomenclature helps us communicate structures without drawing a complete structural formula. Although many names are in common use, the IUPAC system is the standard system used worldwide. Once you learn the basics of the IUPAC system, you can easily name a variety of organic compounds with various functional groups.

Alkanes

To use IUPAC nomenclature, start with the names of the alkyl groups derived from the alkanes by removal of one hydrogen atom, with a general formula C_nH_{2n+1}. The names of the alkyl groups are similar to those of the alkanes given in Table 16.2, except that the *-ane* ending is replaced by *–yl*. Thus, $-CH_3$ is methyl and $-C_2H_5$ is ethyl.

A few rules allow us to achieve consistent nomenclature for the alkanes.

1. Determine the *longest* continuous carbon chain in the hydrocarbon. The name of this alkane constitutes the parent name.
2. Number the carbon atoms in the parent hydrocarbon from the end nearer a branch, so that branch will have the lowest possible position number.
3. Branches attached to the parent chain are named as alkyl groups.
4. The name consists of the position number, a hyphen, the branch name, and the straight-chain name. If there is more than one branch, name them in alphabetical order. If the branches are the same alkyl group, precede the name with a prefix (*di–, tri–*, etc.) indicating the total number of that alkyl group in the molecule, and separate the position numbers with commas.

The cycloalkanes are named in the same way as the straight-chain alkanes, but with the prefix *cyclo–*. Thus, C_6H_{12}, in which six carbons are bonded in a ring, is named cyclohexane.

Alkenes and Alkynes

The names of the alkenes use the alkane root, with the ending *-ene* replacing the *-ane* ending. If a double bond has more than one possible position, a number indicates its location. The carbon atoms are numbered from the end nearer the double bond. For the following compound, the carbon chain is numbered from left to right, which places the C=C bond between carbon atoms 1 and 2:

$$\overset{1}{H_2}C=\overset{2}{C}H-\overset{3}{C}H_2-\overset{4}{C}H_3$$

This molecule has four carbon atoms, so it is derived from butane. It has a double bond, so the *-ane* ending is replaced by *-ene,* giving it the name butene. The position of the double bond is designated by the lowest-numbered carbon involved in that bond, so this molecule is 1-butene. Similarly the molecule $CH_3-CH=CH-CH_3$ is named 2-butene.

The names for side chains on an alkene follow the rules for naming alkyl groups. Side chain names have the *-yl* ending, and a number designates the position of attachment. Again, numbering starts at the end nearest the double bond.

A molecule may have more than one double bond. In such a case, an appropriate prefix (such as *di-*) is placed just before the *-ene* ending to indicate the number of double bonds. For example, $H_2C=CH-CH=CH_2$ is named 1,3-butadiene.

Alkynes are named the same as alkenes, but using a *-yne* ending instead of an *-ene* ending.

> If the double bond is closest to the right side of the molecule, as written, then we start counting from that end of the molecule:
>
> $$\overset{4}{H_3}C-\overset{3}{C}H_2-\overset{2}{C}H=\overset{1}{C}H_2$$

> Some alkenes, such as $CH_3-CH=CH-CH_3$, have geometric isomers. This molecule would be named *cis*-2-butene or *trans*-2-butene.

Aromatic Hydrocarbons

An aromatic group formed by removal of a hydrogen atom from a benzene ring is called an **aryl group,** which is the aromatic analog of an alkyl group. The aryl group formed from benzene, C_6H_5, is called phenyl. For example, $C_6H_5C\equiv CH$ is commonly known as phenylacetylene.

If two or more positions on a benzene ring have substituted groups, the positions are indicated by numbers, counted from the location of the first substituted group. If two methyl groups are substituted on benzene, the molecule is dimethylbenzene or xylene. There are three isomers: 1,2-dimethylbenzene, 1,3-dimethylbenzene, and 1,4-dimethylbenzene. These and several other examples are summarized here:

methylbenzene (toluene)   hydroxybenzene (phenol)   aminobenzene (aniline)

1,2-dimethylbenzene   1,3-dimethylbenzene   1,4-dimethylbenzene

TABLE 16.8	Rules for Naming Other Organic Molecules
Class of Compound	**Rules for Naming**
Alcohols	The common names of the alcohols consist of the alkyl group name followed by the word *alcohol*. The IUPAC names use the alkane root and replace the *-e* ending with *-ol*. Thus CH_3OH is methyl alcohol or methanol, C_2H_5OH is ethyl alcohol or ethanol, and C_3H_7OH is propyl alcohol or propanol.
Ethers	The common names of the ethers consist of the names of the substituted alkyl groups, listed alphabetically, and followed by the word *ether,* as in diethyl ether ($C_2H_5OC_2H_5$) or ethylmethyl ether ($C_2H_5OCH_3$).
Aldehydes	Common names use the ending *-aldehyde,* such as in formaldehyde (HCHO) or acetaldehyde (CH_3CHO). IUPAC names use the alkane name with the ending *-al,* such as in methanal (HCHO) or ethanal (CH_3CHO).
Ketones	Common names use the names of the two alkyl groups attached to the $C=O$ group, followed by the word *ketone*. IUPAC names use the alkane name with the ending *-one*. The solvent commonly called acetone, $(CH_3)_2C=O$, is dimethyl ketone or propanone.
Carboxylic acids	Carboxylic acids often have common names, such as formic acid (HCO_2H) and acetic acid (CH_3CO_2H). The IUPAC names use the alkane name with the ending *-oic acid*. The two examples are methanoic acid and ethanoic acid, respectively.
Esters	Esters have an alkyl group replacing the acidic hydrogen in a carboxylic acid. They are named in the same way as the carboxylic acids from which they are derived, using an *-oate* ending. Thus, $HCO_2CH_2CH_3$ is ethyl methanoate (or ethyl formate), and $CH_3CO_2CH_3$ is methyl ethanoate (or methyl acetate).
Amines	Amines are commonly named by listing the alkyl groups attached to nitrogen, followed by the word *amine*. For example, $NH_2CH_2CH_3$ is ethylamine and $NH(CH_3)_2$ is dimethylamine.

Other Naming Conventions

The other major classes of organic compounds are named by similar rules, which are summarized in Table 16.8.

SUMMARY

Organic compounds contain carbon, hydrogen, and sometimes include other elements. Carbon makes up great numbers of compounds because its atoms form four strong bonds. Carbon can bond to other carbon atoms to form chains or rings. It can have single, double, or triple bonds, and it bonds easily to oxygen and nitrogen. Many organic compounds exist as isomers with the same molecular formula but different structures, physical properties, and reactivities. For this reason, we usually use structural formulas, condensed structural formulas, or line structures when describing organic compounds.

Organic compounds are classified by their shape, the kinds of bonds they contain, and the presence of functional groups. Functional groups impart specific properties to organic compounds; they are generally the part of the molecule that undergoes characteristic reactions.

Hydrocarbons contain only carbon and hydrogen, but can have single, double, or triple bonds between carbon atoms. Hydrocarbons are generally obtained from petroleum. The simplest hydrocarbons are the alkanes, which have only single bonds between carbon atoms. Alkanes are sometimes called saturated hydrocarbons because they have the maximum possible number of hydrogen atoms (C_nH_{2n+2} for chain alkanes). Because of the lack of functional groups, alkanes are fairly unreactive, with the exception of combustion reactions and other reactions that must be initiated by heat or light.

Alkenes contain double bonds and alkynes contain triple bonds. Alkenes and alkynes undergo reactions, such as polymerization and addition reactions, which result in the loss of multiple bonds. The most common aromatic hydrocarbons contain a benzene ring, C_6H_6. The delocalization of electrons in benzene makes it less reactive than alkenes.

Alcohols ($R-OH$) and ethers ($R-O-R'$) contain carbon-oxygen single bonds. Alcohols are liquids or solids at room temperature because of hydrogen bonding. Reactions of alcohols and ethers occur at the oxygen part of the molecule. Alcohols can be oxidized to aldehydes, and then further to carboxylic acids or to ketones.

Aldehydes and ketones contain a carbonyl group ($C=O$). In aldehydes the carbonyl group has at least one hydrogen atom bonded to the carbon atom of the carbonyl group, so it is found at the end of carbon chains.

Carboxylic acids (RCO_2H) are the "organic acids," because their reactivity often involves the loss of H^+. Esters (RCO_2R') form from carboxylic acids and alcohols. Esters are responsible for many of the pleasant odors of plants.

Amines contain one to three carbon atoms with single bonds to a nitrogen atom (RNH_2, R_2NH, or R_3N). Amines generally act as bases in reactions, and are found in drugs such as caffeine and nicotine. Amino acids contain both the amine group and the carboxylic acid group. Amino acids combine to form polypeptides. Proteins are very large polypeptides.

Organic nomenclature helps us communicate without drawing structures. Although many common names are used frequently, the IUPAC system is the international standard.

KEY TERMS

acyclic hydrocarbon (16.2)	alkyl group (16.3)	carboxylic acid (16.7)	hydrocarbon (16.1)
alcohol (16.5)	alkyne (16.3)	cyclic hydrocarbon (16.2)	isomer (16.1)
aldehyde (16.6)	amine (16.8)	cycloalkane (16.4)	ketone (16.6)
aliphatic hydrocarbon (16.2)	amino acid (16.9)	ester (16.7)	organic chemistry (16.1)
alkane (16.3)	aromatic hydrocarbon (16.2)	ether (16.5)	saturated hydrocarbon (16.2)
alkene (16.3)	aryl group (16.10)	functional group (16.1)	unsaturated hydrocarbon (16.2)

QUESTIONS AND PROBLEMS

The following questions and problems, except for those in the *Additional Questions* section, are paired. Questions in a pair focus on the same concept. Answers to the odd-numbered questions and problems are in Appendix D.

Matching Definitions with Key Terms

16.1   Match the key terms with the descriptions provided.
  (a) a compound having the same composition and formula as another compound, but a different structure
  (b) an organic compound containing the functional group $-OH$
  (c) a condensation product of carboxylic acids and alcohols, having the general formula RCO_2R'
  (d) an organic compound containing the functional group $-CHO$
  (e) a hydrocarbon that has a ring or closed-chain structure
  (f) a hydrocarbon that contains only carbon-carbon single bonds
  (g) a group derived from an alkane by removal of one hydrogen atom
  (h) an alkane that contains a ring structure
  (i) a hydrocarbon that contains localized single, double, or triple bonds
  (j) a small molecule that contains both the $-NH_2$ and the $-CO_2H$ functional groups
  (k) the study of organic compounds
  (l) a hydrocarbon that does not contain all the hydrogen atoms that could possibly be included

16.2   Match the key terms with the descriptions provided.
  (a) a hydrocarbon that contains one or more localized carbon-carbon double bonds
  (b) an organic compound containing an oxygen double-bonded to a carbon, which is also bonded to two other carbons: $R_2C=O$
  (c) an organic compound containing the functional group $-CO_2H$
  (d) a reactive part of a molecule that undergoes characteristic reactions
  (e) a hydrocarbon containing one or more triple bonds
  (f) an organic compound that contains the $-O-$ functional group attached to two organic groups: $R-O-R'$
  (g) an aromatic group formed by removal of a hydrogen from a cyclic aromatic hydrocarbon
  (h) a cyclic hydrocarbon with delocalized bonding, which imparts special stability
  (i) a compound that contains only carbon and hydrogen
  (j) a hydrocarbon that contains all the hydrogens it can hold
  (k) a hydrocarbon that has a straight- or branched-chain structure with only single bonds
  (l) an organic base with one or more alkyl groups bonded to a nitrogen atom

Representations of Organic Molecules

16.3   Describe the type of bonds in each of the following molecules.
  (a) CH_3CH_3
  (b) CH_2CH_2
  (c) $CHCH$

16.4   Describe the type of bonds in each of the following molecules.
  (a) $CH_3CH_2CH_3$
  (b) CH_3CHCH_2
  (c) $CHCCH_3$

16.5   Identify and classify the functional group in each of the following molecules.
  (a) CH_3NH_2
  (b) $H_3C-\overset{\overset{\displaystyle O}{\|}}{C}-CH_3$
  (c) CH_3CH_2OH
  (d) $CH_2=CHCH_3$
  (e) $CH_3CH_2CH_2CO_2H$
  (f) $CH_3C\equiv CH$
  (g) $H_3C-CH_2-\overset{\overset{\displaystyle O}{\|}}{C}-H$
  (h) $CH_3CH_2OCH_3$
  (i) $H_3C-CH_2-CH_2-\overset{\overset{\displaystyle O}{\|}}{C}-O-CH_3$

16.6   Identify and classify the functional group in each of the following molecules.
  (a) $CH_3CH_2C\equiv CH$
  (b) $H_3C-\overset{\overset{\displaystyle O}{\|}}{C}-O-CH_3$
  (c) $CH_3CH_2NH_2$
  (d) $H_3C-CH_2-\overset{\overset{\displaystyle O}{\|}}{C}-CH_3$
  (e) $CH_3CH_2OCH_2CH_3$
  (f) $H_3C-\overset{\overset{\displaystyle OH}{|}}{CH}-CH_3$
  (g) CH_3CO_2H
  (h) $CH_3CH_2CH=CH_2$
  (i) $H_3C-\overset{\overset{\displaystyle O}{\|}}{C}-H$

16.7   From the ball-and-stick model, determine the molecular formula and draw the structural formula and the line structure for this molecule.

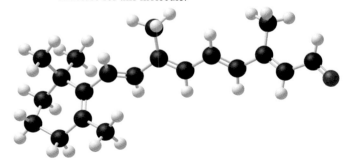

16.8 From the ball-and-stick model, determine the molecular formula and draw the structural formula and the line structure for this molecule.

16.9 Write molecular formulas for each of the following.

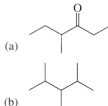

(a)

(b)

16.10 Write molecular formulas for each of the following.

(a)

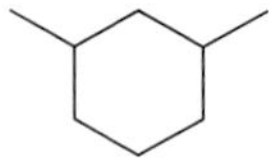

(b)

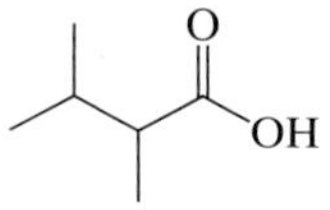

16.11 Draw line structures for each of the following.
   (a) $CH_3(CH_2)_6CH_3$
   (b) $CH_3CH_2COCH_2CH_3$
   (c) $(CH_3)_2CHCH_2CH_2CHCH_2$
   (d) $CH_3(CH_2)_5CHO$

16.12 Draw line structures for each of the following.
   (a) $CH_3(CH_2)_8CH_3$
   (b) $CH_3CH_2CH_2COCH_2CH_3$
   (c) $CH_3CH_2CHCHCH(CH_3)_2$
   (d) $CH_3(CH_2)_5CO_2H$

16.13 Write structural formulas and draw the line structures for each of the following.

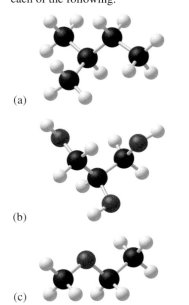

(a)

(b)

(c)

16.14 Write structural formulas and draw the line structures for each of the following.

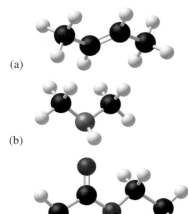

(a)

(b)

(c)

16.15 Draw a perspective drawing for the following molecules.
   (a) CH_4
   (b) CH_3CH_3
   (c) CH_2ClCH_2Cl

16.16 Draw a perspective drawing for the following molecules.
   (a) CH_3Cl
   (b) CCl_3CH_3
   (c) CH_3CHCl_2

Hydrocarbons

16.17 Distinguish between alkanes, alkenes, and alkynes.

16.18 Write line structures for an alkane, an alkene, and an alkyne, each of which contain four carbon atoms.

16.19 Label each of the following molecules as saturated or unsaturated.
   (a) 
   (b) C_4H_6
   (c) $CH_3CHCHCH_2CH_3$

16.20 Label each of the following molecules as saturated or unsaturated.
   (a) $CH_3CH_2CHCH_2$
   (b) C_3H_8

   (c)

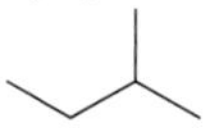

16.21 What distinguishes an aromatic hydrocarbon from other hydrocarbons?

16.22 What feature is common to all aromatic hydrocarbons?

Acyclic Hydrocarbons

16.23 Explain the difference in boiling points of the two molecules represented by the following line structures.

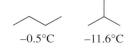

−0.5°C   −11.6°C

16.24 Explain the difference in boiling points of the three molecules represented by the following line structures.

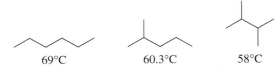

69°C   60.3°C   58°C

16.25 Which compound in each group is expected to have the strongest intermolecular forces?
  (a) CH_4, CH_3CH_3, $CH_3CH_2CH_3$
  (b) $CH_3(CH_2)_4CH_3$, $CH_3CH(CH_3)CH(CH_3)CH_3$, $CH_3CH_2C(CH_3)_2CH_3$

16.26 Which compound in each group is expected to have the strongest intermolecular forces?
  (a) CH_2CH_2, CH_3CHCH_2, $CH_3CHCHCH_3$
  (b) $CH_3CH_2CH_2CH_2CH_3$, $CH_3C(CH_3)_3$, $CH_3CH_2CH(CH_3)_2$

16.27 What is the name of each of the following straight-chain alkanes?
  (a) $CH_3CH_2CH_2CH_2CH_3$
  (b) CH_3CH_3
  (c) $CH_3CH_2CH_2CH_2CH_2CH_2CH_2CH_3$
  (d) $CH_3CH_2CH_3$
  (e) $CH_3CH_2CH_2CH_2CH_2CH_2CH_2CH_2CH_3$

16.28 What is the name of each of the following straight-chain alkanes?
  (a) $CH_3CH_2CH_2CH_3$
  (b) $CH_3CH_2CH_2CH_2CH_3$
  (c) $CH_3CH_2CH_2CH_2CH_2CH_2CH_2CH_2CH_2CH_3$
  (d) $CH_3CH_2CH_2CH_2CH_2CH_2CH_3$
  (e) $CH_3CH_2CH_2CH_2CH_2CH_3$

16.29 Draw all the possible isomers with the formula C_6H_{14}.

16.30 Draw all the possible isomers with the formula C_5H_{12}.

16.31 Write IUPAC names for the following compounds.
  (a) $CH_3CH_2CH_2CH(CH_3)CH_3$
  (b) $CH_3CH(CH_3)CH_2CH_3$
  (c) $CH_3CH_2CH(CH_3)CH_3$

16.32 Write IUPAC names for the following compounds.
  (a) $CH_3CH(CH_3)_2$
  (b) $CH_3CH_2CH(CH_3)CH_2CH_2CH_3$
  (c) $CH_3CH_2CH(CH_2CH_3)CH_2CH_2CH_3$

16.33 Write IUPAC names for the following compounds.
  (a) $CH_3CH_2C(CH_3)_2CH_2CH_3$
  (b) $CH_3C(CH_3)_2CH_3$
  (c) $CH_3CH(CH_3)CH(CH_3)CH_3$

16.34 Write IUPAC names for the following compounds.
  (a) $CH_3CH_2C(CH_3)_3$
  (b) $CH_3C(CH_3)_2C(CH_3)_2CH_3$
  (c) $CH_3C(CH_3)_2CH(CH_3)CH_3$

16.35 Write structural formulas for the following compounds.
  (a) 2-methylpropane
  (b) 3,3-diethyl-2-methylpentane
  (c) 2,4-dimethyloctane
  (d) 3-ethyl-3-methylhexane

16.36 Write structural formulas for the following compounds.
  (a) 2-methylbutane
  (b) 3,3-diethyl-2-methylhexane
  (c) 3,5-dimethylnonane
  (d) 3-ethyl-2,2-dimethylhexane

16.37 Predict the products of the following reactions.
  (a) $CH_3CH_2CH_3 + Cl_2 \xrightarrow{\text{heat}}$
  (b) $(CH_3)_3CCH_2CH_2CH_3 + O_2 \xrightarrow{\text{heat}}$
  (c) $CH_3CH_3 + Br_2 \xrightarrow{\text{light}}$

16.38 Predict the products of the following reactions.
  (a) $CH_4 + Cl_2 \xrightarrow{\text{heat}}$
  (b) $CH_3CH_2CH_3 + O_2 \xrightarrow{\text{heat}}$
  (c) $CH_3CH_2CH_3 + Br_2 \xrightarrow{\text{light}}$

16.39 What is an alkene?

16.40 What is an alkyne?

16.41 Write IUPAC names for the following compounds.
  (a) $CH_3CH_2CH{=}CHCH(CH_3)CH_2CH_3$
  (b)
$$\begin{matrix} H_3C \\ & C{=}C \\ H_3C \end{matrix} \begin{matrix} CH_3 \\ \\ CH_2{-}CH_2{-}CH_3 \end{matrix}$$
  (c)
$$\begin{matrix} H_3C{-}CH_2 \\ & C{=}C \\ H_3C \end{matrix} \begin{matrix} CH_3 \\ \\ CH_2{-}CH_3 \end{matrix}$$

16.42 Write IUPAC names for the following compounds.
  (a) $(CH_3)_2C{=}CHCH_3$
  (b) $CH_3CH{=}CHC(CH_3)_3$
  (c)
$$\begin{matrix} H_3C{-}CH_2 \\ & C{=}C \\ H_3C \end{matrix} \begin{matrix} CH_2{-}CH_3 \\ \\ CH_3 \end{matrix}$$

16.43 Which of the following compounds can exist as *cis* and *trans* isomers?
  (a) $CH_2{=}CH_2$
  (b) $(CH_3)_2C{=}CHCH_3$
  (c) $CH_3CH_2C(CH_3){=}C(CH_3)CH_2CH_3$

16.44 Which of the following compounds can exist as *cis* and *trans* isomers?
  (a) $CH_3CH{=}CHCH_3$
  (b) $CH_3CH{=}CH_2$
  (c) $(CH_3)_2C{=}C(CH_3)_2$

16.45 Write line structures for the following compounds.
  (a) 3-methyl-1-butene
  (b) propyne
  (c) *trans*-2-butene
  (d) *cis*-3-hexene

16.46 Write line structures for the following compounds.
  (a) 2,3,3-trimethyl-1-hexene
  (b) 1-butyne
  (c) *cis*-2-butene
  (d) *trans*-3-hexene

16.47 Complete the following reactions.
  (a) $CH_3CH{=}CHCH_3 + HBr \longrightarrow$
  (b) $CH_3CH_2CH{=}CHCH_3 + Br_2 \longrightarrow$
  (c) $CH_3CH{=}CHCH_3 + H_2 \xrightarrow{\text{Pd}}$

16.48 Complete the following reactions.

(a) $CH_3CH_2CH=CHCH_2CH_3 + HBr \longrightarrow$

(b) $CH_2=CHCH_2CH_3 + Br_2 \longrightarrow$

(c) $CH_3CH_2CH=CHCH_2CH_3 + H_2 \xrightarrow{\text{Pd}}$

Cyclic Hydrocarbons

16.49 What feature is common to all aromatic hydrocarbons?

16.50 Which of the following are aromatic?

(a) $CH_2=CHCH=CH_2$

(b)

(c)

16.51 Name the following compounds.

(a)

(b)

16.52 Name the following compounds.

(a)

(b)

16.53 Write structural formulas for the following compounds.
(a) 1,2,4-trimethylbenzene
(b) chlorobenzene

16.54 Write structural formulas for the following compounds.
(a) 2,5-dichloro-1-ethylbenzene
(b) 2-ethyl-1,3-dimethylbenzene

16.55 Write an equation for the reaction of benzene (C_6H_6) with chlorine (Cl_2) in the presence of an $FeCl_3$ catalyst.

16.56 Write an equation for the reaction of benzene (C_6H_6) with chloromethane (CH_3Cl) in the presence of an $AlCl_3$ catalyst.

Alcohols and Ethers

16.57 Which of the following compounds are alcohols?
(a) CH_3OCH_3
(b) CH_3CH_2OH
(c) $CH_3CH(OH)CH_3$
(d) CH_3COCH_3
(e) CH_3CH_2CHO

16.58 Which of the following compounds are alcohols?
(a) CH_3CHO
(b) $(CH_3)_2C(OH)CH_3$
(c) $CH_3CH_2COCH_3$
(d) $CH_3CH_2OCH_2CH_3$
(e) $CH_3CH_2CH_2CH_2OH$

16.59 An alcohol has the formula C_4H_9OH. How many isomers can exist for this alcohol?

16.60 An alcohol has the formula $C_5H_{11}OH$. How many isomers can exist for this alcohol?

16.61 Name the following compounds using the IUPAC system.
(a) $(CH_3)_2CHCH_2OH$
(b) $CH_3CH(OH)CH_2CH_3$

16.62 Name the following alcohols using the IUPAC system.
(a) $CH_3CH_2CH(OH)CH_2CH_3$
(b) $(CH_3)_2CHCH_2CH_2CH_2OH$

16.63 Write line structures for the following compounds.
(a) 3-hexanol
(b) 2,2-dimethyl-3-pentanol

16.64 Write line structures for the following compounds.
(a) 2-pentanol
(b) 3-ethyl-2-hexanol

16.65 Explain why CH_3CH_2OH has a boiling point of 78°C, but $CH_3CH_2CH_3$ has a boiling point of –42°C.

16.66 Explain why $CH_3CH_2CH_3$ is insoluble in water, but CH_3CH_2OH is soluble.

16.67 What is an ether?

16.68 How do ethers differ from alcohols?

16.69 Which of the following compounds are ethers?
(a) CH_3OCH_3
(b) CH_3CH_2OH
(c) $CH_3CH(OH)CH_3$
(d) CH_3COCH_3
(e) CH_3CH_2CHO

16.70 Which of the following compounds are ethers?
(a) CH_3CHO
(b) $(CH_3)_2C(OH)CH_3$
(c) $CH_3CH_2COCH_3$
(d) $CH_3CH_2OCH_2CH_3$
(e) $CH_3CH_2CH_2CH_2OH$

16.71 Name the following compounds.
(a) $CH_3CH_2-O-CH_2CH_2CH_3$
(b) CH_3-O-CH_3

16.72 Name the following compounds.
(a) $CH_3CH_2-O-CH_2CH_3$
(b) $CH_3CH_2-O-CH_3$

16.73 Write line structures for the following compounds.
(a) ethyl methyl ether
(b) ethyl phenyl ether

16.74 Write line structures for the following compounds.
  (a) methyl propyl ether
  (b) diethyl ether

16.75 Which would have the greater solubility in water, CH_3CH_2OH or CH_3OCH_3? Explain your answer.

16.76 Which would have the higher boiling point, $CH_3CH_2OCH_2CH_3$ or $CH_3CH_2CH_2CH_2OH$? Explain your answer.

Aldehydes and Ketones

16.77 What is an aldehyde?

16.78 What is a ketone?

16.79 Which of the following compounds are aldehydes?
  (a) CH_3CHO
  (b) CH_3COCH_3
  (c) CH_3CH_2CHO
  (d) $HCHO$

16.80 Which of the following compounds are ketones?
  (a) CH_3CHO
  (b) CH_3COCH_3
  (c) $CH_3COCH_2CH_3$
  (d) $HCHO$

16.81 Explain the relative boiling points of these two compounds: HCHO at –21°C and CH_3CHO at 20°C.

16.82 Explain why CH_3CHO is soluble in water, but CH_3CH_2CHO has only limited solubility.

Carboxylic Acids and Esters

16.83 What is a carboxylic acid?

16.84 What is an ester?

16.85 What reagents can be used to prepare the ester $CH_3CH_2CO_2CH_3$?

16.86 What reagents can be used to prepare the ester $CH_3CH_2CH_2CO_2CH_2CH_3$?

16.87 What is a saturated fatty acid?

16.88 What is an unsaturated fatty acid?

16.89 Predict and explain the relative boiling points of CH_3CO_2H and CH_3CH_2CHO.

16.90 Predict and explain the relative solubilities in water of $CH_3(CH_2)_2CO_2H$ and $CH_3(CH_2)_5CO_2H$.

Amines

16.91 What is an amine?

16.92 What makes an amine basic?

16.93 Draw line structures for the following amines.
  (a) $(CH_3)_2NH$
  (b) $(CH_3)_2NC_2H_5$
  (c) $C_6H_5NH_2$

16.94 Draw line structures for the following amines.
  (a) $CH_3CH_2CH_2NH_2$
  (b) $CH_3N(C_2H_5)_2$
  (c) $(C_6H_5)_2NH$

16.95 Explain why CH_3CH_3 has a boiling point of –88.6°C, but the boiling point of CH_3NH_2 is –6.3°C.

16.96 Explain why the boiling point of CH_3NH_2 is –6.3°C, but the boiling point of $CH_3CH_2NH_2$ is 17°C.

Additional Questions

16.97 Outline the different types of hydrocarbons.

16.98 What is an alkane?

16.99 How do alkenes differ from alkynes?

16.100 How do ethers differ from alcohols?

16.101 How do aldehydes differ from ketones?

16.102 How do carboxylic acids differ from esters?

16.103 Name the following compounds.
  (a) $CH_3CH(CH_3)CH_2CH(CH_3)_2$

  (b)
$$\begin{array}{ccccc} H_3C & & CH_2 & & CH_3 \\ & \diagdown & & \diagup & \\ & CH & & CH & \\ & | & & | & \\ & H_2C & \!\!\!\!\!\text{------}\!\!\!\!\! & CH_2 & \end{array}$$

  (c) $(CH_3)_3C-C(CH_3)_3$

16.104 Suggest a simple test to distinguish $CH_3CH_2CH_2CH_3$ from $CH_2{=}CHCH_2CH_3$.

16.105 Suggest a simple test to distinguish $CH_3CH_2CH{=}CHCH_3$ from $CH_3CH_2CHOHCH_2CH_3$.

16.106 Suggest a method to distinguish cyclohexane from cyclo-hexanol.

16.107 Propane, dimethyl ether, and ethanol have boiling points of –42.1°C, –23.6°C, and 78.5°C. Explain the differences in boiling points.

16.108 Name the following compounds.
  (a) $CH_3CH_2CH_2CHO$
  (b) $CH_3CH_2COCH_2CH_3$
  (c) $(CH_3)_2CO$
  (d) H_2CO
  (e) CH_3CHO

16.109 Write structural formulas for the following compounds.
 (a) acetone (propanone)
 (b) butanone
 (c) propanal
 (d) 2-butenal
 (e) acetaldehyde (ethanal)

16.110 Name the following compounds.
 (a) $CH_3CH_2CO_2H$
 (b) $CH_3CH_2CO_2CH_3$
 (c) HCO_2CH_3
 (c) $C_6H_5CO_2CH_2CH_3$
 (d) $(CH_3)_2CHCO_2H$

16.111 Write structural formulas for the following compounds.
 (a) ethyl butanoate
 (b) pentanoic acid
 (c) methyl acetate
 (d) formic acid
 (d) propanoic acid

16.112 Draw the structure and name the ester that is formed in the reaction of 2-propanol with acetic acid.

16.113 Give the common name of the following compounds.
 (a) $(CH_3)_2NH$
 (b) $(CH_3)_2N(C_2H_5)$
 (c) $C_6H_5NH_2$

16.114 Write structural formulas for the following compounds.
 (a) dimethylpropylamine
 (b) dipropylamine
 (c) methylethylamine

16.115 Give the structural formulas of the following compounds.
 (a) 2,2,4-trimethylpentane
 (b) 3-ethyl-2-pentene
 (c) 1,3-butadiene
 (d) cyclobutane
 (e) chlorobenzene

16.116 Give an example of each of the following classes of compounds.
 (a) ketone
 (b) carboxylic acid
 (c) ether
 (d) aldehyde
 (e) alcohol
 (f) alkene
 (g) ester

16.117 Identify the class of each of the following compounds.
 (a) $CH_3CH_2CH_2CH_3$
 (b) $(CH_3)_2CH-O-CH_3$
 (c) CH_3CH_2COOH
 (d) $(CH_3)_2CHCHO$
 (e) $(CH_3)_3CCOOCH_2CH_3$
 (f) $CH_3CH_2NHCH_3$
 (g) $(CH_3CH_2)_2CO$

16.118 Write structural formulas for four isomers of C_4H_9Cl.

16.119 Which of the following compounds can exist as *cis* and *trans* isomers? Write line structures for those that can.
 (a) propene
 (b) 1-butene
 (c) 2-butene
 (d) 2-pentene
 (e) 3-pentyne

Appendix A

Useful Reference Information

Description	Table/Figure Reference	Page
Densities of Some Common Substances	Table 1.4	19
Monatomic Anions and Oxoanions	Figure 3.17	90
Important Polyatomic Ions	Table 3.4	90
Naming Ionic Compounds in Which the Cation Exhibits a Single Charge	Figure 3.24	97
Naming Ionic Compounds in Which the Cation Exhibits Variable Charge	Figure 3.26	99
Common Greek Prefixes	Table 3.9	102
Naming Molecular Compounds	Figure 3.29	102
Naming Acids	Figure 3.36	106
Naming Compounds	Figure 3.37	107
Classes of Chemical Reactions	Table 5.1	167
Decomposition Reactions That Occur When Compounds Are Heated	Table 5.2	168
Activity Series	Figure 5.21	173
Rules Used to Predict the Solubility of Ionic Compounds	Table 5.3	177
Specific Heats of Some Substances	Table 6.2	219
Heat Changes for the Combustion of Fuels	Table 6.3	226
Energy Values of Food Components	Table 6.4	226
Electromagnetic Spectrum	Figure 7.6	240
Electron Configurations of Elements	Figure 7.23	260
Ionization Energies for the Elements	Figure 7.26	267
Successive Ionization Energies for Period 2 Elements	Table 7.1	269
Atomic Radii Values	Figure 7.30	270
Ionic Radii Values	Figure 7.33	271
Ionic versus Covalent Bonding	Figure 8.3	283
Electronegativity Values	Figure 8.5	284

Description	Table/Figure Reference	Page
Geometric Structures Arising from Different Numbers of Atoms	Table 8.5	303
Arrangement of Electron Pairs and Molecular Shapes	Table 8.6	305
Volume Percent of Gases in the Atmosphere	Table 9.1	318
Vapor Pressure of Water at Various Temperatures	Table 9.2	341
Intermolecular Forces and Bonds	Table 10.2	377
Like Dissolves Like	Table 11.2	412
Solubilities of Substances as They Vary with Temperature	Figure 11.19	419
Concentration Units	Table 11.3	422
Common Strong Acids	Table 13.1	495
Common Strong Bases	Table 13.2	496
Common Weak Acids	Table 13.3	497
Common Weak Bases	Table 13.4	498
Weak Acids, Their K_a Values at 25°C, and Their Conjugate Bases	Table 13.5	501
Acid Ionization Constants for Some Polyprotic Acids	Table 13.6	502
pH Scale	Figure 13.12	508
Acid-Base Indicators	Figure 13.16	512
Rules for Assigning Oxidation Numbers	Table 14.1	532
Half-Lives for the Nuclides in the Uranium Decay Series	Table 15.3	586
Functional Groups in Organic Molecules	Table 16.1	610
Classification of Hydrocarbons	Figure 16.2	613
Amino Acids	Figure 16.11	637

Appendix B

Math Toolboxes

Appendix C

Answers to Practice Problems

Chapter 1

1.1 phosphorus, carbon, bromine, sulfur
1.2 (a) Pb   (b) silver
1.3 *Aluminum can,* heterogeneous mixture (metal can with inner plastic liner and paint on the outside).   *Stone fountain,* heterogeneous mixture.   *Helium balloon,* the helium in the balloon is not a mixture. The helium and balloon together are a heterogeneous mixture.   *Pizza,* heterogeneous mixture.   *Copper coin,* if made of pure copper, not a mixture. Newer pennies are a heterogeneous mixture with a zinc core and coating of copper. Older pennies are a homogeneous mixture of copper and zinc.
1.4 (a) A   (b) SO_3
1.5 0.035 g; 3.5×10^{-5} kg
1.6 4.60×10^3 mL; 4.60×10^3 cm^3
1.7 (a) 0.0518 cm^3 (b) 16.0 g
1.8 rises
1.9 (a) 245.1 K   (b) –269°C   (c) 37.0°C; 310.1 K
1.10 (a) physical property   (b) chemical property   (c) chemical property   (d) physical property
1.11 physical change
1.12 the image with atoms moving slower (B)
1.13 light, heat, and chemical energy
1.14 9.83×10^5 J

Chapter 2

2.1 (a) 7 protons; 6 neutrons   (b) atomic number, 7; nitrogen   (c) 13
2.2 6 protons; 6 electons; 8 neutrons
2.3 He-3; $^3_2$He; ^{3}He
2.4 (a) 78   (b) 20   (c) 7
2.5 (a) 9 protons; 10 electrons; anion   (b) 12 protons; 10 electrons; cation   (c) 7 protons; 10 electrons; anion
2.6 (a) $^{34}_{16}$S^{2-} (b) $^{23}_{11}$Na$^+$ (c) $^{40}_{20}$Ca^{2+}
2.7 ^{63}Cu
2.8 6.940 amu which is only slightly lower than the periodic table value of 6.941
2.9 (a) nitrogen   (b) argon   (c) calcium   (d) silver
2.10 (a) Li$^+$ (b) S^{2-} (c) Al^{3+}

Chapter 3

3.1 (b) CO_2 and (d) CCl_4
3.2 KCl
3.3 (a) 2+; Ba^{2+}; barium ion   (b) 1–; Br$^-$; bromide ion
3.4 $PO_4{}^{3-}$

3.5

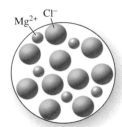

(water molecules not shown)

3.6 KBr; K_2SO_4; FeBr$_3$; Fe$_2$(SO$_4$)$_3$
3.7 potassium oxide; magnesium sulfite
3.8 chromium(III) sulfide; iron(II) sulfite
3.9 (a) Cu$_2$SO$_4$ (b) FeO
3.10 (a) Sn(SO$_4$)$_2$, stannic sulfate; SnSO$_4$, stannous sulfate   (b) ferric sulfide, Fe$_2$S$_3$; ferrous sulfide, FeS
3.11 tetraphosphorus hexoxide; dinitrogen pentoxide
3.12 N$_2$O$_3$
3.13 (a) hydroselenic acid   (b) phosphoric acid
3.14 phosphorus tribromide; magnesium chloride; sulfuric acid

Chapter 4

4.1 6.022×10^{23} molecules N$_2$O$_5$; 1.204×10^{24} atoms N; 3.011×10^{24} atoms O
4.2 60.09 g/mol
4.3 30.5% Fe; 34.9% S
4.4 0.0399 mol
4.5 1.4×10^{-4} mol; 3.53 g; 88 packets
4.6 1.08×10^{21} molecules
4.7 (a) CH$_2$Cl$_2$ (b) CH$_2$O   (c) P$_2$O$_3$
4.8 CuS
4.9 CuSiH$_4$O$_5$
4.10 Cu$_5$Si$_4$O$_{14}$H$_2$
4.11 KSO$_4$; K$_2$S$_2$O$_8$
4.12 66.46%
4.13 solute, H$_2$S; solvent, H$_2$O
4.14 2.64 *M*
4.15 280,000 g or 280 kg
4.16 0.0365 L or 36.5 mL
4.17 0.259 *M*

Chapter 5

5.1 reactants, Fe(*s*) and CuSO$_4$(*aq*); products, Cu(*s*) and FeSO$_4$(*aq*)
5.2 2Na(*s*) + 2H$_2$O(*l*) $\longrightarrow$ 2NaOH(*aq*) + H$_2$(*g*)

5.3 $2C_4H_{10}(g) + 13O_2(g) \longrightarrow 8CO_2(g) + 10H_2O(g)$

5.4 $Zn(s) + 2HNO_3(aq) \longrightarrow Zn(NO_3)_2(aq) + H_2(g)$

5.5 (a) combination   (b) double-displacement   (c) decomposition   (d) single-displacement

5.6 $MgSO_3 \cdot 3H_2O(s) \xrightarrow{\text{heat}} MgSO_3(s) + 3H_2O(g)$ [If enough heat is applied, $MgSO_3(s)$ can further decompose to magnesium oxide and sulfur dioxide.]

5.7 $2Mg(s) + O_2(g) \longrightarrow 2MgO(s)$;
$3Mg(s) + 2N_2(g) \longrightarrow Mg_3N_2(s)$

5.8 Zinc reacts with the acid in tomatoes.

5.9 Co; $SnSO_4(aq)$;
$Co(s) + SnSO_4(aq) \longrightarrow CoSO_4(aq) + Sn(s)$

5.10 $Cd(NO_3)_2(aq) + Na_2S(aq) \longrightarrow CdS(s) + 2NaNO_3(aq)$

5.11 $BaS(aq) + H_2SO_4(aq) \longrightarrow BaSO_4(aq) + H_2S(g)$

5.12 $Ca(OH)_2(aq) + H_2SO_4(aq) \longrightarrow CaSO_4(s) + 2H_2O(l)$

5.13 $2CH_3OH(l) + 3O_2(g) \longrightarrow 2CO_2(g) + 4H_2O(g)$

5.14 $BaCl_2(aq) + Na_2SO_4(aq) \longrightarrow BaSO_4(s) + 2NaCl(aq)$;
$Ba^{2+}(aq) + 2Cl^-(aq) + 2Na^+(aq) + SO_4^{2-}(aq) \longrightarrow$
$BaSO_4(s) + 2Na^+(aq) + 2Cl^-(aq)$;
$Ba^{2+}(aq) + SO_4^{2-}(aq) \longrightarrow BaSO_4(s)$

Chapter 6

6.1 142 mol

6.2 (a) 1.4×10^9 g or 1.4×10^6 kg   (b) 1.6×10^9 g or 1.6×10^6 kg   (c) yes; the law of conservation of mass

6.3 The piece of iron metal would disappear and the blue color of $CuSO_4$ would remain in the solution, though it would be less intense.

6.4 (a) solar cells   (b) 5 cars   (c) 1 motor, 9 frames, 80 wheels

6.5 (a) H_2
(b)

(c) N_2

6.6 $AgNO_3$; 1.0 mol

6.7 Na_2CO_3; 4.57 g

6.8 68%

6.9 endothermic; when the battery runs energy is released. Recharging a battery is the opposite reaction, so recharging must be endothermic.

6.10 A greater mass of water transfers more heat.

6.11 –6910 J; negative; the reaction is exothermic.

6.12 (a) greater; the rise in temperature of the water indicates that the alloy was hotter.   (b) –3400 J

6.13 87.5 kJ

Chapter 7

7.1 (a) infrared   (b) UV   (c) UV

7.2 2.61×10^9/s or 2.61×10^9 Hz; 1.73×10^{-24} J

7.3 (a) 4.85×10^{-19} J   (b) red line at 657 nm

7.4 (a) incorrect; aufbau principle   (b) incorrect; Pauli exclusion principle   (c) correct

7.5 (a) $1s^2 2s^2 2p^6 3s^2 3p^1$ (b) $1s^2 2s^2 2p^6 3s^2 3p^6 4s^2 3d^1$
(c) $1s^2 2s^2 2p^6 3s^2 3p^6 4s^1$

7.6 (a) $1s^2 2s^2 2p^6 3s^2 3p^4$ (b) $1s^2 2s^2 2p^6 3s^2 3p^6 4s^2 3d^{10} 4p^6 5s^1$
(c) $1s^2 2s^2 2p^6 3s^2 3p^6 4s^2 3d^{10} 4p^6 5s^2 4d^{10}$

7.7 (a) 2   (b) 4   (c) 4

7.8 (a) $[Ar]4s^2 3d^{10} 4p^6$ or $[Kr]$; Sr^{2+} and Rb^+ among others
(b) $[He]2s^2 2p^6$ or $[Ne]$; F^- and Na^+ among others
(c) $[Ar]$; S^{2-} and Ca^{2+} among others

7.9 (a) lithium   (b) chlorine

7.10 (a) potassium; valence electron in energy level farther from the nucleus   (b) silicon; increased positive charge in the nucleus is more effective at drawing in valence electrons of chlorine.

7.11 $Y^{3+} < Sr^{2+} < Rb^+ < Br^- < Se^{2-}$

Chapter 8

8.1 (a) ionic; (b) covalent; (c) ionic and covalent

8.2 most polar bond, S–O; partial negative on the oxygen in SO_2 and the chlorine in CCl_4; N_2 and PH_3 consist of nonpolar bonds.

8.3 $\cdot\overset{\displaystyle\cdot}{Si}\cdot$

8.4 $\cdot Be\cdot$; $\cdot\overset{\displaystyle\cdot}{\underset{\displaystyle\cdot}{N}}\cdot$; $(Be^{2+})_3(:\overset{\displaystyle\cdot\cdot}{\underset{\displaystyle\cdot\cdot}{N}}:^{3-})_2$

8.5 $:\overset{\cdot\cdot}{F}:\overset{\cdot\cdot}{N}:\overset{\cdot\cdot}{F}:$
$\quad :\overset{\cdot\cdot}{\underset{\cdot\cdot}{N}}:$

8.6 $\quad:\overset{\cdot\cdot}{\underset{\cdot\cdot}{O}}:$
$\quad\,\Vert$
$H:C:H$

8.7 $:N:::N:\overset{\cdot\cdot}{\underset{\cdot\cdot}{O}}: \longleftrightarrow :\overset{\cdot\cdot}{N}::N::\overset{\cdot\cdot}{O}: \longleftrightarrow :\overset{\cdot\cdot}{\underset{\cdot\cdot}{N}}:N:::O:$

8.8 $:\overset{\cdot\cdot}{\underset{\cdot\cdot}{O}}:\overset{\cdot\cdot}{\underset{\cdot\cdot}{Cl}}:\overset{\cdot\cdot}{\underset{\cdot\cdot}{O}}:$

8.9
```
    H   O   H
    |   ||  |
H — C — C — C — H
    |       |
    H       H
```

8.10 trigonal planar

8.11 bent; about 120°

8.12 C_2H_6, nonpolar; NO_2, polar; SO_2, polar; SO_3, nonpolar

Chapter 9

9.1 0.814 atom; 619 torr

9.2

9.3 4.03 atm

9.4

9.5   −74.4°C
9.6   0.892 L
9.7   28.1 L
9.8   168 L
9.9   10.6 L
9.10   0.0776 mol
9.11   2.46 g
9.12   0.0876 mol
9.13   221 L
9.14   0.0901 mol; 8.83 g

Chapter 10

10.1   −44°C; −62°C
10.2   deposition
10.3   EF; CD; AB
10.4 2.82×10^5 J
10.5 SiH_4
10.6   (a) yes   (b) yes   (c) yes   (d) no
10.7 CH_3CO_2H and HOCl
10.8   (a) Cl_2, larger electron cloud   (b) NH_3, hydrogen bonding   (c) SO_2, dipole-dipole force between polar molecules   (d) H_2Te, larger electron cloud
10.9 $CHCl_3 < CHBr_3 < CHI_3$
10.10 $O_2(l)$; weaker intermolecular forces due to smaller electron cloud
10.11   HF; HF; HF; CO
10.12   SAE 15
10.13   the intermolecular forces between the new material and water molecules
10.14   network solid
10.15   molecular solid; covalent bonds hold the atoms together within the molecules and London dispersion forces hold molecules together.

Chapter 11

11.1   (a) $Zn^{2+}(aq)$, $Cl^-(aq)$,
          $H_2O(l)$; $ZnCl_2(s) \longrightarrow Zn^{2+}(aq) + 2Cl^-(aq)$
       (b) $CH_3OH(aq)$, $H_2O(l)$;
          $CH_3OH(l) \longrightarrow CH_3OH(aq)$
       (c) $K^+(aq)$, $NO_3^-(aq)$, $H_2O(l)$;
          $KNO_3(s) \longrightarrow K^+(aq) + NO_3^-(aq)$
11.2   London dispersion forces between I_2 molecules and between C_6H_{14} molecules
11.3   2.15%; 0.538 g
11.4   87.2%
11.5   1.04 M
11.6   5.34 m
11.7   114 mL; 3.88 g
11.8   0.1373 M
11.9   −2.8°C
11.10   0.5 m $Ca(NO_3)_2$

Chapter 12

12.1

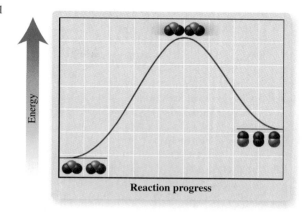

12.2   (a) both decrease; (b) the total number of collisions increases, and the fraction of effective collisions stays the same.
12.3   increase surface area of iron to increase collision frequency; increase oxygen concentration to increase collision frequency; increase temperature to increase collision frequency and fraction of effective collisions.

12.4

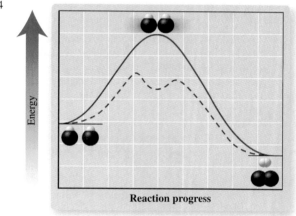

12.5 H_3O^+, catalyst; $H_3C-CH_2^+$, H_2O, $CH_3CH_2OH_2^+$, intermediates
12.6   (a) $K_{eq} = \dfrac{[N_2O_4]}{[NO_2]^2}$ (b) 222   (c) product favored
12.7   (a) not at equilibrium; reaction proceeds in reverse direction   (b) increase
12.8   (a) $K_{eq} = \dfrac{[CO]}{[CO_2]}$ (b) $K_{eq} = [Pb^{2+}][Cl^-]^2$
12.9   (a) reaction shifts to the right; more AgI dissolves
       (b) reaction shifts to the left; AgI precipitates
       (c) reaction shifts to the left; AgI precipitates
12.10   (a) reaction shifts left   (b) nothing
12.11   endothermic
12.12   (a) no; reaction shifts left decreasing products.   (b) yes; reaction shifts right increasing products.   (c) yes; reaction shifts right increasing products.   (d) no; reaction shifts left decreasing products.   (e) no; catalysts do not change the position of equilibrium.

Chapter 13

13.1 (a) OCl^-, base; H_2O, acid (b) F^-, base; H_2SO_4, acid
(c) H_2O, base; NH_4^+, acid

13.2 (a) HF (b) H_2CO_3 (c) H_3O^+

13.3 Glycine has one acidic hydrogen, the one attached to the oxygen.

13.4 (a) strong acid (b) weak base (c) weak acid (d) weak base

13.5 $HClO_4$ is shown in A because it is a strong acid and is completely ionized to $H_3O^+(aq)$ and $ClO_4^-(aq)$.

13.6 HOCl

13.7 (a) $H_2C_2O_4(aq) + H_2O(l) \longrightarrow HC_2O_4^-(aq) + H_3O^+(aq)$
$HC_2O_4^-(aq) + H_2O(l) \longrightarrow C_2O_4^{2-}(aq) + H_3O^+(aq)$
(b) $H_2C_2O_4$

13.8 (a) $1.0 \times 10^{-6}\ M$; acidic (b) $1.0 \times 10^{-12}\ M$; basic

13.9 $[OH^-] = 0.85\ M$; $[H_3O^+] = 1.2 \times 10^{-14}\ M$

13.10 (a) 3.07 (b) 13.00 (c) 0.00

13.11 5.00; no

13.12 $H_2PO_4^-(aq) + H_2O(l) \longrightarrow HPO_4^{2-}(aq) + H_3O^+(aq)$;
A base produces OH^- ions which react with H_3O^+ to produce water. The reaction shifts to the right to reestablish equilibrium, replacing H_3O^+ that was removed.

13.13 (a) no (b) yes; $CH_3CO_2H(aq) + H_2O(l) \longrightarrow$
$CH_3CO_2^-(aq) + H_3O^+(aq)$ (c) no

Chapter 14

14.1 (a) $2Fe(s) + 3Cu(NO_3)_2(aq) \longrightarrow 3Cu(s) +$
$2Fe(NO_3)_3(aq)$ (b) Fe, oxidized; Cu^{2+}, reduced
(c) $Cu(NO_3)_2$, oxidizing agent; Fe, reducing agent
(d) A coating of copper metal appears on the surface of the iron.

14.2 (a) 0 (b) Mn, +3; O, –2 (c) Na, +1; S, +6, S; O, –2
(d) Cr, +6; O, –2

14.3 (a) yes; H_2O acts as both oxidizing and reducing agent.
(b) no (c) yes; Al, reducing agent; O_2, oxidizing agent
(d) yes; Mg, reducing agent; $Zn(NO_3)_2$, oxidizing agent

14.4 (a) oxidation, $Cd(s) \longrightarrow Cd^{2+}(aq) + 2e^-$;
reduction, $Ni^{2+}(aq) + 2e^- \longrightarrow Ni(s)$
(b) $Ni(NO_3)_2$, oxidizing agent; Cd, reducing agent
(c)

As drawn, oxidation will occur in the left half-cell, reduction in the right; $Cd(s)$ is the anode and $Ni(s)$ is the cathode; anions will flow from the salt bridge into the oxidation half-cell and cations will flow into the reduction half-cell; electrons will flow through the wire toward the cathode; the mass of the anode will decrease and the mass of the cathode will increase.

14.5 Ag_2O, oxidizing agent; Zn, reducing agent

14.6 $3Mg(s) + 2Cr(NO_3)_3(aq) \longrightarrow$
$3Mg(NO_3)_2(aq) + 2Cr(s)$

14.7 $6OH^-(aq) + 3Cl_2(aq) \longrightarrow$
$5Cl^-(aq) + ClO_3^-(aq) + 3H_2O(l)$

14.8 Place a solid Cr electrode in a $CrCl_3$ solution to construct one half-cell. In the other half-cell place a Zn electrode in a $ZnCl_2$ solution. Connect the two half-cells with a salt bridge and external wire.

14.9 (a) Ca^{2+} ions are reduced to form Ca atoms. (b) Cl^- ions are oxidized to form Cl atoms. Two of these atoms combine to form a Cl_2 molecule. (c) oxidation,
$2Cl^-(aq) \longrightarrow Cl_2(g) + 2e^-$; reduction, $Ca^{2+}(aq) +$
$2e^- \longrightarrow Ca(l)$ (d) $CaCl_2(l) \longrightarrow Ca(l) + Cl_2(g)$

Chapter 15

15.1 $^{218}_{84}Po$

15.2 $^{234}_{91}Pa$

15.3 $^{11}_{9}C$

15.4 $^{54}_{24}Cr$; $^{209}_{83}Bi + ^{54}_{24}Cr \longrightarrow ^{262}_{107}Bh + ^{1}_{0}n$

15.5 positron ($^{0}_{1}\beta^+$) emission

15.6 $^{208}_{82}Pb$

15.7 12.5%

15.8 11,460 years

15.9 α and $^{0}_{-1}\beta^-$ radiation

15.10 2; $^{235}_{92}U \longrightarrow ^{137}_{52}Te + ^{57}_{40}Zr + 2^{1}_{0}n$

15.11 $2^{16}_{8}O \longrightarrow ^{32}_{16}S$

Chapter 16

16.1

16.2 This molecule has more branching than the other two, so it has less surface area and, therefore, weaker intermolecular forces.

16.3 2-methylpropane

16.4 2,2,5-trimethylheptane

16.5

$$H_3C-\underset{\underset{CH_3}{|}}{\overset{\overset{CH_3}{|}}{C}}-\underset{\underset{CH_3}{|}}{CH}-CH_2-CH_3$$

16.6 C_4H_9Br and HBr

16.7 2-methyl-1-butene

16.8 (a) no (b) yes;

16.9 $CH_3CH{=}CHCH_3 + HBr \longrightarrow CH_3CH_2CHBrCH_3$

16.10 1,3,5-trimethylbenzene

16.11 2,4-dimethyl-3-pentanol

Appendix D

Answers to Selected Questions and Problems

Chapter 1 – Matter and Energy

1.1 (a) mass; (b) chemical property; (c) mixture; (d) element; (e) energy; (f) physical property; (g) liquid; (h) density; (i) homogenous mixture; (j) solid

1.3 (a) homogenous mixture if the dye is evenly mixed into the water; (b) element; (c) homogenous mixture; (d) heterogeneous mixture

1.5 Gasoline, automobile exhaust, oxygen gas, and the iron pipe are matter. Sunlight is energy.

1.7 Elements are composed of only one type of atom. Compounds are made up of two or more different elements.

1.9 Metals are lustrous (shiny) and conduct heat and electricity. In addition you can form wires with metals (ductile) and you can make foil out of them by hitting them with a hammer (malleable).

1.11 (a) titanium; (b) tantalum; (c) thorium; (d) technetium; (e) thallium

1.13 (a) boron; (b) barium; (c) beryllium; (d) bromine; (e) bismuth

1.15 (a) nitrogen; (b) iron; (c) manganese; (d) magnesium; (e) aluminum; (f) chlorine

1.17 (a) Fe; (b) Pb; (c) Ag; (d) Au; (e) Sb

1.19 Ir is the symbol for the element iridium. Iron's symbol is Fe.

1.21 The correct symbol is No.

1.23 The only pure substance is the salt in the salt shaker (if it is not iodized salt). Hamburger: heterogenous mixture. Salt: pure substance. Soft drink: heterogeneous mixture. Ketchup: heterogenous mixture.

1.25 H_2

1.27 The chemical formula is N_2O_4.

1.29 Image D represents a mixture of an element and a compound.

1.31 The elements are O_2, P_4, and He. Fe_2O_3, NaCl, and H_2O are compounds.

1.33

| Liquid state | Solid state |

1.35 gas

1.37 (a) gas; (b) liquid; (c) solid

1.39 Solid

1.41 $O_2(aq)$

1.43 physical properties

1.45 (a) 0.045 g; (b) 1.6×10^{-3} oz; (c) 9.9×10^{-5} lb

1.47 (a) 0.10 mg; (b) 1.0×10^2 µg; (c) 1.0×10^{-7} kg

1.49 (a) 1.2×10^3 mL; (b) 1.2×10^3 cm³; (c) 1.2×10^{-3} m³

1.51 1.6×10^2 mL; 0.16 L.

1.53 1.3 g/cm³; 1.3 g/mL.

1.55 38.5 mL

1.57 The molecules in the liquid state are closer together than molecules in the gas state. More matter in the same volume means that the density is higher.

1.59 oil (lowest) < plastic < water (highest)

1.61 329 K

1.63

Scale	Freezing	Boiling	Difference
Celsius	0°C	100°C	100°C
Kelvin	273.15 K	373.15 K	100 K
Fahrenheit	32 °F	212 °F	180 °F

1.65 No

1.67 Physical properties are (a) mass, (b) density, and (e) melting point. Chemical properties are (c) flammability, (d) resistance to corrosion, and (f) reactivity with water.

1.69 Physical changes are (a) boiling acetone, (b) dissolving oxygen gas in water, and (e) screening rocks from sand. Chemical changes are (c) combining hydrogen and oxygen to make water, (d) burning gasoline, and (f) conversion of ozone to oxygen.

1.71 Symbolic: $Cl_2(g) \longrightarrow Cl_2(l)$. A molecular-level diagram is shown below.

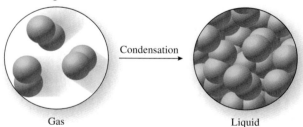

Gas   Condensation   Liquid

1.73 chemical change

1.75 Anna and Bill would have observed kinetic energy from the movement of the welder and the motion of the sparks. The sparks would have glowed, indicating heat, light, and chemical energy.

1.77 The molecules in image A have greater kinetic energy because they are moving faster.

1.79 Any object that would move if allowed has potential energy (e.g., a picture hanging on a wall).

1.81 The falling and rising water, the people walking, and the wind (if it's blowing the leaves) all have kinetic energy. The water, the people, the trees, and the papers on the kiosk all have potential energy. Many objects have potential and kinetic energy.

1.83 Consider as an example a car going down a road. A car going up a hill is converting chemical energy in the fuel to mechanical energy to reach the top. At the top of the hill, the car has potential energy. If it rolls down the hill, it gains kinetic energy but loses potential energy. Other energy conversions can be found while driving a car.

1.85 As the water leaves the top of the fountain, it possesses kinetic energy (going up). That energy is converted to potential energy. As the water falls the potential energy converts back into kinetic energy.

1.87 2.20×10^3 J

1.89 3.47×10^4 cal

1.91 0.209 Cal

1.93 114 Cal

1.95 It is used to explain and predict scientific results.

1.97 (a) hypothesis; (b) observation; (c) theory; (d) observation

1.99 You could do some library research and find the densities of many different woods. By placing samples of the different woods in water, you can determine if the theory is correct.

1.101 At high altitude, the air pressure is lower. As a result, when a balloon rises, it expands. Since the mass of air in the balloon has not changed, but the volume has increased, the density of the balloon is lower.

1.103 zinc

1.105 (a) He; (b) Ne; (c) Ar; (d) Kr; (e) Xe; (f) Rn

Chapter 2 – Atoms, Ions, and the Periodic Table

2.1 (a) neutron; (b) law of conservation of mass; (c) proton; (d) main-group element; (e) relative atomic mass; (f) mass number; (g) isotope; (h) cation; (i) subatomic particle; (j) alkali metal; (k) periodic table

2.3 Dalton used the laws of conservation of mass (Lavoisier) and definite proportions (Proust).

2.5 They differ in their atomic masses and chemical properties.

2.7 Compounds contain discrete numbers of atoms of each element that form them. Because all the atoms of an element have the same relative atomic mass, the mass ratio of the elements in a compound is always the same (law of definite proportions).

2.9 No. Hydrogen atoms are not conserved.

2.11 Thomson's cathode ray experiment

2.13 electrons

2.15 The nucleus of helium has two protons and two neutrons. Two electrons can be found outside the nucleus.

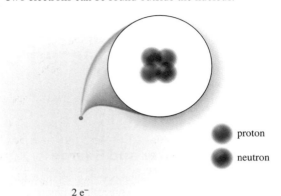

2 e⁻

proton

neutron

2.17 neutron

2.19 Carbon has 6 protons. The relative atomic mass of a carbon atom is 12.01 amu indicating the presence of 6 neutrons. Protons and neutrons have approximately equal masses so the nuclear mass is approximately two times the mass of the protons.

2.21 (a) 1; (b) 7; (c) 80

2.23 protons

2.25 the number of protons

2.27 (b) atomic number

2.29 The following table displays the atomic, neutron, and mass numbers for the isotopes of hydrogen.

	$^{1}_{1}H$	$^{2}_{1}H$	$^{3}_{1}H$
Atomic number	1	1	1
Neutron number	0	1	2
Mass number	1	2	3

2.31

	Protons	Neutrons	Electrons
(a) $^{15}_{8}O$	8	7	8
(b) $^{109}_{47}Ag$	47	62	47
(c) $^{35}_{17}Cl$	17	18	17

2.33 (a) $^{3}_{1}H$; (b) $^{9}_{4}Be$; (c) $^{31}_{15}P$

2.35

	Protons	Neutrons
(a) $^{56}_{26}Fe$	26	30
(b) $^{39}_{19}K$	19	20
(c) Copper-65 or $^{65}_{29}Cu$	29	36

2.37 They differ in the number of electrons.

2.39 (a) an anion with a 1– charge is formed; (b) a cation with a 2+ charge is formed.

2.41 (a) Cu^{2+}, cation; (b) I^-, anion

2.43

	Protons	Electrons
(a) Zn^{2+}	30	28
(b) F^-	9	10
(c) H^+	1	0

2.45

Isotope Symbol	Protons	Neutrons	Electrons
$^{37}_{17}Cl^-$	17	20	18
$^{25}_{12}Mg^{2+}$	12	13	10
$^{13}_{7}N^{3-}$	7	6	10
$^{40}_{20}Ca^{2+}$	20	20	18

2.47   sodium, Na

2.49   copper, Cu

2.51 ^{7}Li has three protons, three electrons, and four neutrons. ^{7}Li$^+$ has only two electrons, and ^{6}Li has only three neutrons. Otherwise they are the same as ^{7}Li. Lithium-6 differs the most in mass.

2.53   The mass of D$_2$ is two times the mass of H$_2$.

2.55   about 40 amu

2.57   The atomic mass unit is defined as 1/12th the mass of one carbon-12 atom.

2.59   (a) 2 amu; (b) 238 amu

2.61   The numerical values of masses of individual atoms are very small when measured on the gram scale. The size of the atomic mass unit allows us to make easier comparisons and calculations of masses of molecules.

2.63   The mass number is the sum of the number of protons and neutrons and is always an integer value. In contrast, the mass of an atom is the actual measurement of how much matter is in the atom and is never exactly an integer value (except carbon-12).

2.65   A mass spectrometer is used to determine the mass of an atom. The mass number of an atom is the sum of the number of protons and neutrons.

2.67   calcium-40

2.69   21.50 amu

2.71   10,810 amu or 1.081×10^4 amu

2.73   2500 amu of boron

2.75   (a) K; (b) Br; (c) Mn; (d) Mg; (e) Ar; (f) Br, K, Mg, Al, Ar

2.77   fluorine

2.79   titanium, Ti

2.81   (a) metal; (b) nonmetal; (c) metal; (d) metalloid

2.83   (a) main group; (b) main group; (c) main group; (d) actinide; (e) transition metal

2.85   group VIIA, the halogen group

2.87   neon

2.89   group VIIIA (18), the noble gas group

2.91   electrons

2.93   (a) group IA (1); (b) group IIA (2); (c) group VIIA (17); (d) group VIA (16)

2.95   (a) Na$^+$; (b) O^{2-}; (c) S^{2-}; (d) Cl$^-$; (e) Br$^-$

2.97   The mass of oxygen added to form Fe$_2$O$_3$ causes an increase in the mass.

2.99   The mass ratios of Zn/S are the same for the two samples (within the significant figures given). The mass ratio is approximately 2.0:1.0.

2.101 Electrons have charge and were readily studied in cathode ray tubes.

2.103 nickel-60

2.105 19 protons and 20 neutrons

2.107 The most abundant isotopes of cobalt have masses greater than the masses of the most abundant isotopes of nickel. As a result, the relative atomic mass of cobalt is greater than that of nickel.

2.109 When there are many isotopes, some of the isotopes can be present in very low abundance. As a result, their masses cannot be determined as accurately and their percentage contributions are also less well known. Both factors result in a decrease in the precision of the calculated relative atomic mass.

2.111 127

2.113 carbon

2.115 20% boron-10 and 80% boron-11; the relative atomic mass (10.81 amu) is about 80% of the difference between the two isotopes.

2.117 Br$_2$(l)

2.119 hydrogen

Chapter 3 – Chemical Compounds

3.1   (a) formula unit; (b) strong electrolyte; (c) molecular compound; (d) acid; (e) nonelectrolyte; (f) oxoanion

3.3   (a) ionic; (b) ionic; (c) molecular

3.5   (a) molecular; (b) ionic; (c) molecular; (d) ionic

3.7   (a) molecular; (b) ionic; (c) ionic; (d) molecular

3.9   LiF

3.11 (a) Na$^+$, sodium ion; (b) K$^+$, potassium ion; (c) Rb$^+$, rubidium ion

3.13 (a) Ca^{2+}, calcium ion; (b) N^{3-}, nitride ion; (c) S^{2-}, sulfide ion

3.15 NO$_2^-$, nitrite ion

3.17 (a) sulfate ion; (b) hydroxide ion; (c) perchlorate ion

3.19 (a) N^{3-}; (b) NO$_3^-$; (c) NO$_2^-$

3.21 (a) CO$_3^{2-}$; (b) NH$_4^+$; (c) AsO$_4^{3-}$; (d) MnO$_4^-$

3.23 SO$_3^{2-}$, sulfite ion

3.25 IO$_3^-$

3.27

3.29 (a) BaCl$_2$; (b) FeBr$_3$; (c) Ca$_3$(PO$_4$)$_2$; (d) Cr$_2$(SO$_4$)$_3$

3.31 (a) K$^+$ and Br$^-$; (b) Ba^{2+} and Cl$^-$; (c) Mg^{2+} and PO$_4^{3-}$; (d) Co^{2+} and NO$_3^-$

3.33 (a) Fe^{2+} will form FeO and Fe^{3+} will form Fe$_2$O$_3$. (b) Fe^{2+} will form FeCl$_2$ and Fe^{3+} will form FeCl$_3$.

3.35 (a) 2–; (b) 2+

3.37 For each "compound" written, the charges do not balance (compounds must have no net charge). (a) Too many chloride ions, NaCl; (b) Not enough potassium ions, K$_2$SO$_4$; (c) There should be three nitrate ions and one aluminum ion, Al(NO$_3$)$_3$.

3.39 (a) magnesium chloride; (b) aluminum oxide; (c) sodium sulfide; (c) potassium bromide; (e) sodium nitrate; (f) sodium perchlorate

3.41 Mn and Co

3.43 (a) copper(I) oxide or cuprous oxide; (b) chromium(II) chloride; (c) iron(III) phosphate or ferric phosphate; (d) copper(II) sulfide or cupric sulfide

3.45 (a) CaSO$_4$; (b) BaO; (c) (NH$_4$)$_2$SO$_4$; (d) BaCO$_3$; (e) NaClO$_3$

3.47 (a) Co^{2+}, cobalt(II) chloride; (b) Pb^{4+}, lead(IV) oxide; (c) Cr^{3+}, chromium(III) nitrate; (d) Fe^{3+}, iron(III) sulfate or ferric sulfate

3.49   (a) $CoCl_2$; (b) $Mn(NO_3)_2$; (c) Cr_2O_3; (d) $Cu_3(PO_4)_2$
3.51   ferrous nitrate and ferric nitrate
3.53

	Ca^{2+}	Fe^{2+}	K^+
Cl^-	calcium chloride $CaCl_2$	iron(II) chloride $FeCl_2$	potassium chloride KCl
O^{2-}	calcium oxide CaO	iron(II) oxide FeO	potassium oxide K_2O
NO_3^-	calcium nitrate $Ca(NO_3)_2$	iron(II) nitrate $Fe(NO_3)_2$	potassium nitrate KNO_3
SO_3^{2-}	calcium sulfite $CaSO_3$	iron(II) sulfite $FeSO_3$	potassium sulfite K_2SO_3
OH^-	calcium hydroxide $Ca(OH)_2$	iron(II) hydroxide $Fe(OH)_2$	potassium hydroxide KOH
ClO_3^-	calcium chlorate $Ca(ClO_3)_2$	iron(II) chlorate $Fe(ClO_3)_2$	potassium chlorate $KClO_3$

	Mn^{2+}	Al^{3+}	NH_4^+
Cl^-	manganese(II) chloride $MnCl_2$	aluminum chloride $AlCl_3$	ammonium chloride NH_4Cl
O^{2-}	manganese(II) oxide MnO	aluminum oxide Al_2O_3	ammonium oxide $(NH_4)_2O$
NO_3^-	manganese(II) nitrate $Mn(NO_3)_2$	aluminum nitrate $Al(NO_3)_3$	ammonium nitrate NH_4NO_3
SO_3^{2-}	manganese(II) sulfite $MnSO_3$	aluminum sulfite $Al_2(SO_3)_3$	ammonium sulfite $(NH_4)_2SO_3$
OH^-	manganese(II) hydroxide $Mn(OH)_2$	aluminum hydroxide $Al(OH)_3$	ammonium hydroxide NH_4OH
ClO_3^-	manganese(II) chlorate $Mn(ClO_3)_2$	aluminum chlorate $Al(ClO_3)_3$	ammonium chlorate NH_4ClO_3

3.55

	potassium	**iron(III)**	**strontium**
iodide	KI	FeI_3	SrI_2
oxide	K_2O	Fe_2O_3	SrO
sulfate	K_2SO_4	$Fe_2(SO_4)_3$	$SrSO_4$
nitrite	KNO_2	$Fe(NO_2)_3$	$Sr(NO_2)_2$
acetate	KCH_3CO_2	$Fe(CH_3CO_2)_3$	$Sr(CH_3CO_2)_2$
hypochlorite	$KClO$	$Fe(ClO)_3$	$Sr(ClO)_2$

	aluminum	**cobalt(II)**	**lead(IV)**
iodide	AlI_3	CoI_2	PbI_4
oxide	Al_2O_3	CoO	PbO_2
sulfate	$Al_2(SO_4)_3$	$CoSO_4$	$Pb(SO_4)_2$
nitrite	$Al(NO_2)_3$	$Co(NO_2)_2$	$Pb(NO_2)_4$
acetate	$Al(CH_3CO_2)_3$	$Co(CH_3CO_2)_2$	$Pb(CH_3CO_2)_4$
hypochlorite	$Al(ClO)_3$	$Co(ClO)_2$	$Pb(ClO)_4$

3.57 $AgCl$
3.59 NF_3, P_4O_{10}, $C_2H_4Cl_2$
3.61   (a) phosphorus pentafluoride; (b) phosphorus trifluoride; (c) carbon monoxide; (d) sulfur dioxide
3.63   (a) SF_4; (b) C_3O_2; (c) ClO_2; (d) SO_2
3.65   The central image represents H_3PO_4.
3.67   (a) hydrofluoric acid; (b) nitric acid; (c) phosphorous acid
3.69   (a) $HF(aq)$; (b) $H_2SO_3(aq)$; (c) $HClO_4(aq)$
3.71   hydrogen ions and nitrate ions
3.73   (a) one sodium ion, Na^+, and one chloride ion, Cl^-
      (b) one magnesium ion, Mg^{2+}, and two chloride ions, Cl^-
      (c) two sodium ions, Na^+, and one sulfate ion, SO_4^{2-}
      (d) one calcium ion, Ca^{2+}, and two nitrate ions, NO_3^-
3.75   (a) electrolyte; (b) electrolyte; (c) electrolyte; (d) nonelectrolyte
3.77   The ions of silver, zinc, and cadmium can each only have one possible charge. They are assumed to have these charges in the compounds they form.
3.79   (a) NO_3^-; (b) SO_3^{2-}; (c) NH_4^+; (d) CO_3^{2-}; (e) SO_4^{2-}; (f) NO_2^-; (g) ClO_4^-
3.81   (a) magnesium bromide; (b) hydrogen sulfide; (c) hydrosulfuric acid; (d) cobalt(III) chloride; (e) potassium hydroxide; (f) silver bromide
3.83   (a) $PbCl_2$; (b) $Mg_3(PO_4)_2$; (c) NI_3; (d) Fe_2O_3; (e) Ca_3N_2; (f) $Ba(OH)_2$; (g) Cl_2O_5; (h) NH_4Cl
3.85 $NaHCO_3$
3.87 $Ca(ClO)_2$
3.89 $H_2O(l)$
3.91 $Cu(s)$, $AgNO_3(aq)$, $Ag(s)$, $Cu(NO_3)_2(aq)$
3.93   (a) NH_3; (b) $HNO_3(aq)$; (c) $HNO_2(aq)$
3.95   Hydrogen atoms attached to oxygen atoms are responsible for the acidic properties.
3.97   All contain oxygen. Magnesium oxide, MgO, has metal and nonmetal ions which makes it an ionic compound. It is a solid at room temperature. Oxygen and carbon dioxide (O_2 and CO_2) are both molecular compounds and gases at room temperature.

Chapter 4 – Chemical Composition

4.1   (a) mole; (b) Avogadro's number; (c) empirical formula; (d) solute; (e) molarity; (f) concentrated solution

4.3   Chemical formulas must have whole number subscripts. (a) H_2S; (b) N_2O_3; (c) $CaCl_2$

4.5   (a) H_2SO_4; (b) SCl_4; (c) C_2H_4

4.7 CO_2

4.9 3.011×10^{23} formula units

4.11 1×10^{23} atoms

4.13 6.022×10^{23} calcium ions

4.15 NaCl, 58.44 g/mol; CH_4, 16.04 g/mol; Cl_2, 70.90 g/mol; SiO_2, 60.09 g/mol

4.17 (a) 472.1 g/mol; (b) 172.18 g/mol; (c) 150.90 g/mol; (d) 120.07 g/mol

4.19 To measure out a useful number of atoms by counting would not be possible because atoms are too small for us to see and manipulate.

4.21 42.394 amu

4.23 (a) 253.8 g/mol; (b) 158.4 g/mol; (c) 56.10 g/mol

4.25 35.9 g/mol

4.27 40.0% calcium

4.29 (a) 0.0999 mol; (b) 0.293 mol; (c) 0.127 mol; (d) 0.0831 mol

4.31 Na

4.33 1.0×10^3 g

4.35 (a) 1.47 mol; (b) 8.84×10^{23} molecules; (c) 8.84×10^{23} N atoms; (d) 4.40 mol of H atoms

4.37 H_2SO_4 has the most atoms per mole. Na has the least atoms per mole.

4.39 1.7×10^{21} molecules

4.41 2.6 g SO_2

4.43 NH_3

4.45 No

4.47 The molecular formula shows the exact numbers of each atom present in a compound. The empirical formula shows the relative amounts of each atom in a compound expressed as small whole numbers.

4.49 H_2O_2 (empirical formula, HO); N_2O_4 (empirical formula, NO_2)

4.51 (a) P_2O_5; (b) same as molecular; (c) same as molecular; (d) $C_3H_5O_2$

4.53 (a) C_3H_2Cl; (b) same as molecular; (c) same as molecular

4.55 NO_2 and N_2O_4

4.57 (a) Fe_3O_4; (b) $C_6H_5NO_2$

4.59 C_5H_6O

4.61 $C_7H_5N_3O_6$

4.63 percent composition by mass and the molar mass

4.65 $C_3H_6O_3$

4.67 $C_6H_{12}O_6$

4.69 (a) 82.25% N, 17.75% H
       (b) 44.06% Fe, 55.94% Cl
       (c) 42.07% Na, 18.89% P, 39.04% O
       (d) 52.45% K, 47.55% Cl

4.71 Chalcocite (Cu_2S) and cuprite (CuO) are both relatively high in copper content (79.85% and 79.88% copper, respectively).

4.73 A solution is a homogenous mixture of two or more substances. Some common solutions: clear drinks (coffee and tea), window cleaner, soapy water, tap water, air, brass (a homogenous mixture of copper and zinc).

4.75 Water is the solvent. Calcium chloride, $CaCl_2$, is the solute.

4.77 Concentrated means high in solute concentration and dilute means low in solute concentration.

4.79 Concentration describes the quantity of solute in a given amount of solvent or solution.

4.81 solution A

4.83 (a) 2.03 M; (b) 0.540 M; (c) 3.20 M; (d) 3.21 M

4.85 (a) 0.375 moles, 28.0 g; (b) 0.513 mol, 72.8 g

4.87 (a) 1.00 L; (b) 0.0833 L

4.89 40.0 mL

4.91 69 mL

4.93 0.02607 M

4.95 9.76 g

4.97 39.95 amu, 6.63×10^{-23} g

4.99 (a) 0.904 mol, 5.44×10^{23} atoms; (b) 0.741 mol, 4.46×10^{23} atoms; (c) 0.169 mol, 1.02×10^{23} atoms

4.101 357 g CH_3OH

4.103 $C_6H_{11}OBr$

4.105 2.458×10^{23} oxygen atoms

4.107 22 mol

4.109 (a) $C_8H_8O_3$; (b) $C_8H_8O_3$

4.111 Al_2O_3

Chapter 5 – Chemical Reactions and Equations

5.1   (a) single-displacement reaction; (b) anhydrous; (c) molecular equation; (d) decomposition reaction; (e) balanced equation; (f) reactant; (g) spectator ion; (h) combustion; (i) precipitate

5.3   The reactants are sodium metal and chlorine gas. The product is sodium chloride.

5.5   Image A represents the reactants and image C represents the products.

5.7   The numbers of hydrogen atoms do not match. One hydrogen molecule should be added to the reactant image.

5.9   The product image should show 5 $XeF_2(g)$ molecules and one unreacted $Xe(g)$ atom.

5.11

5.13 Three major signs are visible (1) a brown gas is formed, (2) bubbles, and (3) color change.

5.15 It is not a chemical reaction because no new substance is formed.

5.17 New substances are formed, so a chemical reaction has taken place.

5.19 No new chemical substances have formed, so no chemical reaction has taken place.

5.21 A chemical equation is a chemist's way of showing what happens during a chemical reaction. It identifies the formulas for the reactants and products and demonstrates how mass is conserved during the reaction.

5.23 (a) chemical reaction; (b) physical change; (c) chemical reaction

5.25   Balancing a chemical equation demonstrates how mass is conserved during a reaction. This makes the equation a quantitative tool for determining the amount of reactant used and product produced.

5.27   (a) $NaH(s) + H_2O(l) \longrightarrow H_2(g) + NaOH(aq)$
(b) $2Mg(s) + SiO_2(s) \longrightarrow Si(s) + 2MgO(s)$

5.29 $2N_2(g) + 6Cl_2(g) \longrightarrow 4NCl_3(g)$

5.31   Image B; $2Mg(s) + O_2(g) \longrightarrow 2MgO(s)$

5.33

5.35

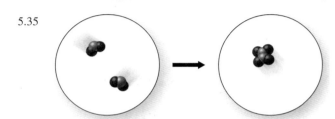

5.37   (a) $2C_2H_6(g) + 7O_2(g) \longrightarrow 4CO_2(g) + 6H_2O(g)$
(b) $4NH_3(g) + 7O_2(g) \longrightarrow 4NO_2(g) + 6H_2O(g)$
(c) $4NO_2(g) + 2H_2O(l) + O_2(g) \longrightarrow 4HNO_3(aq)$

5.39 $Cu(s) + 2AgNO_3(aq) \longrightarrow Cu(NO_3)_2(aq) + 2Ag(s)$

5.41   (a) $CuCl_2(aq) + 2AgNO_3(aq) \longrightarrow$
    $Cu(NO_3)_2(aq) + 2AgCl(s)$
(b) $S_8(s) + 8O_2(g) \longrightarrow 8SO_2(g)$
(c) $C_3H_8(g) + 5O_2(g) \longrightarrow 3CO_2(g) + 4H_2O(g)$

5.43   decomposition: one reactant, more than one product
combination: more than one reactant, one product
single-displacement: element and compound as reactants, different element and compound as products
double-displacement: 2 compounds as reactants, 2 compounds as products

5.45   (a) combination
(b) single-displacement
(c) decomposition

5.47   double-displacement

5.49   (a) combination
(b) single-displacement

5.51   (a) $CaCl_2(aq) + Na_2SO_4(aq) \longrightarrow$
    $CaSO_4(s) + 2NaCl(aq)$: double-displacement
(b) $Ba(s) + 2HCl(aq) \longrightarrow BaCl_2(aq) + H_2(g)$:
    single-displacement
(c) $4Al(s) + 3O_2(g) \longrightarrow 2Al_2O_3(s)$: combination
(d) $FeO(s) + CO(g) \longrightarrow Fe(s) + CO_2(g)$: not classified
(e) $CaO(s) + H_2O(l) \longrightarrow Ca(OH)_2(aq)$: combination
(f) $Na_2CrO_4(aq) + Pb(NO_3)_2(aq) \longrightarrow$
    $PbCrO_4(s) + 2NaNO_3(aq)$: double-displacement
(g) $2KI(aq) + Cl_2(g) \longrightarrow 2KCl(aq) + I_2(aq)$:
    single-displacement
(h) $2NaHCO_3(s) \longrightarrow Na_2CO_3(s) + CO_2(g) + H_2O(g)$:
    decomposition

5.53 $NiCO_3(s) \longrightarrow NiO(s) + CO_2(g)$

5.55   (a) $CaCO_3(s) \longrightarrow CaO(s) + CO_2(g)$
(b) $CuSO_4 \cdot 5H_2O(s) \longrightarrow CuSO_4(s) + 5H_2O(g)$

5.57 $2Mg(s) + O_2(g) \longrightarrow 2MgO(s)$

5.59   (a) $3Ca(s) + N_2(g) \longrightarrow Ca_3N_2(s)$
(b) $SO_3(g) + H_2O(l) \longrightarrow H_2SO_4(aq)$
(c) $4Al(s) + 3O_2(g) \longrightarrow 2Al_2O_3(s)$

5.61 $Zn(s) + SnCl_2(aq) \longrightarrow ZnCl_2(aq) + Sn(s)$

5.63   (a) Ca reacts with water: $Ca(s) + 2H_2O(l) \longrightarrow$
    $Ca(OH)_2(aq) + H_2(g)$
    Ca reacts with HCl: $Ca(s) + 2HCl(aq) \longrightarrow$
    $CaCl_2(aq) + H_2(g)$
(b) Fe reacts with HCl: $Fe(s) + 2HCl(aq) \longrightarrow$
    $FeCl_2(aq) + H_2(g)$
(c) no reaction

5.65   (a) $Zn(s) + 2AgNO_3(aq) \longrightarrow 2Ag(s) + Zn(NO_3)_2(aq)$
(b) $2Na(s) + FeCl_2(s) \longrightarrow 2NaCl(s) + Fe(s)$

5.67   (a) yes; (b) no; (c) no; (d) yes

5.69 $CaCl_2(aq) + K_2CO_3(aq) \longrightarrow CaCO_3(s) + 2KCl(aq)$

5.71   (a) $CaCO_3(s) + H_2SO_4(aq) \longrightarrow CaSO_4(s) + H_2CO_3(aq)$
(b) $SnCl_2(aq) + 2AgNO_3(aq) \longrightarrow$
    $Sn(NO_3)_2(aq) + 2AgCl(s)$

5.73   (a) soluble; (b) soluble; (c) insoluble; (d) somewhat soluble

5.75   lead(II) sulfate

5.77   (a) $K_2CO_3(aq) + BaCl_2(aq) \longrightarrow BaCO_3(s) + 2KCl(aq)$
(b) $CaS(s) + Hg(NO_3)_2(aq) \longrightarrow Ca(NO_3)_2(aq) + HgS(s)$
(c) $Pb(NO_3)_2(aq) + K_2SO_4(aq) \longrightarrow$
    $PbSO_4(s) + 2KNO_3(aq)$

5.79   (a) precipitation of $HgS(s)$
(b) formation of the gas H_2S
(c) formation of the stable molecular compound $H_2O(l)$
    and precipitate $BaSO_4(s)$

5.81   (a) $H_2S(aq) + Cu(OH)_2(s) \longrightarrow CuS(s) + 2H_2O(l)$
(b) no reaction
(c) $KHSO_4(aq) + KOH(aq) \longrightarrow K_2SO_4(aq) + H_2O(l)$

5.83 $2SO_2(g) + O_2(g) \longrightarrow 2SO_3(g)$

5.85   (a) $Cs_2O(s)$; (b) $PbO(s)$ or $PbO_2(s)$; (c) $Al_2O_3(s)$;
(d) $H_2O(g)$; (e) $CO(g)$ or $CO_2(g)$

5.87   (a) $CO(g)$ or $CO_2(g)$ and $H_2O(g)$; (b) $CO_2(g)$; (c) $CuO(s)$
and $SO_2(g)$; (d) $CO(g)$ or $CO_2(g)$ and $H_2O(g)$

5.89   An electrolyte produces ions when dissolved in water. A nonelectrolyte does not produce ions in water.

5.91   (a) and (b) are electrolytes; (c) nonelectrolyte

5.93   nonelectrolyte

5.95   (a) no ions; (b) no ions; (c) $Na^+(aq)$ and $Cl^-(aq)$

5.97   molecular: all substances written as complete chemical formulas or atoms
ionic: all soluble salts, strong acids and bases are written as ions
net ionic: all spectator ions are removed from the ionic equation

5.99   ions that do not participate in a reaction

5.101 $Ag^+(aq) + Cl^-(aq) \longrightarrow AgCl(s)$

5.103 $2Na(s) + 2H_2O(l) \longrightarrow 2Na^+(aq) + 2OH^-(aq) + H_2(g)$

5.105   (a) $Cu(s) + 2AgNO_3(aq) \longrightarrow Cu(NO_3)_2(aq) + 2Ag(s)$
    $Cu(s) + 2Ag^+(aq) \longrightarrow Cu^{2+}(aq) + 2Ag(s)$
(b) $FeO(s) + 2HCl(aq) \longrightarrow FeCl_2(aq) + H_2O(l)$
    $FeO(s) + 2H^+(aq) \longrightarrow Fe^{2+}(aq) + H_2O(l)$

5.107 (a) $2NaCl(aq) + Ag_2SO_4(s) \longrightarrow Na_2SO_4(aq) + 2AgCl(s)$
$2Cl^-(aq) + Ag_2SO_4(s) \longrightarrow 2AgCl(s) + SO_4^{2-}(aq)$
(b) $Cu(OH)_2(s) + 2HCl(aq) \longrightarrow CuCl_2(aq) + 2H_2O(l)$
$Cu(OH)_2(s) + 2H^+(aq) \longrightarrow Cu^{2+}(aq) + 2H_2O(l)$
(c) $BaCl_2(aq) + Ag_2SO_4(s) \longrightarrow BaSO_4(s) + 2AgCl(s)$
$Ba^{2+}(aq) + 2Cl^-(aq) + Ag_2SO_4(s) \longrightarrow$
$BaSO_4(s) + 2AgCl(s)$

5.109 (a) calcium carbonate, $CaCO_3$
(b) $Na_2CO_3(aq) + CaBr_2(aq) \longrightarrow$
$CaCO_3(s) + 2NaBr(aq)$
(c) $2Na^+(aq) + CO_3^{2-}(aq) + Ca^{2+}(aq) + 2Br^-(aq) \longrightarrow$
$CaCO_3(s) + 2Na^+(aq) + 2Br^-(aq)$
(d) $Na^+(aq)$ and $Br^-(aq)$
(e) $CO_3^{2-}(aq) + Ca^{2+}(aq) \longrightarrow CaCO_3(s)$

5.111 (a) $Sr(NO_3)_2(aq) + H_2SO_4(aq) \longrightarrow$
$SrSO_4(s) + HNO_3(aq)$
$Sr^{2+}(aq) + SO_4^{2-}(aq) \longrightarrow SrSO_4(s)$
(b) no reaction
(c) $CuSO_4(aq) + BaS(aq) \longrightarrow CuS(s) + BaSO_4(s)$
$Cu^{2+}(aq) + SO_4^{2-}(aq) + Ba^{2+}(aq) + S^{2-}(aq) \longrightarrow$
$CuS(s) + BaSO_4(s)$

5.113 combination

5.115 (a) $ZnSO_4(aq) + Ba(NO_3)_2(aq) \longrightarrow$
$BaSO_4(s) + Zn(NO_3)_2(aq)$
(b) $3Ca(NO_3)_2(aq) + 2K_3PO_4(aq) \longrightarrow$
$Ca_3(PO_4)_2(s) + 6KNO_3(aq)$
(c) $ZnSO_4(aq) + BaCl_2(aq) \longrightarrow BaSO_4(s) + ZnCl_2(aq)$
(d) $2KOH(aq) + MgCl_2(aq) \longrightarrow$
$2KCl(aq) + Mg(OH)_2(s)$
(e) $CuSO_4(aq) + BaS(aq) \longrightarrow CuS(s) + BaSO_4(s)$

5.117 $2K(s) + 2H_2O(l) \longrightarrow 2KOH(aq) + H_2(g)$

5.119 (b)

5.121 Aqueous solutions of HCl and NaOH, HNO_3 and KOH, HCl and KOH will give the desired net ionic equation.

5.123 No reaction occurs.

5.125

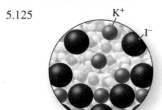

Solution of KI

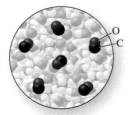

Solution of CO

5.127 Any metal higher in activity can be used. (a) Al; (b) Zn; (c) Mg; (d) Sn

5.129 (a) potassium chromate precipitates iron(III) chromate leaving Al^{3+} in solution; (b) sodium sulfate precipitates barium sulfate leaving Mg^{2+} in solution; (c) silver perchlorate precipitates silver chloride leaving NO_3^- in

solution; (d) barium sulfate is insoluble in water, $MgSO_4$ will dissolve in water.

Chapter 6 – Quantities in Chemical Reactions

6.1 (a) stoichiometry; (b) heat; (c) endothermic reaction; (d) specific heat; (e) actual yield

6.3 (a) $C_6H_{12}O_6 + 6O_2 \longrightarrow 6CO_2 + 6H_2O$
(b) 72 CO_2 molecules
(c) 5 $C_6H_{12}O_6$ molecules

6.5 The coefficients represent the relative number of reactants used and products formed. They can represent the actual numbers of molecules or moles of each substance.

6.7 (a) 1 $Mg^{2+}(aq)$ and 2 $NO_3^-(aq)$ ions
(b) 1 mole of $Mg^{2+}(aq)$ and 2 moles of $NO_3^-(aq)$ ions

6.9 The best representation is (d), however (c) could also be chosen because it has the reactants and products in the correct proportions, though not with the smallest whole numbers. In reaction (a), the reactants are not diatomic. In reaction (b), mass is not conserved (not balanced).

6.11 (a) $\dfrac{15\,mol\,O_2}{2\,mol\,C_6H_6}$; (b) 7.5 mol O_2; (c) 2.9 mol C_6H_6

6.13 (a) $\dfrac{6\,mol\,HNO_3}{2\,mol\,Al}$; (b) $\dfrac{3\,mol\,H_2}{2\,mol\,Al}$; (c) $\dfrac{2\,mol\,Al}{3\,mol\,H_2}$

6.15 (a) not conserved; (b) conserved; (c) conserved; (d) conserved; (e) moles of molecules may be conserved in some cases. All others are conserved in all cases.

6.17 (a) $\dfrac{7\,mol\,C}{2\,mol\,C_7H_5(NO_2)_3}$, $\dfrac{7\,mol\,CO}{2\,mol\,C_7H_5(NO_2)_3}$,
$\dfrac{3\,mol\,N_2}{2\,mol\,C_7H_5(NO_2)_3}$, $\dfrac{5\,mol\,H_2O}{2\,mol\,C_7H_5(NO_2)_3}$
(b) 3.50 mol C, 3.50 mol CO, 1.50 mol N_2, 2.50 mol H_2O
(c) 21.9 mol C, 21.9 mol CO, 9.38 mol N_2, 15.6 mol H_2O

6.19 (a) 249.70 g/mol; (b) 159.62 g/mol; (c) 0.639 g $CuSO_4$

6.21 (a) 2.05 g; (b) 5.89 g; (c) 7.94 g

6.23 (a) 21.0 g CO; (b) 38.0 g I_2

6.25 (a) 1.14 g Cl_2; (b) 0.0760 g H_2O; (c) 2.13 g O_2; (d) 0.544 g Cl_2

6.27 1.0 g $CaCO_3$

6.29 the fuel

6.31 (a) 12 sandwiches; (b) bread; (c) 3 pieces of turkey

6.33 (a) H_2; (b) N_2; (c) H_2 and N_2 (both react completely)

6.35 (a)

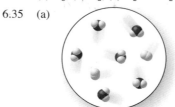

(b) oxygen
(c) hydrogen

6.37

$$2C_2H_2(g) + 7O_2(g) \longrightarrow 4CO_2(g) + 6H_2O(g)$$

Initially mixed	6 molecules	18 molecules	0 molecules	0 molecules
How much reacts	4 molecules	14 molecules	—	—
Composition of mixture	2 molecules	4 molecules	8 molecules	12 molecules

6.39

$$2C_2H_{10}(g) + 13O_2(g) \longrightarrow 8CO_2(g) + 10H_2O(g)$$

Initially mixed	3.10 mol	13.0 mol	0.00 mol	0.00 mol
How much reacts	2.00 mol	13.0 mol	—	—
Composition of mixture	1.10 mol	0.0 mol	8.00 mol	10.0 mol

6.41 (a) P_4; (b) both are consumed completely; (c) O_2

6.43 (a) N_2; (b) 12.2 g NH_3

6.45 limiting reactant, $NaHCO_3$; 4.2 g CO_2 produced

6.47

$$2C_2H_6(g) + 7O_2(g) \longrightarrow 4CO_2(g) + 6H_2O(g)$$

Initially mixed	0.260 g	1.00 g	0.00 g	0.00 g
How much reacts	0.260 g	0.968 g	—	—
Composition of mixture	0.000 g	0.03 g	0.761 g	0.467 g

6.49 (a) $NaOH(aq) + HCl(aq) \longrightarrow H_2O(l) + NaCl(aq)$
(b) NaOH
(c) acidic

6.51 actual yield

6.53 The actual yield greater than the theoretical yield can be caused by contamination. The solid may have not been dry, causing the measured mass to be higher than expected.

6.55 (a) 7.44 g; (b) 8.95 g; (c) 83.1%

6.57 90.5%

6.59 0.47 mol I_2

6.61 84%

6.63 The potential energy is converted to kinetic energy, some of which is transferred to the ground as heat.

6.65 The potential energy of hydrogen and oxygen is used to provide electrical energy which runs electic motors and produces kinetic energy.

6.67 The reaction is endothermic because the decrease in temperature of the surroundings indicates energy is absorbed by the reaction.

6.69 As the liquid evaporates, the molecules going into the gas state must absorb some energy from their surroundings. As a result, the surroundings (water and your skin) lose energy and feel colder.

6.71 The products are lower in potential energy.

6.73 lead

6.75 5470 J or 1310 cal

6.77 −210 J or −2.10 × 10² J

6.79 A calorimeter should have good insulation and a way to accurately and precisely measure the temperature.

6.81 (a) The rock lost heat. (b) $q_{rock} = -1.3 \times 10^3$ J

6.83 Because a reaction is a process, you can only measure the effect it has on the surroundings.

6.85 The chemical reaction is the system and the water and calorimeter are the surroundings.

6.87 (a) exothermic; (b) −525 kJ/mol

6.89 (a) 147 kJ; (b) −24.5 kJ/g

6.91 (a) $q_{nut} = -2.6 \times 10^4$ J so 2.6×10^4 J is released; (b) 6.3×10^3 cal or 6.3 Cal; (c) 3.2 Cal/g

6.93 −256 kJ

6.95 (a) 3.00 mol CO_2, 1.50 mol N_2, 0.250 mol O_2, 2.50 mol H_2O
(b) 7.50 mol CO_2, 3.75 mol N_2, 0.625 mol O_2, 6.25 mol H_2O

6.97 (a) 35.7 g $AgNO_3$; (b) 30.1 g AgCl

6.99 9330 g O_2

6.101 (a) $2K(s) + Cl_2(g) \longrightarrow 2KCl(s)$
(b) Cl_2
(c) There should be some gray solid (K) and a white solid (KCl) inside the container.
(d) 11 g
(e) 4.5 g K

6.103 (a) $\dfrac{1\,mol\,O_2}{2\,mol\,H_2}$; (b) 7.9 g O_2 and 8.9 g H_2O; (c) 4.5 g H_2O

6.105 31.8°C

6.107 0.450 J/g °C; chromium

6.109 2.80×10^3 kJ/mol, 6.70×10^2 Cal/mol

6.111 3 kJ/g

6.113 54.4°C

6.115 0.450 J/g °C; chromium

Chapter 7 – Electron Structure of the Atom

7.1 (a) electromagnetic radiation; (b) frequency; (c) ionization energy; (d) Hund's rule; (e) electron configuration; (f) core electron; (g) orbital; (h) continuous spectrum; (i) isoelectronic

7.3 infrared, microwave, and radio frequency

7.5

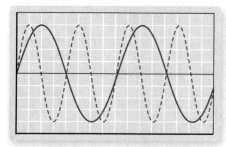

Lower frequency (——) with higher frequency (– – –) superimposed

7.7 blue, yellow, orange, red

7.9 As wavelength increases, the frequency decreases. As wavelength decreases, frequency increases.

7.11 infrared

7.13 Gamma photons have the highest energy and highest frequency. Radio frequency waves are the longest.

7.15 4.00×10^{15} Hz, ultraviolet

7.17 4.27×10^{-19} J, visible light (blue)

7.19   white light

7.21   No.

7.23   No. If they did, line spectra would not exist. The energy of the electron is quantized—it can only have certain values.

7.25   To move to a higher orbit, the electron would have to absorb energy. To go to a lower orbit, it would have to release energy (e.g., a photon).

7.27   A single photon is released.

7.29   n = 6 to n = 3

7.31   n = 5 to n = 3

7.33   The four lines are a result of four different transitions in the hydrogen atom. These transitions are

   n = 6 to n = 2   violet (highest energy)
   n = 5 to n = 2   blue
   n = 4 to n = 2   green
   n = 3 to n = 2   red (lowest energy)

7.35 4.58×10^{-19} J

7.37   Bohr's orbits required that the orbits be a fixed distance from the nucleus with the electron following a specific pathway. Modern orbitals describe the region of space surrounding the nucleus where we are most likely to find the electron.

7.39   Their primary differences are size and energy. The $3p$ orbital is larger and higher in energy.

   2p   3p

7.41   (a) 1; (b) 3; (c) 5; (d) 7; (e) 1; (f) 3

7.43
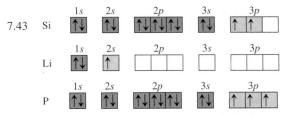

7.45   (a) $1s^2 2s^2 2p^6 3s^2 3p^2$; (b) $1s^2 2s^1$; (c) $1s^2 2s^2 2p^6 3s^2$

7.47   All orbitals can hold 2 electrons (including a $2p$ orbital).

7.49   8 electrons

7.51   3

7.53   elements in group VIIA (17)

7.55   transition metals

7.57   aluminum

7.59   (a) $1s^2 2s^2 2p^6 3s^1$
      (b) $1s^2 2s^2 2p^6 3s^2 3p^6 4s^2 3d^5$
      (c) $1s^2 2s^2 2p^6 3s^2 3p^6 4s^2 3d^{10} 4p^4$

7.61   Bromine should be $1s^2 2s^2 2p^6 3s^2 3p^6 4s^2 3d^{10} 4p^5$. The $4p$ orbital given in the problem has one too many electrons and the 10 electrons in the d sublevel are in the third principle energy level.

7.63   (a) sulfur; (b) nickel; (c) barium

7.65   (a) [Ne]$3s^1$; (b) [Ar]$4s^2 3d^5$; (c) [Ar]$4s^2 3d^{10} 4p^4$

7.67   Bromine ($1s^2 2s^2 \mathbf{2p^6} 3s^2 \mathbf{3p^6} 4s^2 3d^{10} \mathbf{4p^5}$)

7.69   Valence electrons are the electrons in the highest principle energy level.

7.71   No, the d-orbital electrons are always one energy level lower than the valence electrons.

7.73   As you go left to right on the periodic table, one electron is added each time the group number increases. The group number is the number of electrons in the s and p orbitals of the highest energy level. Electrons in the d-orbitals are not included since they are always one energy level below the highest energy level.

7.75   (a) 4th valence level; 2 valence electrons
      (b) 2nd valence level; 7 valence electrons
      (c) 6th valence level; 5 valence electrons

7.77   Cations always have fewer electrons than the element from which they are formed. They are similar in that they possess the same core of electrons.

7.79   (a) $1s^2 2s^2 2p^6$; [Ne]; F^-
      (b) $1s^2 2s^2 2p^6$; [Ne]; N^{3-}
      (c) $1s^2 2s^2 2p^6 3s^2 3p^6 3d^{10}$; [Ar]$3d^{10}$; Zn^{2+}

7.81 O^{2-}, N^{3-}, Na^+

7.83   Valence electrons farther from the attraction of the nucleus are easier to remove. This means that potassium ($4s$ valence electrons) has a lower ionization energy than sodium ($3s$ valence electrons). Since less energy is expended to remove potassium's electron, more energy is left to drive the reaction.

7.85   The first ionization energy of calcium is higher than that of potassium. Since less energy is expended to remove potassium's electron, more energy is left to drive the reaction. In addition, calcium must lose a second electron before it becomes stable. Even more energy is expended to remove this electron.

7.87   P < S < O

7.89   Sodium's valence electron is in a higher principle energy level. This means it is not as close to the nucleus and is therefore easier to remove.

7.91   Fluorine has more protons in its nucleus than oxygen. As a result, the electrons of fluorine are more strongly attracted to the nucleus; the ionization energy is higher.

7.93   First ionization, IE_1: Mg(g) $\longrightarrow$ Mg$^+$(g) + e$^-$
      Second ionization, IE_2: Mg$^+$(g) $\longrightarrow$ Mg^{2+}(g) + e$^-$

7.95   Magnesium because the third electron is removed from the core electrons.

7.97   (a) fluorine has a very high ionization energy; (b) elements do not lose core electrons when forming ions (under normal conditions); (c) high ionization energy and because Ne has a core configuration

7.99   O < S < P

7.101 The valence electrons of chlorine and sulfur are both in principal energy level 3, but chlorine has more protons in its nucleus to attract the electrons.

7.103 (a) Al; (b) S^{2-}

7.105 Both ions are isoelectronic with argon, but the potassium ion is larger because it has fewer protons in its nucleus.

7.107 For hydrogen the energies of the sublevels within a given principal energy level are all the same.

7.109 (a) [Xe]$6s^2 4f^{14} 5d^{10} 6p^3$
      (b) [Rn] or [Xe]$6s^2 4f^{14} 5d^{10} 6p^6$
      (c) [Rn]$7s^2$

7.111 12 filled p orbitals

7.113 10 filled d orbitals

7.115 Both are related to the attraction of electrons for the nucleus. As attraction gets higher, electrons are more tightly held. This results in greater ionization energy and smaller radius.

7.117 The radius of the sodium ion is smaller and the chloride ion greater than for their respective elements in the reactants.

Chapter 8 – Chemical Bonding

8.1 (a) single bond; (b) alkane; (c) covalent bonding; (d) ionic crystal; (e) octet rule; (f) polar covalent bond; (g) alkyne; (h) electronegativity; (i) triple bond; (j) crystal lattice; (k) Lewis formula; (l) chemical bond

8.3 A chemical bond is an attractive force between atoms or ions in a substance.

8.5 Metals form ionic bonds with nonmetals.

8.7 HF, NCl_3, CF_4

8.9 (a) ionic; (b) covalent; (c) covalent; (d) ionic

8.11 A and D

8.13 CH_4, SF_4, HCl

8.15 (a) high; (b) high; (c) low; (d) high; (e) low

8.17 Linus Pauling proposed that the increase in bond strength between unlike atoms was a result of increased ionic character (uneven sharing of electrons). The strength of attraction of an element for electrons in a bond is termed electronegativity.

8.19 Bond polarity occurs when electrons are not equally shared by the two atoms that are bonded together.

8.21 (a) N = Br < Cl < O < F (highest electronegativity)
(b) H < C < N < O < F

8.23
Polar	Nonpolar
(a) $^{\delta+}H-F^{\delta-}$	H—H
(b) $^{\delta+}I-Cl^{\delta-}$	I—I
(c) $^{\delta+}H-I^{\delta-}$	H—H

8.25 (a) H—H < C—H < O—H < F—H
(b) O—Cl < C—Cl < F—Cl < H—Cl

8.27 The Lewis symbol can be used to determine the number of valence electrons and the number of electrons gained or lost to reach an octet configuration.

8.29 (a) $\cdot \ddot{A}s \cdot$; (b) $: \dot{\ddot{I}} :$; (c) $: \dot{\ddot{Se}} \cdot$; (d) $\dot{S}r \cdot$; (e) $\dot{C}s$; (f) $: \ddot{\ddot{A}}r :$

8.31 (a) $: \ddot{\ddot{C}}l :^-$; (b) Sc^{3+}; (c) $: \ddot{\ddot{S}} :^{2-}$; (d) Ba^{2+}; (e) B^{3+}

8.33 (a) $\left[Li^+ \right]\left[: \ddot{\ddot{C}}l :^- \right]$

(b) $\left[: \ddot{\ddot{C}}l :^- \right]\left[Ba^{2+} \right]\left[: \ddot{\ddot{C}}l :^- \right]$

(c) $\left[Ba^{2+} \right]\left[: \ddot{\ddot{S}} :^{2-} \right]$

8.35 The second electron removed from K, potassium, would be taken from the core. These electrons are very strongly held by the atom and are not easily removed.

8.37 When fluorine gains an electron and sodium loses one electron, both have satisfied the octet rule. The ionic bond formed during the process is very stable and the compound is electrically neutral.

8.39 A crystal lattice is the repeating pattern observed in all crystal structures. An ionic crystal is the structure that forms when ions form crystals by minimizing the repulsive energies of like-charged ions and maximizing the contact of oppositely charged ions.

8.41 Six chloride ions surround each sodium ion.

8.43 The ratio of the cations and anions is different, so different ionic crystals are formed.

8.45 NaCl; size of Li^+ is closer to that of Na^+ than of Cs^+

8.47 Each hydrogen needs only one electron to satisfy its valence shell. It does not have another electron to form another bond.

8.49 (a) 2; (b) 3; (c) 1; (d) none

8.51 single bond = 1 shared electron pair
double bond = 2 shared electron pairs
triple bond = 3 shared electron pairs

8.53 (a) IIIA; (b) IVA; (c) VIA; (d) IVA

8.55 (a) H—C≡N:

(b) H—C—C≡N: (with two H on the first C, one above and one below)

(c) H—C≡C—H

(d) C=C structure with two H on each carbon

(e) H—C—C—H structure with two H on each carbon

8.57 (a) $\left[\ddot{O} \begin{smallmatrix} \\ \diagdown \end{smallmatrix} N \begin{smallmatrix} \diagup \\ \end{smallmatrix} \ddot{O} : \\ | \\ : \ddot{O} : \end{smallmatrix} \right]^-$

(b) $\left[\begin{smallmatrix} : \ddot{O} : \\ | \\ : \ddot{O} - S - \ddot{O} : \\ | \\ : \ddot{O} : \end{smallmatrix} \right]^{2-}$

(c) $\left[\begin{smallmatrix} : \ddot{O} : \\ | \\ : \ddot{O} - S - \ddot{O} : \end{smallmatrix} \right]^{2-}$

(d) $\left[\ddot{O} \begin{smallmatrix} \\ \diagdown \end{smallmatrix} N \begin{smallmatrix} \diagup \\ \end{smallmatrix} \ddot{O} : \right]^-$

(e) $\left[: N≡O : \right]^+$

8.59 Resonance descriptions are required when a molecule or ion can be represented by two or more reasonable Lewis structures that differ only in the positions of the bonding and lone pair electrons. The positions of the atomic nuclei do not change.

8.61 Resonance is exhibited in (c) and (d).

8.63 (a) $:\!\overset{..}{S}\!=\!\overset{..}{O}: \longleftrightarrow \overset{..}{S}\!-\!\overset{..}{O}:$ (with $:\!\overset{..}{O}:$ below) resonance structures

(b) resonance structures of SO_3^{2-} type

(c) $:O\!\equiv\!C\!-\!\overset{..}{O}: \longleftrightarrow :\!\overset{..}{O}\!=\!C\!=\!\overset{..}{O}: \longleftrightarrow :\!\overset{..}{O}\!-\!C\!\equiv\!O:$

(d) CO_3^{2-} resonance structures (three, each with 2− charge)

(e) NO_2^{-}/nitrate resonance structures (three, each with − charge)

8.65 Hydrogen's valence shell can only hold two electrons.

8.67 (a) octet obeyed; (b) expanded octet; (c) octet obeyed; (d) octet obeyed

8.69 (a) S, expanded octet; (b) B, incomplete octet; (c) Xe, expanded octet; (d) Cl, odd electron species

8.71 Alkanes with only single bonds, alkenes with double bonds, and alkynes with triple bonds.

8.73 (benzene ring structure drawn with H atoms)

8.75 (a) alcohol; (b) alkane; (c) ether; (d) alkene; (e) ketone

8.77 (a) alcohol; (b) alkene; (c) aldehyde; (d) ether

8.79 (a) alkane; (b) amine; (c) ester

8.81 $H\!-\!\overset{H}{\underset{H}{C}}\!-\!\overset{H}{\underset{H}{C}}\!-\!\overset{H}{\underset{H}{C}}\!-\!\overset{\overset{..}{O}}{C}\!-\!H$ (structure drawn)

8.83 The geometric arrangement of electron pairs and atoms is predicted by finding the geometry that gives the greatest distance between the electron groups and atoms (Table 8.5). The shape is determined by the arrangement of the bonded atoms (Table 8.6).

8.85 Shape is determined by first knowing the number of bonded atoms and unshared electron pairs (Table 8.5). Once this is determined, the shape is given by the arrangement of atoms (Table 8.6).

8.87 (a) (structure drawn)

(b) (structure drawn)

(c) (structure drawn)

8.89 (a) linear; (b) tetrahedral; (c) tetrahedral; (d) trigonal planar; (e) tetrahedral; (f) tetrahedral; (g) trigonal planar; (h) tetrahedral

8.91 (a) linear, 180°; (b) trigonal pyramidal, 109.5°; (c) bent, 109.5°(tetrahedral); (d) bent, 120°(trigonal planar); (e) bent, 109.5°(tetrahedral); (f) tetrahedral, 109.5°; (g) trigonal planar, 120°; (h) bent, 109.5°(tetrahedral)

8.93 (a) 109.5°; (b) 109.5°; (c) not defined; (d) 180°; (e) 120°; (f) 120°; (g) 109.5°

8.95 (a) BF_3; (b) NH_3; (c) SCl_2

8.97 (a) BCl_3; (b) H_2O; (c) $BeCl_2$; (d) NH_4^+

8.99 ClO_3^{-}

8.101 image B

8.103 Each nitrogen has one unshared electron pair.

8.105 The bond angles on the carbon are 109.5°. The bond angles on the nitrogen are 120°.

8.107 A bond is polar if there is a difference in electronegativity between the two atoms that are bonded. A molecule is polar if the polarity of the bonds to a central atom do not cancel out.

8.109 The chlorines are all attracting the electrons of the carbon with equal force. Because the forces are in opposite direction and the molecule geometrically symmetric, the polarity of the bonds cancels in the molecule.

8.111 (a) polar; (b) polar; (c) polar; (d) polar

8.113 (a) Both are tetrahedral, but in the case of CH_2Cl_2, the bond polarities do not cancel and the molecule is polar. The bonds in CCl_4 are all equivalent so the bond polarities cancel.

(b) PF_3 has an unshared electron pair and is not symmetrical so bond polarities do not cancel. BF_3 is trigonal planar (no unshared pair, three atoms) so the polarities of the bonds cancel.

(c) BF_2Cl is trigonal planar like BF_3, but the electronegativities of F and Cl are not the same. As a result, BF_2Cl is polar.

(d) SO_2 has an unshared electron pair so it is not symmetrical. The bond polarities do not cancel. CO_2 is symmetrical so the bond polarities cancel out.

8.115 image B

8.117 The molecule CCl_2F_2 should be soluble in water.

8.119 image B

8.121 (a) $:\!\overset{..}{B}r\!\cdot$; (b) $\cdot\overset{..}{P}b\cdot$; (c) $:\!\overset{..}{S}\!\cdot$; (d) $\overset{.}{C}a\cdot$; (e) $\overset{.}{B}e\cdot$; (f) $:\!\overset{..}{X}e:$

8.123 F > Cl > Br > I

8.125 (a) covalent; (b) ionic; (c) covalent; (d) covalent; (e) ionic

8.127 (a) $H\!-\!N\!=\!N\!-\!H$

(b) $:\!\overset{..}{S}\!=\!C\!=\!\overset{..}{S}:$

(c) $:\!\overset{..}{F}\!-\!\overset{..}{A}s\!-\!\overset{..}{F}:$ (with $:\!\overset{..}{F}:$ below)

(d) $:\!\overset{..}{O}\!=\!C\!=\!\overset{..}{O}:$

(e) $:C\!\equiv\!O:$

8.129 (a)

$$\left[\begin{array}{c} :\ddot{O}: \\ | \\ :\ddot{O}-Cl-\ddot{O}: \\ | \\ :\ddot{O}: \end{array} \right]^{-}$$

(b)

$$\left[\begin{array}{c} \ddot{N}=\ddot{O}: \\ | \\ :\ddot{O}: \end{array} \right]^{-} \longleftrightarrow \left[\begin{array}{c} \ddot{N}-\ddot{O}: \\ \| \\ \ddot{O}: \end{array} \right]^{-}$$

(c)

$$\left[:N\equiv C-\ddot{O}: \right]^{-} \longleftrightarrow \left[:\ddot{N}=C=O: \right]^{-} \longleftrightarrow \left[:\ddot{N}-C\equiv O: \right]^{-}$$

(d)

$$\left[\begin{array}{c} H-C=\ddot{O} \\ | \\ :\ddot{O}: \end{array} \right]^{-} \longleftrightarrow \left[\begin{array}{c} H-C-\ddot{O}: \\ \| \\ \ddot{O}: \end{array} \right]^{-}$$

(e)

$$\begin{array}{c} :\ddot{F}: \\ | \\ :\ddot{F}-B-\ddot{F}: \end{array}$$

8.131 (a) tetrahedral; (b) trigonal planar; (c) bent; (d) trigonal
pyramidal; (e) tetrahedral; (f) bent; (g) tetrahedral;
(h) trigonal planar; (i) bent; (j) bent

8.133 (a) $BeCl_2$ is nonpolar because the molecule is symmetric.
OCl_2 is polar. The molecule has polar bonds and is
bent (not symmetric).

(b) PH_3 is polar. The bonds are polar and the molecule is
not symmetric. BH_3 is nonpolar because the molecule
is symmetric.

(c) BCl_3 is nonpolar because the molecule is symmetric.
$AsCl_3$ is polar because the bonds are polar and the
molecule is not symmetric.

(d) SiH_4 is nonpolar because the molecule is symmetric.
NH_3 is polar because the bonds are polar the molecule
is not symmetric.

Chapter 9 – The Gaseous State

9.1   (a) effusion; (b) Boyle's law; (c) combined gas law;
(d) ideal gas; (e) pressure; (f) Dalton's law of partial pressures; (g) molar volume; (h) ideal gas constant, R

9.3   When compared to other states of matter gases have low
densities and are very compressible. They also take the
shape of any container they are put in.

9.5   The density of warm air is lower than the density of cold
air.

9.7

9.9   Gas pressure is the amount of force exerted by the gas
divided by the area over which the force is exerted.

9.11   Absolute pressure is measured with a device called a
barometer. A tire gauge measures the pressure above
atmospheric pressure.

9.13

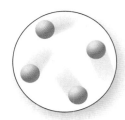

9.15   (a) 0.980 atm; (b) 935 torr; (c) 0.119 atm; (d) 643 Pa;
(e) 1.01×10^3 mm Hg; (f) 563 torr; (g) 1.06×10^5 Pa;
(h) 293 mm Hg

9.17 1.024×10^5 Pa

9.19   As pressure increases, the volume decreases at constant
temperature.

9.21   The particles collide more frequently with the container
walls causing an increase in pressure. Since the number of
moles has not changed, we expect to find twice as many
molecules in the same space. The velocity of the
molecules should not change.

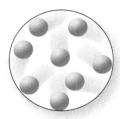

9.23   (a) 1.20 L; (b) 20.0 mL; (c) 0.196 mL

9.25   (a) 81.6 torr; (b) 0.100 torr; (c) 27.4 atm

9.27   182 atm

9.29   0.472 L

9.31   Charles's law states that if the pressure is constant, the
volume and temperature (in Kelvin) are directly proportional to each other (volume increases when temperature
increases; volume decreases when temperature
decreases).

9.33   The velocity of the gas molecules increases as temperature increases. The volume of the container will increase
as the temperature increases.

9.35   (a) 4.51 L; (b) 7.70×10^2 mL; (c) 77.3 L

9.37   (a) 273°C; (b) −266.2°C; (c) −49°C

9.39   0.407 mL

9.41   Nothing happens to the particles if the temperature, pressure, and volume are constant. If the tank is sealed so that
no gas molecules are allowed to escape, then nothing happens to the pressure. If the tank is opened, then the molecules of gas in the tank will leave the tank. Particles leave
the tank to maintain an equilibrium pressure with the air
outside the tank.

9.43   (a) 418 torr; (b) 673 torr; (c) 4.44 atm

9.45   (a) −71.4°C; (b) 1.2×10^3°C; (c) 237°C

9.47   7.71 atm

9.49   (a) 1.2 L; (b) −195.3°C; (c) 9.10×10^3 torr

9.51   1.61 L

9.53   Guy-Lussac's Law states that volumes of gas react in simple whole number ratios when the volumes of reactants
and products are measured at the same temperature and
pressure.

9.55   22.414 L/mol.

9.57   (a) 0.340 mol, 5.45 g; (b) 6.692×10^{-3} mol, 0.8787 g;
(c) 1.70 mol, 47.6 g

9.59   The number of particles is the same. The balloon containing argon has the greater mass therefore greater density since the volumes are the same.

9.61   (a) 0.40 mol, 8.9L; (b) 1.5 mol, 34 L; (c) 2.2 mol, 49 L

9.63   2380 balloons

9.65   11.0 L

9.67   Any gas whose behavior is described by the five postulates of kinetic molecular theory is an ideal gas.

9.69   (a) 0.40 mol, 0.82 L; (b) 1.5 mol, 3.1 L; (c) 2.2 mol, 4.5 L

9.71   (a) 0.245 mol, 3.92 g; (b) 4.33×10^{-3} mol, 8.73×10^{-3} g;
(c) 0.288 mol, 8.07 g

9.73   The balloon filled with CO_2 sinks because the density of CO_2 is greater than the density of air.

9.75   (a) 0.7600 g/L; (b) 1.250 g/L; (c) 1.964 g/L

9.77   (a) 0.673 g/L; (b) 1.11 g/L; (c) 1.74 g/L

9.79   There are 0.208 moles of each gas. (a) 0.42 g; (b) 3.3 g; (c) 13 g

9.81   Dalton's law of partial pressures states that the total pressure in a container is the sum of the pressures exerted by each of the individual gases in a mixture of gases.

9.83   708 torr

9.85   (a) 2.78 mol; (b) 2.08 mol; (c) 2.15 atm; (d) 1.60 atm

9.87   (a) 40.01 g/mol; (b) 30.03 g/mol; (c) 45.99 g/mol;
(d) 20.3 g/mol; (e) 16.0 g/mol

9.89   The five postulates of kinetic molecular theory are
1) Gases are composed of small particles widely separated.
2) Particles behave independently of each other.
3) Gas particles move rapidly in straight lines.
4) Gas pressure results from force exerted by the molecules in the container. The force is the sum of the forces exerted by each molecule as it bounces off the walls of the container.
5) The average kinetic energy of the gas particles depends only on the absolute temperature.

9.91   As the temperature is increased, the kinetic energy of the particles increases. Recall that kinetic energy is the energy of motion. Increasing temperature increases molecular velocity. The particles strike the walls of the container with greater force and, therefore, the pressure is higher.

9.93   Pressure depends on the force of collisions and the frequency of collisions in a given area. The density of gas particles is inversely proportional to the volume. If the volume decreases, the density increases. Since the density of gas particles increases with decreasing volume, the frequency of collisions in a given area has increased. This means that the pressure has increased.

9.95   (lowest velocity) $CO_2 < CH_4 < He < H_2$ (Highest velocity)

9.97   Molecules with lower masses have higher average velocities and as a result diffuse faster.

9.99   neon

9.101   24 L H_2

9.103   54.0 L CO_2 and 85.5 L O_2

9.105 1.68×10^3 g

9.107   The balloon expands or contracts so that the external pressure and the internal pressure are the same.

9.109   The density of a gas decreases if it is heated and the volume is allowed to expand. The air in the hot-air balloon is less dense than the surrounding air, so the balloon floats.

9.111   1100 K

9.113

What will happen if...	Macroscopic view	Molecular-level view
temperature increases and pressure remains constant	balloon gets bigger	molecules move faster molecules collide harder collision frequency increases molecules are further apart
temperature decreases and pressure remains constant	balloon gets smaller	molecules move slower less energetic collisions collision frequency decreases molecules are closer together
10,000 ft (temperature decreases and pressure decreases)	balloon probably gets a little bigger	molecules move slower less energetic collisions lower collision frequency

9.115   6.64 g

9.117   1.75 L

9.119   2.88 L

9.121   1.21 L

9.123   99 K or $-174°C$

9.125   31.0 L

9.127   (a) 10.0 L; (b) 6.67 L; (c) 3.33 L

9.129   229 K

9.131   (a) If the pressure and temperature are constant, increasing the moles of gas increases the volume occupied by the gas. (b) Pressure and temperature

9.133   0.718 L

9.135   Pressure and volume are inversely related.

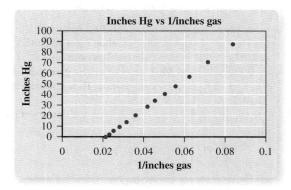

Chapter 10 – The Liquid and Solid States

10.1   (a) vapor pressure; (b) melting point; (c) equilibrium;
(d) evaporation; (e) induced dipole; (f) alloy; (g) boiling point; (h) sublimation; (i) molecular solid; (j) intermolecular force; (k) London dispersion force; (l) melting point;
(m) crystal

10.3

	Liquid	Gas
Particle spacing	dense or closely spaced	molecules far apart
Intermolecular attraction	moderate	weak
Kinetic energy	low	high

10.5 A liquid is fluid because the molecules can move past each other. Attractive forces hold the particles in solids together in a rigid, three-dimensional array.

10.7 The substance is in the solid state. If the substance were a liquid or gas, it would not maintain its shape. In addition, if the substance was not a solid, the images on the surface would not maintain their shape.

10.9 The image is a molecular-level representation of the liquid state. The liquid state and solid state both have dense particle spacing. Basically, the molecules are always touching each other. However, in the image, the molecules do not show any clear pattern. This is a characteristic of the liquid state.

10.11 The solid, liquid, and gas states are represented in circles 1, 3, and 2, respectively. The phase changes are deposition, vaporization, and melting for phase changes A, B, and C, respectively.

10.13 The vapor pressure of a liquid increases as the temperature is increased. A liquid boils when its vapor pressure is equal to the pressure above the liquid. The temperature at which this occurs is the boiling point. The normal boiling point is the temperature at which the vapor pressure of the liquid is equal to 1 atm.

10.15 In order to convert a substance from the liquid to the gas state, the energy of the molecules must increase enough to allow them to overcome their attractions for each other and move about independently. Because the molecules must absorb energy to go into the gas state, the process is endothermic (i.e., absorbs energy from the surroundings).

10.17

10.19 solid state to the gas state; endothermic

10.21 All molecules and atoms have some degree of attraction for each other. As a gas is cooled, the molecules are losing kinetic energy. Eventually the attractive forces are greater than the kinetic energy of the molecules and the molecules begin to coalesce (stick together) to form the liquid state.

10.23 Evaporation of water is endothermic. The energy water needs to evaporate comes from the surrounding air, cooling it. The cooled air can be used to cool a house (if it is circulated by a fan).

10.25 Water boils when it has been heated to the point that its vapor pressure equals the atmospheric pressure. At high elevation, the atmospheric pressure can be significantly lower than 1 atm (about 0.6 to 0.7 atm at 12,000 ft). Since the water boils at lower temperature, the pasta cooks slower.

10.27 As a liquid is heated the molecules of the liquid acquire more energy. The percent of molecules on the liquid surface with enough energy to go into the gas phase increases. This causes an increase in the vapor pressure.

10.29 Gallium is a liquid at 100°C. At 15°C gallium is a solid.

10.31 *BC*

10.33 237 kJ

10.35 41.0 kJ

10.37 An attractive force between individual particles (molecules or atoms) of a substance is called an intermolecular force.

10.39 the gas state

10.41 Ar

10.43 NO, NF_3, and CH_3Cl

10.45 A bond is an attractive force between two atoms in a molecule. A hydrogen bond is not a true bond because it usually takes place between two atoms on different molecules. In comparison to covalent bonds, the strength of attraction is much weaker and the electrons are not shared.

10.47 Only B can hydrogen bond.

10.49 (a) and (b)

10.51 (b)

10.53 The vapor pressures of pure liquids depend on the strength of the intermolecular forces between the particles of the liquids.

10.55 When a molecule is polar, it means that some portion of the molecule appears permanently slightly negative to other molecules and another portion is slightly positive. This comes from the unequal sharing of electrons (Chapter 8). The opposite signed charges on different molecules attract each other (dipole-dipole interaction), so the molecules stick together. The instantaneous and induced dipoles in nonpolar molecules are not permanent.

10.57 As molecular weight increases the importance of London dispersion forces increases.

10.59

	Dispersion Force	Dipole-dipole	Hydrogen Bonding
(a) C_6H_6	x		
(b) NH_3	x	x	x
(c) CS_2	x		
(d) $CHCl_3$	x	x	

10.61

	Dispersion Force	Dipole-dipole	Hydrogen Bonding
(a) Kr	x		
(b) CO	x	x	
(c) CH_4	x		
(d) NH_3	x	x	x

10.63 (lowest boiling point) $CH_3OH < CH_3CH_2OH < CH_3CH_2CH_2OH$ (highest boiling point)

10.65 NO_2

10.67 Hydrogen (H_2) is larger than (but not more massive than) helium. The size of the hydrogen molecule allows the electrons to be moved around more easily and therefore the dispersion force is greater. More energy is required to vaporize liquid hydrogen than helium.

10.69 HCl is polar while F_2 is nonpolar.

10.71 (a) CH_3OH; (b) CH_3OH; (c) NH_3

10.73 (lowest) $He < N_2 < H_2O < I_2$ (highest)

10.75   Both molecules are polar, but water has very strong hydrogen bonding while acetone has dipole-dipole interactions. As a result, water has a higher boiling point.

10.77   On a molecular level, the particles of a substance in the liquid state are higher in energy than for the same substance in the solid state and the molecules have freedom to move apart from each other. In addition, the liquid state does not have a long range structure like the crystal lattice of many solids. On a macroscopic level, liquids flow and can take the shapes of their containers. Liquids tend to be less dense than solids.

10.79   Viscosity is the resistance of a liquid to flow. The more viscous a liquid is the slower it will flow when it is poured.

10.81   All molecules of a liquid experience attractive forces to other molecules in the liquid. Molecules in the middle experience forces that are equal in all directions. On the surface, the molecules only experience attractive forces to the sides and "downward" (toward the center of the liquid). Surface tension causes liquids to bead up on solid surfaces and is also why liquids tend to form spheres. In addition, surface tension is related to capillary action and the formation of a meniscus (curved surface) of a liquid in a container.

10.83   There is an imbalance of forces on the molecules at the surface of a liquid. As the surface area of the liquid is minimized, the imbalance of energy is also minimized. Substances assume the shape with the least surface area for the given volume. As a result, liquids have a tendency to form spherical shapes to minimize this energy difference.

10.85   When ice forms on bodies of water such as lakes and rivers, it stays at the surface and provides an insulating layer. Energy loss through evaporation is minimized and less water freezes than would if the ice did not form at the top. Ice would continually form and sink. Many bodies of water would freeze completely making life in them impossible during the coldest months.

10.87   Crystals are defined as solids that, on a molecular or atomic level, have a regular repeating pattern of atoms. See Figures 10.31 and 10.37.

10.89   Crystal structures help us understand the macroscopic properties of the solid state (i.e., conductivity, hardness, brittleness).

10.91   Close packing is a common structure of metals.

10.93   Amorphous solids have a tendency to flow because the molecules are not locked into a crystal lattice.

10.95   Solids are classified by the strength of attraction between the subunits of the solid. This allows us to broadly classify the characteristics of the solid substances.

10.97   ionic solid

10.99   There are two reasons. First, the attractive forces between metal atoms are relatively weak. This allows them to be separated from each other. Secondly, the electrons in metal atoms are delocalized amongst all their neighboring atoms. Atoms can be moved with relatively little effect on the charge distribution. In ionic solids, cations and anions cannot be separated easily, so particles do not move apart easily. This causes ionic solids to be brittle.

10.101   Both ionic and metallic solids form crystal lattices. Ionic crystals resemble metal crystals with smaller ions fitting into holes between larger ions.

10.103   Molecular solids are held together either by dipole-dipole interactions, hydrogen bonding, or London forces (dispersion forces). Ionic solids are held together by interionic attractions (i.e., forces of attraction between oppositely charged ions). Because ionic interactions are much stronger than intermolecular forces, the ionic solids tend to melt at much higher temperatures.

10.105   (a) covalent bonds; (b) London forces; (c) London forces, dipole-dipole interactions, and hydrogen bonding; (d) covalent bonds

10.107   (a) ionic; (b) molecular; (c) metallic; (d) network; (e) molecular

10.109   (b) and (e) are molecular solids

10.111   molecular solid

10.113 SiF_4

10.115   Below its freezing point, ice is lost by sublimation.

10.117 CH_4 ($-162°C$), C_2H_6 ($-88°C$), C_3H_8 ($-42°C$), C_4H_{10} ($0°C$); London dispersion forces increase with molecular size

10.119   A glass plate (SiO_2) is very polar. Polar substances are attracted to polar substances. Water is attracted to glass, so it spreads out, but mercury is repelled by the glass and tends to bead up.

10.121   (c) is an ionic solid

10.123   (a) Cl_2; (b) CH_3SH; (c) NF_3

10.125

Substance (solid type)	Electrical Conductivity	Hardness	Melting Point	Vapor Pressure
NaCl (ionic)	In molten state	Hard	High	Low
SiC (network)	Not conductive	Hardest	Highest	Lowest
SiCl₄ (covalent)	Not conductive	Softest	Lowest	Highest
Fe (metallic)	Always conductive	Soft*	High	Low

*metals are soft relative to network or ionic solids

10.127   melting; freezing; sublimation; deposition; vaporization; condensation

10.129

Chapter 11 – Solutions

11.1   (a) saturated solution; (b) aqueous solution; (c) molarity; (d) solvent; (e) entropy; (f) solubility; (g) parts per million; (h) miscible; (i) osmosis; (j) colligative property; (k) percent by volume; (l) mass/volume percent

11.3

	Solute	Solvent
(a)	sodium chloride (and other salts)	water
(b)	carbon	iron
(c)	oxygen (O_2)	water

11.5   Strong electrolytes are substances that fully ionize when placed in solution. Strong acids, strong bases, and soluble salts are all strong electrolytes.

11.7   nonelectrolyte

11.9   (a) $H^+(aq)$ and $Br^-(aq)$; (b) $NH_4^+(aq)$ and $Cl^-(aq)$; (c) $CH_3CH_2CH_2CH_2OH(aq)$

11.11   A

11.13   (a) $Ca(OH)_2(s) \xrightarrow{H_2O} Ca^{2+}(aq) + 2OH^-(aq)$

        (b) $N_2(g) \xrightarrow{H_2O} N_2(aq)$

        (c) $CH_3OH(l) \xrightarrow{H_2O} CH_3OH(aq)$

11.15   The intermolecular attractions between water molecules and oil molecules are not the same (water is polar and cooking oil is nonpolar). Since they are not "like" they are immiscible.

11.17   Both molecules are polar and can engage in hydrogen bonding.

11.19   (a) insoluble; (b) soluble; (c) soluble

11.21   In the solvent hydrogen-bonding and dipole-dipole interactions are broken and in the solute ionic bonds are broken. Ion-dipole interactions are formed in the solution.

11.23   In both the solute and the solvent, London (dispersion) forces must be overcome. The new forces are also London (dispersion) forces.

11.25   The attraction between the products (dissolved ions) and water molecules is weaker than the forces of attraction in the pure substances.

11.27   Energy and entropy drive the formation of solutions.

11.29   Ion-ion interactions are very strong. There is not enough energy released or entropy gained in the formation of ion-solvent interactions to make the dissolution process favorable.

11.31   Salts are held together by ionic bonding. Water is held together by dipole-dipole and hydrogen bonding. Although a great deal of energy must be spent in breaking these forces of attraction within the solute and solvent, the formation of the ion-dipole interactions releases enough energy so that the salt can dissolve.

11.33   The solubility of gases such as O_2 decreases as temperature increases.

11.35   The solubility of oxygen and nitrogen gases should be lower at high altitude. The solubility of these gases is directly proportional to their pressures above the solution and these pressures are lower at high altitude.

11.37   The insoluble gases that cause the bends can be redissolved into the blood by putting the divers in pressurized chambers. This increases the solubility of the gas in the blood. Decreasing the chamber pressure slowly (decompression) allows the gases (mostly excess nitrogen) to leave through the lungs.

11.39

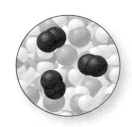

11.41   17.0 g KCl

11.43   On a macroscopic level, you can't dissolve more solid in a saturated solution, but you can dissolve more in an unsaturated solution. On a molecular level, the ions in a saturated solution are in equilibrium with the solid.

11.45   You could add a little solid at a time until no more solid dissolves and you see traces of undissolved solid in the solution.

11.47   The solution will be saturated and 0.85 g $Ca(OH)_2(s)$ will be left undissolved.

11.49   14.3%

11.51   60.0 g KI

11.53   5.423%

11.55   Percent by volume is often used for mixtures of liquids. This is done because liquids are usually measured by volume.

11.57   30.4%

11.59   0.015 mol

11.61   2.41 *m*

11.63   3.0 *M; 3.0 m.*

11.65 9.6×10^{-4} %; 9.6 ppm; 9600 ppb

11.67   (a) 9.0 ppb; yes, the concentration is below the EPA standard (b) 9.0×10^{-6} mg/mL; (c) 9.0×10^{-4} mg; (d) 4.3×10^{-9} mol

11.69   (a) 0.30 L KI; (b) 6.9 g PbI_2

11.71   0.030 mol Na_2CO_3

11.73   57.14 mL

11.75   0.8812 *M* H_2SO_4

11.77   (a) 10.00 mL; (b) 52.50 mL; (c) 37.5 mL

11.79   A colligative property is a physical property of a solvent that varies with the number of solute particles dissolved and not on the identity of those solute particles.

11.81   Osmosis occurs when two solutions of different solute particle concentrations are separated by a membrane that allows the solvent but not the solute molecules to pass (semipermeable membrane). The volume of the more concentrated solution increases because solvent molecules migrate through the membrane to the more concentrated solution.

11.83   Reverse osmosis occurs when solvent is forced to travel in a direction opposite of the direction it would travel during osmosis.

11.85   Water will flow out of the blood cell and the cell will collapse or shrink.

11.87   (a) decrease; (b) increase; (c) decrease

11.89   The boiling point of the water is elevated slightly because of the presence of the sugar.

11.91   (a) 1.3°C (b) 101.3°C

11.93   The sodium chloride solution will have the highest boiling point because NaCl produces two solute particles when it dissolves in water.

11.95   Solution B has the highest concentration of dissolved particles and therefore has the lowest freezing point. Glucose is molecular and images B and C show molecular compounds. Sodium chloride produces two ions of slightly different size as shown in image A.

11.97   (a) 2.0 mol; (b) 5.0 mol; (c) 4.0 mol

11.99   The osmotic pressure of $MgCl_2$ will be greater than that of $NaNO_3$ because each formula unit of $MgCl_2$ produces three ions when it dissolves while $NaNO_3$ produces only two ions.

11.101 Vitamin D is nonpolar.

11.103 The breaking of the crystal lattice and disruption of the intermolecular attractions between water molecules require energy. The formation of ion-dipole interactions between water and the ions releases energy.

11.105 182 g

11.107 8.90 M

11.109 Volume changes with temperature. If the concentration units are based only on mass or number of moles, then they will not change with temperature. Percent by mass (a) and molality (d) do not change with temperature.

11.111 The density of a dilute solution is approximately the density of the solvent. In aqueous solutions, for example, one liter of water has a mass of one kilogram. So the values for molarity (moles/liter) and molality (moles/kilogram) will be very similar.

11.113 (a) 10.5 g $BaSO_4$ (b) 2.33 g $BaSO_4$ (c) 2.33 g $BaSO_4$

11.115 2.0×10^3 g

11.117 solution B

Chapter 12 – Reaction Rates and Chemical Equilibrium

12.1   (a) activation energy, E_a; (b) intermediate; (c) homogenous equilibrium; (d) Le Chatelier's principle; (e) collision theory; (f) equilibrium constant expression

12.3   increasing reactant concentrations, increasing temperature, or providing a catalyst.

12.5   grind the sodium carbonate to a fine powder; increase the temperature; increase the concentration of HCl

12.7   catalyst

12.9   No. If the reactants do not collide with sufficient energy or with the proper orientation, then the activated complex cannot be formed.

12.11 (a) improper orientation; (b) proper orientation; (c) improper orientation

12.13 an activated complex

12.15 the minimum energy required to break bonds in reactants before reactants can be converted into products

12.17 For similar reactions, the higher the activation energy, the slower the reaction.

12.19

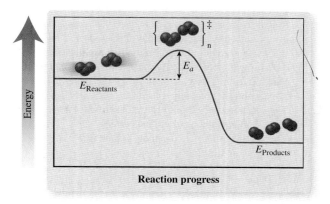

Reaction progress

12.21 Increasing temperature increases the kinetic energy and velocity of the molecules reacting. The fraction of colliding molecules with energy greater than the activation energy increases when temperature is increased.

12.23 increasing temperature

12.25 At room temperature, molecules of propane and oxygen do not collide with sufficient energy to cause a reaction. However, by providing a spark (heat), the kinetic energy of the molecules is increased. Combustion reactions are exothermic, so the reaction is sustained by the heat produced by the reaction.

12.27 Roasts are generally thicker and more massive than steaks. For the same amount of heat, the steak gets hotter faster and therefore cooks faster.

12.29 A catalyst provides a lower energy route (pathway) to the same products.

12.31 Catalytic converters use palladium and platinum catalysts (see Figure 12.11).

12.33 the active site

12.35 Catalysts are only temporarily changed during the course of a reaction (e.g., during the binding of the substrate).

12.37 Chlorine atoms in the upper atmosphere act as catalysts for the decomposition of ozone. Since they act catalytically, they are not lost until they diffuse into outer space (which is a slow process) or are removed by some other reaction.

12.39 an intermediate

12.41 (a) NO_3 an intermediate; (b) $N_2O_5 + NO \longrightarrow 3NO_2$

12.43 (a) H^+ and Br^- are catalysts and Br_2 is an intermediate. (b) $2H_2O_2 \longrightarrow 2H_2O + O_2$

12.45 When a reaction reaches equilibrium, the concentrations of reactants and products stop changing.

12.47 A chemical equilibrium is a state of a reaction where the rate of the forward reaction is equal to the rate of the reverse reaction.

12.49 The system has reached equilibrium in images D and E.

12.51 (a) yes; (b) no; (c) yes; (d) yes

12.53 You could measure the mass of the bromine liquid, but that would be tricky because it is very reactive and volatile. A better option is to measure the pressure of the gas produced. When the partial pressure has stabilized, the system is at equilibrium. Finally, bromine gas has a dark brown appearance. By measuring how much light the gas absorbs, we can determine if its concentration remains constant and the system is at equilibrium.

12.55 The position of the equilibrium tells us whether a reaction is product-favored or reactant-favored.

12.57 The coefficients of the reactants and products are used as exponents in the equilibrium constant expression.

12.59 The brackets represent molar concentrations.

12.61 (a) $K_{eq} = \dfrac{\left[HF\right]^2}{\left[H_2\right]\left[F_2\right]}$

(b) $K_{eq} = \dfrac{\left[CS_2\right]\left[H_2\right]^4}{\left[CH_4\right]\left[H_2S\right]^2}$

(c) $K_{eq} = \dfrac{\left[NO_2\right]^2}{\left[N_2O_4\right]}$

12.63 (a) $C(g) \rightleftharpoons A(g) + B(g)$
(b) $2A(g) \rightleftharpoons 4B(g) + C(g) + D(g)$

12.65 (a) Mathematically, the equations are inverses of each other.
(b) $2NO_2 \rightleftharpoons N_2O_4$; $N_2O_4 \rightleftharpoons 2NO_2$

12.67 0.25

12.69 An equilibrium constant only changes with temperature.

12.71 (a) 3.1×10^{-3}; (b) Because the value of K_{eq} is much less than one, the position of equilibrium is described as "reactant-favored."

12.73 (a) 5.4×10^{-9}; (b) The reaction is reactant-favored since $K_{eq} < 1$.

12.75 The reaction is at equilibrium, so shifting will not take place.

12.77 The reaction will proceed in the reverse direction.

12.79 The reaction will proceed in the forward direction.

12.81 Heterogeneous reactions are those in which the products and reactants are not all in the same physical state.

12.83 The concentrations of pure liquids and solids do not change during chemical reactions, only the amount of the pure substance changes. Because the concentrations of pure substances do not change during reactions, their values are essentially incorporated into the equilibrium constant.

12.85 (a) $K_{eq} = [NH_3][HCl]$
(b) $K_{eq} = [Ca^{2+}][CO_3^{2-}]$

(c) $K_{eq} = \dfrac{\left[H_3O^+\right]\left[F^-\right]}{\left[HF\right]}$

12.87 0.13 M

12.89 Le Chatelier's principle describes how reactions that are at equilibrium respond to changes in reaction conditions. If stress is put on a reaction, the reaction responds to counteract that stress.

12.91 (a) The reaction shifts to the right. (b) The concentration of the products, CO_2 and H_2, will increase and the reactant, H_2O, will decrease in concentration.

12.93 In the following table, the larger arrows ($\Uparrow$) indicate the change made to the equilibrium and the smaller arrows indicate what the concentrations of the products and reactants do following the change. Finally, the central arrow indicates which direction the equilibrium shifts:

	NO(g)	SO₃(g)	$\rightleftharpoons$	NO₂(g)	SO₂(g)
(a)	$\Uparrow$	$\downarrow$	$\rightarrow$	$\uparrow$	$\uparrow$
(b)	$\downarrow$	$\Uparrow$	$\rightarrow$	$\uparrow$	$\uparrow$
(c)	$\downarrow$	$\downarrow$	$\rightarrow$	$\Downarrow$	$\uparrow$
(d)	$\uparrow$	$\uparrow$	$\leftarrow$	$\Uparrow$	$\downarrow$

12.95 (a) This has no effect since the concentration of the solid does not change when more solid is added. (b) This decreases the concentration of Pb^{2+} by shifting the equilibrium left. (c) This causes an increase in Pb^{2+} concentration. When the concentration of lead ion is increased, the equilibrium shifts to the left. However, it never shifts enough to completely remove all the added material.

12.97 (a) right; (b) right; (c) no effect

12.99 (a) no; (b) yes

12.101 (a) left; (b) right

12.103 If the equilibrium shifts to the right, the concentration of products at equilibrium is higher. The value of K_{eq} increases.

12.105 (a) decrease; (b) increase

12.107 exothermic

12.109 An increase in the concentration of NO requires that the equilibrium shift to the right. (a) Water is in the liquid state; removing it does not affect the equilibrium. (b) There are fewer moles of gas on the right. The reaction shifts right increasing the concentration of NO. (c) Since the reaction is exothermic, the reaction will shift right with decreasing temperature producing a higher concentration of NO. (d) Adding oxygen causes the equilibrium to shift right producing a higher concentration of NO. (e) Catalysts do not have an effect on equilibrium concentrations.

12.111 The reaction that produces light slows down and the reactants take longer to be consumed. The lower light output of the cold light sticks is a result of fewer photons being produced per second (lower reaction rate).

12.113 It provides a different reaction mechanism with lower activation energy. In general it splits one difficult step (high activation energy) into two or more steps that have lower activation energies.

12.115 (a) $K_{eq} = \dfrac{\left[H_2\right]\left[I_2\right]}{\left[HI\right]^2}$

(b) $K_{eq} = 0.0199$
(c) The position of the equilibrium favors the reactants.

12.117 The effect of temperature depends on whether the reaction is endothermic or exothermic.

12.119 The reaction must be exothermic since the reaction shifted to the right to form the powder.

12.121 The simplest treatment would be to increase the concentration of oxygen that the patient is breathing. By increasing the O_2 concentration, the second equilibrium will shift to the left and the carbon monoxide can be displaced and exhaled through the blood.

Chapter 13 – Acids and Bases

13.1   (a) strong base; (b) Brønsted-Lowry theory; (c) conjugate acid; (d) polyprotic acid; (e) acidic solution; (f) weak base; (g) amphoteric substance; (h) ion-product constant of water, K_w; (i) self-ionization

13.3   Acids have a sour taste (but you should never taste anything to see if it's an acid!), are corrosive to many metals, turn blue litmus red, and neutralize bases.

13.5   Common "household" bases and their uses

Formula	Name	Use
$Mg(OH)_2$	Magnesium hydroxide	Antacid
$Al(OH)_3$	Aluminum hydroxide	Antacid
NH_3	Ammonia	Cleaning solutions Smelling salts
$Ca(OH)_2$	Calcium hydroxide	Concrete

13.7   An Arrhenius acid produces hydrogen ions (H^+) when dissolved in water. An Arrhenius base produces hydroxide ion (OH^-) when dissolved in water.

13.9   Brønsted-Lowry theory refers to the transfer of hydrogen ions from an acid to a base. Arrhenius similarly describes the release of hydrogen ions into the solution, but identifies OH^- as a component of the base. In Brønsted-Lowry theory, a base is any substance that can accept a hydrogen ion (including OH^-).

13.11 (a) acid; (b) base; (c) acid

13.13 (a) NO_2^-; (b) F^-; (c) $H_2BO_3^-$

13.15 (a) H_2O; (b) $C_6H_5NH_3^+$; (c) H_2CO_3

13.17 (a) $HCl(g) + H_2O(l) \longrightarrow H_3O^+(aq) + Cl^-(aq)$
       (b) $HClO_4(l) + H_2O(l) \longrightarrow H_3O^+(aq) + ClO_4^-(aq)$
       (c) $CH_3CO_2^-(aq) + H_2O(l) \longrightarrow CH_3CO_2H(aq) + OH^-(aq)$

13.19 Both water (a) and hydrogen sulfite (b) are amphoteric.

13.21 Carbonic acid has two acidic hydrogen atoms.

13.23 Strong acids and bases ionize or dissociate completely (100%). For weak acids and bases, the percent of molecules that ionize is usually much lower (typically a few percent).

13.25 (a) strong acid; $H_2SO_4(aq) + H_2O(l) \longrightarrow$
       $HSO_4^-(aq) + H_3O^+(aq)$
       (b) strong base; $Ca(OH)_2(aq) \longrightarrow Ca^{2+}(aq) + 2OH^-(aq)$
       (c) weak base; $Na_2CO_3(aq) + H_2O(l) \rightleftharpoons$
       $HCO_3^-(aq) + 2Na^+(aq) + OH^-(aq)$
       (d) weak acid; $H_3C_6H_5O_7(aq) + H_2O(l) \rightleftharpoons$
       $H_3O^+(aq) + H_2C_5O_7^-(aq)$
       (e) weak base; $C_6H_5NH_2(aq) + H_2O(l) \rightleftharpoons$
       $C_6H_5NH_3^+(aq) + OH^-(aq)$

13.27 A is NH_3, B is HF, and C is HCl.

13.29 Acid A is a stronger acid. The stronger the acid the greater the extent of ionization.

13.31 $0.50 M HCO_2H$

13.33 In a strong acid solution the acid molecules are fully ionized. It does not necessarily mean that the concentration is high. A concentrated acid solution has a high percentage of acid, but this does not necessarily mean that the acid is ionized to any great extent.

13.35 A polyprotic acid is an acid that contains more than one acidic hydrogen atom.

13.37 Hydrosulfuric acid is a weak acid that has two acidic hydrogen atoms. The corresponding acid-base reactions are:
       $H_2S(aq) + H_2O(l) \longrightarrow HS^-(aq) + H_3O^+(aq)$
       $$K_{a1} = 8.9 \times 10^{-8}$$
       $HS^-(aq) + H_2O(l) \longrightarrow S^{2-}(aq) + H_3O^+(aq)$
       $$K_{a2} = 1.0 \times 10^{-19}$$

13.39 The following substances can all be found in a solution of oxalic acid: $H_2C_2O_4$, $HC_2O_4^-$, $C_2O_4^{2-}$, H_2O^+, H_2O. The substance with the lowest concentration is the oxalate ion $C_2O_4^{2-}$.

13.41 Like other equilibrium constants, it is constant only under conditions of constant temperature.

13.43 In pure water, $[OH^-] = [H_3O^+] = 1.0 \times 10^{-7} M$.

13.45 (a) basic; (b) acidic; (c) basic

13.47 (a) $1.0 \times 10^{-12} M$, basic; (b) $1.0 \times 10^{-2} M$, acidic; (c) $4.0 \times 10^{-7} M$, acidic

13.49 (a) $0.010 M$; (b) $0.075 M$; (c) $1 \times 10^{-12} M$

13.51 A decrease in pH represents an increase in the acidity.

13.53 A ten-fold increase in hydrogen ion concentration is 1 pH unit change. The difference in pH units is 1.

13.55 Practically speaking, most solutions of acids and bases are lower in concentration than 1 M so the pH scale typically ranges between 0 and 14. However, the pH values can exceed this range.

13.57 No. A solution is defined as acidic if the pH is less than 7.

13.59 (a) 4.00, acidic; (b) 12.00, basic; (c) 9.19, basic

13.61 (a) 11.00, basic; (b) 7.00, neutral; (c) 4.78, acidic

13.63 (a) 2.00, acidic; (b) 1.12, acidic; (c) 12.00, basic

13.65 Underlined values were given in the problem.

	$[H_3O^+]$	$[OH^-]$	pH	pOH	Acidic or Basic?
(a)	<u>1.0×10^{-5}</u>	1.0×10^{-9}	5.00	9.00	Acidic
(b)	1.0×10^{-10}	<u>1.0×10^{-4}</u>	10.00	4.00	Basic
(c)	1.0×10^{-6}	1.0×10^{-8}	6.00	<u>8.00</u>	Acidic
(d)	2.9×10^{-9}	3.5×10^{-6}	<u>8.54</u>	5.46	Basic
(e)	1.1×10^{-5}	<u>9.0×10^{-10}</u>	4.95	9.05	Acidic

13.67 pH = 2.30, pOH =11.70. The pH and pOH always add to 14.00 at 25°C.

13.69 (a) $1.0 \times 10^{-6} M$, acidic; (b) $1.0 \times 10^{-13} M$, basic; (c) $1.3 \times 10^{-7} M$, acidic

13.71 (a) $1.0 \times 10^{-11} M$, basic; (b) $4.0 \times 10^{-8} M$, basic; (c) $1.2 \times 10^{-2} M$, acidic

13.73 (a) $1.0 \times 10^{-13} M$; (b) $3.0 \times 10^{-4} M$; (c) $4.0 \times 10^{-11} M$

13.75 Acetic acid is a weak acid, so the concentration of hydronium ion would be less than 0.010 M, and the pH would be higher than 2.0.

13.77 0.1 M

13.79 A pH meter typically gives measurements to two decimal places (Figure 13.15).

13.81 Acid-base indicators are typically colored compounds that are also weak acids or bases. The colors of these compounds depend on whether they are in their acid or base forms.

13.83 The color range of an indicator indicates the pH values over which the indicator changes color. The hue of the indicator can be used to estimate the pH of a solution. The range over which an indicator changes color is about 1.5 pH units.

13.85 A selection of indicators would include thymolphthalein.

13.87 It should be red with a small amount of yellow-orange.

13.89 Red (or the presence of red color) indicates the pH is less than 2.5. Yellow indicates a pH between about 2.5 and 8.0. Blue (or a blue hue) indicates pH values above 8.0. It will be completely blue above pH 9.5.

13.91 Universal indicator is a mixture of different indicators. The color of an unknown solution with the indicator present gives a good estimate of the pH. It is similar to pH paper in that pH paper contains universal indicator.

13.93 A buffer contains both a weak acid and its conjugate base (or a weak base and its conjugate acid). As a result, when small amounts of acid or base are added, the buffer can absorb the excess hydronium or hydroxide with only a small change in pH.

13.95 The buffer system of the blood helps to maintain a pH between 7.35 and 7.45.

13.97 Any salt containing $CH_3CO_2^-$ (i.e., $NaCH_3CO_2$).

13.99 (a) Hydrogen phosphate, HPO_4^{2-}, is the acid and the conjugate base is phosphate, PO_4^{3-}.
$HPO_4^{2-}(aq) + H_2O(l) \rightleftharpoons PO_4^{3-}(aq) + H_3O^+(aq)$
(b) When acid is added, hydronium (H_3O^+) is produced and the pH momentarily drops (more acidic). However, the excess hydronium quickly reacts with the PO_4^{3-} shifting the equilibrium to the left to produce the weak acid HPO_4^{2-}.

13.101 Solutions described in (b) and (c) produce buffered solutions.

13.103 Ammonia (NH_3) has a lone electron pair in its Lewis structure. Methane does not have a lone electron pair, so it cannot accept a hydrogen ion (proton) in an acid-base reaction.

13.105 When NaF dissolves in water, sodium and fluoride ions are produced. The fluoride ions react with water to produce a weak base HF by the reaction:
$F^-(aq) + H_2O(l) \rightleftharpoons HF(aq) + OH^-(aq)$
In general the conjugate base (F^-) of a weak acid will react with water to produce hydroxide ions. The molecular species HF is stable, so its formation drives the formation of hydroxide. This solution will be basic. The sodium ion does not react with water.

13.107 No. A neutral solution has a pH = pOH = 7.00 (at 25°C). If sodium hydroxide is added, the hydroxide ion concentration will have to increase. The solution will have a pH greater than 7.00

13.109 Both H_2SO_4 and HNO_3 are strong acids. Both will dissociate completely *for the first ionization*. However, H_2SO_4 is a polyprotic acid. The additional hydronium ion produced by the second ionization will make the pH of the sulfuric acid lower.

13.111 (a) $HOCl(aq) + H_2O(l) \rightleftharpoons OCl^-(aq) + H_3O^+(aq)$;
(b) The HCl reacts with the water to produce hydronium ion: $HCl(aq) + H_2O(l) \rightleftharpoons Cl^-(aq) + H_3O^+(aq)$. The added H_3O^+ causes the equilibrium to shift to the left and some of the excess hydronium is consumed. The HOCl concentration will increase and the OCl^- concentration will go down. While the pH may not drop below 7.00, *anytime* an acid is added to a solution, the solution becomes more acidic (pH is lowered).

Chapter 14 – Oxidation-Reduction Reactions

14.1 (a) electrolytic cell; (b) cathode; (c) oxidation; (d) half-reaction; (e) electrochemistry; (f) electrode; (g) oxidizing agent; (h) oxidation-reduction reaction; (i) corrosion; (j) oxidation number

14.3 a reaction where electrons are transferred

14.5 when there is an increase in charge (oxidation number)

14.7 (a) $Mg(s) + Cu(NO_3)_2(aq) \longrightarrow Cu(s) + Mg(NO_3)_2(aq)$
(b) The charge of magnesium increases by two; 2 electrons transferred; electrons are lost.
(c) The charge of copper decreases by two; 2 electrons transferred; electrons are gained.

14.9 (a) Mg; (b) Zn^{2+}; (c) Zn^{2+}; (d) Mg

(e)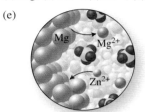

14.11 An oxidation number is a bookkeeping method for tracking where electrons go during a chemical reaction.

14.13 (a) 0; (b) 0; (c) 2+; (d) 1−

14.15 S = 6+, O = 2−

14.17 (a) 5+; (b) 3+; (c) 4+

14.19 (a) 5+; (b) 5+; (c) 5+; (d) 1+; (e) 3−; (f) 3+

14.21 S = 6+, O = 2−

14.23 (a) 7+; (b) 5+; (c) 1−; (d) 3+; (e) 5+

14.25 (a) Cl = 4+, O = 2−; (b) Ca = 2+, F = 1−; (c) H = 1+, Te = 4+, O = 2−; (d) Na = 1+, H = 1−

14.27 (a) N = 3+, O = 2−; (b) Cr = 6+, O = 2−; (c) Ag = 1+, Cl = 1−; (d) S = 4+, O = 2−; (e) C = 4+, O = 2−

14.29 (a) not redox; (b) redox, N_2 is the oxidizing agent, H_2 is the reducing agent; (c) not redox; (d) not redox; (e) redox, O_2 is the oxidizing agent, C_2H_6 is the reducing agent

14.31 (a) $N_2(g) + O_2(g) \longrightarrow 2NO(g)$; (b) The oxidation number increases. Each nitrogen atom loses two electrons to change from an oxidation number of 0 to 2+. (c) The oxidation number decreases. Each oxygen atom gains two electrons to change from an oxidation number of 0 to 2−.

14.33 (a) Fe^{2+}; (b) $Cr_2O_7^{2-}$; (c) $Cr_2O_7^{2-}$; (d) Fe^{2+}

14.35 (a) I_2 is both the oxidizing agent and reducing agent, 1 electron transferred
(b) Cr = reducing agent, H^+ = oxidizing agent, 2 electrons transferred
(c) $Cr_2O_7^{2-}$ is both the oxidizing and reducing agent, 12 electrons transferred
(d) Fe^{3+} = oxidizing agent, Al = reducing agent, 3 electrons transferred

14.37 In the anode half-cell the half-reaction is
$$Fe(s) \longrightarrow Fe^{2+}(aq) + 2e^-$$
In the cathode half-cell the half-reaction is
$$Ni^{2+}(aq) + 2e^- \longrightarrow Ni(s)$$

14.39 The iron electrode will lose mass as Fe is oxidized and becomes $Fe^{2+}(aq)$. The nickel electrode will gain mass as Ni^{2+} is reduced to $Ni(s)$.

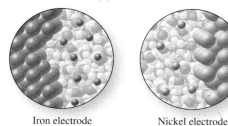

Iron electrode   Nickel electrode

14.41 The $Pb(s)$ gets oxidized and the $PbO_2(s)$ is reduced. The oxidizing agent is $PbO_2(s)$ and the reducing agent is $Pb(s)$.

14.43 oxidation half-reaction $Cd(s) + 2OH^-(aq) \longrightarrow Cd(OH)_2(s) + 2e^-$
reduction half-reaction $2e^- + NiO_2(s) + 2H_2O(l) \longrightarrow Ni(OH)_2(s) + 2OH^-(aq)$

14.45 (a) $3e^- + Fe^{3+}(aq) \longrightarrow Fe(s)$
(b) $Zn(s) \longrightarrow Zn^{2+}(aq) + 2e^-$
(c) $2Cl^-(aq) \longrightarrow Cl_2(g) + 2e^-$
(d) $Fe^{2+}(aq) \longrightarrow Fe^{3+}(aq) + e^-$

14.47 (a) $Zn(s) + 2NO_3^-(aq) \longrightarrow Zn(NO_3)_2(aq) + 2e^-$
$3e^- + Fe(NO_3)_3(aq) \longrightarrow Fe(s) + 3NO_3^-(aq)$
$3Zn(s) + 2Fe(NO_3)_3(aq) \longrightarrow 3Zn(NO_3)_2(aq) + 2Fe(s)$
(b) $Mn(s) + 2Cl^-(aq) \longrightarrow MnCl_2(aq) + 2e^-$
$2e^- + 2HCl(aq) \longrightarrow H_2(g) + 2Cl^-(aq)$
$Mn(s) + 2HCl(aq) \longrightarrow MnCl_2(aq) + H_2(g)$

14.49 $Fe_2(SO_4)_3(aq) + 2KI(aq) \longrightarrow$
$2FeSO_4(aq) + K_2SO_4(aq) + I_2(aq)$

14.51 (a) $Ba(s) \longrightarrow Ba^{2+}(aq) + 2e^-$
(b) $e^- + H^+(aq) + HNO_2(aq) \longrightarrow NO(g) + H_2O$
(c) $2e^- + 2H^+(aq) + H_2O_2(aq) \longrightarrow 2H_2O(l)$
(d) $2Cr^{3+}(aq) + 7H_2O(l) \longrightarrow Cr_2O_7^{2-}(aq) + 14H^+ + 6e^-$

14.53 (a) $3OH^-(aq) + La(s) \longrightarrow La(OH)_3(s) + 3e^-$
(b) $2e^- + H_2O(l) + NO_3^-(aq) \longrightarrow NO_2^-(aq) + 2OH^-(aq)$
(c) $H_2O_2(aq) + 2OH^-(aq) \longrightarrow O_2(g) + 2e^- + 2H_2O(l)$
(d) $8e^- + 3H_2O(l) + Cl_2O_7(aq) \longrightarrow 2ClO_2^-(aq) + 6OH^-(aq)$

14.55 (a) $8H^+(aq) + 3H_2S(aq) + Cr_2O_7^{2-}(aq) \longrightarrow$
$3S(s) + 2Cr^{3+}(aq) + 7H_2O(l)$
(b) $4H^+(aq) + 5V^{2+}(aq) + 3MnO_4^-(aq) \longrightarrow$
$5VO_2^+(aq) + 3Mn^{2+}(aq) + 2H_2O(l)$
(c) $6H^+(aq) + 6Fe^{2+}(aq) + ClO_3^-(aq) \longrightarrow$
$6Fe^{3+}(aq) + Cl^-(aq) + 3H_2O(l)$

14.57 (a) $2NH_3(aq) + ClO^-(aq) \longrightarrow$
$N_2H_4(aq) + Cl^-(aq) + H_2O(l)$
(b) $2Cr(OH)_4^-(aq) + 3HO_2^-(aq) \longrightarrow$
$2CrO_4^{2-}(aq) + 5H_2O(l) + OH^-(aq)$
(c) $2OH^-(aq) + Br_2(aq) \longrightarrow$
$Br^-(aq) + BrO^-(aq) + H_2O(l)$

14.59 $24H^+(aq) + 5C_6H_{12}O_6(aq) + 24NO_3^-(aq) \longrightarrow$
$30CO_2(g) + 12NO_2(g) + 42H_2O(l)$

14.61 The anode could be any metal that is more active than iron (e.g., aluminum or zinc). The supporting electrolyte could be any soluble salt that does not precipitate with Fe^{3+} (e.g., potassium nitrate since all nitrates are soluble). The anodic half-cell would also contain a salt that includes the metal used for the anode (e.g., aluminum or zinc nitrate).

14.63 Au, Ag, Cu, Mn, Mg, Na

14.65 Electrolysis is the process of using electricity to make a nonspontaneous redox reaction occur.

14.67 (a) reduction will begin to take place at the cathode to form sodium metal; (b) oxidation will begin to take place at the anode to form $I_2(g)$
(c) $Na^+(l) + e^- \longrightarrow Na(l)$ (reduction)
$2I^-(l) \longrightarrow I_2(g) + 2e^-$ (oxidation)
(d) $2Na^+ + 2I^-(l) \longrightarrow 2Na(l) + I_2(g)$

14.69

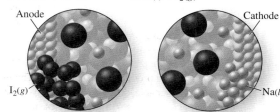

Anode reaction   Cathode reaction

14.71 Magnesium is more active than iron. The magnesium will reduce the $O_2(g)$ and keep the Fe in the reduced state. When the $Mg(s)$ is used up, the iron will begin to corrode.

14.73 One of the methods of preventing corrosion is to prevent contact with oxygen and moisture. The tin coating serves as a barrier to moisture and oxygen. When the tin coating is scratched the more active metal, iron, is more readily oxidized.

14.75 The chrome should corrode first because it is a more active metal than iron.

14.77 (a) 3–; (b) 2–; (c) 3+; (d) 1–; (e) 5+; (f) 3+

14.79 (a) $2H_2O(l) + 2OH^-(aq) + 2CoCl_2(s) + Na_2O_2(aq) \longrightarrow$
$2Co(OH)_3(s) + 4Cl^-(aq) + 2Na^+(aq)$
The oxidizing agent is Na_2O_2 (oxygen is being reduced from 1– to 2–)
The reducing agent is $CoCl_2$ (cobalt is being oxidized from 2+ to 3+)
(b) $2OH^-(aq) + Bi_2O_3(s) + 2ClO^-(aq) \longrightarrow$
$2BiO_3^-(aq) + 2Cl^-(aq) + H_2O(l)$
The oxidizing agent is ClO^- (chlorine is being reduced from 1+ to 1–)
The reducing agent is Bi_2O_3 (bismuth is being oxidized from 3+ to 5+)

14.81 $NH_4^+(aq) + NO_3^-(aq) \longrightarrow N_2O(g) + 2H_2O(l)$
The oxidizing agent is NO_3^- (N is being reduced from 5+ to 1+)
The reducing agent is NH_4^+ (N is being oxidized from 3– to 1+)

14.83 (a) Al; (b) Zn; (c) Mn; (d) Mg

14.85 Brass is an alloy of copper and zinc. Cu doesn't react with acid. There must be a protein in sweat that facilitates the oxidation of the copper to form Cu^{2+}. There are several compounds of copper that are green (including $CuCO_3$).

Chapter 15 – Nuclear Chemistry

15.1 (a) critical mass; (b) fission; (c) alpha particle; (d) radioactive; (e) gamma radiation; (f) nuclear bombardment; (g) nucleon; (h) radiation

15.3 Radioactivity is exhibited by an unstable nucleus which spontaneously decays with the emission of radiation.

15.5 Radioactivity produces photons (i.e., gamma rays) and particulate radiation (i.e., beta rays, positrons, neutrons, and alpha particles).

15.7 Of the elements with atomic number less than 84, technetium (Tc), promethium (Pm), and astatine (At) have no stable isotopes. No isotopes of atomic number 84 or greater are stable.

15.9 Dense materials are better at shielding gamma radiation. Lead, iron, and aluminum (metals) can be used, but concrete and water are effective only if thick layers are used.

15.11 Nuclear decay reactions and bombardment reactions result in the production of radiation. In addition, new isotopes are produced by both.

15.13 During nuclear decay alpha particles, beta particles, gamma rays, and positrons are produced by the nucleus.

15.15 During nuclear reactions, nuclear charge and mass numbers are conserved.

15.17 In a nuclear bombardment reaction a particle accelerator is used to accelerate particles or nuclides to very high velocities. The beam of particles produced is aimed at a target element. If the particle strikes the target with sufficient energy to fuse the target and beam particles, a newly created nucleus is formed. The newly created nucleus spontaneously decays to produce new substances and decay particles.

15.19 $^{222}_{86}Rn$

15.21 $^{99}_{43}Tc$

15.23 $^{15}_{7}N$

15.25 $^{257}_{103}Lr$

15.27 beta particle

15.29 $^{7}_{4}Be$

15.31 (a) $^{238}_{92}U \longrightarrow {}^{234}_{90}Th + {}^{4}_{2}He$

(b) $^{234}_{90}Th \longrightarrow {}^{234}_{91}Th + {}^{0}_{-1}\beta^-$

(c) $^{238}_{92}U \longrightarrow {}^{4}_{2}He + {}^{234}_{90}Th$

(d) $^{10}_{5}B + {}^{2}_{1}H \longrightarrow {}^{11}_{6}C + {}^{1}_{0}n$

(e) $^{210}_{83}Bi + {}^{0}_{-1}\beta^- \longrightarrow {}^{210}_{82}Pb$

15.33 (a) $^{227}_{87}Fr \longrightarrow {}^{4}_{2}He + {}^{223}_{85}At$

(b) $^{18}_{9}F \longrightarrow {}^{0}_{+1}\beta^+ + {}^{18}_{10}Ne$

(c) $^{96}_{42}Mo + {}^{1}_{+1}p \longrightarrow {}^{1}_{0}n + {}^{96}_{43}Tc$

15.35 approximately 1:1

15.37 If the mass numbers of all the stable isotopes are graphed versus their atomic number, a pattern emerges (Figure 15.14). The band of stability is the outline of that pattern.

15.39 The isotope $^{16}_{8}O$ should be the most stable. It has a N/Z ratio of 1.

15.41 (a) beta; (b) alpha and beta; (c) alpha and beta; (d) positron; (e) beta

15.43 $^{234}_{91}Pa \longrightarrow 5\,^{4}_{2}He + 2\,^{0}_{-1}\beta^- + {}^{214}_{83}Bi$

15.45 In a sample of radioactive material, the half-life is the time it takes for half of the radioactive nuclei to decay. For nuclear decay, each half-life has exactly the same duration.

15.47 Film, Geiger-Müller counter, and a scintillation counter can be used to detect radiation.

15.49 One-sixteenth (6.25%) of the original material will be present.

15.51 One-quarter (25%) of the original material will be present.

15.53 22.5 hours (three half-lives)

15.55 5730 years

15.57 Twenty-four hours represents four half-lives. The half-life is 6 hours.

15.59 Radiation powered devices can last much longer than battery powered devices.

15.61 ^{99m}Tc is used because it produces radiation that can be used to image tumors and its short half-life ensures that it won't stay in the body for a long period of time.

15.63 Positron emission occurs for N/Z ratios < 1. $^{11}_{6}C$ has a low N/Z ratio whereas $^{14}_{6}C$ has a high N/Z ratio.

15.65 Radiation can 1) have no effect, 2) cause reparable damage, 3) cause damage leading to formation of mutant cells (cancer cells), 4) kill cells (subsequently absorbed by the body). Mutant cells can also be recognized as foreign entities and be destroyed by an immune response.

15.67 This is a difficult question to answer. How you answer the question stems from your view of our governmental system and your perception of the risk of radon and the benefit of regulations. Clearly, the best protection against unnecessary radon exposure is knowledge of what causes enhanced levels of radon and what measures can be used to reduce radon exposure. The Internet can be a good source of information if you use it carefully (i.e., http://www.epa.gov/iedweb00/radon/index.html).

15.69 You would most likely find $^{210}_{82}Pb$.

15.71 Fission occurs when unstable nuclei decay into smaller nuclei and decay products. Fusion results when smaller particles combine to make larger ("heavier") nuclei.

15.73 three

15.75 In principle, a chain reaction needs the terminating step of the reaction to be a necessary component of the initiating step. For example, if the products of a reaction (initiating step) start a second reaction (terminating step) and the products of the second reaction start the first reaction, you have a chain reaction.

15.77 A fission reactor has fuel rods (containing the radioactive material), control rods (to absorb the excess neutrons), a moderator (i.e., water, heavy water, or graphite), energy converters (convert water to steam which powers electric generators).

15.79   The water is not directly converted to steam by the nuclear reaction as a safety precaution. The steam that would be produced by the nuclear reaction would be highly radioactive and is also made of poisonous heavy water (water with deuterium rather than hydrogen is highly toxic). If there were a leak of the steam, the released products would be very dangerous.

15.81   Isotopes of hydrogen (H-1, H-2, H-3) and lithium are most likely to be used.

15.83 $^2_1H + ^3_1H \longrightarrow ^4_2He + ^1_0n$

15.85 $^1_1H + ^2_1H \longrightarrow ^3_2He$

15.87   In radioactive dating, a sample is taken and analyzed for the quantity of a particular radioisotope. Based on the amount of the isotope, and assuming you know how much of the isotope was initially present in the sample, you can calculate the sample age (using the half-life). For example, if the radioisotope is at $^1/_4$ the original concentration, two half-lives have passed (i.e., $^1/_2 \times ^1/_2$).

15.89   The sample contains about 1/64th less carbon-14 than a modern sample. This is about six half-lives or 34,000 years.

15.91 $^1_1H + ^1_1H \longrightarrow ^2_1H + ^0_{+1}\beta^+$
$^2_1H + ^1_1H \longrightarrow ^3_2He$
$^1_1H + ^2_2He \longrightarrow ^4_2He + ^0_{+1}\beta^+$

15.93   The initial product of the bombardment reaction is carbon-13 ($^9_4Be + ^4_2He \longrightarrow ^{13}_6C$). Loss of a neutron would produce carbon-12 (N/Z =1): $^9_4Be + ^4_2He \longrightarrow ^{12}_6C + ^1_0n$

15.95   With a half-life of 138 days, at the end of one year approximately 3 half-lives have already passed. Only 1/8th of the material would be remaining. You would be better off buying brushes every year (or as you needed them).

15.97   Even though the element chlorine has only been known to exist since 1774, chloride containing compounds were used well before the discovery of chlorine.

Chapter 16 – Organic Chemistry

16.1   (a) isomer; (b) alcohol; (c) ester; (d) aldehyde; (e) cyclic hydrocarbon; (f) saturated hydrocarbon; (g) alkyl group; (h) cycloalkane; (i) aliphatic hydrocarbon; (j) amino acid; (k) organic chemistry; (l) unsaturated hydrocarbon

16.3   (a) seven single bonds
(b) one double bond, four single bonds
(c) one triple bond, two single bonds

16.5   (a) amine; (b) ketone; (c) alcohol; (d) alkene; (e) carboxylic acid; (f) alkyne; (g) aldehyde; (h) ether; (i) ester

16.7   molecular formula: $C_{20}H_{28}O$
structural formula:
$C_6H_6(CH_3)_3CHCHC(CH_3)CHCHCHC(CH_3)CHCHO$
line structure

16.9   (a) $C_7H_{14}O$; (b) C_8H_{18}

16.11   (a)

(b)

(c)

(d)

16.13   (a) $(CH_3)_2CHCH_2CH_3$; (b) $(CH_2OH)_2CHOH$;
(c) $CH_3OCH_2CH_3$
Line structures

(a)

(b)

(c)

16.15   (a)

(b)

(c)

16.17   An alkane hydrocarbon that has only carbon-carbon single bonds, an alkene contains at least one carbon-carbon double bond, and an alkyne has at least one carbon-carbon triple bond.

16.19   (a) unsaturated; (b) unsaturated; (c) unsaturated

16.21   Aromatic hydrocarbons are cyclic hydrocarbons that can be drawn with alternating double and single bonds between the carbon atoms. When the bonds are distributed in this way, the electrons are said to be delocalized.

16.23   The London dispersion forces must be stronger in butane than in isobutane (2-methyl propane). London dispersion forces increase with the length of the molecule because the molecules have more opportunities to interact. Because the London forces are stronger in butane, it boils at a higher temperature.

16.25   (a) $CH_3CH_2CH_3$; (b) $CH_3(CH_2)_4CH_3$

16.27   (a) n-pentane or pentane; (b) ethane; (c) n-octane or octane; (d) propane; (e) n-nonane or nonane

16.29 There are five isomers of C_6H_{14}

16.31 (a) 2-methylpentane; (b) 2-methylbutane;
(c) 2-methylbutane (same as previous structure)

16.33 (a) 3,3-dimethylpentane; (b) 2,2-dimethylpropane;
(c) 2,3-dimethylbutane

16.35 (a) $CH_3CH(CH_3)CH_3$;
(b) $CH_3CH(CH_3)C(CH_2CH_3)_2CH_2CH_3$;
(c) $CH_3CH(CH_3)CH_2CH(CH_3)CH_2CH_2CH_2CH_3$;
(d) $CH_3CH_2C(CH_2CH_3)(CH_3)CH_2CH_2CH_3$

16.37 (a) $CH_3CH_2CH_3 + Cl_2 \xrightarrow{\text{heat}} CH_3CHClCH_3 + HCl$
or $CH_3CH_2CH_3 + Cl_2 \xrightarrow{\text{heat}} CH_3CH_2CH_2Cl + HCl$
(b) $(CH_3)_3CCH_2CH_2CH_3 + 11O_2 \xrightarrow{\text{heat}} 7CO_2 + 8H_2O$
(c) $CH_3CH_3 + Br_2 \xrightarrow{\text{heat}} CH_3CH_2Br + HBr$

16.39 An alkene is a hydrocarbon containing one or more carbon-carbon double bonds.

16.41 (a) 5-methyl-3-heptene; (b) 2,3-dimethyl-2-hexene;
(c) *trans*-3,4-dimethyl-3-hexene

16.43 only (c) because it can be made to be asymmetric along the carbon-carbon double bond axis

16.45 (a)
3-methyl-1-butene

(b)
propyne

(c)
trans-2-butene

(d)
cis-3-hexene

16.47 (a) $CH_3CH{=}CHCH_3 + HBr \longrightarrow CH_3CHBrCH_2CH_3$
(b) $CH_3CH_2CH{=}CHCH_3 + Br_2 \longrightarrow$
$CH_3CH_2CHBrCHBrCH_3$
(c) $CH_3CH{=}CHCH_3 + H_2 \xrightarrow{\text{Pd}} CH_3CH_2CH_2CH_3$

16.49 Aromatic hydrocarbons are cyclic and have alternating single and double bonds (delocalized electrons).

16.51 (a) 1,2-dibromobenzene; (b) 1-chloro-4-ethylbenzene

16.53 (a)
1,2,4-trimethylbenzene

(b)
chlorobenzene

16.55 $C_6H_6 + Cl_2 \xrightarrow{\text{FeCl}_3} C_6H_5Cl + HCl$

16.57 Both (b) and (c) are alcohols.

16.59 There are four isomers.

16.61 (a) 2-methyl-1-propanol; (b) 2-butanol

16.63 (a)
3-hexanol

(b)
2,2-dimethyl-3-pentanol

16.65 The primary difference is that the intermolecular forces between ethanol (CH_3CH_2OH) molecules are hydrogen bonds while those between propane molecules are London dispersion forces. The stronger intermolecular forces between ethanol molecules means that a higher temperature (energy) is needed to raise the vapor pressure of ethanol to 1 atm.

16.67 Ethers contain the functional group $-O-$ attached to two alkyl groups $\left[R^{\diagup O \diagdown} R \right]$.

16.69 Only (a) CH_3OCH_3 is an ether.

16.71 (a) ethylpropyl ether; (b) dimethyl ether

16.73 (a)
ethyl methyl ether

(b)
ethyl phenyl ether

16.75 Water is a hydrogen-bonding solvent, so substances that are hydrogen bonding, polar molecules will be more soluble in water. Ethers (CH_3OCH_3) do not directly hydrogen bond because the oxygen is attached to carbon atoms on either side. Alcohols are more water soluble because the alcohol group $-OH$ can hydrogen bond.

16.77 An aldehyde has the functional group $-CHO$ $\left[\begin{array}{c} O \\ \| \\ R^{\diagup C \diagdown} H \end{array} \right]$ where R can be an organic group or a hydrogen atom.

16.79 Compounds (a), (c), and (d) are aldehydes.

16.81 Both compounds are polar because of the $C{=}O$ bond, but acetaldehyde (CH_3CHO) has stronger London dispersion forces because it has a larger molecular mass. As a result acetaldehyde has a higher boiling point.

16.83 A carboxylic acid contains the functional group $-COOH$
or $-CO_2H$ $\left[\begin{array}{c} O \\ \| \\ R^{\diagup C \diagdown}O{-}H \end{array} \right]$ where R is an organic group.

16.85 One route to the formation of esters is the condensation of a carboxylic acid and an alcohol. Methyl propionate ($CH_3CH_2CO_2CH_3$) can be formed from methanol (CH_3OH) and propionic acid ($CH_3CH_2CO_2H$).

16.87 Saturated fatty acids are high molecular mass carboxylic acids which do not have double or triple bonds in the hydrocarbon chain. As a result, it is no longer possible to add any more hydrogen atoms to the hydrocarbon chain and they are said to be saturated with hydrogen.

16.89 Acetic acid (CH_3CO_2H) should have a higher boiling point than propanal (CH_3CH_2CHO) because acetic acid can hydrogen bond. Aldehydes are polar but do not hydrogen bond.

16.91 An amine is a derivative of ammonia (NH_3) with one or more of the H atoms replaced by organic groups.

$$\left[\overset{\displaystyle \cdot\cdot}{\underset{\displaystyle \underset{H}{|}}{R-N-H}} \right]$$

16.93 (a) $H-\overset{\cdot\cdot}{\underset{|}{N}}\diagdown$ or $\overset{\cdot\cdot}{\underset{|}{N}}\diagdown$

(b) $\diagdown\diagup\overset{}{\underset{|}{N}}\diagdown$

(c) a benzene ring with NH_2 group

16.95 Methylamine (CH_3NH_2) can hydrogen bond so it has relatively strong intermolecular forces resulting in a higher boiling point. Ethane (CH_3CH_3) is nonpolar and cannot hydrogen bond.

16.97 A simple organization of hydrocarbons is given in the chart below.

16.99 Alkenes and alkynes differ primarily in three ways: carbon-carbon bond type, bond angles, and number of hydrogen atoms in the structure.

	Alkanes	Alkenes	Alkynes
bond types	single	double bonds	triple bonds
bond angles	109.5°	120°	180°
general formula (straight or branched)	#H = 2n + 2	#H = 2n	#H = 2n – 2

#H = number of hydrogens
n = number of carbons. The formula assumes only one double or triple bond per structure.

16.101 Aldehydes and ketones are similar functional groups.

A ketone has the general formula: $\left[\underset{R}{\overset{O}{\underset{\displaystyle \diagdown C \diagup}{||}}} R \right]$. The

aldehyde functional group is $\left[\underset{R}{\overset{O}{\underset{\displaystyle \diagdown C \diagup}{||}}} H \right]$.

16.103 (a) 2,4-dimethylpentane; (b) 1,3-dimethylcyclopentane; (c) 2,2,3,3-tetramethylbutane

16.105 One possible test is to compare the solubility of each substance in an alcohol (such as ethanol). However, this test is not very specific. Addition of bromine, Br_2, which is a reddish brown solution would be a better alternative. Bromine will react with an alkene but not with an alcohol. If an alkene is present, the bromine solution will lose color as the bromine reacts with the alkene.

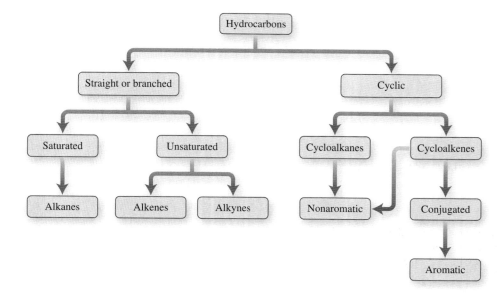

16.107 The trend in boiling points is directly related to the strength of the intermolecular forces in each substance. The weakest attractive forces are found in propane which is nonpolar. Dimethyl ether is polar, but ethanol is both polar and has hydrogen bonding. Because hydrogen bonds are so strong, ethanol has the highest boiling point.

16.109 (a) CH_3COCH_3; (b) $CH_3CH_2COCH_3$; (c) CH_3CH_2CHO; (d) $CHOCHCHCH_3$; (e) CH_3CHO

16.111 (a) $CH_3CH_2CH_2CO_2CH_2CH_3$; (b) $CH_3CH_2CH_2CH_2CO_2H$; (c) $CH_3CO_2CH_3$; (d) HCO_2H; (e) $CH_3CH_2CO_2H$

16.113 (a) dimethylamine; (b) ethyldimethylamine (the official IUPAC name is N,N-dimethyl-1-aminoethane, which is why we use the common name); (c) aniline

16.115 (a) $CH_3C(CH_3)_2CH_2CH(CH_3)_2$; (b) $CH_3CHC(CH_2CH_3)_2$; (c) $CH_2CHCHCH_2$; (d) C_4H_8; (e) C_6H_5Cl

16.117 (a) alkane; (b) ether; (c) carboxylic acid; (d) aldehyde; (e) ester; (f) amine; (g) ketone

16.119 Only (c) 2-butene and (d) 2-pentene can exist as *cis* and *trans* isomers.

Glossary

A

acid a compound that releases hydrogen ions when dissolved in water or donates H^+ to another substance. (3.6)

acid ionization constant, K_a an equilibrium constant for the ionization of an acid in water; a value that expresses the strength of a weak acid. (13.3)

acidic solution a solution in which the H_3O^+ concentration is greater than the OH^- concentration; a solution with a pH less than 7. (13.4)

activated complex an unstable, high-energy chemical species that must be formed as reactants convert to products. (12.2)

activation energy, E_a the minimum amount of energy that reactants must have before they can be converted to products; the difference between the energy of the activated complex and the average energy of the reactants. (12.2)

activity series a list of metals in order of activity. (5.4)

actual yield the amount of product actually obtained in the laboratory from a reaction. (6.5)

acyclic hydrocarbon a hydrocarbon that has a straight- or branched-chain structure with localized bonds. (16.2)

alcohol an organic compound containing the functional group $-OH$. (8.3, 16.5)

aldehyde an organic compound containing the functional group $-CHO$. (16.6)

aliphatic hydrocarbon a hydrocarbon that contains localized single, double, or triple bonds. (8.3, 16.2)

alkali metal any group IA (1) element except hydrogen; Li, Na, K, Rb, Cs, and Fr are alkali metals. (2.5)

alkaline earth metal any group IIA (2) element; Be, Mg, Ca, Sr, Ba, and Ra. (2.5)

alkane a hydrocarbon that has a straight-chain or branched-chain structure with only single bonds. (8.3, 16.3)

alkene a hydrocarbon that contains one or more localized carbon-carbon double bonds. (8.3, 16.3)

alkyl group a group derived from an alkane by removal of one hydrogen atom. (16.3)

alkyne a hydrocarbon that contains one or more triple bonds. (8.3, 16.3)

alloy a mixture of a metal with one or more additional metallic or nonmetallic elements. (10.4)

alpha particle a particle containing two protons and two neutrons; the nucleus of a helium-4 atom. (15.1)

amine an organic base with one or more alkyl groups bonded to a nitrogen atom. (16.8)

amino acid small molecules that contain both the $-NH_2$ and the $-CO_2H$ functional groups. (16.9)

amorphous solid a solid that is not crystalline but has its particles arranged in a random, nonrepeating fashion. (10.4)

amphoteric substance a substance that can act as either an acid or a base. (13.1)

anhydrous a compound free of molecular water. (5.4)

anion a negatively charged ion, such as the chloride ion (Cl^-) or nitrate ion (NO_3^-). (2.3)

anode the electrode that attracts anions and at which oxidation occurs. (14.3)

aqueous solution a homogeneous mixture of two or more substances in which the solvent is water. (1.1, 11.1)

aromatic hydrocarbon a hydrocarbon that contains delocalized bonds made up of alternating single and double bonds that impart special stability. (8.3, 16.2)

Arrhenius model of acids and bases a simple model that describes acids as substances that generate H^+ ions in solution and bases as substances that generate OH^- ions in solution. (13.1)

aryl group an aromatic group formed by removal of hydrogen from a cyclic aromatic hydrocarbon. (16.10)

atom the smallest particle of an element that retains the characteristic chemical properties of that element. (1.1)

atomic mass unit (amu) the basic unit of mass of atoms and molecules; exactly one-twelfth the mass of one ^{12}C atom; 1 amu = 1.6606×10^{-24} g. (2.4)

atomic number the number of protons in the nucleus of an atom of a given element. (2.2)

atomic radius a measure of atomic size; the distance from the nucleus to the outer edge of the atom. (7.7)

aufbau principle an approach to writing ground-state electron configurations of atoms by filling orbitals that are lowest in energy first; means "building up" in German. (7.3)

Avogadro's hypothesis a proposal that equal volumes of all gases contain equal numbers of particles (at constant temperature and pressure); $V = \text{constant} \times n$. (9.2)

Avogadro's number the number of fundamental particles (atoms, molecules, or ions) in 1 mol of any substance; 6.022×10^{23} particles/mol. (4.1)

B

balanced equation a chemical equation that uses coefficients in front of each formula to make the number of atoms of each element the same on both sides of the equation. (5.3)

barometer a closed tube filled with mercury and inverted into a pool of mercury; instrument used to measure atmospheric pressure. (9.1)

base a substance that is capable of neutralizing an acid in a chemical reaction, forming water; a substance that accepts an H^+ from another substance. (3.6)

basic solution a solution in which the OH^- concentration is greater than the H_3O^+ concentration; a solution with a pH greater than 7. (13.4)

battery one or more interconnected voltaic cells used as a portable source of electricity. (14.3)

beta particle a small charged particle that can be emitted from atoms at speeds approaching the speed of light; either an electron or a positron. (15.1)

binary compound   a compound containing atoms or ions of two elements. (3.1)

boiling point   the temperature at which the vapor pressure of a liquid equals the external pressure. (10.1)

bond angle   the angle between the two lines defined by a central atom attached to two surrounding atoms. (8.4)

Boyle's law   a law stating that at constant temperature, the volume occupied by a fixed amount of a gas is inversely proportional to its pressure; $V = \dfrac{\text{constant}}{P}$. (9.2)

Brønsted-Lowry theory   a theory that defines an acid as a substance that donates an H^+ to another substance in solution and defines a base as a substance that accepts an H^+ in solution. (13.1)

buffer   a combination of a weak acid and its conjugate base (or a weak base and its conjugate acid) in similar amounts; a buffered solution resists changes in pH when small amounts of H_3O^+ or OH^- are added. (13.6)

C

carboxylic acid   an organic compound containing the functional group $-CO_2H$. (16.7)

catalyst   a substance that changes the rate of the reaction, but can be recovered unchanged after the reaction. (12.3)

cathode   the electrode that attracts cations and at which reduction occurs. (14.3)

cation   a positively charged ion, such as the sodium ion (Na^+); an ion that contains fewer electrons than protons. (2.3)

chain reaction   a reaction in which the product of one step is a reactant in another step. (15.6)

Charles's law   a law stating that at constant pressure, the volume occupied by a fixed amount of gas is directly proportional to its absolute temperature; $V = \text{constant} \times T$. (9.2)

chemical bond   the force that holds atoms together in a molecule or compound. (8.1)

chemical change   a process in which one or more substances are converted into new substances that have compositions and properties different from those of the original substances. (1.2)

chemical equation   an abbreviated representation of a chemical reaction consisting of chemical symbols and formulas. (5.3)

chemical equilibrium   a state where the rates of the forward reaction and the reverse reaction are equal so the concentrations of reactants and products do not change. (12.4)

chemical formula   a representation of the composition of a compound using symbols of the compound's elements. (1.1)

chemical property   a characteristic of a substance defined by its composition and the chemical changes it can undergo. (1.2)

chemical reaction   a process in which one or more substances are converted into new substances that have compositions and properties different from those of the original substances. (1.2)

colligative property   a solution property that depends only on the concentration of solute particles, not on the identity of the solute. (11.6)

collision theory   a theory that explains the temperature and concentration dependence of reaction rate on the basis of molecular collisions, molecular orientation, and kinetic energy. (12.2)

combination reaction   a reaction in which two elements, an element and a compound, or two compounds join to form a new compound. (5.4)

combined gas law   a law that describes the relation between pressure, volume, and temperature for a fixed amount of a gas; $\dfrac{P_1 V_1}{T_1} = \dfrac{P_2 V_2}{T_2}$. (9.2)

combustion reaction   a reaction that involves oxygen molecules as a reactant and that rapidly produces heat and flames. (5.4)

compound   a substance composed of two or more elements combined in definite proportions. (1.1)

concentrated solution   a solution that contains a relatively high concentration of solute. (4.4)

concentration   the relative amounts of solute and solvent in a solution. (4.4, 11.4)

condensation   the conversion of gas-state molecules to the liquid state. (10.1)

conjugate acid   a substance that forms after a base gains an H^+ ion. (13.1)

conjugate base   a substance that forms after an acid loses an H^+ ion. (13.1)

continuous spectrum   a spectrum in which all the visible wavelengths are present. (7.1)

core electron   an electron in an atom that is not a valence electron; an inner electron. (7.5)

corrosion   a slow deterioration of metals due to interaction with the environment. (5.4, 14.7)

covalent bonding   the bonding between two atoms resulting from the sharing of electrons. (8.1)

critical mass   the smallest amount of fissionable material necessary to support a continuing chain reaction. (15.6)

crystal   a solid having a shape bounded by planes intersecting at characteristic angles. (10.4)

crystal lattice   the repeating pattern of atoms or ions in a crystal. (8.2, 10.4)

crystalline solid   a substance in which the particles are arranged in a regular, repeating geometric structure. (10.4)

cyclic hydrocarbon   a hydrocarbon that has a ring or closed-chain structure. (16.2)

cycloalkane   an alkane that contains a ring structure. (16.4)

D

Dalton's atomic theory   the first widely accepted theory that describes matter as being composed of atoms; published in 1807. (2.1)

Dalton's law of partial pressures   a law stating that gases in a mixture behave independently and exert the same pressure they would if they were in the container alone; $P_{total} = P_A + P_B + P_C + \dots$. (9.3)

decomposition reaction   a chemical reaction in which a substance breaks down into simpler compounds or elements. (5.4)

density   the ratio of the mass of a substance to its volume. (1.2)

deposition   the conversion of gas-state molecules to the solid state. (10.1)

diatomic molecule   a neutral particle consisting of two bound atoms. (2.5)

diffusion   the movement of gas particles through the atmosphere from high concentration to low concentration. (9.4)

dilute solution   a solution that contains a relatively low concentration of solute. (4.4)

dilution   the process of adding more solvent to give a solution of lower concentration. (4.4)

dipole-dipole force   the attraction between the partially negative end of one polar molecule and the partially positive end of another polar molecule. (10.2)

double bond   a covalent bond that involves the sharing of two pairs of electrons. (8.3)

double-displacement reaction   a reaction in which two compounds exchange ions or elements to form new compounds. (5.4)

E

effusion   the movement of gas particles through a small opening. (9.4)

electrochemistry   the study of the relationship between chemical reactions and electrical work. (14.6)

electrode   a solid material that conducts electrical charge and is used as part of an electrochemical cell; one electrode (the anode) acts as the site of oxidation, and the other (the cathode) acts as the site of reduction in electrochemical reactions. (14.3)

electrolysis   a process in which a nonspontaneous oxidation-reduction reaction is forced to occur upon passage of electric current through a sample. (14.6)

electrolyte   a substance that dissociates or ionizes into aqueous ions upon dissolving and conducts electricity when dissolved in water. (3.1)

electrolytic cell   an electrochemical cell through which electric current is passed to cause a nonspontaneous oxidation-reduction reaction to occur. (14.6)

electromagnetic radiation   a form of energy that can be described in terms of oscillating waves that move through space at the speed of light; also called light or radiant energy. (7.1)

electron   a negatively charged subatomic particle. (2.2)

electron capture   the incorporation of a $1s$ electron into the nucleus, resulting in the conversion of a proton into a neutron. (15.2)

electron configuration   a description of the distribution of electrons in atomic orbitals or sublevels. (7.3)

electron dot formula   see *Lewis formula*. (8.3)

electron dot symbol   see *Lewis symbol*. (8.2)

electronegativity   a measure of the ability of an atom to attract electrons within a bond to itself. (8.1)

element   a pure substance that cannot be broken down into simpler stable substances in a chemical reaction. (1.1)

element symbol   a shorthand version of an element's name consisting of one or two letters, the first of which is uppercase. (1.1)

empirical formula   a formula written with the simplest ratios of atoms or ions (the smallest whole-number subscripts). (4.3)

endothermic reaction   a chemical change that absorbs heat. (6.6)

energy   the capacity to do work or to transfer heat. (1.1, 1.3)

entropy   a measure of the tendency for matter to become randomly distributed. (11.2)

enzyme   a protein molecule that acts as a catalyst to a specific reaction or to a narrow range of reactions. (12.3)

equilibrium   a state of balance between opposing processes where both processes occur at the same rate. (10.1)

equilibrium constant expression   the expression obtained from a reaction's balanced equation that relates the relative amount of products to reactants at equilibrium. (12.5)

ester   a condensation product of carboxylic acids and alcohols, having the general formula RCO_2R'. (16.7)

ether   an organic compound that contains the $-O-$ functional group attached to two organic groups: $R-O-R'$. (16.5)

evaporation   a process by which molecules pass from the liquid state to the gaseous state. (10.1)

exothermic reaction   a chemical change that releases heat. (6.6)

F

family   a vertical column of related elements in the periodic table; another name for group. (2.5)

fission   the splitting of a heavy atom into two or more lighter atoms and some number of neutrons with the release of energy. (15.6)

fluid   a substance that can flow (a gas or a liquid). (10.3)

formula unit   the smallest repeating unit in an ionic compound. (3.3)

freezing point   the temperature at which the liquid and solid states of a substance are in equilibrium (identical to the melting point). (10.1)

frequency   the number of wave cycles passing a stationary point in 1 s. (7.1)

functional group   a group of atoms substituted for hydrogen in the formula of a hydrocarbon that gives the compound its characteristic properties; a reactive part of a molecule that undergoes characteristic reactions. (8.3, 16.1)

fusion   the combination of light nuclei to form heavier nuclei. (15.6)

G

galvanic cell   an electrochemical cell in which an oxidation-reduction reaction causes electrons to flow through an external circuit; also known as a *voltaic cell*. (14.3)

gamma radiation   the highest-energy and most-penetrating type of electromagnetic radiation. (15.1)

gas   the physical state in which matter has no fixed shape or volume but expands to fill its container completely and can be compressed easily. (1.1)

Gay-Lussac's law of combining volumes   a law stating that in chemical reactions occurring at constant pressure and temperature, the volumes of gaseous reactants and products are in the ratio of small whole numbers. (9.2)

ground state   the state of lowest energy in an atom where the electrons occupy orbitals that are lowest in energy. (7.3)

group   a vertical column of related elements in the periodic table; another name for family. (2.5)

H

half-cell   the individual compartments in which oxidation and reduction occur. (14.3)

half-life   the time it takes for half of a sample of a nuclide to decay to a different nuclide. (15.3)

half-reaction   the oxidation and reduction processes that occur in the individual cell compartments of an electrochemical cell. (14.3)

halogen any group VIIA (17) element; the nonmetals F, Cl, Br, I, and At are halogens. (2.5)

heat the energy that is transferred between objects due to a difference in their temperatures. (6.6)

hertz (Hz) a unit of frequency equal to 1/s, or cycles per second. (7.1)

heterogeneous equilibrium an equilibrium for a reaction in which reactants and products are not all in the same physical state. (12.5)

heterogeneous mixture a combination of two or more substances that is not uniform throughout. (1.1)

homogeneous equilibrium an equilibrium for a reaction in which all reactants and products are in the same physical state. (12.5)

homogeneous mixture a combination of two or more substances that has uniform composition. (1.1)

Hund's rule a rule stating that electrons are distributed in a set of orbitals of identical energy in such a way as to give the maximum number of unpaired electrons. (7.3)

hydrocarbon a compound that contains only carbon and hydrogen. (8.3, 16.1)

hydrogen bond an especially strong dipole-dipole force between polar molecules that contain hydrogen attached to a highly electronegative element. (10.2)

hydronium ion an aqueous hydrogen ion; $H_3O^+(aq)$. (13.1)

hypothesis a tentative explanation for the properties or behavior of matter that accounts for a set of observations and can be tested. (1.4)

G

ideal gas a gas that follows predicted behavior, as represented by the ideal gas law. (9.2)

ideal gas constant, R a constant used in the ideal gas law that relates pressure, volume, amount of gas, and temperature. (9.3)

ideal gas law a law stating the relationship between pressure, volume, temperature and amount of an ideal gas; $PV = nRT$. (9.3)

induced dipole a temporary dipole formed by attraction of the electrons in the molecule to a nearby dipole. (10.2)

inner-transition metal an element that is a lanthanide (atomic numbers 58 to 71) or an actinide (atomic numbers 90 to 103). (2.5)

instantaneous dipole a temporary dipole formed by the movement of electrons within a molecule. (10.2)

intermediate a relatively unstable chemical species that is produced in a step during a reaction, and then reacts in a subsequent step. (12.3)

intermolecular force a force of attraction that occurs between molecules. (10.2)

ion an atom or molecule bearing an electrical charge. (2.3)

ion-dipole force an intermolecular force between an ion and a polar molecule. (11.2)

ionic bonding the bonding between cations and anions resulting from electrostatic attractions of opposite charges. (8.1)

ionic compound a compound consisting of cations and anions in proportions that give electrical neutrality. (3.1)

ionic crystal a solid structure in which the ions are arranged in a regular repeating pattern. (8.2)

ionic equation a form of a chemical equation in which ionic species are represented as the separated ions. (5.5)

ionization energy the minimum energy required to remove a valence electron from a gaseous atom or ion; usually reported in kJ/mol. (7.7)

ion-product constant of water, K_w the equilibrium constant for the self-ionization of water; the product of the H_3O^+ ion concentration and the OH^- ion concentration in any aqueous solution; $K_w = [H_3O^+][OH^-] = 1.0 \times 10^{-14}$ at 25°C. (13.4)

isoelectronic having the same number of electrons. (7.6)

isomer a compound having the same composition and formula as another compound, but a different structure. (16.1)

isotope an atom of an element with a specific number of neutrons. (2.2)

isotope symbol an element symbol with the atomic number (Z) as the left subscript and a mass number (A) as the left superscript. (2.2)

K

ketone an organic compound containing an oxygen double-bonded to a carbon, which is also bonded to two other carbons: $R_2C{=}O$. (16.6)

kinetic energy the energy possessed by an object because of its motion. (1.3)

kinetic-molecular theory of gases a model of ideal gas behavior; assumes that gas particles occupy no volume and that there are no intermolecular forces between them. (9.4)

L

law of conservation of energy a law stating that energy can be converted from one form to another, or transferred, but it cannot be created or destroyed. (6.6)

law of conservation of mass a law stating that the mass of the substances produced in a chemical reaction equals the mass of the substances that reacted. (2.1)

law of definite proportions a law stating that all samples of the same pure compound will always contain the same proportions by mass of the component elements. (2.1)

Le Chatelier's principle a principle stating that a system at equilibrium reacts to a disruption such as a change in concentration, volume, or temperature in such a way as to counteract the disruption and reestablish equilibrium. (12.6)

Lewis formula a representation of a compound consisting of the symbols of the component elements, each surrounded by dots representing shared and unshared electrons. (8.3)

Lewis symbol a representation of an atom consisting of the symbol for the element surrounded by a number of dots equal to the number of valence electrons. (8.2)

limiting reactant the reactant that is completely used up in the reaction and therefore determines the amount of other reactants that react and the amount of products that should form. (6.4)

line spectrum a spectrum produced by a substance in which only certain wavelengths are present. (7.1)

liquid the physical state in which matter has no characteristic shape but takes the shape of the filled portion of its container and can be poured. (1.1)

London dispersion force the attractive force that results from formation of temporary dipoles in molecules. (10.2)

M

main-group element   a member of one of the A groups of elements in the periodic table; also called a *representative element*. (2.5)

mass   a measure of the quantity of matter. (1.1)

mass number   the sum of the number of protons and the number of neutrons in the nucleus of an atom or ion. (2.2)

mass/volume percent   mass of solute divided by volume of solution and multiplied by 100%. (11.4)

matter   anything that occupies space and has mass. (1.1)

melting point   the temperature at which the liquid and solid states of a substance are in equilibrium (identical to the freezing point). (10.1)

meniscus   the curved upper surface of a liquid column. (10.3)

metal   an element characterized by luster and the ability to conduct electricity. (1.1)

metalloid   an element that has properties intermediate between those of metals and nonmetals; also called *semimetal;* an element with physical properties resembling a metal but chemical reactivity like that of a nonmetal. (2.5)

miscible   the ability of liquids to mix in all proportions. (11.1)

mixture   a combination of two or more substances that can vary in composition. (1.1)

molality   moles of solute per kilogram of solvent. (11.4)

molar mass   the mass of 1 mol of any substance, in units of grams per mole. (4.1)

molar volume   the volume occupied by 1 mol of a gas, which equals 22.414 L at STP for an ideal gas. (9.2)

molarity, *M*   the number of moles of solute per liter of solution. (4.4, 11.4)

mole   the amount of a substance that contains the same number of fundamental particles (atoms, molecules, or formula units) as there are atoms in exactly 12 g of ^{12}C (Avogadro's number); 6.022×10^{23} particles. (4.1)

molecular compound   a compound composed of two or more nonmetals and existing in discrete units of atoms held together in a molecule; electrons are shared between atoms rather than transferred from one atom to another. (3.1)

molecular equation   a form of a chemical equation in which substances are represented as if they existed as molecules (using formula units), even though the substances may exist in solution as ions. (5.5)

molecular formula   a formula that indicates the actual number of atoms present in a molecule. (3.5)

molecular solid   a solid made up of discrete molecules held together by intermolecular forces. (10.4)

molecule   two or more atoms bound together in a discrete arrangement. (1.1)

monatomic ion   an ion of a single atom. (3.2)

N

net ionic equation   a form of a chemical equation in which ionic compounds are represented as the separated ions, and spectator ions are eliminated. (5.5)

network solid   a solid in which all atoms are held together by covalent bonds to make a giant molecule. (10.4)

neutral solution   a solution in which the OH⁻ concentration is equal to the H_3O^+ concentration; a solution with a pH of 7. (13.4)

neutralization reaction   a reaction of an acid and a base to form an ionic compound and water. (5.4)

neutron   an electrically neutral subatomic particle. (2.2)

neutron number   the number of neutrons in the nucleus of an atom. (2.2)

noble gas   any group VIIIA (18) element; He, Ne, Ar, Kr, Xe, and Rn are noble gases. (2.5)

nonelectrolyte   a substance that retains its molecular identity upon dissolving and does not conduct electricity when dissolved in water. (3.1)

nonmetal   an element that typically has a dull appearance and is a poor conductor of electricity. (1.1)

nonpolar covalent bond   a bond resulting from the equal sharing of electron pairs. (8.1)

normal boiling point   the temperature at which the vapor pressure of a liquid equals the standard atmospheric pressure of 1 atm. (10.1)

normal freezing point   the temperature at which the liquid and solid states of a substance are in equilibrium at a pressure of 1 atm. (10.1)

nuclear bombardment   a nuclear reaction that is induced by the impact of a nuclide or a subatomic particle on another nuclide. (15.2)

nucleon   a nuclear particle (neutron or proton). (15.1)

nucleus   the central core of an atom which contains the protons and neutrons; contributes most of the mass of the atom. (2.2)

nuclide   an atom having a specified atomic number, mass number, and nuclear energy. (15.1)

O

octet rule   a rule stating that atoms tend to gain, lose, or share electrons to achieve an electronic configuration with eight electrons in the valence shell. (8.3)

orbital   a three-dimensional region in space around a nucleus where an electron is likely to be found. (7.3)

organic chemistry   the study of organic compounds. (16.1)

osmosis   the passage of solvent molecules, but not solute particles, through a membrane from a solution that is less concentrated in solute to a solution that is more concentrated. (11.6)

oxidation   a chemical process, coupled with reduction, in which an atom increases in oxidation number. (14.1)

oxidation number   a charge assigned to each atom in a compound according to a set of rules used in an electron accounting system. (14.2)

oxidation-reduction reaction   a reaction in which electrons are transferred. (14.1)

oxidizing agent   an atom, ion, or molecule that brings about an increase in the oxidation number of another substance. (14.1)

oxoanion   an anion containing oxygen attached to some other element. (3.2)

P

parts per billion   the mass of solute divided by mass of solution and multiplied by 10^9. (11.4)

parts per million   the mass of solute divided by mass of solution and multiplied by 10^6; in aqueous solutions it is essentially the milligrams of solute per liter of solution. (11.4)

Pauli exclusion principle   a principle stating that a maximum of two electrons can occupy an orbital, and the electrons must have opposite spin. (7.3)

percent by mass   the mass of solute divided by mass of solution and multiplied by 100%. (11.4)

percent by volume   the volume of solute divided by volume of solution and multiplied by 100%. (11.4)

percent yield   the ratio of the amount of product actually formed in a reaction to what is predicted by stoichiometry, multiplied by 100. (6.5)

percentage composition by mass   an expression of the percentage of the total mass due to each element in a compound. (4.2)

period   a horizontal row of elements in the periodic table. (2.5)

periodic table   an arrangement of all the known elements by increasing atomic number and arranged in columns and rows to emphasize periodic properties. (2.5)

pH   a measure of the acidity of aqueous solutions; the negative logarithm of the H_3O^+ concentration: pH = $-\log[H_3O^+]$. (13.5)

photon   the smallest packet of energy that describes the energy of a specific type of electromagnetic radiation. (7.1)

physical change   a process that changes only the physical properties of a substance, not its chemical composition. (1.2)

physical property   a characteristic of a substance that can be observed without changing its composition. (1.2)

physical state   a form that matter can take; possibilities are solid, liquid, or gas. (1.1)

polar covalent bond   a bond resulting from the unequal sharing of electron pairs. (8.1)

polarity   the separation of electronic charge within a bond (or molecule). (8.1)

polyatomic ion   an ion containing two or more atoms, such as SO_4^{2-}. (3.2)

polyprotic acid   an acid containing more than one acidic hydrogen. (13.3)

positron   a particle having the same mass as an electron but the opposite electrical charge. (15.1)

potential energy   the energy possessed by an object because of its position. (1.3)

precipitate   an insoluble solid deposited from a solution. (5.4)

precipitation reaction   a reaction in which the reactants in solution exchange ions to form a solid. (5.4)

pressure   the amount of force applied per unit area. (9.1)

principal energy level   an energy level containing a set of electrons in orbitals of similar sizes. (7.3)

product   a substance formed from another substance during a chemical reaction. (5.1)

proton   a positively charged subatomic particle. (2.2)

pure substance   matter that has the same chemical composition, no matter what its origin. (1.1)

Q

quantized   having only specific allowed values. (7.2)

R

radiation   the emission of particles or electromagnetic waves. (15.1)

radioactive   having the ability to emit ionizing radiation spontaneously. (15.1)

radiocarbon dating   a technique for the establishment of the age of a carbon-containing object by determination of the amount of C-14 present in the object. (15.3)

reactant   a substance converted into another substance during a chemical reaction. (5.1)

reaction rate   a measure of how fast reactants are converted to products; usually measured as the change in concentration of a reactant or product per unit time. (12.1)

reducing agent   an atom, ion, or molecule that brings about a decrease in the oxidation number of another substance. (14.1)

reduction   a chemical process, coupled with oxidation, in which an atom decreases in oxidation number. (14.1)

relative atomic mass   the average mass of an atom of an element, taking into account the masses and abundance of all the naturally occurring isotopes. (2.4)

resonance hybrid   an average or composite Lewis formula derived from two or more valid Lewis formulas that closely represents the bonding in a molecule. (8.3)

S

salt bridge   a device used in electrochemical cells that allows migration of ions to maintain electrical neutrality but prevents mixing of the anode and cathode solutions. (14.3)

saturated hydrocarbon   a hydrocarbon that contains all the hydrogens it can hold as a result of having only carbon-carbon single bonds. (16.2)

saturated solution   a solution that is in equilibrium with excess solute. (11.4)

scientific method   a method of inquiry or investigation that involves cycles of observation and interpretation. (1.4)

self-ionization   a process where one molecule transfers a proton to another molecule of the same substance. (13.4)

semipermeable membrane   a membrane that allows the passage of solvent but not of solute particles. (11.6)

single bond   a covalent bond that involves the sharing of one pair of electrons. (8.3)

single-displacement reaction   a reaction in which a free element displaces another element from a compound to produce a different compound and a different free element. (5.4)

solid   the physical state of matter characterized by a fixed shape and low compressibility. (1.1)

solubility   a ratio that describes the maximum amount of a solute that dissolves in a particular solvent to form an equilibrium solution under specified conditions. (11.4)

solute   the substance being dissolved; usually the component of a solution that is present in the lesser amount. (4.4, 11.1)

solution   a homogeneous mixture of two or more substances uniformly dispersed at a molecular or ionic level. (1.1, 11.1)

solvent   the substance doing the dissolving; usually the component of a solution that is present in the larger amount. (4.4, 11.1)

specific heat   a property that is a measure of the amount of heat required to raise the temperature of 1 g of a substance by 1°C. (6.6)

spectator ion   an ion that is not involved in a chemical reaction. (5.5)

spontaneous   a chemical reaction or physical change that occurs without outside intervention. (14.6)

standard temperature and pressure (STP)   0°C and 1 atm. (9.2)

stoichiometry   the use of quantitative relationships between substances involved in a chemical reaction to determine the amount of a reactant or product. (6.1)

strong acid   an acid that ionizes completely when dissolved in water. (13.2)

strong base   a base that ionizes or dissociates completely when dissolved in water. (13.2)

strong electrolyte   a substance that dissociates or ionizes completely into ions upon dissolving and conducts electricity well. (3.1)

subatomic particle   a particle found within an atom. (2.2)

sublevel   an energy level consisting of only one type of orbital at a specific principal energy level; usually consisting of more than one orbital. (7.3)

sublimation   the vaporization of a solid. (10.1)

supersaturated solution   an unstable solution that contains more dissolved solute than the maximum dictated by the solubility. (11.4)

surface tension   work necessary to expand the surface of a liquid; the property of a liquid surface that causes it to behave like a stretched membrane. (10.3)

T

temperature   a measure of how hot or cold a substance is relative to some standard. (1.2)

theoretical yield   the maximum amount of product that can be obtained in a chemical reaction from known amounts of reactants; the amount of product calculated by assuming all the limiting reactant is consumed. (6.5)

titration   a process of determining the concentration of one substance in solution by reacting it with a solution of another substance of known concentration. (11.5)

transition metal   a member of one of the B groups of elements in the periodic table. (2.5)

triple bond   a covalent bond that involves the sharing of three pairs of electrons. (8.3)

U

unsaturated hydrocarbon   a hydrocarbon that does not contain all the hydrogen atoms that could possibly be included as a result of having carbon-carbon double or triple bonds. (16.2)

unsaturated solution   a solution that contains less than the maximum amount of solute possible in a stable system. (11.4)

V

valence electron   an electron in the highest principal energy level, the valence level. (7.5)

valence level   the principal energy level that is highest in energy and contains the largest filled orbitals. (7.5)

valence-shell electron-pair repulsion (VSEPR) theory   the shapes of molecules result from the tendency for electron pairs to maximize the distance between them to minimize repulsions. (8.4)

vapor pressure   the partial pressure of gas molecules above a liquid (or solid) when the two states are in equilibrium. (10.1)

viscosity   resistance to flow. (10.3)

voltaic cell   an electrochemical cell in which an oxidation-reduction reaction causes electrons to flow through an external circuit; also known as a *galvanic cell*. (14.3)

volume   the amount of space a substance occupies. (1.2)

W

wavelength   the distance between two corresponding points (such as the crests) on a wave. (7.1)

weak acid   an acid in which only a fraction of the molecules ionize when dissolved in water; an acid that forms an equilibrium with its conjugate base. (13.2)

weak base   a base in which only a fraction of the molecules ionize when dissolved in water; a base that forms an equilibrium with its conjugate acid. (13.2)

weak electrolyte   a substance that dissociates or ionizes partially into ions upon dissolving and conducts electricity to a slight extent. (3.1)

work   the action of a force over a distance. (1.3)

Photo Credits

Chapter 1

Chapter 1 Opener: © Robert Holmes/CORBIS; **1.1:** © Photo-Link/Getty Images/McGraw-Hill; **1.2:** © McGraw-Hill Higher Education, Inc./Stephen Frisch, photographer; **1.4 (phosphorus):** © E. R. Degginger/Color Pic, Inc.; **1.4 (carbon):** © Steve Cole/Getty Images/McGraw-Hill; **1.4 (bromine):** © McGraw-Hill Higher Education, Inc./Stephen Frisch, photographer; **1.4 (sulfur):** © McGraw-Hill Higher Education, Inc./Stephen Frisch, photographer; **1.4 (lead):** © Richard Treptow/Photo Researchers, Inc.; **1.4 (gold):** © Erich Schrempp/Photo Researchers, Inc.; **1.4 (copper):** © David Nunuk/Photo Researchers, Inc.; **1.4 (aluminum):** © Edward Kinsman/Photo Researchers, Inc.; **1.4 (nickel):** © McGraw-Hill Higher Education, Inc./Stephen Frisch, photographer; **1.4 (tin):** © Tom Pantages; **1.4j (bottom right):** © Richard Treptow/Photo Researchers, Inc.; **Ex 1.1 (iron):** © E. R. Degginger/Color Pic, Inc.; **Ex 1.1 (iodine):** © Andrew Lambert Photography/Photo Researchers, Inc.; **Ex 1.1 (magnesium):** © McGraw-Hill Higher Education, Inc./Stephen Frisch, photographer; **Ex 1.1 (sulfur):** © Richard Treptow/Photo Researchers, Inc.; **Ex 1.1 (aluminum):** © McGraw-Hill Higher Education, Inc./Stephen Frisch, photographer; **1.5 (left):** © TH Foto-Werbung/Photo Researchers, Inc.; **1.5 (right):** © Richard Trepto /Photo Researchers, Inc.; **1.6:** © Jean Leo Dugast/Peter Arnold, Inc.; **Ex 1.3 (fountain):** © D. Normark/PhotoLink/Getty Images/McGraw-Hill; **Ex 1.3 (pizza):** © Kevin Sanchez/Cole Group/Getty Images/McGraw-Hill; **Ex 1.3 (coin):** © Royalty-Free/Corbis; **Ex 1.3 (balloons):** © Jules Frazier/Getty Images/McGraw-Hill; **Ex 1.3 (can):** © Brian Moeskau/Moeskau Photography; **1.8:** © Coco McCoy/Rainbow; **1.9:** © Jules Frazier/Getty Images/McGraw-Hill; **1.10:** © Robert Holmes/CORBIS; **1.12:** © Royalty-Free/CORBIS/McGraw-Hill; **1.13:** © Charles D. Winters/Photo Researchers, Inc.; **1.14:** © Paul A. Souders/CORBIS; **1.16:** © Brian Moeskau/Moeskau Photography; **1.17 (top):** © Arnold Fisher/Photo Researchers, Inc.; **1.17 (bottom):** © Doug Menuez/Getty Images/McGraw-Hill; **p. 16:** © NASA Johnson Space Center (NASA-JSC); **1.18 (both):** © Brian Moeskau/Moeskau Photography; **1.19:** © Richard Megna/Fundamental Photographs; **p. 18 (bottom):** © James P. Birk; **p. 20:** © Brian Moeskau/Moeskau Photography; **1.26:** © Bettmann/CORBIS; **1.27 (both):** © Brian Moeskau/Moeskau Photography; **1.28:** © Charles D. Winters; **Ex 1.13:** © David Grossman/Photo Researchers, Inc; **1.29:** © Brian Moeskau/Moeskau Photography; **1.30, 1.31, 1.33, 1.35:** © Richard Megna/Fundamental Photographs; **1.34:** © Brian Moeskau/Moeskau Photography; **1.36 (left):** © Gibson Stock Photography; **1.36 (middle):** © Creatas/Picturequest; **1.36 (right):** © David Hills Photography; **Question 1.23, p. 43:** © Patricia Brabant/Cole Group/Getty Images; **Question 1.24, p. 43:** © Gibson Stock Photography; **Question 1.81, p. 46:** © Robert Holmes/CORBIS; **Question 1.82, p. 46:** © Royalty-Free/CORBIS

Chapter 2

Chapter 2 Opener: © Brian Moeskau/Moeskau Photography; **2.1:** © Coco McCoy/Rainbow; **2.1 (inset):** © James P. Birk; **2.2:** © Time Life Pictures/Mansell/Time Life Pictures/Getty Images; **2.3 (all):** © Richard Megna/Fundamental Photographs; **2.5:** © Dr. K.W. Hipps, Washington State University; **2.14:** © Tom Pantages; **2.19 (all), 2.21 (all), 2.22:** © McGraw-Hill Higher Education, Inc./Stephen Frisch, photographer; **p. 68:** Courtesy of Gary J. Schrobilgen, Department of Chemistry, McMaster University

Chapter 3

Chapter 3 Opener: © Reed Kaestner/Old Web Aliases/Image State; **p. 80 left, middle:** © Phil Degginger; **p. 80 right, 3.4 (all), 3.6:** © McGraw-Hill Higher Education, Inc./Stephen Frisch, photographer; **3.8:** © Brian Moeskau/Moeskau Photography; **3.9:** © Charles D. Winters/Photo Researchers, Inc.; **3.15a:** © Brian Moeskau/Moeskau Photography; **3.15b:** © Tony Freeman/PhotoEdit; **3.16 (both):** © Brian Moeskau/Moeskau Photography; **3.18 (left):** © Arnold Fisher/Photo Researchers, Inc.; **3.18 (right):** © Dane S. Johnson/Visuals Unlimited; **3.22, 3.30, 3.33, 3.34:** © Brian Moeskau/Moeskau Photography

Chapter 4

Chapter 4 Opener: © Lester Lefkowitz/CORBIS; **4.1:** © Paul Silverman/Fundamental Photographs; **4.2:** © Paul A. Souders/CORBIS; **4.3:** Courtesy of Kennecott Utah Copper; **4.4. 4.5:** © Charles E. Rotkin/CORBIS; **4.6:** © Tom Pantages; **4.7:** Courtesy of B.L. Ramakrishna/IN-VSEE Project/Arizona State University; **4.8, 4.13, 4.17 (both):** © Richard Megna/Fundamental Photographs; **4.12:** © Siede Preis/Getty Images/McGraw-Hill; **4.18:** © Brian Moeskau/Moeskau Photography; **4.19, 4.20:** © James P. Birk; **4.21 (all):** © McGraw-Hill Higher Education, Inc./Stephen Frisch, photographer; **p. 140, 4.22 (all), 4.23 (all):** © Brian Moeskau/Moeskau Photography

Chapter 5

Chapter 5 Opener: ©Ellen Senisi/The Image Works; **5.1:** © Tom Pantages; **5.2 (both), 5.3:** © E. R. Degginger/Color Pic, Inc.; **5.4:** © Richard Megna/Fundamental Photographs; **5.5:** © Charles D. Winters; **5.7 (all):** © Brian Moeskau/Moeskau Photography; **5.8a:** © Paul Silverman/Fundamental Photographs; **5.8b, c, g:** © Richard Megna/Fundamental Photographs; **5.8d, e:** © Charles D. Winters/Photo Researchers, Inc.; **5.8f:** © E. R. Degginger/Color Pic, Inc.; **5.9:** © Joel Gordon; **5.10:** © Eastcott-Momatiuk/The Image

Works; **5.11:** © Sami Sarkis/Getty Images; **5.12 (lithium):** © Chemistry at Work/Videodisk/Videodiscovery, Inc.; **5.12 (sodium):** © Richard Megna/Fundamental Photographs; **5.12 (potassium):** © Richard Megna/Fundamental Photographs; **5.12 (rubidium):** © Chemistry at Work/Videodisk/Videodiscovery, Inc.; **5.12 (cesium):** Courtesy of Open University; **5.13:** © Richard Megna/Fundamental Photographs; **5.15:** © Royalty-Free/CORBIS; **5.16:** © Tom Pantages; **5.17, 5.18, 5.20:** © Richard Megna/Fundamental Photographs; **5.22:** © E. R. Degginger/Color Pic, Inc.; **5.22 (inset):** © James P. Birk; **5.24:** © Bettmann/CORBIS; **5.25 (both):** © Tom Pantages; **5.26, 5.28:** © E. R. Degginger/Color Pic, Inc.; **5.29 (left):** © Kristen Brochmann/Fundamental Photographs; **5.29 (right):** © NYC Parks Photo Archive/Fundamental Photographs; **5.30 (top left, top right):** © Richard Megna/Fundamental Photographs; **5.30 (bottom left):** © Sami Sarkis/Getty Images; **5.30 (bottom right):** © S. Solum/PhotoLink/Getty Images; **5.32:** © Richard Megna/Fundamental Photographs; **Question 5.13, p. 184:** © Richard Megna/Fundamental Photographs; **Question 5.14, p. 184:** © Richard Megna/Fundamental Photographs

Chapter 6

Chapter 6 Opener: © PhotoLink/Getty Images; **6.1:** © StockTrek/Getty Images; **6.3:** © Tatsuyuki Tayama/Fujifotos/The Image Works; **6.6 (left):** © Richard Megna/Fundamental Photographs; **6.6 (middle):** © E. R. Degginger/Color Pic, Inc.; **6.6 (right):** © Richard Megna/Fundamental Photographs; **6.7 (both):** © Charles D. Winters; **Ex 6.3 (both)** © E. R. Degginger/Color Pic, Inc.; **Ex 6.7, p. 210, Ex 6.7, p. 211:** © Tom Pantages; **6.11:** © E. R. Degginger/Color Pic, Inc.; **6.19:** © Brian Moeskau/Moeskau Photography; **Question 6.19, p. 227:** © Brian Moeskau/Moeskau Photography; **Question 6.20, p. 228:** © Richard Megna/Fundamental Photographs

Chapter 7

Chapter 7 Opener: © Richard Cummins/CORBIS; **7.1:** © PhotoLink/Getty Images; **7.2 (left):** © Tom Pantages; **7.2 (right):** © S. Wanke/PhotoLink/Getty Images; **7.3 (left):** © McGraw-Hill Higher Education, Inc./Stephen Frisch, photographer; **7.3 (right):** © Washington DC Convention & Visitors Bureau; **7.4 (top):** © Lawrence Manning/CORBIS; **7.4 (bottom):** © Mark Maio/Phototake; **p. 240:** © Dr. Arthur Tucker/Photo Researchers, Inc.; **7.25 (lithium):** © Chemistry at Work/Videodisk/Videodiscovery, Inc.; **7.25 (sodium):** © Richard Megna/Fundamental Photographs; **7.25 (potassium):** © Richard Megna/Fundamental Photographs; **7.25(rubidium):** © Chemistry at Work/Videodisk/Videodiscovery, Inc.; **7.25 (cesium):** © Chemistry at Work/Videodisk/Videodiscovery, Inc.

Chapter 8

Chapter 8 Opener: © Steve Prezant/CORBIS; **8.8c:** © Arnold Fisher/Photo Researchers, Inc.; **8.10:** © Herve Berthoule/Photo Researchers, Inc.; **8.11:** © Charles D. Winters/Photo Researchers, Inc.; **p. 290:** © Charles D. Winters/Photo Researchers, Inc.; **8.19**

(both): © Brett Bergherm/AGE Fotostock; **8.25 (all):** © McGraw-Hill Higher Education, Inc./Stephen Frisch, photographer; **8.34:** © Kip Peticolas/Fundamental Photographs

Chapter 9

Chapter 9 Opener: © Royalty-Free/CORBIS/McGraw-Hill; **9.1:** © David Frazier; **9.4 (both):** © Michele Westmorland/Danita Delimont.com; **9.5 (both):** © Brian Moeskau/Moeskau Photography; **9.6:** © BananaStock/Picturequest; **p. 323 (thumbtack):** © Brian Moeskau/Moeskau Photography; **9.10 (both):** © McGraw-Hill Higher Education, Inc./Stephen Frisch, photographer; **9.12:** © David Frazier; **9.13:** © Michael Newman/Photo Edit, Inc; **p. 326:** © Tom Pantages; **9.17 (both):** © Richard Megna/Fundamental Photographs; **9.23:** © Royalty-Free/CORBIS/McGraw-Hill; **Question 9.59, p. 357 (both):** © Royalty-Free/CORBIS/McGraw-Hill; **Question 9.60, p. 357 (both):** © Royalty-Free/CORBIS/McGraw-Hill; **Question 9.73, p. 358:** © Tom Pantages; **Question 9.74, p. 358 (both):** © Royalty-Free/CORBIS/McGraw-Hill; **Question 9.113, p. 359 (both):** © Jules Frazier/Getty Images/McGraw-Hill

Chapter 10

Chapter 10 Opener: © Martin Bond/Photo Researchers, Inc.; **10.1:** © Tom Pantages; **10.2a:** © Phil Degginger/Color-Pic, Inc; **10.2b:** © PhotoDisc/Getty Images/McGraw-Hill; **10.2c:** © Phil Degginger/Color-Pic, Inc; **10.3:** Courtesy of Materials Research Society; **10.4 (left, middle):** © Brian Moeskau/Moeskau Photography; **10.4 (right):** © L. S. Stepanowicz/Visuals Unlimited; **10.5:** © Charles E. Rotkin/CORBIS; **10.6 (left):** © Emanuele Taroni/Getty Images; **10.6 (right):** © L. Hobbs/PhotoLink/Getty Images; **10.8:** © PPP/Popperphoto/Robertstock; **10.11:** © John A. Rizzo/Getty Images; **10.13:** © Matt Meadows/Peter Arnold, Inc.; **10.16:** © E. R. Degginger/Color Pic, Inc.; **10.17a:** Courtesy of Oleg D. Lavrentovich, Liquid Crystal Institute, Kent State University; **10.17b:** Courtesy of B.L. Ramakrishna/IN-VSEE Project/Arizona State University; **10.24a:** © David Young-Wolff/Photo Edit, Inc; **10.24b:** © Arthur S. Aubry/Getty Images; **10.25:** © John A. Rizzo/Getty Images; **10.26 (top):** © PhotoLink/Getty Images; **10.26 (bottom):** © Photodisc/Getty Images/McGraw-Hill; **10.27:** © James P. Birk; **10.28:** © E. R. Degginger/Color Pic, Inc.; **10.29a:** © SPL/Photo Researchers; **10.29b:** © Keith Brofsky/Getty Images/McGraw-Hill; **10.30:** © Richard Megna/Fundamental Photographs; **10.31a:** © Siede Preis/Getty Images/McGraw-Hill; **10.31b:** © Tomi/PhotoLink/Getty Images; **10.32:** © Dr. K.W. Hipps, Washington State University; **10.33 (left):** © PhotoLink/Getty Images/McGraw-Hill; **10.33 (middle-left):** © James P. Birk; **10.33 (middle-right, right):** © Ryan McVay/Getty Images/McGraw-Hill; **10.34:** © Bill Aron/Photo Edit, Inc; **10.35:** © Brian Moeskau/Moeskau Photography; **10.38a:** © Royalty-Free/CORBIS/McGraw-Hill; **10.38b:** © Martin Bond/Photo Researchers, Inc.; **10.41:** © Steve Cole/Getty Images/McGraw-Hill; **10.42:** © C Squared Studios/Getty Images/McGraw-Hill; **10.43:** © Steve Cole/Getty Images; **Question 10.7, p. 400:** © Royalty-Free/CORBIS; **Question 10.8, p. 400:** © James P. Birk

Chapter 11

Chapter 11 Opener: © Stockbyte/Veer; **11.1 (left):** © Royalty-Free/CORBIS; **11.1 (right):** © Randy Allbritton/Getty Images; **11.2:** © Brian Moeskau/Moeskau Photography; **11.3a:** © E. R. Degginger/Color Pic, Inc.; **11.3b:** © Aaron Haupt/Photo Researchers, Inc; **11.3c:** © Peter E. Smith/Natural Sciences Image Library; **11.4:** © Brian Moeskau/Moeskau Photography; **11.5 (left, middle):** © Brian Moeskau/Moeskau Photography; **11.5 (right) :** © Charles D. Winters; **11.6:** © Brian Moeskau/Moeskau Photography; **11.7:** © McGraw-Hill Higher Education, Inc./Stephen Frisch, photographer; **11.8:** © Brian Moeskau/Moeskau Photography; **11.9:** © J.Ledesma/UNEP/Peter Arnold, Inc.; **11.12 (all):** © Brian Moeskau/Moeskau Photography; **11.15 (all):** © Phil Degginger/Color-Pic, Inc; **p. 416 (margin top):** © Brian Moeskau/Moeskau Photography; **p. 416 (margin bottom):** © Brian Moeskau/Moeskau Photography; **11.16 (both):** © Brian Moeskau/Moeskau Photography; **11.17:** © Tom Pantages; **11.18:** © Brian Moeskau/Moeskau Photography; **11.22:** © E. R. Degginger/Color Pic, Inc.; **11.23 (all):** © McGraw-Hill Higher Education, Inc./Stephen Frisch, photographer; **11.24 (top):** © LADA/Photo Researchers, Inc; **11.24 (bottom):** © E. R. Degginger/Color Pic, Inc.; **11.25 (all):** © McGraw-Hill Higher Education, Inc./Stephen Frisch, photographer; **11.29:** © E. R. Degginger/Color Pic, Inc.; **11.30 (all):** © David M. Philips/Photo Researchers; **11.32 (both) :** © Brian Moeskau/Moeskau Photography

Chapter 12

Chapter 12 Opener: © Steve Kaufman/DRK PHOTO; **12.2 (left):** © Yoav Levy/Phototake; **12.2 (right):** © Paul Silverman/Fundamental Photographs; **p. 463 margin (both):** © Brian Moeskau/Moeskau Photography; **12.3 (both):** © Phil Degginger/Color-Pic, Inc; **12.4 (both):** © McGraw-Hill Higher Education, Inc./Stephen Frisch, photographer; **12.5:** © Brian Moeskau/Moeskau Photography; **Ex 12.3:** © Charles D. Winters/Photo Researchers, Inc; **12.11a:** © E. R. Degginger/Color Pic, Inc.; **12.14 (all):** © McGraw-Hill Higher Education, Inc./Stephen Frisch, photographer; **12.18 (both), 12.20 (all), 12.22 (all):** © Tom Pantages; **12.23 (both):** © E. R. Degginger/Color Pic, Inc.

Chapter 13

Chapter 13 Opener: © Brian Moeskau/Moeskau Photography; **13.1, 13.2, 13.3, 13.6, 13.8, 13.9, 13.17 (both):** © Brian Moeskau/Moeskau Photography; **p. 496 (margin):** © Charles D. Winters/Photo Researchers, Inc.; **13.7:** © Charles D. Winters; **p. 500 (margin):** © David Young-Wolff/Photo Edit, Inc; **13.13:** © Leonard Lessin/Peter Arnold, Inc.; **13.15:** © Charles D. Winters/Photo Researchers, Inc.

Chapter 14

Chapter 14 Opener: © Tom Pantages; **14.1:** © Tom Pantages; **14.2:** © Michael Macor/San Francisco Chronicle/Corbis ; **14.3:** © Brian Moeskau/Moeskau Photography; **14.4:** © Chuck Pearson/Whitehead Street Pottery; **14.5 (both):** © McGraw-Hill Higher Education, Inc./Stephen Frisch, photographer; **Ex 14.1 (both):** © Joel Gordon; **Practice Problem 14.1 (both):** © E. R. Degginger/Color Pic, Inc.; **14.7:** © Brian Moeskau/Moeskau Photography; **14.9 (top):** © E. R. Degginger/Color Pic, Inc.; **14.9 (bottom):** © Gary Yeowell/The Image Bank/Getty; **14.13a:** © McGraw-Hill Higher Education Inc./Stephen Frisch, photographer; **14.14:** © Haskell Smith/ScienceTreasures.com; **14.15:** © Chris Sorensen Photography; **14.16 (both):** © Brian Moeskau/Moeskau Photography; **14.19:** © Jim Varney/Photo Researchers, Inc.; **14.24 (all):** © Brian Moeskau/Moeskau Photography; **14.26:** © Jeff Greenberg/The Image Works; **14.28:** © Tony Freeman/PhotoEdit

Chapter 15

Chapter 15 Opener: © Argus/Peter Arnold, Inc.; **15.1a:** © Spencer Grant/Photo Researchers, Inc.; **15.1b:** © Zephyr/Photo Researchers, Inc.; **15.2:** © Gondelon/Photo Researchers, Inc; **15.3:** © ISM/Phototake.com; **15.12:** © David Parker/Photo Researchers, Inc.; **15.13 (both):** © SPL/Photo Researchers; **15.16:** © Tom Pantages; **15.18:** © James P. Birk; **15.20 (left):** © ADEAR/RDF/Visuals Unlimited ; **15.20 (right):** © TIim Beddow/Photo Researchers, Inc; **15.21 (top):** © Tom Pantages; **15.21 (bottom):** © Richard Megna/Fundamental Photographs; **15.25:** © Martin Bond/Photo Researchers, Inc; **15.27:** © Jerry Mason/Photo Researchers, Inc.

Chapter 16

Chapter 16 Opener: © Brian Moeskau/Moeskau Photography; **16.5 (both):** © Tom Pantages; **16.6:** © Phil Degginger; **16.7:** © Tom Pantages; **16.8:** © Tony Freeman/PhotoEdit; **16.9:** © James P. Birk; **16.10:** © Richard Megna/Fundamental Photographs

Index

Page numbers followed by an "f" indicate figures, "mn" indicates marginal notes, and "t" indicates tables.

A

Absolute temperature scale, 22
Absolute zero, 22, 330mn
Accuracy, 37
Acetaldehyde, 300t, 460, 632, 640t
Acetaminophen, 606mn
Acetate ion, 90t
Acetic acid, 103, 104–5f, 490, 633, 640t
   acidic hydrogen atoms in, 494, 494f
   acid ionization constant of, 501, 501t
   aqueous solution of, 496, 497f
   dissociation in water, 81, 82f, 105, 500, 500mn
   Lewis structure of, 494, 494f
   reaction with sodium bicarbonate, 450, 450f
   structure of, 300t
   as weak acid, 496, 497f, 497t, 500–501
Acetone, 300t, 369, 610t, 632, 640t
Acetylene, 613, 623
   combustion of, 182, 226t, 623mn
   molecular and empirical formulas of, 129, 129f
Acetylsalicylic acid. See Aspirin
Acid(s), 103–6
   Arrhenius model of, 491
   Brønsted-Lowry theory of, 491–92
   carboxylic acid, 103, 104f, 300t, 610t, 632–33, 640t
   conjugate acid-base pairs, 492–94
   containing polyatomic ions, 106
   definition of, 103, 490–92, 491f
   formula of, 104
   ionization in water, 409
   naming of, 105–6, 106f, 108
   polyprotic, 502–3, 502f
   reactions with metals, 172–73
   strong, 105, 105f, 105t, 138mn, 494–95, 495f, 495t, 498–500
   weak, 105, 105f, 138mn, 411mn, 494, 496, 497f, 497t, 498–503, 501–2t, 502f
Acid-base neutralization reaction(s), 177f, 180, 427, 431–33, 431f
Acidic hydrogen atom, 494, 494f
Acidic solution(s), 503–7, 504t
Acid ionization constant (K_a), 500–501, 501–2t
Acidosis, 513mn
Acid rain, 103mn, 154, 178–79, 179f, 489, 495
Actinide(s), 5f, 67
   electron configuration of, 260, 260mn
Actinium-227, radioactive decay of, 583
Activated complex, 452–53, 452f
Activation energy, 452–53, 452f, 457–59, 457f
Active site, 458, 458f
Activity series, 172–75, 173f, 264–66, 264f, 266f, 553f, 554–55
Actual yield, 213–14, 213f
Acyclic hydrocarbon(s), 612–13, 613f, 615–25

Addition
   numbers in exponential notation, 36
   significant figures, 38
Addition polymerization, 624
Addition reaction(s), of alkenes and alkynes, 623–25
Age of Materials, 363
Agorca M5640, 127
Agricultural runoff, 80mn
Air. See also Atmosphere
   density of, 21
   as gaseous solution, 8mn, 408
   temperature-density relationships, 320mn
Air pollution, 318, 318mn, 447
   indoor, 592–93
   pollutant concentration, 425mn
-al (suffix), 640t
Alanine, 637f
Alcohol(s), 610t, 629–30, 631t
   boiling point of, 631
   combustion of, 181mn
   denatured, 629
   functional group in, 300, 300t, 610t
   naming of, 630, 631t, 640t
   reaction with carboxylic acids, 633
   structure of, 631t
Aldehyde(s), 610t, 631–32
   functional group in, 300t, 610t
   naming of, 640t
   structure of, 631
   Tollens test for, 632
Algal bloom, 80mn, 419
Algebraic equations, 349
   solving of, 349
Aliphatic hydrocarbon(s), 299, 299f, 612, 612mn
Alizarin yellow, 512f
Alkali, origin of word, 70mn
Alkali metal(s), 67, 70mn
   electron configuration of, 264, 266
   ionization energy of, 266
   reaction with water, 165, 165f, 266, 266f
   reactivity of, 264
   valence electrons of, 264
Alkaline battery, 548, 548f
Alkaline earth metal(s), 67, 69, 70mn
   electron configuration of, 265
   ionization energy of, 266
   reactivity of, 264, 266
Alkaloid, 490
Alkalosis, 513mn
Alkane(s), 299, 299f, 610t, 612, 613f, 615–21, 615mn
   boiling point of, 615–16, 616f, 631
   branched-chain, 617
   combustion reactions of, 621
   continuous-chain, 615mn
   intermolecular forces in, 616
   London dispersion forces in, 616, 616f
   naming of, 616–20, 617–18t, 638–39
   reactions of, 620–21, 621f
   reactions with halogens, 621, 621f
   straight-chain (normal), 616f
   structure of, 616–20, 618t
   unbranched, 615mn
Alkene(s), 299, 299f, 610t, 613, 613f, 621–23
   addition reactions of, 623–24

   isomers of, 622
   naming of, 621–23, 622t, 639
   polymerization of, 624
   reactions of, 623–25
   structure of, 621–23, 622t
Alkyl group, 619
Alkyne(s), 299, 299f, 610t, 613f
   addition reactions of, 623–24
   naming of, 623, 639
   polymerization of, 624
   reactions of, 623–25
   structure of, 623
Alloy, 8mn, 363–64, 392t, 393–95, 393–94f, 394t, 408, 408f
Alpha amino acid, 636, 637f
Alpha decay, 581–83, 581t
Alpha particle(s), 54–55, 55f, 570–72, 571–72f, 580, 580f
Alpha particle emission, 573–74, 573f
Alpha radiation, properties of, 572, 572t
Alternative-fuel vehicle, 195–96, 553
Altitude
   atmospheric pressure and, 323mn
   boiling point and, 369
Aluminum, 4–5, 6f
   density of, 18f, 386
   preparation of metallic form, 526, 526mn
   reaction
      with bromine, 170, 170f, 170mn, 212–14
      with iron(II) oxide, 158–59, 159f, 171, 172mn
      with oxygen, 525
   specific heat of, 219t, 220
Aluminum(III) bromide, 96, 170, 170f, 170mn
   formation of, 212–14
Aluminum chloride, formation of, 287
Aluminum hydroxide, as antacid medication, 490
Aluminum ion, 63, 70, 87t, 96, 263, 287
Aluminum oxide, 96, 96f
   formation of, 158–59, 159f, 171, 172mn, 525
   preparing aluminum metal from, 526
Alundum, 392t
Amine(s), 610t, 635–36
   basic properties of, 491
   functional group in, 300t, 610t
   naming of, 640t
   as weak bases, 497–99
Amino acid(s), 279, 608mn, 636, 638
   structure of, 301, 636, 637f
Aminobenzene. See Aniline
1-Amino-4-chlorobenzene, 627t
Ammonia, 103–4, 105f, 635
   aqueous solution of, 491
   atmospheric, 318t
   bond angle in, 305mn
   bonds in, 292
   formation of, 208–9, 464, 464f, 467, 476–77
   in household products, 490
   hydrogen bonds in, 382f
   intermolecular forces in, 386
   ionization in water, 498f
   melting point of, 85
   as molecular solid, 398

   reaction with hydrogen chloride gas, 345, 345f, 492mn
   reaction with water, 104, 104f
   solid, 84–85, 85f
   solubility of, 178
   as weak base, 497, 498f, 498t
Ammonium chloride, formation of, 345, 345f, 492mn
Ammonium compound(s), decomposition of, 168t
Ammonium ion, 88, 90t, 104, 104f, 501t
Ammonium nitrate, in cold packs, 416mn
Amorphous solid(s), 390, 390f
Ampere (A), 39
Amphetamine, 607, 607mn, 611, 626, 635
Amphoteric substance(s), 493
amu. See Atomic mass unit
-ane (suffix), 616, 638–39
Anhydrous compound(s), 169
Aniline, 627t, 635, 639
Animal fat, 634–35
Anion(s), 61–62. See also Ionic bond(s)
   electron configuration of, 262–63
   formation from neutral atoms, 61–62, 62f, 285
   Lewis symbol for, 287
   oxoanion, 88, 89t, 90f, 164
   size of, 271, 271f
Anode, 538–39, 538–39f, 542
Antacid, 180, 280, 407, 407mn, 417, 490, 496, 498t
Ant bite, 632
Antifreeze, 18f, 437, 629
Antimony, physical properties of, 69f
Aqueous solution(s), 15, 83, 409, 409f, 427–33
   acid-base neutralization reaction, 427, 431–33, 432f
   chemical reactions in, 182–83, 182f
   electrical conductivity of, 81, 81f
   electrolyte solution, 409–12, 409–11f, 410t
   in human body, 415mn
   precipitation reactions, 427–30, 428f
   solution process, 413–17, 413–17f
   symbol for, 15t
   temperature effects on, 419, 419f, 426
Archeological dating, 587–88
Arginine, 637f
Argon
   atmospheric, 318t
   electron configuration of, 254–55, 259, 265
   intermolecular forces in, 384
   kinetic energy of gas, 28
   London dispersion forces in, 378
   stable compounds of, 70mn
   volume, pressure, and temperature relationship, 333, 336
"Aromatic," 616mn
Aromatic character, 626
Aromatic hydrocarbon(s), 299, 299f, 612mn, 613–14, 613f, 626–29, 627t
   Lewis formula for, 300f
   naming of, 626–28, 627t, 639–40
   reactions of, 628–29

Important Polyatomic Ions

1– Ions

nitrate	NO_3^-
nitrite	NO_2^-
bicarbonate (hydrogen carbonate)	HCO_3^-
perchlorate	ClO_4^-
chlorate	ClO_3^-
chlorite	ClO_2^-
hypochlorite	ClO^-
cyanide	CN^-
hydroxide	OH^-
acetate	$CH_3CO_2^-$
permanganate	MnO_4^-

2– Ions

chromate	CrO_4^{2-}
dichromate	$Cr_2O_7^{2-}$
sulfate	SO_4^{2-}
sulfite	SO_3^{2-}
carbonate	CO_3^{2-}
oxalate	$C_2O_4^{2-}$
peroxide	O_2^{2-}

3– Ions

phosphate	PO_4^{3-}
phosphite	PO_3^{3-}
borate	BO_3^{3-}

1+ Ion

ammonium	NH_4^+

Flowchart for Naming Compounds

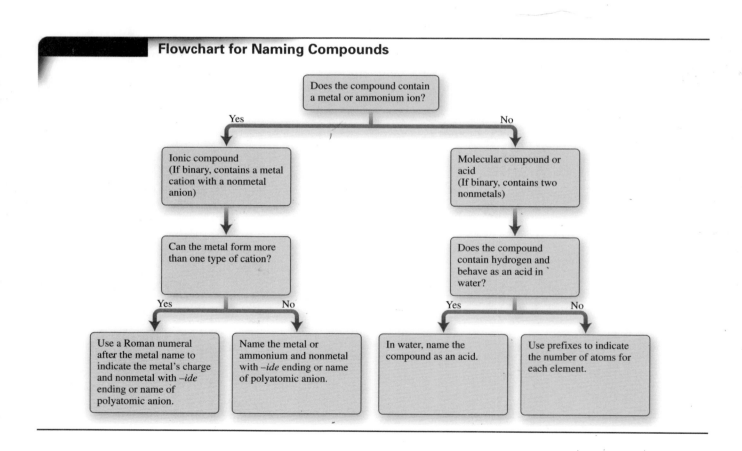